Handbook of Automotive Body & Systems Design

To Paul, Peter and Denise

Handbook of Automotive Body & Systems Design

by

John Fenton

Professional Engineering Publishing

Professional Engineering Publishing Ltd.
London and Bury St Edmunds, UK

First published 1998

ISBN 1–86058–076–X

A CIP catalogue record for this book is available from the British Library.

Printed in Great Britain by Redwood Books, Trowbridge, Wiltshire

CONTENTS: AUTOMOTIVE BODY & SYSTEMS DESIGN

Chapter 1:
Creating and prototyping

Chapter 2:
Ergonomics, seating and packaging

Chapter 3:
Aerodynamics and air-conditioning

Chapter 4:
Body trim and fittings

Chapter 5:
Body electrical/electronic controls

Chapter 6:
Body-electronics and telematics

PREFACE

This literature survey is aimed at providing the vehicle body designer and engineer with a broad as possible introduction to the factors affecting body aesthetic design and body systems. Because body design and engineering stands somewhere between the disciplines of product design and mechanical engineering, and now includes very considerable electronic input in both body design automation and control of vehicle systems, a considerble degree of cross-discipinary work is involved. The refenced leads provided in the text are intended to help designers and engineers from whatever background discipline. The widespread availability of computing power to designers and engineers has created the possibility of considerably shortening the lead-times between design conception and prototype manufacture. Some guidance is therefore also provided for individuals who may not have enjoyed background training in these areas.

The book is an introduction to body design and body systems in the sense of providing readers with an insight into analytical methods given in a wide variety of published sources such as technical journals, conference papers, and proceedings of engineering institutions, for which a comprehensive list of references is provided. The analyses are therefore not necessarily fully developed or rigorously evaluated; recourse to the original references is necessary particularly in order to understand the limiting assumptions on which the analyses are based.

Some of the earlier works involved analysis techniques developed from empirical experimentation and predate the widespread introduction of SI units. In such cases conversion to SI units, and reconstruction of the related graphical illustrations, was felt to be inappropriate. Much of the analytical work is centred around impending legislation and, where this is quoted in the text, it is for illustration only and it is, of course, important to examine the latest statutes in the locality concerned. The list of references given at the end of each specialist chapter is a key element of the publication, providing where possible a link to the original publication source. Where the original publication is no longer in print many of the sources can be found in such libraries as those of the Institution of Mechanical Engineers, London, or the Motor Industry Research Association, in Nuneaton, as well as the British Library

JOHN FENTON

Chapter 1: Creating and prototyping

Facets of design; 'technik' and design; distancing of design and manufacture; design training; aesthetics and industrial design; formal aesthetics and shape; evolution of body shapes; computer-aided-drafting; C.A.D. and surface development; 3D modelling; computer-aided-manufacture; surface geometry and definition; rapid prototyping; rapid tooling

A merging of engineering and design disciplines is necessary for successful execution of vehicle body shaping and packaging. This can involve severe clashes, in traditional societies, and even the most progressive manufacturers have sharp divides between product engineering and styling departments. Body fittings and interior trim also need a multicultural approach so as to come up with aesthetically pleasing effects which are also durable and appropriate to the performance behaviour of the vehicle.

Facets of design

Engineering design and aesthetic design are sometimes seen as the only two faces of design. This somewhat narrow interpretation does not give justice to ergonomic design, design for manufacture and design for environmental fitness, for example. In reality design should have even broader implications and be concerned with every conceivable interface that the engineering artefact has with the outside world and its publics.

Huge markets have been lost by companies, indeed countries, who consider products such as vehicles merely as engineering products. Of course one sees the same dangers in over-concern for some other facet of design, without consideration for the remainder at the same time. Vehicles which have little more than showroom sales appeal are now quite likely to fail in the market place if designers have failed to build in the reliability that guarantees a long trouble-free life. Engineers and designers need to be very broad in their views and talents so as to try and gauge the multitudinous external effects on their designs.

Historically, well engineered vehicles had tremendous market appeal in the limited markets into which they were sold. In the UK passenger car field one thinks of the products of companies such as Frazer Nash, Aston Martin and Alvis in the 1920s and 1930s. Well engineered products were crucial to the discerning markets into which these high-class vehicles were sold. But these products had something more than what the disciplines of engineering science had to contribute, not least that difficult to define quality that comes from being built by skilled and professionally-devoted craftsmen. Their engineers and designers were often far from being limited only to a knowledge of engineering science. The breadth of their education and training showed in the 'elegance of line' of their products; the fitness for purpose of their interior layouts also revealed an enthusiasm for the product which they shared with their customers.

In the creative process of writing the specification for an engineering design it is suggested that the design engineer goes through the thought processes of assimilation of information governing the product; this is followed by a period in which conscious solutions are produced based on the information obtained; next follows an incubation period where the subconscious (the mind in a 'dream' state) sends out ideas for further possible solutions. Types of creative thinking have been considered by de Bono[1] who has defined three thought processes. Insight thinking is the first, where there appears to be no definite step leading up to the solution. Next is sequential thinking where the solution follows a progressive sequence of steps such as modification, improvement, mistakes, and new ideas. Final stage is strategic thinking where careful consideration is given to the choice of the most appropriate steps to take along the path to he solution. Here the mathematical/ scientific approach to problems establishes a definite method of working which is followed through to the final result.

Because the contemporary educational environment favours the 'judicial' rather than 'creative' approach to original work the average person's balance has been disturbed and it is necessary to force the discipline of adopting an initial creative stage, when no judgement or adverse comment is allowed, then progress to a separate assessment stage where the ideas are evaluated. While any specific idea is necessarily the product of an individual, it has been shown useful to have group brainstorming session in which one person's ideas generates modifications and better ideas in other individuals.

According to one-time Royal College of Art teacher, L Bruce-Archer, there are five elements of the design activity. He suggested that a key element in designing was the formation of a prescription, or model, of a finished work in advance of its embodiment; furthermore that the embodiment of the design as an artifact was essential. He even suggested that arriving at the solution by strict calculation was not regarded as designing; that originality was essential and that designing implies a purposeful seeking after solutions rather than idle exploration.

What distinguishes engineering design from other design activities is the extent to which technological factors are involved. A sculptor is mainly concerned with form, shape and texture, together with an awareness of the technical factors related to the highly skilled craftsmanship involved. However, the sculpture can be executed with technological knowledge amounting to no more than a routine. In the engineering design of a vehicle (discounting the Madison Avenue type 'stylist') the designer requires considerable knowledge of technology alongside his or her creative and synthesising abilities. When, of course, the appropriate technology for vehicle design becomes complex then the engineering design involves analytical procedures.

The basic functions of engineering design are to determine exactly what is to be made, to define it as clearly and completely as is necessary, to ensure that it will be made exactly as required and then to ensure it will be put to work as intended. The design drawing was the classic expression of engineering design, linking the work of the designer to the manufacturing engineer. Advances in information technology have substituted the CAD perspective for the drawing and in many cases the design database is also the database used to cut the first dies of the body panels. The interaction between designer and manufacturer is thus much closer, in the new age of simultaneous engineering.

The historic Fielden Report[2] summarised the engineering designer's role with this definition '...of being the use of scientific principles, technical information and imagination in the definition of a structure, machine or system to perform pre-specified functions with maximum economy and efficiency'. A further historic definition, for the industrial designer, is `...for the making logical decisions in which engineering necessity and social needs are equated with aesthetic sensitivity — when the social needs and aesthetic sensitivity are conscious activities: it is a specialist sector of engineering where the relationship of the user to the product is predominant'. In recent years, of course, the Design Council and the Engineering Council have tried to move the traditional role of the industrial designer, with the consumer, to engineering in general and a broader based individual is evolving less torn by the rigid cultural divides of arts, sciences and engineering.

Technik and design

English speaking countries seem to confuse science and technology in their publishing and broadcasting media. Is this because the word "technik", discussed in detail later, which appears in a number of other languages has never properly been translated[3] into English? The Germans and Swedes for example seem to define technik as "useful arts of manufacture" whereas perhaps when English translators first con-

Fig 1: An interesting 'meeting of the two cultures of architecture and vehicle design' was seen in an exhibit at London's Design Museum by Tetra Associates. Le Cobusier's vision of the car of the future is seen in this wooden model he made in 1928. He favoured a rear engine and hatchback layout with three front seats and one, occasional, at the rear.

sidered the word, they confused it with "technology" meaning scientific study of the teaching of techniques. It may even be the reason why British politicians think teaching technology will solve manufacturing problems and why many "trained" British engineers seem to be uncomfortable with manufacturing processes. Perhaps if Britons had a greater understanding of the concepts behind the word technic they would see a greater dignity in manufacturing.

CP Snow's "two cultures", of humanities and sciences (Fig 1) might have been too narrow a view of the problems of a declining manufacturing base. In Germany, it is suggested, there is acceptance that technology and science are separate cultural areas with separate work patterns for those involved in them. There would be more likely to be concern at the two cultures of quest for knowledge, generally, on the one hand, and making things, on the other. There would still be a separate regard for the third subculture of "arts", in the sense of fine arts and performing arts.

George Stephenson, who is regarded as a key player in getting the first industrial revolution off the ground, by his application of the steam engine in early locomotives, is often said to have been better endowed with technical rather than scientific knowledge. Probably the same could be said of many founding industrial entrepreneurs. Even the first successful production of the steam engine, at the Boulton and Watt factory in Birmingham, hinged on the manufacturer's advanced skills in metal-cutting.

It is argued that the "age of science" has existed for no longer than two centuries whereas key engineering principles go back to the history of man and his use of tools. The English-language notion of "applied science" could be seen as providing a misleading invitation for natural scientists to tackle elements of manufacturing which they neither understand or are competent to deal with.

In continental Europe it has been suggested that some three-quarters of graduate engineers are concerned with the technic functions of manufacturing. The notion of engineering as applied science is perhaps suspect and even the difference between pure and applied mathematics difficult to define. The idea of applied science suggests something which changes on application, like burning a match. It is interesting to note that economists see science as a "public good" which does not change its character on transaction.

These ideas even throw doubt on traditional teaching of engineering in which first basic theory is taught; then the theory is used in a training environment; finally the fully prepared engineer is ready for work. However, a notion of technik is that there is never a body of knowledge to be used in all technical circumstances. While theoretical propositions are, of course, used in technical work they are not exclusively of the physical sciences. Early eduction of engineers would be perhaps better be spent on the simulation of events in the process of making things.

At this point in the age of science, known as the information revolution, we now find that scientific knowledge and engineering principles combine to be of considerable assistance in the computer modelling of engineering problems It is accepted that industrial designers should understand and practice the skills of manufacturing but strangely there is often a reluctance to accept that engineering science should be a part of their training. Provided they are part of a simultaneous engineering team in which each specialist individual contributes his skill to the common purpose, this may well not matter. If they work alone in a studio without contact with other specialists, there may be a problem.

Design and manufacture distanced

The giant Exhibition Road complex in London includes such as establishments as the Victoria and Albert Museum, the Royal College of Art, Imperial College and the Science Museum. According to AN Wilson[4] "...Prince Albert wanted his 'Albertropolis' to be a vast structure with four institutions devoted to raw materials, machinery, manufacture and the 'plastic' arts...". While such places were founded very much with the industry connection in mind, many beleive that they heve drifted into the control of an artistic elite with limited intersest in the art/design/industry interplay.

An interesting background to the drifting apart of design and manufacture is provided by Penny Sparke[5], who traces the period commencing with the early employment of artists by the textile and ceramics industries through to the influence, at the the turn of the century, of William Morris. This designer's "arts and crafts" movement was intended to reunite artist and craftsman — split apart by the economic "division of labour" in the early nineteenth century; in the event, it is argued, the movement sadly attracted artists away from industrial design.

With modern trends to a reuniting the 'division' exemplified by such factors as pioneering 'group assembly' at Volvo and Saab and 'team management' at Nissan and Toyota, there is a clear movement to integrating the tasks of design, development, manufacture, testing and marketing. In Sparke's view, the over-close uniting of design and marketing, in the US 'industrial design consultancy' movement, has had a negative effect — however — in the cosmetic treatment of products to try and obtain market acceptance long after the normal design stage.

If dishonest design is disguising, or artificially enhancing, the appeal of a product, one has to find a positive definition of honest design. A somewhat extreme view of this is provided by US designer Victor Papanek[6]. He rather outspokenly asserts that "...advertising design, in persuading people to buy things they don't need, with the money they don't have, in order to impress others who don't care, is probably the phoniest field in existance today!" Papanek's cry is to unite the division between form and function as part of the 'total function' — or Yin-Yang as the Eastern sages describe it. His book proposes a structured approach to conceptual designing in which the following seven major 'blocks' are overcome; namely perceptual, emotional, associational, cultural, professional, intellectual and environmental. He points to the success of AMC Jeep, successfully produced since 1943 to the highest quality standards. in debunking the myth that 'quality no longer counts' — a happily out-dated concept as recent events in the volume vehicle industry have shown. He insists, more concretely, that design education should shift from "arts-and-crafts" concepts to interactive learning technques involving a familiarisation with computer-integrated manufacture. The cult of surface newness should go in favour of fundamental functional design — mechanism and function should not be 'relegated to some product engineer', he argues. Strong commercial advantages arise from this, not only in gaining lasting market acceptance, but in a reduction in high development costs associated with traditional 'cut and try' methods. Computer prediction by design calculation, and increasingly development verification using sophisticated computer modelling techniques, to 'test on screen', give a new importance to analytical techniques.

Design training

While many consider the single Euoropean market as one of just the flow of goods backwards and forwards across the national boundaries, there is also the flow of ideas and attitudes to training. How, for example, might the UK benefit from adopting some of the German attitudes to education, work and professional training? The Siemens company has been in the news recently for its link-up with British electronic companies to form a pan-European conglomerate. A copy of the company's house-journal[7] at the beginning of the 1980s carries some far-seeing views by a prominent British consultant who speaks with the experience of both countries attitudes to design engineering education and training. Of course, there is not such a divide between arts/humanities and science/technology in Germany as there is in Britain — but there are many other interesting differences as well, it seems.

Above all, it is argued, the German entrepeneurial spirit is strong, possibly as a result of the aftermath of World War 2 when many companies had to restart from nothing. And another background point to the greater industrial success is the relative immunity from take-overs. The banks play an important role in industry in that their representation on supervisory boards of large firms is a means of much more truly representing, in turn, the interests of the small investors alongside workers and executive directors who have seats on these beards. In the UK the relative influence of any of these people, as individuals, is insignificant at shareholders meetings compared with the united voice of such people through the bank's representation. And as the banks take a long term view of dividend policy, executive directors are free to get on with their jobs without worrying about stock-market vultures.

The author dates the German educational revolution to Baron Rumford's introduction of compulsory elementary education in 1803. But beyond an education system, which is now said to give an 18-19 year old leaving school level pegging with university graduates in some other countries, lies the national tradition in management for hard work — which is seen in Germany not merely as a means of earning money but as something which brings satisfaction. This dates from a proud craft tradition and, highly significant, is the attention paid in Germany to providing part-time education for a senior craftsman when he reaches the point of taking on the responsibilities of management. In the UK, of course, the transition is often without any training whatsoever.

At Siemens, some 12 % of the workforce are said to have a higher engineering qualification and the author maintains that in Germany the concept of advancement by business-school (Harvard-type) training is quite foreign. Management theory is relatively low on the manager's interest level and as line management is strong, and chains of command kept short, there is an organisational clarity largely absent in the UK large firm. Dependency on management experts for advice could lead to longer chains of command, it is considered. Lack of business schools may also help to explain relative lack of interest in take-overs. Hence the German axiom that it is better to put a competitor out of business by better design and manufacture rather than buying him out of the market; it has also been obeserved that there is a healthy distrust of authoritarian management in Germany and many Americans have said to have been surprised, as a result, at the number of directors names appearing on a German 'company's letterhead.

There is a different culture, too, in the higher education establishments and this could lead to surprise, by foreigners, at the number of doctorates on the company letterhead. It is explained that this could well lead back to college days where the young student might well be asked to become his professor's assistant for one or two extra terms added to his college stay. During this period he will help his professsor in lecturing, and perhaps even book-writing, while at the same time working on his doctorial thesis. He may thereafter spend a year or so as research worker but after that, he would be firmly kicked out into industry as there is no precedent, as in the UK, for technology teachers to gently climb up the ladder to professorship. Either you are a full professor with professional industrial experience behind you or you are a dogsbody assisting your professor! The engineering academic world is also dependent for each teacher on part-time professors who are in senior positions in industry — a task which in Germany is taken seriouly and with pride. Many such people are found to have preceded their academic life with apprenticeship training and a background of craft perfectionism combined with sound scientific training is often the result. While the UK Finniston Report on 'Engineering our Future' compared the UK position with Germany, the author maintained that it overlooked the fundamental differences in the two countries, technological education systems — further missing the point that 'Britain has plenty of clever and lazy people but inadequate numbers of clever and industrious ones!'

Aesthetics and industrial design

As successful aesthetic design comes from practice rather than conventional teaching, much depends on the commitment to good industrial design by the practioner before he or she puts pencil to paper. Fig 2 reveals some of the results of good industrial design practice while the text below argues that the overall design of vehicles is a difficult mix of engineering-science, constructional skills and aesthetics which can only be described by the non-English word 'technik'. This concept is considered here with other factors which ultimately affect appearance — in an introductory article intended to find the right frame of mind for embarking on successful design.

The market opportunities for well designed vehicles are one of the positive aspects of the worsening market response to the design of conventional volume-built vehicles. A uniform product for all, it is often argued, seems to be the result of aerodynamics and 'clinic' design which boils down a wide sample of individual preferences into a 'something for all but not much for any'. Another prevailing factor could be the high percentage of volume-built vehicles which are purchased by companies either for commercial use in fleets, in the case of CVs, or as 'job' or 'status' cars. This inevitably presents a different set of criteria to the product planners and designers — as the fleet sector will have different market demands from those of the individual-purchase sector.

In the case of cars, the end-user in the company will usually have a 'say' in the choice within a band of offerings made to him or her by senior management or the transport excutive concerned with fleet purchase. The decision to buy might be influenced by something seemingly as trivial as the status of the 'L' pack over the 'GT' pack or how the neighbours will rate the car in the pecking order of consumer-hardware purchases. Strong in the mind of the fleet user will be the knowledge that he or she will not be paying for fuel, service and repairs — and that replacement will occur in as little as one year's time.

In the case of the large CV fleet, the purchasing decision for the volume vehicle is more likely to be made on perceived reliability; running cost based on servicing by commercal garages (it has even been known to make the selection on service-interval multiplied by recommended charge times projected life) and above all by life-end price in the resale market. Fitness for purpose is likely to be relatively low down on the priority list and the scope for good design very limited indeed.

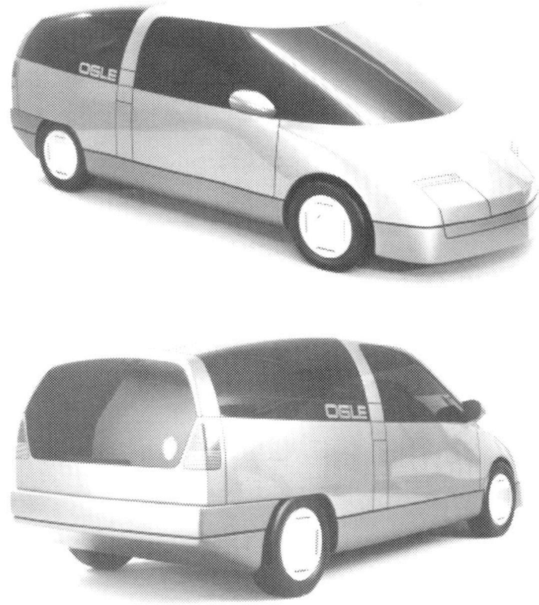

Fig 2: Design studies by Ogle

Product planners in the volume sector seem to have switched from talking about function versus appearance and are now addressing themselves to whether the perceived market demands within a particular known sector should take preference over opening a new niche to widen the sector by appealing to a greater range of customers. In the specialist vehicle sector, by contrast, there seems to be scope for aiming at 'good design' which attracts new customers for its own sake.

Recognition of design The sympathetic use of materials in design and construction, it is often suggested, can favourably affect the design of the end-product and materials, a real consideration for the end-user's enjoyment in driving or even just sitting inside the vehicle. The way the vehicle responds to the controls — and the sound engineering of its functional systems — can also influence a wide range of buyers.

Design has happily entered the political arena in recent years and is no longer left to the design-conscious few who can so often be out of touch with

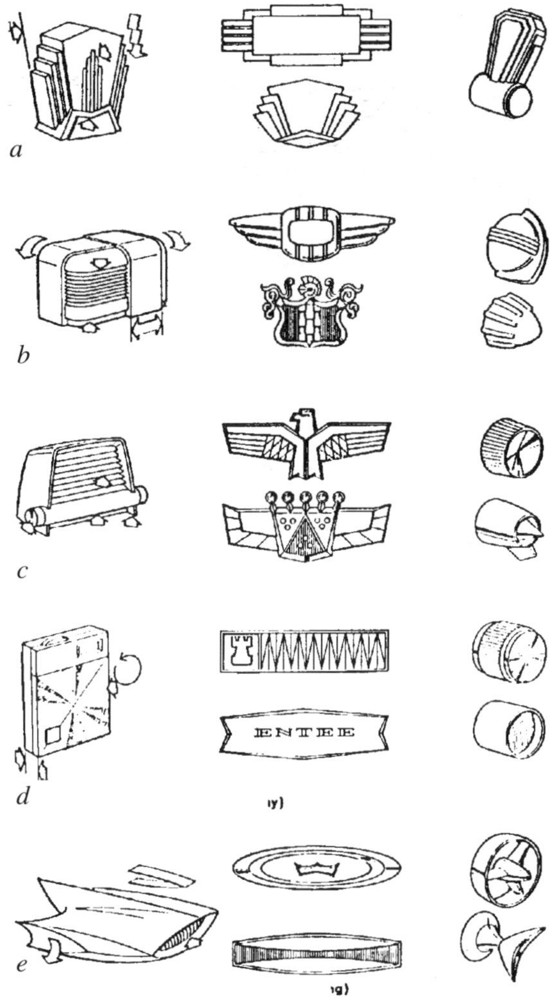

Fig 3: Styling cliches: a. stepform, 1925-35; b. streamform, 1930-40; c. taperform, 1945-55; d. sheerform, 1955-65; e. sculptureform, 1960-70

day-to-day marketing and manufacturing. 'Design or decline' is the catchy slogan of the new thinking. Recent reports show that the UK design consultancy business has been growing at a rate of 30 % per year and is now beginning to get somewhere near in total spend to those leading marketing services — sales-promotion and advertising; it has already overtaken market research and public relations in overall turnover terms. While this is encouraging in one sense, it hides the fact that 20% of companies have no internal design resource. Nor does it reveal that in many smaller companies, good design no longer necessarily flows from good craft skills of the workforce directed by enlightened management who can properly exploit such a resource. The integrated approach to design is now being advocated — in place of merely 'making things look good'. The tendency to use dedicated stylists is losing ground to sharing many job functions, at all levels, in the key aspects of industrial design and engineering for quality.

Market experience shows that specialist commercial and industrial vehicles, unlike their 'volume' counterparts, can often be operated by those who are very much concerned wih fitness for purpose; vehicles are as likely to be serviced in the operator's own workshops as in 'high-street' service stations and will probably be driven by experts who have a real appreciation of the engineering of their functional systems. Car buyers are likely to be enthusiasts who are concerned not only about the performance but also by sculptural shape of the product.

Stylists, how they took control The American-originated styling-for-sales strategy historically grew out of the Madison Avenue-based advertising industry and many of its participants had their offices in that neighbourhood — a world away from the manufacturing base. In the UK, it was argued, there was an additional factor to this geographical divide; according to 1960s design consultant Hugh Clausen[8], organisations like the Institution of Mechanical Engineers separated the traditional drawing office based designer from the so-called 'professional (chartered) engineers; more recently such institutions have been accused of overloading the contact hours in engineering teaching courses so that students miss the wider education necessary for development into good working designers.

The need for good industrial design is an inevitable result of subordinating design to styling, which consciously makes a visual addition. The development of styling cliches corresponding to the changes in fashion over the years, Fig 3, is completely divorced from function or beauty in the classical sense. A prominent design teacher, John Grant, who advocated this view, argued that industrial design had the ultimate aim of producing the formal properties which related the function and structure of the product to what the buyer sees and uses. The designer, in this context, synthesised what the engineer analysed, Fig 4. He lists objectives as function, utility (permanence of usage), simplicity (ease of manufacture and service) and elegance (in both appearance and the scientific sense). Whereas the styled product had to

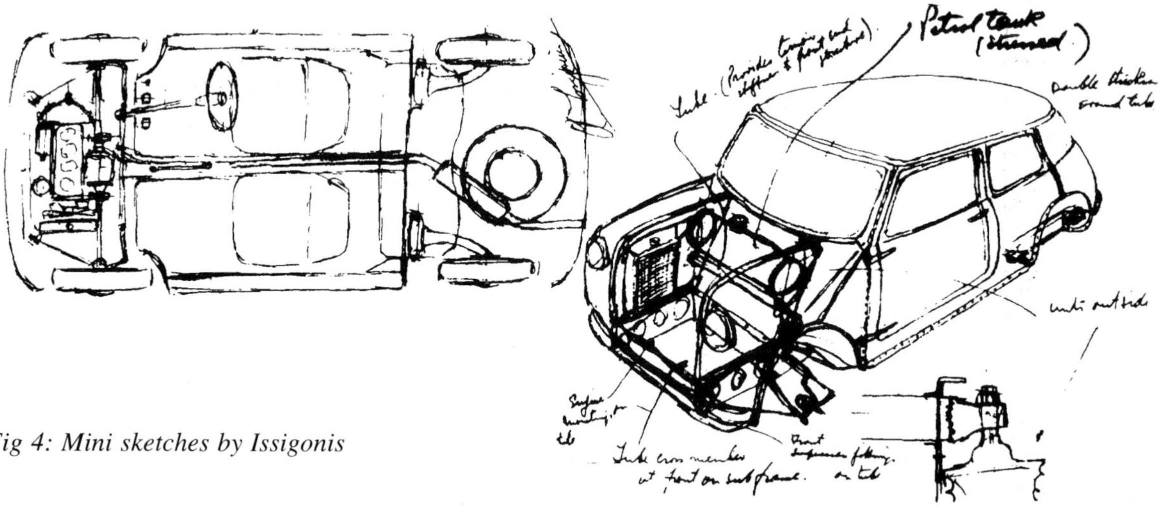

Fig 4: Mini sketches by Issigonis

be changed every time that fashion changed, good design could establish market leadership over a number of years; this was the key message of good industrial design.

In the same way as Prince Charles encourages reverence for the British architectural heyday — the regency era when designers drew on the classical aesthetic proportions of ancient Greece — so also were vehicle design teachers like Carl Olsen reported to have advocated that in car aesthetiscs, one should pay regard to the heyday of UK body design — when sleek knife-edged bodies were built on such chassis as Frazer Nash, Aston Martin, Bentley and Alvis.

Another teacher, Nigel Chapman, was reported as saying 'it is pointless trying to pander to what appears to be public taste — the industrial designer must work with engineers to evolve the most appropriate concept'. Another prominent consultant, Douglas Scott, who had the London Transport Routemaster bus among his design portfolio, stressed the need for the designer to have an intimate knowledge of the construction processes; he is reported as saying, for example, that 'rather than the presswork engineer adding his opinion a the early concept stage, it is far better for the designer to have a real feeling for presstool work and sheet metal assembly in the same way as the sculptor in clay has a feel for his material .

Fit and finish Attention to detail is another important element in the visual impressions made on vehicle users. A designer's overall control is necessary to avoid details being irretrievably changed between design sign-off and production-build as has so commonly occurred at the 'volume' manufacturers. Another pitfall is the tendency for a CAD draughtsman to select and arbitrarily blend between surfaces, during development of a shape, rather than visualise the sort of effect achieved by the sculptor with a few deft strokes of the scalpel. Expert care is necessary, too, in the choice of surface texture, colour and contour, when specifying interior finishes in the immediate vicinity of driver and passenger. And it is vital to understand that detail finishes applying to one material rarely read across to another.

Inter-disciplinary design approach As introduced in an earlier section, 'technik' is an interesting concept taken from the non-English language which seems to have no direct interpretation in modern Britain. The word is used in Germany and Sweden and has different interpretation from 'technology' for which equivalent words also exist in those languages. It is a more realistic way of expressing what in Britain is mistakenly referred to as 'applied science' because it relates much more to engineering technique than the manufactured product.

Thus engineers in Sweden and Germany are seen not so much as technologists, closer to science on the one hand, or technicians close to the craft skills, but as 'contrivers' who devise, make and improve artifacts. The lesson is being learned that it is not enough just to employ a scientist and craftsman, in combination, to achieve a well designed vehicle. There must be a master designer in control who has a properly integrated training in the arts and sciences.

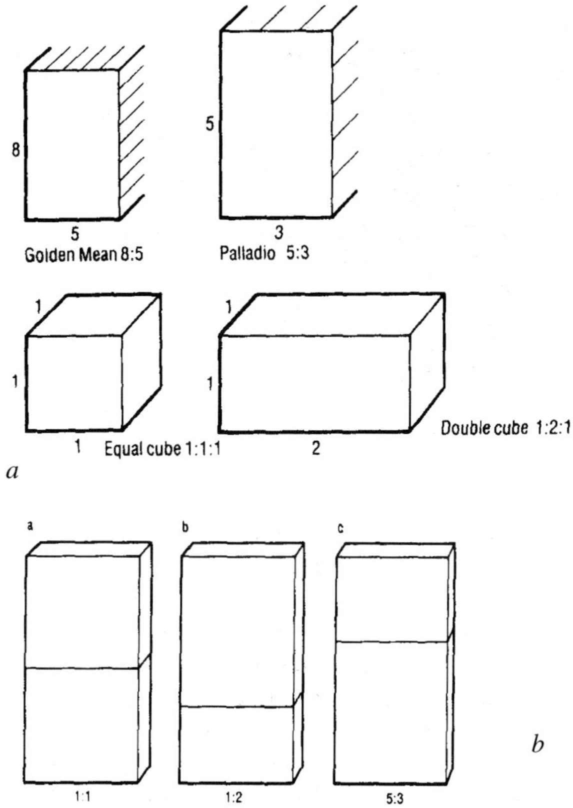

Fig 5: Classic proportions (a, above) and stability in form (b, below)

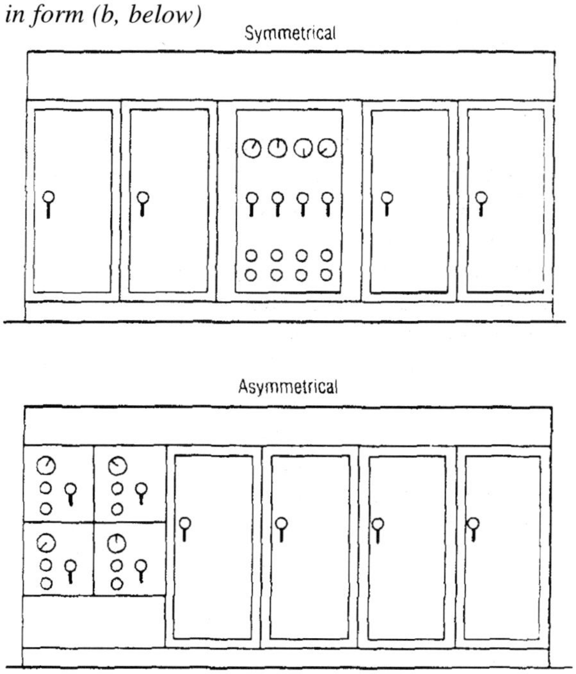

Fig 6: Introducing asymmetry

Formal aesthetics in design

The post-war foundation of UK industrial design was based on the Scandinavian model with an aesthetic based on simple lines, use of natural materials and an overall feeling of clarity and light in interiors. The moving apart of the two design cultures between then and now is attributed to the relative isolation of Scandinavia on the one hand and the heavy influences on UK design of such factors as the 'throw-away' design culture of North America. By the 1970s Scandinavia had lost its earlier design supremacy to countries such as Italy. However, the beginnings of Scandinavian design, much earlier in the century, are said to have been inspired by the German Werkbund example, a derivative of which we examine below.

Balance in form design According to Flurscheim[9], form has been defined as the image presented by the outer surface of an object — but it is also the manner in which it is able to be perceived. It cannot therefore be treated purely as a functional problem and must be viewed from the perspective of the potential user and to its environment — unlike a work of art which is often viewed in isolation. Form is able to convey certain messages about the object through historical association and other learnt attributes.

'Balance' in form is based on historical perceptions of proportion. Euclid's 'golden mean' calls for a ratio in rectangular dimensions of 8 to 5. Palladio is associated with 5:3 and Adam with the double cube, Fig 5. Stability is another aspect of balance, in considering shapes that generate an impression of equilibrium. Instability can be suggested even on simple rectangular shapes, Fig 5b, according to the positioning of their horizontal dividing lines with respect to the mid-height level. Tall free-standing structures are given stability in another way — the introduction of a very slight taper. Symmetry is also a factor in balance and whereas symmetry might suggest a state of order, Fig 6, asymmetry can often create a greater sense of interest. Repetition, too, can be used to enhance balance; the repeating of individually similar shapes can help to create an impression of unified order.

Functional design will be achieved when all the visual components support a tidy and co-ordinated impression of function without the need for any imposed style. Because of their complexity many products could present an untidy and unco-ordinated impression and a styled co-ordinating medium is then necessary. One method is to base the design on straight lines with flat and rectangular forms. Another is to employ curved lines with two or three dimensional curved forms.

German influence in design

The conscious use of formal aesthetics in contemporary design dates from the foundation of the German Bauhaus design school in the 1920s. The history of the school is described by Whitford[10] who speaks of its revolutionary effect on art education, one of the aims of which was to elevate the status of 'crafts' to that of 'fine art.' The Great Exhibition of 1850, held in Paxton's Crystal Palace was earlier to have cast doubts in the minds of some of its thoughtful visitors who contrasted the wonder of the building with the paucity of design in the objects it contained. While this had led to the end-of-century craft revolution with its art-nouveau style, the later Bauhaus movement was to spawn the more design-conscious art-deco style. During its life-time, however, the Bauhaus spurned any sort of styling cliche at all.

One of the most important of the school's industrial design influences was the work done by its prominent teacher, Peter Behrens, in an earlier appointment, when he started to revolutionise the design of AEG products, including the logo of Fig 7 designed as long ago as 1907. Two fundamentals of the Bauhaus approach were: (a) forbidding the students to specialise and (b) that all art education be based on craft training; however, first director Gropius argued, at the end of World War 1, against 'the dangerous worship of might and the machine which led us over the spiritual to the economic abyss' — not surprising therefore that the Nazis closed the school in the 1930s.

A significant milestone in aesthetic design teaching came in 1925 at the Bauhaus when the modern artist Kandinsky introduced a design course which aimed at an art which was 'intellectually controlled as well as emotionally expressive'. A

Fig 7: Gropius logo for AEG

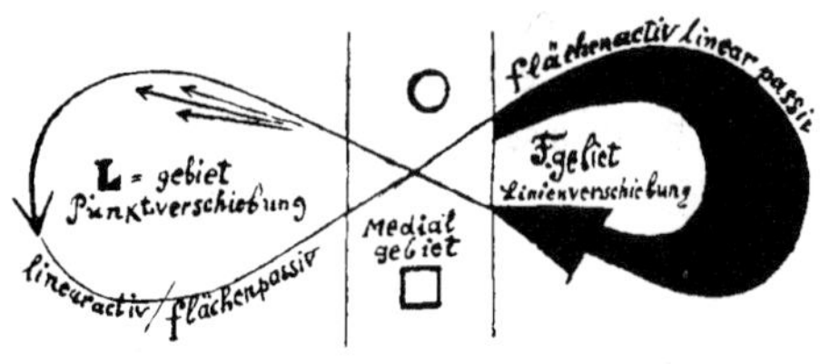

Fig 8: Basic elements of form

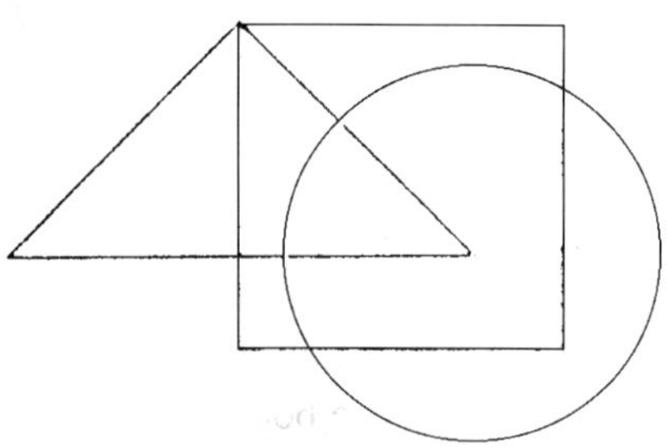

Fig 9: Klee's active and passive lines

Number (of BPAs), both cardinal (depending on the "degree of patterning": 1, 20, 250, . . .) and ordinal (1st, 16th, 39th, . . .)

Position in the field; for example, upper left, far right, . . .

Size relative to the field; for example, large, small, . . .

Shape as determined by the contours; for example, round, oblong, square, . . .

Direction relative to the field; for example, vertical, diagonal, . . .

Texture as a visual surface pattern; for example, striped, dotted, smooth, . . .

Surface quality; for example, specular, glossy, matte, lustrous, iridescent, . . .

Color as a property of the surface; for example, red, light blue, dark green, . . .

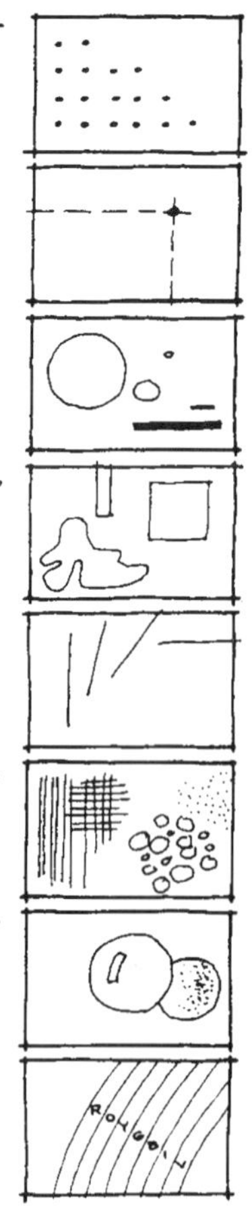

Fig10: Objects and their environment

fellow teacher, Itten, argued that the clear geometric form is the one most easily comprehended, its basic elements being the circle, square and the triangle, Fig 8. Form was also colour, he asserted, those of the spectrum being the ones most easily comprehended.

His colour classes were based on the foundation that there were seven distinct types of colour contrast: a simple contrast between one colour and another, a contrast between a colour of a light and a colour of a dark tone, between a warm colour and a cold, and between complementary colours. In place of the familiar 'colour wheel', with the colours of the spectrum around its periphery, he designed a colour sphere to describe all the relationships.

Kandinsky tried to rationalise the phenomenon of synthaesthesia, where as one sense is stimulated another reacts, and postulated the four major 'sounds' of colour contrasts as: warm and bright; warm and dark; cold and bright; cold and dark. 'Temperature' is the tendency of any colour towards yellow (absolute warmth) or blue (absolute coldness). Colours were also considered to have 'meanings' — yellow was earthly and blue heavenly, the former symbolising opposite feelings to the latter. Since green is made by mixing yellow and blue it creates balance and harmony, in his opinion. Kandinsky also believed in the basic elements of geometric form being the circle, square and triangle. The square he considered symbolised peace and the triangle vehemence.

The creation of a line from a point was thought to have its strength dependent on the number of forces applied to produce an end result, a single regular force producing a straight line while two or more forces produce angular or wavy lines. Vertical lines are 'warm' while horizontal lines are 'cold'. Another teacher, the prominent artist Paul Klee, tried to rationalise the 'movement' and direction of line. He produced the diagram in Fig 9 to compare active and 'passive' lines, between which limits it was said to be 'medial' — when defining a coherent form. He maintained that medial lines define either 'structural' or 'individual, forms. Structure is achieved if the same visual elements can be repeated infinitely whereas an individual form cannot have anything added or subtracted.

Self-training in form design

A valuable course of instruction for students of formalised design is provided by Peter Thiel[11] in an admirable American work of cross-disciplinary teaching. Of the many contributions within it from prominent design teachers, one which puts design into focus will be valuable to those who, for example, approach the subject from an engineering background: 'a design may be very beautiful, but it is not art ... the artist wants to make a visual statement ... the designer needs a client to present him a problem'.

The course sees five clear stages in the design process; preliminary design, detail design, approval, implementation and evaluation. The problem of perception is one which is treated in some depth on the course, starting from the understanding that the'material' of visual perception which enters our eyes is a relatively small spectrum of electro-magnetic radiation, emanating either from light sources, direct, or reflected by illuminated sources in the environment. The light energy is modulated, patterned in both space and time dimension, by interaction of the light sources in reflection, refraction and diffraction with respect to the atmosphere and environmental surfaces, Fig 10.

Without such modulation we would see 'nothing' and visual perception is thus the de-coding of the modulated information. This is illustrated by forms of camouflage, Fig 11, when the units of the patterned field have such a strong visual relationship with one another that we are unable to organise them into an alternative arrangement. The four types shown are termed somatolysis (body dissolution), disruptive colouration (dazzle painting), countershading and the suppression of form-revealing cast shadows.

Another useful concept is the bounded visual field which the human eye can perceive. Generally all the eye perceives is an arrangement of patches of colour, variously shaded. Perception of solid form is an experienced-learning process; we 'see' nothing but flat contours and only by a series of 'experiments' do we find, for example, that a stain of grey indicates the dark side of a solid substance or that a faint hue means that the object is far off. There is also the realisation that the eye's field is roughly oval in shape, extending some 150 degrees vertically and 180 degrees laterally, Fig 12. The construction for an equidistant hemispherical projection to simulate the eye's field of vision is shown in Fig 13. Main points to realise are that although the visual world is unbounded and uniformly clear, the actual visible field is bounded and has a centre-to-periphery gradient of clarity. The hemispherical projection is actually a fish-eye lens photograph, with a 180 degree vertical subtense and a uniformly clear field. Fig 14 shows a fish-eye photograph with a hemispherical projection alongside showing the basic pattern areas (BPAs). In this projection, the BPAs can be described as having different shapes and sizes; being in different positions; having different colours and textures; being vertical or horizontal and being either glossy or matte. This leads to the identification of the primary static visual attributes .

According to Whitford, one of the Bauhaus' greatest teachers was Laslo Moholy-Nagy who considered the camera-less photograph to be a valuable tool in design teaching. Photo-sensitive paper exposed to light will record the varied intensity of light sources in black, white and scales of grey. Laying solid objects on the paper will project shadows in various grey scales. The concept of 'figure' and back-'ground' was also crucial to understanding visual perception. When we attempt to penetrate a camouflage we try to regroup the basic pattern areas in terms of alternate figures on new grounds.

Fig 11: Forms of camouflage

Fig 12: Field of vision

Organic form

Many express the desirability of natural form so it is useful for the designer to rationalise the attributes of such forms. A good exercise is to consider the tideline jetsum which one encounters during a beach walk. The man-made items washed ashore from ship's waste will stand out a mile unless they are the product of an outstanding designer who can make good claim to the natural origin. The difference between the two classes of object, it is considered, is one of growth. As organic forms develop the effects of time on development are clear — the form is defined by the change in form. Environmental forces to which an object, or creature, is subjected in its lifetime eventually reach an equilibrium in the final form. Man-made objects which reflect natural forms are thus seen in designs for traditional ships, Fig 15.

An interesting exercise in design training is the use of grid-lines for the purpose of scaling up arbitrary shapes. The exercise is valuable not for the means of scaling up, which of course can be done

Given point P, a stationary point at eye level; line PF, the horizontal line of sight; and point 0, any point in the forward field of view;
to find 0', the hemispherical projection of point O, in a circle of any convenient radius H; with horizontal axis L'R', vertical axis Z'N', and center P':

1. Establish plane PSOF through point O and line PF. This intersects vertical plane LR-ZN in line PS.
2. Find β, angle APS in right triangle PAS, using true lengths of AP and AS.

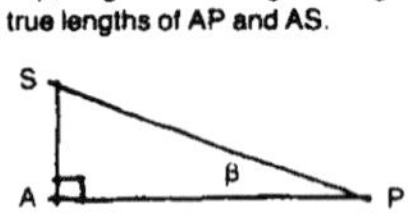

3. Find α, angle FPO in right triangle PFO, lying in plane PSOF; using true lengths of PF and FO. Note that FO = PS.

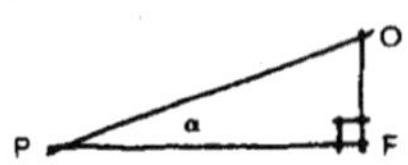

4. Draw radius H from center P' at angle β.
5. Lay out distance r from P' on H, where r = (α/90) H. This locates O'.

Note that when α = 90°, r = H, and that when α = 0°, r = 0. Note also that lines and edges parallel to PF will lie along a radius in the hemispherical projection.

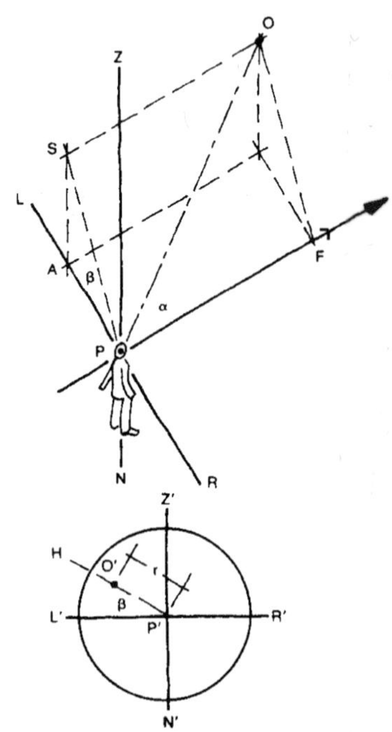

Fig 13: Construction for hemispherical projection

optically, but in helping to develop skills in estimating distances. Freehand 'fairing-in' of the enlarged contour is particularly instructive; reproducing points of inflexion where two curves of opposite slope join together, is also demanding of effort. Where curves join tangentially with straight lines the effect of the curve radius gradually increasing to infinity has to be reproduced and it is helpful for the student to think of the path of an aircraft coming into land as being a good analogy to the shape required. Such flowing transition curves are also those which approximate the ideal for vehicles moving through fluids, traditionally boats through water, but with increasing road speeds, also cars moving through air.

Conveying expression is another task for the designer and a visually aware person has to recognise the key visual features of, let's say, a steep rock, a willow tree or the colours of a sunset and the effects each has in putting over its expressive message. The columns of a temple, Fig 16, are expressive — owing to careful proportioning by their architects in classical times. Location, proportion and shape of the columns are all important in making their image contain the desired impression. The illustration is, in fact, a stored image supplied by Autodesk to purchases of their Autosketch bottom-of-the-range CAD program and also gives the proportions of the 'golden' rectangle upon which so many Georgian reproductions of classical buildings and styles were based.

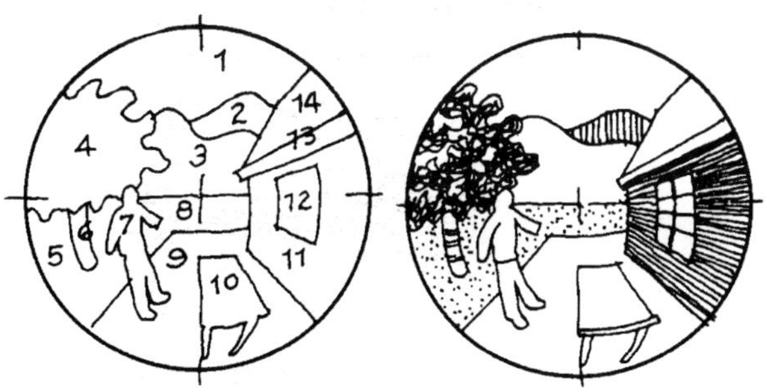

Fig 14: BPAs in fish-eye photograph

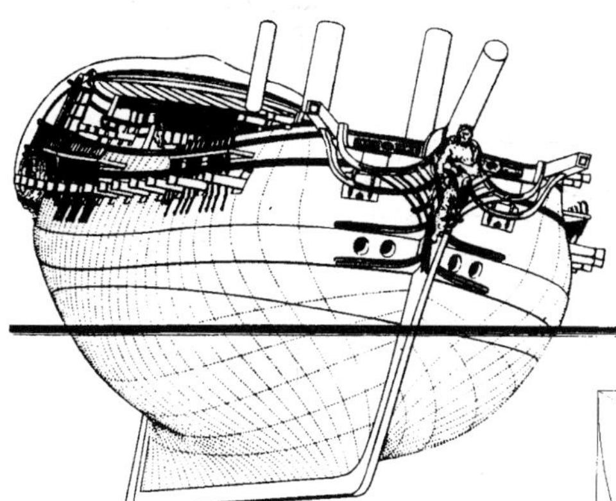

Fig 15: Natural form in classic ship

Gestalt psychologists are the originators of concepts that govern the organisation of objects within a visual field. Four laws are espoused; that of 'proximity' stating that the relative closeness of some units, compared with others at a greater distance, cause the closer elements to be seen together as a separate entity. The law of similarity concerns the tendencies of commonalities in the attributes of a number of separate elements to be related as a group. The law of continuance refers to the relating tendency of similarities in the changes in the attributes of a series of adjacent basic pattern areas. Finally, the law of closure deals with our perceptual tendency to group certain visual elements and, by `filling-in' the gaps between them, to establish one simpler larger form. The four rules are illustrated in Fig 17.

Colour attributes

Colour is said to be perceived in five different ways; the illuminant mode in which it is perceived as belonging to a source of light; the illumination mode based on illuminated objects casting shadows; the

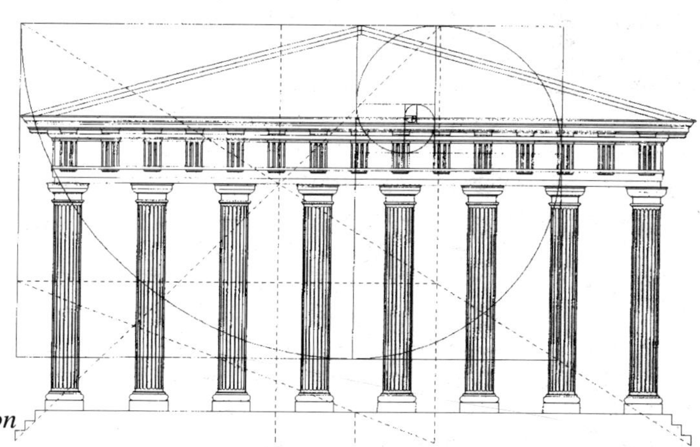

Fig 16: Classic proportion

surface mode in which colour is perceived as belonging to a surface; the volume mode involving passage of light through a semi-opaque mist such as smoke and finally, the film or aperture mode in which a colour is seen through an aperture, as in a mask, isolated from and unrelated to other colours. One of the best known systems of colour notation is due to Albert Munsell whose familiar hue-circle and hue, value and chroma scales are shown as Fig 18. This allows for 100 hues of the five predominant colours of the rainbow spectrum of white light. Extending vertically through the centre of the hue circle is the scale of reflectances, known as 'values'. Radiating out from the central core of colour solid are steps of saturation, known as 'chroma'. The scales are important for the correct specification of colour under standard conditions of illumination and viewing. A contemporary journal[12] gives a step-by-step process for choosing interior colours based on these concepts and taking in those of warmth, spaciousness and excitement level discussed earlier in this chapter. The question of surface texture is also crucial to the perception of colour. The directional pattern of light reflected from a surface depends on the degree of roughness of the surface and on the pattern of that roughness. In the extremes, 'diffuse' or 'specular' light will be reflected from a surface of random micro-facets or of a mirror-like finish. Three dimensional objects illuminated by sunlight, say, will also show a different pattern of illumination at a scale above that of surface facets — due to the chiaroscuro effect. This is because of the overall form of the object and its relationship with the light source. Fig 19 shows typical conditions of highlights and shades for different shaped objects.

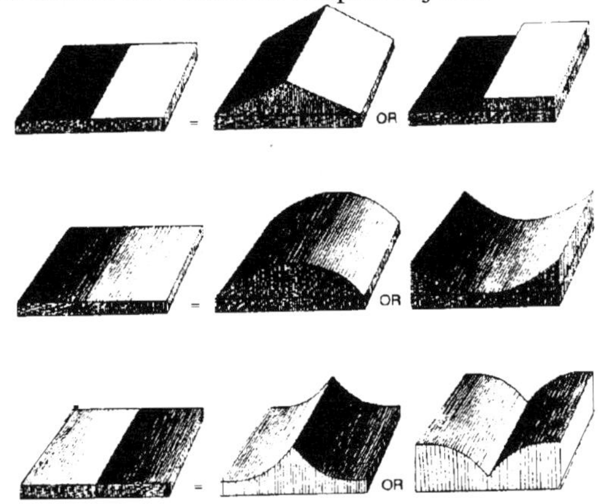

Fig 19: Highlights and shades

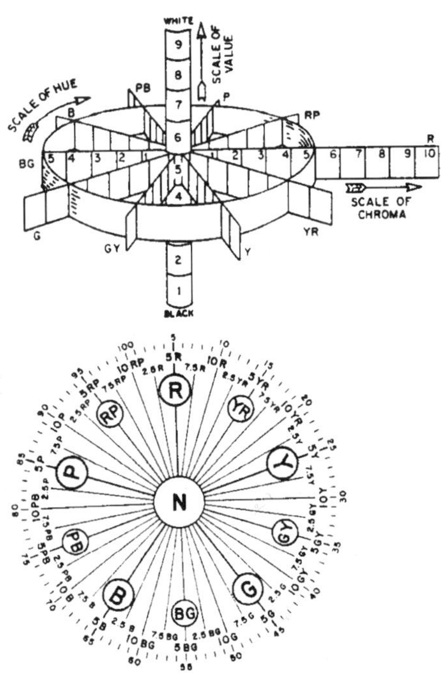

Fig 18: Munsell hue-circle and colour scales

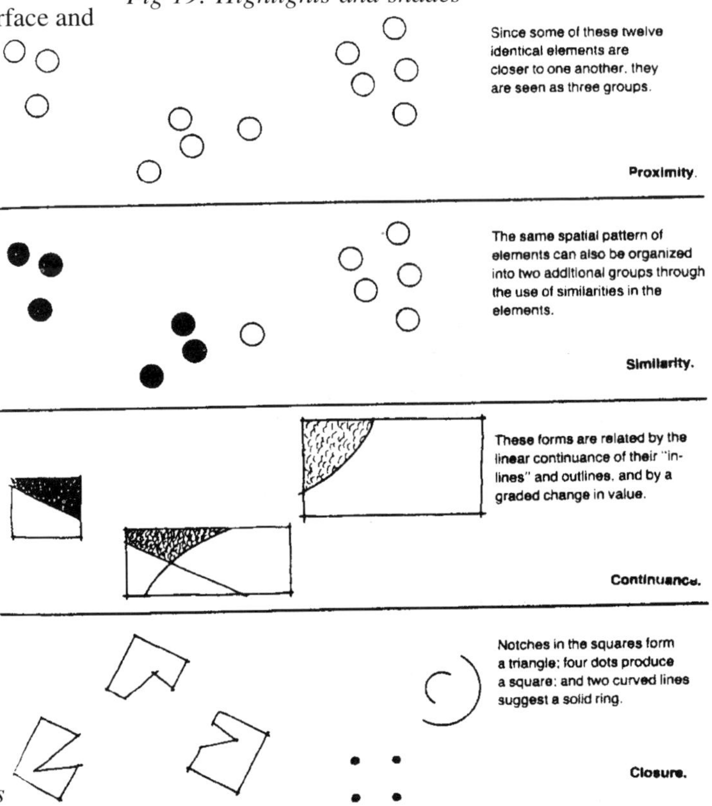

Fig 17: Gestalt laws

Evolution of body shapes

Fig 20: The change from horse to horseless brought barely perceptible changes in body style

Fig 21: Compared with the upper view, the lower one shows the advantage of making key lines intersect at a perspective point

Fig 22: Early attempts at streamlining failed to arrange the bulk of the vehicle towards its leading edge

The relationship between utility and aesthetics in design has plagued automotive designers since the beginnings of the styling revolution in the 1920s. In a paper[13] based on earlier writings in 1932, a UK-resident GM MD talks of the 'fashionable' development of architectural lines of vehicles, during the 1930s.

The author points out that there was surprisingly little evolution in the development from horse-to horseless-carriage, Fig 20, with unequal diameter front/rear wheels, lack of windscreens and absence of front doors, persisting to well into the second decade of automobile production. And it was not until 1919 that the closed saloon predominated over the open-car. Generally the early trends in design were to more simple shapes for the faster-moving bodies that were developed, the same trend being noticeable, too, in design details such as headlamps, which were to develop from historic lantern shapes to near tear-drop designs. Surprise was expressed in 1920s-30s car designs, by the author, that the golden rule of fast-moving objects being designed with horizontal lines predominating, was often overlooked. The rule, that the human eye appreciates repetition to a very considerable degree, was also being broken by such incongruities as rectangular-shaped radiators being grafted on to rounded body shells.

Two further rules which were continually broken, a. that the eye appreciates the appearance of perspective in the arrangement of lines and b. that generally the eye discourages sudden changes in the direction of the major design-lines, became clear in examining contemporary vehicles of the time. In the first case, the top contour line of the car, the window-sill line and the running board line did not meet at a point, Fig 21. In the second case, he points out that the front view of the Rolls-Royce radiator has the required repetition of equal 'columns', like a Greek temple; however, its side view, juxtaposed against a rounded 'streamline' body, looks incongruous.

The author also observed that Nature's law, of having fast-moving objects with most of their bulk near the leading edge, was also frequently broken, noticeable in even the most streamline contemporary body — shown by the author, Fig 22. This lead him to propose the surprisingly advanced layout, Fig 23, for the early 1930s. Here three occupants use the front seats and only two, the rear. Rear wings are fared into the body in an interesting way and use of a rear engine location allows optimum streamlining. Another streamline design by the author, Fig 24, exploits the use of catenary curves which simplified early drafting procedures. Even in those early days, however, the author warned of the inevitable similarity in shapes between future models, as aerodynamic perfection was approached.

The author freely admits the streamlining of cars in the thirties owed more to opportunistic marketing than measured drag reduction or side-wind stability and it is interesting to note how the Americanisation of the UK motor industry had started in earnest well before World War 2. While the author pays tribute to continental design such as Citroen's distinctive aesthetic, Fig 25, in such cars as the Light Fifteen, exploiting the early recognised advantage of front-engine/front-drive, he has little to say about indigenous UK makers of the 1930s.

While the mainstream makers such as Austin and Morris continued to ape the lines of the up-market 'carriage trade', it was left to smaller specialists such as Riley, Fig 26, and Rover to develop more individualistic styles.

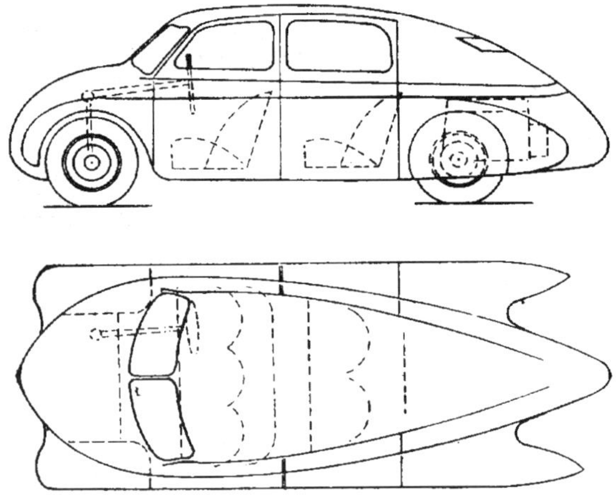

Fig 23: Moving the predominant bulk to the leading edge of the vehicle

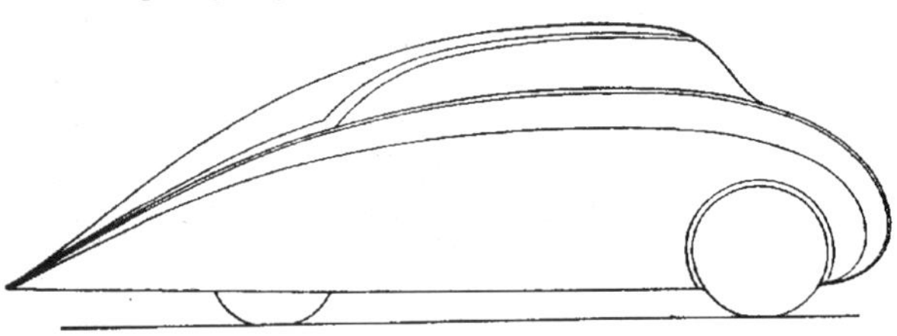

Fig 24: Streamlining using catenary curves

Fig 25: Citroen's use of "traction avant"

1944 onwards

The post-war period industrial design has been examined by Boris Ford[14] who argues that the term 'industrial design' was hardly used before the war and is formed from a fusion of the pre-war disciplines of industrial art and engineering design. Initially there was the belief that, like the Supermarine Spitfire, "if it works right, it should look right" — quality should be perceptible. In the automotive world many would argue that pre-war products from such specialist makers as Aston-Martin, Fig 27, and Frazer-Nash fell

Fig 26: Mass-market Morris (left) compared with up-market Riley

into this category.

In the UK, 1944 saw the establishment of the Council of Industrial Design but the exhibition "Britain can make it" in 1946 was thought by the author to show disappointing results, from an industry not yet re-adjusted from wartime working. The Festival of Britain in 1951 was to prove more optimistic and considerable successes were seen in such fields as furniture manufacture. There was also considerable success in the design consultancy sector, with some world-class UK firms in being: among which the author notes Ogle Design and its work on the Reliant Scimitar GTE and Plaxtons coaches in the 1960s.

However, the author noted that the development of the world market for motor vehicles mitigated against considering auto-design in purely national terms and top British designers were found working for such companies as Citroen and Audi while some British companies adopted the American practice of styling. Austin used Raymond Loewy's US offices and later the Italian company Pininfarina. The author praises Alex Issigonis for his Morris Minor which was quiet and spacious compared with its rival Volkswagen Beetle, but was unreliable in export markets, demonstrating British design's "greatest weakness in commercial terms, namely, an insularity that frequently rendered it incapable of understanding conditions other than its own". Similarly the enormously popular Mini "could not be produced in quantities or at a cost that translated the vehicle's astonishing status into profit". When BL finally began to take automobile design seriously, "its management were still incapable of exploiting superb ideas". The author consider's David Bache's Range Rover combined the best in the new industrial design and older engineering disciplines but was never produced in quantities capable of meeting demand and soon overseas competition produced their own versions, having greater reliability. Fig 28, however, shows there was very considreable design talent at Triumph even before Michelotti was engaged in the BL era.

Fig 27: 1934 Aston-Martin Le Mans

Fig 28: Triumph sports cars before and after BL acquistion and Italian styling. Top: 1938 Dolomite; Centre: 1946 Roadster; Bottom: 1974 Spitfire

CAD and surface development

The advent of CAD has revolutionised the designers workplace. A complete CAD system involves several layers of software between the electronics and the user starting with, at the computing 'machine' itself, the operating system. This does basic 'housework' such as looking after the disc drives, which hold the next layer up, accepting keyed input and arranging screen display. In most micro-computers (PCs) the operating system is incorporated into the basic 'interpreter' circuitry which converts keystroke data into machine code understandable by the central processor unit (CPU).

Other layers of software are the 'programming language' and the 'application program'. Levels of programming language vary according to similarity with the language of the user (high, such as BASIC) or the machine (low, an assembler language). A good programmer is thus a person good at language — not necessarily a conventional engineer. In some cases a programmer may use a 'compiler' to turn the program language into binary instructions for direct access to the operating system.

Graphics are fundamental to both CAD and surface development; they are described in quality terms by the 'resolution' of either screen display or printed image — referring to the numbers of lines in the display times the number of dots per line. A bank of random access memory (RAM) is often reserved for graphics. Normally the graphics display is controlled by a dedicated microprocessor called the 'video controller'; each bit of the RAM corresponds to one dot on the screen which is switched on or off (lit up) by a 1 or an 0 from the CPU — the technique of bit mapping. A more accurate representation, used in 'drawing' rather than 'painting' is vector graphics in which lines are described mathematically and can be accurately scaled.

Only by 1980 had even the mini-computer-based CAD systems acquired the ability to successfully model a part in 3-D. Two of the most capable were considered to be DUCT and Unigraphics. By 1982 the serious commercialisations of desk-top microcomputers, plus the advent of networked engineering workstations, changed the face of CAD. The AutoCAD package has become an important contender with good drafting facilities backed by after-sales support. The Autodesk company have fully exploited the PC's potential and realised a system which offers both 3-D surface and full solid-modelling. Memory limitations in the future for many systems are in prospect of being overcome by new technology introductions such as writeable CD-ROM disks.

Key elements of a CAD system are devices for input, display, storage and output. A 'mouse', joystick controller or light pen might typically be used for input; display screens are usually of the raster-scan or the refresh-tube type; storage is either by machine memory (RAM) or by magnetic (hard or floppy) disc; output may be to an impact (dot-matrix), non-impact (bubble-jet, laser) printer or a plotter. Basic elements of a drawing are points, straight lines, arcs and curves. Positioning is achieved either by using a cursor, in conjunction with an accurate background reference grid to which a point is 'snapped to' — or numerically, using cartesian or polar co-ordinates, Fig 29. Some CAD facilities available with a typical package are shown in Fig 30. It is also possible to construct associated databases of useful specificational information, also libraries of symbols and frequently used detail components. Layering is another useful facility — comparable with a number of transparent printed sheets in conventional drawing — and editing functions are a dramatic improvement over rubber erasers.

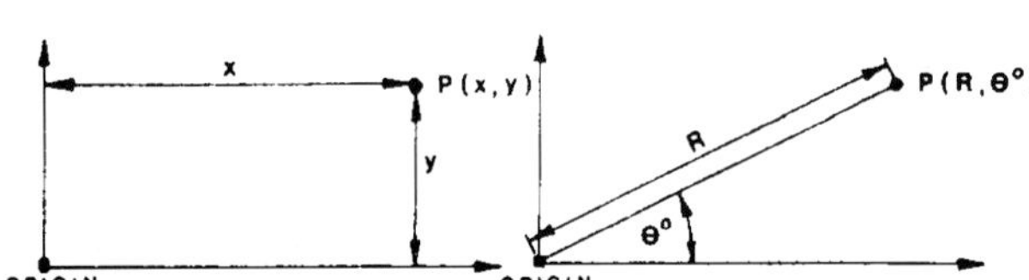

Fig 29: Cartesian and polar co-ordinates

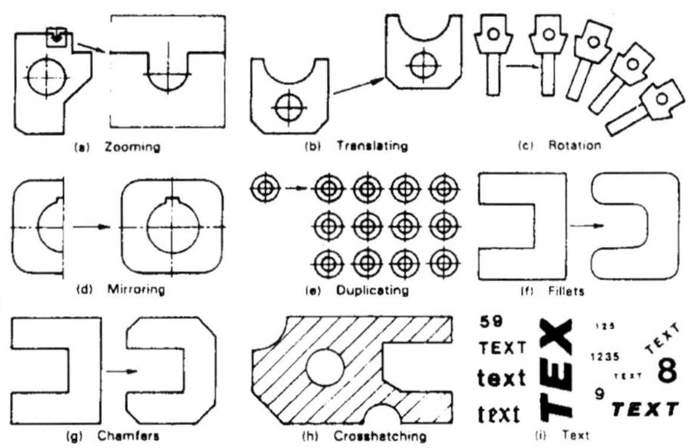

Fig 30: Some facilities available with CAD

With an adept knowledge of co-ordinate geometry, it is possible to produce clear drawings with a very modest desk-top computer provided it is properly interfaced with a suitable plotter. Working would inevitably be slow, however, and happily the computer-aided-drafting programs carry out the co-ordinate geometry automatically so that spatial points can be defined accurately and joined with lines or pre-defined mathematical curves at great speed. The computer performs graphical geometry calculations for tangent lines to curves, optimised fillet radii and all manner of constructions. It is thus only a short step to plot the motion of linkage mechanisms and check large assembies of parts for foul conditions.

CAD has, of course, only a limited use in the creative side of design and so before making a system purchase, it should be realised that the system is going to be of most benefit after the creative work has been carried out. Creative design is seen as a process in which ideas are generated, sorted and then tested — followed by a process of modification until the desired result is achieved. It is estimated that the designer spends 20% of his or her time searching for

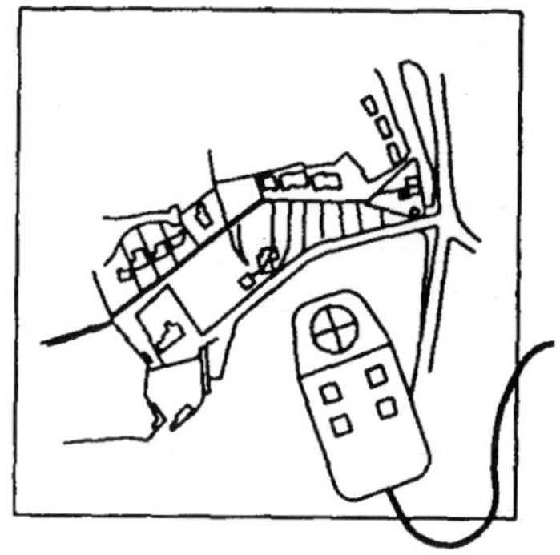

Fig 32: Digitising tablet

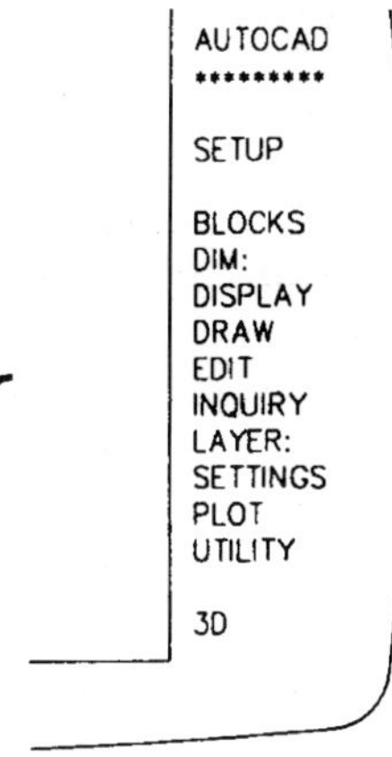

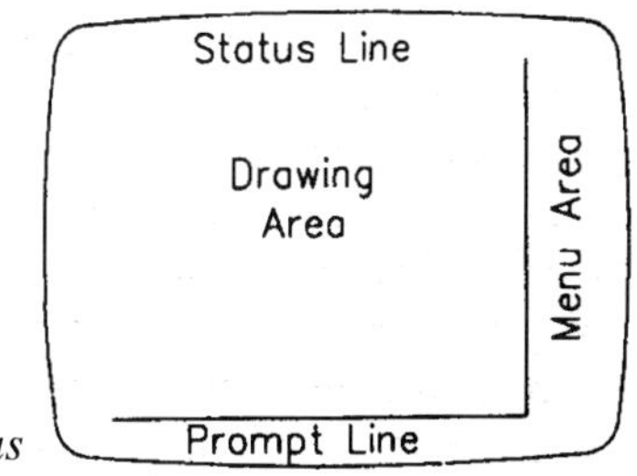

Fig 33: Screen areas

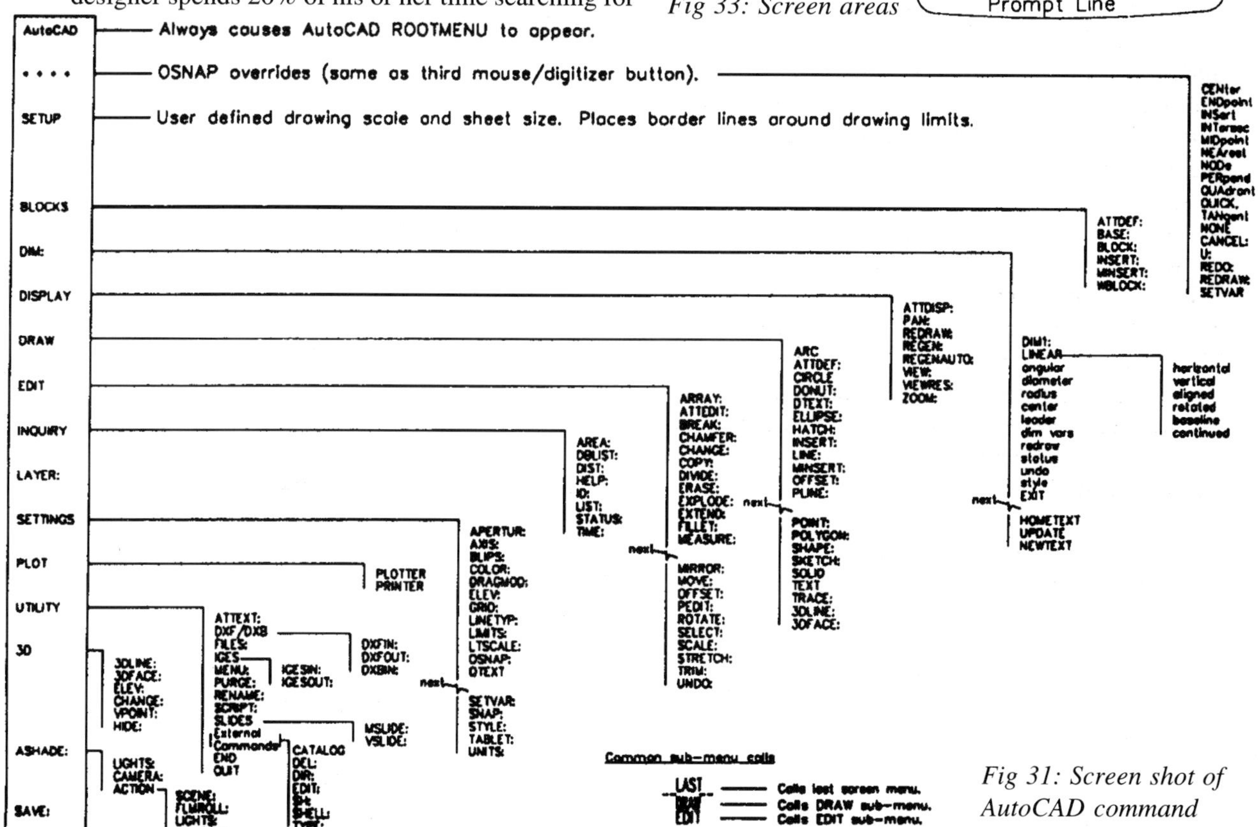

Fig 31: Screen shot of AutoCAD command menu

information, 65 % generating and sketching possible solutions and 5 % carrying out calculations before the stage of carrying out computer-aided-drafting. There is a further 10% of time writing reports to convince others of the viability of the design. The multi-tasking ability of the better desk-top computers does at least allow the more or less simultaneous use of programs for word-processing, calculation and drafting as well as calling on databases of information. Computer-aided-design usually implies carrying out other functions as well as drafting.

A typical drafting-only package uses a rollerball 3-button mouse with 'pick' button to fix a point on the drawing screen, for starting a line. A 'return' button for repeating commands and a third for accessing the command menu completes the trio; Fig 31 shows the AutoCAD comand menu. A digitising tablet, akin to a 2-D scanner, can also be used where points on a 'hard-copy' drawing need to be read into the picture, Fig 32.

The AutoCAD screen has four main areas, Fig 33, the menu being based on a tree structure, root-menu, which is shown inset. To join a new drawing element on to a specific feature of the current drawing, an 'Object SNAP' facility exists whereby a connection can be made to the closest line on a grid underlying the drawing. A feature which is special to the program is Polylines, Fig 34, which enables the drawing of a series of interconnected lines and/or arcs. A further option allows specific widths to be specified. After construction, the combination can be treated as a single object and even a smooth curve fitted to the constructed shape if required. In a similar way to word-processing, 'blocks' of drawings, as with blocks of text, can be separately saved and introduced as required elsewhere. Objects can be measured to obtain areas for use in calculation — merely by going through the kind of sequence shown in Fig 35. There are also facilities for ordering arrays, Fig 36, and for divide, measure, Fig 37 and scaling, Fig 38. Dimensioning, of course, is automatic and the addition of textural annotation can be added in the same way as a word-processing or desk-top publishing system.

The Windows environment has made AutoCAD more accessible to less experienced users. This is well described by Yarwood[15]. The screen is shown in Fig 39 and the use of toolbars obviates the need for working with a draughtsman's tablet, since these

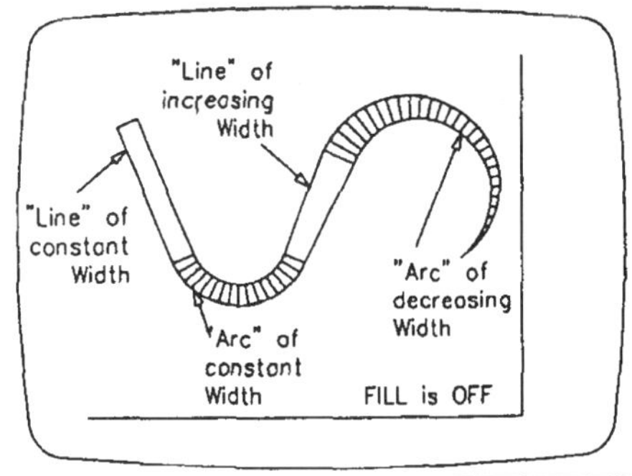

Fig 34: Polylines

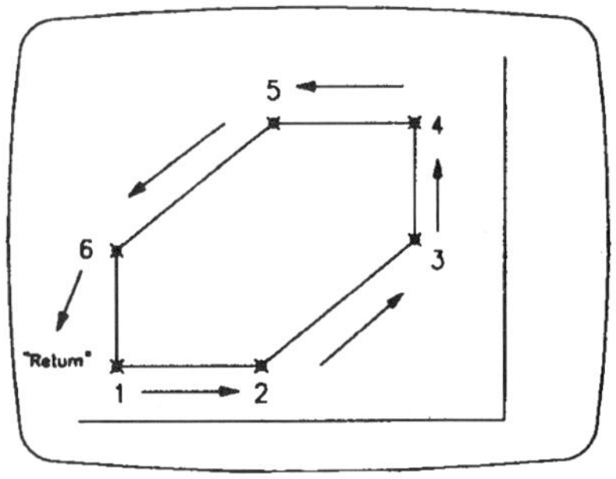

Fig 35: Area measurement

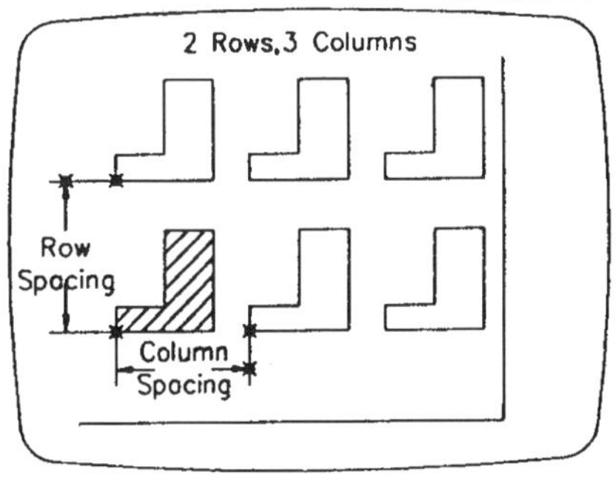

Fig 36: Ordering arrays

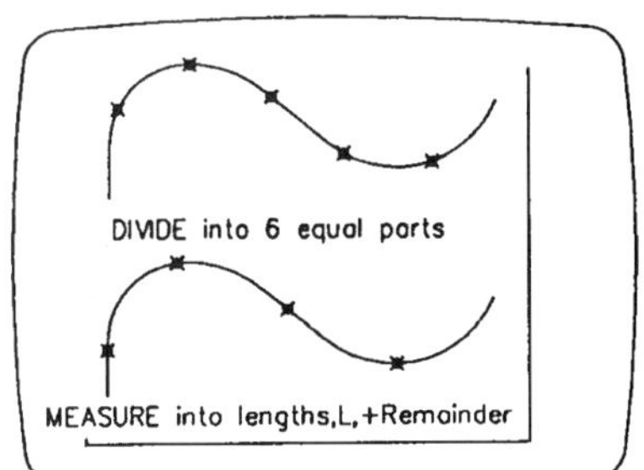

Fig 37: Divide and measure

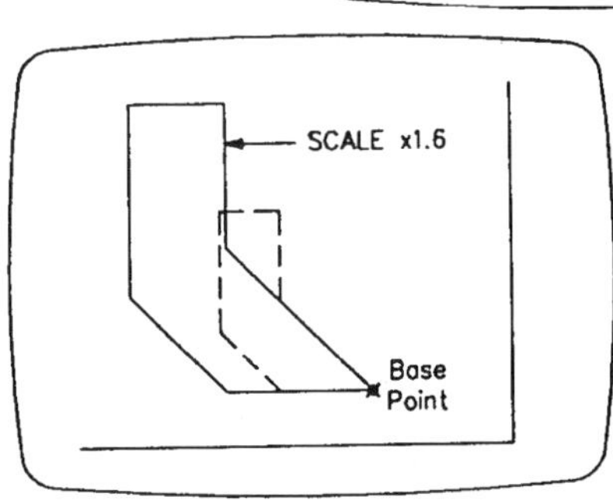

Fig 38: Scaling

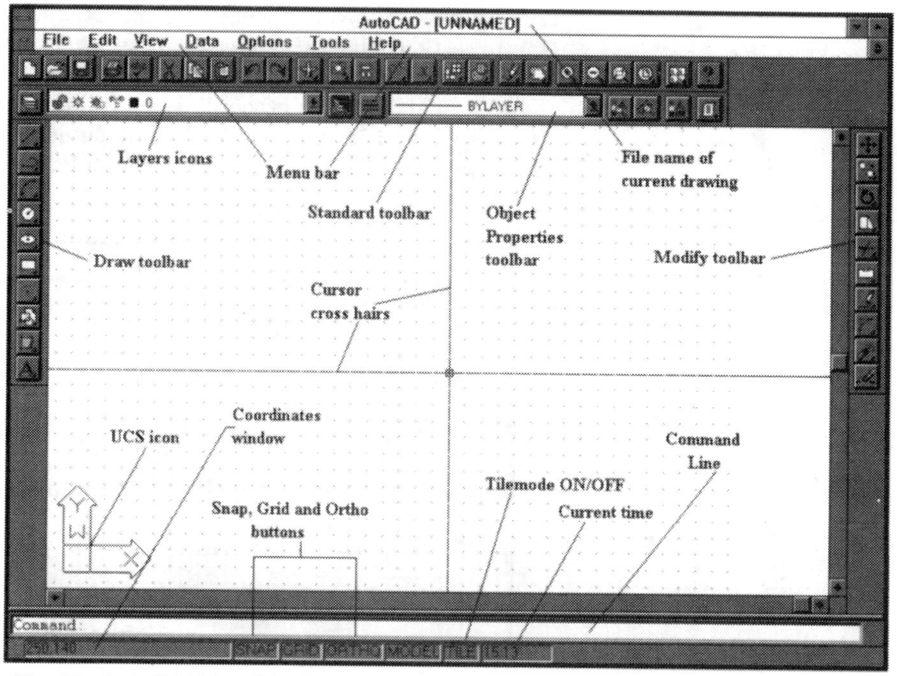

Fig 39: AutoCAD for Windows screen

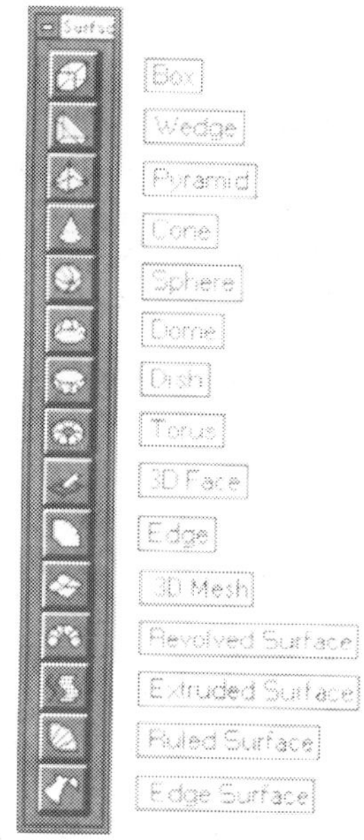

Fig 40: Surfaces toolbar

groups of icons representing tools or commands are now on-screen. The comand line is seen at the base of the screen and prompts the user to give his next instruction, having recorded previous ones, even those applied freehand. There are also pull-down menus and dialogue boxes with further commands, typical of a graphical user interface. Movement of the mouse can be made to shift the axis system for drawing, either orthographic or isometric. A surfaces toolbar, Fig 40, is used for constructing 3-D drawings and standard tools construct cubes, pyramids, cones, spheres, domes and torus, using surface meshes. Either a revolved or extruded surface tool can be used to generate more complex solids. Ruled and edge surface tools allow primative surface-development excercises to be carried out. There are also user-co-ordinate-system (UCS) and solids toolbars, Fig 41, the former allowing the graphics window to be set at any angle to the co-ordinate system. The latter allows almost any solid model to be constructed, with sufficient practice, and a render toolbar can be used to produce realistic, photo-like images.

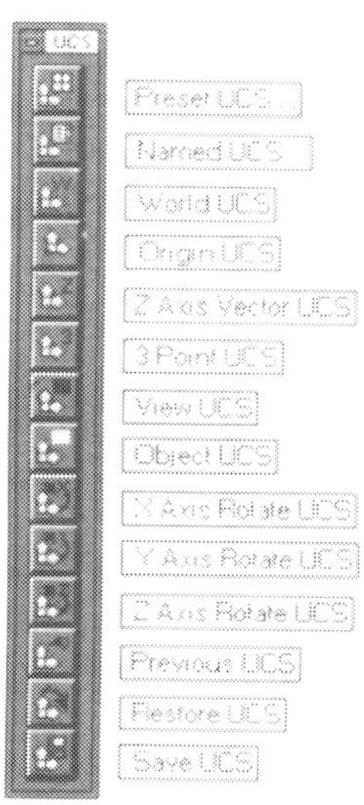

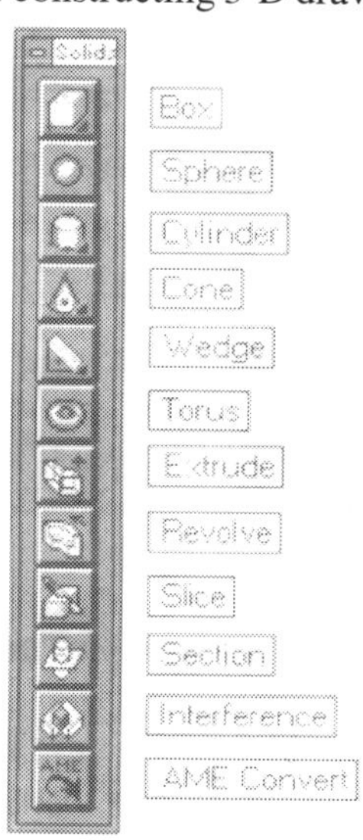

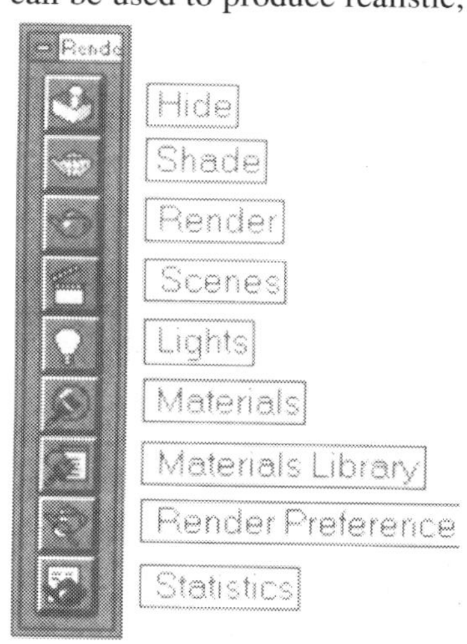

Fig 41: UCS, solids and render toolbars

Add-on, simplified and specialist facilities

A technique under discussion for CAD is the so-called 'plastic modeller', in which relevant dimensions of objects can be specified as variables rather than numerical values. Thus design work can proceed with loosely designed shapes whose parameters would be specified as a result of calculation or 'fit' to another object — necessitating interfacing with another graphical database.

The fact that it is actually more difficult to operate a sophisticated CAD system than to draw comparatively simple shapes on a home-computer — with the aid of a well produced manual — begs the important question as to at what level does a company, of a particular size and complexity of design involvement, enter the CAD market. This is a question almost impossible to answer by the vendors themselves and for which independent advice is very hard to come by. If an office computer exists at the workplace, it could be very worthwhile experimenting with simple programs if the drafting to be done does not extend into the more complex area of surface-development.

A program such as MLD2 CAD has a 'parametric' facility whereby a drawing's standardised elements can be stored and used at different scales and even different key dimensions for different proportions. There is facility, too, for automatic removal or presentation of hidden detail — facilitating construction of assembly drawings from detail ones. There are also window commands which allow magnification or reduction of separate sections in four areas of the screen, Fig 42. It has links to other proprietary software for drawing and data transfer permitting either finite element structural analysis or computer-aided manufacture by the generation of defined cutter paths.

CAD packages differ from drawing ones in several respects and have a clear concept of the end-product in their make up. Working speed is, of course, increased by having the potential to produce libraries of standard components and symbols. However, there is the potential problem that 'screen refresh' can take a long time in view of the time taken for the program to redraw the complex geometric

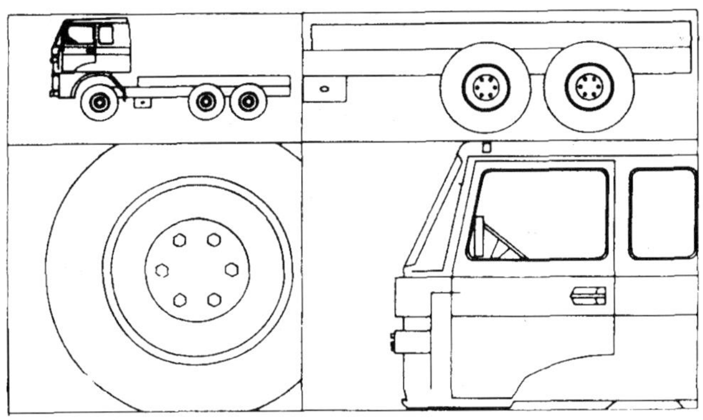

Fig 42: MLD2 windows and parametric design

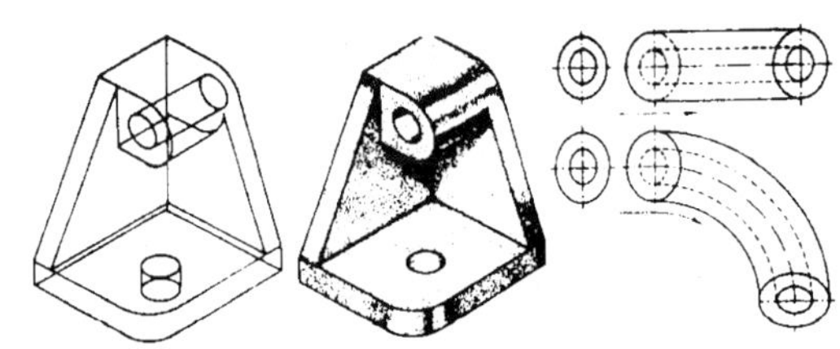

Fig 43: Wire-frame, solid and 2.5D models

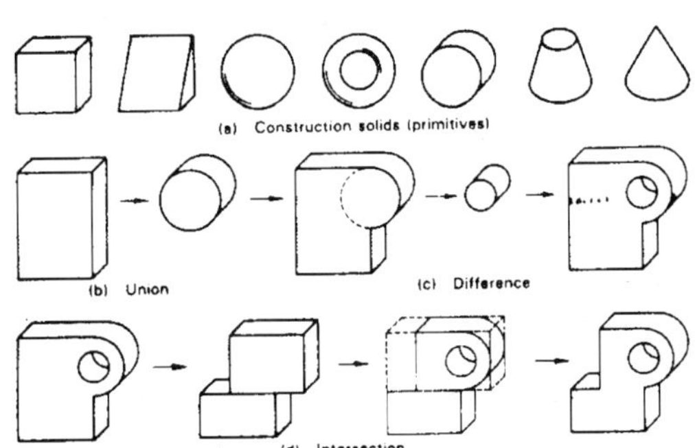

Fig 44: 3D modelling

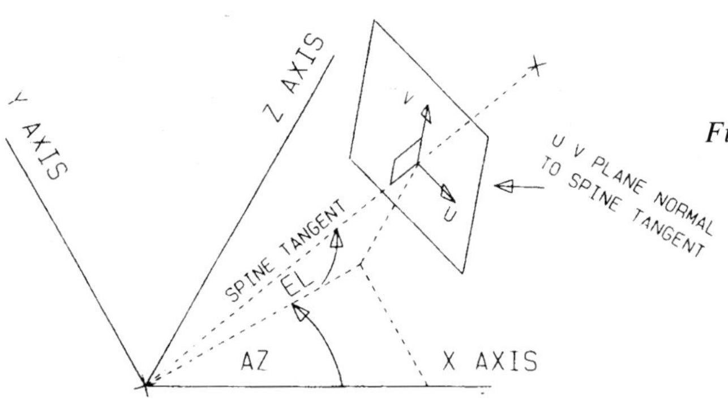

The relationship between the UV and XYZ axes. When looking forwards along the spine, the U axis is to the right of the parallel to the XY plane. The V axis is orthogonal to the U axis and the spine tangent while being as upright as possible.

Fig 45: Spine concept

constructions on each occasion. In advancing from 2- to 3-D facility, the simpler programs have only a wireframe facility and no 'solid' visualisation. Often a rotation feature is provided whereby the current viewing position may be rotated about the azimuth and elevation directions of the drawing in any combination. In the absence of mouse control, various key combinations move the cursor along the three co-ordinate axes. Points can usually also be defined by direct keyboard input.

3-D modelling/surface development

Three-dimensional modelling involves the spatial image being drawn in a pictorial projection using x-y-z co-ordinate geometry — and stored in memory. It can be recalled as either pictorial or orthogonal projection based on several 2-D views. It can also be represented by only its edges (wire-frame) to minimise computer memory as well as in solid form. Further a so-called 2.5 D presentation, Fig 43, can be used in which a wire-frame 2-D shape can be translated and rotated into a 3-D shape. Solid modelling is carried out with the processes of 'union', 'difference' and 'intersection', Fig 44. Surface modelling lies between solid and wire-frame. The initial wire-frame has its gaps filled by flat or rounded surfaces which can then be shaded. Complex surfaces cannot be simply represented by geometrically simple solids and so a surface therefore has to be developed by defining and joining curves. The 'complexity' of the surface refers to its underlying mathematical description. For computerising this process, Delcam's DUCT is a valuable package which was first developed at Cambridge University in the late 1960s. One of its major claims is its ability to direct the cutter-path of a die machining tool with a reliability sometimes not found in rival CADCAM systems.

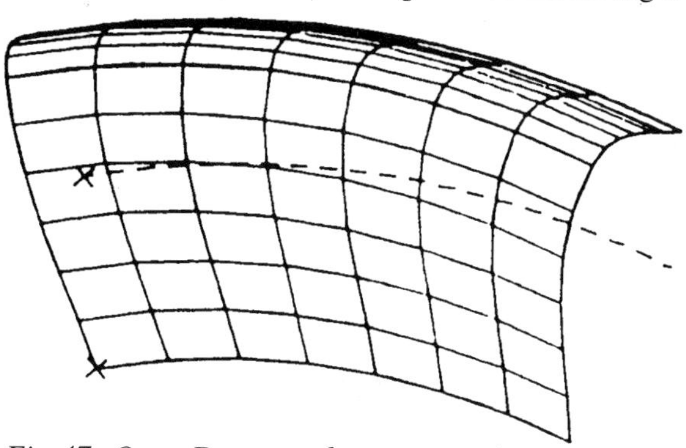

Fig 47: Open Duct panel representation

Basically it is a modeller working in terms of centre lines and sections. A straight or curved reference line, the Spine, has sections normal to it, Fig 45, and is defined at least by any two points in x,y,z co-ordinates and additionally by an angle if required. Successive sections are described in u,v co-ordinates with axes parallel to x,y. Between the data points on the sections, the program uses the Bezier principle whereby the curve is defined in terms of points and associated control points. So-called 'strength factors', Fig 46, are used in conjunction with angles which can be specified in two directions, at the data points, to describe the surface. Open ducts, Fig 47, singly or in combination, allow body-panel type surfaces to be modelled.

The latest version is DUCT 5 which is a direct surface modeller said to be faster than wire-frame based methods. It incorporates automatic blending, trimming and filleting of intersecting surfaces and a relational database is available which allows surfaces to be assembled together. Areas, volumes and masses of completed panels can also be calculated.

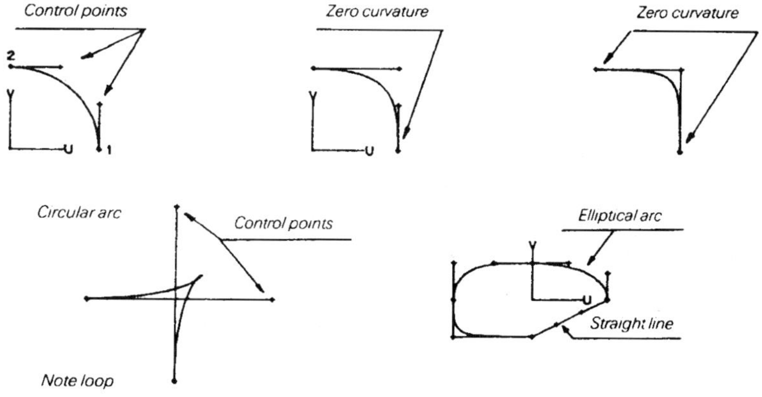

Fig 46: Strength factors

Specialist surface modelling

Surface modelling lies between solid and wire-frame modelling, the surface being developed by defining and joining curves. The use of the DUCT 5 package for this purpose has been extended by derivation 5.3 which has enhanced CAD facilities as well as a refined CAM side of the package. One of its potential applications is for the complex surfaces of moulded rear-lamp lenses and similar body hardware.

Enhancements include Ductshade for colour shaded images, Ductflow for plastics flow analysis, Ductmill and Ductviewmill for machining visualisation. There are also new facilities for parametric wireframe, surface (Fig 48) and solid modelling which allow the designer to develop and fully define almost any shape, within a package particularly famed for its CAM facilities. There is also an ability for interactive surface styling which optimises accuracy, speed and ease of design modification plus an intelligent drafting system based on active dimensions, with full associativity to the surface model. An enhance technique for generation of NC data to provide NC data for up to five axis control of CNC machine tools is also featured.

Design Using Computer Technology (DUCT) works in a similar way to a design draughtsman, in terms of centre lines and sections normal to these. A basic 'duct' consists of a reference line, the spine, and the sections normal to it . The spine can be straight or curved and is defined by at least two points in x,y and z co-ordinates; angles can also define the direction of the spine through any point. Each duct section must have the same number of points, where each point is defined in terms of U-V co-ordinates, the V axis tending towards the zed one. The shape of the section between two points is defined by the tangent angle and strength factor at each point, as illustrated in the earlier description.

Bezier curves are used, in which definition is by data points associated with control points, the form of the curve between data points being defined by control points. Duct surfaces are not sensitive to irregular spacings of patch corner points and the designer is free to specify the angles in two directions, at data points. A single surface will model a complex form for subsequent die-cutting or pattern-making. The package can also import surfaces from other CAD systems, Fig 49.

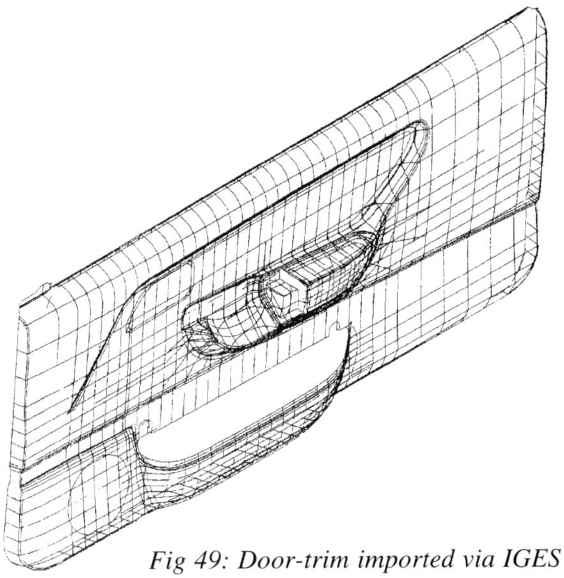

Fig 49: Door-trim imported via IGES

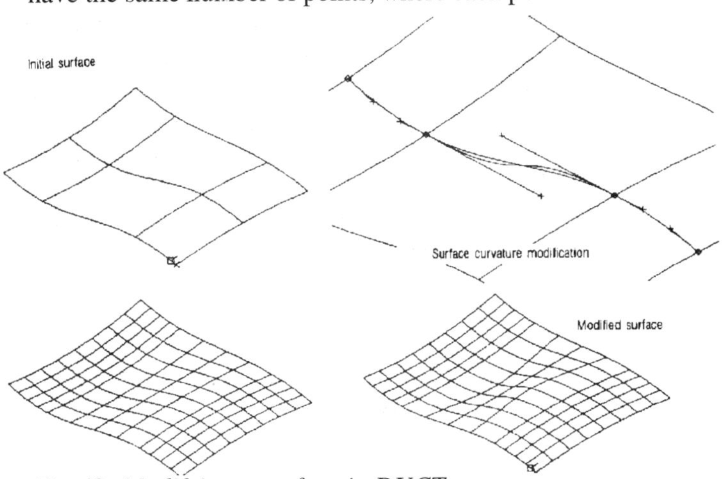

Fig 48: Modifying a surface in DUCT

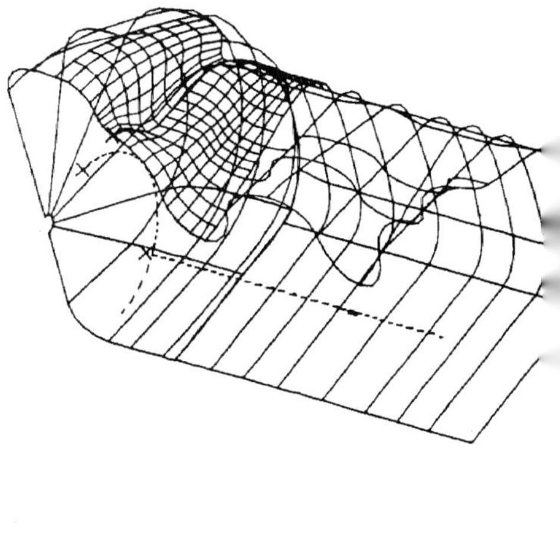

Fig 50: Combination of two open DUCTS

Fig 50 shows a combination of two open ducts which might be used to define a particularly awkward pressing. While single ducts will rarely describe such complex shapes, the ability to have either continuity or discontinuity of angles, both around the sections and along the spine, makes for considerable flexibility. Matching of ducts along their edges is achieved by interrogative routines, perhaps one of the most powerful aspect of the program. Curve fitting is used to join points representing a surface description of a part involving interpenetrating surfaces.

Fig 51: Scooter styled with Duct Designer

EXPAND, to provide contraction allowance

ROTATE, to place it in the right position.

ED VIEW 1 – edit view to give first angle projection.

SET ISometric – to select an appropriate scale and viewing angle to show the best image of the component and the cutter.

LIMit DUCT – to define the boundaries between which machining is to be carried out.

EDit RADius to define the cutter.

Edit TIP RADius to define the cutter tip radius (any tool between a ball nosed cutter and a slot drill can be chosen).

Set feedRATE, plunge RATE and spindle speed (FRAT, PRAT and RPM)

Set TOLerance and CUsp tolerance to define number of cutter moves and passes to provide required finish.

THickness allowance for finishing (by hand or by machine) or for taking into account spark erosion gaps.

Set SAFE Z height to define the clear plane.

Finally, PRINT TOOL to provide a list of current machining parameter settings.

Fig 52: Cutter-path generation instructions

An intelligent wireframe constructor eliminates the need for laborious construction geometry; just by defining a convenient workplane allows geometry to be created in any orientation. A trimming facility allows complex parts to be trimmed to boundaries; there is no limit to the number of surfaces used to define a part and the system automatically ensures continuity across patch boundaries. A true rolling ball fillet can also be automatically generated between surfaces, with variable radius if required. A single shell can be created from several surfaces and it can then be manipulated as an entity so that the advantages of solid modelling are combined with surface design.

Product geometry for a computer model can be created from information from several different sources. Firstly a concept design by free-form drawing of curves, and dynamic manipulation of surfaces, can be produced. Alternatively a standard 2D engineering drawing can have 3D geometry created from it or a physical model can be scanned, or co-ordinately measured, to input sufficient key dimensions. In the first case making use of the direct surface construction technique can be used quickly to define complex forms. For product visualisation a colour image can be quickly rendered and pattern mapping effected.

For analysis work, the integral finite-element mesh generator can be applied for stress calculations and interfaces exist with such packages as NASTRAN and PAFEC. The generator creates plate elements which may be beam with 2 or 3 points, triangular with 3 or 6 points or quadrilateral with 4 or 8 points. Block ones can also be created from the surface elements. For styling work, Duct Designer is especially useful for design of products including curved surfaces which cannot be unambiguously represented in 2D, Fig 51. For relief imaging which allows the raising of 2-D badge and nameplate patterns in 3-D relief, Artcam permits mould-tools to be made with the relief incorporated. Separate degrees of relief can be calculated for each colour of the pattern and parameters such as start height, limit height and wall height can be separately defined.

Computer-aided manufacture

Cutter paths for machining of pressing dies or moulding patterns may be generated automatically to cut either along the spine or around the sections. Typical instructions are listed in Fig 52. A cutter location data file will be produced from this information and the cutter path drawn on screen. Finally this file is processed for the particular machine tool controller, outside the program. Roughing cutter paths require more thought than those for finish machining. Two basic strategies were supplied in the original program which can be combined to form a third. The first is to machine to-and-fro, the cutter lifting as required to avoid obstacles; the second is contouring, Fig 53. According to the material being cut the tool will either plunge or ramp into the material as it changes from one level to the next Split faces are generated automatically along the split lines. The new roughing module Ductrough allows the shape for machining to be loaded and NC data to generated as a picture file.

The wireframe constructor now has a method for parametric modelling based on active dimensions. Thus the geometry of any part is determined by a set of parameters which are updated by the change of any one dimension. Also, primitive surfaces created from the wireframe will be modified by altering the dimensions on the wireframe geometry from which they were created. Methods have also been added to create surfaces directly from a wireframe, useful to tooling engineers who may receive information from clients as just wireframe data and partially surfaced models, Fig 54. The shells solid modelling facility is more robust in that it is more tolerant of gaps between surfaces and of those which are not actually trimmed. Unlike conventional solid models the shell does not have to enclose the volume completely. On the CAM side, 5.3 can undertake more standard tasks automatically such as calculation of area, volume and mass; and determining split lines and draft angles. Improved methods for interactive filleting help combine surface features; it will fillet around a model continuously until it arrives at a junction requiring a decision. When this is made it continues automatically and filleting between sur-

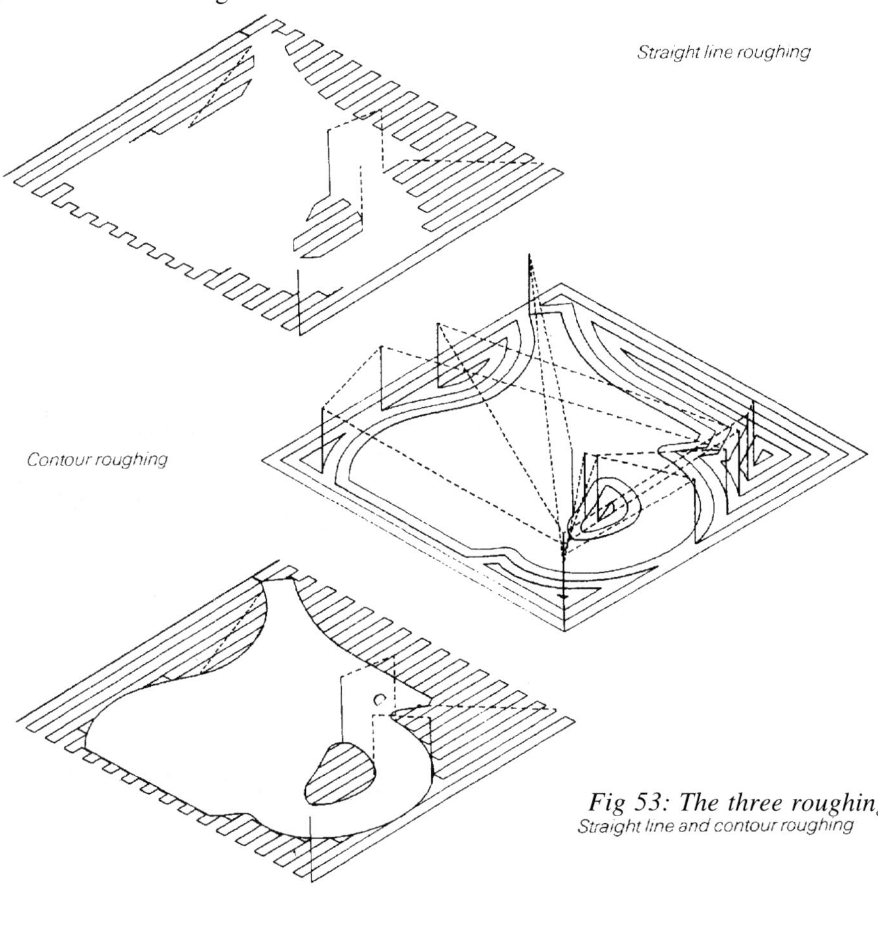

Fig 53: The three roughing strategies
Straight line and contour roughing

faces separated by gaps is now possible. Milling strategies include raster, multiple offsetting, user defined z-height or climb milling, Fig 55. The program can also automatically pre-drill blocks where plunge moves are to take place or holes may be cleared by ramping between levels. Pocket milling has been introduced; this allows individual areas to be cleared in sequence instead of machining the whole job level by level. In finish machining a new shallow region option allows regions with a gradual slope angle to be machined smoothly while, in contrast, the constant z-height option ensures that a good surface finish can be obtained on parts with steep surfaces.

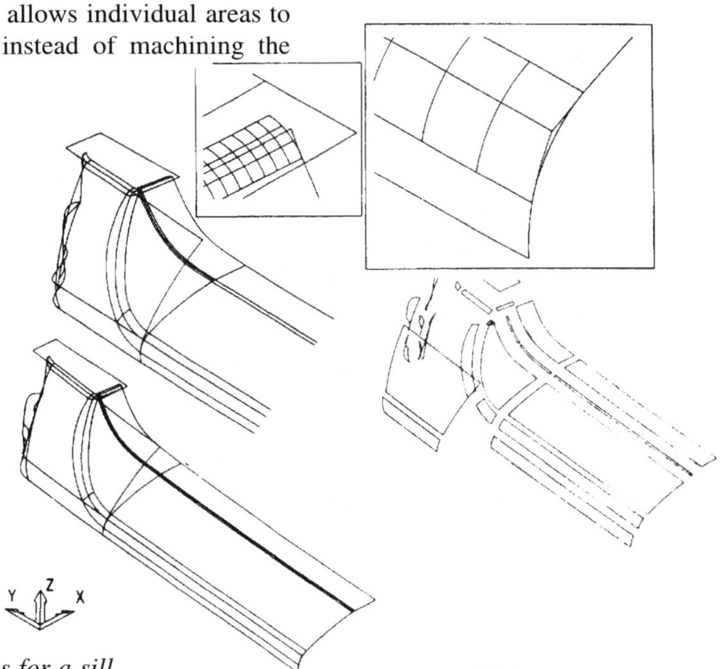

Fig 54: Patching surfaces for a sill panel

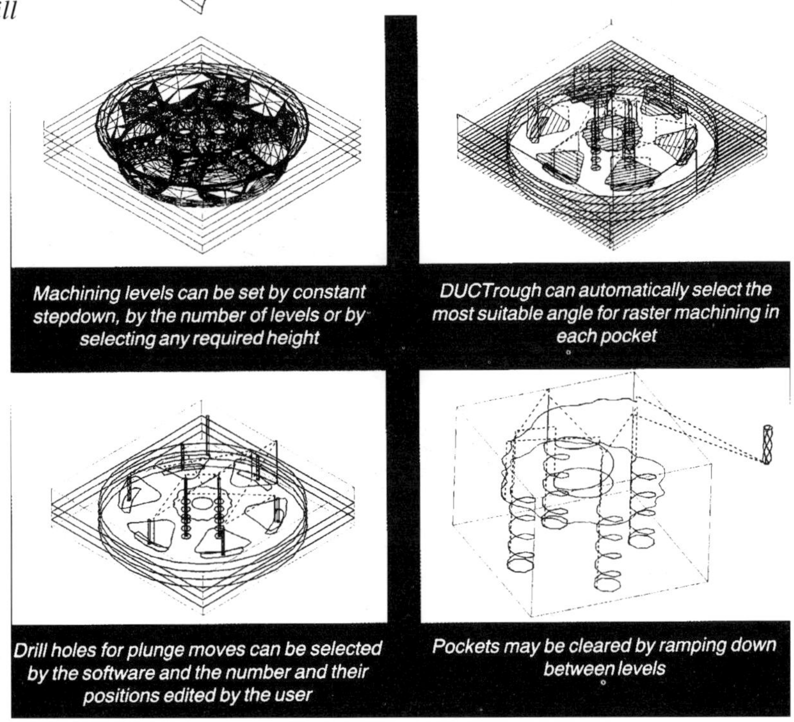

Fig 55: Facilities offered by the DUCT roughing module

Surface geometry

Curved body panels are, of course, preferred to flat ones to avoid distorted light reflections and obtain greater panel stability against vibration. There is also an 'aesthetics discord' in joining a flat panel to a curved one. Even in single curved surfaces it is aesthetically preferable to avoid curves with constant radius since the line joining surfaces generated by different radii is also problematical, Fig 56. Parabolic or elliptical curves are therefore preferred. Better still, double curvature surfaces ensure overall fluidity of shape.

In three-dimensional curvature, equality, similarity and affinity (collineation) are three key parameters (Fig 57). In the first, the generators are identical; in the second they generate curves whose slopes are equal at different positions; in the third, curves I to IV are related by the curves at V and W. The varying degrees of affinity are shown in Fig 58; this shows the effect of different initial curves used for interpolating the shape on subsequent drawings between the boundary curves. Different surfaces are thus obtained: a. continuous edge from the corner; b. fading edge; c. fading corner.

After the shape has been chosen by clay or computer modelling, the surface is split up into sections corresponding to the major boundaries, prior to collineation, Fig 59. Intermediate curves between the boundaries are collineated using trapezoidal, triangular, radial or rectangular divisors, Fig 60. In the first case, boundary curve K1 is represented in the subsidiary diagram by the base line JE drawn to any convenient length. Normal boundary lines AE and FJ are represented by vertical lines (with the same lettering) drawn to a suitable scale. Straight connecting lines like AF and BG are drawn to complete the diagram. Section lines I and II are represented on the diagram by drawing normals on the base line through K and P which are themselves chosen so that AP:AK:AF is a constant relation in the two figures. The intersections L,M,N and R,S,T can then be transferred back to the first figure and these, when connected, give the lines K3, K4 and K5. With triangular divisors, only three boundary curves are given and in both methods linear proportionality of the intermediate curves is the simplest form of dependence. Higher forms of collineation, such as radial and rectangular divisors, are used to obtain a more refined aesthetic.

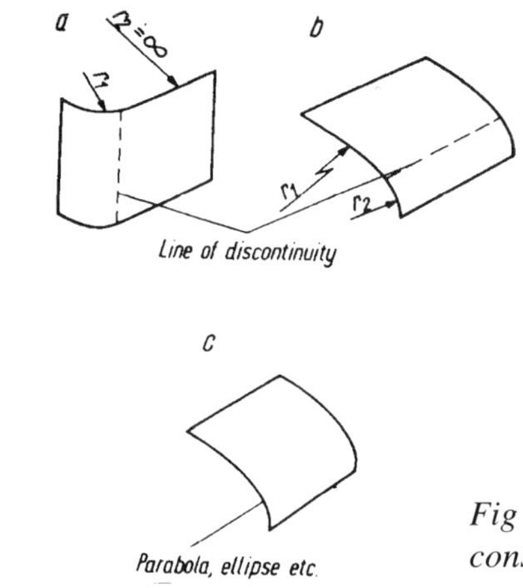

Fig 56: Avoiding constant radius

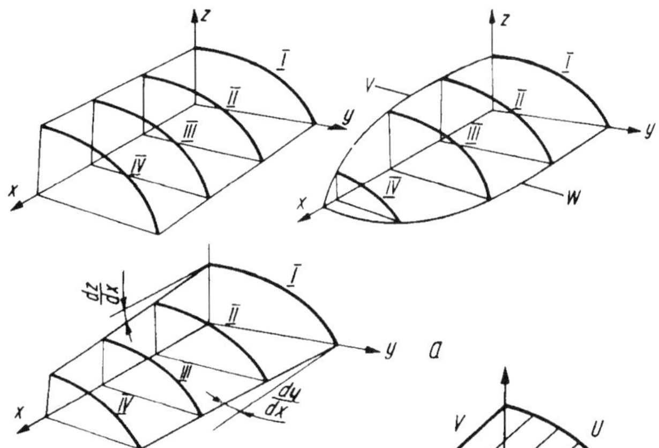

Fig 57: Equality (top-left), similarity (top-right)and affinity

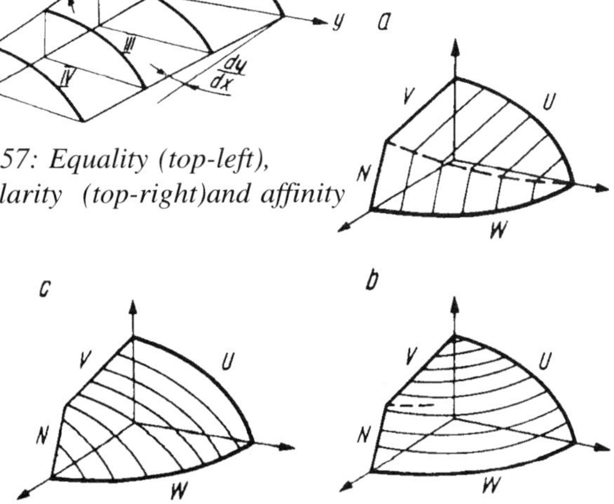

Fig 58: Degrees of affinity

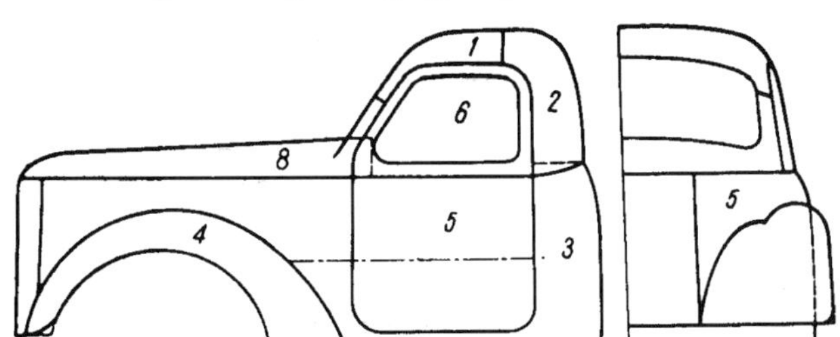

Fig 59: Sections by major boundary

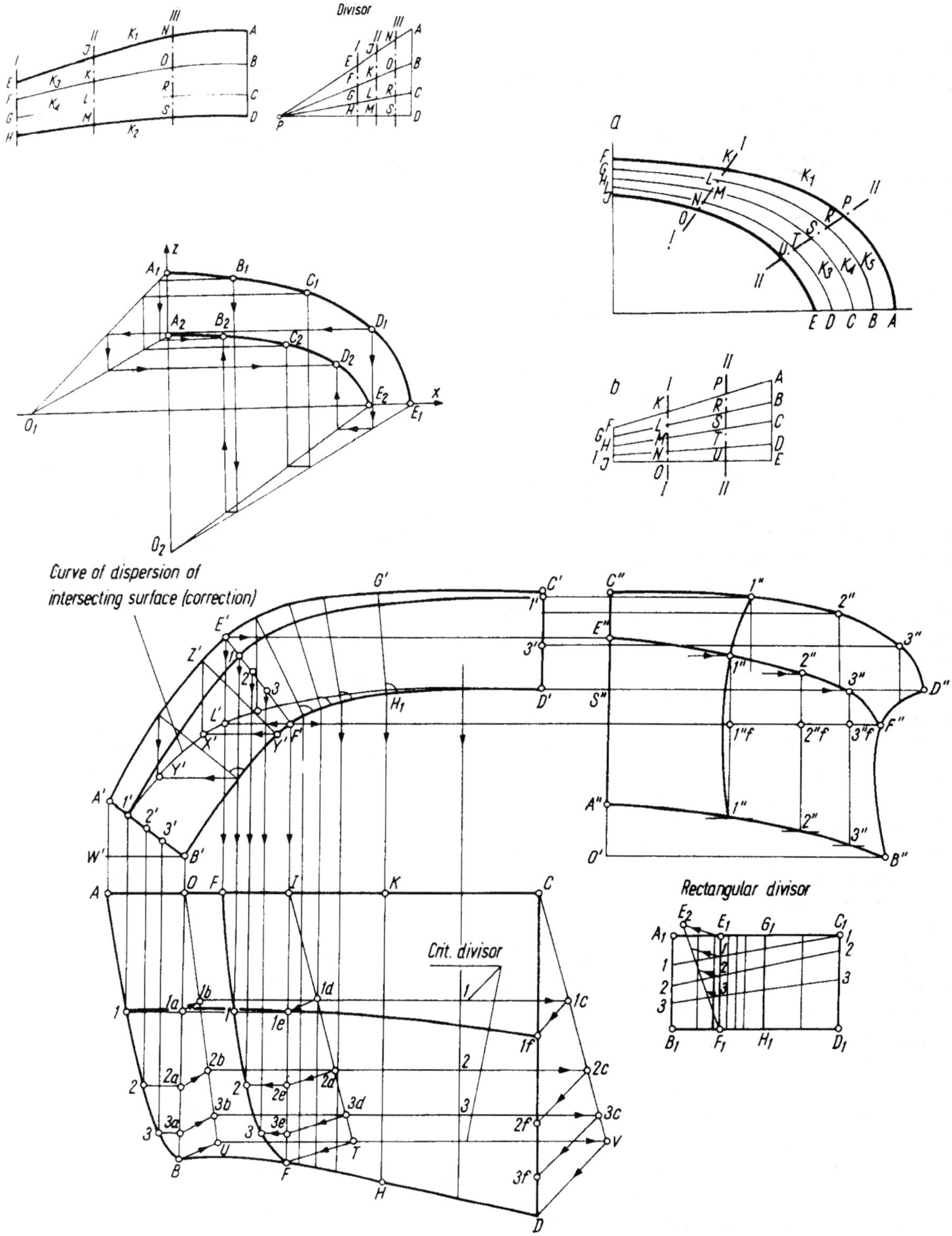

Fig 60: Surface development by rectangular (bottom), trapezoidal (top right), triangular (top left) and radial (top left — lower) divisors

Surface definition techniques

Prior to the introduction of computer techniques, surface definition was by means of descriptive geometry, with the surface being described by a set of curves with enough information to manufacture templates, from which a master model was produced. In computerized methods the three types of surface descriptions are ruled surfaces, surfaces-of-revolution and the more complex sculptured surfaces. The latter are generated by a series of section curves, blends between adjacent ones being called fillet surfaces. Fig 61 due to Paul G Ward[16] shows a single surface generated from four curves, the hole being obtained in an editing operation involving projecting a circular entity on the surface.

Surface geometry in vehicle bodies is influenced considerably by aesthetic considerations. Curved panels are, of course, preferred to flat ones for avoiding distorted light reflections from non-true surfaces. There is also a 'discord' in joining flat panels to curved ones and even in single-curved surfaces it is preferred to avoid curves with 'constant radius' because of the 'difficult' line between surfaces of different constant radius. Hence the preference for conicsection curves and better still surfaces with double curvature — which are said to ensure greater fluidity of shape.

According to MIRA's vehicle systems engineers, the crux of surface-development is the fitting of curves to co-ordinates; the computer uses polynomial expressions to describe the curves and progressively smooth the surface. Errors on the surface of the full size clay concept model of the vehicle body need correcting by a smoothing line. The surface is described by a grid of coordinates and it is usually smoothed over four coordinate lines using the 'hardpoints' as reference. First the lines are smoothed over the points, and then the surface over the grid; final stage is to join up with surfaces on adjacent grids. In the smoothing process the polynomial equations are solved to give progressively finer amplitude tolerance on the surfaces — to the point where they can be passed to tooling. Lines joining the control points are termed polygonal lines and the representation of a curve involves drawing many short straight lines, involving very large files of data. The differential equations which describe a curve may be expressed as a second-order (Bezier), third-order (B-spline) and fourth-order (NURBS), Fig 62.

Medland[17] helps to distinguish between solid modelling and surface development packages. In

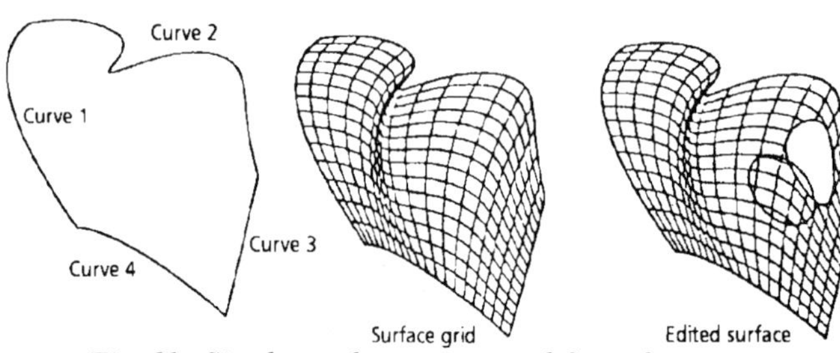

Fig 61: Single surface generated from four curves

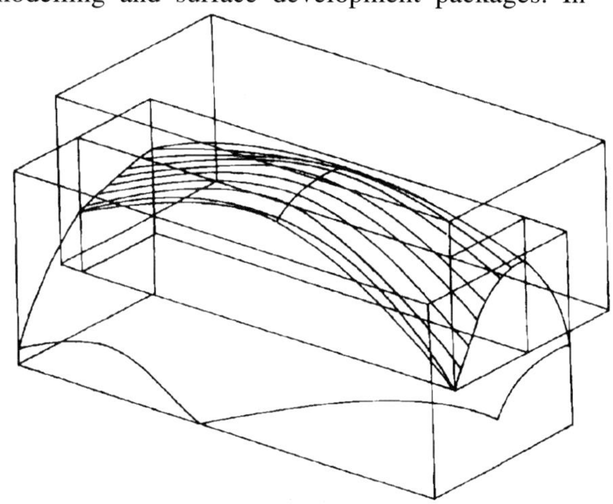

Fig 63: Trimming of surfaces to produce sub-surfaces

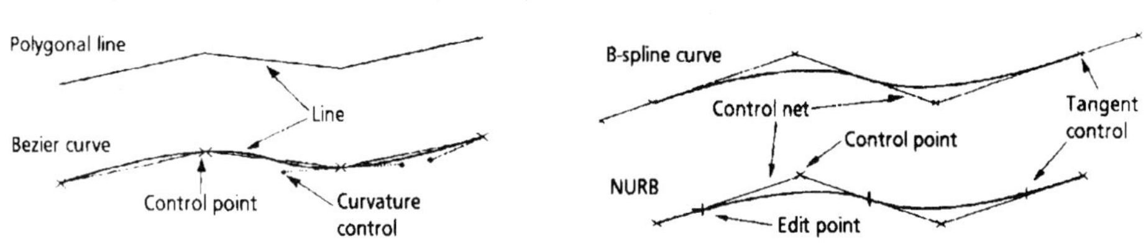

Fig6 2: Different methods of describing the same curve

surface modelling which lies somewhere between the two, a wire frame model has surfaces added to it which may be plane, ruled or surfaces-of-revolution defined between the boundary entities. Surface development packages usually involve complex free-form surfaces using Bezier curves, B-splines, Coons or patches. He points out that a body within a spatial domain is made up from a combination of edge and surface descriptions. The visual appearance of the solid object can thus be generated and the intersection of surfaces calculated. Excursions of domains can thus be considered as simple flat surfaces existing in the local orthogonal plane. Surfaces can thus be 'trimmed' by the specified range in the co-ordinate directions, Fig 63, and sub-surfaces thus produced. The association of two domains can then be interpreted via a hierarchical search extending down from checking geometric intersections, through an edge condition check into one examining the sub-surface relationships.

Whereas, in solid modelling, surfaces are represented by geometrically simple solids, surface modelling involves the development of a surface by defining and joining the curves. With the Bezier principle, the curve is defined in terms of points and associated control points. Strength factors, Fig 64, are typically used in conjunction with angles which can be specified in two directions, at the data points, to describe the surface.

A comparison of the basic techniques of computerised surface-development is provided by Vittorio and Mazzanti[18]. The 'Simple Poles' technique was one of the first solutions for non-interpolating (smooth) curves; these are modified by moving the control points. The tangency condition is equivalent to the co-linearity of three control points but the drawback is the difficulty in defining curves of rapidly changing curvature. The solution can be either to break the curve into segments or increase the number of control points but both these steps can create further problems of complexity.

The next alternative is using cubic uniform B-splines, giving the opportunity for multi-segment curves (multi-patch surfaces) with each described by a cubic polynomial. A requirement for a curvature continuity condition for adjacent segments simplifies the modelling but limits the application with real-life objects. This leads on to considering Rational B-splines which allow unique representation both for analytical and sculptured surfaces. There are, however a greater number of parameters to control.

A further evolution of the Simple Poles is known as Generalized Poles and overcomes the limitations of the first form while maintaining an intuitive approach by the user. It can be thought of as consecutive segments of Simple Poles with continuity constraints. It is thus more general than Uniform B-Splines and less so than nonuniform ones (NURBS). This approach was taken by the authors' companies but with an extension to allow representation of single surfaces with sharp edges, giving it the new title of DG Poles (discontinuous generalised). Fig 65 is an example of a panel obtained with DG Poles, for which Bezier representation would have required over 100 patches.

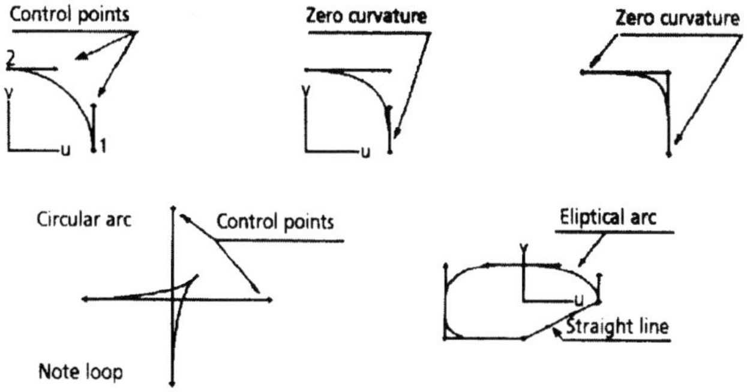

Fig 64: Strength factors, typically used to describe surfaces in conjunction with angles that can be specified in two directions at the data points

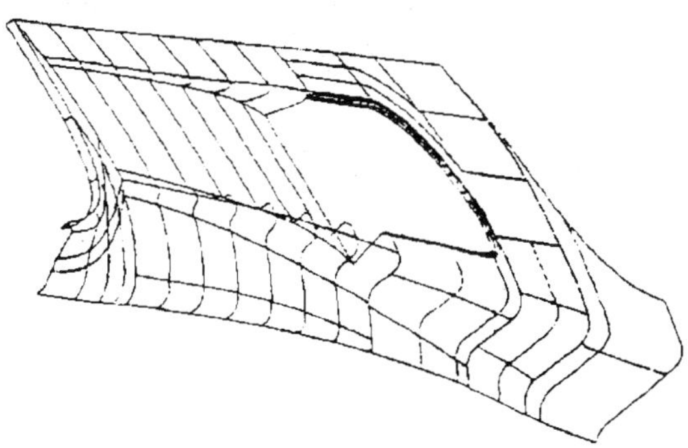

Fig 65: Panel obtained with DG poles

Virtual modelling and 2 to 3D conversion

The ICEM Surf package for concept styling is seen by its suppliers as being akin to "virtual modelling" with the ability to operate without the use of the traditional clay master model. The stylist can either draw freehand in the package or develop an existing design, for face-lifting purposes, and in both cases a high-quality mathematical model will exist very early in the design-to-build process. Control Data Systems were early exponents in mainframe systems devoted to body design, starting with Chrysler. Over four years ago US Ford, in its C3P global program, switched to ICEM Surf for Class A exterior and interior surfaces and the latest Taurus car is one of the first results.

For the redesigned Taurus, rather than digitize data from a clay model, Ford used the computer model to generate surfaces of a quality that was good enough for official "studio release", not possible with other packages, apparently, Fig 66. The so-called Surface Layout is the official delivery of the vehicle's conceptual model from Body Design to Body Engineering. The package was able to combine various elements of earlier competing concept designs in arriving at this final surface solution. In face-lifting operations the package has the facility to take panels from earlier vehicles, which are not mathematically defined, and spread a grid over them to read the surface and tailor it to fit the envisaged design.

In the latest release of the package, ICEM Surf data interfaces are available for CADDS, CATIA and Inventor CAD surface models without special data-translation peripherals. Underbody structure design is also facilitated by a new function within Surf Master called Feature Design which allows features to be defined and applied to different surface geometries. Stereographic large-screen projection has also been introduced; so too has global filleting for intersecting surfaces. The global parametric modelling facility allows large panels such as bonnet lids to be created in their entirety rather than from a series of individual surface elements or patches. After creation, they can be extended or compressed without losing the continuity of patch transitions and the proportions of a complete body can be altered, for instance, to create models for different wheelbases, while retaining overall surface quality.

Another supplier, Radan, also now has a 2D design automation package called New Radraft, with the belief that with most companies, even when they model in 3D, there is usually a major requirement for producing 2D drawings for sub-contract manufacture and so on. Design changes can be implemented quickly with the package, due to the intelligent association of orthogonal views, which automatically update when dimensions are changed. A further tool known as New Radesign also allows drawings created with New Radraft to be transformed into 3D solids, for the benefit of draughtsman who normally only work in 2D. Makers of special purpose vehicles, ML Douglas Equipment have benefited in the design of their aircraft towing tractors, Fig 67, by using the package. The freedom to move from 3D models to 2D drawings and vice versa cut the design time for the latest model from six to two weeks.

Reverse engineering

For automotive companies still working with physical surface models, or existing real-world panels which have not previously been mathematically defined, the Atos Video 3D measuring system, Fig 68, from Newport is a useful tool in non-contact surface probing. While non-contact probing has been possible with large 3D laser installations, the Atos XL is a portable system using a photographic technique. A high-resolution digital still-video camera, with computer interface, is involved. A controller and driver; tripod and two special rules complete the pack, all of which can be stored and transported in three suitcases. So-called topometric sensing can be carried out to a data density of 430,000 object points per view within a few seconds. Point clouds can then be made available for post-processing in CAD and FE packages. Various fringe patterns are projected on to the surface and recorded by two cameras from different angles. Precise 3D co-ordinates are independently calculated for each of the CCD camera pixels, using triangulation methods and digital image processing.

Adhesive retro-reflective points are attached to the surface to be digitized, and to elements, defining its co-ordinate system at a typical separation of 250 mm, for measurement purposes. Two reference rules are also positioned close to the object surface. Initially, without a tripod, multiple shots at various view angles are then taken, the views being selected to overlap one another and include reference targets in

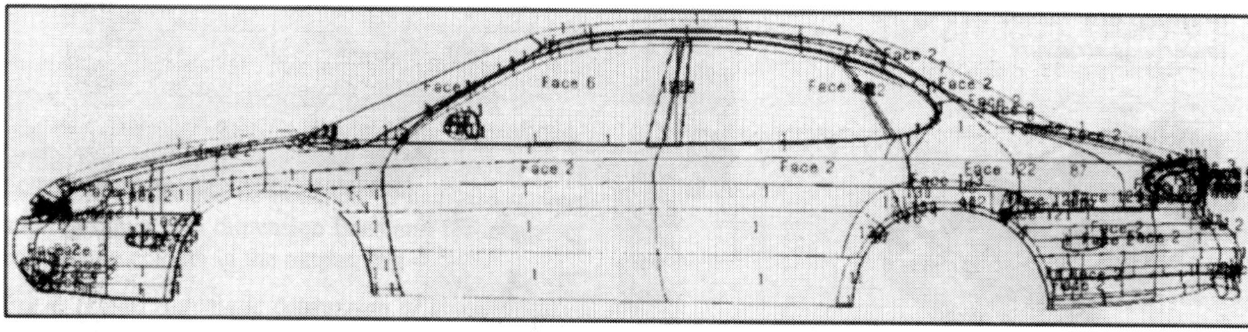

Fig 66: With ICEM Surf , the surface model is defined by a relatively small number of surface patches.

each shot. The data obtained is down-loaded for photogrametric evaluation to compute the 3-D co-ordinates of all the target reference markers over the whole surface of the large object. Then, to obtain surface detail, a tripod is used with probe-head for measuring within a range of 300-400 mm over a section of the surface which includes three reference points. The resulting cloud of 3D data points obtained is then meshed into the previously obtained co-ordinate system. An entire body takes up to 100 ATOS measurements providing data with a separation of 4 points per square mm. About half a day would be required to cover a square metre, a full-size car taking one to two days. The system has already been taken up by Mercedes-Benz and benchmarking evaluation is under way at VW-Audi

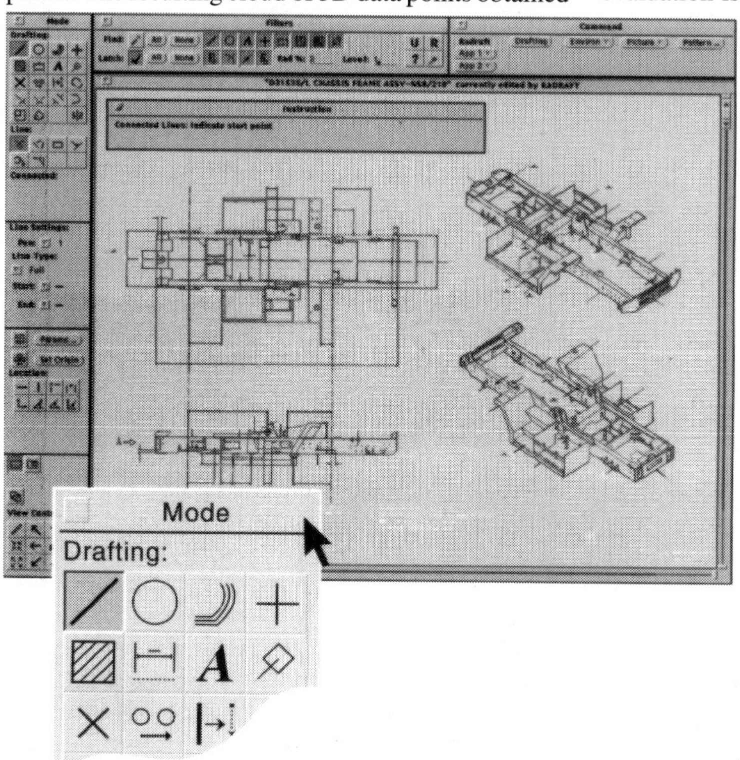

Fig 67: Radan technology in use by Douglas for its aircraft tug for the Swedish market showing 3D and 2D drawings also a solid view of the underbody structure

Fig 68: The Atos sensor and its application to a body clay model

Rapid prototyping

The ability to provide users of CADCAM techniques with tangible prototypes within a few hours, by stereolithography and other related processes, has been a major advance in speeding development times. Most techniques involve building up the model in a series of layers, with data derived from a series of slices through the shape held by the CAD system. Complex shapes are represented by a combination of solid-modelling and surface development methods.

In stereolithography the model is made from a liquid polymer that cures when exposed to ultraviolet light. A laser beam is made to scan the surface of the liquid, in the X-Y plane, by a galvanometer and solidifies it to a measured depth and radius. A string is formed when the laser spot is drawn across the surface and a series of these enables a solidified film to be made which corresponds to the CAD slice. The tank containing the liquid is then raised in the Z-direction by a given amount and a second layer produced; this is repeated to completion of the model. The ability to produce both internal and external surfaces has widened the application into a variety of hollow model shapes.In laminated object modelling (LOM), models are built up from layers of sheet material pre-coated with a heat-sensitive adhesive. Modelling is performed on a movable platform which can be incremented vertically by the thickness of the sheet which is fed from a roll into position and bonded to the stack by a heated roller. An X-Y positioner then guides a laser to cut the sheet to the desired shape.

Selective laser sintering The University of Leeds is offering a service to companies who wish to explore the potential of layered manufacturing (rapid prototyping) using selective laser sintering (SLS). This SLS is a newer process than the more common stereolithography (SLA). SLS (Fig 69) forms solid objects by using a laser to sinter powder layer by layer under control of instructions derived from a CAD representation of the object. The main, currently demonstrable, advantages over other processes are exploitation of polycarbonate in investment casting and a range of non-toxic materials. Also no additional geometry is needed as in SLA where the required shape must be extended to provide the supports needed during manufacture.

The issues with all layered manufacturing processes are repeatability, accuracy and surface finish. All layered manufacturing processes require a layered representation of the object currently derived from the *de facto* standard STL file developed for Stereolithography. This is a triangular faceted representation of a solid object.

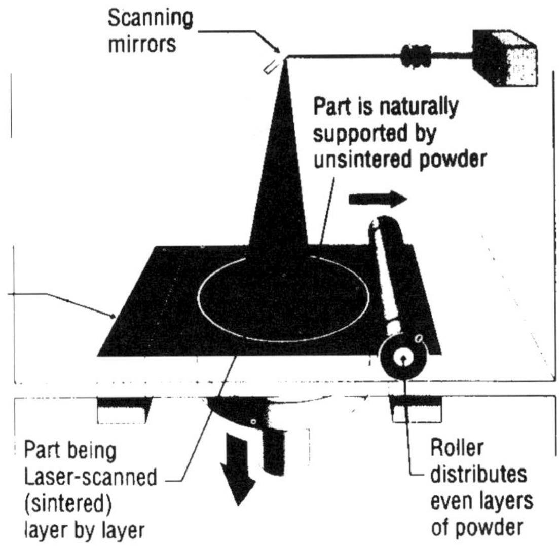

Fig 69: Selective laser sintering

Fig 70: Ford tailgate screenwiper cover

Rapid tooling

As an extension of rapid prototyping, two forms of rapid tooling are currently under automotive-industry scrutiny: the more familiar soft tooling, using secondary processes such as epoxy or Kirksite spray metal tooling to produce plastic parts in 10-1000 quantities. Hard rapid tooling is under an earlier stage of exploration.

Rapid tooling systems such as 3D Systems' QuickCast 1.1 process allow the production of precision castings from RP patterns. The process involves the combination of generation of CAD negatives with stereolithography. Negative patterns can then be shell investment cast, producing core and negative cavity inserts. Ford and GM are said by the company to be among those experimenting with the technique for generation of prototype and, eventually, production tooling. An early application of the technique was used to tool the 30,000 parts per month rear screenwiper cover for the Ford Explorer, Fig 70.

3D Systems' SLA-350 stereolithography machine provides build envelopes up to 350 x 350 x 400 mm and incorporates the Zephyr recoating system, Fig 71. It deposits a layer of resin as it sweeps over the part. Layers as thin as 0.01 mm can be specified for a smoother surface finish than is conventionally available. It also simplifies building parts with trapped volumes particularly SL tool patterns. There are four stages involved: when the laser finishes solidifying a layer, the recoater is ready to sweep the part; the part is lowered only one layer thickness before recoating; next the thin resin coat is applied; after coating the next layer is solidified and the recoater reservoir refilled. Another fascinating application has involved the incorporation of reverse engineering to create a pipework checking fixture, Fig 72.

Mercedes-Benz is a pioneer of applying rapid-prototyping techniques and has recently taken delivery of a machine that allows large moulds to be produced in sand for metal castings. The Zeiss EOSINT S 700 equipment can produce sand moulds from a CAD design in just a few days, within standard tolerances of cast prototypes. Moulds up 720 x 380 x 380 mm can be produced. It claims to be the only RP system to use two lasers and scanners simultaneously, which doubles the building volume compared to single laser systems. It also doubles building speed and has a throughput of 100 kg of sand per day.

Metal moulds for plastics injection moulding can also now be produced by rapid tooling techniques at the Sunderland company Amsys. Using the selective laser sintering (SLS) process it again works from a 3-D CAD file using a mould material with properties similar to P20 steel, or 7505 aluminium, and capable of producing over 50,000 parts. Lead times from initial CAD to finish parts is put at 2-3 weeks, compared to 12-15 weeks by conventional methods. A 'green' mould of loosely sintered metal particles is first produced then, after a polymer infiltration step, it is infiltrated with molten copper in a furnace to form the dense metal mould. Final stage is polishing and integration with the mould base.

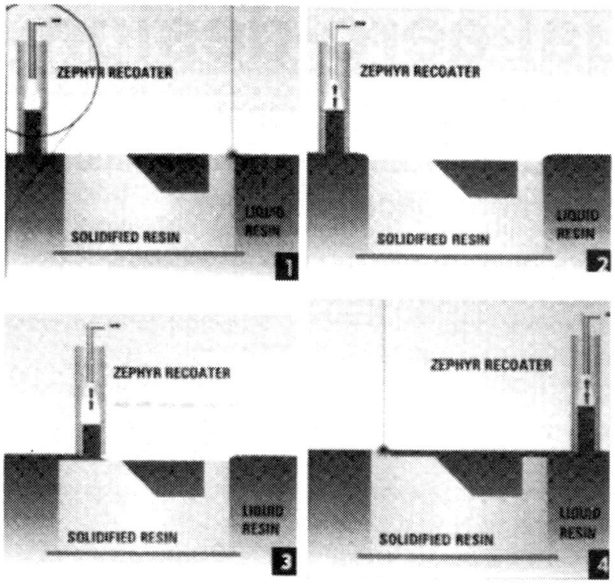

Fig 71: Stages of Zephyr recoating process

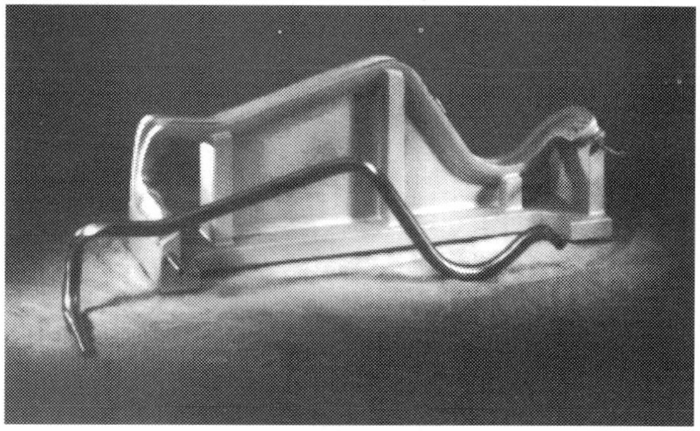

Fig 72: Ford pipe checking fixture

Chapter 2: Ergonomics, seating, and packaging

Dimensional factors in seated comfort; interior (control) ergonomics; ergonomic system design by computer; seat comfort; suspension seats; reducing back pain; split-frame seating; dashboard instruments; advances in electronic displays; CV-cab ergonomics; CV legal dimensions; mechanical package layout

Dimensional factors in seated comfort

Audi researchers[1] have stressed the importance of careful integration of all the requirements of seat design in order to ensure that the driver's perception is one of pleasure rather than extreme disappointment, which can quite easily become the case. Ergonomic considerations must cater for the wide range in human stature, the principal body dimensions for vehicle seat design being shown in Fig 1. Computer-aided procedures allow the designer to investigate a wide range of seat positions, with the help of three-dimensional manikins. Another key consideration is to examine the postures of those who have, for example, long legs and a short trunk, or vice versa, and do not conform to the averaged percentile norms, Fig 2.

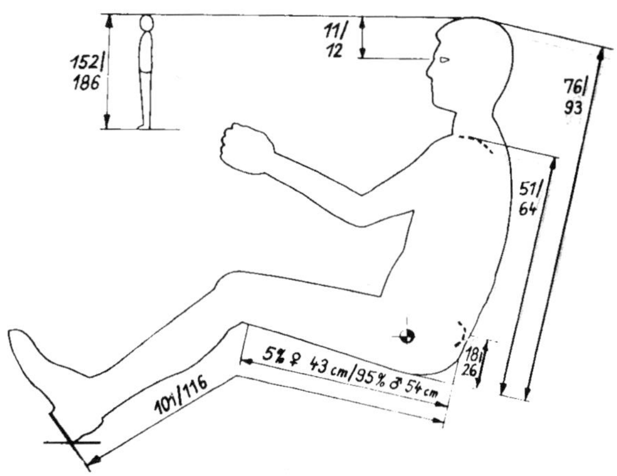

Fig 1: Principal dimensions for seat design, 5th percentile woman and 95th percentile man

The contrast has also to be resolved between the extra relaxation enjoyed by the seated person and the constriction of certain organs and the restriction of blood flow through the veins, from the lower extremities of the body. There is also greater susceptibility to vibration loads due to the absence of damping characteristics of the legs in an otherwise upright posture. The authors suggest the best compromise is obtained when the seated posture minimises the strain on the human organism imposed by external loads as well as equilibrium between the ligaments and the sections of muscle which act in the opposite sense to them. While the body must be 'held', a degree of freedom of movement must be provided. There is also the requirement to give sufficient rearward recline of the seat to displace the body's CG behind the ischial contact points, so that arm and leg movement is free from unwanted feedback effects at the other extremities. Automotive designers have also obtained subjective survey results of comfort angle ranges which have proved useful in configuring the seated person in the vehicle, Fig 3. Cushion compression effects must also be taken into account when transferring the anthropometric data into seat dimensions.

In the seated position the load imposed by the body on the legs rests on the bones of the pelvis which roles to the rear and initially gives a pleasant relaxation sensation, Fig 4. But the natural uniform load on the elements of the spine is departed from, with attendant health risks. Back muscles are elongated and the abdominal area is compressed. The forces imposed by the seated occupant must be reacted where the pelvic bone structure projects the farthest, in an area where blood flow through the muscle is least affected. Lateral location of the pelvis must also be provided. A wide support area should be provided to avoid high pressure areas. The authors suggest 1-3 N/cm^2 at the ischial support areas, falling to 0.2-0.8 at the boundaries. All-foam cushioning systems are believed by the authors to provide the necessary compromise in the spring characteristic distribution over the surface of the seat.

Resonances are stimulated in relaxed seated persons at around 2-6 Hz in the shoulder area, which can affect breathing comfort, also at 4-5 Hz over the entire trunk, leading to pains in the lower body, and at 20-25 Hz in the eyes, leading to vision impairment. They lead to the optimum comfort criteria used by suspension engineers. The seat springing can, of course, cause a dynamic-magnifier effect if appropriate damping is not provided. It is suggested that foam systems should be used in which static seat compression should be 40% of total spring travel.

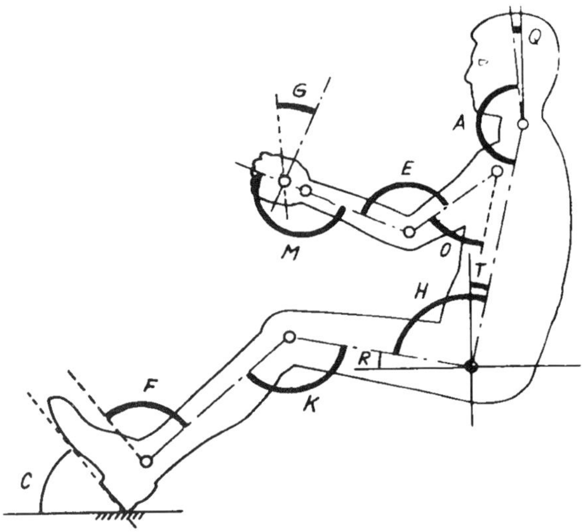

	-5	-4	-3	-2	-1	0	1	2	3	4	5
A:	125	138	150	160	168	175	180	187	195	206	220
E:	30	50	67	80	90	100	116	132	148	164	180
F:	65	69	73	77	81	85	96	105	113	120	125
G:	0	5	10	15	20	25	28	31	34	37	40
H:	60	80	90	96	99	100	103	110	122	145	180
K:	45	60	90	105	115	120	124	132	143	158	180
M:	140	153	163	170	176	180	183	187	190	195	200
O:	-60	-28	-5	6	15	25	27	31	48	87	160
Q:	-60	-48	-36	-24	-12	0	12	24	36	48	60
T:	-30	-5	10	20	23	25	27	28	34	51	85

Fig 3: Recommended comfort angles +/- 5 degrees

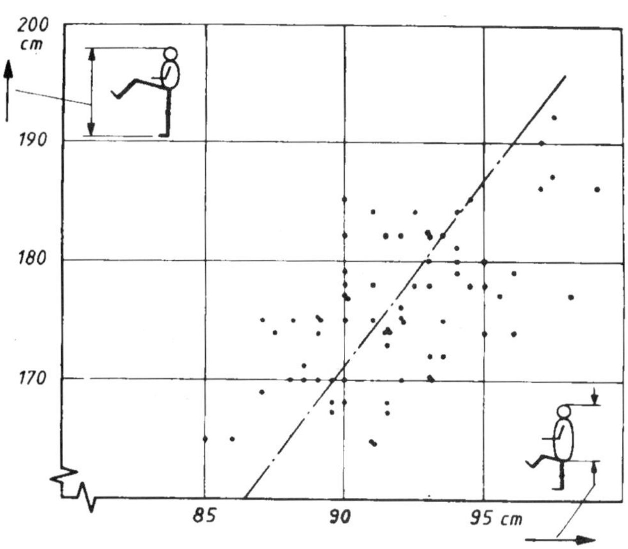

Fig 2: Body stature and seated height of a population of 70 male test subjects

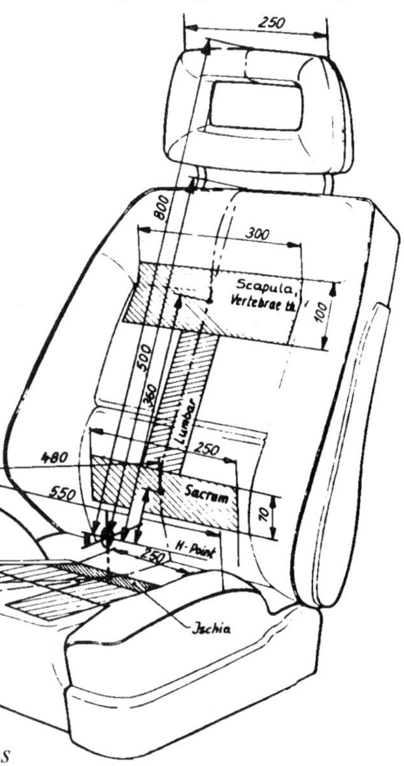

Fig 4: Basic seat dimensions

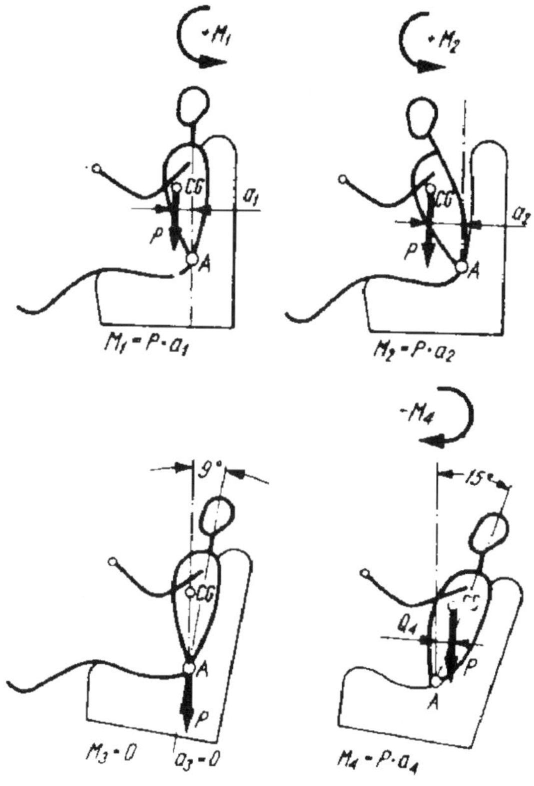

Fig 5: Results of misplaced back-rest

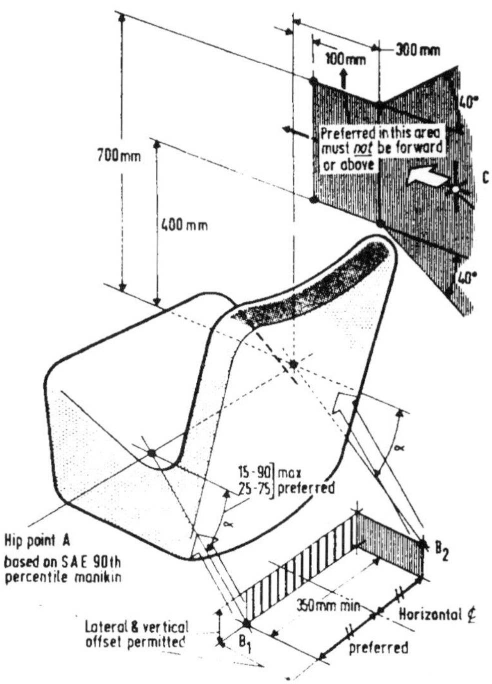

Fig 6: BS seat belt anchorage

Seating design pitfalls

Seats and associated upholstery trim design can have dramatic effect on a potential customer both at the point of sale of the vehicle and at the end of a long journey. Leaving the selection of seat parameters until the 'afterthought' stage in the design of the complete vehicle can easily lead to failure in the market place for most categories of vehicle.

The section which follows this one, on ergonomic layout of vehicle interiors, makes clear the importance of seat dimensioning, and provision of adjustment, to both the comfort of occupants and the adequate control of the vehicle. Fig 5 shows how even the wrong backrest inclination has the potential to cause severe discomfort over a long journey. Equally important is the choice of cushion springing and damping characteristics as well as the contouring of the seat and squab profiles and the choice of upholstery trimcloth.

The importance of seating to vehicle control leads the writer to recall the words of a latter-day automotive wit, commenting on the occupant's anatomical relationship with the primitive vehicle seats of days gone by. 'To seat two spheres upon one other and regard it as stable is foolish...' this is an extreme view of the driver's difficulty in maintaining a position of stability with respect to the controls but it nicely emphasises the risk of losing the essential 'seat of the pants' feedback of vehicle motions necessary to properly involve the driver in the 'control loop'. Medical scientists argue that muscular interaction with vehicle motions, sensed through the seat, is governed by anticipation of (experience) reaction and by a sense of balance. While the interaction becomes a reflex condition, considerably more energy is required with conventional seat systems, to maintain a relationship with the controls, than is necessary to actual control the vehicle.

Fig 6 shows the seat belt anchorage positions defined by British Standard BS AU 4B:1965 which are chosen to provide safety restraint in impact without undue discomfort in normal driving. In fact, much driving discomfort has been traced to vehicle manufacturers providing too great a static deflection for the seat springs in order to enhance showroom appeal. In general, it is argued, this can lead to undesirable overrun-bounce — an undamped motion in phase with the vehicle's wheel suspension frequency.

Interior ergonomics (vehicle controls)

Driver and passenger comfort; ease of entry and exit; driver relationship with the primary controls of the vehicle — all these affect the interior package dimensions of passenger-cars and public service vehicles: also the cab layout of goods and industrial vehicles. While the high volume builders invest heavily in studies to optimise ergonomics of particular designs, the specialist builder often needs to make recourse to more generalised published data to optimise driver and passenger spatial envelopes for a relatively low volume production vehicle.

Ergonomics is strictly the study of matching man to machine and is a surprisingly recent science. Much dimensional data used in the UK for matching drivers to the controls of their vehicles came from studies of World War Two fighter plane cockpits.

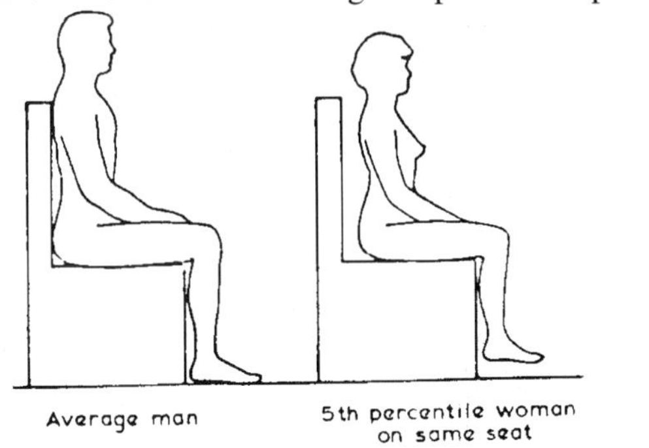

Fig 8: Seat measurements as affected by percentiles

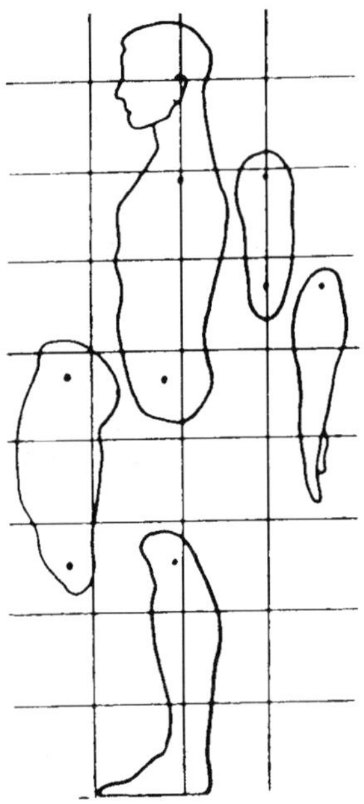

Fig 7: Dimensions of average male on 10 inch grid

1

Dimension	*Percentile male* 5th	50th	95th	*Percentile female* 5th	50th	95th
1 SRP to knee hinge	20	22	23½	18½	20½	22½
2 Knee hinge to sole of foot	18½	20	22	15½	18	20
3 Buttock to patella	21½	23½	25	20½	22	24
4 Buttock to back of knee	17	19	20½	16	18	19½
5 Knee height (no shoes)	20	21½	23½	17	19½	21½

Dimension	*5th %ile female*	*95th %ile male*
6 Shoulder point above seat (sitting shoulder height less 2 in)	17 in	23 in
7 Shoulder point to elbow hinge (shoulder to elbow less 1½ in)	10½ in	14½ in
8 Elbow hinge to closed hand (elbow to fingertip less 1 in for hinge and 4 in for closed hand)	11 in	15 in
9 Sitting eye level above seat	25½ in	33½ in

2

SEAT PARAMETERS

Optimum E = 105° (90% acceptable, 102°–108°)
C = 6° (90% acceptable, 5°–8°)
B ≤ 16 in for small women
= 17 in as compromise (provided knee pivot of small woman > 2½ in above seat front edge, to allow full pedal depression)
A = 18–24 in
D < 15–16 in

* Failure to meet this dimension causes accident risk due to reduced pedal force arising from popliteal pressure. Seat front edge must be rounded and softly padded.

3

Normal and maximum (emergency) pedal efforts given by shortest women and tallest men in 90% range, for three seat heights; M = Males, F = Female.

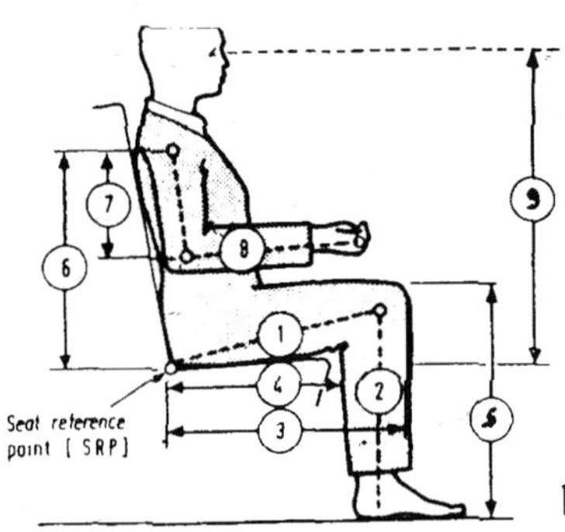

b

Seat height (in) SRP	Front edge	*Knee angle* M	F	*Norm. force (lb)* M	F	*Max. force (lb)* M	F
11	13	133°	126°	75	45	135	80
9	11	140°	142°	80	50	155	95
7	9	145°	135°	90	55	175	120

Fig 10: Seated occupant with tables of seat dimensional data

Hence earlier data sources need treating with care in respect of the special levels of physique in the subjects concerned — and often their above average stature. Buses and taxis, for example, need designing for a user-population of much less fit, and often smaller stature, people. The use of templates in drawing work therefore needs some caution. Fig 7 shows the dimensions of an 'average' man; since it is usual to design within percentile extremes, this could be scaled up and down to represent these extremes

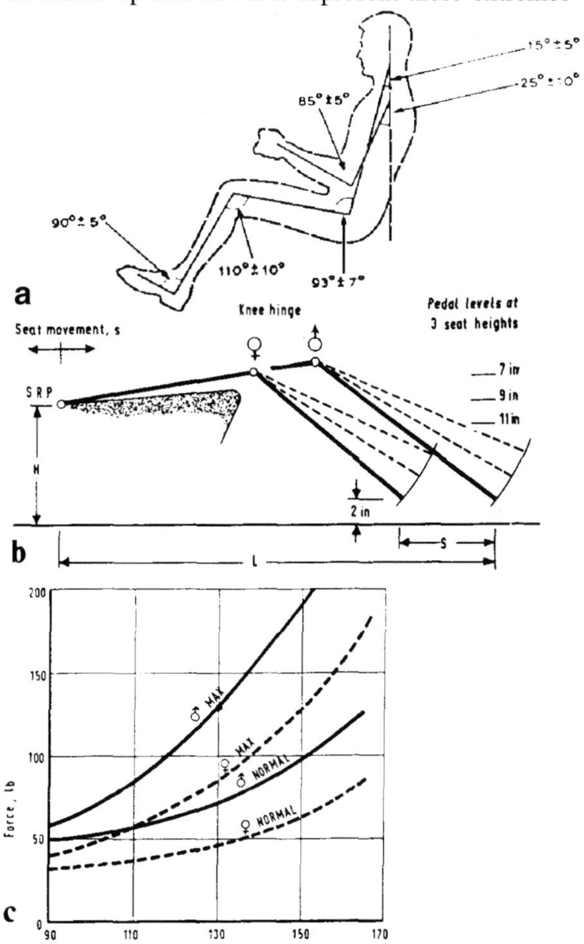

Fig 9: Recommended comfort angles

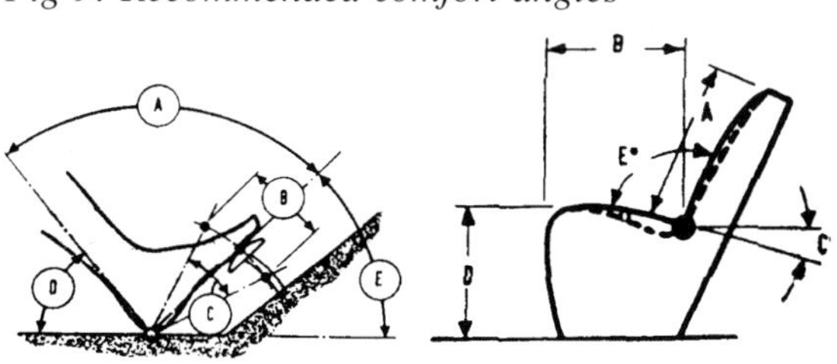

Fig 11: Foot control data

and two jointed models made for fit testing on scale drawings of interiors.

Percentile correction

As there are few women, say, smaller than the 5th percentile and few men larger than the 95th, these correspond to statures of 58 and 72 inches. By definition, percentiles relate to a single dimension for a group of people and it should thus be realised that the 'average man' concept could be invalid for different demographic groups of people, even when factored for stature as described above. Fig 8 shows an unfactored 'average' man alongside a 5th percentile woman sitting on the same seat. Minimum fatigue over long periods in the seated position has led to comfort angles as shown in Fig 9 but another very important consideration is the provision of space for a person to change position during long sustained periods of driving. By defining a seat reference point (SRP) as the intersection of the seat and squab surface, Table 1 in Fig 10 shows the necessary factoring of data to accommodate percentile variations.

Foot controls

Early work by researchers at Cardiff University[2] focused on the different ergonomic requirements for the different controls in a vehicle. He pointed out that foot controls are divided between those like brakes, needing considerable force, for which free leg movement is required, and the 'precise' control of throttle, say, where use of ankle articulation is better suited. For braking and declutching a situation of progressive force with virtually zero movement is ideal. Critical factors for pedal force and movement are: knee angle in leg operation optimum 160-165 degrees (cf. most comfortable relaxed range 120-130 degrees) and ankle angle in leg and ankle operation, within 90-125 degrees absolute minimum. Both angles are obtained from a spatial relationship of seat and control; see Fig 11. Optimum lateral offset of pedals from body midline is 25 in (635 mm); at 6.5 in (165 mm) offset, leg force potential is reduced by 75%. Alternative approaches in calculation are either to determine seating/pedal position, given the force required hence knee angle from Fig 9a; or, given seat/pedal position, to determine maximum allowable force. Optimum seat parameters are shown in Fig 9b and Table 2 of Fig 10. A worked example to find leg operation parameters for 95th percentile man to 5th percentile woman is as follows, using the data from Fig 9b: procedure for calculation is to design for adequate vision of the shortest woman and then use available pedal pressure for this position from Table 3 of Fig 10. Pedal movement is then arranged to be in line with the lower leg. The greatest force can be applied by the instep; it is reduced by 50% if toes are used. Hence it is best to site the pedal 6 in (15.25 mm) above floor level with further 2in (7.6 mm) clearance for toes; pedal travel should be restricted to 2in (7.6 mm) maximum before pressure is required.

Ankle controls, such as throttle, are best with pedals hinged at their base. Optimum and maximum angles for barefoot are given in Fig 11 and Table 3 of Fig 10; when pedal is hinged at heel-to-floor contact point, angle E should be increased, to allow for heels, by 5 degrees for men and 10 degrees for women compared with the no-shoes condition. For example, in Fig 46, E = 45 degrees (55 degrees with shoes) for women and E = 55 degrees (58 degrees with shoes) for men (also E = 71 degrees and 73 degrees for men and women, with 7in seat height) so an acceptable pedal can be arranged.

Designing controls for ankle operation starts with the determination, from Table 1 of Fig 10, of seat height for population extremes and finding corresponding D in Fig 11. A pedal angle is then selected to lie within allowable range. Throttle and brake pedal surfaces are then aligned for quick foot interchange, so the throttle pedal is mounted closer to the SRP than the depressed brake pedal (using smaller knee angles than for the brake pedal when calculating

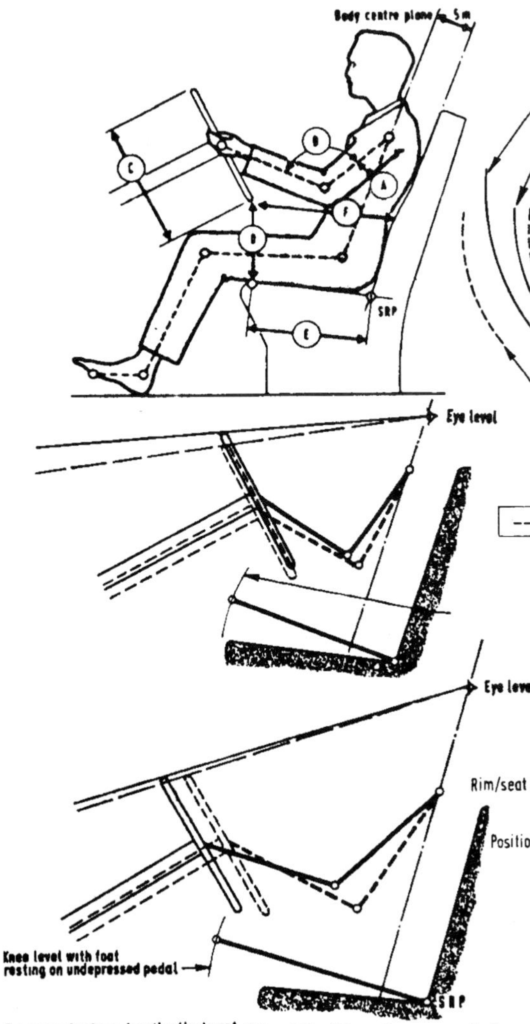

Fig 12: Hand control data

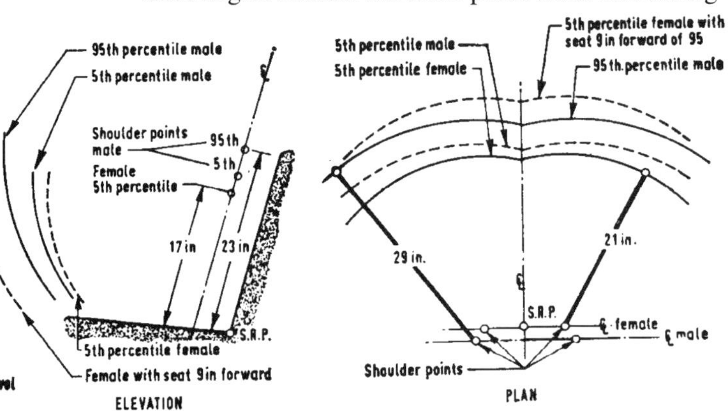

Fig 13: Reach envelopes

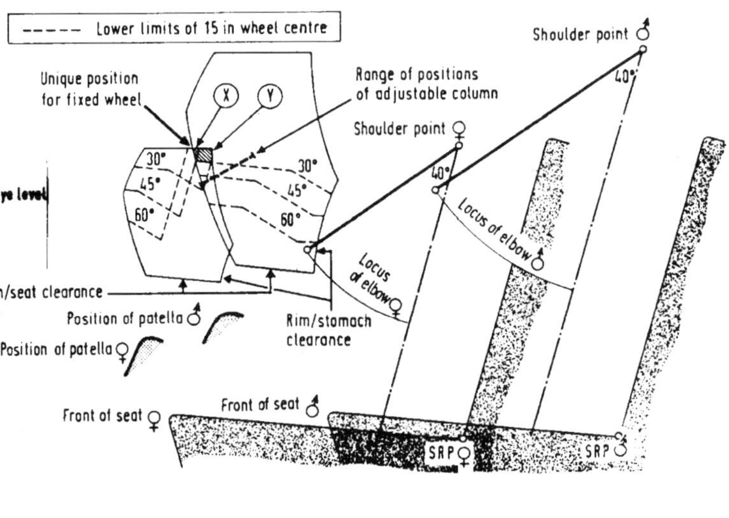

Fig 14: Improved vision with adjustable column

the position). Sufficient throttle spring loading is required to support the foot (48 lb or 22kg). As an example, when E = 55 degrees to suit the smallest woman on an 11 in (280 mm) seat, A is 115 degrees for the largest man on a 7 in (177.5 mm) seat. If depression is 15 degrees, this gives an unacceptable maximum for A of 130 degrees. The remedy is to increase E to 60 degrees and reduce the depression.

Hand controls

Murrell relates the areas of hand reach to the SRP as in Fig 12 — applying to the seat-belt restrained driver. Note that the large stature person seems unusually badly off in relative terms — but this is due to fact that he/she must sit far back to ensure proper use of foot controls.

Several factors affect steering wheel position; applied hand force depends on column rake, perpendicular distance from SRP to wheel centre and seat back inclination. Maximum force is available when pulling longitudinally with fore-arm and shoulder acting together — while faster, but low efficiency operation, motion is obtained with movement across the body. A wheel tilted 30 degrees to the vertical could have a rim force 1.7 times that with a vertical wheel but the rate of turning is halved and the energy expended reduced considerably. A competition car thus has a near vertical wheel and an extended arm position gives maximum speed for a short-term high stress activity. Heavy industrial vehicles, by contrast, favour the flatter wheel.

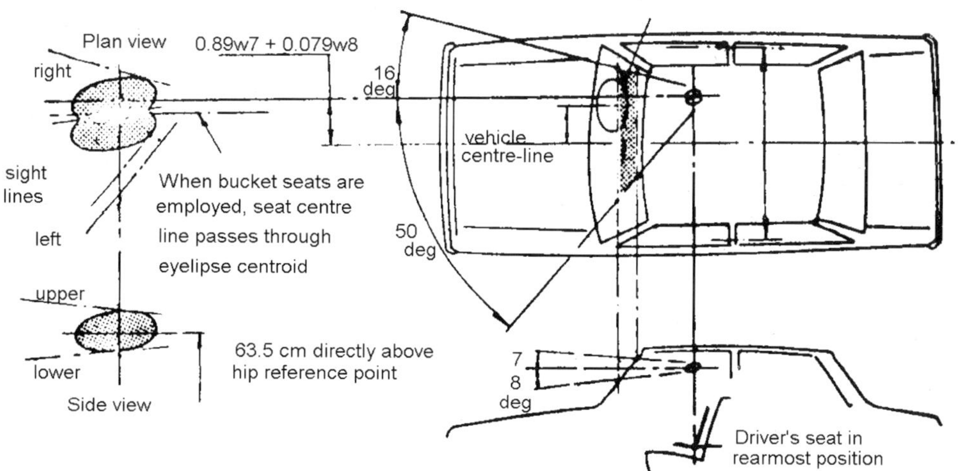

Fig 15: Eyelipse location data

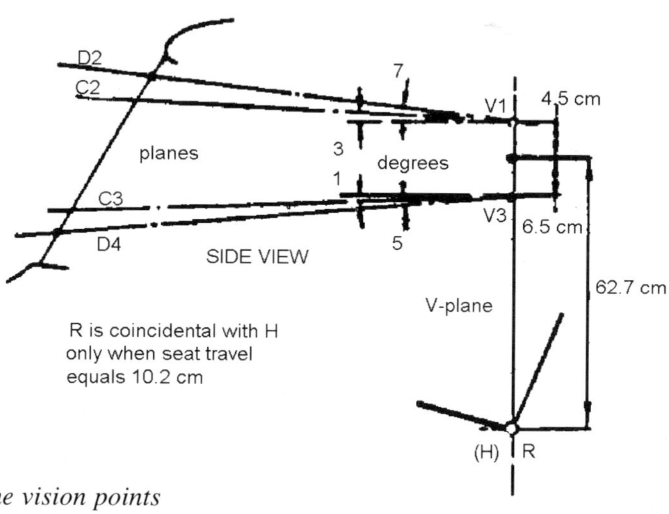

Fig 16: Position of the vision points

Design procedure is to locate wheel centre and inclination for population extremes then determine reach envelopes using dimensions in Table 1 of Fig 10 and Figs 9a and 11. Wheel centre must then fall in the common area shown in Fig 13. If the column length is non-adjustable, no position meets population extremes; for a 45 degree wheel the small area X applies and for a 30 degree wheel the larger area Y is common. If column length is adjustable by 3.5 inches, for this example, then a satisfactory condition exists for both driver extremes — anywhere with a 30 degree wheel and for a small range within the 45 degree inclination. Fig 14 shows the resulting improved woman's vision when an adjustable column is used.

Forward vision

The SAE Eyelipse is a common standard used for optimising forward vision; this is a contour of eye position boundaries for a range of driver statures. Tangents drawn from the contours are constructed to obtain vision areas. Fig 15 shows the method for locating the eyelipse and European practice is to locate the plan-view eyelipses on a longitudinal line passing through the centre of the driving seat. Since the maximum sideways viewing angle of the eye is 30 degrees, some geometric modification to the planview contour is required above this angle. Fig 16 indicates the position of the vision points (V) within the vehicle and it can be seen that a plane in which the V-points are located is required to be positioned 0.4in forward of the hip reference point (joint articulation centre) for each 0.5 inches of longitudinal seat travel in excess of 4 inches, in this example.

Car and CV drivers and passengers

Another USA standard, the AMA measuring system, has been used widely to determine the relative benefits of vehicle space envelopes. This is shown in Fig 17 with standard manikins used in the system and the critical dimensions defining space and comfort for occupants. Comparisons between a sports saloon, bus and limousine, Fig 18, are interesting in this context. In passenger service vehicles, the journey time exerts considerable influence on comfort and

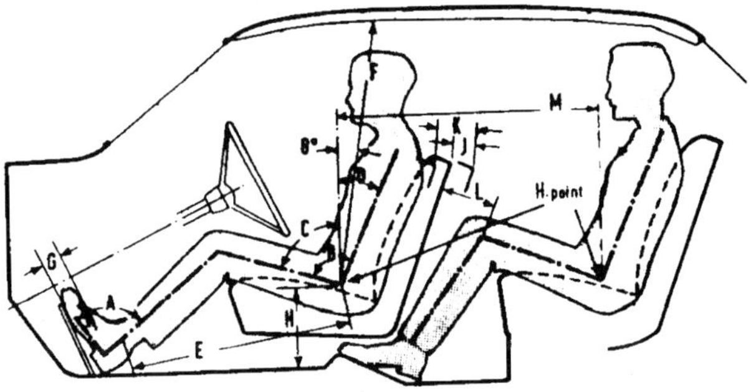

Fig 17: AWA measuring system

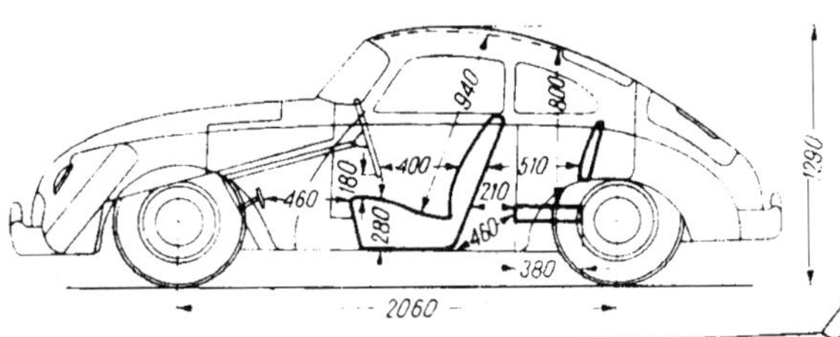

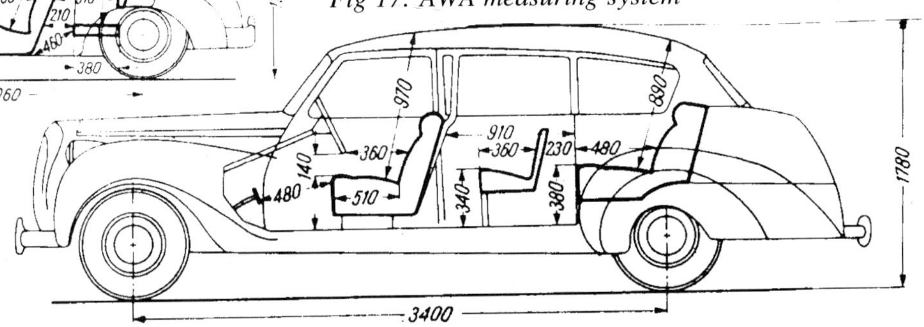

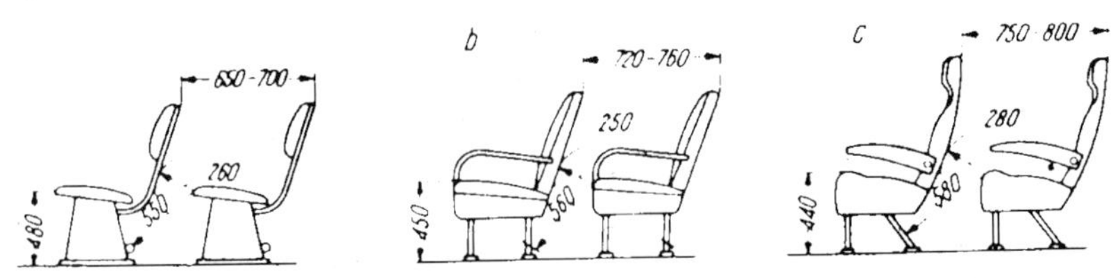

Fig 18: Sports saloon, limousine and bus dimensional comparison

thus seating dimension requirement varies for city-bus, inter-city bus and touring coach. The driver's cabin for a PSV also requires different criteria from a passenger car as shown in the Leyland National bus example of Fig 19. This vehicle was also studied extensively from the interior movement and entry/exit condition for passenger comfort, Fig 20.

Computerised approaches to ergonomic system design

The advance of computer simulation into ergonomic package and driver-control layout has revolutionised the interior design process. Standard codes are now available which enable the package to be readily related to the vehicle structure and the layout of mechanical systems.

The Systems Engineering Dept at GM has been prominent in the use of ergonomics to optimise occupant-packaging design of vehicles. The techniques rely on a human factors database which has been built up over many years. Studies on hundreds of human subjects have resulted in standardized drafting templates, leading to analysis of the occupant's space envelope within the vehicle, such as that in Fig 21. As well as the static anthoprometric measurements, such as those described in the previous section, so-called functional, task-oriented, measurements are taken with the human body at work, in motion, in driver- and passenger-space attitudes. These are expressed in three-dimensional co-ordinates around the major reference points, Fig 22a and b. Key points are H (hip), BOF (ball of foot) and AHP (accelerator heel point); such procedures are now standardised under SAE codings such as J1100.

Selection of design limits is carried out to determine what percentage of the population is to be accommodated and to arrive at the amount of seat adjustment required. When seat position is plotted against percentile, an almost linear portion com-

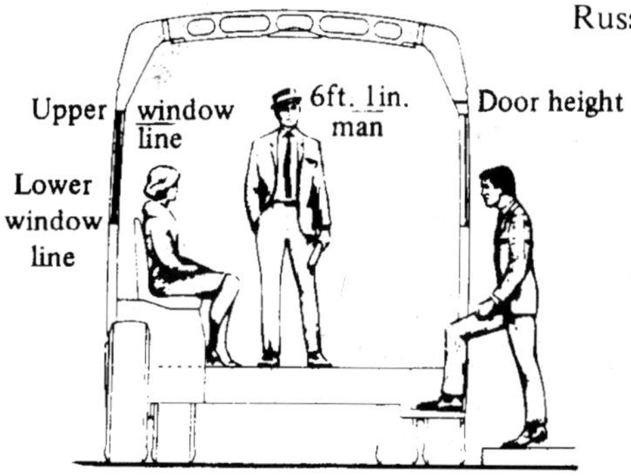

Fig 20: Entry conditions for Leyland National bus

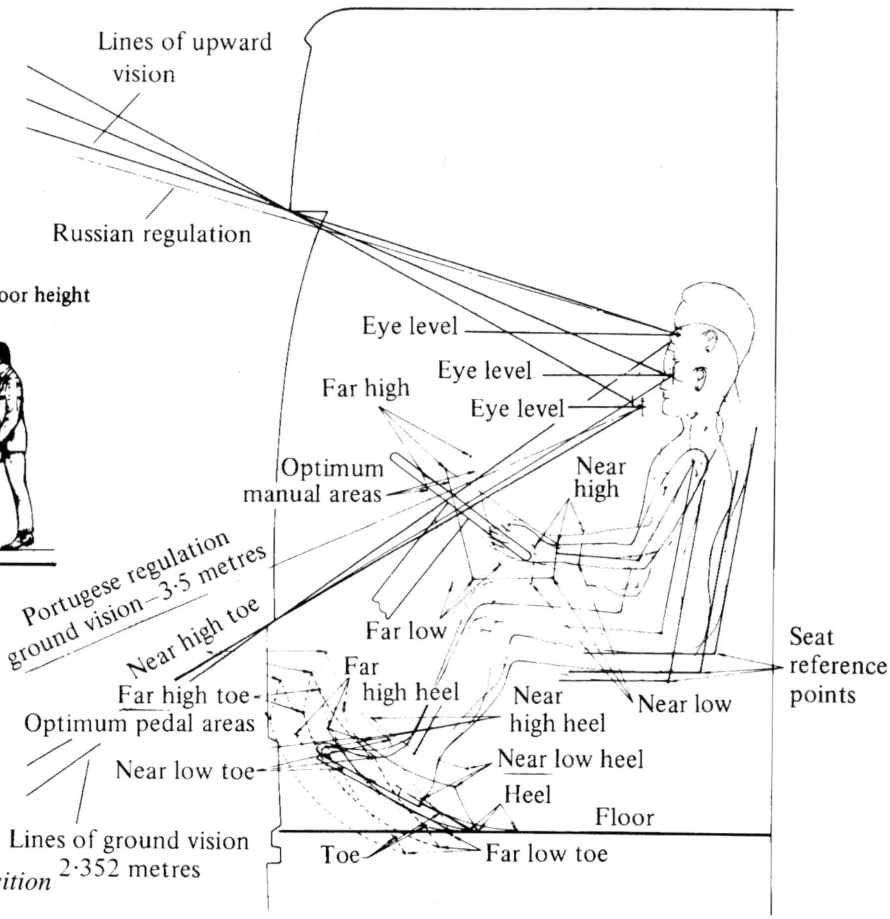

Fig 19: Bus driver seating position

prises the central 90% of the sample, Fig 23, and in this example horizontal adjustment of 168 mm, if properly positioned, will accommodate this range. The difficulty of simply placing a fifth percentile female manikin on a vehicle drawing, in order to obtain the fifth percentile reach diagram, say, is that the interaction of anthoprometrical variables with the workspace cannot always be predicted.

This has led the SAE to a task-oriented statistical approach involving the development of functional anthoprometric models, including computer-aided-design tools. These describe 'probable' location of the reference points, expressing design limits in percentile levels of accommodation, the table in Fig 24 providing an interesting comparison between task-oriented models and manikin ones. Such techniques were used to draw up SAE J287 Driver Hand Controls Reach (1990) — and a typical envelope is seen in Fig 25. As well as statistical models of the main reference points, other tools used include H-point machines such as that in Fig 26 and vehicle dimensioning procedures that provide standard measurements as a database for analysis and comparison of vehicle interior space.

Usually the packaging engineer develops the vehicle from the ground up, the seat height and inclination being variable to vehicle category between, say, sports car (low) and van (high). Ground clearance and underframe structure define the floor level — which locates accelerator heel point and so on. Finally, sight lines are drawn from the eyelipse

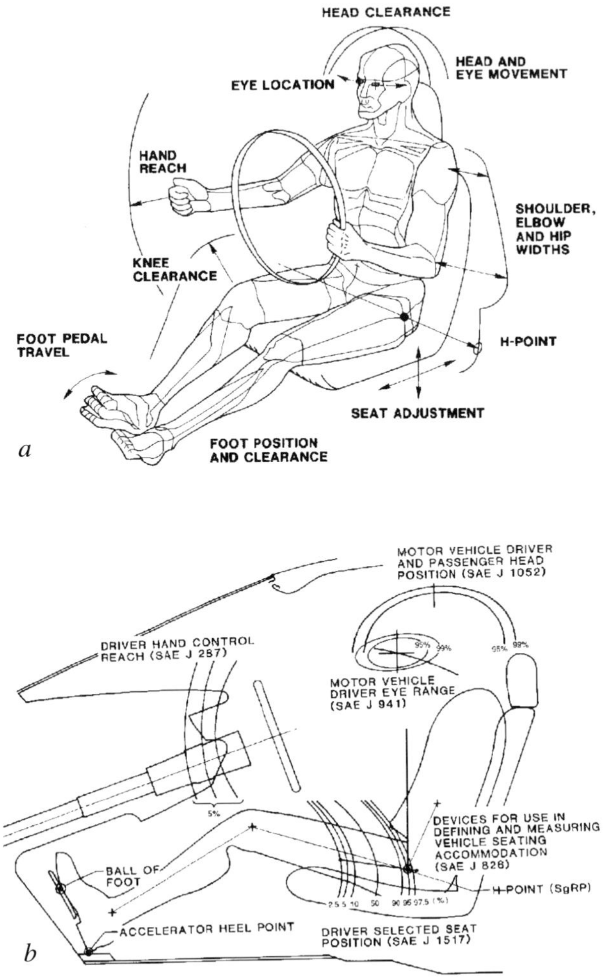

Fig 22: Major reference points

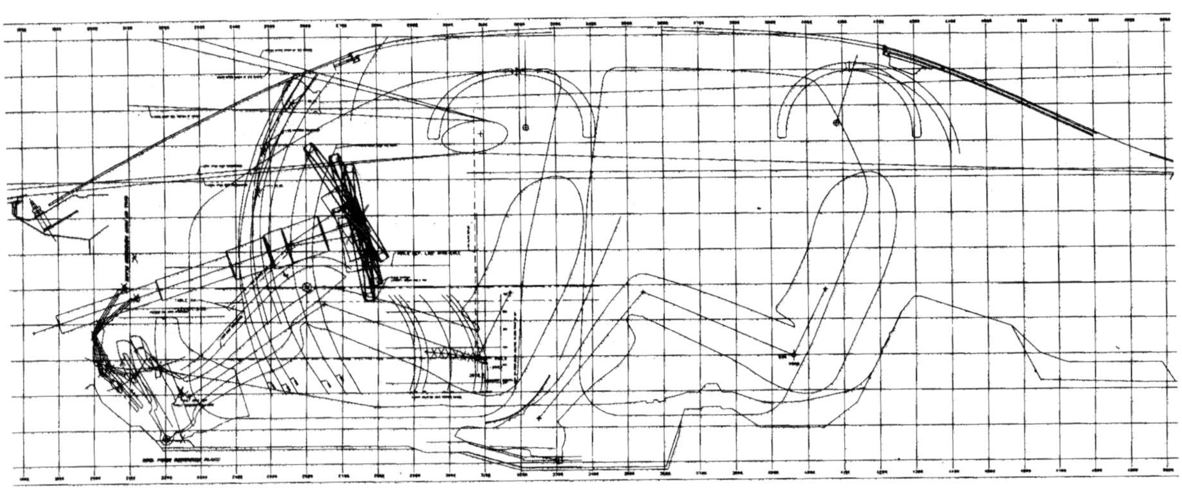

Fig 21: Interior space envelope

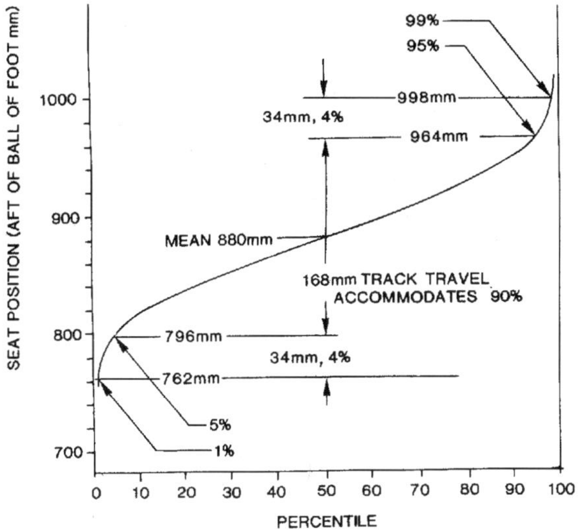

Fig 23: Percentile plots to find seat adjustment

Task oriented percentile models	Manikin oriented models
Uses anthropometric data to define a user population stratified across stature, weight and age.	Uses anthropometric data to define 95 per cent male and fifth percentile female dimensions.
Conducts tests for specific tasks (reach, eye location) in vehicles to develop statistical models defining spatial requirements.	Uses manikins and/or selected large (95 per cent) males and small (5 per cent) females to define spatial requirements.
Assumes that sample populations define user requirements.	Assumes that specified large males and small females comprehend all user requirements.
Assumes user needs are expressed by a central tendency (usually a normal Gaussian distribution) with exclusions at both ends, for example 95 per cent accommodation with 2.5 per cent excluded at each end.	Assumes that 2-D (or 3-D) manikins can predict or model human requirements Assumes that a given percentile person is definable from the sum of the parts.
Many levels of accommodation are described, but only for the task studied.	Many tasks are measured and evaluated, but only for a defined small and large user.
Results in well defined *statistical* model that defines accommodation levels for *specific tasks.*	Results in questionably defined *geometric* manikin models that pedict accommodation for only *two extreme percentile people.*

Fig 24: Task-oriented vs manikin models

and roof clearances measured to the head-position contours, and the previously discussed reach-clearances included.

One of the most sophisticated computer-aided-ergonomics packages is SAMMIE-CAD, now being developed by Loughborough University. Models are constructed from primitive shapes such as cuboids, spheres, cylinders, meshes and polyprisms which can be specified interactively, by name and key dimension. Irregular solids can also be modelled and curves are represented by multi-facets. A reflection facility allows mirror-image copies to be made, which simplifies defining the second side of a vehicle, say. The program's workplace modeller is strong in its ability to specify logical or functional relationships between components, important for the operation of doors, pedals and adjusters — by the man models of Fig 27. The man model is constructed from 17 pin-joints and 21 rigid links encased in solid modules of 'flesh'; link lengths and joint constraints are data-driven and can easily be controlled by the user. The designer interacts with the system via a menu-driven interface, some 30 menus each having up to 20 commands. The vehicle model is built up from data as previously described and can be complex or relatively simple, as in Fig 28.

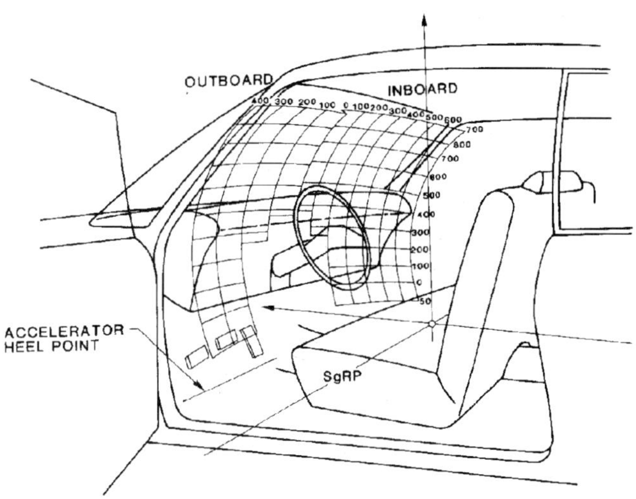

Fig 25: Control reach envelope

Measuring seat comfort

Researchers at Ford have constructed sensitive manikins for measuring seat comfort which are based on a distribution of small load cells over the contacting surface. This leads to pressure distribution 'contour maps' such as that in Fig 29. Such an apparatus is used in conjunction with special seats whose cushion-springing can be adjusted by means of helical springs distributed across the surface which can be adjusted to different rate settings. A computer analysis procedure then co-relates the two distributions for a variety of human test subjects. An optimum comfort distribution, as shown in Fig 30, is then reproduced as closely as practically possible, on the manikin. The manikin can then be used to rate different seating systems for comfort level.

Displacement rather than pressure is also used as a measurement criterion for seat comfort in vehicles such as AFVs and racing cars where the driver needs to be held still by a fairly elaborate support mechanism. In work carried out by Southampton University, a lateral backrest support was constructed from foam blocks for this purpose, Fig 31, and the motion of the driver's head measured by an ultrasonic displacement transducer. Subjects were subjected to 240 seconds of Gaussian random lateral (y-axis) vibration at 1.0 metre/sec/sec rms with a frequency bandwidth of 0-3.15 Hz, in stages, while backrest height was altered in 100 mm increments between 0 and 700 mm. Probability density functions of head displacements were calculated from the time histories given by the ultrasonic transducers and the mean distribution across the 16 subjects for the eight seating conditions is shown in Fig 32.

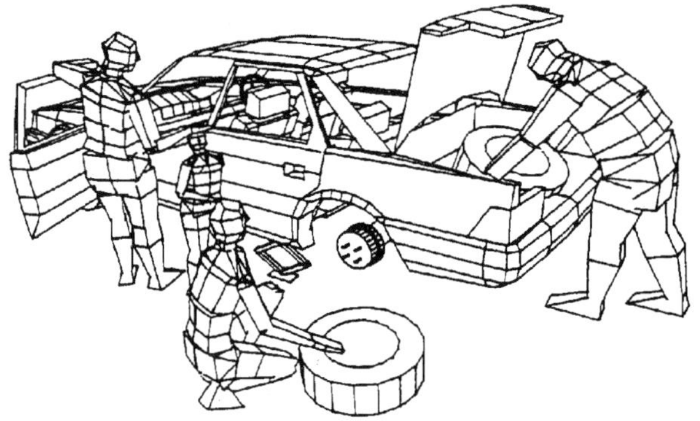

Fig 27: SAMMIE man models

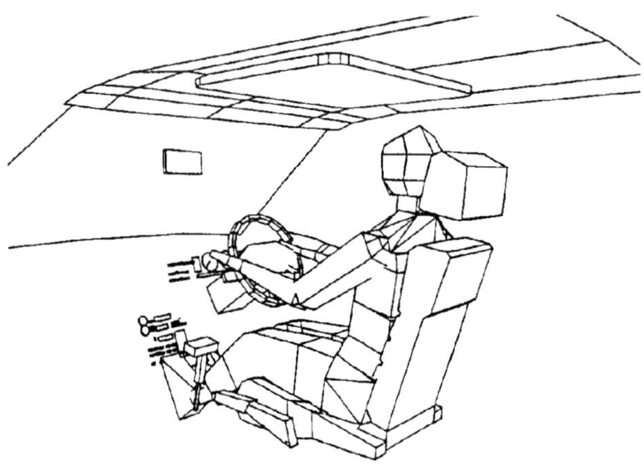

Fig 28: Simple vehicle model

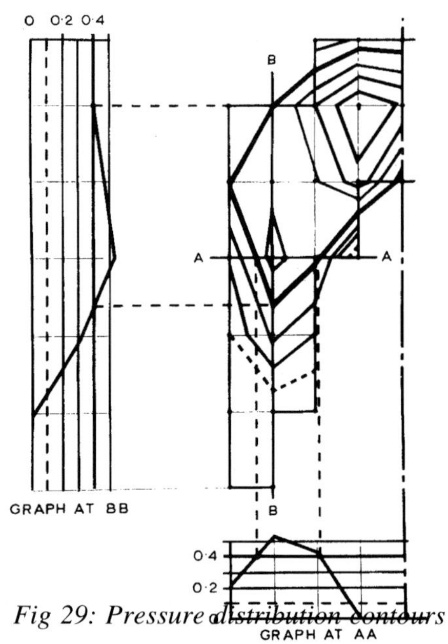

Fig 29: Pressure distribution contours

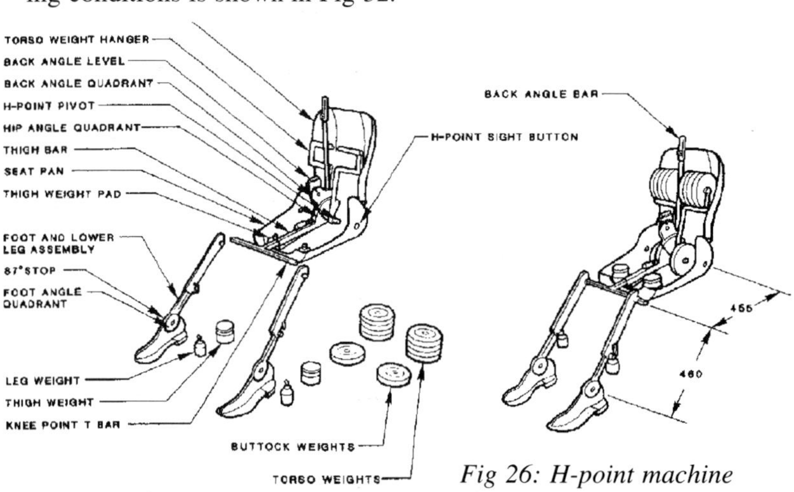

Fig 26: H-point machine

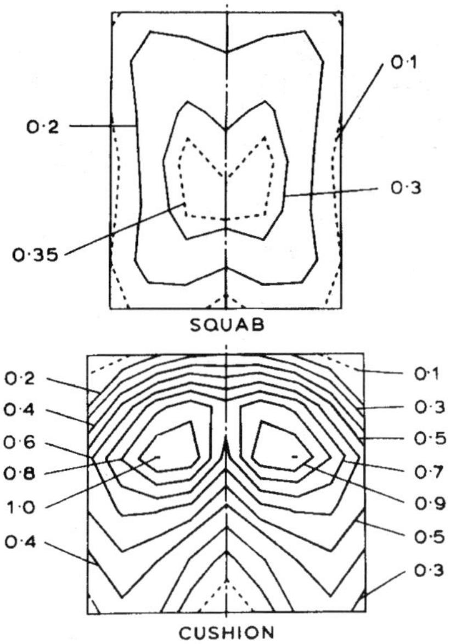

Fig 30: Optimum comfort distribution

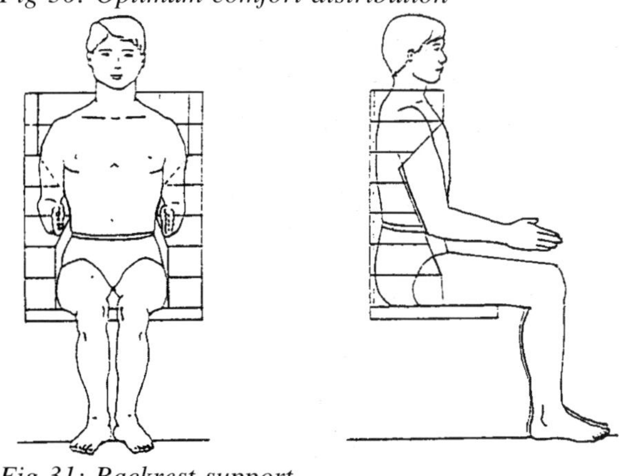

Fig 31: Backrest support

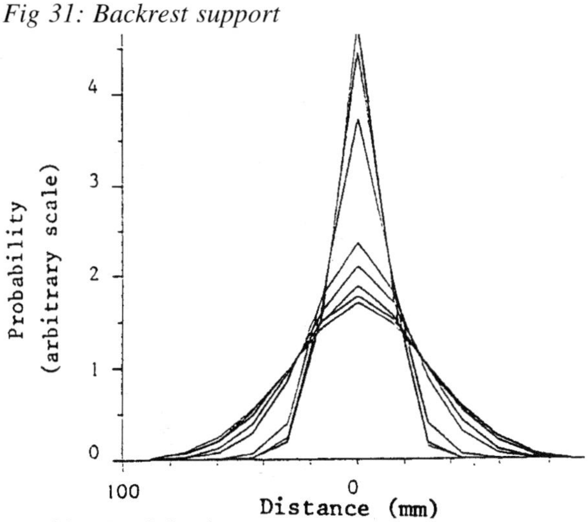

Fig 32: Head displacements measured

Seat cushion deflection and ride comfort

According to Delphi Automotive engineers[3], the dynamic response of vehicle seating has tended to be overlooked in earlier studies. The authors point out that the following seat deflection data is now being utilized for designing and evaluating automotive seats. Indentation load deflection (ILD), D-point penetration (DPP) and thigh point penetration (TPP) for specifying desired cushion deflection characteristics. Design engineers use such data as guidelines during the design and evaluation process, and to meet H-point requirements. Human Factors Engineers utilize deflection data for studying posture, lumbar position, occupant feel, eye position and clearance between human body and hardware. The load-deflection data may also be utilized for competitive benchmarking, assessing stiffness, and to estimate seat natural frequencies.

The local stiffness is considered to influence occupant feel and driving comfort. Similarly, seat natural frequencies are crucial for ride comfort and quality since these frequencies should not interfere with vehicle or occupant body natural frequencies. Furthermore, the stiffness characteristics derived from the deflection data can be utilized for tuning seat dynamic response to specific road vibration inputs. Static deflection of a seat system is one of the objective parameters which can be effectively utilized for improving the quality of seats for comfort, as well as optimizing automotive ride and handling characteristics. Therefore, it is important to study the relationship between seat deflection and occupant comfort under dynamic conditions.

The natural frequency value of the seat is significant for occupant comfort and ride quality of an automobile because, if the seat natural frequency matches with vehicle ride frequencies, the seat may affect vehicle ride handling characteristics, causing squeaks, rattles and high magnitudes of vibration, leading to occupant discomfort. Also, if the seat natural frequency matches with human body resonance frequencies, then the occupant body components may undergo large motions resulting in extreme discomfort.

The tuned natural frequency should not vary with a change in occupant or ride condition. In order to meet this essential design criteria, seat engineers are required to have tools to determine dynamic response characteristics of seat subsystems. Seat cushion deflection is an important parameter, useful to calculate cushion stiffness characteristics and to estimate the undamped natural frequency of a seat system, which may possess different stiffness/spring characteristics. An automotive seat consists primarily of soft foam, trim and suspension and will deflect when an occupant is seated. Therefore, the seat engineer has the task of deciding the type of spring characteristics required to accomplish design objectives for maximum comfort. Seat cushion deflection data can also be utilized to determine the minimum clearance required between the bottom of the seat cushion and body floor pan.

The local stiffness properties of the seat cushion influence the pressure distribution at the cushion-occupant interface and therefore is a significant parameter that influences occupant feel at all times. Emerging and innovative design concepts such as adjustable lumbar and adjustable back essentially provide the occupant with the ability to control the desired local stiffness. Scientific studies have indicated that electromyography (EMG) of the lower back and trunk muscles varies with changes in seat surface stiffness. Therefore, the spring rate or the stiffness coefficient is an important parameter of subjective feeling. This denotes that the seat engineer decides what amount of stiffness is required for the targeted market for maximizing seat comfort.

Fig 33 shows a typical deflection curve the amplitude x representing dynamic deflection with respect to the equilibrium condition, for both rising rate (duffling) and linear rate springs and Fig 34a shows variation in natural frequency with vibration amplitude for such springs plus a falling-rate (softening) spring. This demonstrates that the natural frequency of a non-linear spring varies with changes of amplitude. Fig 34b allows consideration of the response of a hardening spring where the vibration amplitude increases with excitation frequency up to a point where damping limits the motion. At the point of vertical tangency of the response curve, slight increase in frequency causes a jump down in amplitude to the unstable region of the response curve. Thus, the authors conclude that, in order to maintain a constant natural frequency for all levels of ampli-

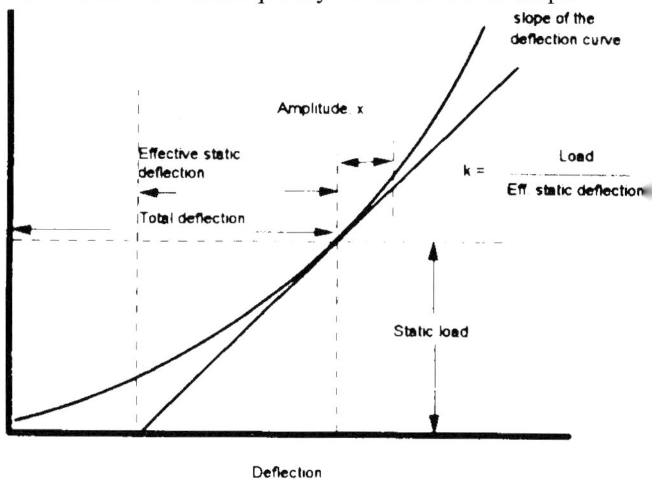

Fig 33: Derivation of stiffness coefficient (load: x-axis)

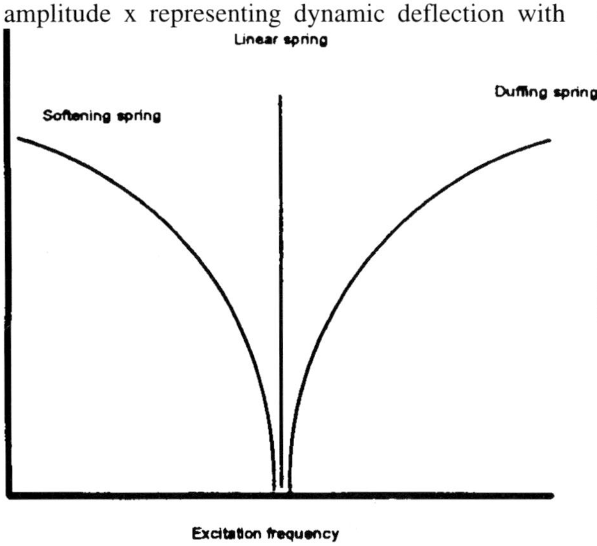

Fig 34a: Amplitude/frequency relationship for different rates

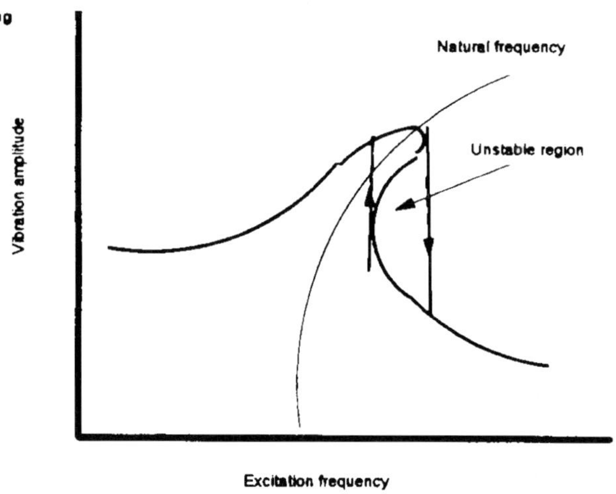

Fig 34b: Unstable response of progressive rate spring

tude, linear spring characteristics are required.

In another presentation Delphi authors[4] consider a further comfort criterion, that of comfort perceived during posture change on the seat, in relieving discomfort. General improvements in vehicle ride comfort have made the study of seat-occupant interface pressures and vibration levels less significant in the authors' view. Perceived comfort relates both to the level of discomfort and to the elimination process of discomfort, it is argued. In Fig 35, at Phase I when discomfort is severe, comfort is at the lowest. At Phase II, discomfort level is being reduced, and comfort increases due to the less severe level and to the reduction rate of discomfort. Phase III is when discomfort is removed but perception of comfort adapts, possibly to an "indifference" saturation level. At Phase IV, some insignificant feature may become the focal stimulus and discomfort creeps up to attract occupation attention, so that perceived comfort drops. Phase V represents another cycle of discomfort removal.

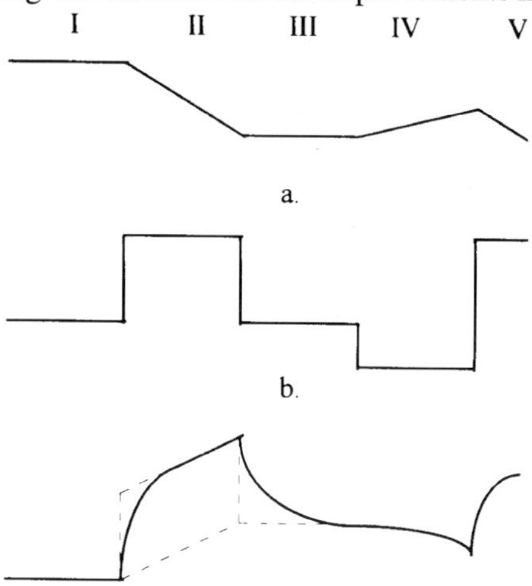

Fig 35a: Time courses of comfort: a. time course of discomfort; b. rate of reduction of discomfort; c. as a delayed response function of a and b

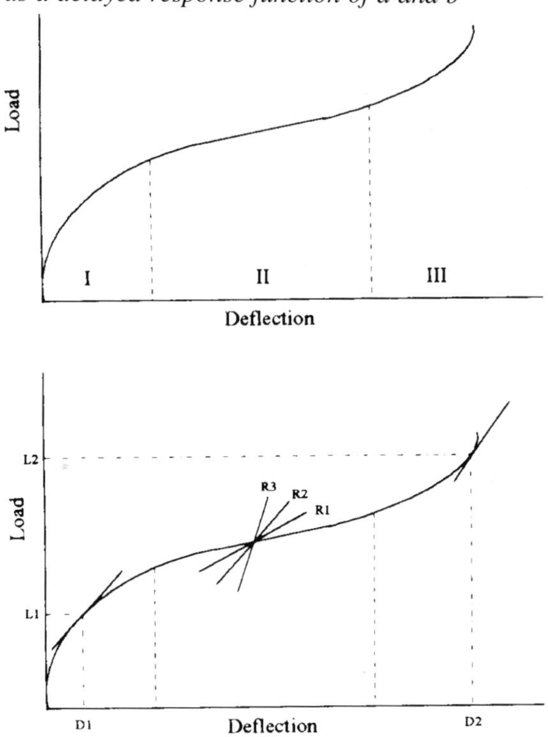

Fig 35b: Seat force-deflection curves (local modules at D1, L1 and D2, L2; dynamic modules at vibration rates R1<R2<R3)

The concept of temporal comfort has its impor-

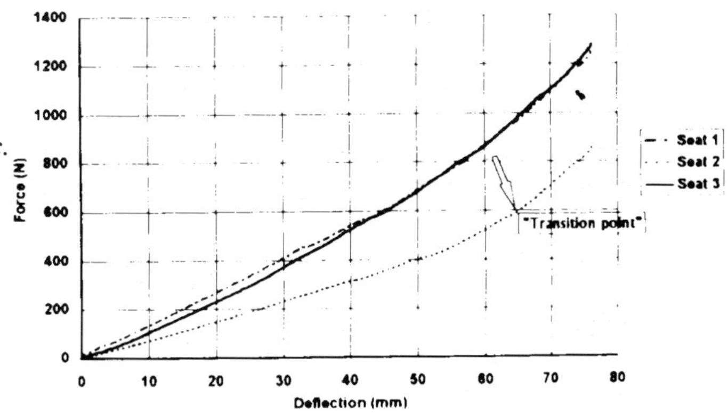

(a) Using buttock-form plate indentor

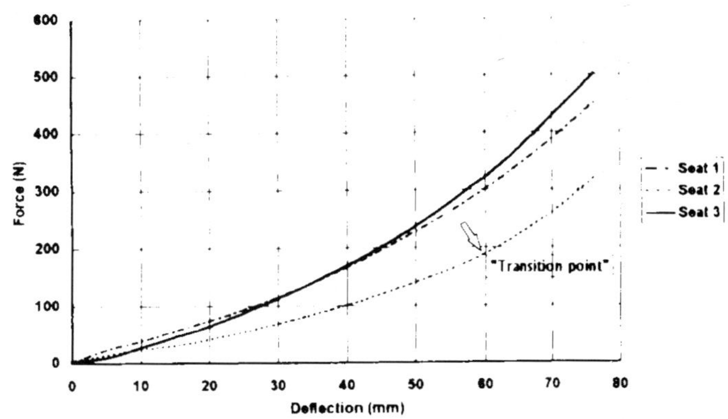

(b) Using a three-inch-diameter indentor at buttock-pocket

Fig 36: Force-deflection test for three seats

tant implication to the design of seats as it tells us, say the authors, that just as with any other tasks, sitting is a naturally fatiguing process. No matter how good a seat or a vehicle is, it cannot maintain an occupant free of discomfort for hours of sitting. And this is a biological mechanism to protect the human tissues from compression damage. If a product is to excel, it has to provide effective ways to minimize the accumulation of discomfort and help remove it. Careful consideration should be given to provide features for soothing and relieving discomfort. With such an active approach towards comfort some of the factors that cause occupant discomfort can be suppressed.

According to Fig 35a, the designer has to evaluate a seat in terms of its capability of relieving or eliminating discomfort, in addition to testing the absence of discomfort. The authors feel that there are some seat features which may promote comfort and, for lack of proper terminology, they call them transient features. If a seat provides proper postural support and occupant protection, the judgment of comfort is often made at the moments when the occupant is being seated and/or when he/she is seeking change of posture by fidgeting or repositioning. He/she may sit on the seat, bouncing around and compressing and twisting the body on the seat, a ubiquitous practice when someone is asked to give a quick rating on seat comfort. To better understand the transient features, the authors classified the process of ride comfort into three modes: initial comfort, transient comfort and dynamic comfort. This is described in Fig 35b, a hypothetical load-deflection curve of a seat cushion assembly.

The seat should provide a gradual, gentle, cosy support to the "landing" or indenting of occupant's body upon seat cushion. The initial softness and sensation of being gently but progressively cushioned is very important to occupant's perceived comfort. To properly support riding postures, the cushion should further produce increasing resilience to the body load indenting the seat, a feature well held in polyurethane foam. This property in foam padding may be expressed by the initial local modulus of the load-deflection curve. In terms of this principle, the local modulus at (Dl, Ll) of Phase 1, Fig 35b, should not exceed that of Phase 11 which is the normal supporting phase. The local modulus is the gradient of the curve at a given deflection under load. Earlier work concluded that the two seats with low initial local modulus were apparently more comfortable and preferred than the two seats with the highest initial local modulus. As an example, the situation depicted in Fig 35b obviously fails this requirement. Of course, the presumption is there are no significant problems with other seat features such as contouring. On seat-occupant interface, body pressure distribution may examine the supporting quality by identifying hard points and inappropriate posture.

Transient comfort is perceived as the subject adjusts posture or fidgets around on the seat. Riding tasks often require frequent change of posture, so do the muscle groups when some relaxation is in need. A comfortable seat should not constitute any local impact or interfere with such changes, but provide gentle damping effect on the transient loading. This again relates to the local modulus on the basis of settlement deflection, the deflection at static body load. After the occupant sits down and the cushion settles, the initial local modulus does no affect perception, but the local modulus at the settled deflection starts to influence the perception of cushion comfort. The amount of settlement deflection depends mainly on occupant's body weight. The majority of automotive seats operate at a settlement deflec-

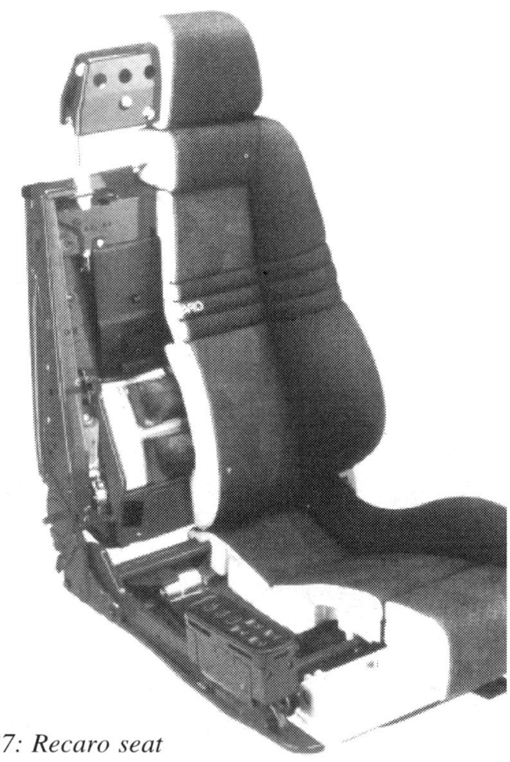

Fig 37: Recaro seat

tion of 15 to 50%, at the second stage on the load-deflection curve. However, a large and heavy occupant may indent the seat more than 50% close to the Phase III when the cushion hardens up or "bottom-out". A small occupant may indent less than 15%, and in the case shown in Fig 35b, the seat would also feel harder.

Fig 36 describes results of a force-deflection test on three automotive seats, using (a) a buttock-form indentor and (b) a three-inch-diameter flat indentor, simulating the vicinity of the right ischial tuberosity. Seats 1 & 3 have no apparent differences between stages and were subjectively perceived as being slightly too hard. Seat 2 has low modulus before 65 mm deflection, the "transition point", and then the modulus increases to a level similar to that of the other two seats. This feature provides excellent initial and transient comfort if the cushion deflection is less than 65 mm under the occupant's body weight. Yet the support is effective without bottoming-out as the modulus increases after 65 mm deflection. Fig 36b indicates that, in order to ensure initial and transient comfort, the weight taken by each ischial tuberosity should be less than 200 N. Since the ischial tuberosities take the majority of body weight in sitting, this load limit to the region is important to avoid the numbing feeling of the buttocks.

Seating comfort under vibration conditions can be considered as the transient comfort when the imposed loading upon static settlement deflection is either cyclic or random. The local modulus or rather, dynamic modulus, depends partially upon the vibration rate, or frequency. The higher the frequency, the higher the dynamic modulus. However, transmissibility measurements of real vehicle seats show high levels of vibration isolation above 20 Hz. Actual vibration sensitive frequency is between 0 and 6 Hz. For this frequency range, reducing seat dynamic modulus is very important for achieving vibration comfort. However, measurement of such modulus may have technical difficulties, the authors explain.

Fig 38: ASTec technology

Proprietary seating systems

The German Recaro company are prominent in sophisticated up-market seating systems and have just set up a UK distribution centre. Features of the company's seats, Fig 37, include ergonomically-shaped upholstery to support the spine; extendable seat cushion to ease stress on the thighs and adjustable height and tilt. There are also shaped side bolsters on the seat cushion, and backrest, also a variable lumbar support system. The latter comprises air-chambers which are individually inflatable. Safety features include a crossmember in the seat frame to counteract submarining effect on impact. The seat cushion is mounted in suspension bushes to further isolate vibration and there is a vent system to optimize the control of seat temperature. The vent system draws moisture through the seat cover and the Gore-tex membrane into air-channels in the upholstery. Fans in the cushion and backrest draw the moisture out of the air channels and out of the seat via slots in the backrest. Optional heating elements are also available.

Another interesting development is the ASTec seat from Textron, Fig 38, which automatically adjusts to the occupants shape and size when sat upon. So called 'active surface control' is achieved by electronically actuated pneumatics to control 10 air cells positioned between upholstery and foam springing. Sensors on the cells measure pressure between the seat and the occupants back/lower-body, sending signals to microprocessors which store a bio-mechanical model in their memory, for comparison. Any discrepancy causes individual air cell pressures to be adjusted.

Suspension seats

For commercial and industrial vehicles, there is a strong case for suspension seats in which controlled levels of springing and damping can be introduced, Fig 39 showing one of the original Bostrom designs. In designing seat systems it is useful to observe measured comfort criteria such as those established by Janeway. A simplified approach is to regard the driver on his or her seat as a single-degree-of-freedom spring/mass/damper system with the familiar response characteristic shown in Fig 40. Maximum mechanical response of different parts of the body of a 170 lb (77 kg) man sitting on a rigid chair, and standing, are shown in Fig 41, due to Bostrom. This company has also measured vibration levels in a range of vehicle categories as shown by Fig 42. (A) represents on-highway trucks, wheeled earthmovers and farm tractors; (B) represents corresponding off-highway working and (C) is the same for non-suspended seats. Cushion design should observe the anatomical necessity of providing firm support only for the bones immediately above the buttocks since the surrounding areas have little pressure capacity. Too deep a cushion deflection would overload the surrounding areas and cause discomfort — an acceptable maximum is put at 3 in (76 mm). Fig 43 shows belt-level response of a man subject to 1/4 in (6.35 mm) amplitude vibration when sitting on three typical cushions, without squabs, which unlike suspension seats cannot have their characteristics altered to isolate offending frequencies. Cushion design has been revolutionised by moulding techniques developed for low density flexible foams. While hot moulding has been established for over 20 years, the recent existence of cold moulding has helped lower costs due to lower energy requirements and cheaper tooling. Two-component dispensing machines are required, with manually opened GRP moulds, as the basic equipment. Materials used are polyethers expanded by isocyanates. Fig 44 shows systems available from ICI Polyurethanes. Hardness of foam can be varied by adjusting the isocyanate component in different parts of the mould — giving the possibility of moulding dual-hardness cushions to obtain a soft centre with stiffer side bolsters.

Techniques are also available for moulding the cushion on to the upholstery system when temperature tolerant coldmoulding is used. For industrial vehicles, composite moulding of slush-moulded PVC

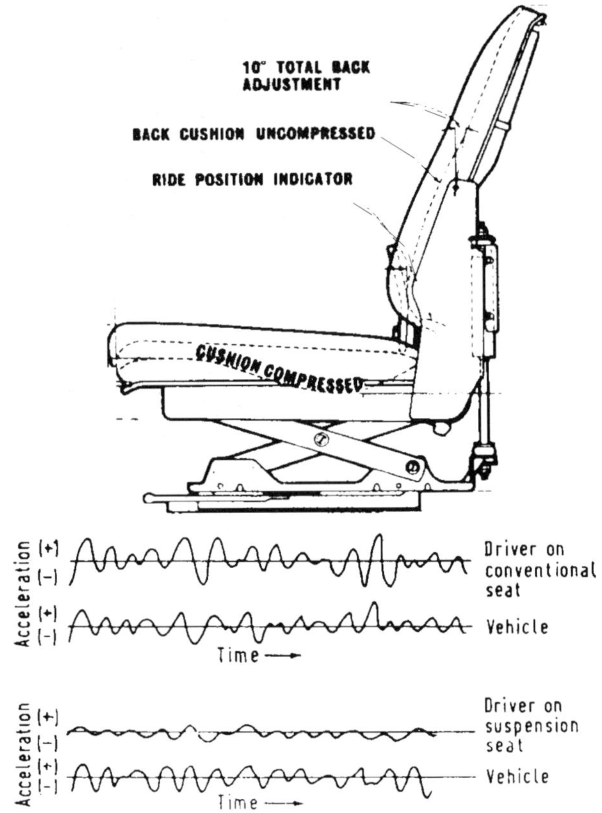

Fig 39: Bostrom suspension seat, with responses

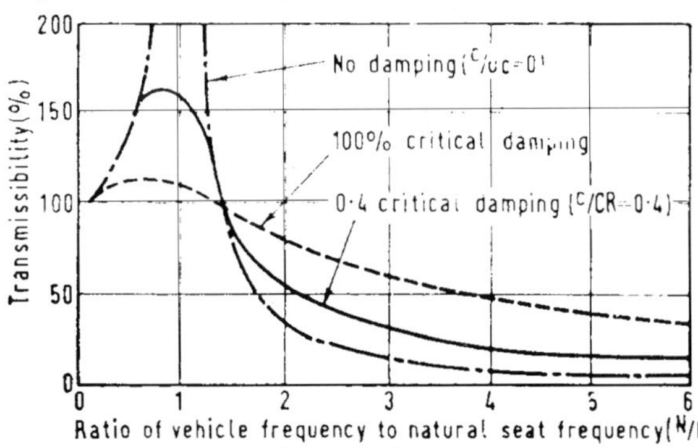

Fig 40: Response of single degree system

plastisol covers, filled with cold-moulded foam, is common. For other vehicles stretch-fabrics can be employed as the top layer of a multilayer laminate — over 23 mm thick peeled flexible foam, a PVC film and further inner layer of 1 mm thick foam. The so-called Skinform process involves such a laminate, being held under a controlled tension which is chosen according to the cushion design, drawn by vacuum into specially designed moulds. The coldcure cushion-foam is then moulded in-situ. The more traditional way of covering a moulded foam cushion pad, for coaches, might involve fabric moquettes. Examples in 85/15% wool/nylon are now available for coach seat covering, with backing materials which allow a thick pile effect to be created. For cars, nylon faced fabrics are typical, backed by foam and a layer of knitted cotton skrim. The exposed top surface is usually proofed against stain marking. PVC coated fabrics are also available which provide excellent simulation of leather — even to the extent of fine-perforation for breathing. Foam backings and two-way stretch backing-cloths allow cooling flutes to be moulded-in and a variety of embossed surface effects are possible.

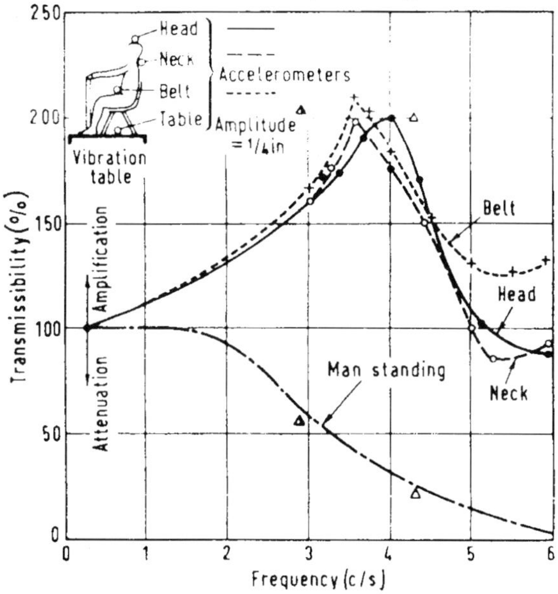

Fig 41: Different body area responses

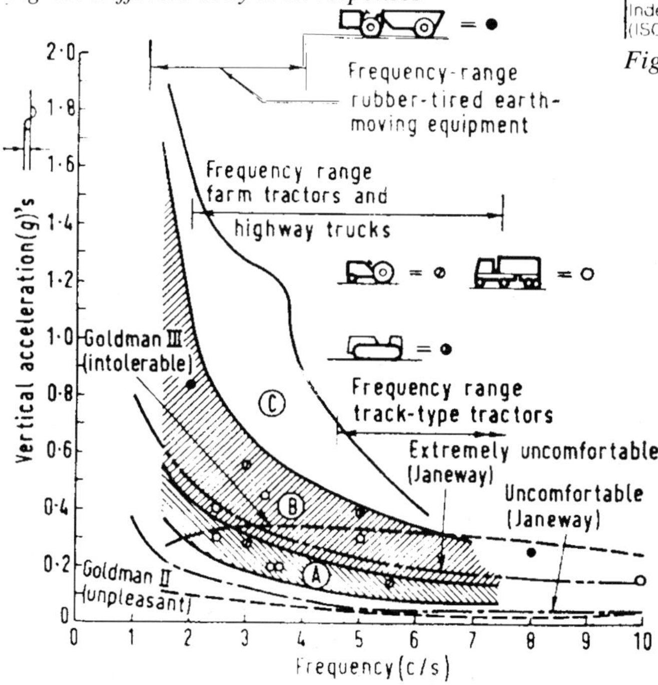

Fig 42: Different vehicle category responses

Formulation	Typical automotive system	Typical furniture system	
'Suprasec' VM 28 (pbw)	–	55.4	47.8
'Daltocel' C 971 (pbw)	–	100	–
'Suprasec' VM 25 (pbw)	63.3	–	–
'Daltocel' C 1565 (pbw)	100	–	–
'Daltocel' C 969 (pbw)	–	–	100
CFM-11 (pbw)	0 to 10	8	–
Isocyanate index	100	88	80
Processing conditions			
Polyol blend temperature (°C)	20 to 25	20 to 30	20 to 30
Isocyanate temperature (°C)	20 to 25	20 to 30	20 to 30
Mould temperature (°C)	40 to 45	ca 30	ca 30
Mould occupation time (minutes)	4 to 6	4 to 6	4 to 6
Post-cure	None	None	None
Foam properties			
Density (kg/m^3)	35 to 55	39	53
Compression set (50% compression 22 h at 70°C) - (% set)	7 to 10	4	7
Tensile strength (KPa)	145	60	80
Elongation at break (%)	140	110	105
Tear strength (N/m)	200 to 300	110	160
Indentation hardness at 40% (N) (ISO 2439)	150 to 300	165	200

Fig 44: Some MDI-based cold-cure moulding systems

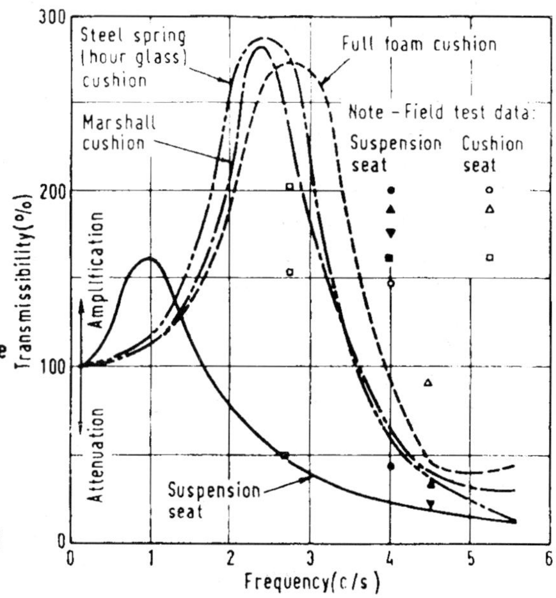

Fig 43: Belt-level response of vibrating subject

Reducing back-pain from seating

Engineers have realised for some years that a beneficial effect on seat occupants can be had by decoupling the motion of the seat back and seat cushion but little has been done to measure this effect and see how it could be exploited in vehicle seat design. Researchers at the Institute of Sound and Vibration Research[5] have now studied the concept in some depth and concluded that there may be real improvement by employing a moving back-rest system (sliding on near-upright rails), both in the higher floor-motion frequency range (4-16 Hz), in receiving more benefit from the seat cushioning, and in the lower frequencies, by reduction in strain on the lumbar spine caused by differential motion of seat and back.

The measurement system used by the Human Factors Research Group at the Institute is seen in Fig 45 and testing was carried out by mounting the complete rig on a one metre stroke vertical electro-hydraulic vibrator. Seat angles were measured using an SAE H-point manikin and seat back angle was set at 20 degrees from the vertical, while maintaining 97.5 degrees between seat back and squab. A foot-rest was used that could be adjusted to keep the occupant's upper legs approximately horizontal while maintaining about 120 degrees knee-angle representative of a typical driving position. Test occupants were in the 25-33 age range, weighing 61-76 kg and measuring 1.75-1.85 metres in height (the 50th percentile male weighs 74.5 kg and is 1.74 metres tall).

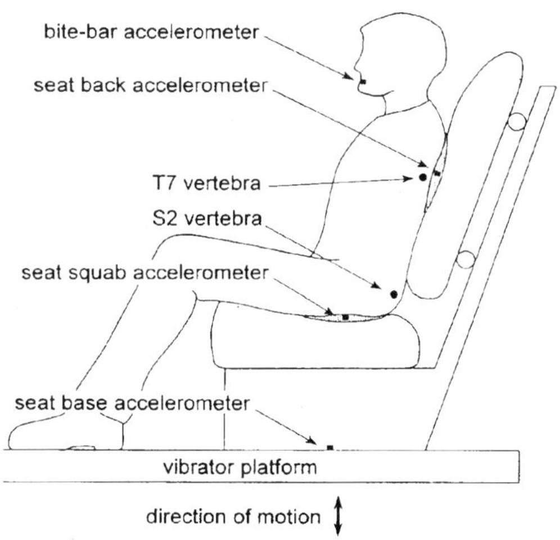

Fig 45: Seat rig set-up

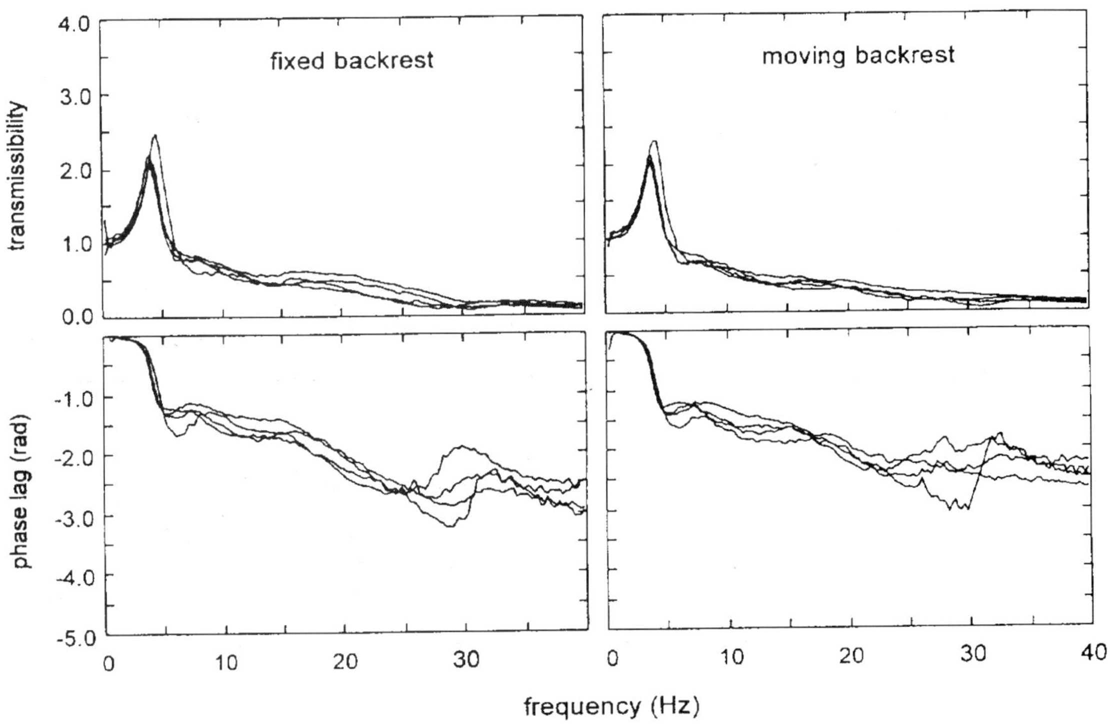

Fig 46: Vertical transmissibilities and phase differences from platform to squab for the four subjects at 1.5 m/s^2 r.m.s.

Input magnitude (m/s² r.m.s.)	Back-rest condition	(Seat surface)/(platform) Frequency of first resonance (Hz)	Transmissibility at first resonance	Phase lag at first resonance (rad)	(Seat back)/(platform) Frequency of first resonance (Hz)	Transmissibility at first resonance	Phase lag at first resonance (rad)
0.5	Fixed	4.75	2.49	−0.86	4.50	1.42	−0.32
	Moving	4.50	2.45	−1.01	4.25	2.63	−1.25
1.5	Fixed	4.00	2.12	−0.68	3.75	1.29	−0.18
	Moving	3.75	2.09	−0.68	3.75	2.17	−0.97

Fig 47: Mean frequencies, transmissibilities and phase lags of resonances between platform and seat surface/back for the four subjects

The slide travel for the seat back-rest was 105 mm and testing was carried out in the slide free and fixed conditions to measure the resulting effects. Input motion to the system was presented for 60 seconds at both 0.5 and 1.5 m/s² r.m.s. Accelerations at the six sites were measured using piezo-resistive accelerometers. Fig 46 shows the vertical transmissibilities and phase differences between the platform and the seat squab surface which shows the frequency of the resonance was greater with the fixed seat-back. The table in Fig 47 shows the numerical values of these frequencies and the transmissibilities at resonance with fixed and moving backrests for the two input magnitudes. With the moving seat back, at most frequencies between 6 and 40 Hz, the vertical motion on the back-rest was attenuated relative to the platform.

Fig 48 shows transmissibilities and phase differences between the platform and the lumbar spine which showed a degree of comfort roughly comparable with the occupant sitting with his back off the seat-back. Similar recordings were also plotted for platform-to-thoracic-spine/head motions and for platform to bite-bar. In most cases resonances occurred at frequencies around 4 Hz but were higher with the fixed back than the moving back, consistent with an increase in stiffness of the seat-body system when the seat-back is fixed. With the moving back-rest, the motion on the seat back was 104% of that on the seat surface for the 1.5 m/s² r.m.s. input while with the fixed back this fell to 60% which led to the conclusion of less lumber strain being imposed with the moving-back system.

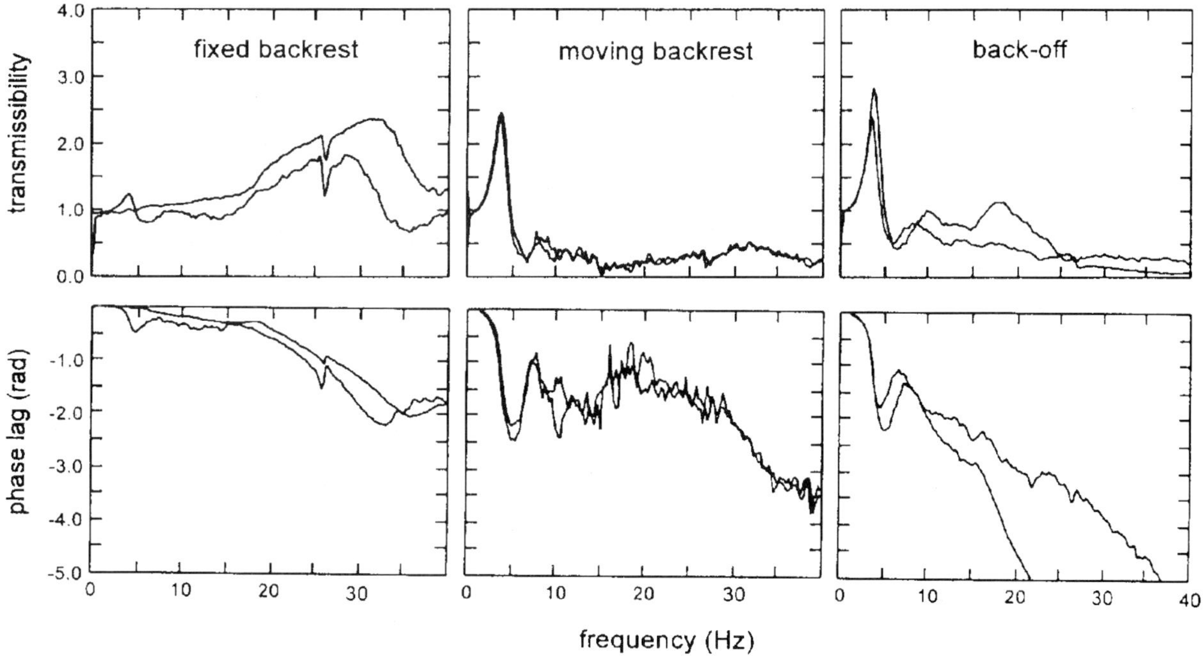

Fig 48: Vertical transmissibilities and phase differences between platform and lumbar back for two subjects at 1.5 m/s² r.m.s.

Split-frame seating

According to ITT researchers[6], with conventional seating when the passenger adjusts the seat bottom cushion, the entire combination, both the cushion and the back, moves together, unlike the split frame sea assembly in which the two components can move independently. With most seat adjusters, the occupant can raise and lower the front of the cushion. They can also independently raise and lower the rear of the cushion. However, with the split frame, the seat-back remains fixed, while with the conventional seat it also tilts. Research has shown that the distance between the lumbar mid-points, Fig 49, from the seated 5th percentile female to the 95th male is approximately 48 mm. In this case, the split frame seat with active lumbar support will work ideally. Since the cushion is adjustable independently of the back (usually about 40 mm or more of vertical adjustment is available) the lumbar support can be positioned precisely where an individual needs it most. In fact, when the seat cushion is positioned at a contact eye point (short and tall) for optimum viewing, the lumbar bladders are automatically placed at the optimum location on the back. Also, in a car with a split frame seat, the headrest can remain at a fixed height on the seat-back. This allows the seat cushion to be raised until the operator's head and eyes move into near optimum position for safe driving, with the head restraint automatically corrected.

In conventional seats, the seat—back is connected to the seat cushion, which in combination is supported on relatively long spindly linkages by the track and motor sub-assemblies (Fig 50a). These can lead to a feel of instability, and an overall perception of lack of precision in the total car package. In split frame seats, the seat back is secured directly to the track sub-assemblies system (Fig 50b) and bypasses all of the motion linkages. The seat pan linkages only have to support the dead weight of the individual and none of the torque applied to the seat-back. The load path for conventional seats requires that particular components must be strengthened and have mechanical robustness (Fig 51a). The components are normally manufactured from high strength heavy gauge steel in order not to fracture during any crash. The spilt frame seat, on the other hand, is arranged for the load path to travel more directly from the seat-back to the floor pan. Here the seat cushion is completely bypassed. Consequently, the split frame design can actually dispense with some components, because there is reduced loading during both normal use and any crash (Fig 51b). Additionally, the integral seat becomes more practical for the average vehicle. The loads are transmitted directly to the floorplan and body cross members, and not by the seatpan and adjustment mechanisms. The configura-

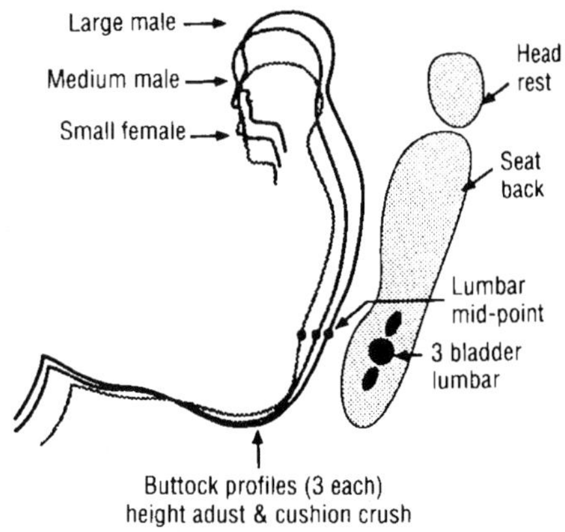

Fig 49: Split-frame lumbar positioning

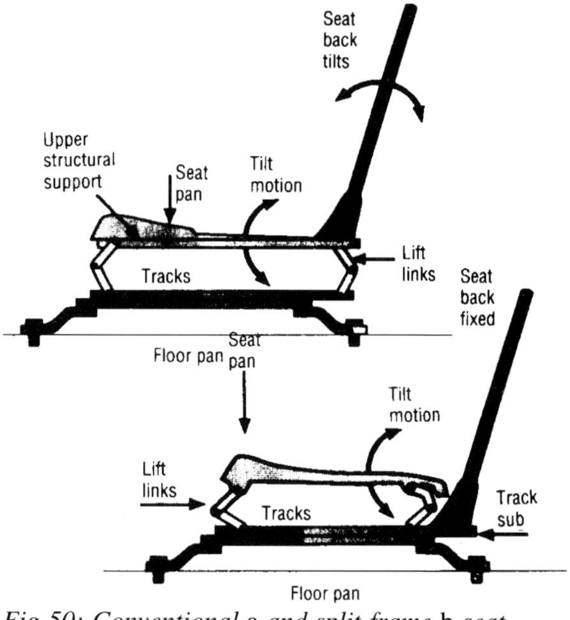

Fig 50: Conventional a *and split-frame* b *seat construction*

a

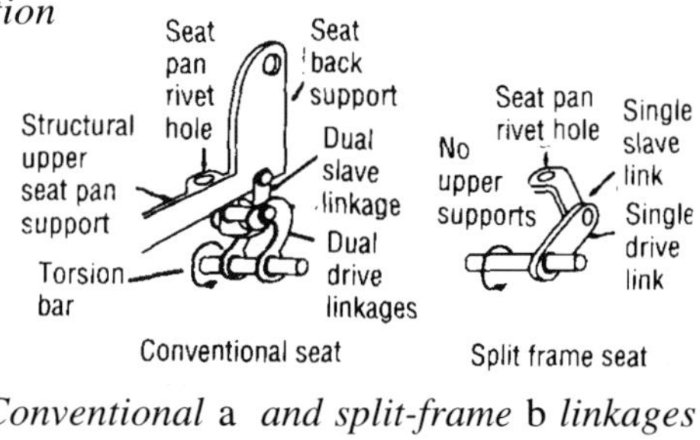

Fig 51: Conventional a *and split-frame* b *linkages*

tion shown in Fig 52 can withstand 10kNm torque about the recliner pivot without failure.

In conventional seating, a small person can have trouble sitting comfortably against the length of the back cushion. Conversely, taller people will have their knees overhanging the cushion, and thus lack good thigh support. The split frame seat can be designed to lengthen the bottom cushion as the cushion is lowered, and shorten the effective length as it is raised, see Fig 53. Since in a split frame design the seat back is at a fixed height, an unobstructed view of the instrument cluster can more easily be achieved. Moreover, since the load path goes directly into the seat track sections, the strength required in the structural pieces of the seat systems can more easily be predicted. In manufacturing the advantage for the split frame is that it can be assembled as a completed frame module and fully validated before it is trimmed out. Conversely, the complete split frame module is more awkward to handle on a trim assembly line which may still use manual trim processes. One advantage for the conventional seat is that the seat cushion and back are normally trimmed independently as separate components. Then they are bolted together and attached to the power seat tracks. However, the entire unit must now be validated, and if there are any problems, then the whole seat may have to be detrimmed in order to be repaired.

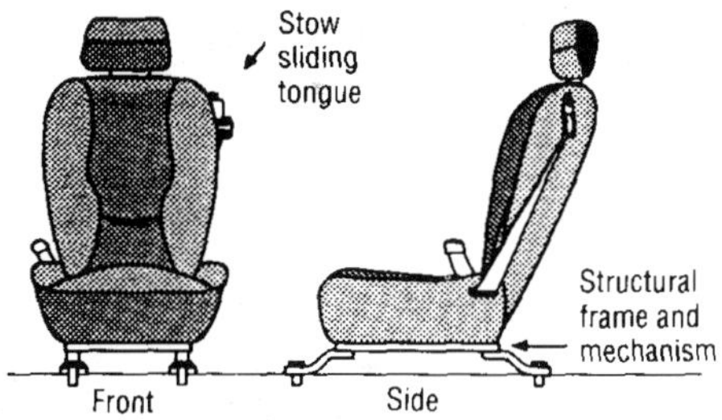

Fig 52: Integral construction

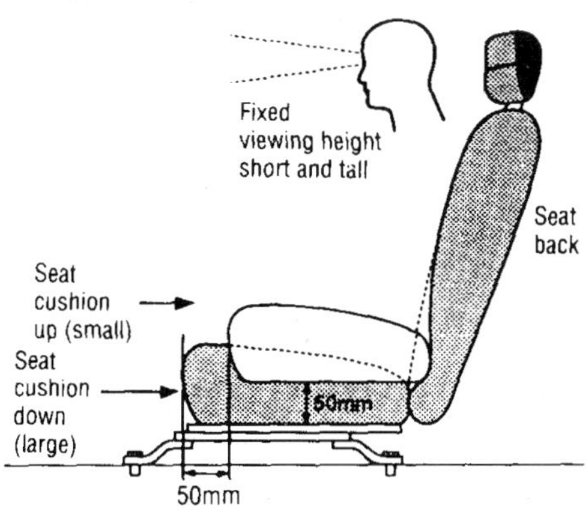

Fig 53: Split-frame cushion extension

Dashboard instruments

The controversy continues as to whether car-customers and CV-operators prefer traditional or 'high-tech', fully or minimally informative, dashboard instrumentation. The answer which the volume vehicle builders continually fail to address is that different customers/operators have different preferences.

Conventional wisdom for the volume builders is provided by theoretical ergonomists and industrial psychologists who seek to generalise human behaviour and present some sort of mythical 'mean' person for product planners to try and satisfy. While the specialist vehicle marketing man will quickly recognise the limitations of such an approach, such research should not be dismissed out of hand because most systematic investigation is likely to yield data which can be selectively exploited in design.

Early ergonomic work was carried out for military purposes in order to optimise the information systems available to those who controlled aircraft and armoured fighting vehicles and this is exemplified by the list of requirements produced by University of Wales researchers[2] in the early 1960s, reproduced as Fig 54. This helps to classify, and rate for performance, all the different information displays on the dashboard. While digital displays are excellent for accurate visualisation of information, their ability to display comparatative values is severely limited.

Speed recording is really best presented by a circular instrument with the pointer angle aiding assimilation. Interestingly, expert studies have shown black figures on a white dial are preferable to the reverse because dark backgrounds tend to impair vision. Otherwise very bold markings are needed and where colours are used, strong contrast is necessary.

'Experts' will argue that oil-pressure, oil-temperature and engine-speed are of value only to those who can properly interpret them. While in itself this is an acceptable view, many would argue that these instruments have immense educational value to the uninitiated — one only has to suffer the expense of engine failure once on a high performance vehicle to learn the importance of maintaining oil pressure.

One of the benefits of re-configurable solid state displays is that they give the opportunity to call up a subsidiary panel that could give more definite information as to whether temperature is too hot. On rear-engine water-cooled engine vehicles in particular, it is important to signal this information as drivers are likely to be unaware of escaping steam in this situation. The argument could be extended, too, to 'battery charging' indication as many specialists might prefer to know, as well, the rate of charge — and even a switch-out system that would register the rate of discharge (current consumption) and perhaps even the state of charge of the battery, on a selectable basis.

Makers of ammeters have pointed out the very wide range of information that can be learned from that instrument for the discerning driver. The ability to check unduly long initial charge rates (indicating wrongly set output controllers) or inadequate running charge rates (indicating impending electrical failure due to slipping fan belt) could be valued by some drivers. It is also sometimes forgotten that the ammeter is also a good check of ignition function — as slow cranking of the engine will register a fluctuating discharge rate to indicate satisfactory contact breaker functioning.

Experts rightly advise of the safety aspects of providing too much information on the main displays. As the single-channel mechanism of the brain has a strictly limited capacity for data transmission, it is wise to relieve it of any checking function — leaving this to either subsidiary or selectable displays. It takes about 1.25 seconds to read a value and 0.75 to make a check depending on the number of excursions the eyes must make. Remember to locate instruments which have to be read, therefore, in the centre of vision and limit peripheral vision of warning signals to within 60 degrees on each side of the visual centre line. Another valuable lesson is to arrange instruments in a logical order of sequence and if all are of the analogue, pointer, type — having their pointers parallel and upright for the most usual operating condition.

Fig 55: Back-lit LCD

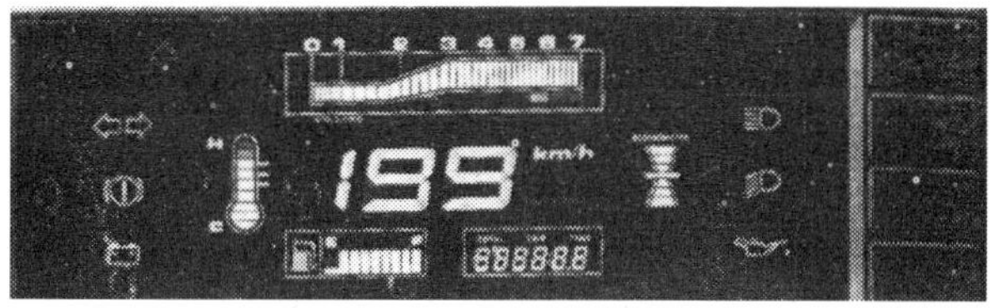

Fig 56: VF/LC hybrid display

Solid state displays

Computer graphic technology has great potential for dashboard display systems if it can be used in sympathy with customer needs — and has particularly revolutionised the availability of selectable information. Liquid crystal displays of the type pioneered for pocket calculators are widening their appeal with the advent of back-lighting which has considerably improved their visibility, Fig 55.

Other developments include the integration of drive circuits on to the displays so that extensive

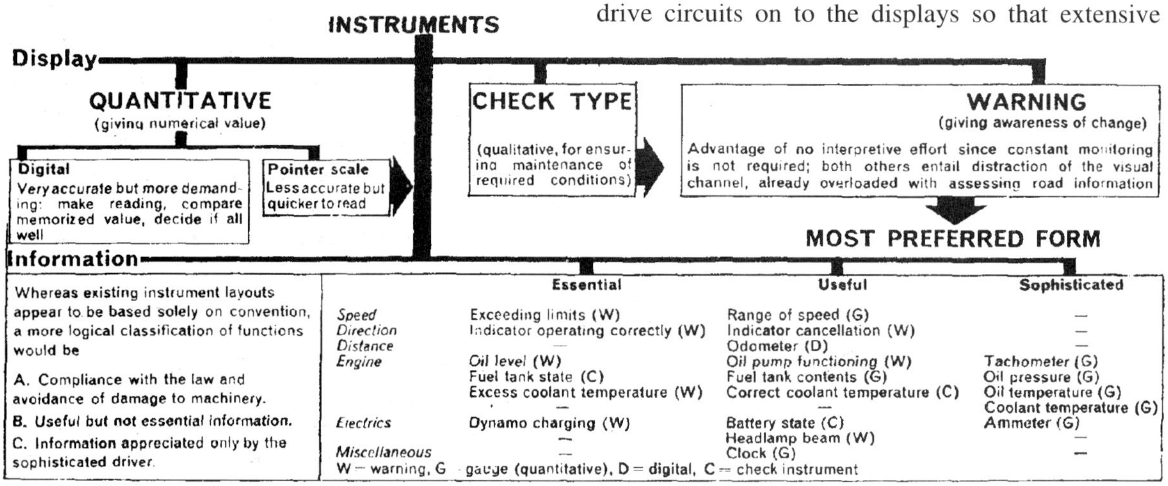

Fig 54: Design rationale for dashboard instruments suggested by Murrell

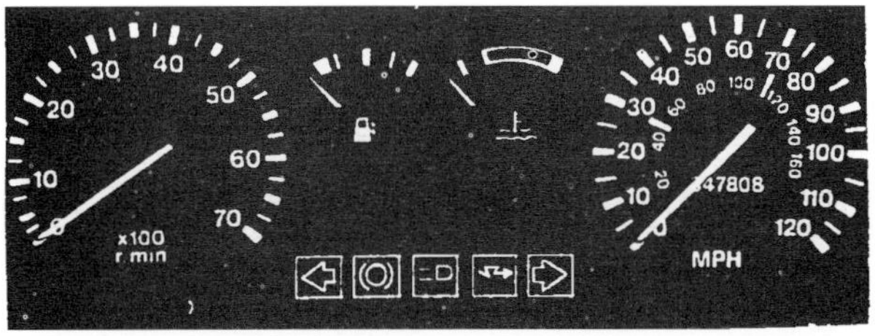

Fig 57:Test display

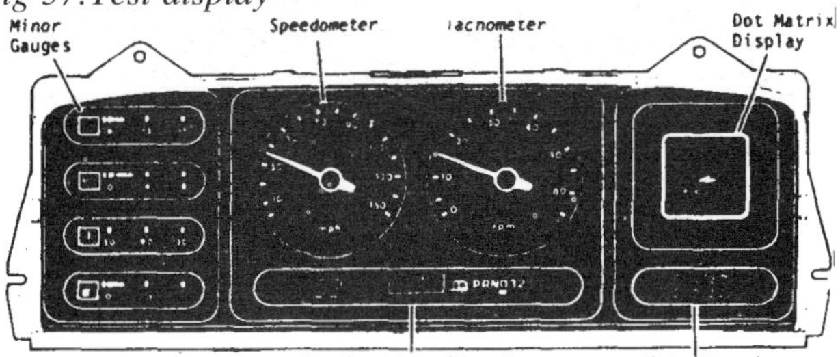

Fig 59: XJ40 display

interconnections between modules will not be required. This technology is now preferred to vacuum fluorescent displays of the type pioneered for the Mark 1 Austin Maestro — though hybrid systems of both technologies are said to be a possibility, Fig 56. Cathode ray tube technology is another important contender and benefits from the considerable development work that has gone into computer monitors. Compact size commensurate with low-cost is one of the problems yet to be solved commercially — and, like LCDs, the advantage of being able to program in the occasional display of less essential information.

The problem of visibility changing on LCDs has been overcome by so called 'transflective' designs using a combination of front and backlighting. Background discolouration with temperature change can be developed out and colour displays can be used without fear of predominant colours changing with variation of ambient conditions. In tests carried out by the Institute for Consumer Ergonomics some idea of subject colour preference has been scientifically evaluated. Using the display in Fig 57 preferences were recorded as shown in Fig 58.

MOST PREFERRED

	Men Age (years)			Women Age (years)			Colour	Bifocal
	17-30	31-50	50+	17-30	31-50	50+	Defect	Wearers
Most Attractive	G	G BG	O Y	G	BG Y	BG Y	G	Y O
Easiest to Read	BG G	BG Y	O Y	BG G	BG Y	BG Y	O	Y
Least Distracting	G BG	G BG	Y	G	BG	BG Y	R	BG Y

LEAST PREFERRED

	Men Age (years)			Women Age (years)			Colour	Bifocal
	17-30	31-50	50+	17-30	31-50	50+	Defect	Wearers
Least Attractive	-	R O	R G	O	R	R	R Y	R G
Most Difficult to Read	R	R O	R G	R O	R	R G	R	O R
Most Distracting	O Y	O Y	R	O Y	R	R O	O Y	-

Fig 58: Display preferences

Vacuum fluorescence has been successfully used on the Jaguar XJ40 for the warning signals which are seen on a 32x32 dot-matrix display, Fig 59. Below these a 20 character caption or ISO symbol is used as an aid. The trip computer is operated by nine keys to give nine displays and has facility for keying in intended journey length. Bar graphs are used to present fuel-in tank, oil pressure, coolant temperature and battery voltage, Fig 60. The odometer display data is held and updated in an EPROM close to the computer.

Advanced instrument displays

Commonly used forms of twisted nematic LCD have dark figures appearing on a light background. Polarisers on the front and back of the display sandwich long rod-like molecules of an organic fluid which are twisted through 90 degrees from one polarizer to another. When no electric power is applied, light will pass from front to rear polariser plate by a light reflection process. On reaching the polariser plate it is reflectcd back towards the front of the cell via a multi-purpose surface which acts as a reflector during daylight conditions. In passage the light is twisted again and emerges on the front polariser as a light spot on a bright back-ground, with no image visible. On energising the panel the molecules are altered such that the light is no longer twisted with the result that the rear polariser absorbs the light and a dark spot appears on the light background of the rear polariser.

Contemporary technology employs colour segments with backlighting added for large displays. Fig 61 shows a chip-in-glass fluorescent indicator panel em-bodying such techniques and Fig 62, a suitable operational block diagram. Methods employed to produce a three-dimensional appearance for the alpha-numeric may involve colours (phosphor materials) and filters. When current is applied to the CIG FIP grid the phosphor coated characters emit light which can be filtered to any colours. White light can be generated by a bar segment anode printed with closely spaced blue green and red phosphors.

Multi-colour cells are an important advance which makes use of separately controlled active elements, within each viewed active cell, in a technique known as 'combe patterning'. Each element has a unique fixed filter which, with alternative or combined activation, will give a perceived colour change. A display can thus be made to adopt a secondary colour when critical levels are reached, Fig 63. According to which basic LCD technology is used, twisted nematic (TN) or super-twisted-nematic (STN), the colour of the base monochrome cells can be created by fixed filters located in a set area to the rear of the cell, Fig 64.

Such a colour cell may comprise, typically, three grouped pixels, each individually filtered, being individually or simultaneously activated, and producing up to seven colours. TN cells, in static drive, are said to offer good contrast, viewing angle, and colour realisation but for larger areas STN cells are specified where multiplex rates in excess of those controllable by TN are required. Molecules may be twisted in the range 180-270 degrees, some 90 degrees more than in a TN cell. STN cells are however limited by colour variance and operating temperature extremes. For driver-controlled switching applications, with status information monitoring, the LCD can be overlaid with a membrane touch-sensitive

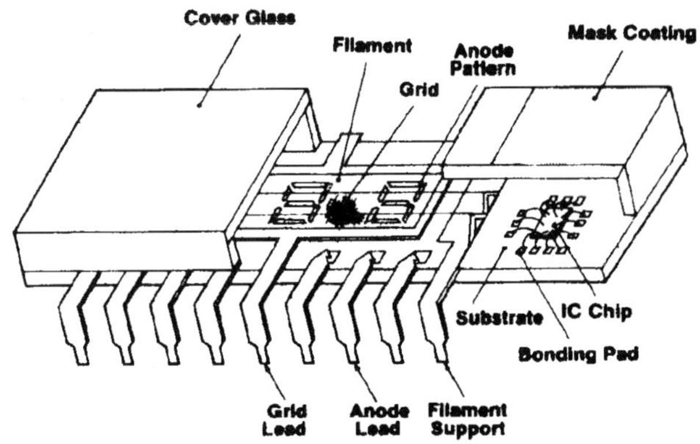

Fig 61: Chip-in-glass fluorescent indicator

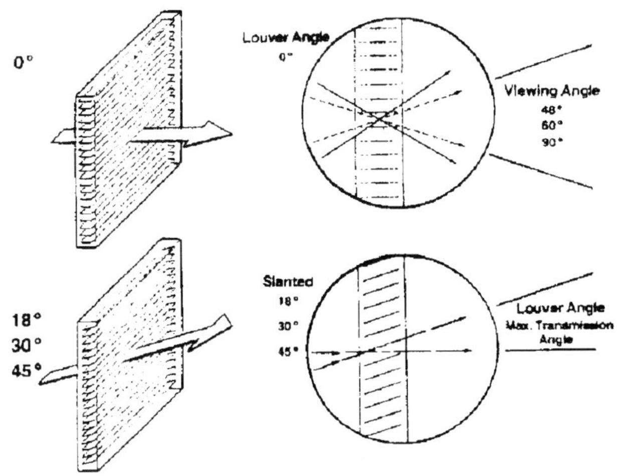

Fig 63: Louvre technique for reducing glare

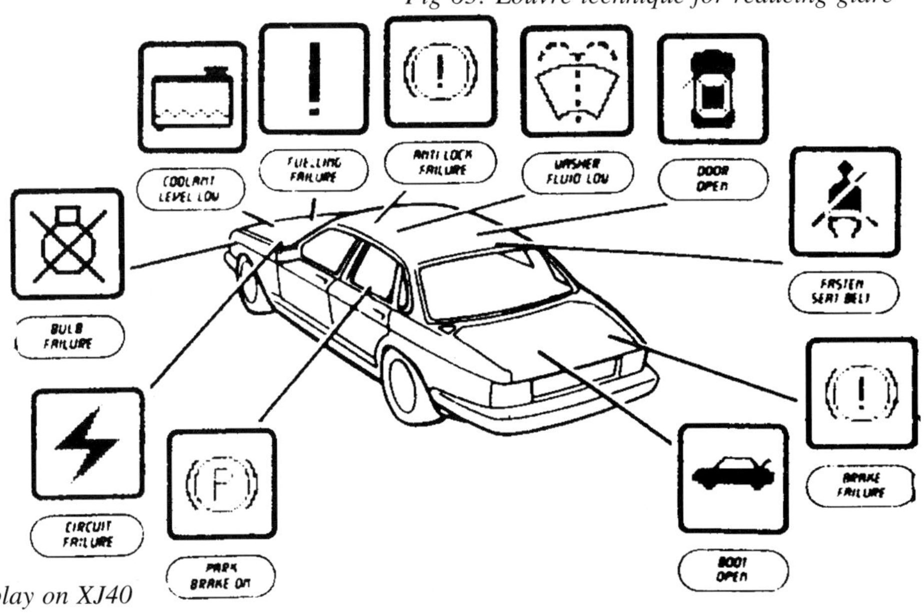

Fig 60: Icon display on XJ40

cell. Local pressure on the membrane would activate the function shown on thc relevant part of the display while a single main polling switch would remap the display and the membrane activatcd zones.

CRTand VF displays

Cathode ray tube displays rcquire primary driver information in a high intensity, high resolution multi-colour image for direct rather than reflected or projected viewing. A suitablc device would be a single-gun CRT with set colour phosphor zones, the primary display driver being a mono graphics controller with a multi-pixel based display driving capability and an integrated serial data link for translating vehicle sensor information. A typical system would involve a display screen energised by a door opening, causing a 'wake-up' signal to display a menu page allowing the selection of all subsequent displays.

Touch technologies available include conductive-membrane, capacitive and scanning infrared beam types. In the resistive membrane system of Fig 65 a resistive coating is applicd to a glass implosion shield in front of the display which has a sheet or metalliscd embossed mylar over its glass. An air gap between thc two layer prevents contact. A grid of electrodes is etched on to respcctivc surfaces in X and Y directions respectivcly. On application of local pressure a short is caused between X and Y electrodes at the contact point and the computer signalled appropriately.

Vacuum fluorescent displays involve a triode, with cathode grid and anode, the filament cathode being a suitably coated tungsten wire. When heated to 600 C, thermo-electrons are emitted towards thc cathode at a rate determined by thc voltage applied to the grid stretched over the anode. The electrons collide with a phosphor layer on the thick film anode to emit light. The technique of 'self-standing grids' allows location of the grid in a position to suit the phosphor pattern. Segments of an analogue bar graph can be progressively illuminatcd by variation of the drive method. When large area displays are required, location and connection of display elements can prove difficult and a solution is to display information on to thc unused portions of the instrument front face by means of a mirrored reflection area in the front and dark toned areas at thc rear of the display lens. Transparent symbol slides may be uscd to projcct appropriate images. These can also be the basis of head-up displays which allow the driver to focus on the road-line. Here a VFD may be hidden within the fascia and information projected on the windscreen, suitably coated with a transparcnt film

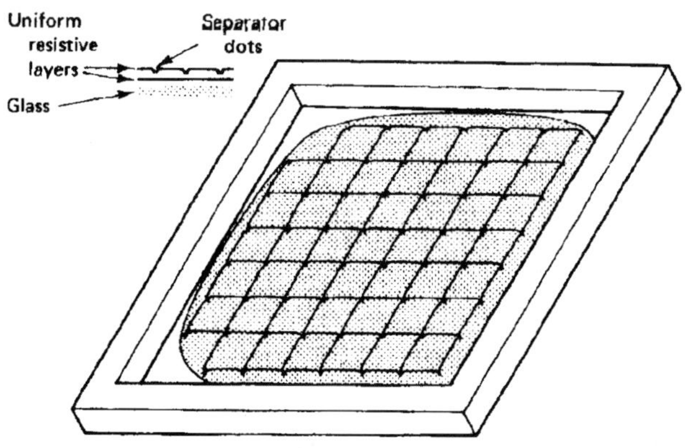

Fig 65: Touch-screen technology

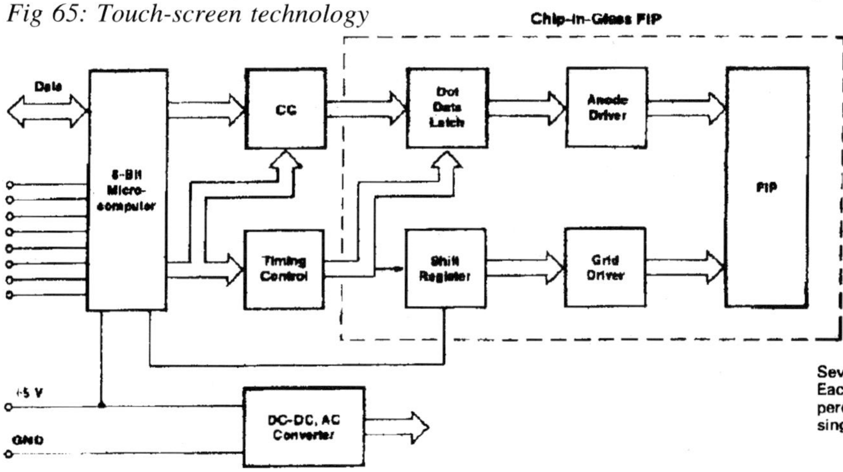

Fig 62: Operational block diagram for panel in Fig 63

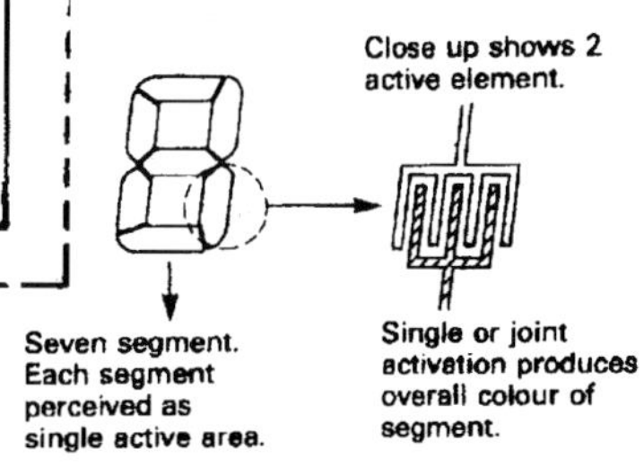

Fig 64: Colour-effect technique

acting as a half-mirror.

Illumination of displays in future could well utilise new light-emitting plastics, exploiting the technology of electro-luminescence. This permits the emittancc of light when an applied AC voltage excites chemical compounds placed betwcen metal electrodes. Reducing glare from electronic instruments has been carried out by light control film (LCF) techniques. This is an optically transparent film having uniformly-spaced micro-louvres, Fig 66. Louvres shield the unwanted light from the driver's eyes. VDO are probably now the main manufacturer of automotivc instrument displays. The company has been active in liquid crystal displays with negative contrast — bright segments on a dark background — with light provided by transflective design in which a combination of front and back lighting is involved. The company have developed a technique whereby the cell appears neutral in colour, under transmitted back light, due to a normally grey background coloration. Contrast of the display is almost independent of the indicator colour chosen. This makes it an attractivc proposition for multi-colour displays without the necessity for black-masking outside the active driven areas. As seen in Fig 66, the display can also be supplied in non-rectangular formats to improve viewing through the steering wheel.

In providing secondary displays such as warning messages, the company can meet the requirement for alpha-numeric characters. When symbols are required, a dot-matrix presentation is usually necessary. Fig 67 shows examplcs from the available range which may include arbitrary graphics as well as symbols. Pulsating displays can also be used to animate the action of a windscreen wash. Long texts may appear running vertically in a roll-mode or horizontally as a moving screen. Warning symbols can also be made to appear alternatcly in orange or red, say, to signify the urgcncy of the warning. Dot-matrix formats also open up thc possibility of providing further displays in thc periphcral view of the driver in a different form than is possible on the central speedometer and associated display, Fig 68.

CRTs, EL and VF displays are luminous (active) and so need a sufficient supply of electrical energy. While EL and VF displays typically rcquire an installation depth less than 40 mm, CRTs require a depth equivalent to their screen diagonal and contrast can be impaired by bright ambient light. However, they offer the greatest information density due to their high resolution. By comparison LCDs are passive displays and thc liquid crystal really acts like an electrically controllable shutter for external light, thus dictating the need for backlighting. Power consumption is comparatively low and involves 1 to 10 microwatts per square centimetre. Negative displays, dark segments against a light background, while providing a more attractive optical effect, cost more to manufacture than positive ones. Contrast and range of viewing angles of an LCD abatc with increasing multiplex rates, required for increasing resolution.Displays driven at multiplexing rates up to 1:16 are currently used in vehicle applications, providing 80x200 image dots for adequate readability. Siemens now offer the TTD display which offers three levels of information prcscntation. Two outer zones display time and outside temperature while the date is shown in the central zone, with an 8-figurc LCD screen; when the radio is switched on, this is over-laid with the transmitter frequency.The size of the display is 132 x 26 mm, with 180 individual segments, to provide clear legibility. Thc central zone display allows the radio to bc remote mountcd and at night its brightncss is automatically dimmed to reduce brightness, Fig 69.

Fig 66: Neutral-background display

Fig 67: Symbols available from VDO

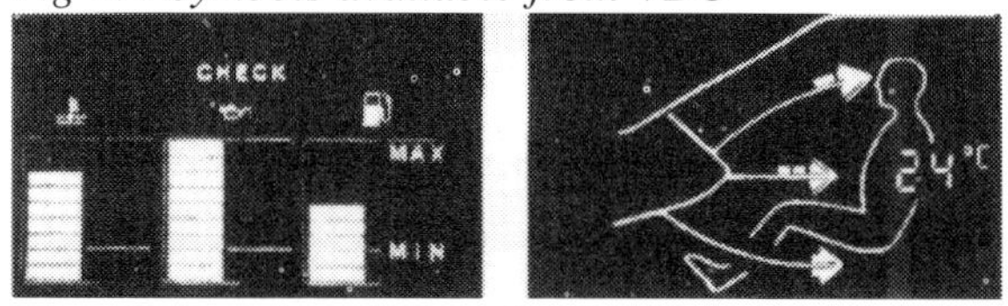

Fig 68: Peripheral-view displays

Fig 69: Siemens TID system

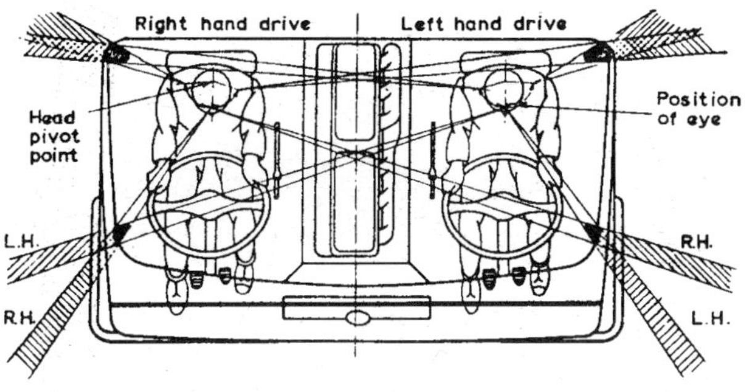

Fig 70: Handing of cab

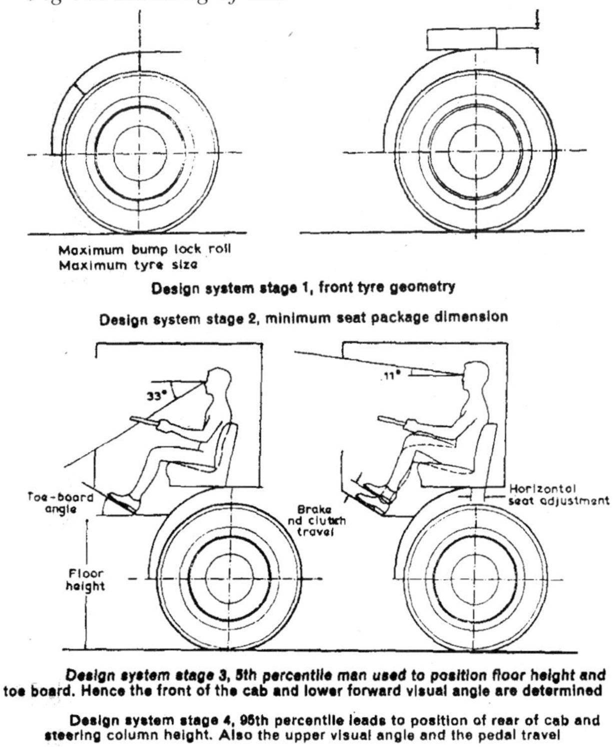

Fig 71: Design scheme for basic layout

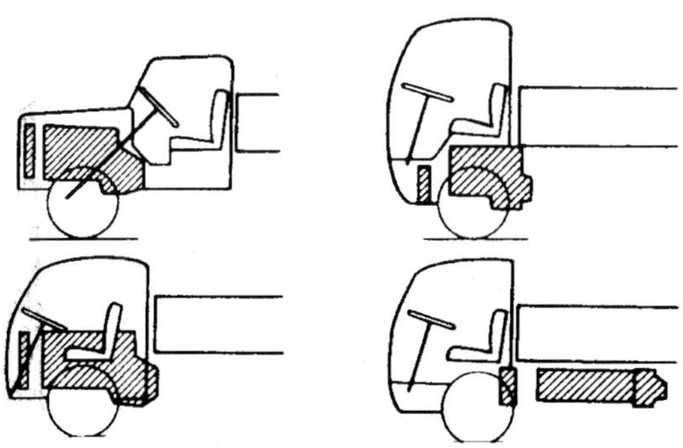

Fig 72: Effect of engine position

Truck cab ergonomics

Given the constraints of legal dimensions on cab layout discussed in the following section, the approach to ergonomics is similar to that on cars. However, while very necessary improvements have been made to cab comfort and convenience in recent years to provide the standards of a private car, it is important to realise that there are some fundamental differences in ergonomic design required of the two workplaces. For the long-distance CV driver in particular it is imperative that proper dimensional relationship between seat and controls are obtained to minimise fatigue and make the working environment as pleasant as possible.

According to Loughborough's HUSAT Research Institute, advances in driver information systems are likely soon to revolutionise the ergonomic design of vehicle cabs. The ability to travel at higher speeds in an increasingly more complex road and traffic network is also an important engine of change. Systems for route navigation, guidance, anti-collision, driver-status (drowsiness) monitoring, parking aids and auto-policing systems are all on the horizon. Research into head-up and head-down displays is well underway in helping to alleviate these problems. In considering driver information systems, it is essential to configure these so that they are a real help to drivers and not overload them with redundant information. Tests on design concepts should of course be made with real drivers and not information scientists! It is also vital not to test the systems in isolation but as an integral part of the driving task.

The earlier section defined the basic criteria for posture in the driver environment — with the accent on car layouts. The use of servo-assistance for steering and braking has, of course, meant that the CV cab environment is tending towards the car one as the control efforts between the categories tend towards one another. This has been reflected in the steepening of steering wheel rake and postural relationship with pedals based on improved positional control rather than the application of heavy pedal efforts. However the fundamental differences in construction between cars and heavy commercials often means the CV driver is often placed much higher with relation to the road and yet has to make frequent dismounts in the case of delivery operations. Much larger planform areas, bulkier bodies and restricted rear vision also differentiate the CV cab — not to mention the intrusion of a large engine and the need to operate a multi-speed gearbox.

The need to meet 'handing' requirements of international operation (Fig 70) affects both categories of course but the physical size of the engine can make the distance between LH and RH driving positions greater. However, the picture shows how careful siting of the pillars can maximise the field of view. A design procedure established by Leyland engineers for fixing the basic dimensions of a cab is shown in Fig 71. The procedure would, of course, have to be modified according to the engine position (Fig 72). Sometimes a modular construction, Fig 73, is used to build different versions of a basic design without serious cost penalty. The dimensions relating driver to his/her seat and the vehicle controls have been recommended as in Fig 74.

A new approach to cab vision problems is in the use of three-dimensional graphics computer programs such as SAMMIE. This allows considerable evaluation resource, Fig 75, prior to committing to metal. It can involve constructing a grid on up to all six sides of a cube around the cab design in 3-D so as to plot out exactly what can be seen. Legislation for mirror performance, Fig 76, is defined in terms of the areas to be seen on the ground and these can also be modelled by the computer in evaluating mirror sizes and positions. As speeds and traffic densities increase, greater concern is being shown for cab safety design. One suggestion has been to make the cab as effectively a detachable capsule from the chassis frame which itself would take the impact of a collision while the cab and its occupants would be partially cushioned by its support springs, Fig 77. There were safety considerations, too, in mounting and dismounting from cabs and appropriate dimensions needed to apply to widely ranging cab sizes and types, Figs 78 and 79. The real challenge in cab

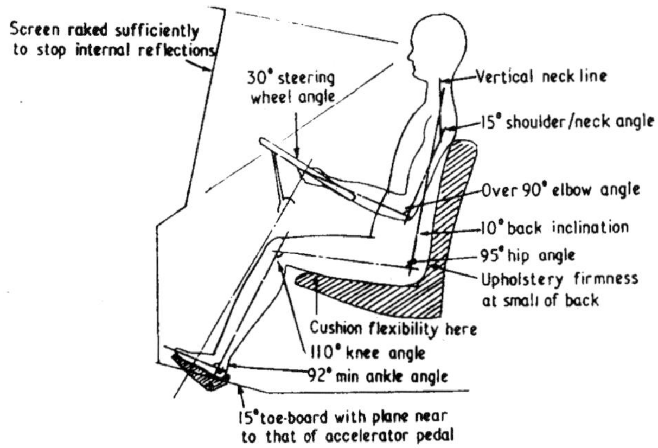

Fig 74: Driver/seat/control relationship

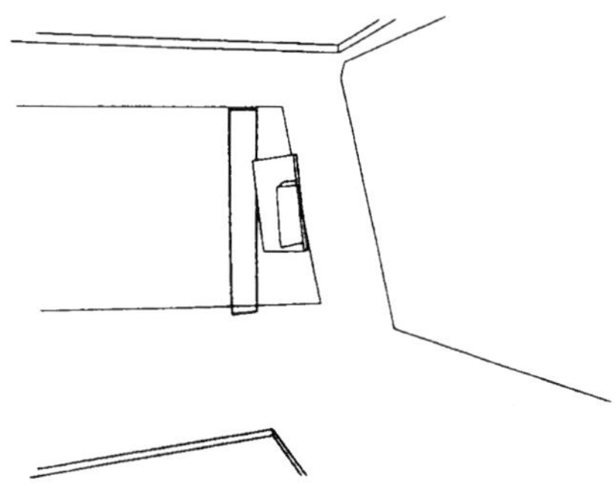

Fig 75: Computer-modelling the interior

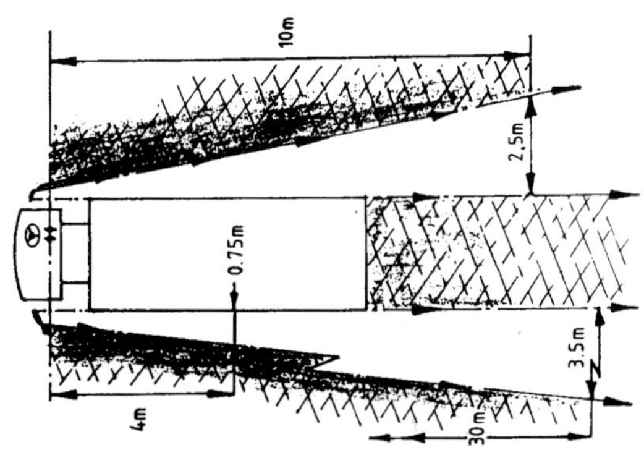

Fig 76: Simulating mirror performance

Segments of the module 1—segments common for all cabs; 2—segments of normal control cab only; 3—segment of forward control; 4—segments of forward control cab; 5—segment of long-distance forward control cab

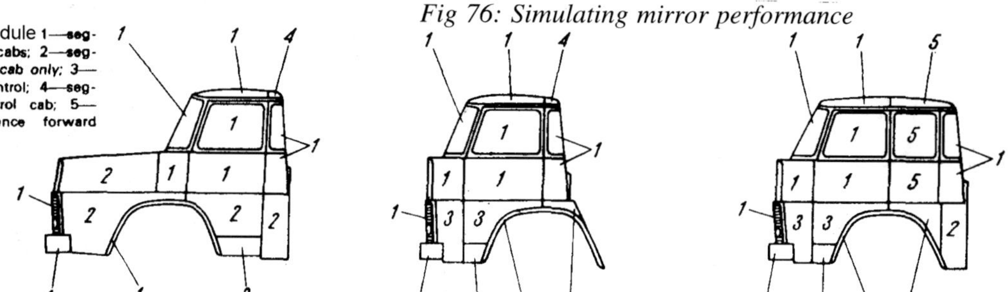

Fig 73: Modular layout

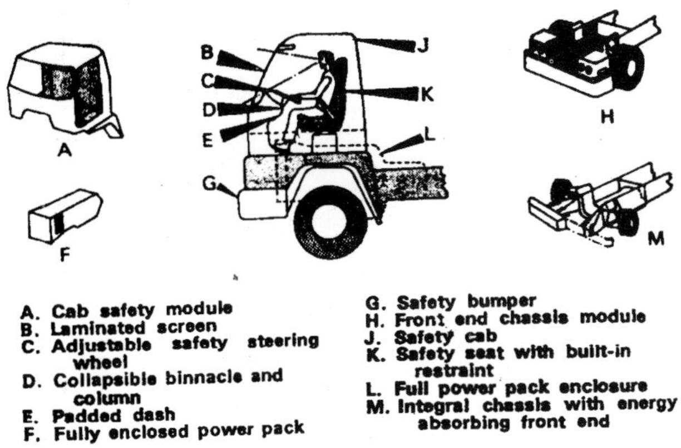

Fig 77: Safety considerations

ergonomics is to provide a workspace that suits a wide range of body statures. For this there is no substitute for a properly executed package drawing layout using largest and smallest percentile anthropometrical data. Pioneering work by Rebiffe[7] portrays the body as a bio-mechanical model consisting of eight links as in Fig 80. To establish control comfort areas the hip reference point (H) for a small person is positioned directly above the contact point between buttock and seat. Hip segment HE is drawn at an angle corresponding to the seat back inclination (25 deg to the vertical), followed by the upper limb links with angles A5 and A6 at 45 and 80 deg respectively. By rotating links EC and CM in turn, the shape Es, Fs, Gs, Hs is traced out in which the hand controls can be positioned. A similar procedure is adopted for the foot controls and the process repeated for the other population extreme, using similar constructions as those given earlier for cars.

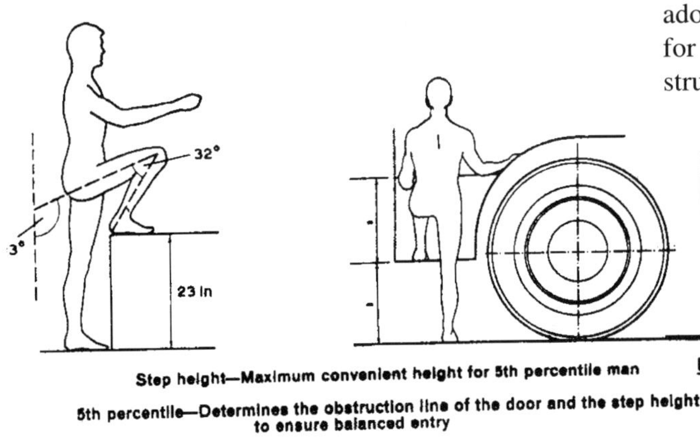

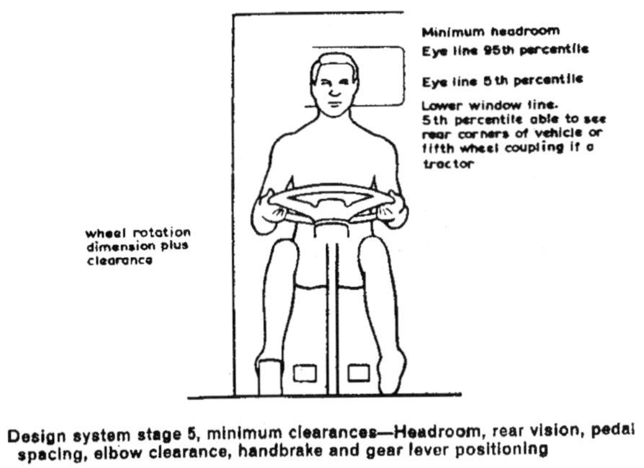

Fig 78: Dismounting dimensions

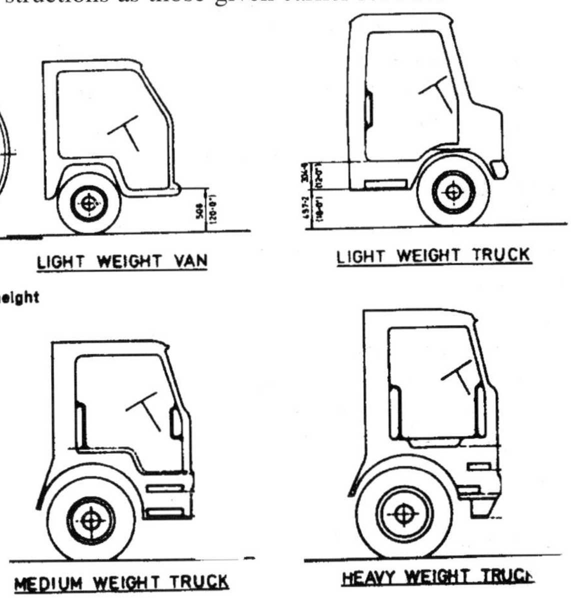

Fig 79: Effect of vehicle type

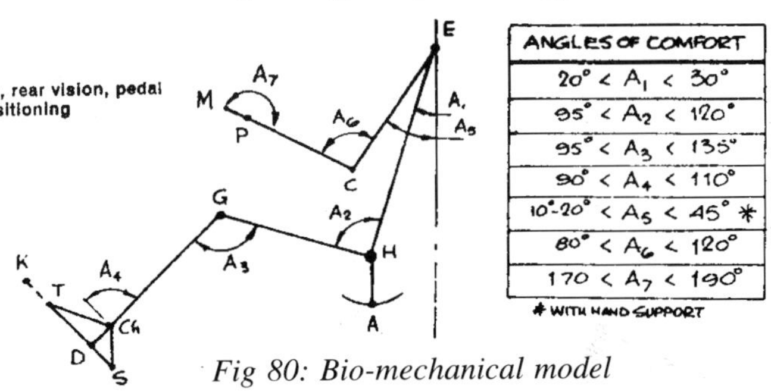

Fig 80: Bio-mechanical model

CV packaging affected by legal weights and dimensions

Trucks differ from cars and buses in that their size and shape is very much dictated by commercial and legal considerations. Construction and Use regulations define the UK legal dimensions of goods vehicles. While this section may help the interpretation of statutory instruments published by HMSO, it is important to refer to the original instruments as frequently as possible to record the frequent changes that take place.

The UK and Ireland were provided with a derogation by the EEC which exempts them from EC directives for heavy trucks until the end of 1998. However, weight increases have already been allowed on such categories as four-axled articulated vehicles and drawbar units.

L = 12 metres B = 2.5 metres (with 2.75 concession for heavy tractors and 2.6 for fridge vehicles whose side walls are at least 45 mm thick)
R = 0.6wb (for refuse vehicles, works trucks and tippers R = 1.15 maximum, regardless of wb)
W_A = 4.6 tonnes for one wheeled axles (or 5.09 with wide/twin tyres)
W_A = 9.2 tonnes for two wheeled axles (or 7.12 for steered axles without load compensation)
W_A = 10.17 tonnes for twin-tyred two wheeled axles (or 10.5 for single wide tyres)
W_G = 14.23 tonnes for vehicles less than 2.65 metres wb
W_G = 16.26 tonnes for vehicles at least 2.65 metres wb
W_G = 17.00 tonnes for vehicles at least 3.0 metres wb
For three and four axled rigid vehicles:
W_G = 16.26 tonnes for 3-axled vehicles (increasing with wb as in Table 1)
W_G = 18.29 tonnes for 4-axled vehicles (increasing with wb as in Table 2)
$R_{6/8}$ = 0.6 $wb_{6/8}$ (overhang measured from 110 mm rear of bogie mid-span)

Fig 81b: Rigid vehicle dimensions

Tractor/trailer combinations

For trailers generally:
B = 2.5 metres (or 2.3 for tractor GVW < 3.5 tonnes)
B = 2.6 metres for fridge trailers with walls over 45 mm thick
For drawbar trailers:
L = 12 metres for W_G>/= 3.5 tonnes (or 14.04 metres for composite — dolly + semi-trailer)
L_C = 18 metres W_G = 18 tonnes for wheelbase above 3 metres
W_{GT} = 24.39 tonnes (or 32.52 tonnes with power-assisted brakes)
W_{GT} = 40.65 tonnes when drawn by 'locomotive'
For articulated combinations:
L = 15.5 metres (or 16.5 if unit can turn within concentric circles of 12.5 and 5.3 metres radii)
L = minimum defined by Table 3 for artics with W_{GT}>/= 32.5 tonnes
W_G = 14.23 tonnes for wb>/=2.0 metres (20.33 for wb>/=3.0 with 3 axles)
W_G = 16.26 tonnes for wb>/=2.4 metres (22.50 for wb>/=4.0 with 3 axles)
W_G = 17.00 tonnes for wb>/=2.7 metres (24.39 for wb>/=4.9 with 3 axles)
W_{GT} = 20.33 tonnes (or upwards according to Tables 3 and 4 for three or more axled tractive units).
H = 4.2 metres for W_{GT}>/=32.52 tonnes

Fig 81c: Tractor/trailer combinations

Referring to the diagrams in Fig 81a, for rigid vehicles generally, dimensions in Fig 81b apply and for tractor/trailer combinations, those of Fig 81c.

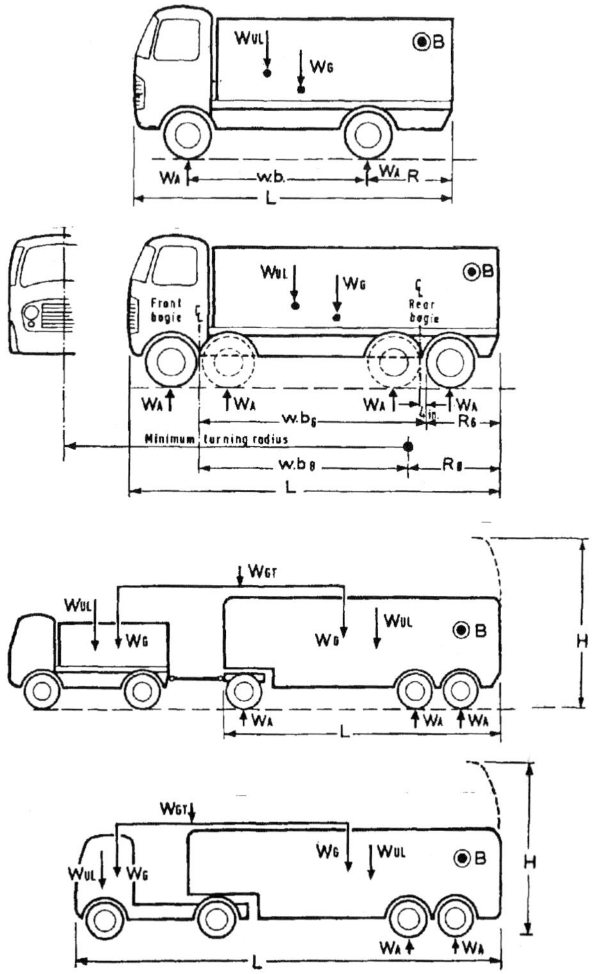

Fig 81a: Truck dimensions

Min. distance between axles (m)	*Max weight:* - when plated weight of neither axle exceeds half of the specified weight (kg)	- in cases not within previous column when plated weight of neither axle exceeds 10170 kg	- in cases not within either column on left (kg)
1.02	16,260	12,200	10,500
1.05	17,280	15,260	10,500
1.20	18,300	16,270	15,260
1.35	18,800	17,280	16,500
1.50	19,320	18,300	18,000
1.80	20,000	19,000	19,000
1.85	20,340	19,320	19,320

Axle weights alter for closely spaced tandem axle arrangements according to the data in Fig 82 and the tables of Fig 83.

2 ***Vehicles with four or more axles:***

Must not exceed 18,290kg except as shown below:

distance between front & rear axle at least 3.7m and max. axle weight not more than 8,640kg	**20,330**
distance between front & rear axle at least 4.6m and max. axle weight not more than 8,640kg	**22,360**
distance between front & rear axle at least 4.7m and max. axle weight not more than 8,640kg	**24,390**
distance between front & rear axle at least 5m and max. axle weight not more than 9,150kg	**24,390**
distance between front & rear axle at least 5.6m and max. axle weight not more than 9,150kg	**26,420**
distance between front & rear axle at least 6m and max. axle weight not more than 9,660kg	**26,420**
distance between front & rear axle at least 5.9m and max. axle weight not more than 9,150kg	**28,450**
distance between front & rear axle at least 6.3m and max. axle weight not more than 9,660kg	**28,450**
distance between front & rear axle at least 6.3m and max. axle weight not more than 9,400kg	**30,490**
distance between front & rear axle at least 6.5m and max. axle weight not more than 9,660kg	**30,490**

1 ***Three-axled vehicles:***

Must not exceed 16,260kg except as shown below:

	kg
distance between front & rear axle at least 3m	**18,290**
distance between front & rear axle at least 3.2m and max. axle weight not more than 8,130kg	**20,330**
distance between front & rear axle at least 3.9m and max. axle weight more than 10,170kg	**20,330**
distance between front & rear axle at least 3.9m and max. axle weight not more than 8,640kg	**22,360**
distance between front & rear axle at least 4.6m and max. axle weight more than 10,170kg	**22,360**
distance between front & rear axle at least 4.9m and max. axle weight not more than 9,400kg	**24,390**
distance between front & rear axle at least 5.1m and max. axle weight more than 10,170kg	**24,390**

5

Relevant axle spacing (m): where motor vehicle has 2 axles	*where motor vehicle has at least 3 axles*	*Max. weight (kg)*	*Min. overall length (m)*
less than 2.2	less than 2.2	20,330	-
at least 2.2	at least 2.2	22,360	-
at least 2.6	at least 2.6	23,370	-
at least 2.9	at least 2.9	24,390	-
at least 3.2	at least 3.2	25,410	-
at least 3.5	at least 3.5	26,420	-
at least 3.8	at least 3.8	27,440	-
at least 4.1	at least 4.1	28,450	-
at least 4.4	at least 4.4	29,470	-
at least 4.7	at least 4.7	30,490	-
at least 5.0	at least 5.0	31,500	-
at least 5.3	at least 5.3	32,520	-
at least 5.5	at least 5.4	33,000	10.0
at least 5.8	at least 5.6	34,000	10.3
at least 6.2	at least 5.8	35,000	10.5
at least 6.5	at least 6.0	36,000	11.0
at least 6.7	at least 6.2	37,000	11.5
at least 6.9	at least 6.3	38,000	12.0

3

10.0m = 33,000kg pmw
10.3m = 34,000kg pmw
10.5m = 35,000kg pmw
11.0m = 36,000kg pmw
11.5m = 37,000kg pmw
12.0m = 38,000kg pmw

4

Number of axles of motor vehicle	*Intermediate axle weight (kg)*	*Spacing between rear-most axle and foremost axle (m)*	*Max weight transmitted to road by all wheels of vehicle (kg)*
2	-	at least 2.0	14,230
2	-	at least 2.4	16,260
2	-	at least 2.7	17,000*
3 or more	not over 8,390	at least 3.0	20,330
3 or more	not over 8,640	at least 3.8	22,360
3 or more	not over 10,500	at least 4.0	22,500
3 or more	not over 9,150	at least 4.3	24,390
3 or more	not over 10,500	at least 4.9	24,390

Fig 83. Dimension tables

Vehicle mechanical package layout

Important to the body designer is the packaging of the mechanical components in relation to the main operational function of the vehicle. Some of the factors affecting the choice of power train and running gear layout are considered here in relation to the different categories of special vehicle.

An early breakthrough in sports car occupant and mechanical packaging was by the Lotus Europa, shown in Fig 84, whose imaginative package helped to achieve remarkably low frontal area and drag coefficient at the time of introduction. Vehicle weight of around 1350 lb was split 45/55 % front/rear. The central box backbone chassis was a key feature of the design — together with front suspension beam and splayed chassis rear-extensions supporting the power train. The rearward 'mid' engine configuration does, of course, compromise the space required for rear occupants in the conventional 'passenger' car. One of the main debates of mechanical packaging is therefore that of the desirability or otherwise of grouped engine and drives, front or rear, against the use of front and rear drive. Some argue that because of the large clearances swept out by the front wheels in steering, space between them is of little use except for the engine. Basic stability in handling also comes from a concentration of weight at the front — as well as easier engine access, in the case of 'normal control' vehicles. Easier engine cooling and inherently better gear shift linkage is also argued in favour of front-engine/rear-drive. However, front 'grouping' is preferred for proximity of driver to controls as well as better rear-end structural integrity for most body types. Supporters of rear grouping of engine/drive argue that there is no restriction on turning circle as front wheels are free of the engine package between them. Better traction on gradients is also claimed — arising from the greater weight on the driving wheels — and there is said to be a minimum change in centre-of-gravity of the vehicle with change in payload and or passengers. Noise, fumes, heat and exhaust are also more easily removed and there is no problem of having to steer driven wheels. Any grouped system is claimed to be more efficient, arising from the proximity of engine and drive; also the lack of any intrusion by propeller shafting between the front and rear 'axle' positions obviously eases pressure on the passenger space. Fig 85 compares front and rear grouped engines of two former passenger cars, the Morris 1100 and NSU Prinz. The transverse engine layouts are also seen to advantage against the Alpine Renault in-line engine, seen in Fig 86, which en-

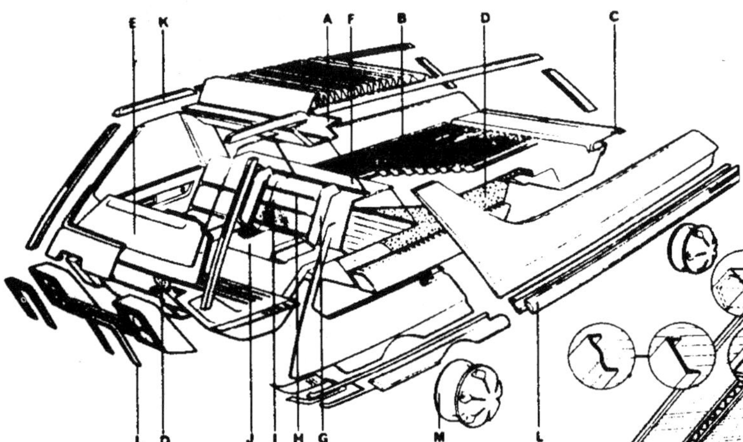

Fig 88: ALCOA concept van

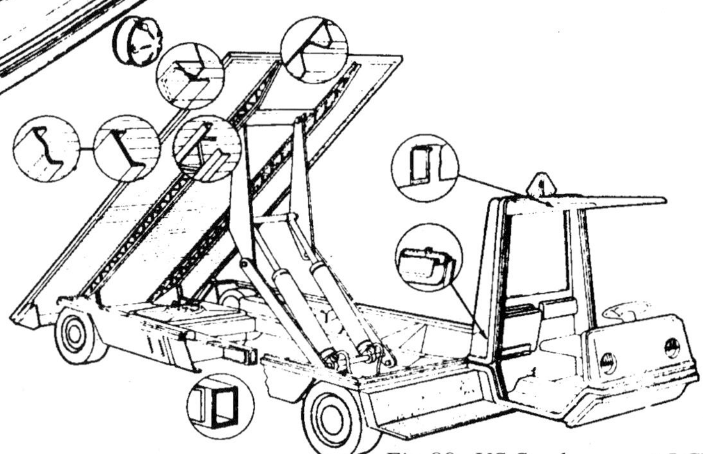

Fig 89: US Steel concept LCV

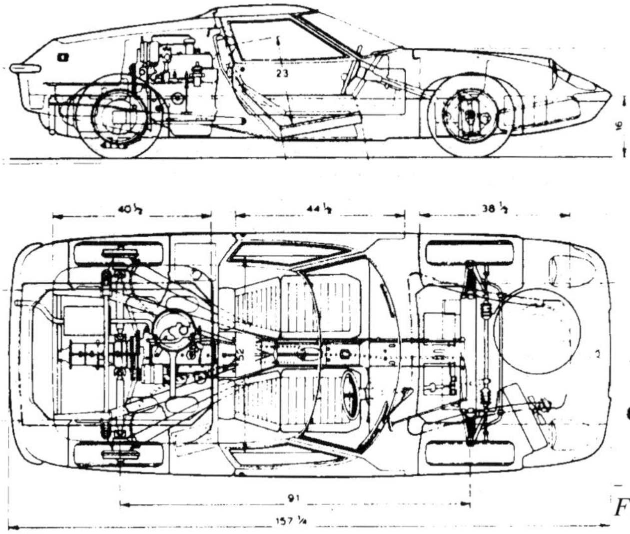

Fig 84: Lotus Europa

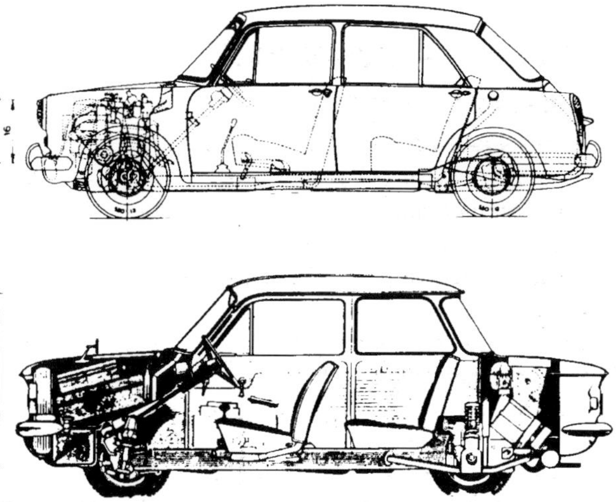

Fig 85: BMC 1100/ NSU Prinz comparison

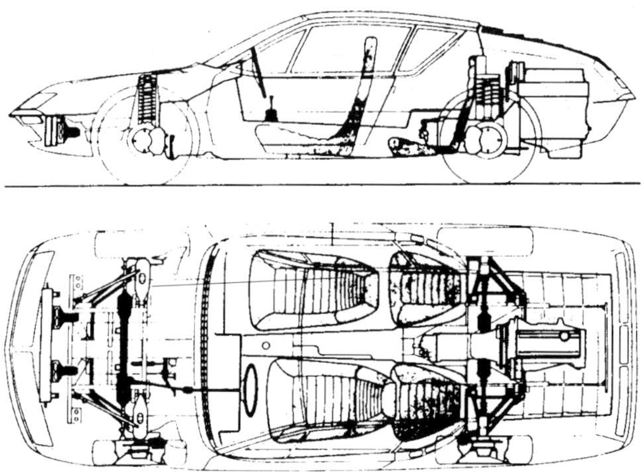

Fig 86: Alpine Renault

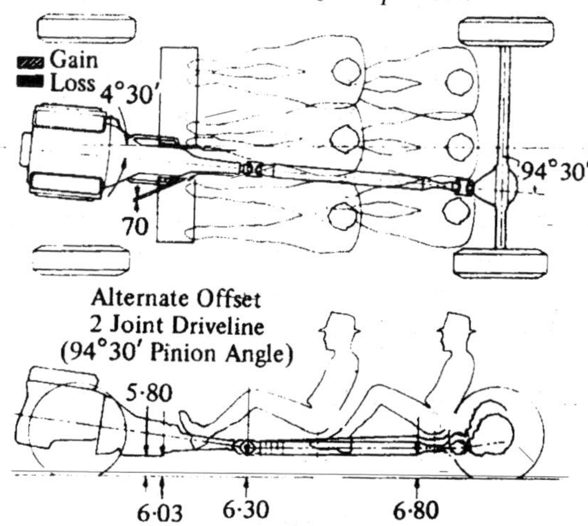

Fig 87: Ford Techna concept

croaches to some extent into boot space. In the Ford Techna concept car of the late 1960s, a variant on the front-engine/rear-drive theme shows that a driveline angled in plan view can alleviate passenger-to-mechanicals clearance problems, Fig 87. In panel vans and LCVs, the conventional layouts are sometimes challenged by new concepts such as those in Figs 88 and 89, proposed respectively by aluminium suppliers ALCOA and steelmakers US Steel. In the first, the chassis (D) is of foam-filled roll-form construction and an extruded tailboard (C) folds flush with the deck for easy loading. (A) is storage area for a sliding tilt while (B) is the deck floor made from interlocking extruded planks. A tubular seat structure (H) abuts a cast frame (G) and supports an aluminium screen mesh (I) around which foam and fabric is bonded. (F) is a storage space, (E) a stamped instrument panel and (K) an extruded windscreen frame. Forged wheels are at (M) and rubber inserts in critical contact areas at (L). The US Steel Venture Wagon uses a flat-four engine in unit with the rear drive to achieve its pleasing low profile. The offset cab mounted on an outrigger permits the carriage of long and awkward shapes.

Goods vehicle layout

Concept commercial vehicles have also been used to show the possibilities for improving the packaging of larger goods vehicles such as the light/medium distribution van and the inter-urban supply vehicle of Fig 90. The Leyland TX450 has unusually low slung power and drive trains to give exceptionally low deck height. It is notable also for the inclusion of a compactly packaged system of independent wheel suspension, front and rear, using low rate air springs. An automatic transmission is also accommodated within the depth of the new-departure fabricated aluminium alloy chassis frame. In general, however, the mechanical packaging of the heavier on-road goods vehicles in current production is unlikely to alter because of the major OEMs' need to keep up volumes to maintain economy and hence the need to retain standardised layouts so as to appeal to as wide a range of end-users as possible.

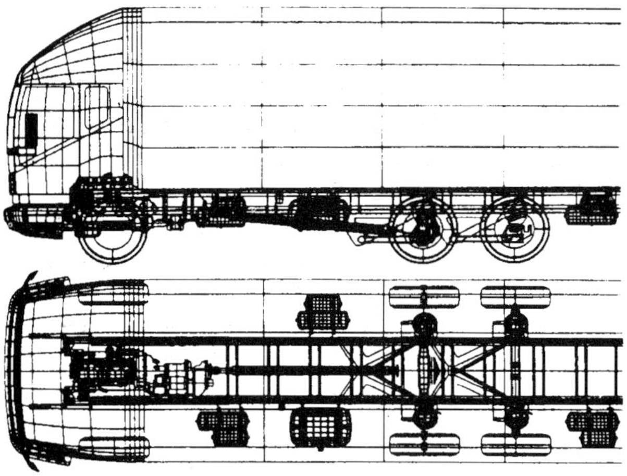

Fig 90: Leyland concept truck

Fig 91: Citroen H-van

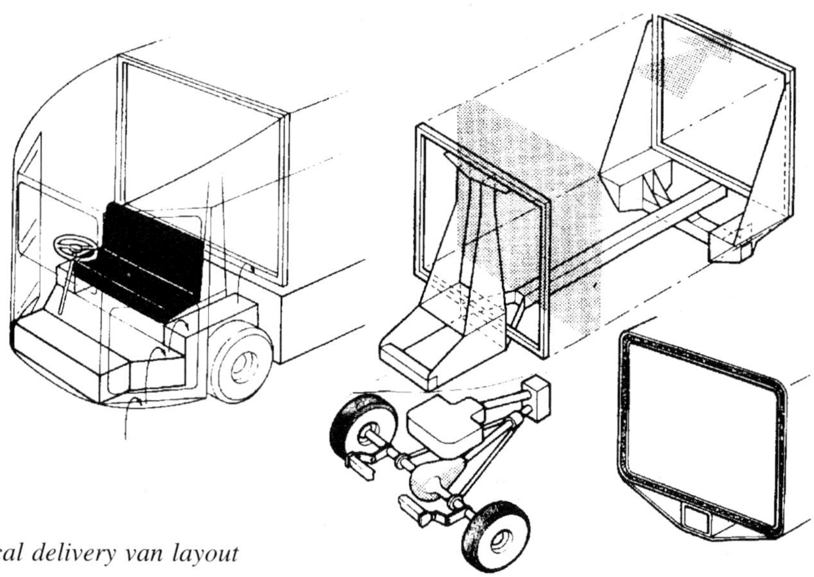

Fig 92: Radical delivery van layout

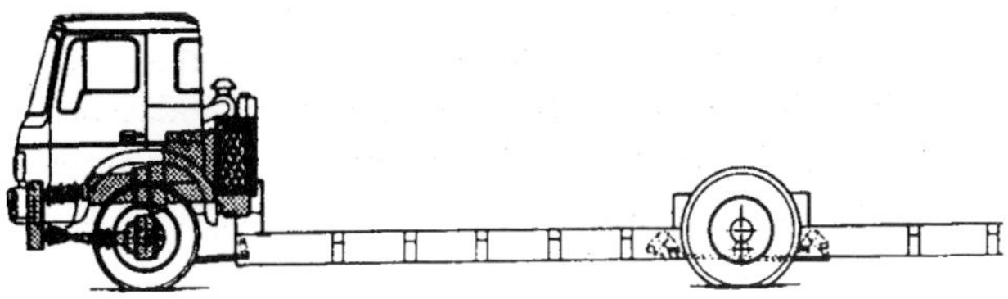

Fig 93:Bruce delivery truck conversion

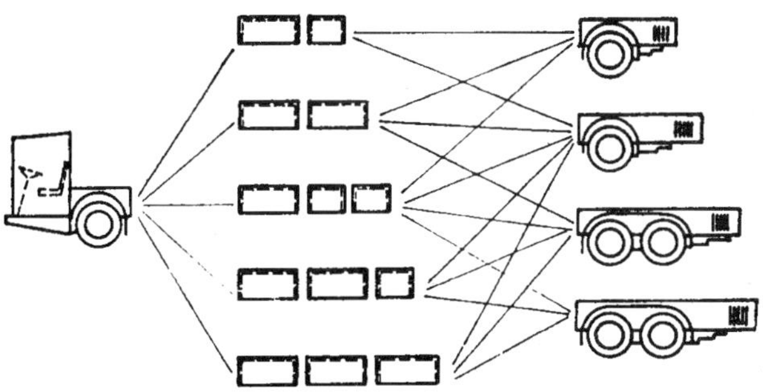

Fig 94: Pawlowski modular system

Scope for specialist delivery vehicles

Considerable opportunities therefore exist for special-purpose-vehicle builders to develop designs for specific operations. The large-capacity delivery vehicle is a case in point where, already, imaginative cab designs by the major OEMs, as well ingeniously thought out box-van body designs by bodybuilders have achieved good compromise solutions. The optimum solution, however, might await the commitment of that special purpose vehicle builder who is able to configure the mechanicals, and integrate body and cab, in a way that maximises the efficiency of the overall delivery operation.

Among the needs of the delivery vehicle operator are an across-cab and front-rear (cab to load-space) walk-through facilities. Car standards of comfort and convenience in the driving controls and minimum vehicle overall dimensions, for ease of parking, must come with best use of interior space for payload and access compatible with ISO pallet dimensions. A level-control on the suspension system would also be a valuable attribute for street deliveries, as would the minimum possible rear deck dimension. A move in this direction was made by the Citroen H-van, Fig 91, but presumably volumes were not at that time high enough to amortise the tooling costs of high volume production.

A more radical proposal is shown in Fig 92 based on the silhouette of the engine used in an earlier delivery truck, the horizontal flat-four of the Albion Claymore. The main body is made from a sandwich panel tube structure, using structural panels of the type described in the section on sandwich panels. The well-aft disposition of both axles gives acceptable weight distribution for front-drive. The expedient of using a 4WD transfer gearbox, for an easily made-available front-drive arrangement, has already been demonstrated by the Pitt Trailers conversion of the Bedford Rtype truck for the CEGB in the 1960s. It could possibly give a more compact layout than the Bruce Leyland conversion for Scottish and Newcastle Breweries shown in Fig 93 for which the overriding factor was the achievement of a 20 in (508 mm) platform deck height, to ease the unloading of drinks crates from an open platform body. A yet more radical proposal for specialised goods vehicles is due to the late Dr Ing Janusz Pawlowski from Poland's Institute of Industrial Design. Here the inter-axle space occupied by the chassis frame is replaced by structural compartments which slightly increase deck height but result in extra lower-level accommodation for freight. The whole leads to a modular system of construction in which different overall lengths and drive systems can be achieved with minimal modification, Fig 94.

Chapter 3: Aerodynamics and air-conditioning

Aerodynamic basics; racing car aerodynamics; bluff-body aerodynamics; local air flows; CFD for aerodynamic performance prediction; body heat transfer; air-conditioning systems; HVAC sophistication and modularisation

Aerodynamic basics

Widening realisation of the importance of aerodynamics, prompted by the fuel crises of the 1970s, has focused interest by designers in most of the different vehicle sectors. The subject is one in which lessons learned for one vehicle category, in one generation, can very often be applied elsewhere in another generation. As well as aerodynamic drag reduction, improvements in handling stability and noise prevention can follow careful design in this area.

According to Costin[1], The effects of changes in drag co-efficient on engine power requirement are shown in Fig 1, from which its importance over rolling drag can also be seen. Even the effect of removing a rubber windscreen surround, standing proud by 0.5 inch, and taking off two wing mirrors, measuring 5 x 3 inches, is quite significant.

Key to drag reduction is in the basic expression for air resistance:

$$= C_d A \rho V^2 /2g$$

where *A* is vehicle frontal area, *r* the air density and *V* the velocity through the air. This expression defines the much discussed drag coefficient C_d which is mostly a function of the shape of the vehicle. With the massive attention taken by the volume builders in recent years, textbook figures for 'typical' vehicle shapes quickly become outdated. Figures for the most basic shapes are useful and some of these are given in Fig 2. A 'teardrop' shape, adopted by water as it falls through the air, is as low as 0.1. The value is much affected by objects which project into the streamlines around a moving vehicle and by rear end shape affecting the turbulence in the wake of the vehicle. Air density varies with atmospheric temperature *t*, and pressure *p*, according to the law $\rho = 144p/Rt$ where *R* is the gas constant, equal to 53.3 for

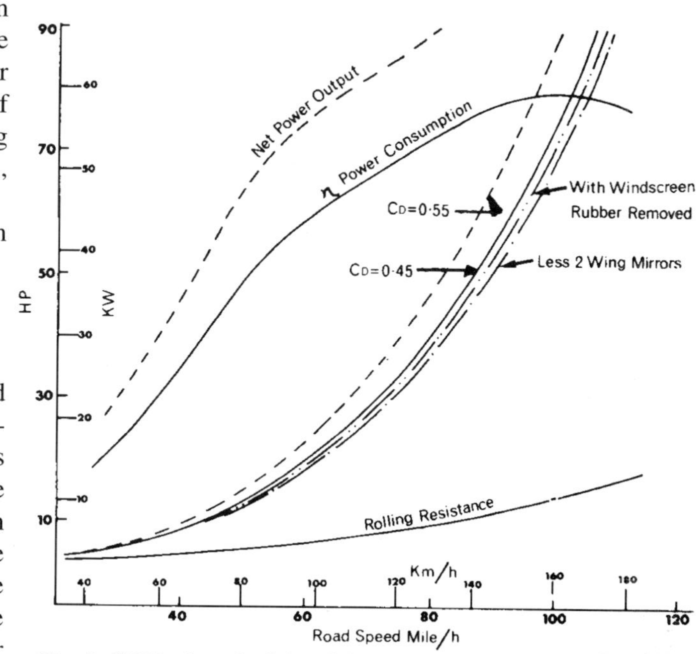

Fig 1: BHP absorbed for different drag co-efficients

air. In standard atmospheric conditions, $D = 0.0763$ lbf/ft^3. Total drag comprises air resistance, defined above, plus skin friction (depending on surface roughness; this could amount to 10% of air resistance) plus resistance to airflow through cooling and interior ventilation systems.

Product of drag coefficient and frontal area is a term frequently used to compare vehicles aerodynamically. This product can actually decrease with a rise in frontal area if, for example, a protruding part is covered with a streamline fairing. Values for A in square feet (and square metres) are typically 14-20 (1.3-1.85) for small cars, 22-30 (2.04-2.70) for large cars and 60-80 (5.6-7.4) for buses and trucks. Areas can be estimated as 0.9 x body height x wheel track. Engine power required to overcome air resistance is $0.0026C_dAV_r^{\,2}V/375$ where V_r is the geared road speed. This shows that in still air conditions the power required to overcome air drag is proportional to the cube of vehicle speed — indicating the importance of aerodynamics at higher vehicle speeds.

Fig 3 illustrates air-flow around a car body to show how the flow separates and rejoins as a body passes through the air. As the air approaches the body it has to divide, the point of diversion being termed the stagnation point — at which the dynamic pressure head is $\rho V_r^2/2$. As air flows around the body a pressure distribution is built up as shown; over the roof, increase in air speed causes a suction zone to develop due to the pressure drop. Conversely, at the base of the windscreen, the streamlines separate and a pressure zone results from the low velocity. The air resistance, overall, is the summation of the horizontal components of the pressure distribution vectors. This shows why nose shape is also critical to drag.

Reducing the wake size, by maintaining unseparated air flow further back along the body, is also helpful — as in an aerofoil wing section. While the flow remains streamline, air adheres to the vehicle over a thin boundary layer whereafter, moving away

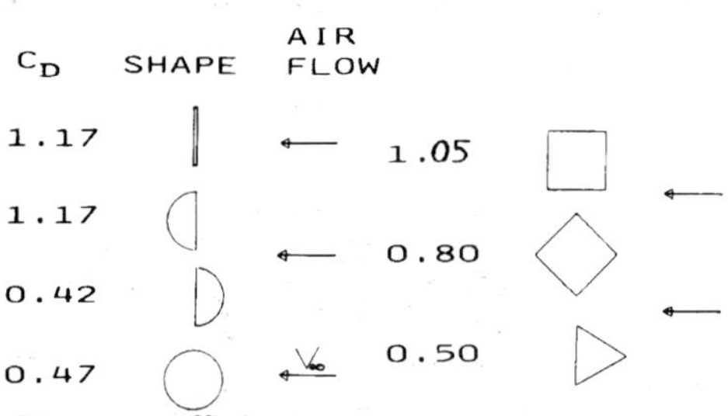

Fig 2: Drag co-efficients

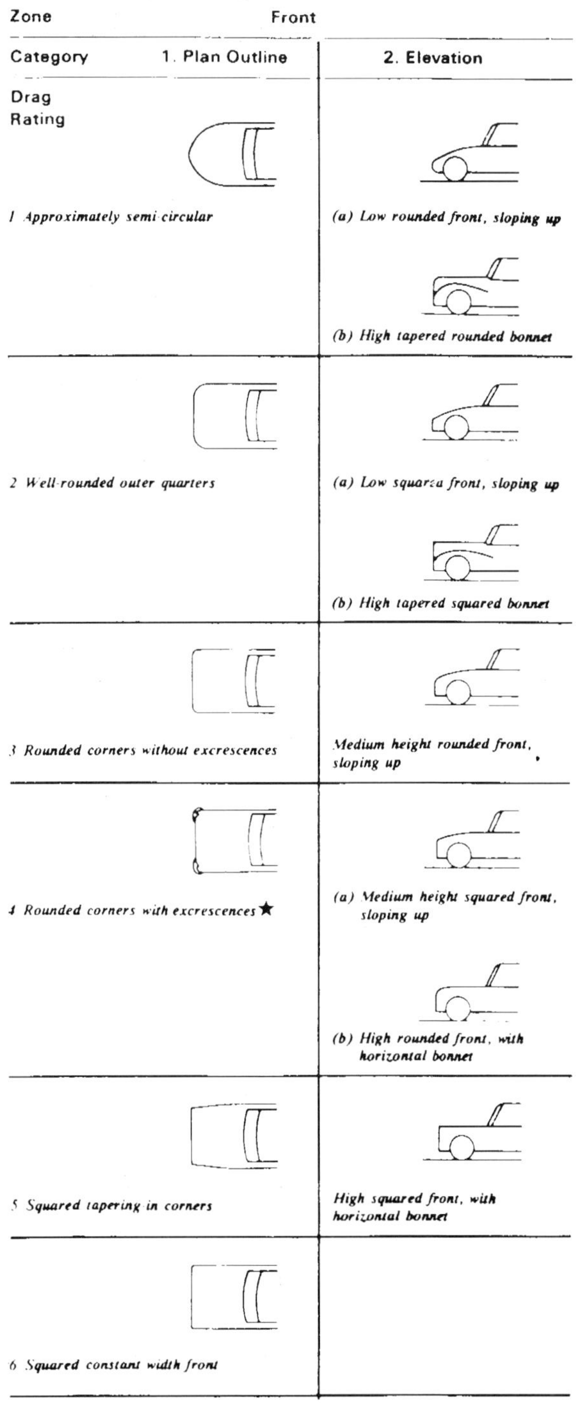

Fig 4: Components of drag

Zone	Windscreen/Canopy junction	
Category	3. Scuttle and wing cross-section	4. Windscreen plan
Drag Rating		
1	 *Flush bonnet and wings, well-rounded body sides*	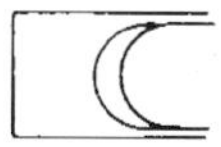 *Full wrap-round (approximately semi circular)*
2	 *High scuttle, low wings*	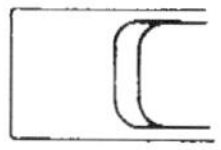 *Wrapped round ends*
3	 *(a) Bonnet flush with rounded top wings*  *(b) High scuttle, with rounded-top wings*	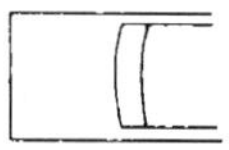 *Bowed*
4	 *Bonnet flush with squared-edge wings*	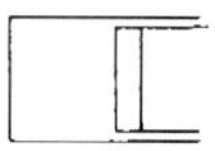 *Flat*
5	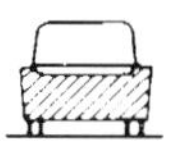 *Depressed bonnet, with high squared edged wings*	

Zone	Canopy	
Category	5. Windscreen peak	6. Canopy plan
Drag Rating		
1	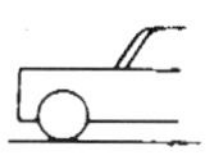 ***Rounded***	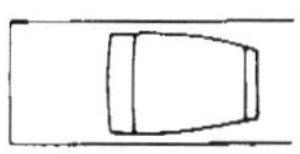 *Well- or medium-tapered to rear*
2	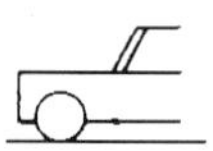 ***Squared (including flanges or gutters)***	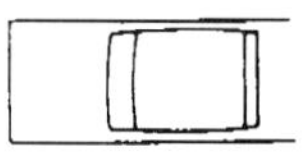 ***Tapering to front and rear (max. width at BC post) or approximately constant width***
3	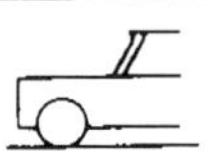 ***Forward-projecting peak***	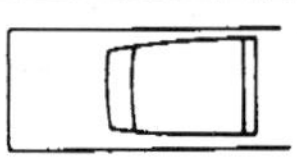 ***Tapering to front (max. width at rear)***

Zone	Rear canopy/boot	Lower rear quarters	Underbody
Category	7	8	9
Drag Rating			
1	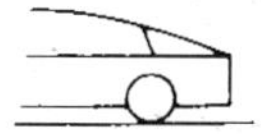 ***Fast-back, i.e., roof line continuous to tail***	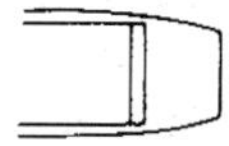 ***Well-or medium-tapered to rear***	***Integral, flush floor, little projecting mechanism***
2	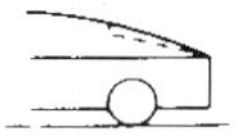 ***Semi-fast back i.e., with discontinuity in line to tail***	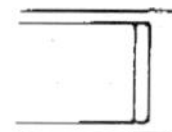 ***Small taper to rear or constant width***	***Intermediate***
3	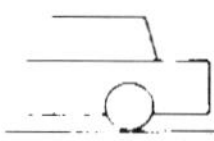 *Squared canopy with boot rear edge squared*	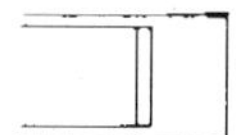 *Outward taper (or flared-out fins)*	*Integral, projecting structure and mechanism*
4	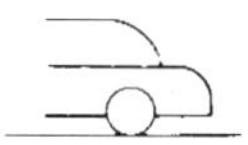 *(a) Rounded canopy with rounded boot* 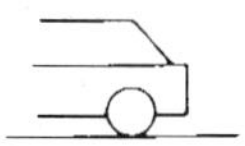 *(b) Squared canopy with short or no boot*		*Intermediate*
5	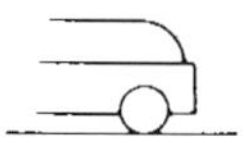 *Rounded canopy with short or no boot*		*Deep chassis*

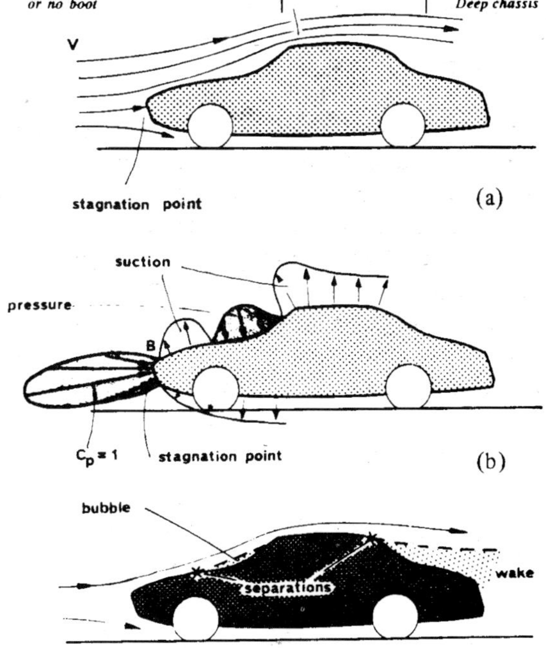

Fig 3: Air-flow around a car body

from the surface, it is sheared until the air velocity builds up to that of the main stream. Surface roughness gives rise to tangential viscous forces, skin friction, while severe surface discontinuities cause separation of the flow which is a starting point for the turbulent wake. At vertical corners, like the windscreen edge, flow slows down causing the pressure to rise and the possibility of a separation. Flow would probably re-attach itself further down stream along the side of the vehicle. The design aim is thus to maintain the boundary layer as far back along the body as possible by using very gentle curvatures. The advantage of sloping sides becomes clear, in this respect, as the blending in of the windscreen will be easier — and air will make its way around the sides to sweep the separation point further towards the tail of the vehicle.

Much valuable work has been carried out by the Motor Industry Research Association at Nuneaton. A number of their earlier reports are now publicly available. Fig 4 provides[2] a method of adding up the effects of the shapes of different areas of a car body to predict drag coefficient at the conceptual stage in design. Ratings shown for each configuration indicated are summed to a total drag coefficient being obtained from:

$$C_d = 0.16 + 0.0095 \times \text{drag rating}$$

Lift and pitch

The near teardrop shape shown in Fig 5 was tested some years ago at Imperial College in London[3]. It provided a drag coefficient of just 0.17 with ground clearance at 1/5th the depth of the body. Such an idealised shape would require modification to reduce the effects of changing angle of incidence to the oncoming airstream in a practical road vehicle. With constant incidence and decreasing ground clearance, negative lift begins to appear as a result of the venturi effect beneath the body. Tests showed the pitching moment increased with decreasing ground clearance while drag increased due to negative camber effects. The effect on drag co-efficient of varying incidence angle is shown in Fig 6. Blunting the trailing edge increases both drag coefficient and the lift curve slope.

When using aerofoil wings, such as that sectioned in Fig 7, to induce negative lift, it is usual to define the wing plan area S and chord length C in calculation: *Pitching moment* = $\rho V^2 SC/2$ *x pitching moment co-efficient*. The angle of incidence of the

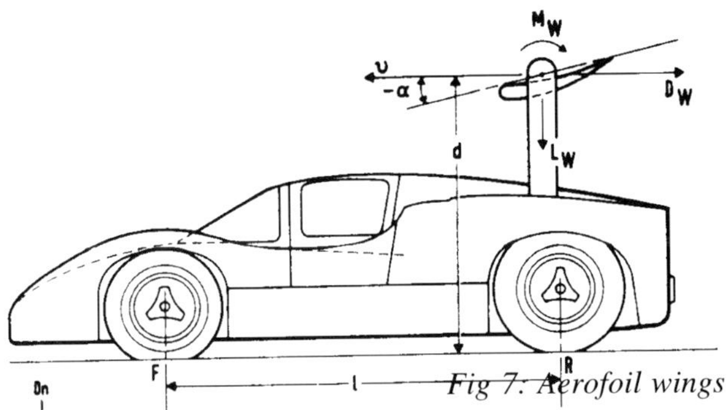

Fig 7: Aerofoil wings

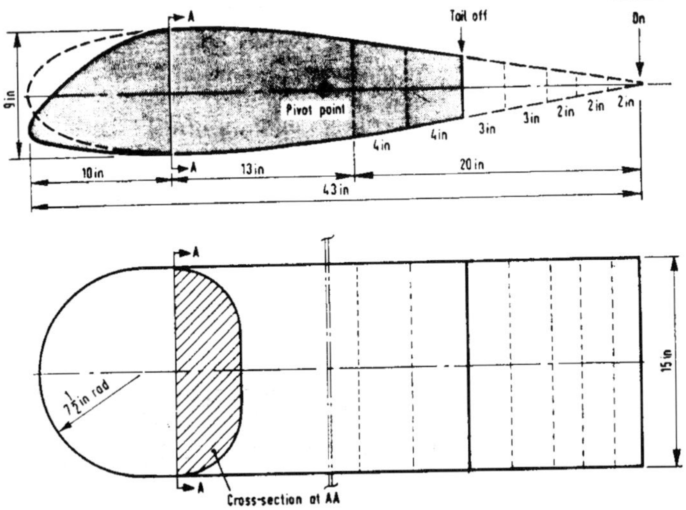

Fig 5: Near tear-drop shape

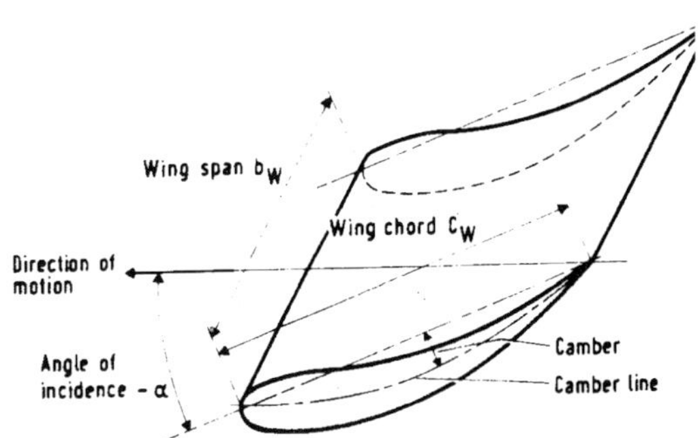

Fig 8: Cambering an aerofoil

wing is measured from the zero-lift attitude of the wing as a datum. Direction of resultant force on a wing is usually downwards and slightly backwards with angle to the vertical called down-wash. Lift per unit span = $\rho V^2 C L_c/2$ where L_c is lift co-efficient. The characteristics of the aerofoil can also be changed by cambering as shown in Fig 8.

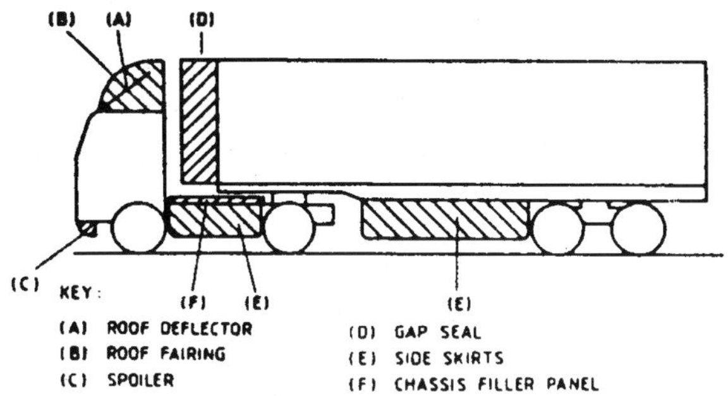

Fig 9: Aero-devices on artic container

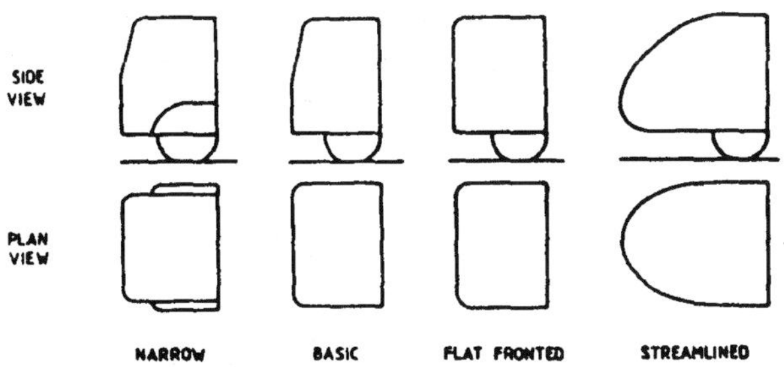

Fig 11: Cab-shape effect

Bluff vehicles

The effect of prevailing wind on drag has led to the use of wind-averaged drag coefficients. For a large box-shaped goods vehicle, side winds can substantially increase the drag coefficient from its 'head-wind' value. Tests at MIRA on an articulated container van model, Fig 9, have shown that a range of add-on devices can combine to effect drag reductions up to nearly 40%. Cab leading edge design is critical but streamlining the remainder of the cab is found to provide no benefit unless attached flow is maintained. Of the add-on devices, apart from roof-deflectors for high bodied vehicles, side-skirts were shown as particularly beneficial. Fig 10 shows proportion of drag due to aerodynamic effects — the dominant element at 90 km/h and above for this shape of vehicle. Fig 11 shows the effect of cab shape for different trailer body configurations.

From the US, engineers at the Kenworth truck company have compiled a table, from different sources, of the effects of individual changes on drag reduction; see the table in Fig 12.

Aerodynamic stability

Roll, pitch and yaw modes need to be considered in respect of vehicle stability. Roll stability can be achieved by ensuring that centre of pressure (point of application of wind sideforce) is close to the roll axis of the car; small deviations are tolerable if the vehicle does not have roll-steer characteristics. Distribution

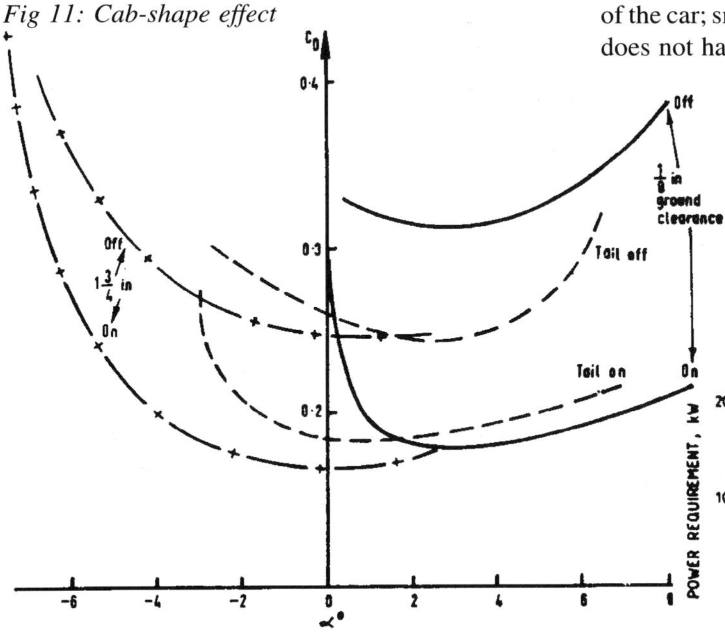

Fig 6:Incidence angle affecting drag

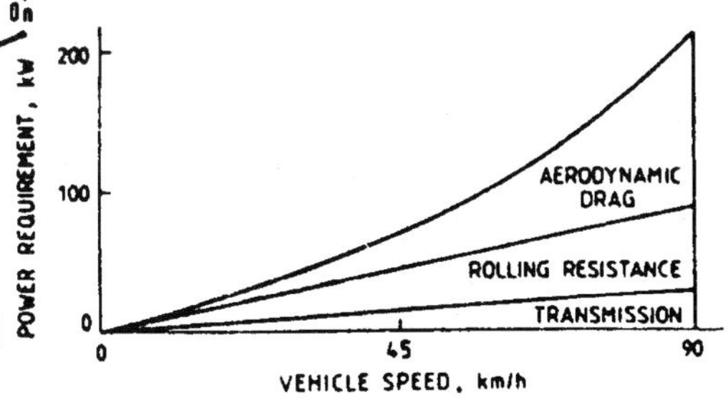

Fig 10: Contributors to motion resistance

of lift forces acting on a vehicle determines the magnitude of the pitching moment — which tends to vary the wheel loading with speed. Tail-end spoilers used to form the top edge of a drag-reducing cut-off tail can also induce downward stabilising forces on the road-wheels. Light-weight vehicles with centre-of-gravity well aft are less easy to control and are susceptible to zero or small negative lift at the front. Drooping the nose of the vehicle can have a stabilising effect in this respect — equivalent to reducing the

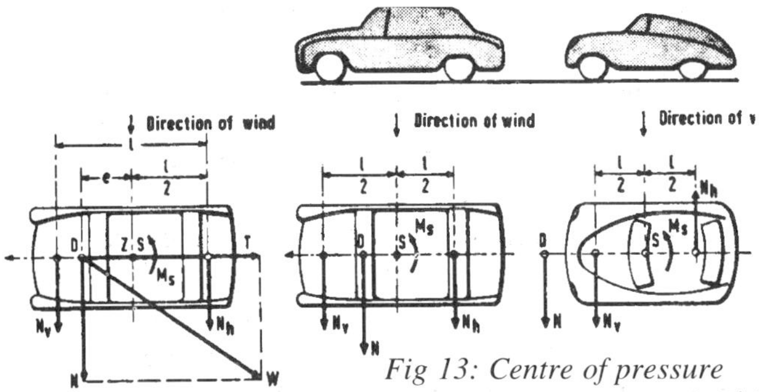

Fig 13: Centre of pressure

Modification	Tractor-Trailer Low	Tractor-Trailer High	Straight-Truck Low	Straight-Truck High
1. Tractor or Cab				
a) Deflectors	-0.01	0.15	-0.02	0.12
b) Roof fairings	-0.10	0.24	N/A	
c) Front edge rounding	0.00	0.06	N/A	
d) Aero Bumper	-0.05	0.05	N/A	
e) Rear cab extenders	0.04	0.08	N/A	
f) Sunvisor	-0.01	0.005	N/A	
2. Van Trailer or Body				
a) Front face fairing	0.09	0.16	0.14	0.22
b) Edge rounding (vs. square)				
- front posts	0.08	0.10	0.12	
- top front rail	0.05	0.08	0.10	
- posts and rail	0.13	0.18	0.22	
- top longitudinal	0.00	0.02	0.00	
- rear posts and header	0.02	0.10	N/A	
c) Edge beveling				
- 45 deg. front post & rail	0.04	0.05	0.10	
- shallow bevel on rear posts and header	0.03	0.09	0.03	0.05
d) Refrigeration unit	0.03	0.10	0.04	
e) Edge fairings	0.0	0.06	0.12	
f) Front turning vanes	0.0	0.08	0.20	
g) Smooth sides (vs. ext. rib)	0.03	0.09	N/A	
h) Skirts	0.02	0.07	0.02	
i) Bogie fairings	-0.02	0.02	N/A	
j) Streamlined tail (with streamlined front)	0.0	0.13	N/A	
3. Miscellaneous				
a) Reduced gap (1.5m to 1.0m)	0.02	0.10	N/A	
b) Gap seal	0.04	0.10	0.03	
c) Fill gap	0.06	0.12	N/A	
d) Deflector & body fairing	0.0	0.15	N/A	
e) Streamlined truck	0.45	0.55	0.40	

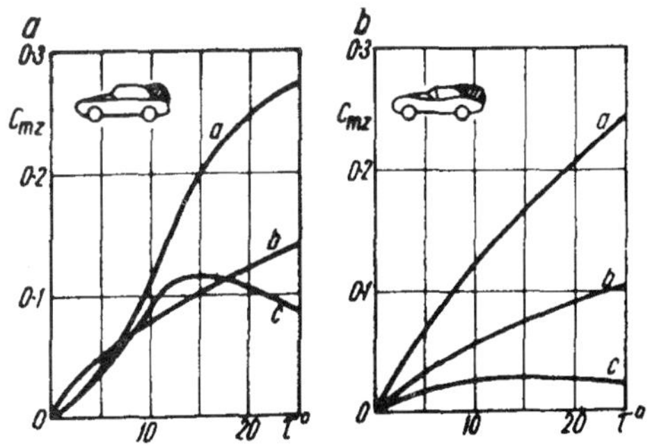

Fig 14: Effect of stabilisers

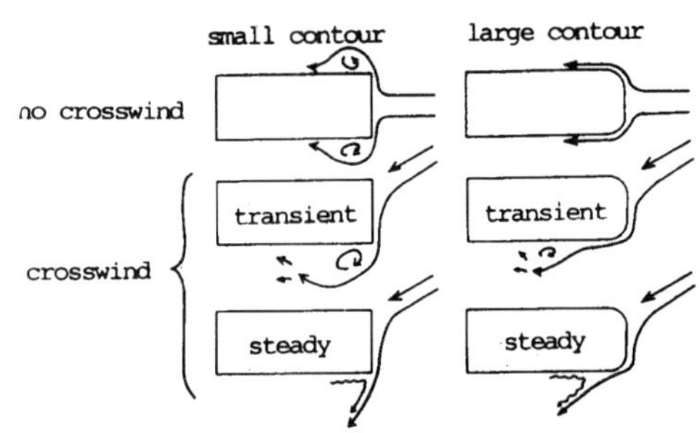

Fig 15: Cross-wind effect on yaw

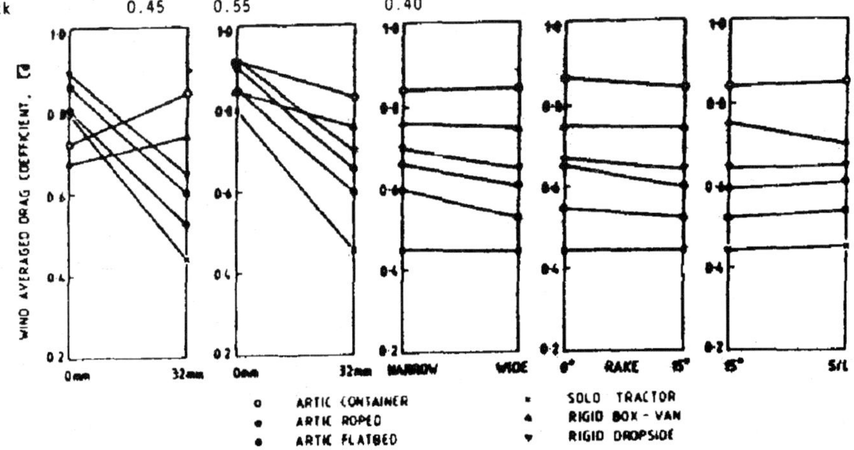

Fig 12: Constituent effect on drag reduction: from base of C_D = 55%

angle of incidence of an aerofoil. The top surface of wedge-shaped bodies also provides a negative angle of incidence resulting in a downforce, as does an underside venturi. Care must, of course, be exercised in reducing ground clearance — as the venturi will stall in the limit with the result that reduced mean velocities can cause destabilising lift forces.

Stability in the yawing plane governs the amount of steering correction required in cross winds. Only when the centre of pressure corresponds with the neutral steer point will the vehicle maintain its heading and merely drift sideways. Ahead of this point instability occurs and it is preferable to arrange the centre-of-pressure just behind the neutral steer point. Sharp corners on a body should be avoided in the interests of reducing sharp pressure changes when the vehicle is subject to side gusts — with consequent changes in yawing moment. Fig 13 shows the force pattern, with centre of pressure at *D*, at which air resistance *T* and normal wind force *N* act; a torque $M_s = Ne$ develops around the vertical axis Z through the centre of gravity. Streamline vehicles for which $Ne = (N_v + N_h)/2$ have a greater turning moment than bluff ones for which $Ne = (N_v - N_h)/2$. Use of stabilisers, or fins, at the rear of the vehicle can give good results. Fig 14 shows measured yawing moment co-efficients of production cars modified by the addition of stabilisers: a saloon car; b coupe; c twin stabilisers.

Care must be taken in relating results measured in wind tunnels with those expected on the open road. To account for viscous and inertia effects in flows around models in tunnels and full sized vehicles on the road, the Reynolds Number: $R_e = rVW_n n/m$ where W_n is nominal model size and n is viscosity coefficient. The expression needs to be evaluated for both conditions. For equal Reynolds numbers, flow conditions are thought to be identical for model and vehicle. Thus forces can be scaled up in proportion to scale ratio. Normally the nominal model size is defined by the width dimension but if the height is very different from the width, the square root of the frontal area should be used. A fullsize truck at 55 mph has a Reynolds number, based on width, of 4.1×10^6. The SAE standard (J1252) states that a minimum of 0.7×10^6 is an acceptable test condition provided wind speed is less than 300 ft (91.5 m)/sec to avoid air compressibility effects.

An interesting use of wind tunnel testing has been carried out by Toyota to study vehicle motion in a cross-wind. Effect on yawing moment co-efficient was investigated — of altering front end corner contours using 1/10 th scale models. Fig 15 shows flow pattern variation. The company devised a method of inducing flow separation in order to more nearly equalise flow in steady and transient conditions. Ducting is used to connect high and low pressure areas on the vehicle in order to blow air out of slits in the front corners, Fig 16, in order to induce the desired separation.

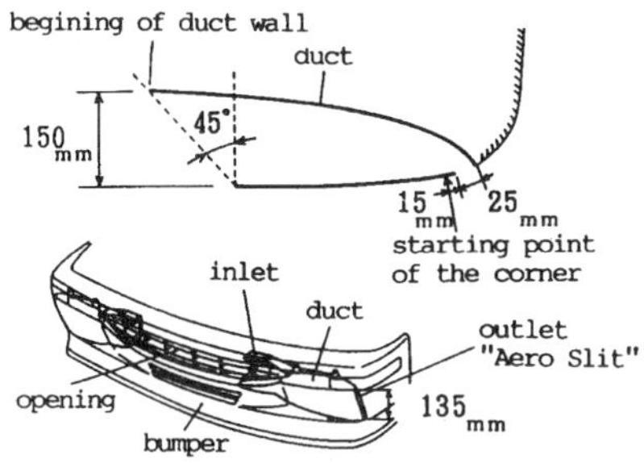

Fig 16: Use of aero slits

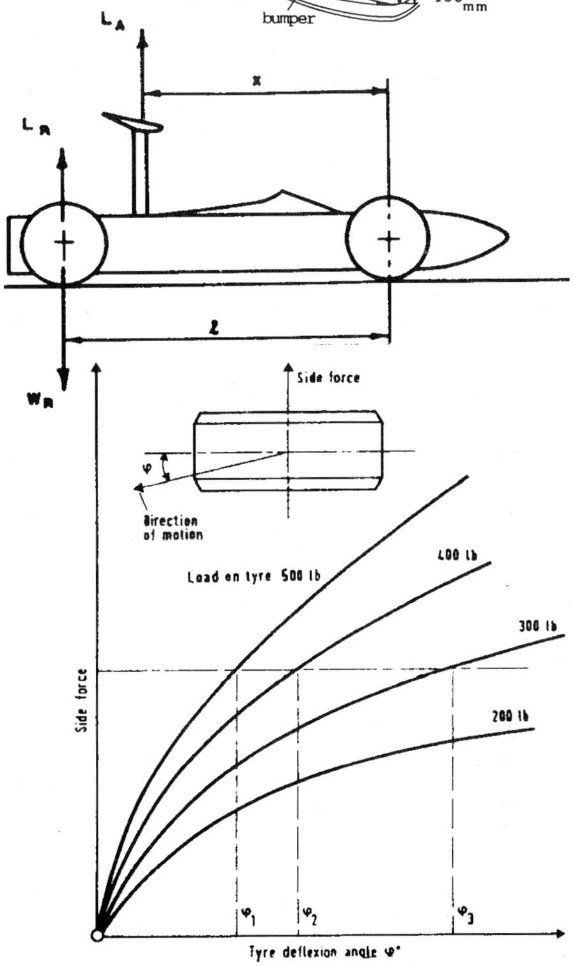

Fig 17: Effect of down-force on handling

Race-car aerodynamics

The aerodynamic design of racing car bodies varies with the rules allowable by different race formulae as well as periodic changes made by the authorities in search of better and safer sport. About the only common theme between models is the basic subject discipline which yields the essential principles for making design improvements whether by add-on devices or changes in body shape.

Serious work on racing car aerodynamics commenced, some would say, with the introduction of negative lift devices in the 1960s. According to City University researchers[4] such devices put very high loads on suspension systems and can radically affect cornering characteristics of the tyres as shown in Fig 17. Choice of wing aerofoil sections is thus crucial to overall handling of the racing car. While the aerospace industry has much data available on the performance of aerofoils, it has to be understood that much of this is based on wind tunnel measurements inappropriate to the race-car application. Dynamic similarity is crucial to the extraction of meaningful data in design. This is the consideration of air flows around two similar bodies, but of different size, to ensure that inertia and viscous forces at equivalent points in the flow path are in the same ratio. This is ensured by calculating Reynolds Number for each flow case, as defined in the earlier article. As theoretical relation between lift/drag co-efficients and Reynolds number is mathematically complex, it is

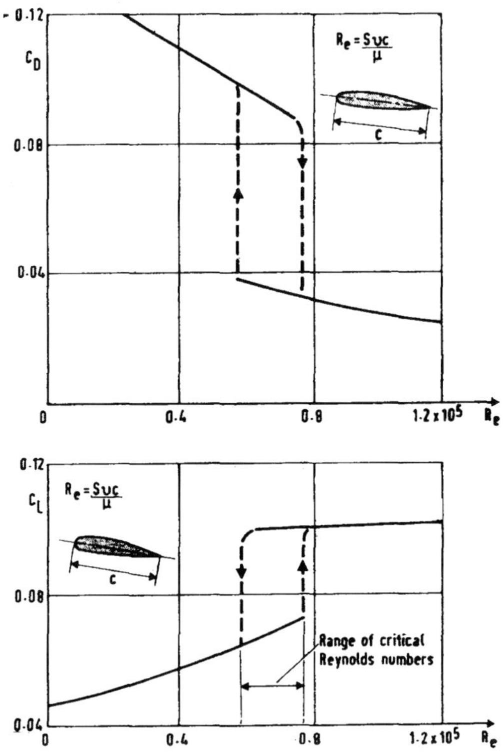

Fig 19: Characteristic curves for aerofoils

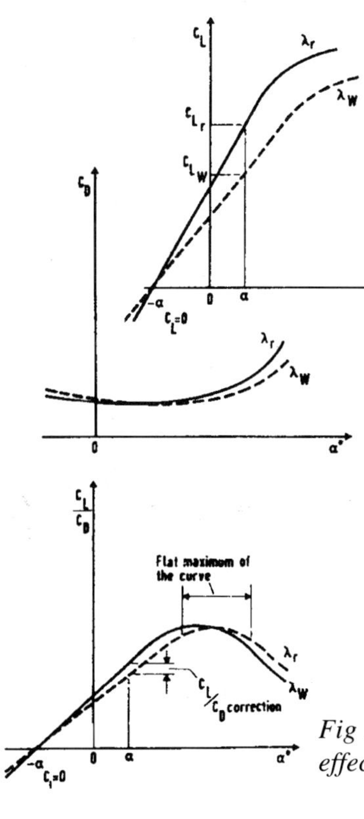

Fig 18: Reynolds number effect on lift/drag co-efficients

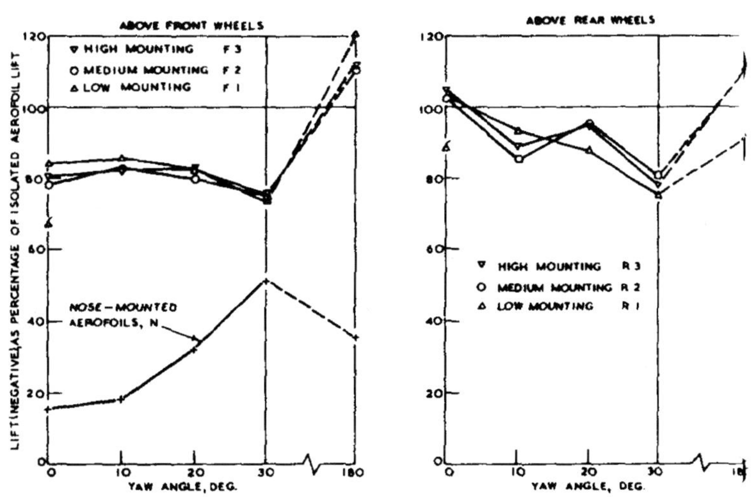

Fig 23: Negative lifts obtained in tests

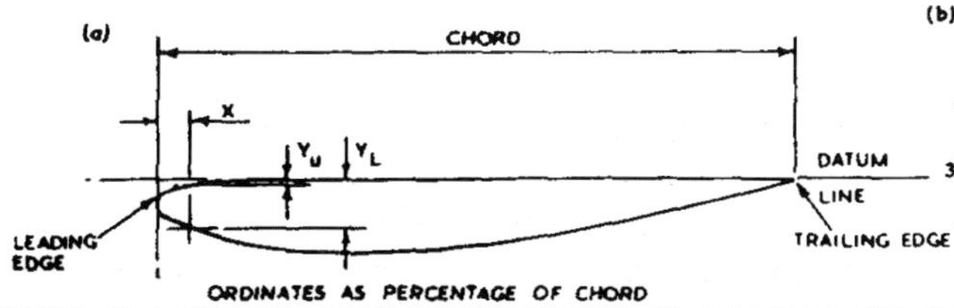

ORDINATES AS PERCENTAGE OF CHORD

X	0	1·25	2·5	5	10	15	20	30	40	50	60	70	80	90	100
Y_U	3·50	1·93	1·47	0·93	0·42	0·15	0·03	0	0	0	0	0	0	0	0
Y_L	3·50	5·45	6·50	7·90	9·60	10·68	11·36	11·70	11·40	10·52	9·15	7·35	5·22	2·80	0·12

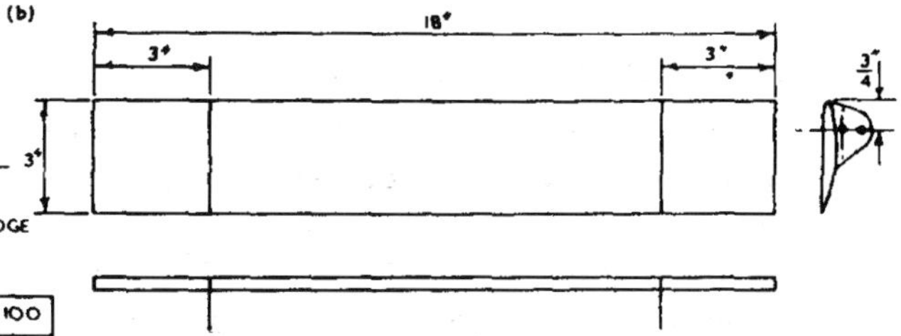

Fig 20: Test section for aerofoil evaluation

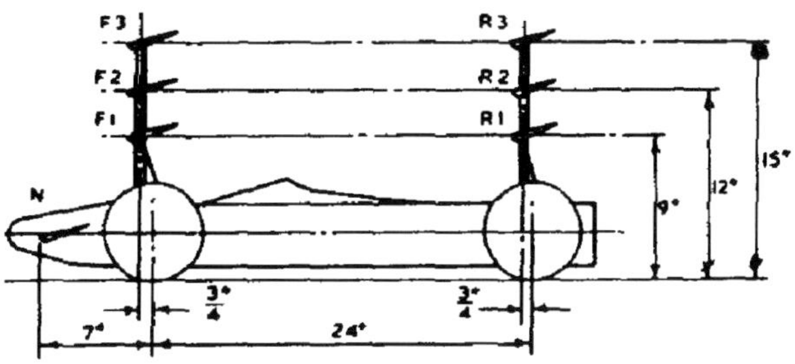

Fig 22: Wing mounting positions

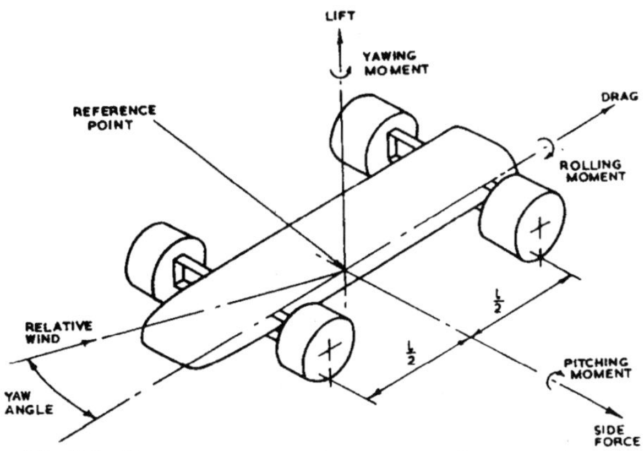

Fig 21: Force components measured

usual to determine this experimentally for all but the simplest of shapes.

Fig 18 shows that for aerofoil shapes, only at low *Re* flows is there a significant jump in *Cl* and *Cd* values. The 'characteristic dimension' for the *Re* expression is in this case the chord of the aerofoil. Flows corresponding to higher speeds of the vehicle aerofoil would behave predictably; see the data in Fig 18. Design procedure is thus to find the range of angles of incidence for the wing at the design operational speed of the vehicle. Selection of the aerofoil section should be on the basis of highest possible lift force with the minimum of drag at the angles of incidence under consideration. Fig 19 shows characteristic curves for a selection of different aerofoils based on work carried out at the City University.

A comparatively flat maximum for the *Cl/Cd* characteristic shows of course that drag is minimum over a wide range of incidence angles. Because the rectangular planforms of the wing used on the car may be different from those of the wind-tunnel testpiece, aspect ratio $\lambda = b2/S$ should be noted in calculation: the ratio of wingspan to constant chord in the case of a rectangular planform. Compared with a

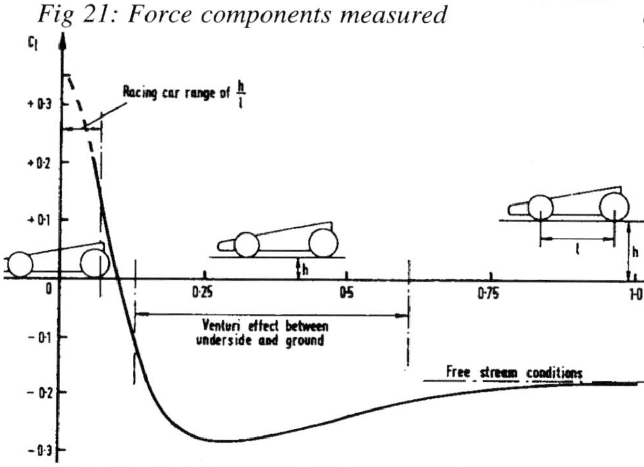

Fig 24: Model for race body aero testing

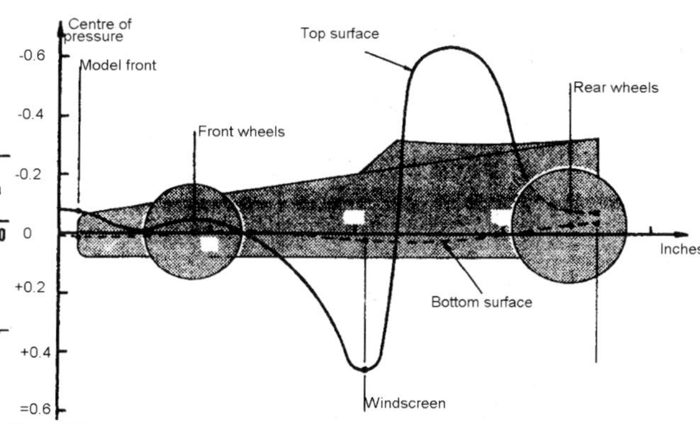

Fig 25: Pressure distribution on test model

wing of infinite span, a typical short wing would suffer additional induced drag (as seen in Fig 18 when comparing lamda values for wind-tunnel (w) and real (r) devices); conversion tables must therefore be used. The change in lift and drag co-efficients can also be calculated from:

$$dC_l = (C^2\omega t57.3/a\pi)[(1/L\omega t) - (1/L)]$$
$$dC_d = (C^2\omega t/\pi)[\{1 + 2dC/C_l t/L\omega t\} - 1/L]$$

for lift and drag change. Wing positioning must be such as to operate high enough above the car to be in a clean airstream while avoiding an excessive negative pitching moment from the drag of the wing.

Early work sponsored by the Jim Clark foundation[5] was valuable in optimising race-car aerofoils, based on the section shown in Fig 20. Test work involved measuring the force components shown in Fig 21 at a wind tunnel speed of 100 mph. For the wing mounting positions shown in Fig 22, negative lifts produced are shown in Fig 23 expressed as a percentage of the force produced by an isolated aerofoil without the car. Note that when mounted

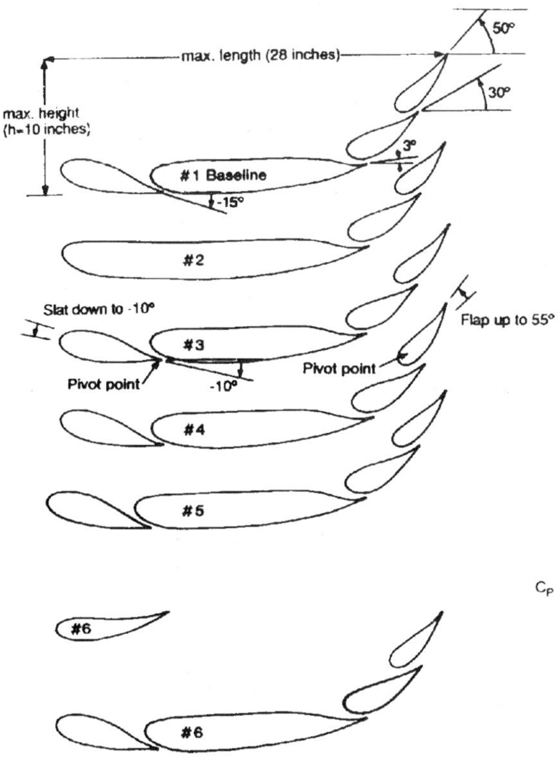

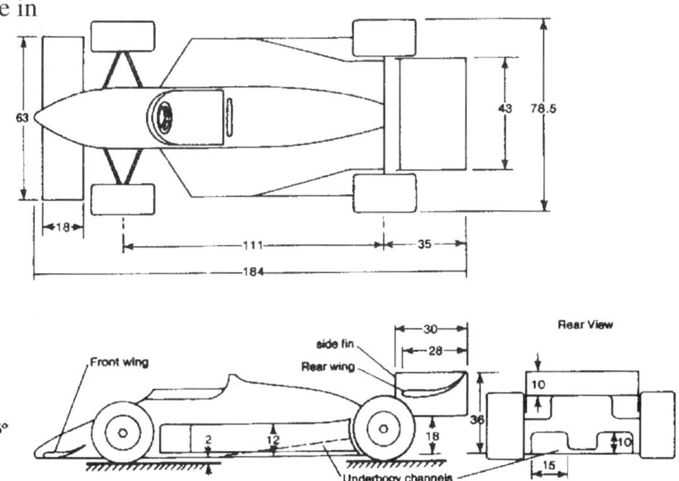

Fig 26: CART regulations

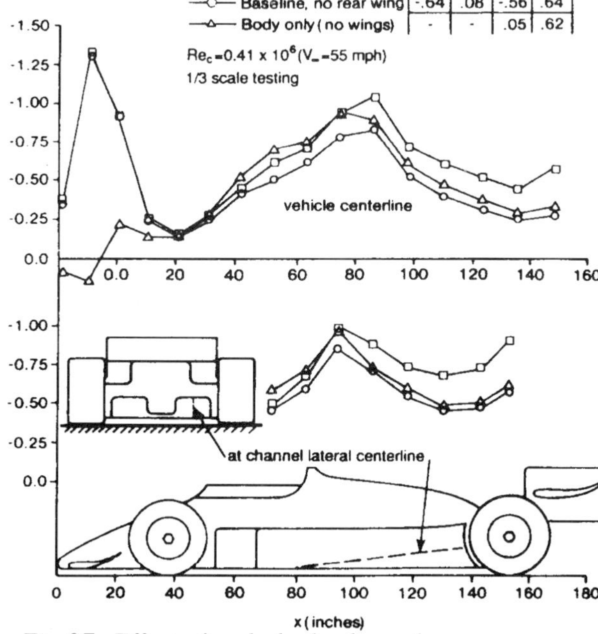

Estimated coefficients	C_{L_f}	C_{L_r}	C_L	C_D
Baseline	-.39	-1.32	-1.71	1.06
Baseline, no rear wing	-.64	.08	-.56	.64
Body only (no wings)	-	-	.05	.62

Fig 27: Effect of underbody channels

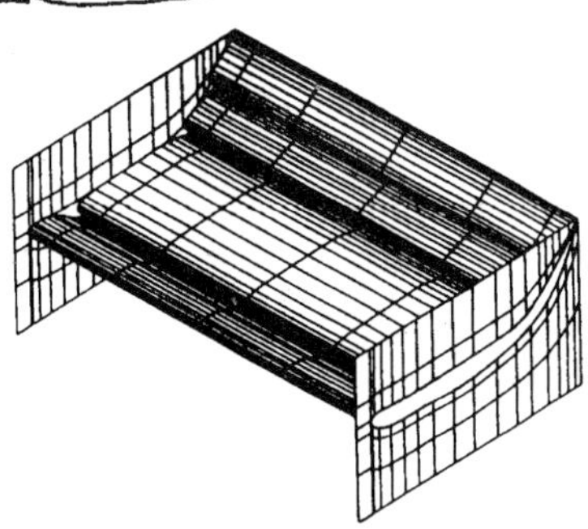

Fig 28 Computer-optimisation of wing

above the front wheels the aerofoil lost 20% effectiveness at all yaw angles up to 30 degrees. When placed above the rear wheels it only fell to this level at 30 degrees. Also shown is the variation of lift of the stub aerofoils (N) mounted on the nose of the car. Aerodynamic characteristics of the typical wedge shape of racing car bodies was also the subject of pioneering work at City University[6]. The basic form of such vehicles also has a substantial effect on reducing lift. One eighth scale model tests were used to demonstrate the ground effect in Fig 24. When the underside is parallel to the ground there is effectively a negative angle of incidence with consequent downforce — in the high, 'free-flight' condition. With reduction in h there is an increasing venturi effect beneath the car creating a suction which adds to the negative lift. At very low clearances an upward lift force occurs when the airflow stalls beneath the body. In the pressure distribution plot of Fig 25 the nett upward lift resultant can be seen — with a lift coefficient in this case of 0.23. More recent work at San Diego University[7] in the USA showed that cars designed to CART regulations, Fig 26, could have rear wings developing -2.2 negative lift co-efficient. The small front wing positioned far forward near the ground operates in the undisturbed air stream and works at fairly low loading to trim downforce distribution on the vehicle. Underbody channels are also permitted in the body configuration, by the regulations, to enable further increase in downforce. The interacting effect of these with the rear wing is shown in Fig 27 found in one-third scale model tests. Optimising a high lift rear wing of very modest aspect ratio was achieved by computer modelling of multi-element aerofoil systems, Fig 28. Using a finite-element model of the optimum section computation was carried out and checked against an experimental model made from 664 panels. Test results are shown in Fig 29.

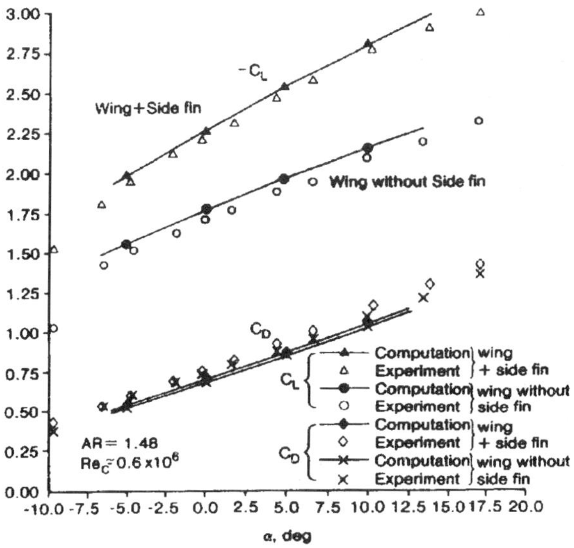

Fig 29: Test results from computer model (left)

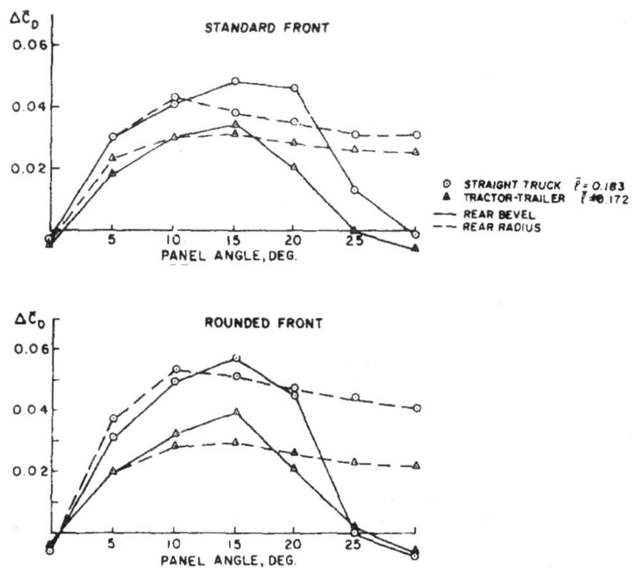

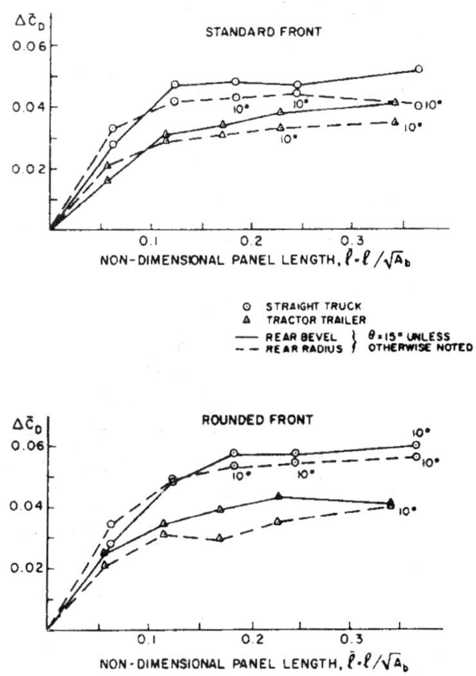

Fig 30: Effect of bevel angle (left) and length (right) on drag

Truck aerodynamics

The bluff rather than streamline bodies which apply to many special vehicles require a different aerodynamic analysis from that applying to the more streamline shapes. This section examines both the basic principals involved and the effects of some add-on devices which can be used, given a proper understanding, to effect quite substantial fuel-economy savings.

The previous section was based generally on the assumption of streamline air flow over the body surfaces. Here the effect of turbulence caused by bluff body shapes is considered so that the flow over typical box-cube shapes can be understood.

The viscosity and compressibility of the air becomes significant when air is flowing over a bluff body which grossly disturbs the air stream and drag co-efficient depends upon Reynolds Numbers:

$$= \rho VL/\nu$$

where r is air density, V relative velocity, ν viscosity and L is a major dimension of the bluff body with respect to the air stream.

The objective in aerodynamic design of bluff bodies is to try and avoid separation of airflow by using devices which attach the flow to the surface as far as possible along the vehicle towards its wake. SAE Recommended Practice J1252 suggests using the square root of the frontal area for the 'significant dimension' L in the Reynolds Number equation — if vehicle height is very different from vehicle width. A large truck might have a Reynolds Number of 4.1 x 10^6 and generally speaking the high drag of truck container-bodies arises from pressure reduction over the rear face of the vehicle due to the entrainment of air by the shear layers separating from the rear edges.

Either bevelling or radiusing the rear edges help to reduce this effect. Tests made by the National Research Council of Canada[8] show the effect of rear-panel bevel angle and length on drag for both square and radius fronted vehicles in Fig 30. The danger of applying streamline flow theory directly to a bluff body is shown in Fig 31 which compares calculation with experimental result for flow around a circular cylinder. Only around the stagnation point where the flow separates at the front of the body is the flow similar to the theoretical 'streamline' case.

The shearing of fluids can be visualised with

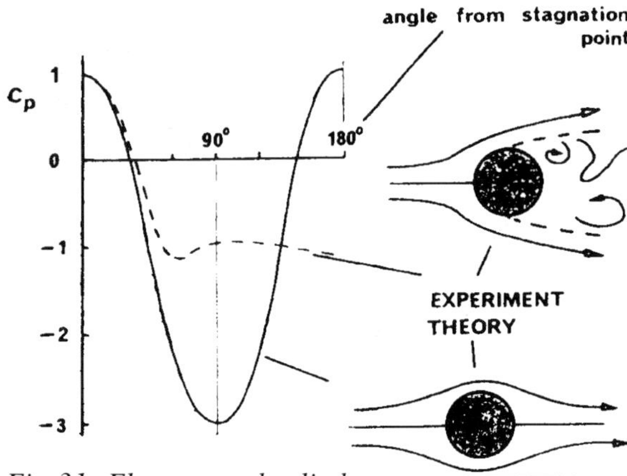

Fig 31: Flow around cylinder

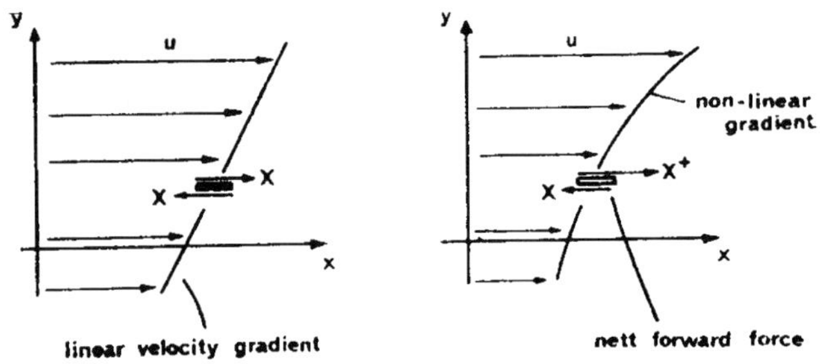

Fig 32: Shearing of a fluid

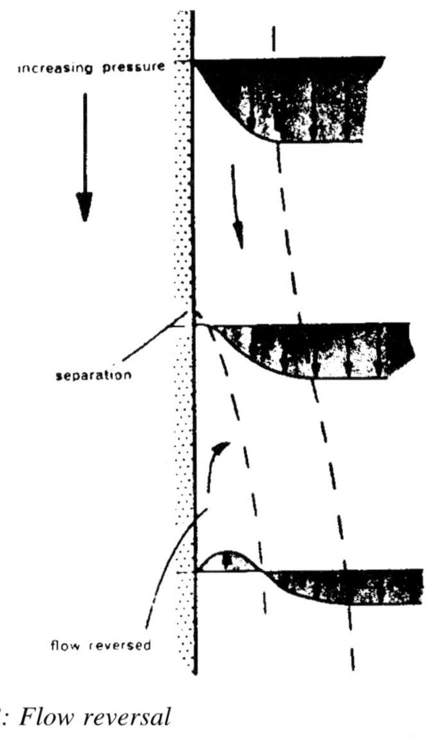

Fig 33: Flow reversal

the help of the diagram in Fig 32. Here flow is considered to be in the x-direction and its velocity increasing with the profile shown in the y-direction. For any element a shear stress X exists because the fluid tries to reduce the relative motions of adjacent layers. When the velocity profile is linear, tangential force each side of the element is zero; otherwise there is a nett forward force due to pressure gradient in addition to the viscous force. This can be represented mathematically and simplified to produce the Reynolds Number equation stated above.

If, due to the shape of the vehicle body, the pressure outside the boundary layer, between air and body-surface, increases in the flow direction, the gradient effect will also be felt within the layer. However, as within the layer, the velocity has already been reduced because of the viscous stresses, there is possibility of the fluid being brought to rest or reversed, Fig 33. In such cases the layer separates from the surface and is no longer under direct control of the body. This explains the formation of the wake. The occurrence of separation depends on several factors including the position of the boundary layer upstream of the area being considered. Various flow patterns can arise from separations caused by different body shape effects, Fig 34. Boundary layer control is principally a matter of moving separation points as far aft of the body as possible, either by arranging for very gradual adverse pressure gradients, using curved front-ends, or using layer control devices. Such devices either speed up or remove

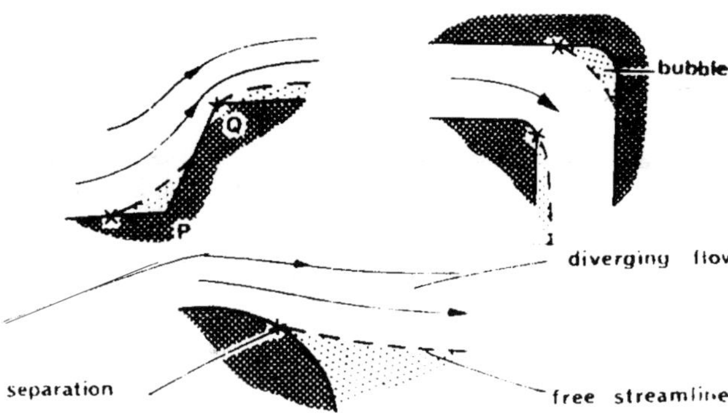

Fig 34: Body shape effects

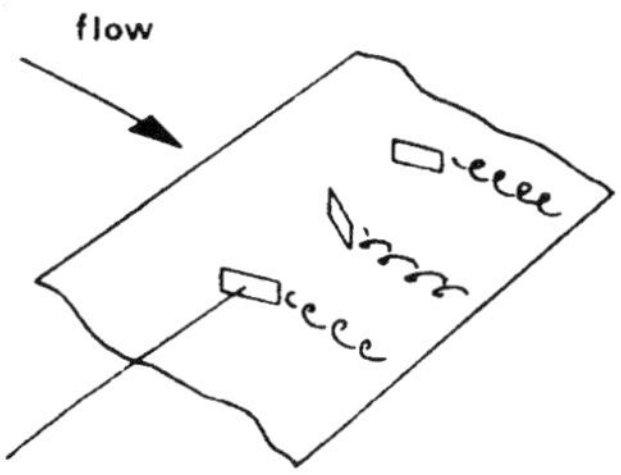

Fig 35: Vortex generators

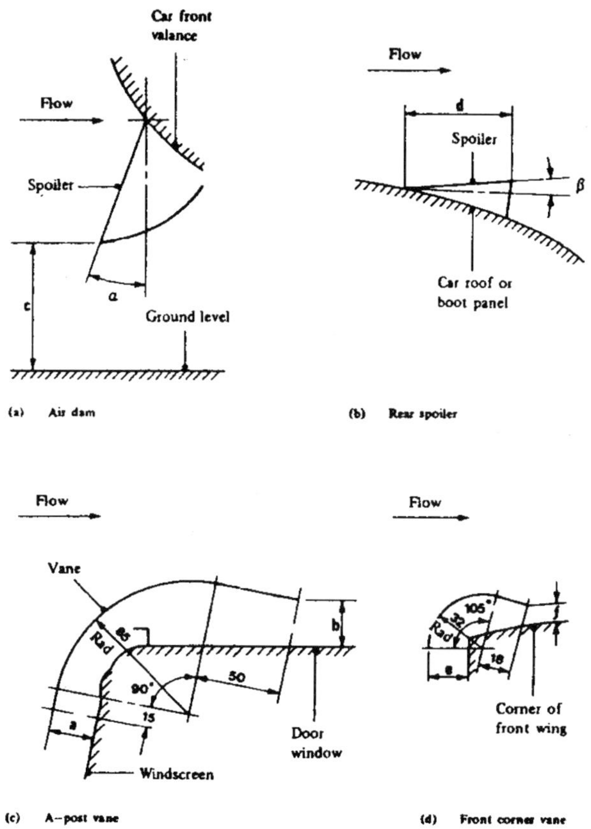

Fig 36: Turning vanes

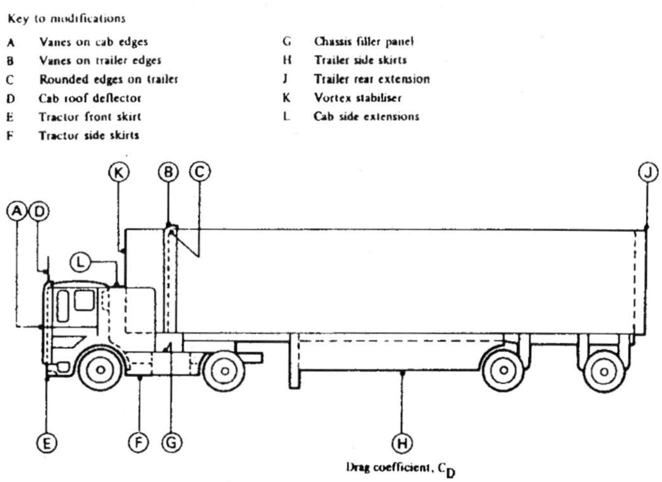

Fig 37: Drag reduction devices

stagnant air near the surface; sometimes this can be done by sucking air through slots in the surface which might double as air inlet ducts for another vehicle system. Small inclined slats, or vortex generators, are also used — the small vortices produced having the effect of breaking up the stagnant layer, Fig 35.

Experimental work carried out at MIRA[9] has provided a useful evaluation of add-on devices. Fig 36 shows turning vanes used for attaching flow around relatively sharp corners such as windscreen pillars with which some 3% drag reduction was measured. Their use on the leading edges of box van

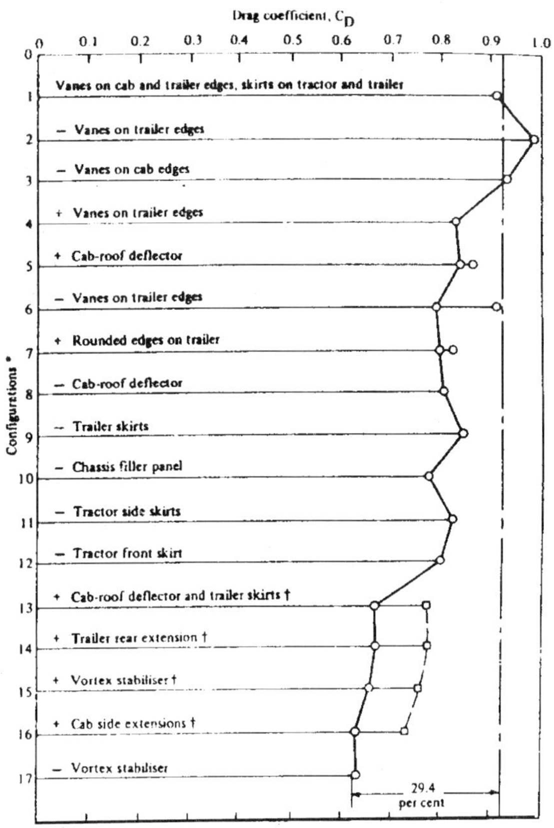

Fig 38: Incremental drag reductions

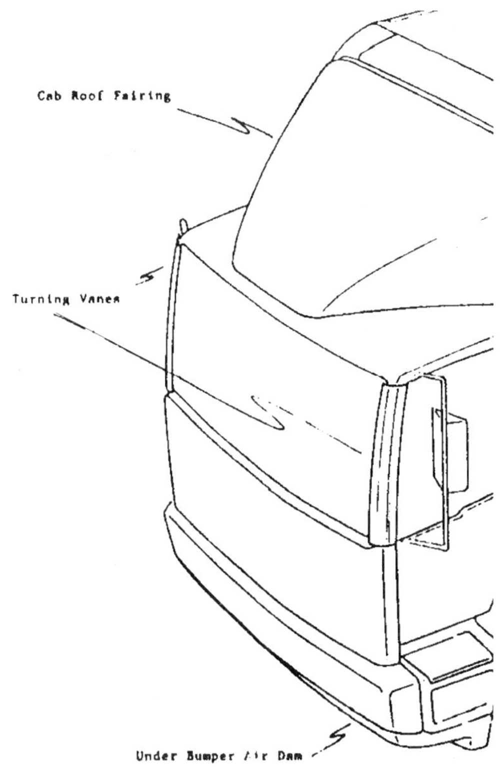

Fig 39: Leyland Freighter mods

Roof deflector)with)high)load/	10 to 12%
Roof fairing)box body	10 to 25%
Air dam (rigid)	up to 4%
(articulated)	up to 2%
Cab side deflectors	up to 2%
Chassis skirts (rigid)	5 to 12%
(articulated on tractor only)	up to 4%
Leading edge turning vanes	up to 1%
Typical wind averaged drag coefficient reduction,	

Fuel Consumption (mile/gallon)

Test Speed and Gear	Standard at 10 000kg	Prototype at 10 273kg	Improvement
Test Track			
40 mile/h, 5th gear	14.69	17.86	22%
50 mile/h, 6th gear	12.61	16.49	31%
60 mile/h, 6th gear	9.70	13.56	40%
Motorway (maximum gradient 1:1000)			
40 mile/h, 5th gear	15.60	19.85	27%
50 mile/h, 6th gear	13.01	17.83	37%
60 mile/h, 6th gear	10.18	14.38	41%
Leyland to Kendal (M6 and A6, return) Test Route			
M6	11.42	14.64	28%
A6	13.98	16.35	17%
Overall	12.63	15.46	22%

Fig 40: Fuel consumption reductions for different add-on devices

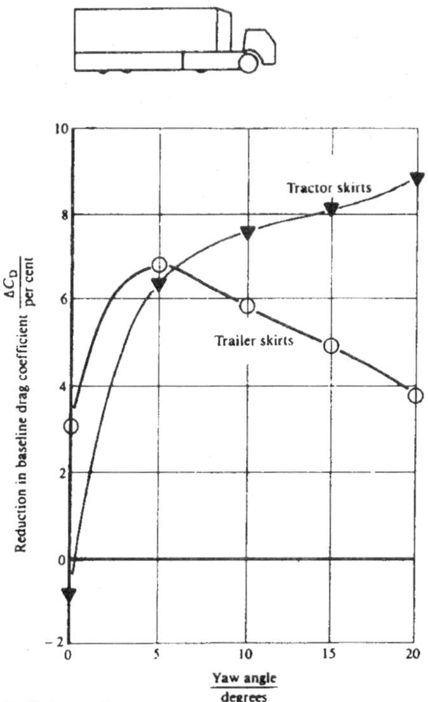

Fig 41:Skirt advantages

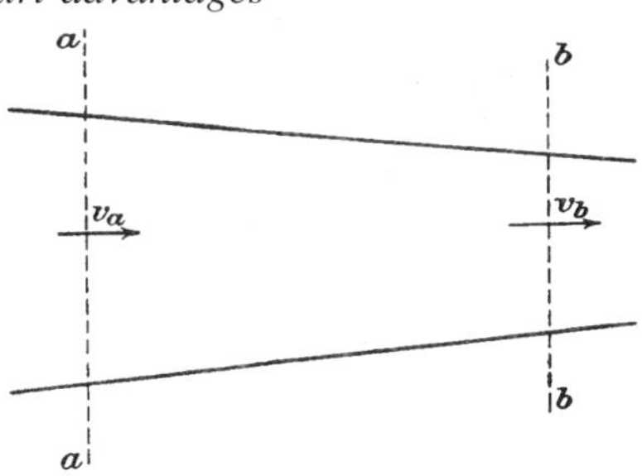

Fig 42: Flow in a duct

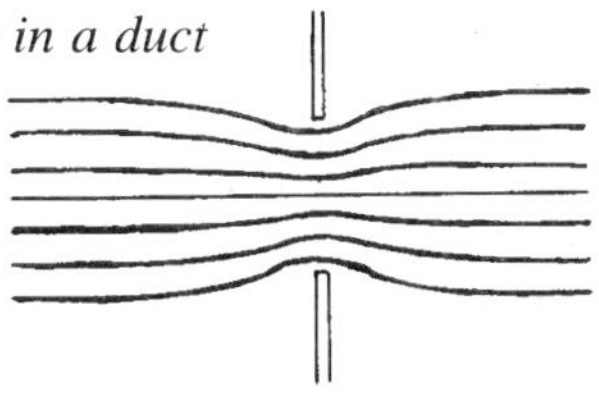

Fig 43: Constriction in a duct

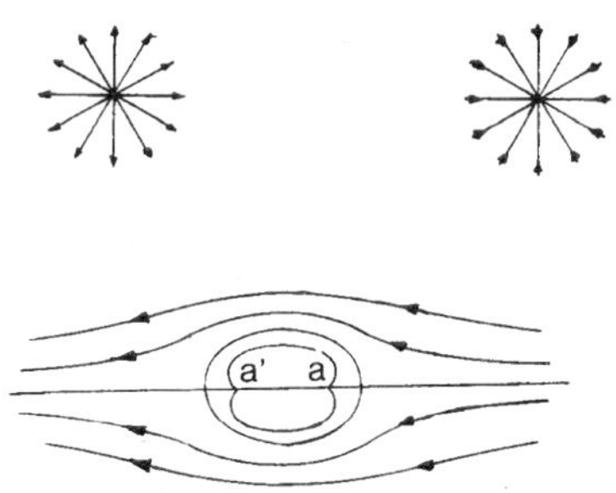

Fig 44: Flow around oval section

bodies, however, is reported to have produced reductions up to 30%. The degree of reduction depends very much on the shape of the cab ahead of the box container and in this context it is important to note that tests have shown vanes mounted to the windscreen corners of a cab have actually increased the overall drag of the vehicle. Rounded leading edges for box bodies have resulted in 12-13% reductions rising to 15% in conjunction with cab-roof deflectors. Effects of tractor and trailer skirts, on articulated vehicles, have also been recorded, Fig 37. In combination, a 30% reduction in drag of articulated vehicles is achievable with such modifications, Fig 38. The Leyland Freighter vehicle is shown in detail in Fig 39 with the modifications and their contribution highlighted. Fig 40 shows some of the fuel consumption benefits in tabular form. Model testing carried out at the Cranfield Institute of Technology[10] has also thrown light on the effect of skirts and cab-to-body seals. The cab to body gap can be particularly detrimental in cross winds when there is a tendency for crosswinds within it to separate fore-aft flow prematurely on the leeward side of the vehicle. The same problem applies with crosswinds blowing through the underside of the vehicle. The advantages of skirts in this context are shown in Fig 41.

Local airflows

Local airflows in the vicinity of the engine cooling system are examined next, which are of interest both in determining the contribution of the system to overall vehicle drag and in determining the airflow available for cooling heat transfer.

Detailed studies of local airflows require a knowledge of some of the basics of fluid mechanics, starting with the definition of a fluid as a material substance which cannot sustain a shearing stress when at rest. In fact, real fluids are not completely inviscid but often the viscosity effects are negligible, as is the fluid's incompressibility under normal conditions. Crucial to the understanding of forces imposed on bodies in a fluid stream is the significance of relative body between body and stream — as opposed simply to the absolute velocity of the stream. Forces on any point on the inclined surface of the body can be resolved into a pressure normal to the surface and a shearing stress tangential to it. The 'total pressure force' is the summation of all the elemental point forces normal to the surface and the 'skin-friction' is the corresponding summation of tangential forces.

In flow through ducts such as that defined by the air-inlet system, an equation describing the continuity of flow can be written which is useful in calculation. Between two stations (1,2) of the duct, Fig 42, with the assumption of continuity of flow (no fluid added or removed at any point), and with the fluid assumed to be incompressible:

$$s_1 v_1 = s_2 v_2$$

where s is the spacing of the streamlines.

Since v is inversely proportional to s, the effects of constrictions in the ducts are shown as Fig 43. These show the classic increase in velocity at the throat of a venturi, as the streamlines come close together, and on the outer edges of a cylinder crossing a wide, rectangular-section, duct. Corresponding reductions in velocity occur upstream and downstream of the cylinder.

The Bernoulli equation relates fluid velocity to pressure, p such that:

$$p + \rho v^2/2 = B$$

where B is the Bernoulli constant and the second term is the dynamic pressure, or the increment above the static pressure (at a stagnation point in the flow) that describes the freestream pressure at any point in the flow. In air at sea level its effect is fairly minimal, and drag forces are the result of pressure changes rather than air-density changes.

A rotating fluid is termed a vortex, and a free vortex if no external forces are applied to it. For tangential flow velocities v_1 and v_2 at radii r_1 and r_2 from its centre, a centrifugal head of pressure is imposed:

$$= (v_2^2/2g) - (v_1^2/2g)$$

the basis of the power developed by a centrifugal blower. Applying Bernoulli's thereom, with constant pressure head, the static pressure decreases rapidly towards the centre of the vortex. When the fluid flow is filled with vortices, a so-called shearing flow occurs as in the case of the shallow layer (boundary layer) close to the surface of a body in the fluid stream or in its wake. If the shearing flow takes place in a very thin layer it is called a vortex sheet.

Sources and sinks are useful concepts in visualizing fluid flow, the former being a point in the fluid from which flow issues symmetrically in radial direction — and the latter is its converse. An oval shaped section body in a flow could be represented by a source and sink pair, Fig 44, the two being analogous to the poles in a bar magnet and their associated magnetic field; a and a' are two stagnation points where the uniform flow velocity is equal and opposite

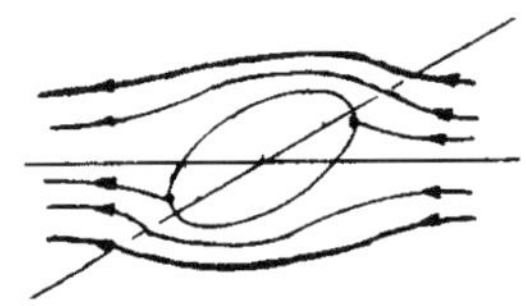

Fig 46a: Flow around elongate section

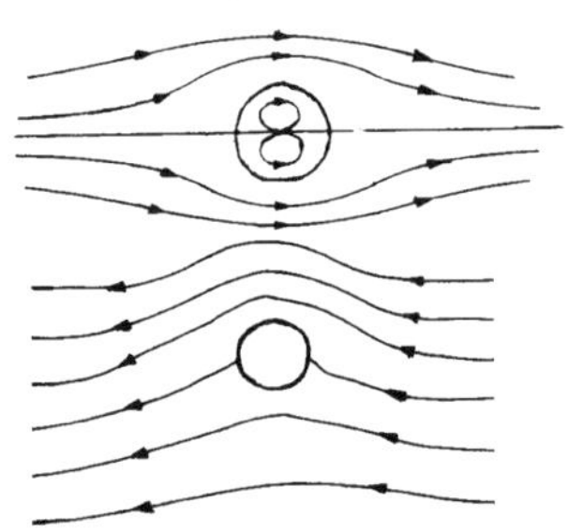

Fig 45: Vortex and doublet

Fig 46b: Flow around flat plate

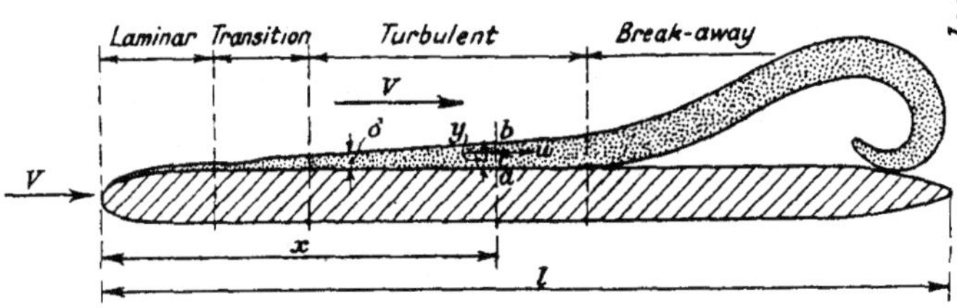

Fig 47: Velocity variation in boundary layer

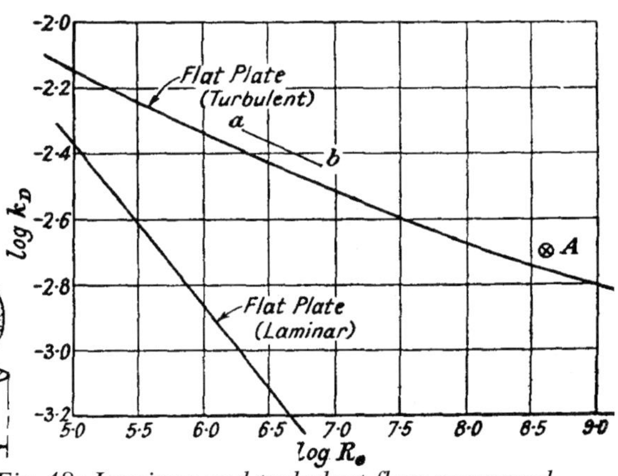

Fig 48: Laminar and turbulent flow compared

to the velocity caused by the source-sink pair. A circular section body in a flow can be represented by so-called doublet in which the distance between source and sink shrinks to nothing. When a vortex and doublet combine in a uniform stream, Fig 45, the stagnation points are displaced from the vortex field away from the axis of symmetry. The effect of the vortex is thus said to give the flow a camber or curvature as a symmetrical body in the stream which was cambered at an an angle to the stream would give a similar flow field.

If there is no circulation around a body of elongated section, Fig 46a, inclined to the free-flow the streamline pattern is symmetrical with upstream and downstream stagnation points — and for a flat plate as in Fig 46b. Variation in velocity within the boundary layer, thickness δ, can be visualised with the help of Fig 47 representing a surface past which fluid is flowing with velocity V. The body immersed in the fluid will have a Reynolds number of $R_e = Vlv$ where l is its linear dimension and n the fluid viscosity. For low R_e, say less than 500,000, the flow within the layer is streamline and for high values, above 1,000,000, turbulent. In practice different combinations as shown apply. In the laminar flow region, thickness of the boundary layer can be approximated as:

$$=5.83(lx)^{1/2}/R_e^{\ 1/2}$$

Drag co-efficient for a flat plate can be found from the Karman equation to be:

$= 1.327(1/R_e)^{1/2}$ for laminar flow and

$=0.072(1/R_e)^{1/5}$ for turbulent flow, Fig 48.

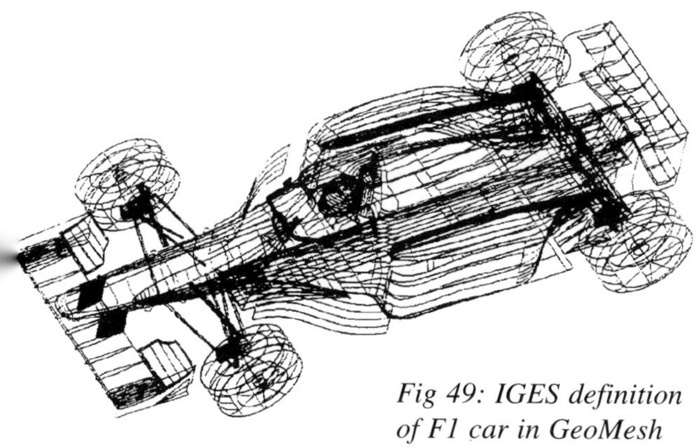

Fig 49: IGES definition of F1 car in GeoMesh

Predicting aerodynamic behaviour with computational fluid dynamics (CFD)

A good deal of the high cost of moving-ground wind-tunnel testing of racing cars can now be avoided with the use of computional fluid dynamics according to researchers at Fluent Europe[11]. Hitherto, the long time required to set up grids, and turbulence model limitations, precluded the widespread use of CFD, but most of the teething problems have now been overcome and the use of RAMPANT code by Benetton is seen here.

Benetton, like most of the major Formula 1 teams, run a number of CAD workstations interactively during the day, and potentially this computational power is wasted overnight. However, unstructured codes such as RAMPANT parallelise to homogenous and heterogeneous networks of work stations and have been shown to have solution efficiencies around 90%. Thus, with up to 10 HP9000/700 series work stations in parallel, Benetton engineers have the computational power of a small supercomputer available overnight.

All new predictive computational technologies, CFD included, must be validated and cross-referenced with known experimental data. After an extensive period of testing and verification for various types of flows around an F1 car, Benetton started to integrate RAMPANT into its design philosophy.

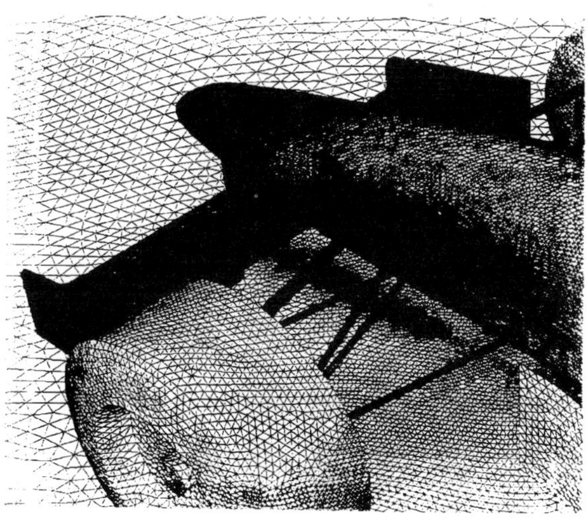

Fig 50: Surface mesh at front corner of car

Initially, because of hardware limitations, certain parts of the car were only modelled in two dimensions, three dimensional calculations coming later. As always with CFD, useful information can be gained from 2-D simulations and key trends may be picked up in 2-D or even from coarse 3-D simulations by careful choice of boundary conditions for parametric flow studies.

The Benetton aerodynamicists found CFD to be a useful design tool in association with their wind tunnel experiments, especially for optimising configurations at the start of the design cycle. Typically, three or four times as many configurations could be considered than with the old approach and for certain F1 components a modified aerodynamic development process was devised involving the following steps: i) The aerodynamic idea. ii) Simulation and preliminary investigation of the idea using CFD. iii) Design and manufacture of the necessary parts for the wind tunnel model. iv) Extensive evaluation of the new concept in the wind tunnel. v) Design of the components for the real car and their manufacture. vi) Real car test of the new components.

A range of aerodynamic factors and allied considerations have to be addressed: i) Downforce characteristics. ii) Drag characteristics. iii) Stability in cornering and iv) How does it affect the car's balance? The car's driveability at different speeds, over bumps and under acceleration and braking conditions v) Practicality in relation to the rest of the car.

The CFD process for a typical Formula 1 simulation using RAMPANT involves four stages: i) Construction of a CAD definition of the car's geometry. ii) Creation of a computational grid that defines both the car body, the moving road plane and the fluid phase around the car, usually extended to some fictitious far-field boundary. iii) Choice of boundary conditions for the CFD code, turbulence model(s) and properties of the fluid to be modelled. iv) Solution of the governing fluid flow equations and post-processing, or interpreting, the CFD predictions.

RAMPANT's pre-processor GeoMeshTM, can perform stages (i) and (ii) in association with an

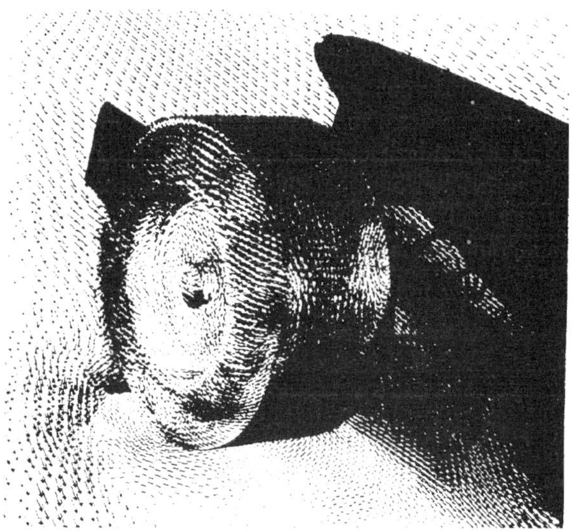

Fig 52: Air flow velocity vectors near front wheel:

Fig 51: Airflow velocity vectors near the moving ground and outlet planes

automatic mesher called TGridTM. The solving and postprocessing represented by stages (iii) and (iv) are performed within RAMPANT itself. Many companies, like Benetton, have other CAD packages like EDS Unigraphics for producing their geometries. The output from these CAD packages can be pulled into GeoMesh directly via standard IGES files (Fig 49).

Three key flow areas for race-cars are external flow over the front quarter, radiator side pod internal flow and external flow over the rear downforce wings. Fig 50 shows a surface mesh for a 300,000 cell CFD simulation for a front-end. Fig 51 shows the way oncoming airflow accelerates under the front wing and beneath the aerofoils whilst stagnating at the front wheels and nose regions. In flow passing around and over the wheel, Fig 52, it produces a complex set of separations emanating from the sides and crown of the wheel. The wake can be seen to be skewed towards the body and horseshoe vortices are visible, rolling off the rear of the tyre. Oil and water radiators are kept separate by aerofoils in the side pods of the car, Fig 53, cooling air being pulled into the pod by motion of the car. Flow through the duct is attached to the walls and non-uniform heating of the airflow is apparent on the wall of the outer radiator duct but seems to disappear with turbulent mixing in the exhaust duct as the flows recombine. Fig 54 shows a 2-D simulation of the triple element rear wing with oncoming airflow of 60 m/s. CFD helps narrow down the best combinations of angles of attack and wing positions by pointing out possible flow interactions prior to tunnel-testing.

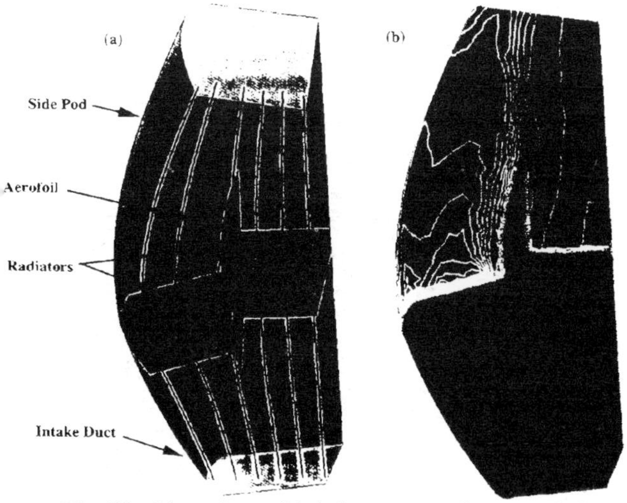

Fig 53: Plan view of (a) flow streamlines and (b) surface temperatures in radiator side pod

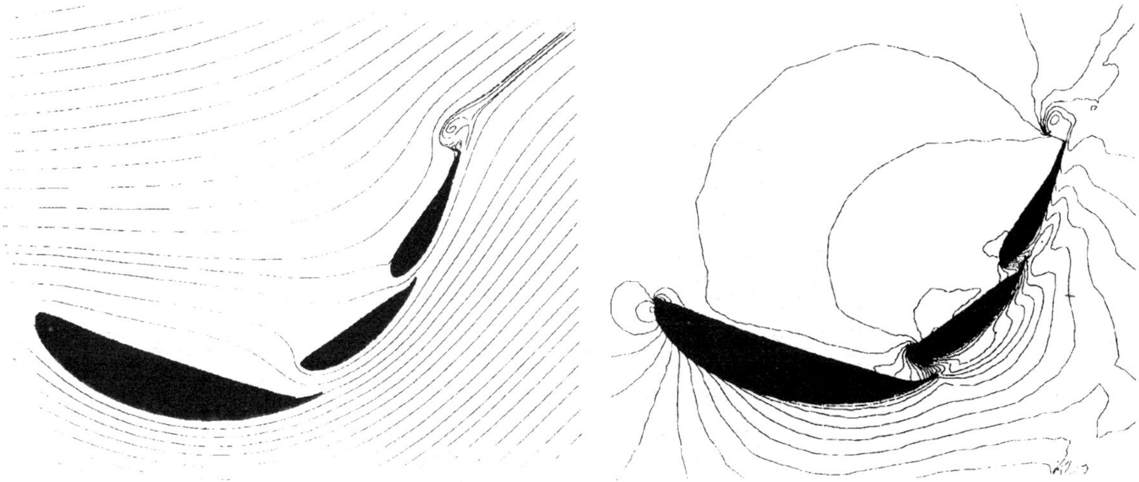

Fig 54 (below): Air flow path lines (left) and pressure contours (right) around wing assembly

Rear surface flows

Researchers at Durham University[12] have undertaken an important project which it is hoped will allow predictions of the turbulent flow regimes at the rear of vehicle bodies. They adopted the relatively simple Ahmed idealised vehicle form, used in verifying CFD codes, seen in Fig 55 in conjunction with surface mounted hot film gauges and liquid crystal films for flow measurement. Even at the front of the model, some flow separation was shown to exist at the corners, as seen by the oil flow visualisation technique of Fig 56. The unsteady component of the surface static pressure co-efficient is found to change along the length of the model, Fig 57. In this figure, transition and reattachment of the separation bubble at the front of the model appear only as a small rise in the measured RMS unsteadiness, at $x/L = 0.1$. The first evidence of a large and rapid rise in unsteadiness at the rear of the car is at $x/l = 0.79$. This point lies upstream of the sharp corner between the flat roof and the backlight at $x/L = 0.82$. Surface oil flow visualisation over the backlight shows a closed separation bubble, Fig 58, with the flow detaching at the top corner but re-attaching under the influence of the longitudinal vortex filament either side of the backlight. In general, the author found CFD predictions for this test case to show flow either to be fully separated or fully attached, with strong dependence upon the structure of the computational grid.

Refining CFD codes Researchers at MIRA[13] have considered how CFD meshes can be refined and turbuence models introduced to improve CFD analysis, there work again being based on the Ahmed model in studying the front-end separations. A base model of 348,000 cells (mesh A) is seen in Fig 59 having refined meshes of 385,696 and 388,896 in regions B and C. A mesh embedding technique is used in the refinement which gives the necessary localised density without increasing cell numbers to the level that would be inviable for industrial computing. This upstream refinements were used to test some unexpected pressure co-efficient peaks which had occurred in earlier work using mesh A only, Fig 60. The figure shows a relatively smooth upstream curve for mesh C which also predicted a C_p max value of 1.079 for the stagnation point, not far from the theoretical 1.00. The paper also shows the effects of numerical differencing schemes and turbulence models on the results.

Moving ground effect Rover researchers[14] explain their attention to moving-ground effect as being due to the interest in improving aerodynamic performance of vehicle undersides. Again the authors used an idealised test model like the Ahmed one to which "roughness strips" were attached beneath the floor. They were also able to change the back-light

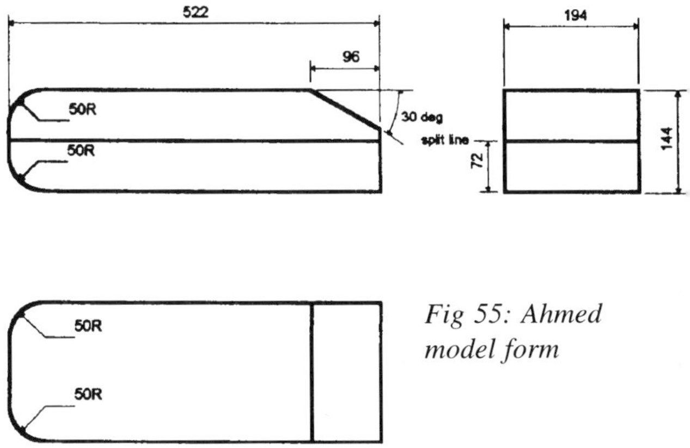

Fig 55: Ahmed model form

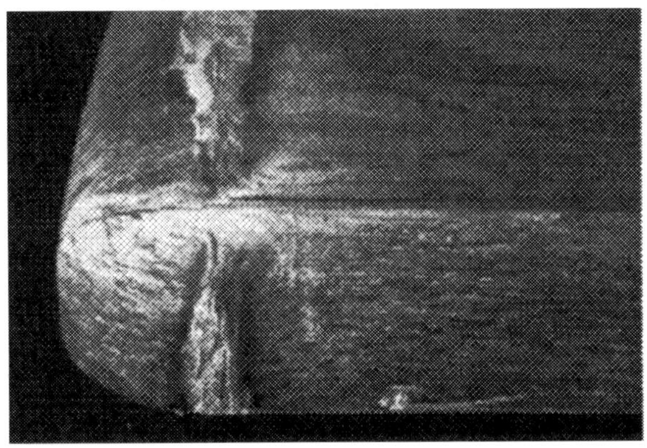

Fig 56: Oil flow visualisation for nose of model

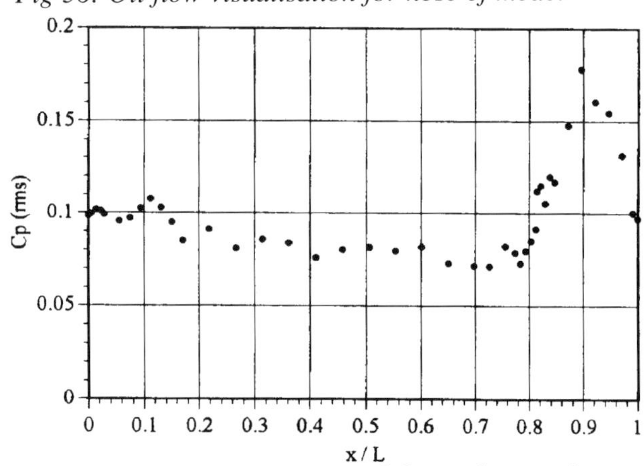

Fig 57: Unsteady component of the surface static pressure co-efficient along the upper body surface

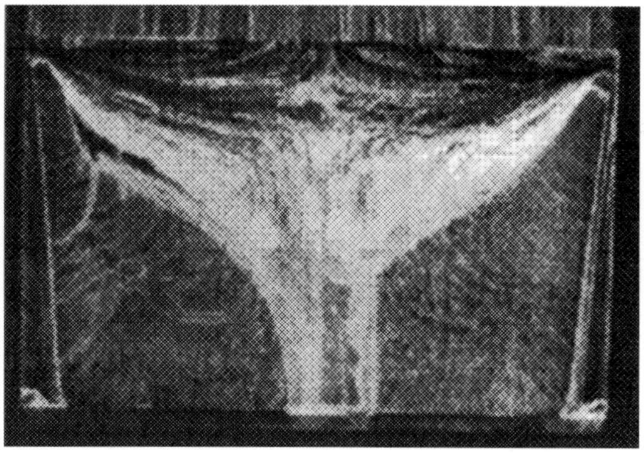

Fig 58: Oil flow visualisation on backlight

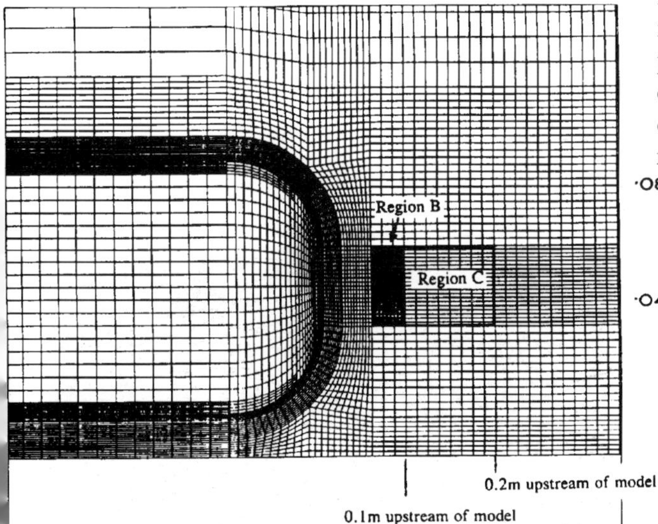

Fig 59: Mesh A and cell refinement regions

angle to observe the effect of a wide range of lift and drag characteristics as well as a variety of wake structures. In the first test, at 50 mm ground clearance on the 1.044 metre long model, underfloor drag was found to be almost independent of ground simulation but overall drag was higher with a moving belt. When the ground clearance was halved, different results were obtained as seen by the graphs presented in the paper. Fig 61 shows underfloor drag plotted as a function of backlight angle for different roughness conditions at the two ground clearances. The results at the higher ride height conflict with the results from full-scale tests which the authors attributed are due to interference effects on real vehicles which were not present on the model. At the lower ride height, considerably less than on normal cars, substantial differences between stationary and moving ground results are seen.

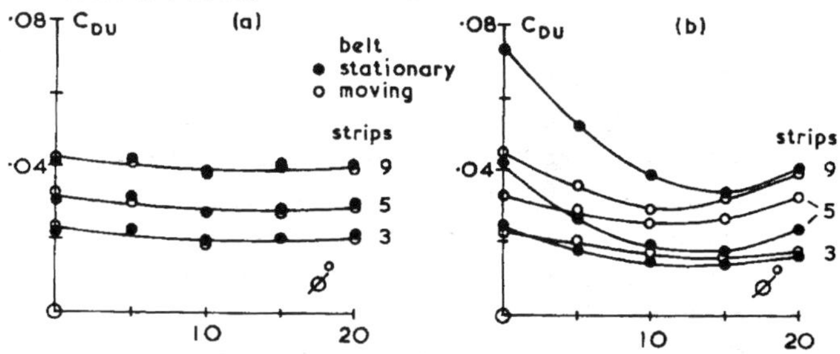

Fig 61: Underfloor drag vs backlight angle for ground clearance (a) 50 mm and (b) 25 mm

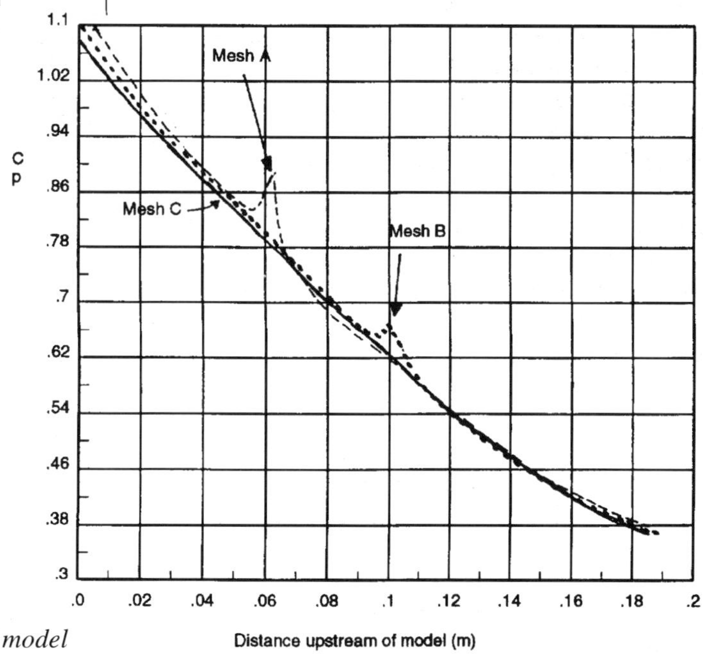

Fig 60: Pressure co-efficient upstream of model

What type of CFD package?

Barnard[15] sees CFD as the possibility of calculating aerodynamic forces, so far not been possible with accuracy by conventional methods. Unlike aircraft in which airflow generally stays attached to the surface and flow is either axisymmetrical, over the fuselage, or largely 2-dimensional, over the wings and tail-plane, road vehicle airflows are strongly 3-dimensional and dominated by separation effects, Fig 61. With CFD, numerical methods can be used to look at small elements of the surface (as in finite element analysis of stress) and averaging of turbulence effects can be used to obtain reasonable computational answers. So as to minimise the number of cells (elements) used, differential cell sizes are now being used, a very fine mesh of cells being specified in the sensitive areas. In order to keep a constant check on the accuracy of CFD modelling, using Ahmed forms, complete wind-tunnel measured flow data having been obtained for this shape and tested against CFD modelling. Also the shape, for which the hatchback backlight angle can be varied in increments, is demanding — in that drag of a hatchback falls initially, with increasing backlight angle, but then rises as strong vortices form over the backlight, reaching a peak at around 30-35 degrees. In general the author recommends using CFD for low-accuracy work and more expensive physical model testing in wind tunnels for more accurate analysis.

MIRA and DRA are reported[16] to be co-operating in the evaluation of aerospace CFD techniques for automotive use using the SAUNA (Structured and Unstructured, Figs 63, 64, Numerical Analysis) code which has an automated grid generator and an advanced algorithm for solving flow equations which will, say the authors, provide more accurate solutions than is possible with the commercial CFD packages so far used for automotive work on external body airflows. Some of these commercial codes have tended to use unstructured grids; however, the algorithms used have required tetrahedral cells to be approximately equal sided to achieve acceptable numerical solution — which imposes a considerable limitation. SAUNA's Multiblock suite has recently been extended to accept a number of unstructured blocks within a structured grid which is seen as a way forward. However, a long programme of validation against wind tunnel testing is required.

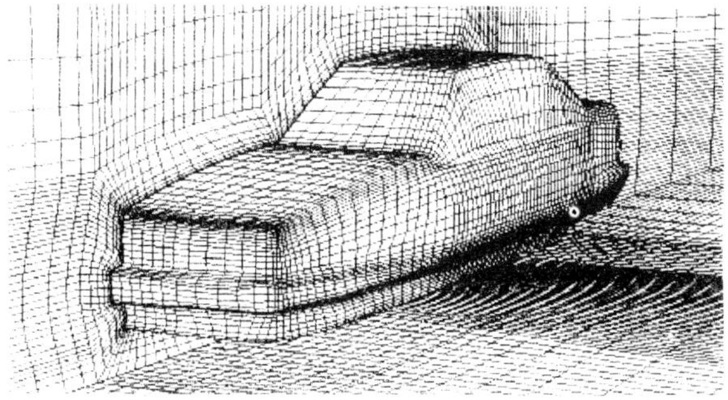

Fig 61: Grid of 98,000 cells on Volvo 850 model

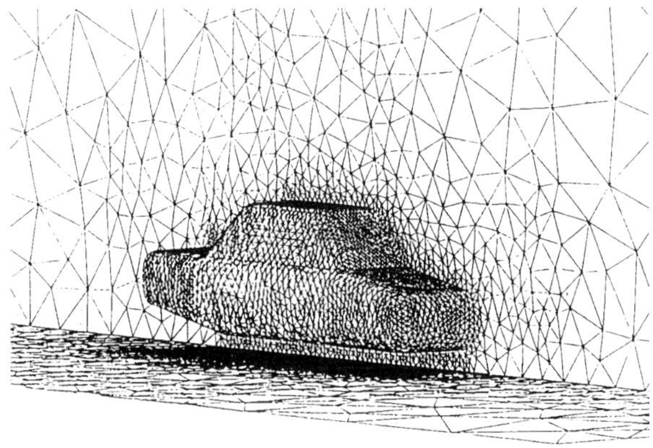

Fig 62: Unstructured grid around MIRA reference car

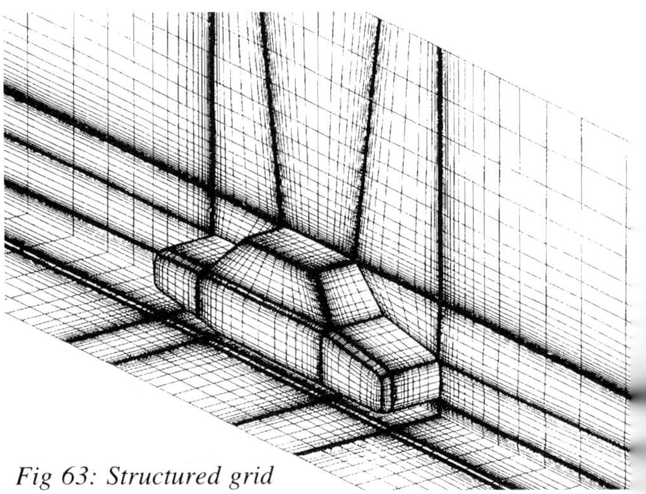

Fig 63: Structured grid around MIRA reference car

Body heat transfer

Good insulation of a car saloon or CV cab is possible by careful design and should precede the design of interior climate control systems. This requires an understanding of the heat flow mechanisms through and from the skins as well as consideration of the properties of the different alternative insulants. Overall heat transfer co-efficients can then be built up so that materials and layouts can be chosen at the concept stage and related to their predicted heat losses.

Creating a comfortable vehicle environment requires the same rules as used for the achievement of comfortable buildings. In this respect human comfort has been found to be optimised when a person's natural heat loss equals the heat required to be dissipated. The liberation of heat from the body has been found to be 30% by convection, 25% by evaporation and 45% by radiation and this should be the aim to optimise comfort. Different internal temperatures suit different levels of human activity — sedentary work such as driving requires a temperature of 19C. Heat is lost through the 'fabric' of the vehicle, by conduction and subsequently being dissipated by radiation and convection on the outside. Heat is also lost by 'ventilation' due to intake air from the vehicle's heater air intakes and around the weatherstrips of the doors and windows. The diurnal cycle of daytime and nighttime temperatures obviously affects the 'fabric', and 'ventilation', heat losses. The fabric loss is calculated by estimating the heat loss for each surface enclosing the space and then summating the total:

$$Q = UA(T1 - T2)$$

where Q is rate of heat transfer in watts; A is area of element losing heat in square metres; $T1$ is internal environment temperature in C: $T2$ is the outside temperature in C and U is the overall thermal transmittance co-efficient in W/m^2C and:

$$Qf = Qa + Qb + Qc \text{ etc.}$$

Typical U values are 5.6 for glass and 1.5 for well trimmed and insulated panels. U values can also be calculated by summing the conductivities and resistances of the respective elements of the panels:

$$U = L1/k1 + L2/k2 + Ra + Rb \text{ etc.}$$

with heat path lengths L in mm; thermal conductivity k in W/mC and elemental resistances Rn in m^2C/W. Air infiltration losses can be calculated from the general formula:

$$Qv = 0.33NV(T1 - T2)$$

where Qv is the rate of heat transfer to ventilating air; $0.33N$ the ventilation allowance based on the assumption of one complete air change per hour; V is the vehicle internal volume in m^3.

In practice the surface area of the vehicle saloon or cab acts as both reflector and emitter — by re-radiating some of the heat absorbed. The basic mechanisms of heat transfer are described in a later section on refrigerated bodies where it is noted that the conduction mechanism is similar to the conduction of electricity. With heat, however, there are no specific insulators except vacuum since the energy

MATERIAL	THERMAL CONDUCTIVITY (Btu. in/ft². ° F.h)
METALS	
Copper	2670
Aluminium	1460
Steel	330
MINERALS	
Granite	20·3
Limestone	10·6
Glass	7·3
PLASTICS (UNEXPANDED)	
Polythene	2·5
PVC	1·2
Polystyrene	0·9
GASES	
Hydrogen	1·18
Air	0·18
Trichlorofluoromethane	0·057

Fig 64: Thermal conductivities

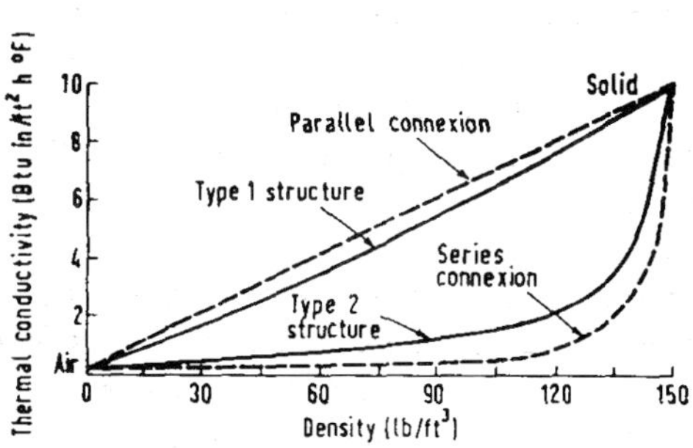

Fig 65: Effect of structure

moves not by electrons but by atomic vibration. There are thus differences in temperature profile along a cooling fin to that expected from an electrical analogue. Similarly heat radiation should not be confused with atomic radiation generally where electrons are also involved. In heat radiation, electromagnetic radiation is involved at various wavelengths.

Radiant heat transfer is not stopped by vacuum and is generally transmitted in relation to the fourth power of the surface temperature of a body and its emissivity. Radiation is also received by the body from heat flows in the reverse direction — depending on the fourth power of the surrounding temperatures, which may be far from constant, and on the absorbivity of the body. It is important to realise, in calculation, that absorbivity only equals emissivity at one particular wavelength.

In natural convection, flow is generated by the buoyancy of warm fluid relative to cold fluid and if speeds are low, flow can be laminar rather than turbulent. Factors affecting fluid motion are thermal expansion co-efficient and density of the surrounding fluid — and the relative position of the heat transfer surface and its temperature difference with the fluid. There is no exact proportionality with temperature difference so a heat transfer constant co-efficient cannot be defined; generally flow is related to between the 1.25 and 1.33 power of the temperature difference resulting in co-efficients between 5 and 20 W/m^2. In the case of forced convection there is a proportional relationship but the temperature gradient depends on the thickness of the sluggish 'boundary' layer of fluid between the heat-transmitting surface and the forced draught of fluid. This thickness varies with the inverse square root of the draught velocity, approximately.

The method of calculating heat loss by the combined effect of different mechanisms is best understood by example. A metal cab might be assumed to have an internal volume of 1000 litres and weigh 100 kg, its steel panels having specific heat of 2000 J/kg K and their surface covered by insulation 30 mm thick having conductivity of 1.1 W/mC and surface area 15 square metres. A reasonable assumption would be for the convective and radiative 'co-efficients' to average $12W/m^2C$. Overall resistance of the insulation = 0.03/(1.1 x 15) = 0.0018 K/W. Surface resistance (convection plus radiation) = 1/(12 x 15) = 0.0055 K/W. Overall resistance = 0.0018 + 0.0055 = 0.0073 K/W. Heat loss per θ temperature difference with surroundings: = θ/0.0073 W and heat capacity = 100 x 2000 = 200,000 J/K for the metal cab. So for every watt, the temperature falls at 1/200,000 K/s — so the rate of temperature loss with time *t* is = $-\theta/(0.0073 \times 200{,}000)$ so that:

$$\log_e \theta = -t(0.0073 \times 200{,}000).$$

Thus for a surrounding temperature of 10 C, cab temperature will drop from 80 to 40 C in time:

$$= \log_e(80 - 10)/(40 - 10) \times 200{,}000 \times 0.0073 \times 1/3600$$
$$= 9.03 \text{ hours}$$

Fig 64 shows the thermal conductivities of a number of media at ambient temperature. In the case of gases the conduction mechanism involves the loss of energy by molecular collision, rather than vibration as is the case with solids. The lower conductivity of gases is an important factor in the choice of insulating materials and underlies the reason for gas expanded, or 'blown', plastics insulants of various types. In blown insulants, assumed to have a regular cubic lattice, two extremes can be considered, type-1 having solid walls with cubical pores and type-2 comprising cubes of solid material surrounded by air passages. Values given in Fig 65 are based on a pore-free solid phase having density of 150 lb/ft^3 (2430 kg/m^3) and thermal conductivity of 10 BTU (2.52 kgcals) in/ft^2F. Fibrous insulants tend towards one extreme and cellular materials towards the other. Solid phase and gas phase are essentially in parallel in the type-1 structure and in series for the type-2 structure, the formulae for calculating these two extremes being shown in Fig 66. Fig 67 shows the heat loss to atmosphere from a plane vertical surface by convection and radiation where:

$$Qc = 0.3(T1 - T2)^{5/4}$$
$$Qr = 1.73 \times 10^{-9} E(T1^4 T2^4)$$

E is the surface emissivity taken as 1.0 for the example in the figure which applies to an ambient temperature of 65 F (18.3 C). At low temperatures, thermal radiation wavelengths are in the infra-red region; at elevated temperatures it is the predominant form of heat transmission as shown by the figure.

If a material absorbs all the radiation falling upon it, and reflects and transmits none, it is known as a 'black body'. Such a body also has maximum radiation when it is heated and is thus taken as the standard against which the emission from other bodies is compared. Surface emissivity is thus the ratio of its radiation to that emitted from a black body at the same temperature. A black body is defined as a surface with absorbivity of unity. While non-metallics might have a value of 0.9, that of aluminium is only 0.05 which accounts for the use of foil forms of the material in insulation. The radiation flux per unit area of any body is called the emissive power = 5.67 x 10^8T^4 where T is the thermodynamic temperature of the body, this radiation being spread over a wide spectrum of wavelengths (Fig 68).

Real surfaces are not 'black' and emit less radiation and their monochromatic emissivity is the ratio of the emissive power of a black body at the same wavelength and temperature. It varies considerably with wavelength and such surfaces are called 'selective emitters'. A 'grey body' is thus defined as one having constant emissivity at all temperatures and wavelengths — in order to obtain solutions to real problems by treating selective emitters in small increments of wavelength of constant emissivity. Considering a grey body in black surroundings is a convenient way to reach a solution, Fig 69, for which heat transfer rate:

$$=\varepsilon A_1 \sigma (T_2^4 - T_1^4)$$

suffix 1 referring to the grey body and suffix 2 the black surroundings. This is a useful approximation for real bodies for which a realistic estimate of constant emissivity can be made in 'large' surroundings, implying that the reflected energy is unlikely to be re-incident on the relatively small grey body — and will by successive reflections to the surroundings become less and less. Fig 70 shows the heat transfer by solid conduction, air conduction and radiation, as a function of density, for an insulating material at mean temperatures of 0, 200 and 400 F, values being representative of a fine-fibred mineral wool material for which solid conduction has been taken as *0.005D* and $kr = 1.08 \times 10^{-9} T^3 m/D$; convection effects are neglected. In solids with a greater degree of parallel connection than mineral wool, conduction increases more rapidly with density — with consequent effect on the overall conductivity. In Fig 71 conductivities of two mineral wool samples are compared, at a density of 7 lb/ft^3, with mean fibre diameters of 5 and 15 microns respectively. From this it is seen that by reducing fibre diameter it is possible to obtain thermal conductivity values not much more than still air.

Hence the trend to reduce this limiting value by replacing air in expanded plastics insulants by such gases as trichlorofluoromethane having thermal conductivity of only 0.057 BTU in/ft^2 h F. Substitutes which are kinder to the ozone layer are, of course, now being sought. In the case of a vehicle body which

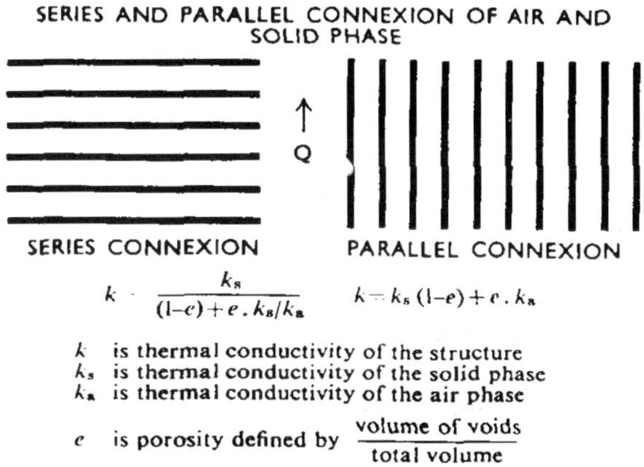

Fig 66: Calculating conductivity

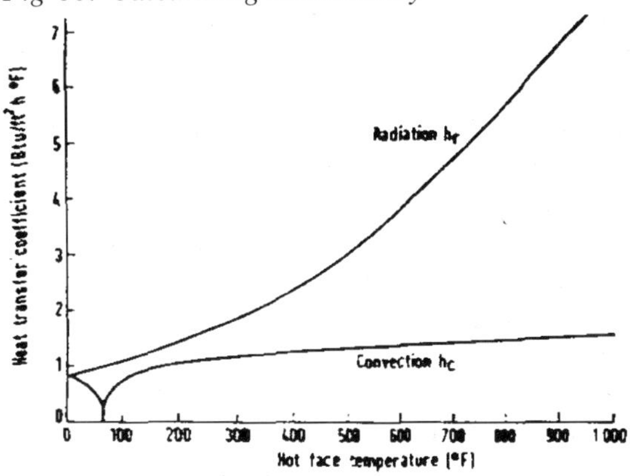

Fig67: Heat loss to atmosphere

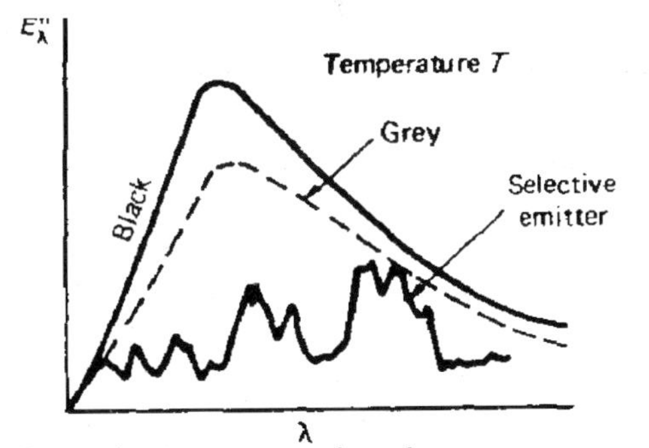

Fig 68: Radiation vs wavelength

can be likened to two two fluids being separated by a solid wall, the resistances for conduction and convection in series may be added and the heat transfer rate expressed in terms of the fluid temperatures at either side. Putting *(Tf1 - Tf2) = 0*, the overall temperature difference, then for a plane wall, an overall heat transfer co-efficient can be defined by:

$$1/U = 1/h1 + \sigma(\delta x/k) + 1/h2$$

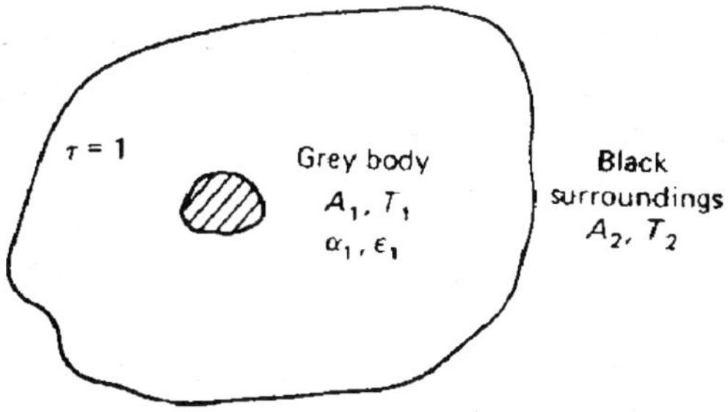

Fig 69: Grey body in black

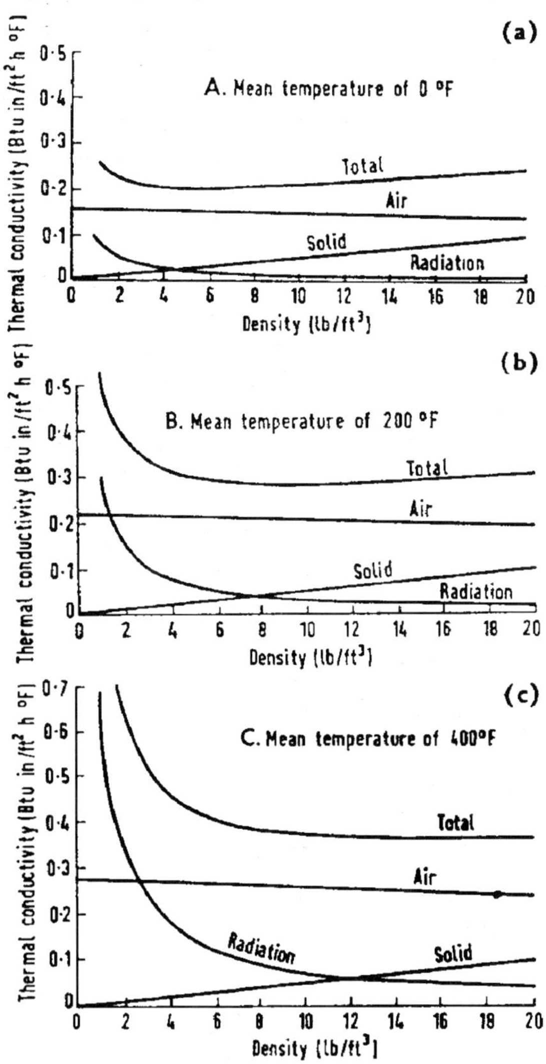

Fig 70: Heat transfer at different mean temperatures

Air-conditioning systems

The combination of heating, cooling, airchange and dehumidification that constitutes air-conditioning can be successfully and economically applied to vehicles only if performance requirement is carefully calculated.

The requirements for full air conditioning of a confined space cover control of air heating, humidity, air movement, air changes, odour preventatives and air distribution — not just the cooling for which the load is calculated below. The cooling is usually achieved by passing air over the evaporator plates of a refrigeration system. To avoid a maze of control knobs that would be required for all these facilities, integrated automatic controls are essential in the vehicle application.

Research done on studying the discomfort of close, crowded rooms within buildings has shown it to be due to the physical rather than chemical properties of air — its temperature, humidity and stillness. The presence of a variable quantity, normally around 1%, of aqueous vapour in air is of especial importance and is generally defined by either the absolute or relative humidity. Absolute humidity denotes the weight of moisture in a given volume of moist air. Amount of moisture which air can hold is temperature dependent and is 'saturated' at any temperature

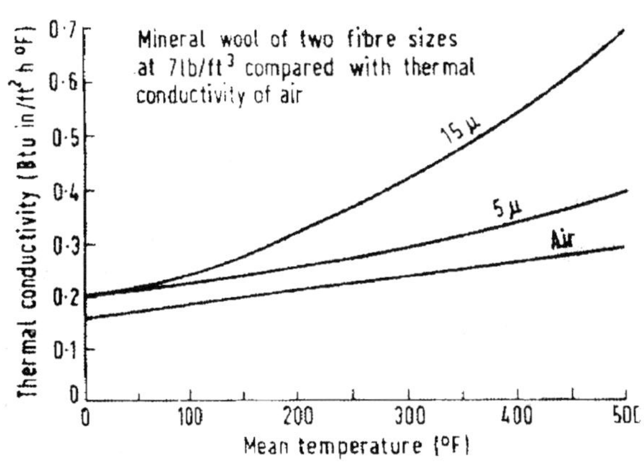

Fig 71: Effect of fibre size

according to the maximum amount of moisture it can hold — i.e. 100% humidity. Relative humidity is the proportion to that 100 expressed as a percentage. On cooling moist air, a temperature is reached (the dew point) below which condensation occurs. A guide to preferred air change can be gained from building, and even marine, practice. It has been stated that a seaman's shipboard accommodation should provide a minimum of 120 ft^3 (3.36 m^3) per occupant and traditional building standards suggest that each person requires 1000 to 1200 ft^3 (28-33.6 m^3) of fresh air per hour for satisfactory comfort. A rather larger fresh air requirement has been stated for vehicle drivers — of 1500-1800 ft^3 (42-50.4 m^3)/hour — typically 125 ft^3 (3.5 m^3)/minute for a five-seater car. This allows an increment for reducing inward draughts. Dehumidification can be obtained by cooling the incoming air.

Successful application of vehicle interior air-conditioning depends on a reasonably accurate cooling load calculation for the vehicle interior. Three principal constituents of the cooling load are solar radiation on the vehicle, heat input due to passengers and heat input from the fresh air supply into the vehicle. Up to 300 btus (75.6 kgcals)/hour can be transmitted into a typical car by the sun's rays and it useful therefore to assess the different transmissilities of glass and inner-trimmed steel surfaces. An example may be 6 x 4 ft (1.83 x1.22 m) for the roof area of a saloon car and twice 2.5 x 6 ft (0.76x1.83 m) for the side panels (less windows) also 2.5 x 4ft (0.76x1.22 m) for the rear panel — giving a total of 64 ft^2 (2.42 m^2). Glass area is computed in the same way and in this case is taken as 5ft^2 (0.465 m^2) for the windscreen, 11 ft^2 (1.02 m^2) for the side windows and 3ft^2 (0.28 m^2) for the rear-light — totalling 19 ft^2 (1.77 m^2). The conductivity of trimmed steel panels is quoted as 14 btu (3.53 kgcal)/ft^2/h so that in this example the heat flow through them due to solar radiation on the outside is 895 btu (225.5 kgcal)/hour. The equivalent figures for the glass area (whose quoted heat-transmissibility is 160 btu (40.3 kgcal)/ ft^2/h) are 3040 btu (766 kgcal)/h for the glass alone and a combined panel-and-glass total of 3935 btu (992 kgcal)/h.

The figure above could be substantially reduced by the use of tinted glass. The reduction could be of the order of the quoted heat transmission per person of 400 btu (100 kgcal)/h — which for a 5 seater makes the total of 2000 btu (504 kgcal)/h. Such figures give a clue to the temperature difference on a full and empty bus! Total heat to be removed from the fresh supply can be calculated from the assumption of given climatic conditions which in this case are taken as 100 F (37.8 C) dry bulb temperature, 79 F (26.1 C) dry bulb and outside-air relative humidity of 40%. Given also the interior requirements, of say 80 F ((26.7 C) dry, 57F (13.9 C) wet and 50% RH, sensible heat to be removed from 1 lb of fresh air is 0.238 x 20 = 4.76 btu (1.2 kgcal)/lb of dry air. Heat to be removed from moisture per pound of ambient air for the given reduction in wet bulb temperature (reducing vapour pressure from 0.375 to 0.246 lb/in^2 [0.026-0.017 bar]) is obtained by subtracting the product of lbs of moisture per lb of interior air and heat content per lb of moisture at the interior wet bulb temperature

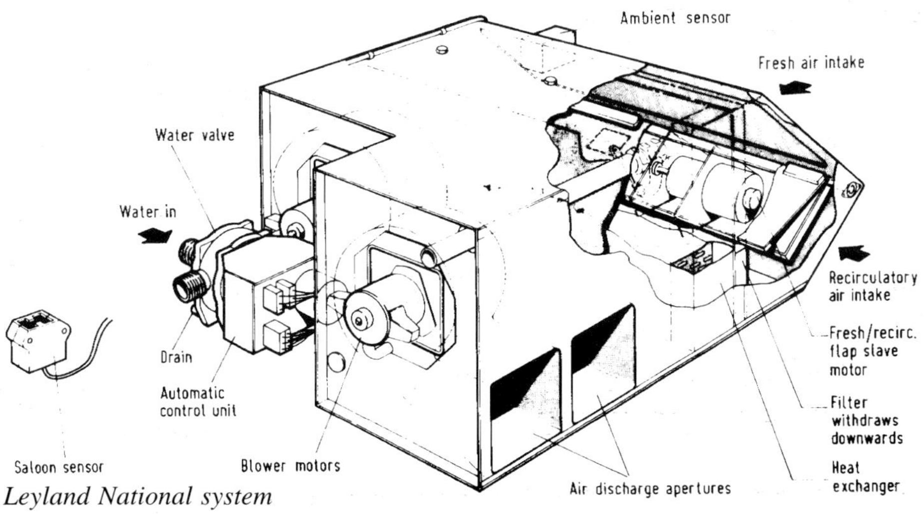

Fig 73: Leyland National system

from the equivalent product for the exterior air. In this case it amounts to 0.01645 lb/lb x 1095 btu/lb - 0.0111 lb/lb x 1085 btu/lb = 5.95 btu (1.5 kgcal)/lb of air to be extracted. Total heat to be extracted is thus 10.71 btu (2.7 kgcal)/lb for the air and vapour mixture. Quantity of dry air in the incoming stream of fresh air at 125 ft^3 (3.5 m^3)/min, $w = Pa.Q/R.T$ where Pa is dry air atmospheric pressure, Q is quantity of fresh air supplied, R is gas constant for air and T is the absolute temperature of incoming air. Thus w = 8.67 lb (3.94 kg)/min and total heat to be extracted = 5575 btu (1405 kgcal)/hour with moisture to be removed totalling 0.0464 lb (0.02 kg)/min. Total cooling load thus amounts to 10,575 btu (2665 kgcal)/h.

Cadillac was one of the first car models for which a full fresh-air air-conditioning system was developed, Fig 72, complete with automatic control. After all incoming air is cooled and dried by the evaporator a variable amount is diverted and reheated sufficiently to provide the required duct air temperature where it is remixed with the remainder. Flow is then automatically proportioned between the outlets, recirculation being used only during initial cool down from high ambient temperatures. Blower speed is varied automatically to suit demand.

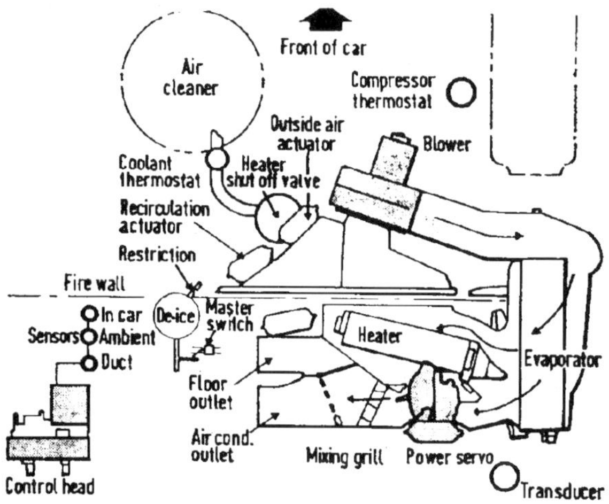

Fig 72: Early Cadillac system

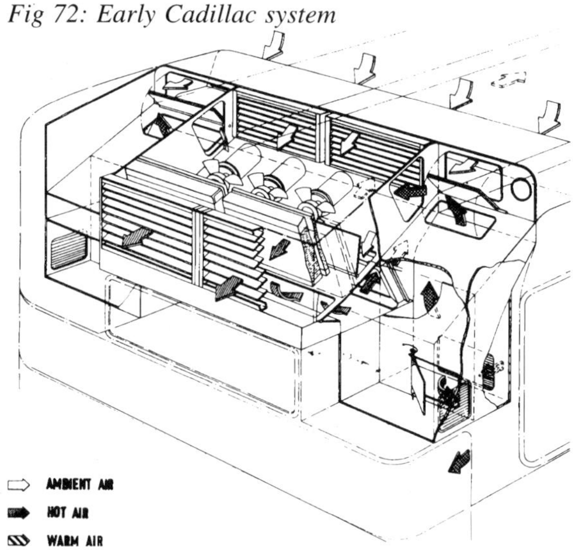

Fig 74: Patented system for buses

In the case of buses and coaches, the cooling loads are sufficiently high to warrant the size of industrial systems used for room air-conditioning. There is a more frequent requirement with buses than with cars for the unit to be run while the vehicle is at rest and therefore separate motor systems are often specified. The Leyland National bus with its familiar rear-end roof-pod represented an important step forward in bus air-conditioning which grew out on an interesting development aimed at a combined engine cooling/saloon heating system. In the event, the system realised, Fig 73, is a unitised heater which feeds air for conditioning from a built-in duct in the vehicle roof — which can be either cooled or heated as required by the operation. The saloon temperature controller has a transistorised amplifier which responds to three thermistors, sensing ambient, heater-outlet and saloon temperatures respectively. The amplified signal energises a servo-motor driving a water regulating valve. Each heater unit has an output of 60,000 btu (15120 kgcal)/hour/100 C temperature difference and supplies 750 ft^3 (21 m^3)/minute airflow against a duct resistance of 0.75 inches water gauge (0.188 bar) — and providing 25 air changes per hour in the single-deck vehicle. The design allowed either tandem heater units or a side-by-side refrigeration unit to be installed according to operating climate. The original development which was not productionised culminated in a patent specification (GB1205941), Fig 74.

Truck cabs can be fitted with roof-top systems, of which the Kysor one shown in Fig 75 was a pioneer. An engine driven compressor absorbs 4.25 bhp when operating in a 100 F (37.8 C) ambient with 50% RH. The largest model had a thermal capacity of 16,000 btu (4032 kgcal)/h. Some of the latest systems under consideration are based on electric motor rather than engine driven arrangements. Refrigeration efficiency would be increased by redesign of the refrigeration condenser unit, Fig 76, giving 75% higher co-efficient of heat transfer and 20% reduction in pressure drop over the conventional serpentine type. Fig 77 shows the trade off between improved efficiency and larger area required. These improvements have been shown in tests to result in substantial reductions in electrical power demand.

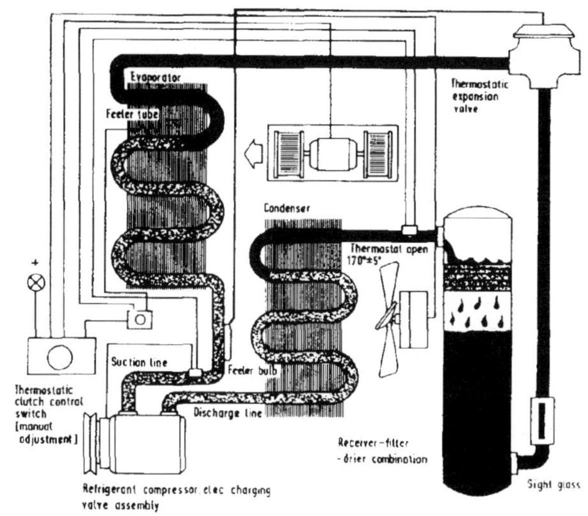

Fig 75: Kysor truck-cab unit

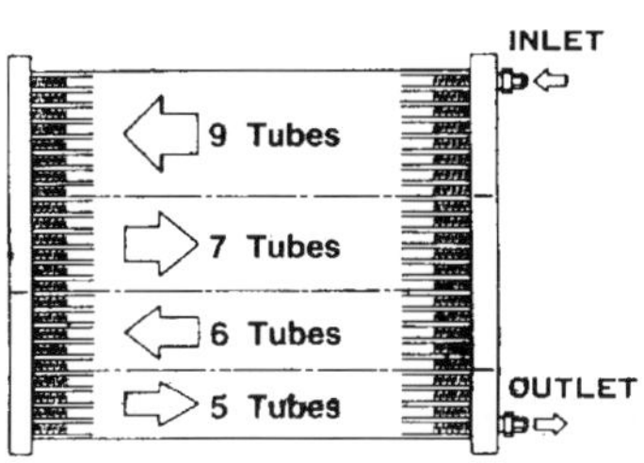

Fig 76: Optimising the condenser

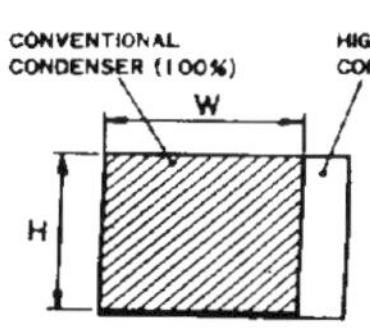

	W (mm)	H (mm)	T (mm)
CONVENTIONAL CONDENSER (SERPENTINE TYPE)	410	355	22
HIGH PERFORMANCE CONDENSER (PARALLEL FLOW TYPE)	500	360	16

Fig 77: Efficiency vs area

Sophistication in climate control

According to Delphi Automotive, Heating, Ventilation and Air Conditioning (HVAC) is one of the fastest growing areas of European vehicle content. The advantages of air-conditioning include: increased occupant comfort and driver concentration; reduction in air humidity and hence a reduction in window fogging; filtering of cabin air to remove pollen and other irritants and the fact that driving with windows-up increases security and reduces noise.

The company point out that between 1965 and 1985, the penetration of air-conditioning in the Japanese new car market rose from under 5% to more than 90%. Market penetration had reached a similar figure in the USA several years earlier, whilst even Canada, which has a similar or colder climate than Europe, now has 60% of new cars fitted with air-conditioning. Europe is the next major growth area, with fitment of original equipment air-conditioning systems increasing rapidly from just 9% in 1991 to 32% in 1996. Very high levels of repeat purchase suggest that this trend is likely to continue, reaching a predicted 54% by 2000. Much of this growth is expected in smaller cars and monospace vehicles, where the proportion of glass to metal is high and the steeply raked screens make the interior temperature particularly vulnerable to solar loading.

Rivalling air-conditioning for rapid growth is particle filtration. Today's high-performance cabin air filters can remove many of the airborne particles that can cause irritation, such as pollen and even some bacteria, and can include a carbon filter to remove smells. These filters are now widely fitted to cars that do not have air-conditioning, as well as to those that do, and have become a point of differentiation for vehicle manufacturers. Rapid market growth is also expected in other areas that provide additional functionality and control. These include electronic systems to check the filter condition, devices to provide fast warm-up and multi-zone control systems that allow each occupant to set his own preferred temperature and fan speed.

In Delphi designs, the evaporator is integrated with the HVAC module, Fig 78, which adds a fan, a switchable heat exchanger (the heater core) for heating the air, an air filtration system, a distribution system and a control system. The refrigerant evaporates in the evaporator, absorbing heat from the fins over which the ventilation air is flowing. The compressor then compresses the refrigerant so that it can be returned to the liquid state in the condenser, releasing the absorbed energy. A filter drier then removes any moisture from the refrigerant before it is passed back through the expansion valve into the evaporator, where the cycle starts again. Manual air-conditioning systems cool the air entering the cabin without reference to the cabin temperature. Automatic climate control systems allow the car interior temperature to be set by the user and then maintained automatically.

Compressor technology for Europe Until recently European vehicles fitted with air-conditioning were almost exclusively the large, powerful cars and commercial vehicles to which earlier US technology was well suited. For smaller European vehicles, where there is currently rapid growth in demand for air-conditioning, the requirements are significantly different, Fig 79. To accommodate the tighter packaging and lower power outputs of the engines, air-conditioning systems must be more compact, lighter weight, more energy efficient and smoother in operation. Delphi's new Compact Variable Compressor (CVC) has been developed for such applications. The six cylinder CVC can displace between 8 and 125 cc. The output of variable displacement compressors is adjusted by varying the displacement of the compressor cylinders by using a swash plate to increase or reduce the piston stroke. Because they run continuously without clutch cycling, variable displacement compressors provide significantly greater refinement, improved dehumidification and superior fuel economy compared with traditional fixed displacement compressors. The company manufactures VDCs in five, six and seven cylinder configurations to suit a wide range of vehicle applications. They can be used in both Thermostatic Expansion Valve (TXV) and Orifice Tube (OT) systems, and mounting, porting and pulleys can be designed to meet customer requirements. The units are of light-weight, die-cast aluminium construction. While the largest size weighs 6.8 kg, the CVC weighs just 5.1 kg. The next generation of CVC is expected to weigh around 4.2 kg.

Fig 78: HVAC module

Electronic and multi-zone controls The traditional method for controlling the wobble plate position and hence the capacity of a variable displacement compressor uses a pneumatic system that senses the input and output pressures. Delphi's new electronic control system replaces the pneumatic control valve with a pulse width modulated solenoid. Closed loop control inputs are provided by sensors for condenser outlet pressure, evaporator air outlet temperature and/or evaporator outlet pressure. This provides much more sophisticated control and hence introduces new opportunities to modulate pump output to suit operating requirements. The benefits of electronic control include fuel savings, reduced frosting (and hence a more stable cabin temperature), soft start (to eliminate the clutch engagement disturbance of traditional systems) and enhanced diagnostics. Control inputs may also be received from the engine management system, so that when the accelerator is fully depressed or when the engine temperature is high, the compressor displacement (and hence the power drain) is reduced to a minimum.

Some current production vehicles, such as limousines, are already available with HVAC systems that provide independent control of temperature and airflow for the front and rear compartments. This is achieved by fitting two HVAC modules, each with a blower motor, heater core and evaporator. Clearly these systems bring significant cost, weight and packaging penalties — which are said to be overcome by Delphi's Quad Zone concept which provides individual control of up to four independent zones from a single HVAC module. Internal partitions effectively divide the module into four segments. A single, electronically controlled blower is used and air flows to individual zones are controlled by electrically operated restrictor flaps. A logic-based electronic control unit ensures that all settings are optimized for the requirements of each zone. In the vehicle, each occupant may have his own controls for temperature, air flow and delivery location, mounted in the door console. A modified seatbelt sensor may be used to deactivate unoccupied zones and all zones are deactivated when the driver selects the defrost setting. An automatic climate control facility could be added if required.

Solar, air-quality and filter sensors As the glass area in cars increases and windscreens become more horizontal (most notably in Multi Purpose Vehicles), so the temperature in the cabin becomes more sensitive to solar loading. Delphi's Solar Sensor uses a light-sensitive device to detect increased solar loading and to adjust the output of the air-conditioning system (which must be electronically controlled) accordingly. The ability to predict demand allows a more stable cabin environment, increases occupant comfort and can reduce fuel consumption. The company's air quality sensor will automatically operate interior air recirculation when the quality of external air entering the HVAC system falls below a predetermined level. The system can detect oxidising agents (NO, NO_2, SO_2) and reducing agents (CO, C_x, H_y, O_3, CO_2). To enhance safety, the maximum allowable period of continuous recirculation can be predetermined. The sensor reacts very quickly, greatly reducing the ingress of polluted air compared with the level that can be achieved with manual operation. Switching to recirculate in highly polluted ambient atmospheres will also help to protect the filter and hence increase its service life. The Filter Check system notifies the driver when the HVAC air filter element should be replaced. The system works by passing an infrared light beam through the filter element. When the element's opacity drops to a predetermined level, indicating that the filter is approaching the end of its useful life, the system illuminates an indicator light on the instrument panel. The system enhances safety and comfort by ensuring that the driver is notified when the effectiveness of the filter is impaired.

Future technologies Cabin heater systems have traditionally been entirely dependent on the engine for the generation of heat. As engines become more efficient, less heat is available for the passenger compartment. This results in very long cabin warm-up periods and an inability to provide sufficient warmth when external temperatures are very low. Delphi has a number of technologies that can provide solutions both for vehicles powered by high efficiency conventional engines, such as direct injection diesels, and for vehicles powered by propulsion systems that create virtually no usable heat, such as electric motors. One solution is the Positive Temperature coefficient Ceramic device (PTC), is effectively an electrically powered, resistive ceramic heater placed in the heat exchanger airflow. A PTC can provide very fast cabin warm-up and may also be

Fig 79: Ferrari 456GT climate control system

configured to provide pre-heat, activated remotely, before the driver enters the vehicle. A similar effect can be achieved by using modified diesel glow plugs to pre-heat the engine cooling water.

The company's computer-controlled electric heat pump can provide heating or refrigeration from a single, electrically-driven unit. The technology is already a standard fitment on the General Motors EV1 electric car. The heat pump is, in principle, an air-conditioning system that can be run in reverse. In an air-conditioning system, compressed refrigerant passes through the evaporator (a heat exchanger) where it absorbs heat as it turns from a liquid to a gas. The refrigerant is then pumped to a condenser (also a heat exchanger) where it returns to the liquid state and in so doing deposits the heat into the vehicle airstream. An air-conditioner can only pump heat one way. By using the system in both directions, a heat pump can pump heat either way using the heat exchangers to either absorb heat or reject heat depending on the direction of operation. On the EV1, the warm powertrain coolant can also be passed through the heater core.

System design The traditional approach to HVAC system design has been for the vehicle manufacturer to source components and subsystems for each application and then to design the vehicle structure with the appropriate spaces available for assembly. System tuning and calibration would be largely carried out with the resulting system installed in the vehicle. Considerable time and cost savings, as well as superior refinement, packaging and performance, can now be achieved by designing HVAC as a single, integrated system. This also allows reductions in power consumption, simplified manufacture and easier recycling. Opportunities for modularization can then be fully developed and the system can be supplied for vehicle prototypes with most of the calibration work already complete.

During development, finite element analysis is used, Fig 80, to assess structural strength, vibration characteristics and heat conduction. Computational Fluid Dynamics (CFD), Fig 81, provides information on temperature distribution and the velocities in air and fluids. CFD can also be used to identify possible sources of turbulence, which can be a source of noise.

Modularization Computer modelling is proving particularly valuable for the development of HVAC modules. Delphi is already able to supply highly integrated units for the cockpit mounted subsystems (the HVAC Module) and the front-end subsystems (the Condenser Radiator Fan Module - CRFM). The CRFM integrates the air-conditioning condenser, the engine coolant radiator and the oil coolers in a single unit. It consists of a fan-motor-shroud assembly, a radiator with available in-tank engine and transmission oil cooling, and one of two condenser types: either headered tube and centre, or tube and fin. The next stage of simplification is to integrate elements of the HVAC system within the

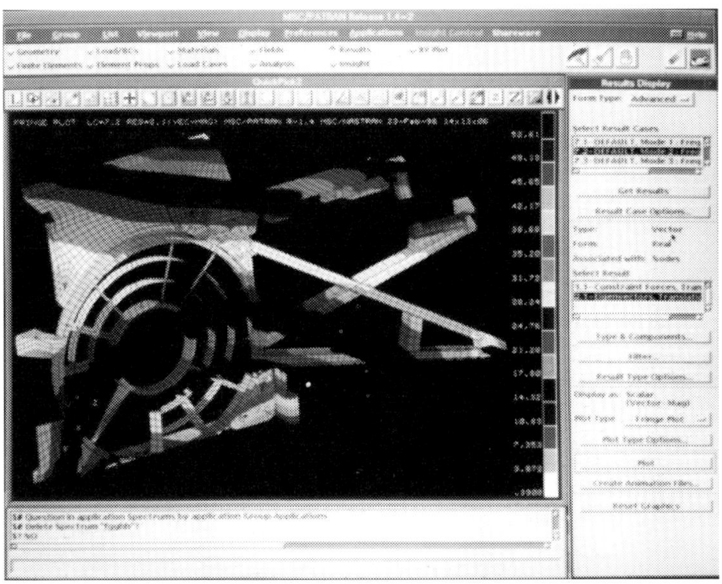

Fig 80: Finite element analysis

cockpit module (as discussed in the next section) and to integrate non-HVAC components with the CRFM. The cross-car beam and the cockpit HVAC ducting, for example, could be formed as a single component.

Integrating HVAC with the instrument panel

According to Delphi the focus of next-generation modularisation is to use systems integration techniques to reduce the number of interfaces and to develop the synergistic benefits between components and sub-systems. Substantial benefits can be achieved by directly replacing a conventional dashboard with a cockpit module, but the greater opportunity is to reassess the entire area so that neither the physical nor the functional extent of the module is greatly constrained by the need to interface with existing systems and structures. Quality is enhanced by a reduction in components and interfaces and by the easier access provided by out-of-vehicle assembly. Modules can be tested more quickly and more thoroughly.

The area directly in front of the driver and front seat passenger, commonly called the cockpit, is one of the most complicated and tightly packaged assemblies in any vehicle. It is also one of the most inaccessible. The cockpit may contain 60% of the total vehicle wiring system, most of the vehicle functions are controlled from this area and, with several hundred individual components, it is also one of the most common sources of rattles and squeaks. The quality of design and manufacture is therefore critical. The space occupied by the cockpit is restricted by the requirement to provide good visibility and leg room for front seat occupants and by the intrusion allowed into the engine bay.

Fig 81: CFD applied to air flow in HVAC module

This trend is particularly challenging in smaller cars, where airbags, air-conditioning and driver information systems are becoming rapidly more popular. Even in larger vehicles, new technologies such as navigation are adding to the number of systems requiring mounting space and wiring routes. Packaging, however, can not be allowed to detract from crashworthiness. The cockpit is supported on a cross-car beam that reinforces the vehicle structure and provides mounting points for safety systems such as airbags and the collapsible steering column. In an impact, it will help control intrusion, absorb energy and ensure that these systems maintain the correct alignment relative to the occupants. Traditionally the cross-beam is of welded steel sheet, but other designs currently in production include tubular weldments, aluminium and magnesium castings and, in a few applications, plastic composite structures.

Traditional manufacturing practice is to assemble the cockpit from individual components and sub-systems, often many hundreds of individual items, in what is known as a 'layered build'. The layers are built up within the vehicle, with many operations having to be accomplished 'blind'. Some of these problems can be overcome by assembling the cockpit off-line, out of the vehicle. The completed assembly is then transported to the vehicle assembly line and inserted into the vehicle either through the door or windscreen opening.

The Delphi Cockpit Module, Fig 82, can combine elements of the ventilation system, instrumentation, dash panel, electrical/electronic systems, steering column and wheel, airbags, pedal assembly, entertainment and other systems. The traditional cross-beam can be replaced by a one-piece structural member in a lightweight material such as die-cast magnesium, aluminium or a composite. Components and sub-assemblies can be integrated into this carrier. This will dampen structural noise and vibration; improve serviceability (Fig 83), assembly and disassembly; also reduce weight, component and manufacturing costs. The cross-beam, Fig 84, can also be designed to shield sensitive electronics such as a databus linking the various sub-systems incorporated in the module. A light-weight flexible circuit, with electrical connections and mounting features, can be designed into the beam to replace conventional wiring assemblies, clips and channels. Up to 50% of the wiring content may be replaced by an integral data communications bus. Ducting for the ventilation system may also be included in the crossbeam, often making the system more efficient, less noisy and much more space-efficient.

Next-generation cockpit modules, Fig 85, may encompass even more sub-systems. Delphi is already investigating opportunities for integrating the vehicle fire wall, including sub-systems such as the brake servo, that are traditionally mounted there. Further simplification can be achieved by extending the rationalisation of interfaces. The throttle pedal, for example, could be an integrated electronic unit communicating with the Engine Management System via a central high-speed databus. Electronic throttle control would also allow the addition of cruise control and traction control with virtually no additional cost or complication. The company has already developed an electric/electronic architecture that uses a serial data communication system based on the widely adopted CAN bus. This operates from a body computer module to provide a high-speed link to the engine management system and a low-speed link to systems such as instrumentation, entertainment and driver communications. A separate network can be used for safety systems such as airbag controls.

Such a system has many benefits. A central databus will, for example, allow the replacement of HVAC (Heating Ventilation Air Conditioning) actuation cables, which present packaging, user feel and manufacturing issues, with electronic actuation. This will also allow the addition of automatic climate control features with little additional cost. Further weight and cost savings are being introduced by the use of new materials, including various glass-filled thermoplastics such as polypropylene and PET. These can allow, for example, ventilation ducts, wiring supports and column and airbag mounts to be integral with the cross-beam. Because the cockpit module is a single unit from a single supplier, materials can be matched for ease of recycling.

Fig 82: Delphi cockpit module

Fig 83: Service accessibility

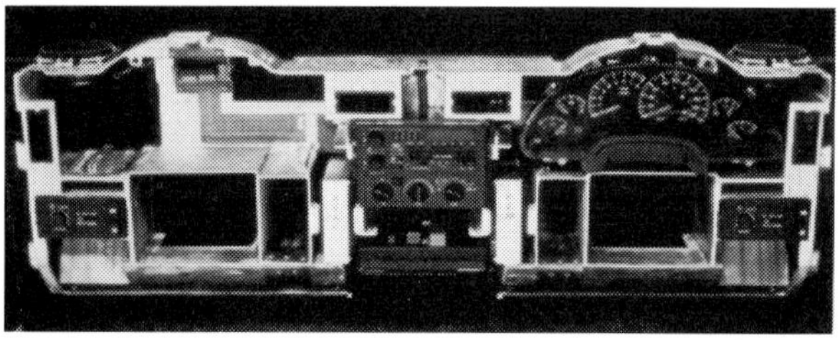

Fig 84: Cross-beam

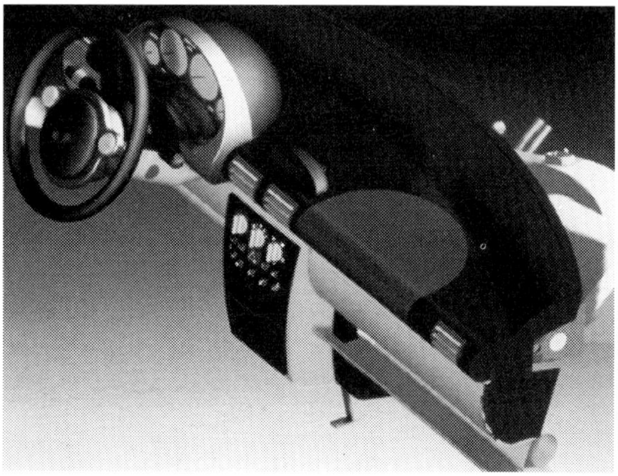

Fig 85: Next-generation module

Calculating airflow for HVAC and engine cooling

According to Cranfield Univerity researchers[17] stimating the quantity of excess heat to be dissipated for a given vehicle speed is a useful first step. This can be obtained from:

$$K_l A(T_w - T_a)V^{0.8}$$

where A is radiator area, K_l the radiator constant, T_w the radiator wall temperature and T_a the air temperature; V is airflow velocity at the entrance to the radiator. Rate of air mass flow required is $=\rho A V^2$.

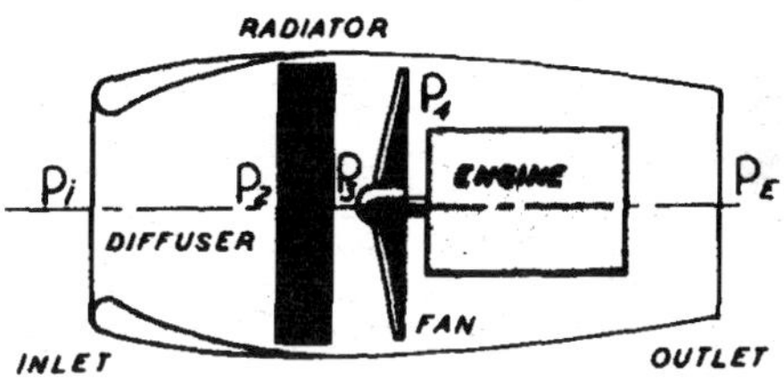

Fig 86: Ideal cooling system

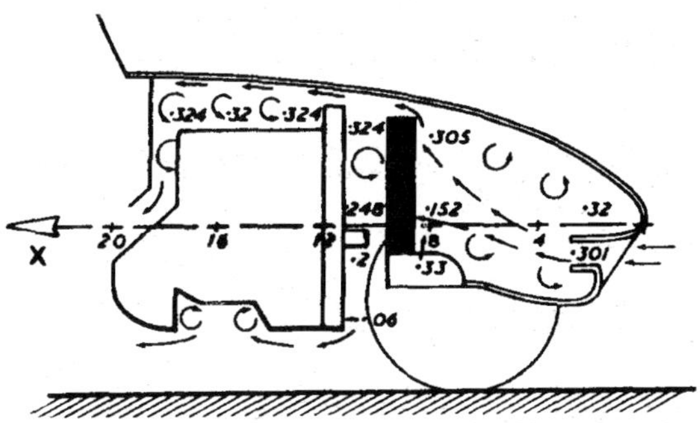

Fig 87: Actual cooling system

The ideal cooling system, Fig 86, comprises a straight sealed duct, the inlet of which leads to the diffuser and then through to the radiator, followed by an outlet at the rear. In a less idealised system such as Fig 87, with typical internal pressure distribution shown, the aerodynamic efficiency of the radiator can be estimated from a formula due to Skibor-Rylski as:

$$2K(T_w - T_a)/\rho V^{1/2} C_d$$

where C_d is its drag co-efficient. As the efficiency increases with airflow velocity, a long duct in front of the radiator is to be preferred with a gradually increasing cross-section. The diffuser should regulate the airflow to obtain the optimum speed. Efficiency is maximised by having the airflow sufficiently turbulent for the air particles to spend the most time collecting heat from the walls. The front grille can be designed to produce the right conditions of turbulence.

Experiments have shown that the diffuser duct should not decline more than 7 degrees to its centreline at any point, to avoid flow separation from its surface. Total drag due to the internal airflow can be expressed as the loss in momentum of the air between inlet and outlet, the product of air mass flow and velocity differential between inlet and outlet. This needs to be added to the pressure drop over the same path times the exit area.

Chapter 4: Body trim and fittings

Interior trim basics; process advances in body trim; PSV interior/exterior body hardware; smart glass windows; extended uses of glass; lighting technology; electro magnetic compatibility; wiper/winding/locking mechanisms; other interior mechanisms

Interior trim

Polymer-based products continue to compete with pressed woodfibre board as a trim base, Fig 1 showing a panel produced by British Moulded Fibre in the latter material. Because of the relatively inexpensive tooling, the process is also attractive for the moulding on of fabrics and trimcloth to produce 'integrated' trim panels. Soft feel can be obtained with foam interlayers beneath the trim cloth which is usually vacuum-drawn on to the fibreboard.

An alternative trim-base uses wood-flour rather than woodfibre. This fine sawdust-like material is blended with polypropylene and moulded in a press with somewhat more expensive tooling but at least having potential for medium volumes. Trimcloth covering is also possible in a single closing of the mould, with such materials as vinyls, textile fabrics or carpet. Fig 2 shows a selection of items made with the process by John Cotton (Colne).

Materials used for seat upholstery, such as those described in the later section, are also used extensively in keying the door and side panels of car saloons and CV cabs to the texture of the seat. Particularly good leather simulations have been developed in PVC by such firms as Weston Hyde. Vynide is a solid PVC-coated fabric which was developed about 20 years ago. Perforated forms of the material make it attractive for headlinings. When used with knitted backcloths, allowing two-way stretch, quite sharply curved surfaces can be trimmed. Surfaces also can be embossed to obtain leather or other 'grain' effects. Flutes, too, can be moulded-in to obtain 'traditional' effects. Thermoforming plas-

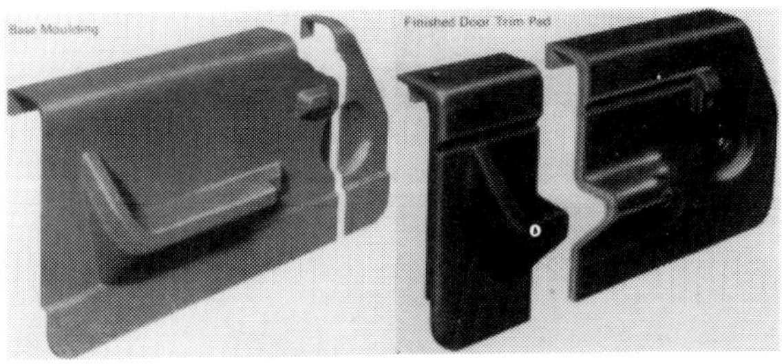

Fig 1: Woodfibre based panel

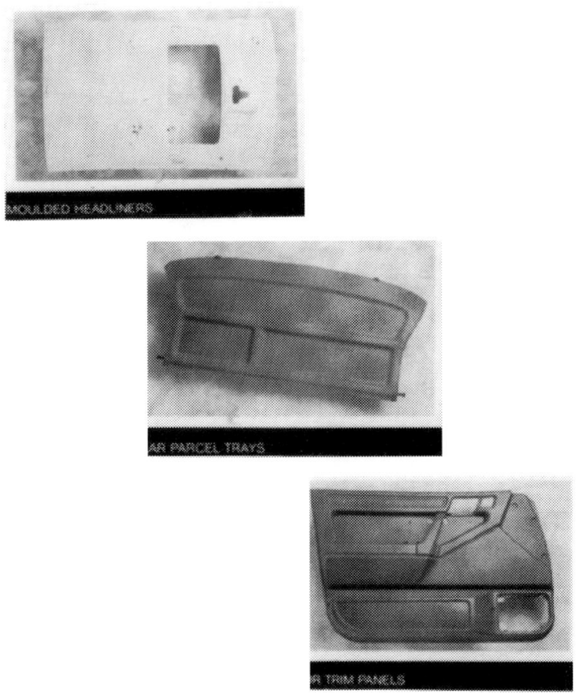

Fig 2: Polypropylene/wood-flour panels (top)

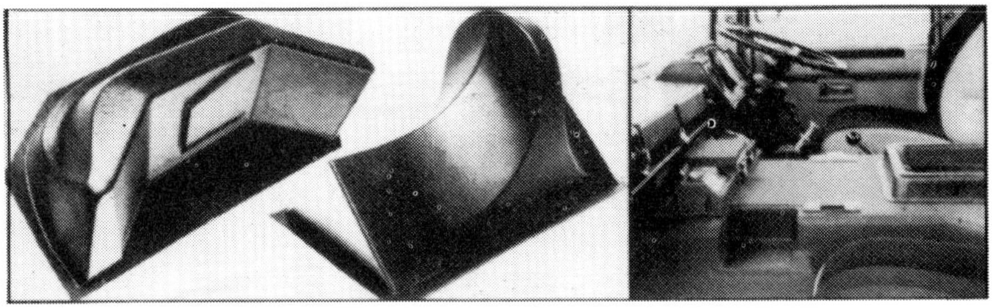

Fig 4: CV panels in Cycolac ABS

tics are another alternative to 'cloth' covered boards. Fig 3 shows a process developed by Borg Warner for thermoforming ABS. The same company has produced design rules relating thicknesses and draw depths available with the material. Well seasoned timber or composite tools can be used for very small runs — or to make prototype parts while metal powder filled heat-resistant epoxides can be used for medium runs. Draft angle should be 2-3 degrees and depth of draw should be less than half the finish width of the panel. Fig 4 shows Cycolac ABS parts and assemblies for CV cabs. Polyurethanes in various forms are also valuable as trim-base materials with their ability to exist in rigid, semi-rigid and flexible states. Hot and cold cure foams are both used, the former requiring aluminium moulds. The table in Fig 5 shows some of the properties which are typically obtainable with cold and hot-cure foams. Cold-cure foams can be produced in wood or GRP moulds but normally it is difficult to produce foams less dense than 45 kg/m^3. They do however have intrinsic flame-retardent qualities. Semi-rigid and self-skinning foams may be produced either by the so-called prepolymer or one-shot processes. In the former, excess toluene diisocyanate is reacted with polyol as a first step and the foam reaction is completed subsequently, with addition of water and catalysts during moulding. With the one shot system all the chemicals are mixed together at the time of moulding. Adjustment of the chemical formulation results in typical properties for a crash pad panel as shown in the table in Fig 6. Pads may be produced with a variety of cover materials such as vac-formed ABS or PVC; blow-moulded, blow-formed, rotationally-cast or slush-moulded PVC. The last two methods particularly suit arm-rest production. Flooring systems are also produced in PUR by spray-forming a PVC skin into an electroform,

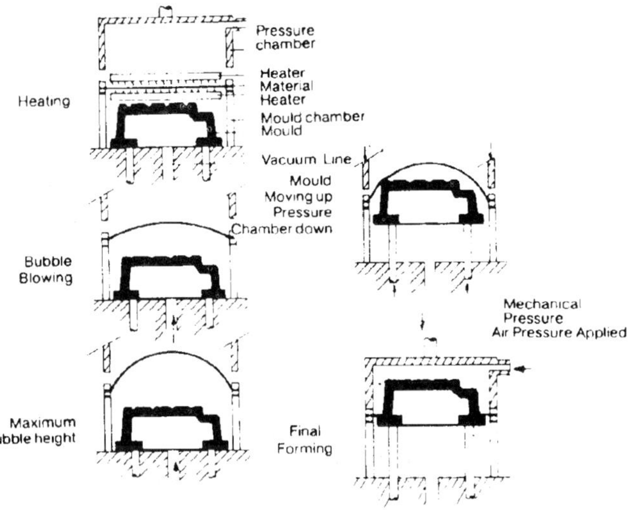

Fig 3: Thermoforming of ABS

Property (BS3379)	Typ hot cure	Typ cold cure	32 kg/m^2 cold cure
Density (kg/m^3)	24-45	45-70	32
Tensile strength (kN/m^2)	110-165	69-97	62-67
Tear strength (N/m) ASTM-D1564	350-610	157-200	157-200
% elongation at break	250-350	90-110	80-90
75% compression set	6-9	5-7	5-7
90% compression set	6-9	5-7	5-7
Hardness (12 in ind 40% comp),kg	18-35	15-45	20-24
Isoprene coating (% elong at break)		600	
Dynamic fatigue (% height loss)	4-12	1-4	1-2
Resistance (% hardness loss)	15-35	5-10	15-20
(to ISO draft spec TC45/WGH-N84)			

Fig 5: Properties available with cold- and hot-cure foams

Density kg/m^3	48-160
Tensile Strength kN/m^2 (BS3379)	138-414
Tear Strength N/m ASTM-D 1564	157-525
Compression Set 75%	25-35
(BS3379) 90%	80-90
Elongation at Break % (BS3379)	50-90

Fig 6: Properties of semi-rigid polyether polyurethane foam

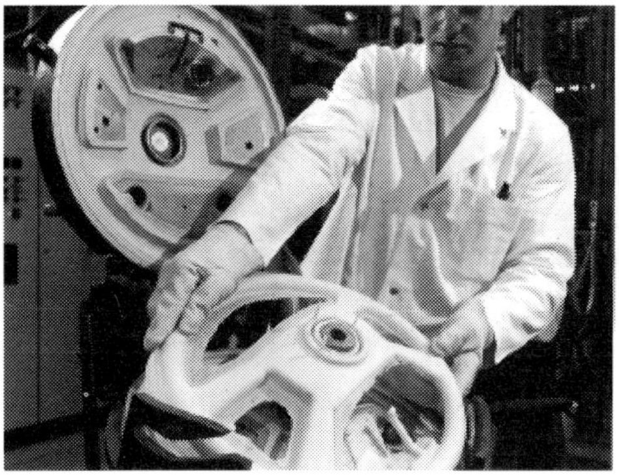

Fig 7: De-moulding a steering wheel cover

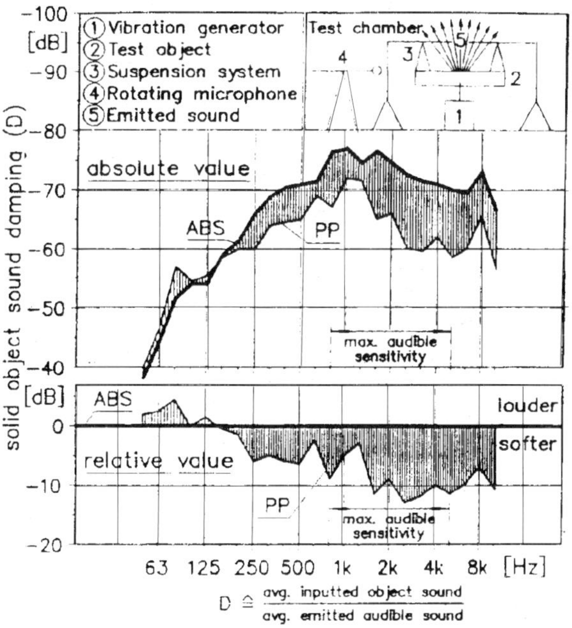

Fig 10: Noise characteristics of B-post trim panels

the skin, after curing, being filled with semi-rigid foam backed in the foaming operation by an 0.08 mm thick sheet of PVC. Door inner panels by contrast are usually made in the material either by vac- or blow-forming the covers in ABS. Self-skinning foams find applications in steering wheel trim and both acrylic and PUR colour coatings can be applied. An interesting development for in-mould coating systems for PUR panels is Trimbond by MacPherson Industrial Coatings. The paint formulation contains its own mould release agent and dries in a few seconds on the mould surface prior to injecting the PUR. After three minutes the coated panel can be withdrawn from the mould; a variety of surface textures are possible. A steering wheel cover by BP Chemicals is shown in demoulding in Fig 7. Trends towards softer touch trim finishes has led to the use of PVC compounds and the dry-powder slush-moulding process, Fig 8. Claims for the process include greater design freedom than that allowed by vac-forming as well as a softer skin and deeper grain. It is said to be less costly than wet slush moulding and a variation is called double slush moulding, where skin and PVC foam are formed integrally in one tool. The table in Fig 9 lists rival processes competing for trimmed instrument panel construction with advantages and disadvantages, due to US consultants Phillip Townsend Associates. In door panels, there is sign of a trend away from ABS towards polypropylene due to lower cost and better acoustic performance. Fig 10 due to Hoescht shows noise test results carried out on B-post trim panels for an executive car.

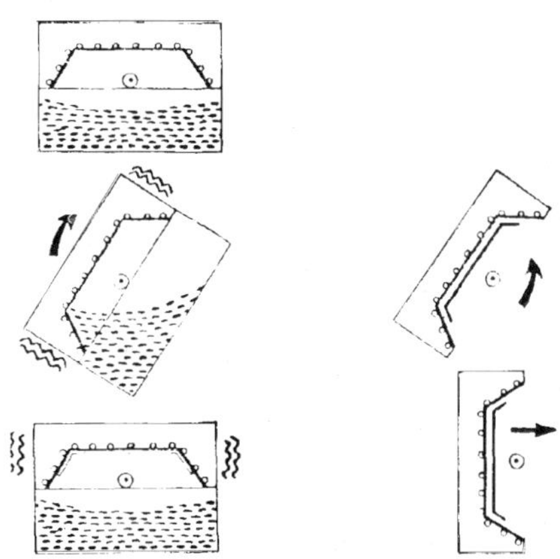

Fig 8: Dry slush-moulding

PROCESS:	DP-PVC	DP-PVC	Wet slush	Negative thermoform	Positive thermoform	Thermoform	Spray
PROCESS VARIANT:	Single slush	Double slush					
SKIN MATERIAL:	PVC	PVC	PVC	PVC/ABS alloy	PVC/ABS alloy	EPVC	PU
SKIN THICKNESS (MM):	1.5	1.5-2	1.8	1.1	1.1	1.1	1.1
GRAIN REPRODUCTION:	Excellent	Excellent	Fair	Fair	Excellent	Fair	Fair
FOAM:	PUF	PVC	PUF	PUF	PUF	PVC	PUF
FOAM PROCESS:	Injected	Integral	Injected	Injected	Injected	Integral	Injected
MOLD COST ($'000):	200	20	20	50	50	50	40
CYCLE TIME (SECONDS):	315	320	420	45	45	45	150
SOFTNESS:	Excellent	Excellent	Fair	Fair	Excellent	Fair	Fair
RELATIVE COST INDEX (UNITS/ SKIN):	200	?	260	100	120[e]	100	157

Fig 9: Rival processes for instrument-panel trimming

Process advances in body trim

The need to reduce costs by consolidation of parts in such assemblies as instrument panels, door cases and seats, both for cars and CVs — Figs 11/12, is leading to one-shot moulding techniques. Environmental legislation is also leading to the use of common materials.

For door-rim panels a principal technique in use by major suppliers John Cotton is a trim base panel of moulded Woodstock. This material is bought by the company as an extruded blend of polypropylene and woodflour and then processed on a machine which permits forming, covering and trimming in one operation. Its properties are in the table of Fig 13. Coverings commonly in use are supported expanded vinyls (SEV), textile fabrics or carpeting. The company can also offer a wide range of alternative trim-base materials as seen in the table of Fig 14, which is an excellent survey of materials currently applied to vehicle trim. In self-supporting moulded headliners the company uses an extensive range of substrates including resin-bonded and glass reinforced polyurethanes (500-610) and felts; FSL Dielark styrene (580-640), PPO foam and Eften corrugated board (975); figures in brackets being the weight in gms/metre[2]. Coverings include non-woven fabrics of polyester, nylon or viscose in spun-bonded, point-bonded or needle-punched construction; polyester or nylon knitted fabrics, normally as a raised tricotte knit to give a velour effect; other options are woven fabrics, artificial suede, PVC washables and flock coatings. With the popular knitted fabrics, the material is flame-laminated or sinter-bonded on to the substrate

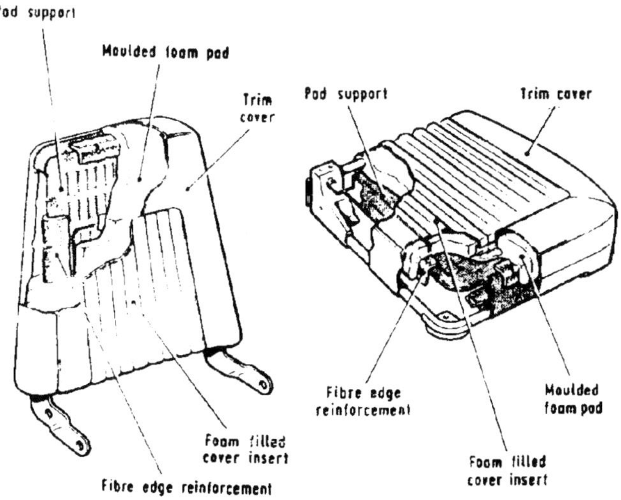

Fig 11: Car-seat cushioning/upholstery

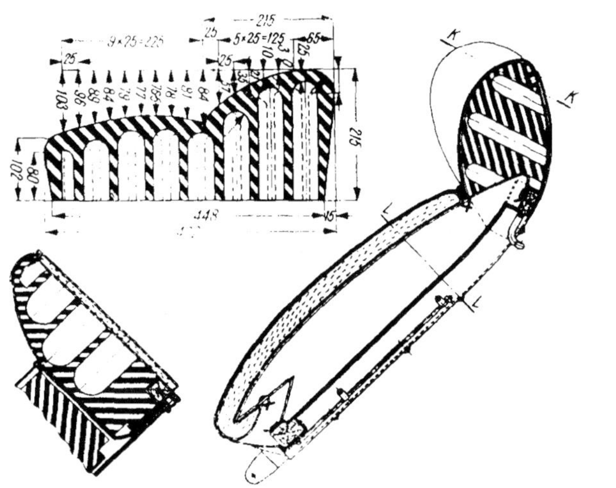

Fig 12: Coach-seat upholstery

Material	Fimit Lignatok	BMF	Mip Van Dresser	Northern Fibres	Lignatok	Cekazell	Rover 200 Woodstock	Woodstock Equivalant HMPP	Strepazzini Resine	Fimit Felt	Glass Mat Pu Rim Exd 113/1	Empe-worke Pelliform
Application	Fiat Tray	Montego	Maestro		Sierra	Parcel Tray	Rover 200	Matl. Data	Lancia	Fiat	Matl. Data	
Description	Wood Fibre Pehnolic Resin	Wood Fibre Aromatic Polyester	Wood Fibre Phenolic Resin	Wood Fibre Phenolic Resin	Wood Fibre Aromatic Polyester	Waste Fibre	Woodflour Filled Polypropylene	40% Wood Filled Polypropylene	20% Milled Glass Filled Resin	Felt Phenolic Resin	450 g/m^2 Glass Mat PU	100% Jute Thermoplastic Resin
Density (kg/m^3)	1050	650	656	760	820	860	890	1000	720	1000	480	770
Thickness mm	2.8	2.2	3.5	3.3	2.4	4.5	2.4	3.0	3.8	3.0	3.5	2.5
Mass per unit area (kg/m^2)	2.9	1.4	2.3	2.5	2.0	3.9	1.7	3.0	2.7	3.0	1.7	1.9
Flexural Strength (MN/m^2)	56	20	24	34	46	34	>27	38	24 41	76	31	46
Flexural Modulus (MN/m^2)	–	–	–	–	–	1700	–	4000	2500	–	1200	1500
Flammability	Self Ext 30 secs.	Burns Slowly	Self Ext 45 secs.	Self Ext 45 secs.	Burns Readily	Self Ext	Burns Slowly	Burns	Self Ext 30 seconds	Self Ext 30 seconds		Burns Slowly
Hardness (double thickness) (N)	810 (> 2000)	200 (360)	(970)	190 (1200)	550 (1020)	790 (1200)	(> 2000)	1350 (> 2000)	370 (> 690)	900 (> 2000)		500-1000
Wicking Length (mm)	0	0	45	35	55	–	0	0	0	30	0	–

Fig 14: Range of alternative trim systems

Thickness	2.5 ± 1 mm
Density	1.0 gm/cm³ min.
Moisture Content	2.0% max.
Water Absorption	3% max.
Flexural Strength	Long 35 N/mm² min. Trans. 30 N/mm² min.
Flexutal Modulus	Long 1400 N/mm² min. Trans. 900 N/mm² min.
Tensile Strength	Long 16 N/mm² min. Trans. 12 N/mm² min.
Shrinkage on Drying	0.5% max

Fig 13: Woodstock properties

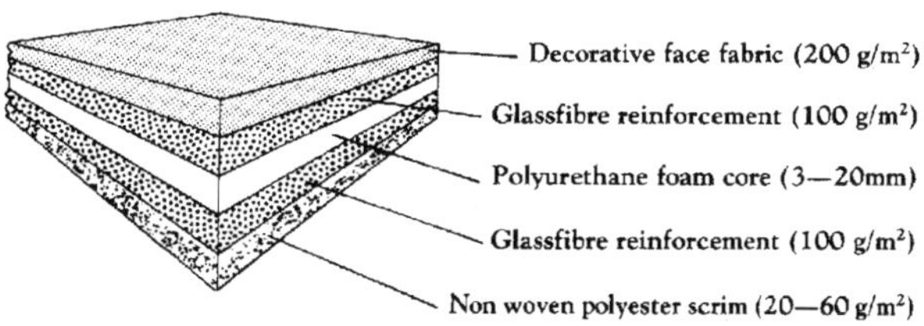

Fig 17: PUR headliner construction

foam of 1-4 mm thickness which gives a cloth of appealing surface texture and soft-touch feel. Knitted fabrics can also be drawn into most shapes and are readily dyed to colour key with other parts of the vehicle interior. A variety of edge conditions is offered, Fig 15; With the cropped edge, a finisher or an overlapping adjacent panel is normally necessary. The cloth-wrapped edge removes the need for a

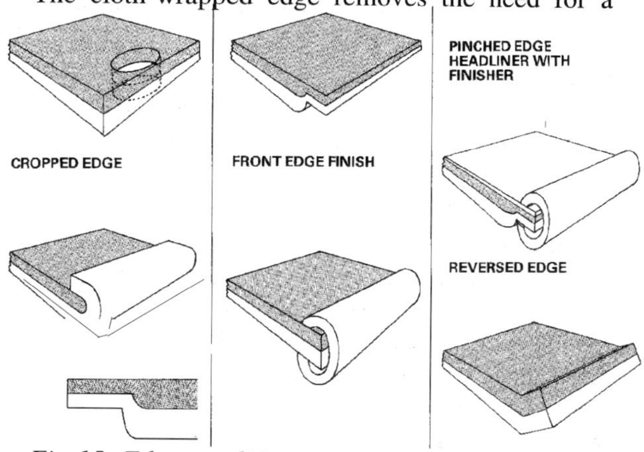

Fig 15: Edge conditions

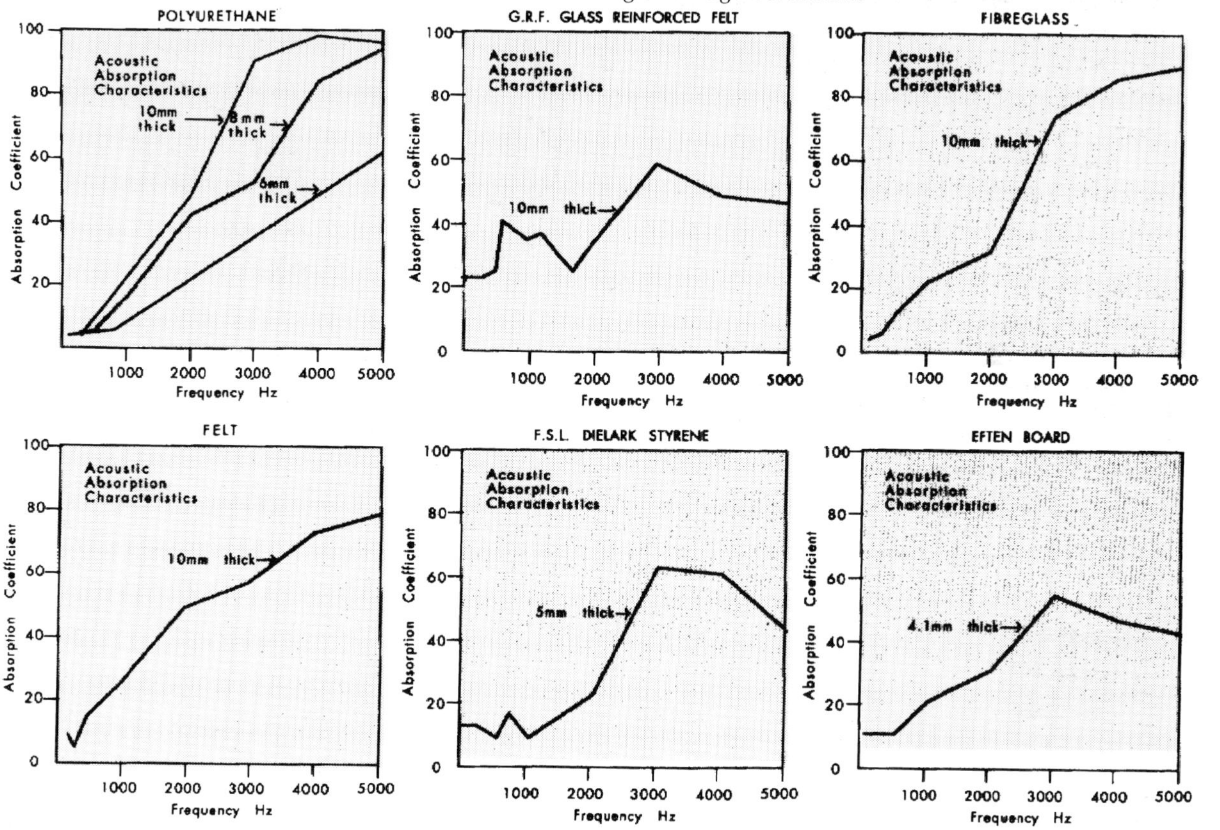

Fig 16: Acoustic performance

finisher but as the substrate has to be trimmed to size first two separate operations are involved. With the pinched edge, a compressed edge is designed into the moulding which is subsequently cropped to size, having first moulded the substrate and covering as an integral unit. Sometimes the so-called reversed edge is acceptable, as usually only the corner bend is visible. Acoustic performance of headliners made from the different substrates is seen in Fig 16.

Polyurethane headliners have been developed by the company which can be made in one piece for a car or cab roof. Construction is shown in Fig 17; the PUR core is based on an open-celled BASF polyester foam which is thermoformable and has unusually low density and good sound-absorbent characteristics. Hardness can be varied right through from flexible to rigid state. It has good dimensional stability up to 100 C and has survived simulation testing for tropical conditions (200 hours at 40 C and 90% relative humidity). It meets FMVSS flammability rating, and with all covering layers in place, weighs 1 kg/metre^2. Glass-reinforced felt has also recently been developed by the company, for headliners, and has strength characteristics which suit applications involving the suspending of sunroof frames, light fittings or speaker grilles. Traditional resin-bonded felts are enhanced by a glass reinforcement to provide twice the rigidity for a given density. Fixing clips and brackets can be integrally moulded and weight is in the order of 1300 gm/metre^2. Fig 18 shows typical weatherstripping employed at the junctions with closure panels.

Noise insulation and recycling Recent developments by ICI involve a scientific approach being made to vehicle interior noise reduction by exact specification of trim material properties and dimensions. Heavy covering layers of carpeting and similar materials are mounted on flexible foams of known stiffness so that the spring/mass can be exploited in noise isolation, Fig 19. The company co-operated with German supplier GNM in the production of flexible foam backed acoustic mats for Audi and Mercedes Benz. Fig 20 shows the press used in form-fit bonding the material. PUR foam densities from ICI are now down to 48 kg/m^2 and mould cycle time is down from five to two minutes. So-called tactile foams have been developed for the formed carpet application which ensures the absence of voids between foam and floor-pan.

With the prospect of future legislation specifying fixed quantities of recyclate in plastics systems used in vehicle trim, there is an onus on designers to select suitable systems for future vehicles. The trend is to reduce the number of material types used in such items as instrument panels and door casings. Polypropylene is being favoured to replace a number of current mixed material systems. Also, despite

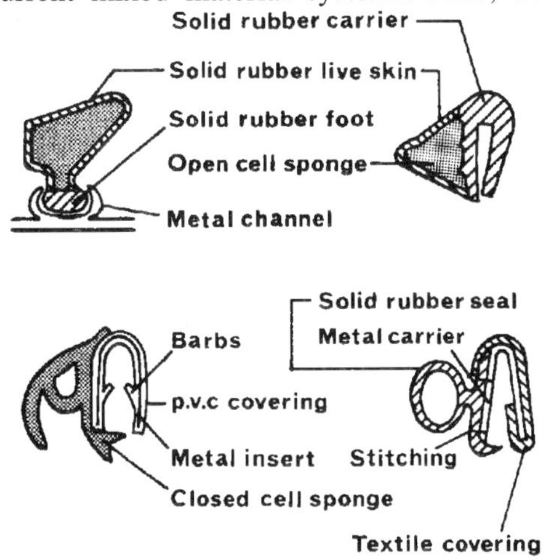

Fig 18: Weatherstripping

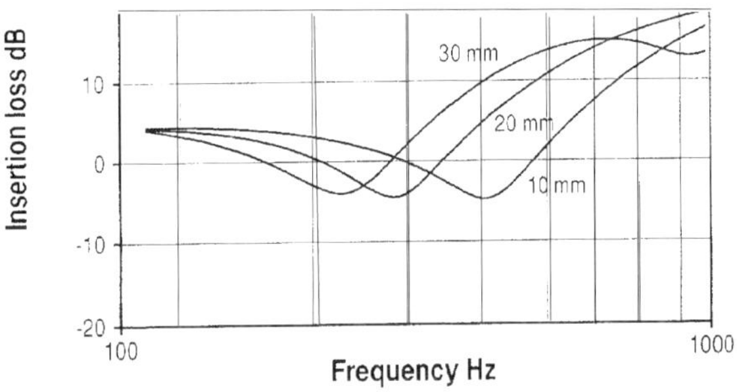

Fig 19: Using spring-mass system for noise isolation

Fig 20: Form-fit bonding press

warnings of problems with fogging and halogen emission in manufacture, PVC coverings have been developed which are now more environmentally acceptable. Soft feel versions are now available for covering skins of instrument panels and for substrates the trend is to talc-filled polypropylene. Eventually the plastics suppliers hope to be able to produce PP skins of equal acceptance in order to commonise the material usage. As a substrate the material claims to have an advantage over ABS in eliminating squeaks and rattles. Adhesion and cost problems have yet to be overcome with PP covering skins.

PP fabrics are available for upholstering of seats, however, and these are said to be strong, durable, light-fast and have good heat tolerance. Development work is still underway on improving abrasion resistance and proneness to electrostatic attraction of dust. Seating in single material systems is however leading some companies to develop polyester fabrics which 'conform' with expanded polyester cushion foams. Du Pont recently launched Ecofil, a compression-mouldable fibre cluster material which can be formed into seat cushions and a 20% weight saving was claimed over expanded PUR foams. Polyester matrix composites are being mooted for the associated seat frames and Dacron polyester fabric for the upholstery. Woollen fabrics are being promoted for top-class vehicles and for some time it has been offered by Jaguar and a number of continental manufacturers. Claims for the material are good tensile strength, tear strength and abrasion resistance. Lack of static 'dust' problems, and the resilience which allows moulding around complex shapes are also claimed. Comfort is enhanced by low thermal conductivity and safety by its flame retardancy.

Fig 21: Double-decker test mock-up

PSV interior/exterior hardware

Unique to the PSV category, of course, is the effect on vehicle design of passenger movement on and off the vehicle, and between saloons in the case of a double-decker. Ogle Design made an important study in this area, on behalf of London transport and Leyland bus, using the full-scale mock-up shown in Fig 21. With this equipment it was possible too infinitely adjust the staircase and exit positions with respect to the entrance and so test the validity of ticketing systems and entrance step configurations. Extensive user trials were carried out with test subjects from all quarters of the population. These lead to the favouring of designs which permitted dual stream boarding and optimum handrail layouts were determined as well as a design of angled split steps.

Fig 22 shows the effect of varying step to kerb distance; the split step reduces the exclusion of ageing and infirm sections of the population from the buses from a 26% to around 6%. The angled split step avoids the usual split-step problem of squeezing two steps into the space normally occupied by one and increases the tread area by 50%. Staircase layout trials, Fig 23, showed the ideal arrangement was one in which the bottom of the staircase is aligned opposite the front half of the exit doors so that dual-stream alighting is possible. Both benefit from being as close to the front of the vehicle as possible so as to improve driver-supervision and enhancing passenger flow by minimising the number of longitudinal seats forward of the exit. Staircase design evaluation brought requests for straight steps and absence of any protrusion of the bottom step into the lower saloon. The longer staircase shown in Fig 24 was shown in the user-trials to be considered an improvement over conventional layouts along with its developed handrail profile. Elliptical sections were preferred for the recommended handrail layout for the entrance — as shown in Fig 25.

Studies by Leyland Bus into air-conditioning for double deckers also showed the considerable extra capacity required of air-conditioning systems for occupant comfort. Passenger loadings as high as 102 seated and 30 standee passengers for such vehicles operating in Hong Kong amounted to a high cooling load as did high solar gain arising from large surface areas, a large proportion of which is glass — also the frequent door-openings involved in multi-stop duties. High idle speeds are required of the prime-moving engine to drive the refrigeration plant through the gearbox power-take-off. Because roof-mounted units are not possible, substantial saloon space is required for both plant and associated distributive ducting. Saloon temperature range for this particular application was 22-28 C for which a cooling capacity requirement in the upper saloon was 27kW and 18.5 kW in the lower saloon — with corresponding evaporator airflows of 3000 cubic metres per hour for each saloon. There was a dedicated compressor, condenser, evaporator and control system for each saloon. An engine idle speed of 700 rpm was require to drive the two compressors which together absorbed 30 bhp. In the associated blower system a 240 amp alternator is required to provide 24 volts DC to all the system motors.

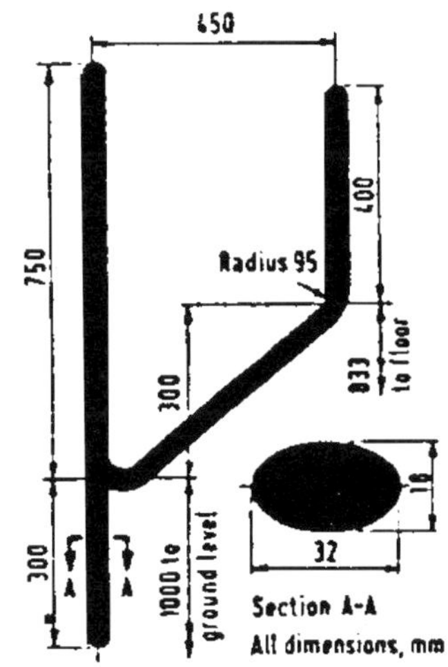

Fig 26: Hand-rail recommendations

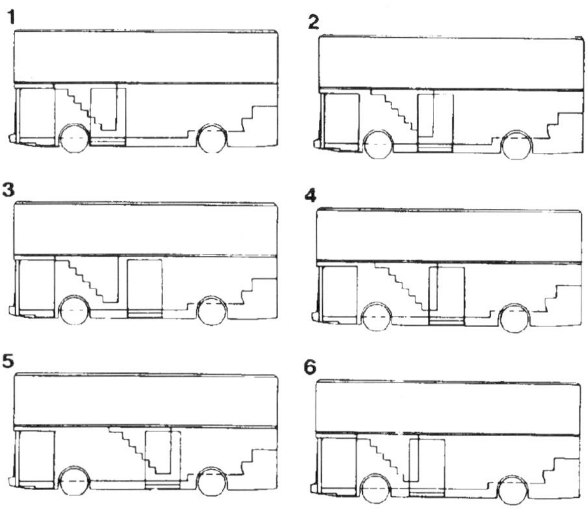

Fig 23: Staircase position alternatives

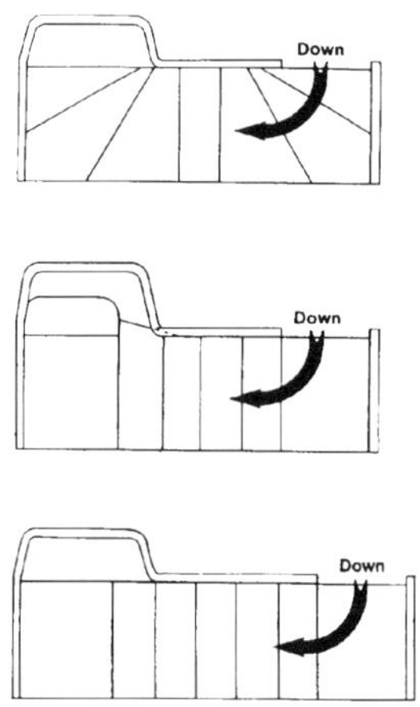

Fig 24: Staircase types: top, angled; centre, MCW Mark 1; bottom, Ogle's design

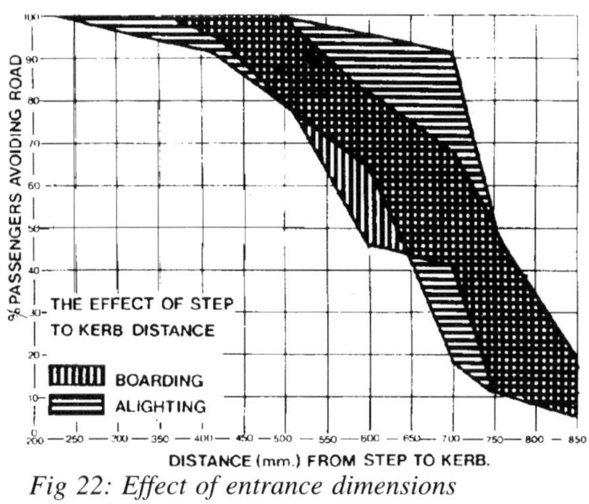

Fig 22: Effect of entrance dimensions

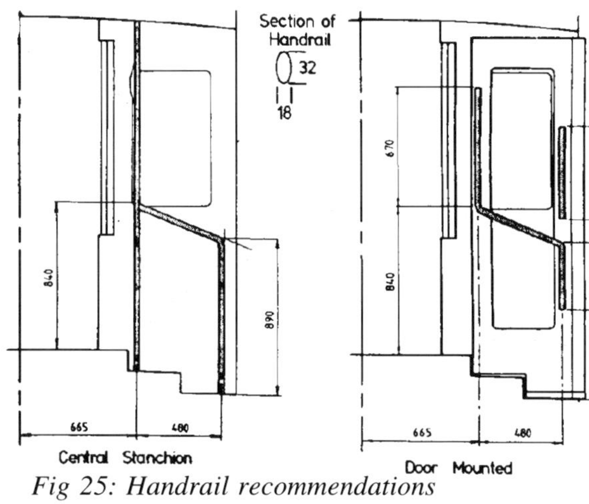

Fig 25: Handrail recommendations

Catering for the disabled

Detail recommendations such as those in Fig 26 have been made on the design of handrail systems (in this case for a two step entrance involving rises of 311 and 216 mm) as well as those for levels of braking and acceleration tolerable to infirm standing passengers. This has led to derivative of acceleration values, 'jerk', being defined: 1.5 is prescribed as a maximum tolerable level. The greatest incidence of accidents was found to be during entry/exit situations and the researchers designed the retractable step system in Fig 27 to help alleviate this problem. The mechanism was designed at Cambridge University and the step retracts into the side of the vehicle without affecting ground clearance. A sensitive strip on the edge of the step prevents injury during extension and sensitive matting on top prevents the step rising when anything comes in contact.

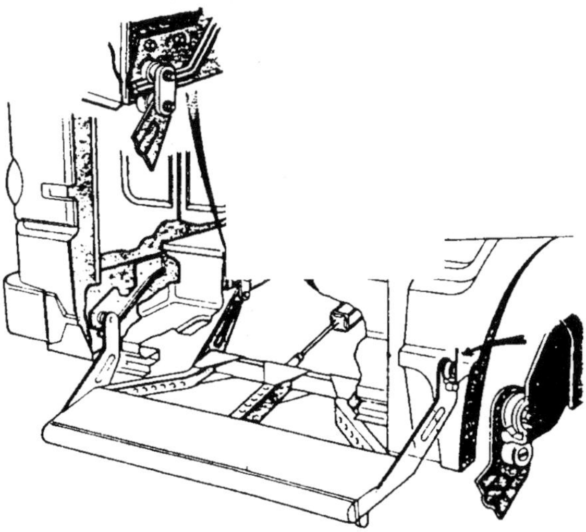

Fig 27: Retractable step mechanism

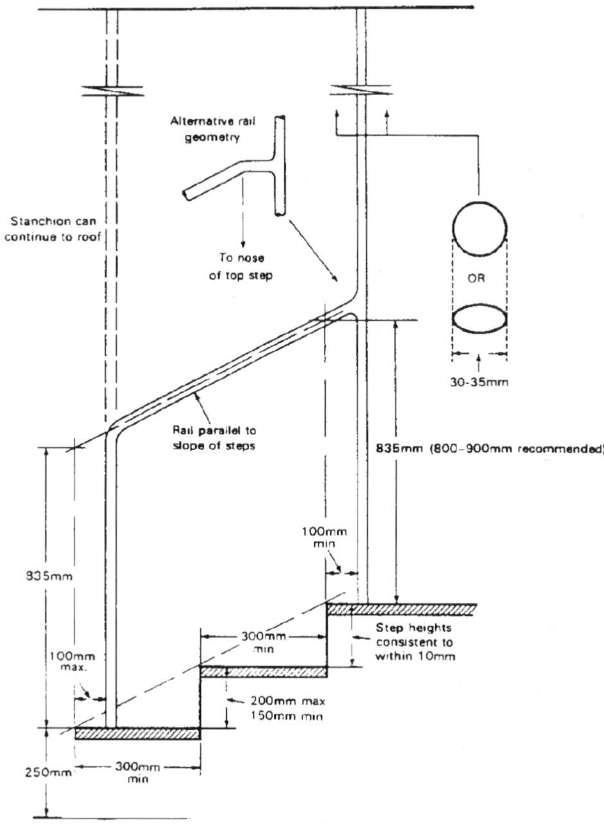

Fig 30: Preferred step dimensions

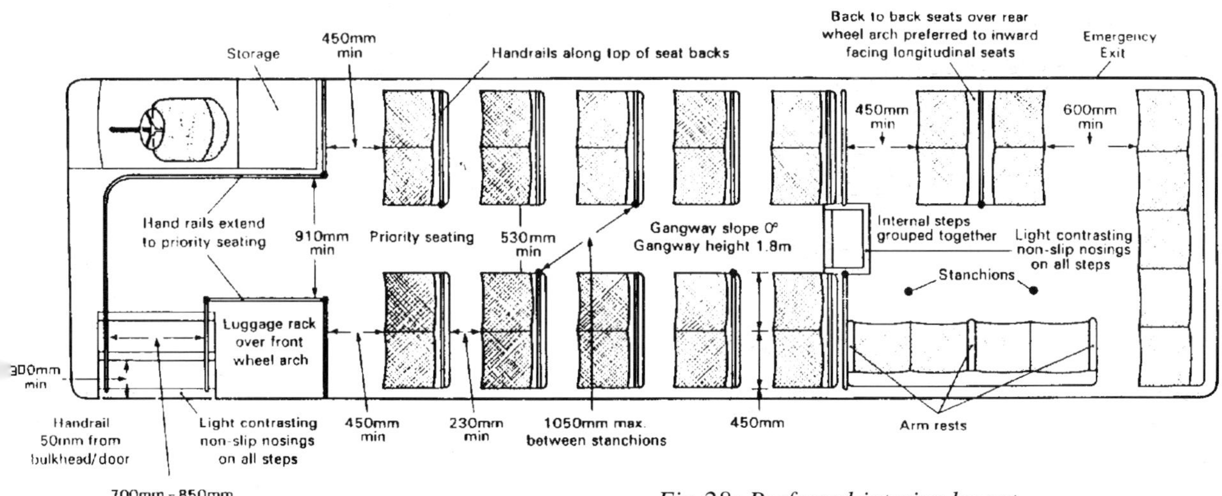

Fig 28: Preferred interior layout

The Department of Transport have now issued a recommended specification, in TRRL Report 137, for buses used on local services to improve accessibility for passengers with mobility disorders. A preferred interior arrangement is shown in Fig 28 and designers are asked to improve on PSV (Conditions of Fitness, Equipment, Use and Certification) Regulations 1981. Headroom should be more than 1.8 metres in the gangway and stanchions/handrails should have circular sections of between 30 and 35 mm diameter, with 50 mm clearance to adjacent surfaces. They should be non-slip in surface texture and coloured to contrast sharply with surroundings. The design of centre handrails for a dual stream entrance/exit should follow the layout in Figs 29a,b and step dimensions in Fig 30. Seat cushion height at the front edge should be between 430 and 460 mm and cushions should not slope more than 30 mm over their depth. Headroom above the cushion top surface should be over 1 metre.

Gradually, bus/coach bodybuilders are availaing themeselves of computer-aided-design packages such as SAMMIE for modelling the optimum ergonomic solutions for passengers. Fig 31 shows its use for the interior layout of passenger vehicles.

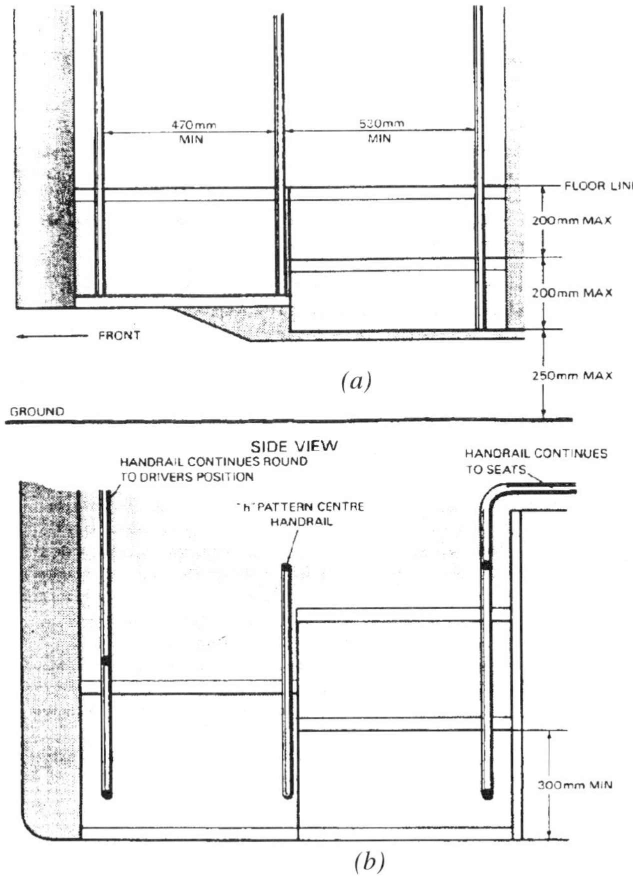

Figs 29: Dual stream access layout

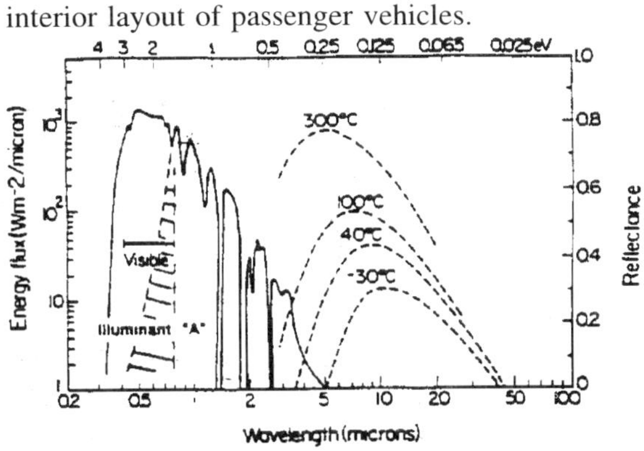

Fig 32: Solar load spectrum

Fig 31: Interior layout using SAMMIE

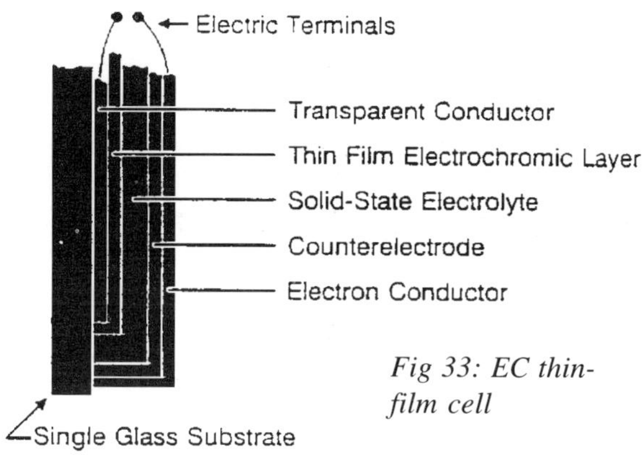

Fig 33: EC thin-film cell

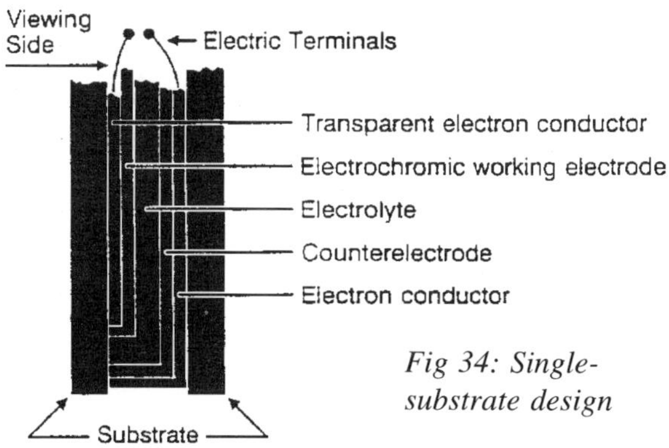

Fig 34: Single-substrate design

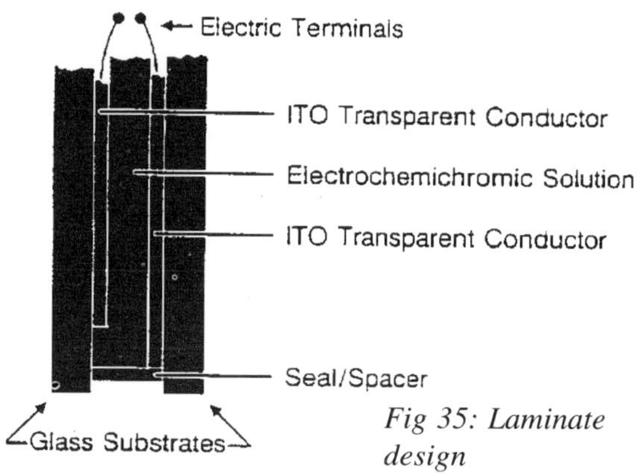

Fig 35: Laminate design

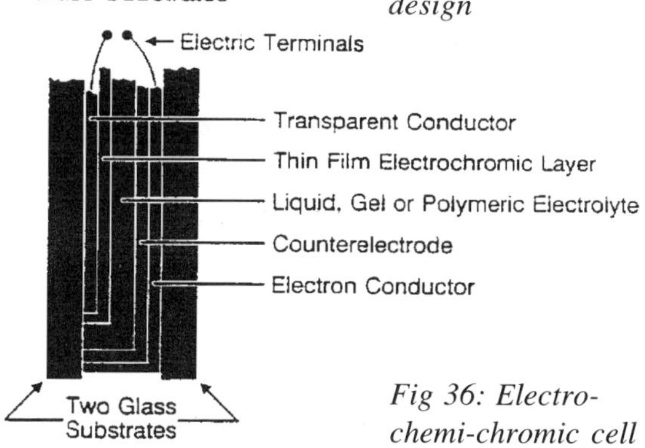

Fig 36: Electro-chemi-chromic cell

Smart-glass: more than just glazing

Solid-state techniques for tinting of glass are leading to smart window systems which have the potential to reduce considerably vehicle interior heat build-up by solar radiation.

Donnelly Corporation[1] have pioneered several systems for the control of glass transmissibility. Chromogenic technologies include Liquid Crystal and Electrochromic, they report. A variety of designs include laminated constructions in which two glass sheets enclose the electrochromically active medium and solid-state devices made by deposition of thin films into a single-layer glass substrate.

Systems must, of course, meet safety legislation for auto-glass governing luminous transmission and shatter protection. The projected increase in glass usage on vehicles, for styling and aerodynamic considerations is stimulating the use of smart glass systems for controlling the solar heat loading inside the vehicle as well as making possible special stylistic effects.

Fig 32 shows the solar load spectrum incident on a vehicle from which it can be noted that if all infra-red radiation could be effectively filtered, half of the solar radiation could be rejected without any loss in visibility. A number of existing tinted glass systems are partially effective in this respect. Donnelly feel that more effective glazing systems could result in the downsizing of air-conditioners by 50% or more. Smart windows would have the objective of transmitting the full solar spectrum in cold winter months while being able to reflect most of the infra-red in summer.

Electrochromism is a reversible colour change in a material induced by an applied electric field. Devices may be EC thin-film or chemi-chromic, the latter involving solutions of the materials, subject to redox reactions, rather than films. Fig 33 shows the first category, comprising a working electrode that colours and a counter-electrode that is a sink and source for the ions used during colouration. These are separated by an electrolyte and there are conducting layers of a transparent material such as indium tin oxide which provide electrical contact to the electro-active layers.

When a solid-state electrolyte is used the device can be a stack of films deposited on to a single substrate, Fig 34, and for liquid or gel systems a laminate design is usual, Fig 35. Thin-film devices retain colour when voltage is removed, important in automotive glazing for long-term parking without power. The material also has the property of maintaining a degree of colouration according either to the magnitude or the duration of the applied voltage.

Construction for electro-chemichromic technology is shown in Fig 36. The effect of a small applied voltage through the transparent conductors causes the solution to colour. Donnelly point out that the gel variant has been manufactured by the Japanese Asahi Glass Company of the type described in Fig 37. Spectral transmission and reflectance as a function of applied voltage is in the graph and table of Fig 38 a and b. Speed of response is dependent on window size and using 10 ohm/ft^2 (100 ohm/m^2) resistant ITO film, a 30 x 40 cm window takes about 120 seconds to dim to 30% transmission, whereas one tenth of the area would take about five seconds.

In an electrochromic sun-roof developed for a Nissan concept car the construction shown in Fig 39 was used. Here a third auxiliary electrode was required because, initially, both of the electrochromic layers (tungsten oxide and prussian blue) are in the non-transparent, oxidized, state. Its purpose is to reduce the blue to a colourless white which forms a redox couple with the tungsten oxide. Panels measuring 40 x 40 cm have been demonstrated which dim from 60% to 5% visible transmission.

Encapsulated liquid crystal technology also has potential automotive application. Devices usually consist of LC droplets, of 1-10 microns, within a polymer matrix. These are further sandwiched between two conductive electrodes. In the off state the LC molecules are aligned so that their refractive index differs from that of the matrix.

	Transmission Characteristics	
	% of Solar Spectrum*	%. Weighted Illuminant A
Bleached	59.2	79.6
Colored @ 0.50 V	26.8	42.8
Colored @ 0.75 V	5.6	10.0

	Reflection Characteristics	
	% of Solar Spectrum*	%. Weighted Illuminant A
Bleached	15.4	12.7
Colored @ 0.50 V	26.8	42.8
Colored @ 0.75 V	5.6	10.0

* With reference to air mass 2 (AM2) between 300 nm to 2500 nm

Fig 37: Gel variant of electro-chemichromic

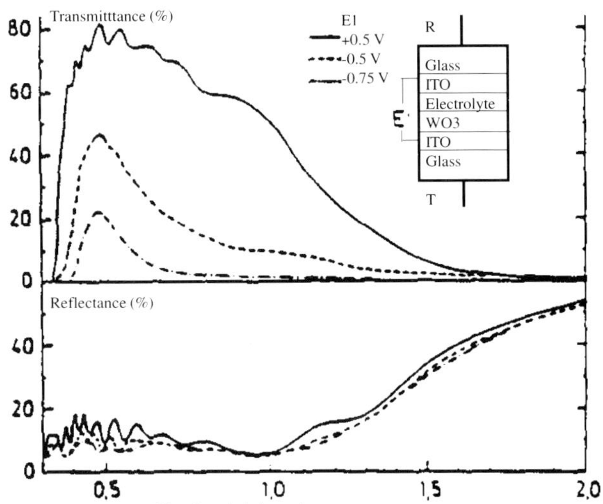

Fig 38: Performance characteristics

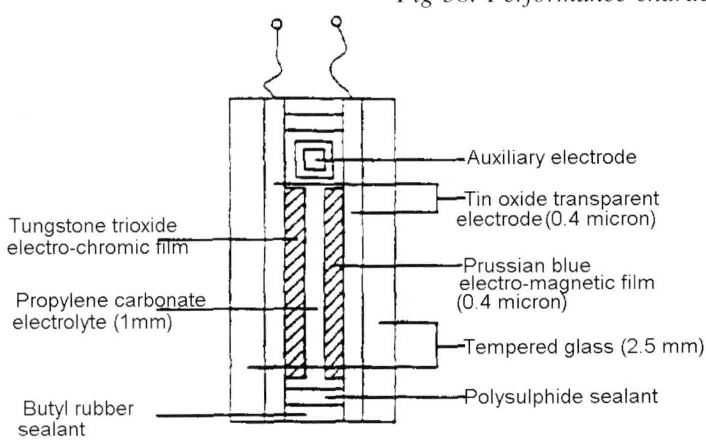

Fig 39: Cell with gel electrolyte

Other extended uses for glass

The Triplex prize-winning RCA-concept Locust combine harvester, Fig 40, has a virtually all-glass cab from which an unusually wide angle of view can be obtained. However, an opaque roof and negatively raked side panels reduces the greenhouse effect, aided by graduating tinting of the blue-green polarised glass. Three convex, and one concave, surfaces are involved and double-curvature used is said to minimise the "sphereing" effect. The traditional horizontal chassis frame is replaced by a four-post and cap backbone structure from which curved panels can be conveniently hung to give a clean exterior appearance.

The Triplex Hotscreen system for electrically heating a laminated glass windscreen is seen in its latest form as the optional Quickclear screen offered by Ford on its Mondeo model. The system is considerably improved over the early versions of three decades ago which were poorly perceived because of high current consumption. Now the current drawn is much less of a drain on the battery and improvements have been made in crimping to nullify the effect of haze that could arise from interference patterns caused by a combination of refraction and reflection of the wires. A low melting point solder-coated copper bus bar has also been recently introduced to eliminate any risk of corrosion. The latter replaces the earlier pressure contact to a silver-coated copper bar.

The current system, Fig 41, comprises a grid of tungsten filaments with a diameter of about 0.2 mm, sandwiched by the glass laminate. Power rating is 600 W/metre^2 and recent MIRA testing has confirmed that defrosting from -5 C occurs in less than two minutes. From -18 C, under 78/317/EEC conditions, 100 % windscreen clearance time is five minutes (compared with 80% in 40 minutes required of the regulation). The regulation is framed against typical defrost blowers built into conventional heating systems and really is a very poor incentive to improved safety, according to Triplex Product Development Manager, Peter Pennels. Typical heater blowers rate only 4kW at full blast. In large windscreen PSV applications blow-defrosting would be unacceptable because of the "dry-eyeball syndrome" and this vehicle category has used electrically heated systems for some time.

The company[2] has also been examining Head-Up-Displays for night-time vision enhancement and a system installed in a demonstrator vehicle allows a driver to see the image derived from an infrared camera reflected on the windscreen and superimposed on the view ahead, Fig 42. An additional set of headlamps is involved which emit near infrared and illuminate the area normally covered by the main beam lamps, without dazzling oncomers. A camera then senses the forward view and projects it on to the windscreen via an LCD display and head-up reflector. Another possibility is the use of far infrared, a camera which would sense thermal rather than visual radiation. A Sobel edge detection algorithm would be required to process the camera output before the image was viewed.

For the HUD, the three elements are the electronic display, optical projector and a combiner for superimposing the images. Normally the combiner is a freestanding glass element with either a reflective coating or containing a reflective holographic element; obviously the windscreen is a simpler and cheaper alternative. The image luminance depends

Fig 40: Locust combine harvester concept

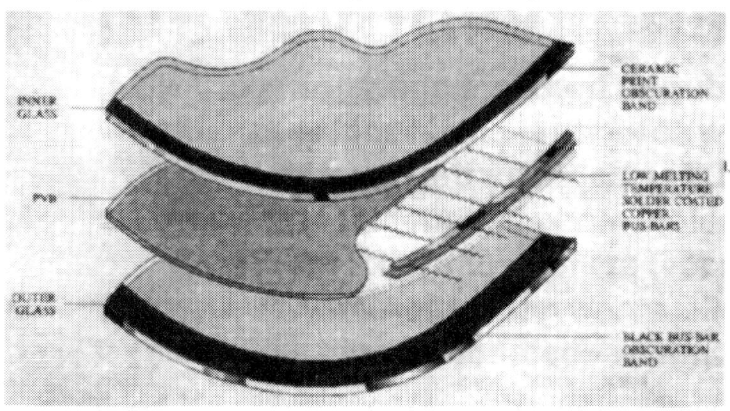

Fig 41: Electrically-heated laminated windscreen

on the windscreen's reflectivity which can be enhanced by reflective coatings, Fig 43. When a holographic element is used for enhancement, in the form of a hologram laminated in the windscreen, it possesses different focusing characteristics to the windscreen surface and has a different optical axis. In one system demonstrated, a CRT was used for the display, with a reflective optical projection system located in the dashboard area. The focusing mirror was aspheric and the instantaneous field of view of the HUD image was 14 degrees. The reflectivity of the holographic optical element is dependent on both light wavelength and the angle at which it is used.

The latest Range Rover's Triplex wind screen is a key feature of the vehicle's new look, being uniquely curved at either end to a minimum radius of curvature less than half that found in conventional vehicle windscreens. This was achieved by new glass shaping process developed by Pilkington Glass Products Developments. Land Rover wanted to reshape the windscreen to improve aerodynamic drag; however, the packaging limitations of the vehicle did not allow for a generous planform curvature. Conventional laminated windscreens are shaped by a sag bending technique where a pair of flat glass sheets are supported around the periphery on a mould. The glass and mould pass through several heating and cooling stages in a furnace and, by application of the appropriate heat distribution over time, the glass forms under gravity to the required shape. However, there are limitations on the complexity of shape that can be formed. The Range Rover windscreen has a relatively deep cross-curvature (from top edge to bottom edge), a small minimum radius towards the top of the wing and required close control of the edge shape to blend into the vehicle contours seamlessly.

The die-assisted concept was therefore conceived, whereby the windscreen would be sag bent close to the design shape, then the glass pushed with a die to control the additional shape required. The die introduces control over edge shape (fit-to-car), cross-curvature and minimum radius and pushes out the inverse curvature in the corners, which is a feature in many deeply curved sag bent glasses. This ensures a minimum radius of curvature of 230 mm compared with a figure of 500 mm on conventional windscreens, Fig 44. The new Range Rover also has the largest single curvature door glass of any volume production vehicle - 602 mm (h) x 946 mm (w) compared with 590 mm (h) x 605 mm (w), making it more than 1.5 times the size of the previous model's door glass — while its rear quarterlights incorporate a number of glass-based products including an aerial for the in-car entertainment and a sensor for the remote central locking system. Triplex achieved another new departure by undertaking complete sub-assembly work for Land Rover and supplying glass parts with bonded attachments. In addition, the glass in the new Range Rover model is 40% thinner than the glass used previously as many of the glazing developments centred around saving weight for improved performance and fuel economy.

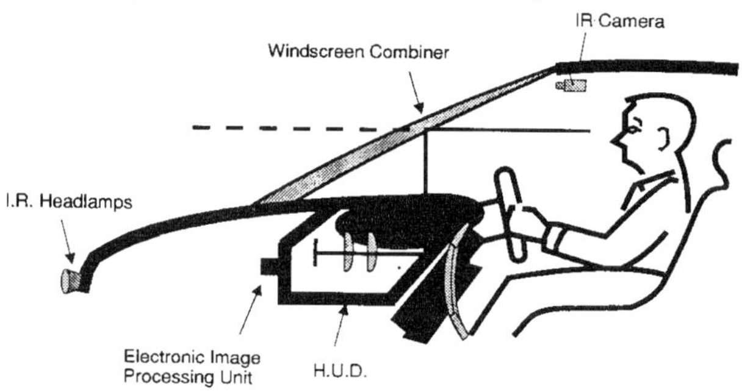

Fig 42: Vision enhancement system

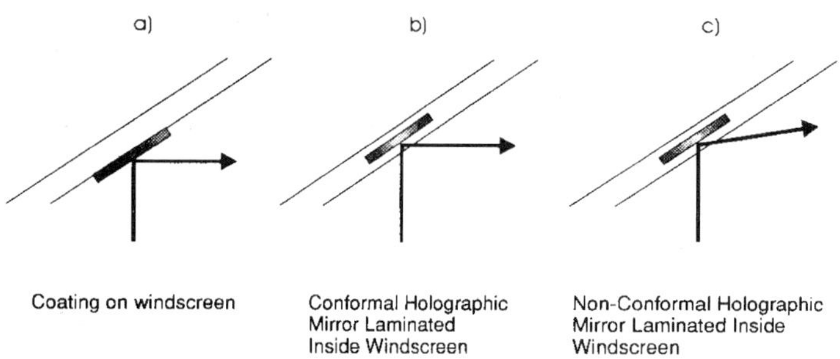

Fig 43: HUD combiner alternatives

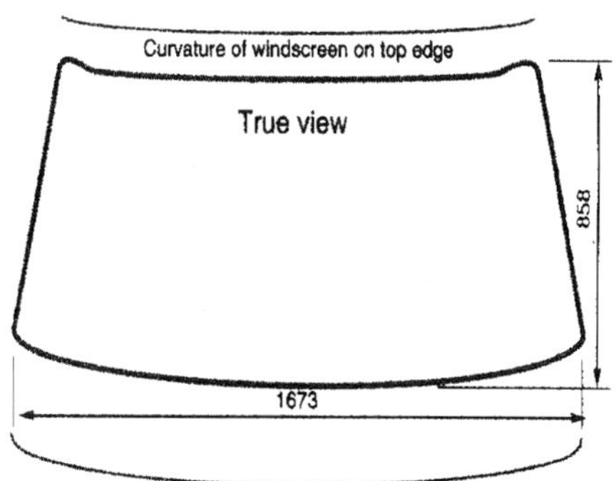

Fig 44: Range Rover windscreen

Refraction and Reflection.
1 Incident beam. 2 Refracted beam.
3 Reflected beam.

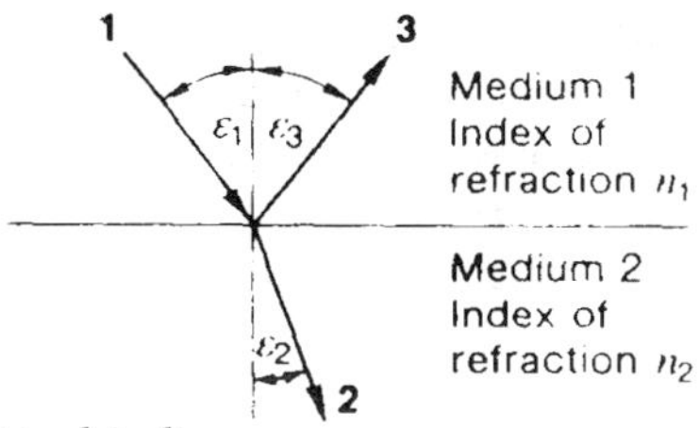

Fig 45: Optical indices

Cylindrical lens element of a headlamp lens.

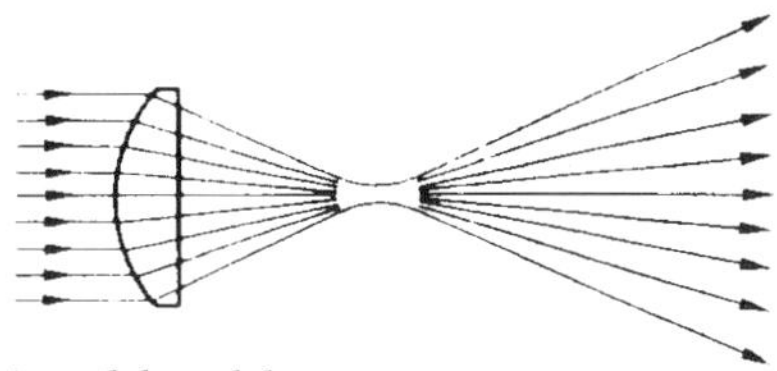

Fig 46: Conventional head-lamp

Parabolic reflector.
F Focal point, S Vertex, f Focal length

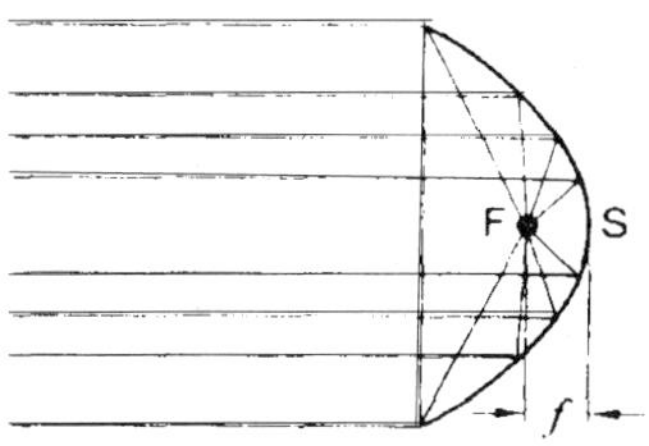

Fig 47: Parallel beam

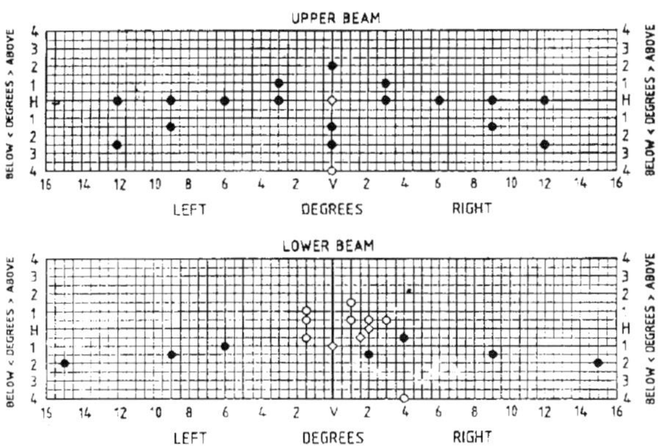

POINTS MARKED ○ DESIGNATE MAXIMUM CANDELA
POINTS MARKED ● DESIGNATE MINIMUM CANDELA
POINTS MARKED ◇ DESIGNATE MAX. & MIN. CANDELA

Fig 48: Beam assessment diagram

Lighting design/development

The specification of headlamp systems can be helped by an appreciation of the basic optics of lighting and an understanding of development trends past and present. Bosch have done much to disseminate knowledge of the theory, developed here, as well as being prominent in developing improved systems.

The optics of lighting has to be understood for a proper appreciation of existing and developing systems. Simple geometric laws can be used to trace the path of 'light rays' which are refracted as they pass from one medium (air) to another (glass). The law of refraction gives the relationship:

$$n_1 sin\ e_1 = n_2 sin\ e_2$$

where n_1 and n_2 are the indices of refraction, Fig 45. These are a function of the wavelength dispersion. The reflected ray is diverted from the surface at equal angles to the normal and the degree of reflection depends both on the angle of incidence and the refractive indices of the air and glass. Total reflection occurs if the angle of incidence is greater than or equal to the angle of total reflection. Degree of reflection varies from 70% for polished chrome to 89% for vacuum-deposited aluminium.

Conventional headlamp lenses comprise cylindrical elements, parallel rays being made to converge in a focal line, Fig 46. Other prismatic elements are also used to aid the process, to direct the rays into an acceptable beam. Incident parallel rays remain parallel after deflection by the prisms. From the 'point' light source, such as a bulb filament, rays emanate radially and are conventionally gathered by a parabolic reflector to be reflected as an approximately parallel beam. The point source is thus enlarged and its luminous intensity is a product of the light source luminance, the degree of reflection and the aperture area of the reflector, Fig 47.

As visible radiation sensitivity varies from person to person, a generalised function has been established to describe it. This is the spectral brightness response Φ. Luminous flux is the radiant power emitted by a source, assessed on the basis of Φ according to:

$$\Phi = K_m \int P_\lambda V\lambda . d\lambda$$

where K_m is known as the maximum luminous efficacy (683 lumens/watt). Values of F for a 2 degree field of vision are listed in European standards such as DIN 5031, Part 3. P_l is known as the spectral distribution of radiant energy.

Procedures for measuring photometric properties of automotive lamps are covered in SAE J1330. Other SAE standards cover headlamp beam switching, manual (J564) and automatic (J565). Performance requirements for headlamps is covered in J1383, with the system of beam assessment being shown as Fig 48. Conventional headlamp reflectors with focal length about one quarter of the diameter have provided a traditionally accepted solid angle. A reflector diameter D and source area A_s would cause an area illuminated at distance L of:

$$A_s(L/f)^2 \text{ or } A_s(4L/D)^2$$

and the intensity of illumination is proportional to:

$$f^2/L^2 \text{ or } D^2/16L^2$$

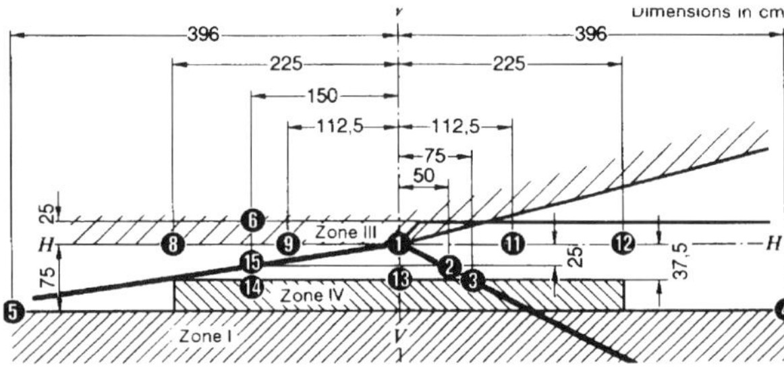

Lower beam				Upper beam	
Measurement points	Specified lx values	Measurement points	Specified lx values	Measurement points	Specified lx values
1. E_{HV}	≤ 0.7	Max. value in Zone I	≤ 2 E_{50R}	7. E_{max}	> 48 < 16 × E_{75R} (< 240)
2. E_{75R}	≥ 12	Max. value in Zone III	≤ 0.7	8. $E_{H-5.15°}$	≥ 6
3. E_{50R}	≥ 12	Min. value in Zone IV	≥ 3	9. $E_{H-2.55°}$	≥ 24
4. E_{25R}	≥ 2	13. E_{50V}	≥ 6	10. E_{HV}	≥ 0.8 E_{max}
5. E_{25L}	≥ 2	14. E_{50L}	≤ 15	11. $E_{H+2.55°}$	≥ 24
6. E_{B50L}	≤ 0.4	15. E_{75L}	≤ 12	12. $E_{H+5.15°}$	≥ 6

Fig 50: Measurement points for illuminance

Eye sensitivity

Fig 49 shows the relative spectral sensitivity $V(v)$ of the light-adapted eye. Other definitions which affect the physiology of vision are luminous energy, being the spectral radiation energy assessed on the basis of V_l; the luminous intensity, being the ratio of luminous flux to the irradiated solid angle; the illuminance, being the ratio of incident luminous flux to the area of illuminated surface; luminance, being the ratio of luminous intensity to the apparent projected area of the surface from which it is emitted; luminous emittance, being the ratio of the emitted luminous flux to the power absorbed and, finally, the solid

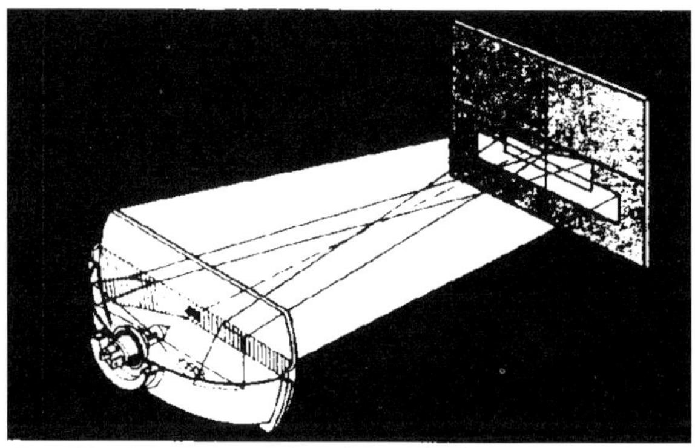

Fig 51: Asymmetric low-beam light

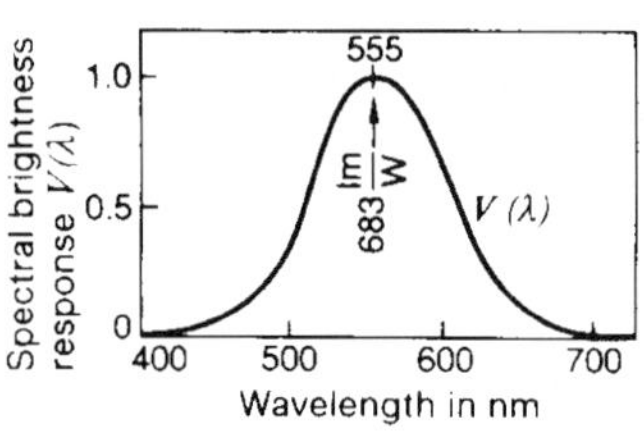

Fig 49: Spectral density of light-adapted eye

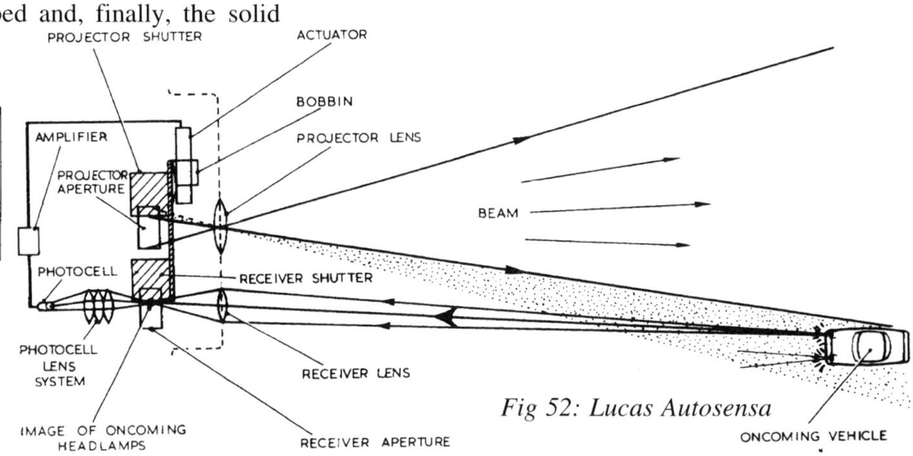

Fig 52: Lucas Autosensa

angle which is the ratio of the irradiated portion of the surface of a sphere (concentric with the light source) to the square of the sphere's radius. While a number of these may not be used in design, they are of value when checking the specifications of proprietary lamps which may be expressed in different ways.

It is also useful to differentiate between direct glare, experienced in the line of sight, and indirect glare seen at the extremities of the angle of sight. Luminance values above about 20 cd/cm^2 produce absolute glare from a source which cannot be reduced in its effect by additional illumination of its background. Good lighting depends on three eye sensitivities: that for discrimination, that for shape, and that for speed of perception. In the case of glare, one European country stipulates that glare is effectively eliminated if each headlamp generates no more than 0.7 lx, at headlamp height, at a distance of 25 metres. Specified values are given in Fig 50 which apply to a screen at that distance with point (1) lying in the headlamp axis.

Key parameters for conventional parabolic reflectors, Fig 47, are focal point F, vertex S and focal length *f*. The range of the lamp is defined as that distance at which the luminous flux has a specified value, say 1 lux. Then a geometric range is considered to be the distance of the horizontal component of the light/dark cut-off on the road surface with 1% low beam inclination. Typically focal lengths lie between 15 and 40 mm for reflectors whose forward projected areas are referred to as the 'illuminated area'. The lamps 'effective luminous flux' is that portion of the filament's flux which is beamed by the reflector, via the refractive lens, to the road surface. This will be higher for short focal lengths because the reflector converts more of the flux into a light beam.

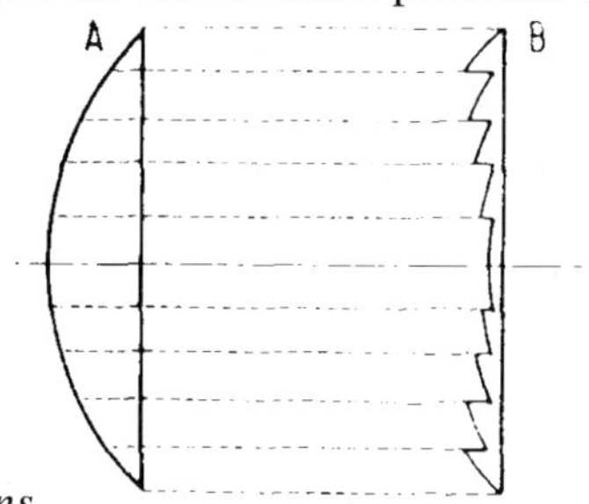

Fig 53: Fresnel lens

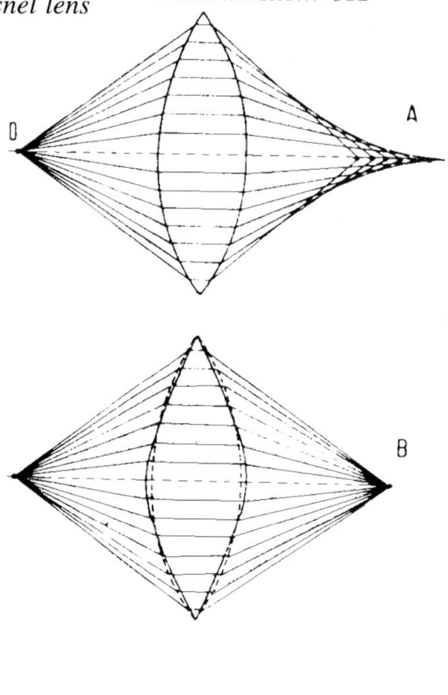

Fig 54: Overcoming aberration

With increased traffic densities high-beam driving is more of a rarity and thus much development has gone into improving low-beam performance. Asymmetric low-beam lamps have extended visual range to the near-side while halogen bulbs have substantially raised luminous intensity. Low-beam is conventionally achieved by the appropriate filament being ahead of the focal point while a cap keeps the lower area of the reflector unlit; the cap's edge defines the cut-off, Fig 51. Departures from convention have included stepped reflectors with parabaloid sections of different focal lengths. Another type is the Homofocal reflector in which different quadrants of the reflector have the same focal point but have differing focal lengths. These quadrants aid near-field and lateral illumination but do not increase range and its economic production was made possible by plastics moulding.

Another ingenious development was the Lucas Autosensa, Fig 52, in which a receiver and projector was combined in one unit. An amplified signal from a photocell operated a linear actuator to move a shutter across the aperture which 'homed' on the image of an approaching vehicle; a second shutter moved in unison to cut-off the off-side edge of the beam. The important contribution of the lens glass can be visualised by imagining the collapse of a conventional lens into a Fresnel type, Fig 53. It is possible to achieve such shapes in plastic that were hard to obtain in glass. Also it is possible to overcome the aberration effects of glass lenses with spherical curvature, Fig 54. The spherical lens cannot focus all incident rays at the same point, resulting in the aberration of the image, whereas a non spherical surface can be more easily moulded in plastic.

Recent developments

Complex Surface headlamps are a relatively new departure, in which, instead of using a parabolic reflector, Valeo have one which is developed by computer analysis of some 10,000 surface points, Fig 55. Luminous flux is increased by the more efficient use of reflector surface and some 80% more light output is claimed, in low-beam configuration, than for a conventional lamp of equivalent reflector surface area. The latest versions require no inner lens and so can use a clear outer glass.

Both Valeo and Bosch are also developing gas-discharge lamps using electric arcs in lieu of tungsten filaments. A small quartz bulb contains zenon under high pressure as well as mercury and solid-state halogenures. After an initial arc is struck in the zenon, on turning on the lamp, a high current is applied which evaporates the other constituents — increasing luminous flux. An initial over-voltage ignites the light quickly; thereafter a ballast device

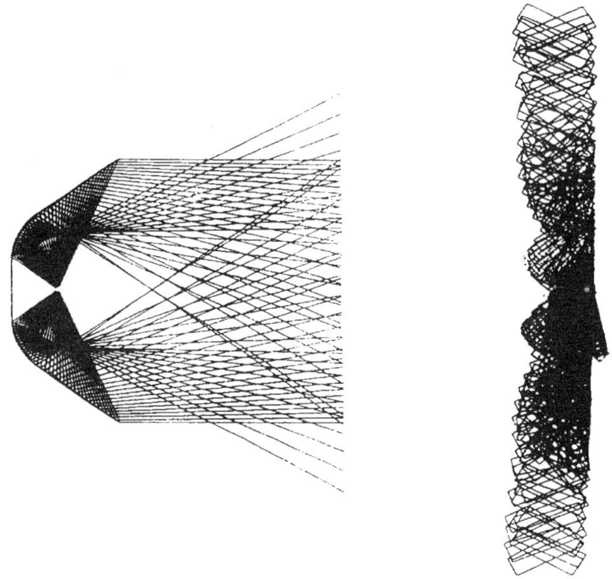

Fig 55: Complex surface system with beam pattern (below)

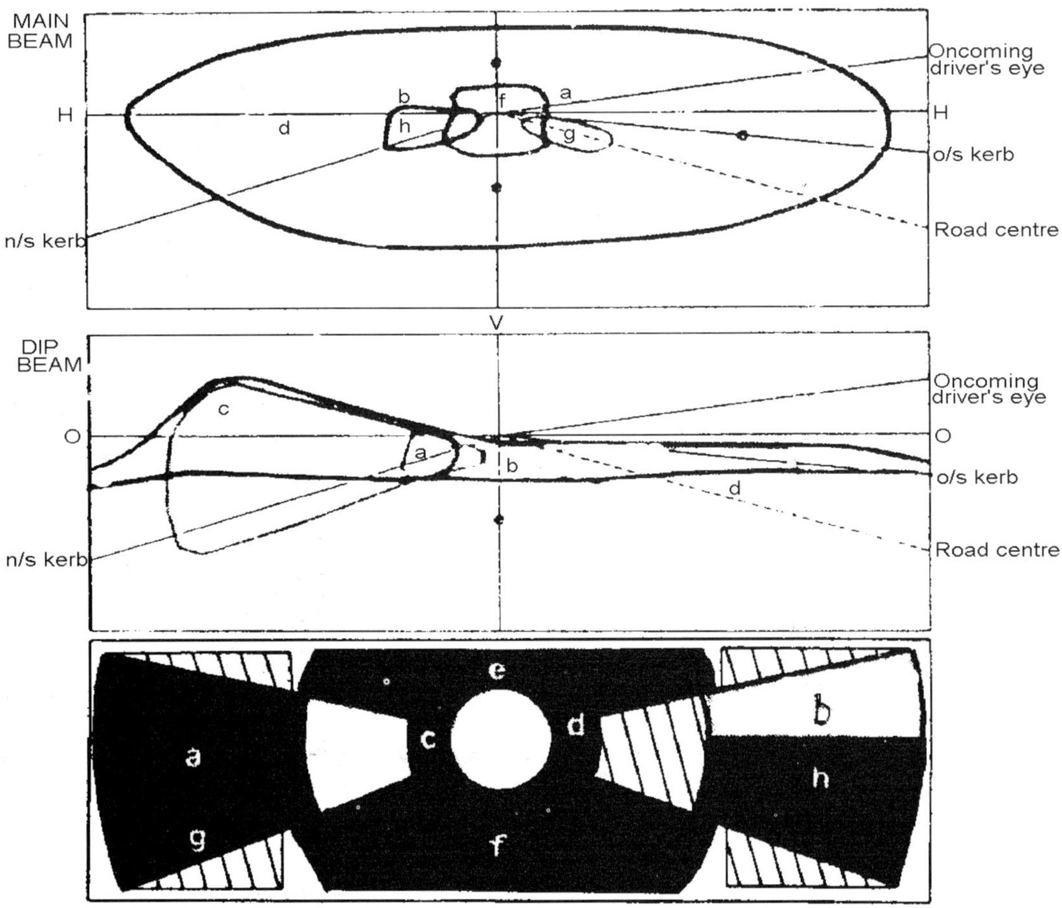

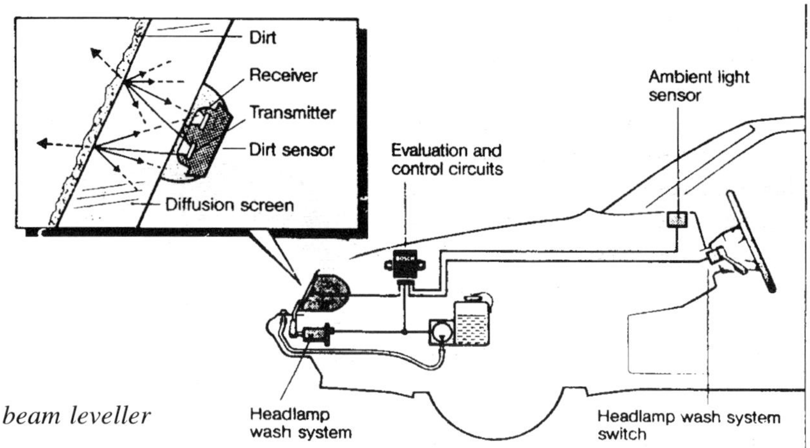

Fig 56: Automatic beam leveller

optimises power — reducing from the initial level of 75 W to a stable state of 35 W (initial strike voltage is 6-12 kV for the arc, made possible by a high step-up transformer and an AC supply). A doubling in field of vision is claimed over comparable lamps with a conventional halogen bulb of 55-65 W. The gas-discharge lamp is also 2.5 times brighter and gives a bluer light. Bulb life is also substantially enhanced with the prospects of lasting the whole vehicle life, for certain categories, by virtue of the lower power requirement and cooler running. About 2-3 seconds will be required to reach full luminosity for the automotive versions — compared with the one minute for street lamps of this type. First applications are expected on performance cars using four-headlamp systems. They will be used as low-beam lamps in pair with conventional high beam ones. The higher intensity beam would require an automatic levelling device.

The Bosch automatic beam leveller is another significant advance; it uses servo-motors to keep beam-aim within controlled limits. Another likely requirement, for which Bosch has also produced a solution, is a means for automatic washing after dirt build-up on the lens. A sensor for this purpose uses infra-red to gauge the equivalent amount of white light reflected back inside a lamp, from a dirty lens, Fig 56. The company's polyellipsoidal PES headlamp for low profile applications, in which an aperture of only 60 mm diameter provides normal levels of distribution using intense focusing of light, has also been upgraded. The PES Plus has had its cutoff mask modified and additional reflector elements incorporated so that 50% greater light intensity has been achieved for the 520 metre (near) field of vision, Fig 57.

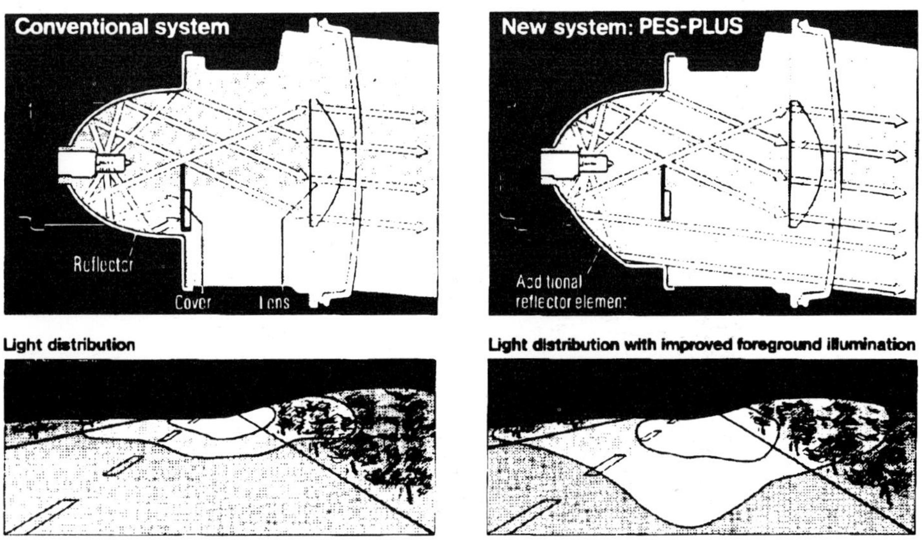

Fig 57: Conventional vs PES lamp

Electro magnetic compatibility

The successful use of vehicle on-board electrics and electronic controls depends upon the ability of individual systems to share the electro-magnetic spectrum of frequencies without degradation of performance. The systems also need protection from electromagnetic interference (EMI) from road-side and other sources of radio-frequencies.

According to RFI Shielding Ltd[3], the beginning of 1992 saw the introduction of European Directive 89/336/EEC on EMC compliance for electronic goods sold on the market and a legal stimulus is added to the need to ensure good protection. There is also the safety influence of strong EM fields possibly causing the malfunction of anti-lock braking or engine-management systems. Most electronic devices involve rapid changes in voltage and/or current levels with the resultant development of electro-magnetic energy over a range of frequencies. This will be radiated in the space surrounding the circuit in question and be conducted into the wiring which communicates with the circuit.

Fig 58 shows the effect of providing a metal barrier against the incoming wave, the latter having two main component fields, one electrical and one magnetic — each at right-angles to one another and the direction of travel. A degree of reflection occurs at each surface of the barrier. The effectiveness of the screening is often quoted in decibels — such that the reduction in strength of the emergent field to one tenth of the incident value would be given an effectiveness of 20 dB; see table in Fig 59.

Absorption loss depends on the thickness of the barrier as shown in Fig 60 depending on the

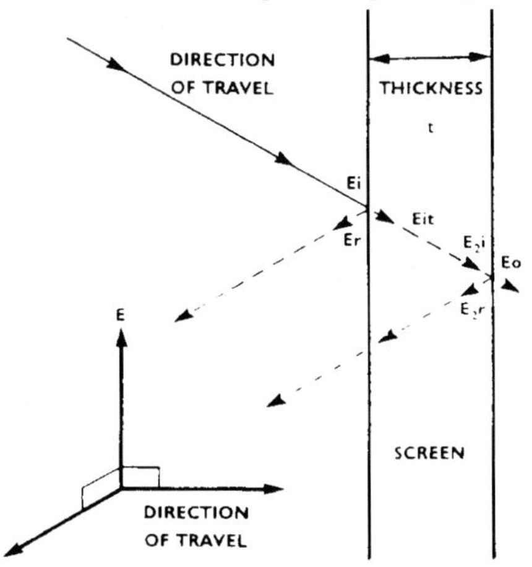

Fig 58: Reflection in metal barrier

S = SCREENING EFFECTIVENESS (dB)
A = Absorption loss (dB)
R = Reflection loss (dB)
B = Correction factor (dB) (for multiple reflections in thin screens)
S = A + R + B

dB	Percentage reduction
0	0
20	90
40	99
60	99.9
80	99.99
100	99.999, etc

Fig 59:Screen decibel values

A = $0.1315\,t\sqrt{f\,\sigma\,\mu}$ (dB)
t = screen thickness (mm)
f = frequency (Hz)
σ = conductivity relative to copper
μ = permeability relative to copper

Note: For screen thickness (t) in inches replace the constant 0.1315 with 3.34

Material	σ	μ
Copper	1.00	1
Aluminium	0.61	1
Brass	0.61	1
Tin	0.15	1
Steel (SAE 1045)	0.10	1000
Monel	0.04	1
Stainless steel	0.02	500
Electroless nickel	0.02	1

Fig 60: Thickness/permeability effect

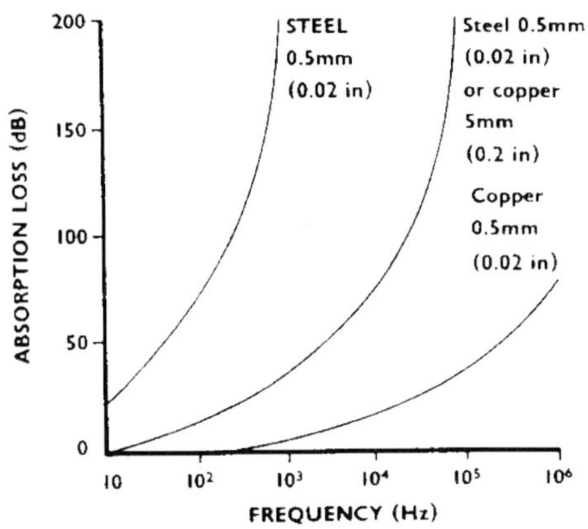

Fig 61: Frequency effect

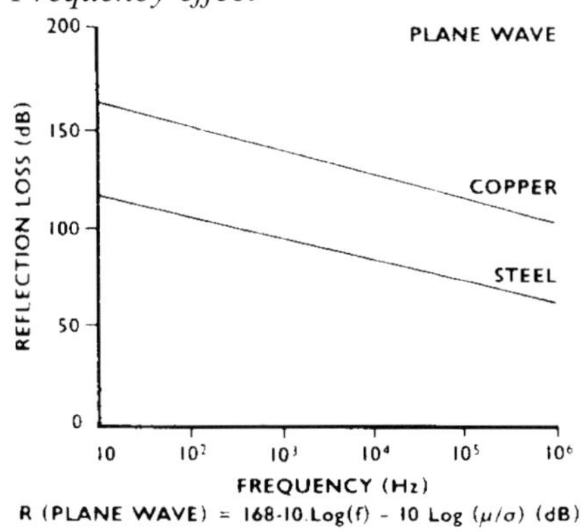

Fig 62: Total effectiveness

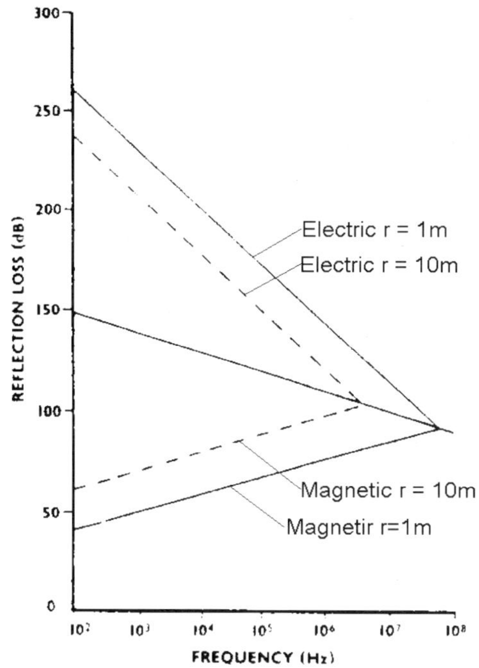

Fig 63: Straight antenna radiation (r is distance form source to screen m)

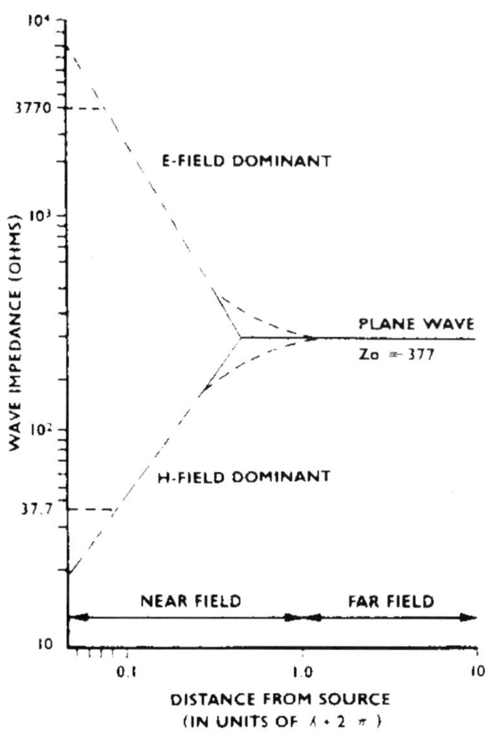

Fig 64: Screen impedance

conductivity and permeability of the materials from which it is made. Variation of absorption loss with frequency for two screening materials is shown in Fig 61. Reflection loss increases with ratio of impedance of the incident wave to the impedance of the screen material. Beyond a distance of one sixth of a wavelength from the source wave impedance is constant at about 377 ohms — while the impedance of the screen material is proportional to the square root of the frequency times the permeability divided by the conductivity.

Total effectiveness for a shield, allowing for absorption/reflection loss of plane waves, is seen in Fig 62. This refers to a copper 0.5 mm thick screen in the 'far-field' where the wave-front is plane.

Near and far fields

Below the one sixth wavelength distance from the source, 'near-field' conditions are said to exist. For a straight rod-antenna emitting radiation, wave impedance is high near the source, Fig 63. However, if the emitting antenna is a small wire loop, the field is predominantely magnetic and the wave impedance is low near the source, Fig 64.

For small screened devices, working in the near-field is the norm and formulae for calculation are shown in Fig 65. the correction factor B is normally only computed for the near-field case and then only if the absorption loss is less than 10 dB. Variation of reflection loss with distance and frequency, for a copper screen, is shown in Fig 66

REFLECTION LOSS IN THE NEAR FIELD

R (Electric) = 321.8 − 20.Log(r) − 30.Log(f) − 10.Log(μ/σ) (dB)

R (Magnetic) = 14.6 + 20.Log(r) + 10.Log(f) + 10.Log(σ/μ) (dB)

r = distance from source to screen (m)

f = frequency (Hz)

μ = permeability relative to copper

σ = conductivity relative to copper

CORRECTION FACTOR B

B = 20.Log(1-exp(−2.t/δ)) (dB)

t = screen thickness (mm)

δ = skin depth = 0.102 ÷ √(f.μ.σ) (mm)

For (t/δ) = 0.1, B = −15dB

= 0.5, = −4dB

= 1.0, = −1dB

Fig 65: Calculation of losses

indicating the closer the screen is to the source, reflection loss for electrical fields is higher and the reverse applies to magnetic fields.

EMC materials and construction At very low frequencies only high-permeability materials are effective for screening; cheaper materials are suitable for high-frequencies. Joints or holes in the screen will, of course, limit its effectiveness. By considering the hole in the screen as a wave-guide for the emitting radiation, its size can be optimised for screening purposes. If wavelength of the EM energy is too long relative to the gap dimension of the hole, little energy passes through and the 'cut-off' condition for the wave guide is said to to exist. Fig 67 shows formulae for cut-off frequencies in round and rectangular wave-guides as well as effectiveness indications at well below cut-off — a function of the length to diameter ratio.

An alternative way to obtain a good length to diameter ratio is to bond a small tube on the aperture for those cases where penetration of the enclosure is essential. These considerations are also used in the design of multi-hole perforated screens.

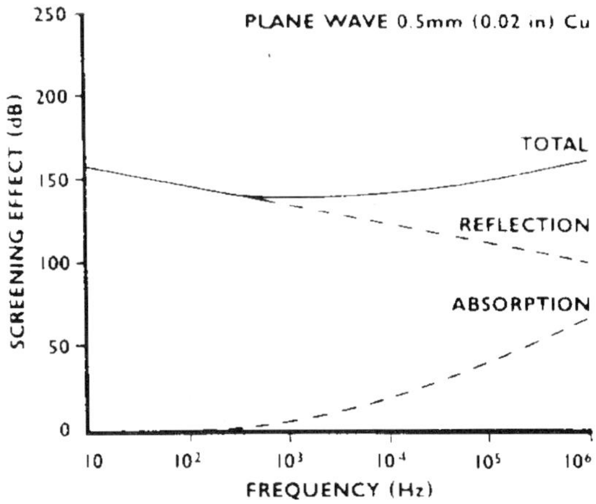

Fig 66: Loss variation with distance and frequency

WAVEGUIDE CUT-OFF FREQUENCY (fc)

In round guide, fc	= 175.26/d GHz (6.9/d in.)
d	= waveguide diameter (mm)
In rectangular guide, fc	= 149.86/a GHz (5.9/a in.)
a	= largest dimension of waveguide cross-section (mm)

SHIELDING EFFECTIVENESS (S) OF WAVEGUIDE

For operating frequencies well below cut-off

S (round)	= 32.t/d (dB)
S (rectangular)	= 27.2.t/a (dB)
t	= Screen thickness

LAT

Fig 67: Cut-off frequencies

Screen-wiper and door-latching mechanisms

The fitment of body mechanisms to vehicles has sometimes been regarded almost as an afterthought within the design process — and understandably there has been a desire to utilise where possible proven products from past applications. Often these units are designed for a restrictive specification imposed by a particular vehicle and their perfomance may be a poor compromise for the application. As computer resources have made possible a simplification of hitherto lengthy design procedures, it is well worthwhile considering a wide choice of drive linkages and using a mechanism design package to optimise the choice for a particular application.

International safety standards are having a profound effect on the configuration of window and door mechanisms. Among others, FMVSS and ECE legislators lay down rules for wiped areas and wiper speeds, in the case of screen-wipers, also fully and secondary latch positions, with integrity under 10/12 kN transverse/longitudinal loads, for door-latches.

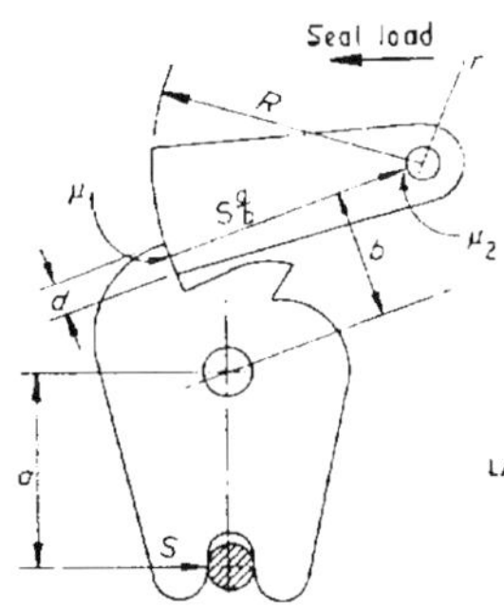

Fig 68: Latching mechanism

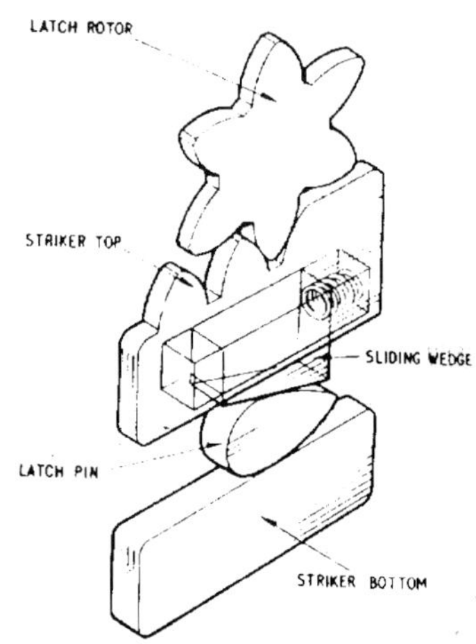

Fig 69: Rotating bolt latch

Established design for a latching device is shown in Fig 68. Rotating bolt latches such as the Wilmot Breeden example shown in Fig 69 solve the problem with sliding latches, of overcoming the frictional force on the slider created by the door seals. Effort for withdrawing the pawl of this claw latch is:

$$(dsa/b)m_1 + m_2r/R)$$

where $m_{1/2}$ are the friction co-efficients at the positions shown. Sometimes a mechanical stop, or detent is applied to the rotating element — in this case a striker pin is engaged with the arm of the claw in the closed position. Further complication of the basic linkage results from the need to hold the door rattle-free over rough terrain where forces of about 50 lbf (1.1 kN) can occur between door and pillar due to body torsional deflection. Means for resisting this separation range between moving-wedge devices to elastically deforming members. The first type is illustrated in Fig 69 which a sprung wedge moves in to take up any gap between door and pillar members. Shown in Fig 70 is the total latch and lock assembly in plant-on form, with staple which forms the striker.

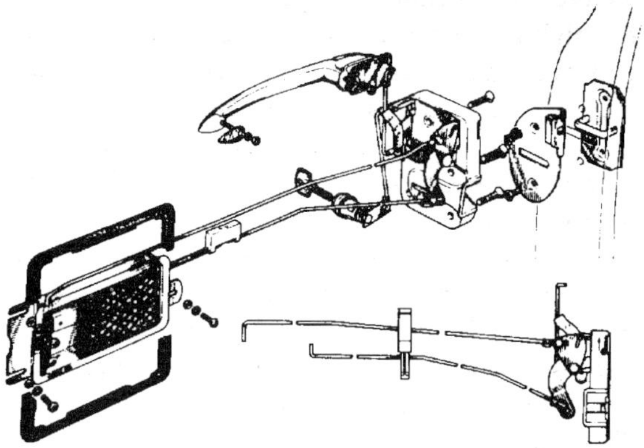

Fig 70: Latch and lock assembly

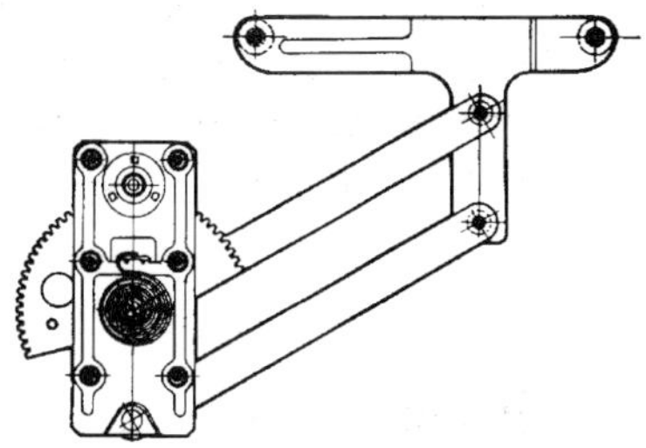

Fig 71: Arm-and-segment regulator

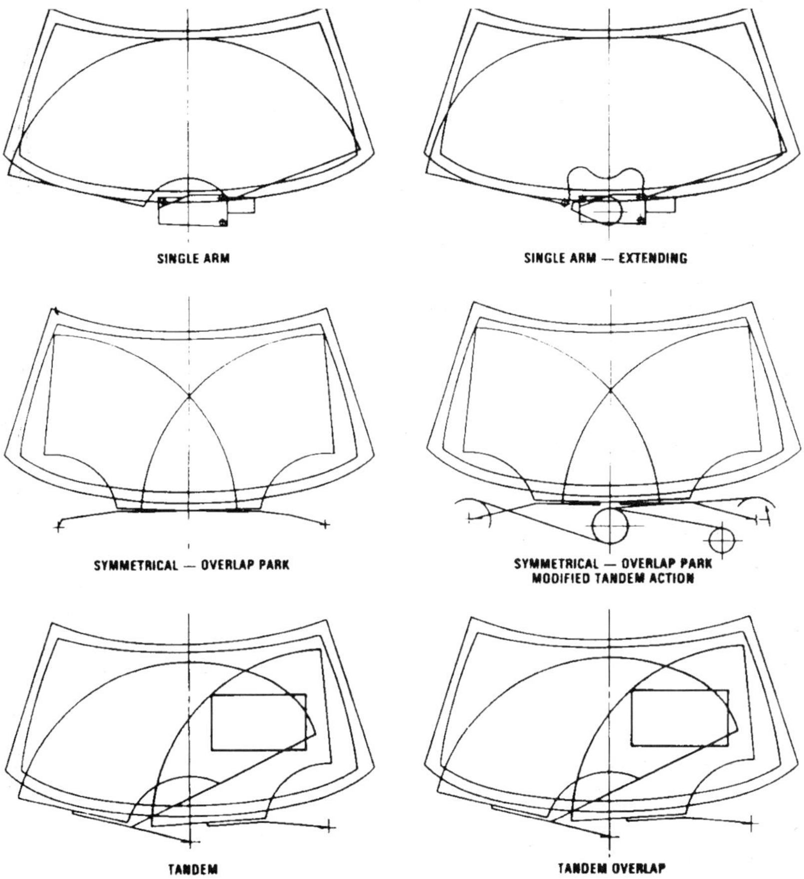

Fig 72: Screenwiper linkage variations

Window-regulators/screen-wipers

In window regulators, the arm and segment type (Fig 71) was the traditional pattern for volume cars and trucks; in recent years there is evidence of a switch to the use of flexible plastic perforated tapes which act as racks to engage pinion gears on the winder. With the first type, a number of multi-arm variants are available for extra large and curved glasses.

Windscreen wiper mechanisms are governed by wiper-area patterns quoted in such standards as SAE J903A and J941. Some wiper patterns, available from Delco Dynamics linkages, are shown in Fig 72. Linkage types, like that in Fig 73 from Wilmot Breeden, are said to be more efficient than the flexible rack type but the latter is less demanding in space. A number of software packages on mechanism linkage design particularly suit the screenwiper. One due to Singh of Ford (USA) used the model in Fig 74 to represent the two basic four-bar linkages of the mechanism, for obtaining the characteristics shown in Fig 75 — which can be analysed in respect of proneness to stall, unduly high velocity/ acceleration peaks or poor load distribution on the motor between forward and return strokes.

Synthesis of mechanisms

The rational rather than intuitive design of mechanisms involves both finding the best type of mechanism and the dimensions of the mechanism. Elements of a linkage which connect together are called kinematic pairs, as exemplified by those in Fig 76. So-called lower and higher pairs are those with less or more than two degrees of freedom while 'mobility' of a mechanism is the residual number of degrees of freedom which can be found from:

$$3(n-1) - j_h - 2j_i$$

where j_h and j_i are the numbers of higher and lower pairs respectively and n the number of links. This will govern the way a number of links can be put together to form different linkages.

A kinematic chain is a combination of kinematic pairs in which each link forms part of two pairs and in which relative motion is completely constrained. Fig 77 shows three arrangements in which each link forms part of two turning pairs, so that the first part of the definition of a kinematic chain is satisfied by each arrangement. At (a), a rigid frame is formed, with no relative motion between links, whereas in (b) relative motion is completely constrained. However, if five bars are connected together

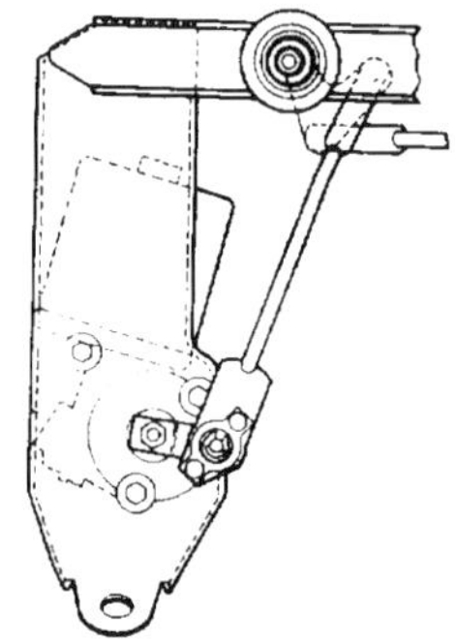

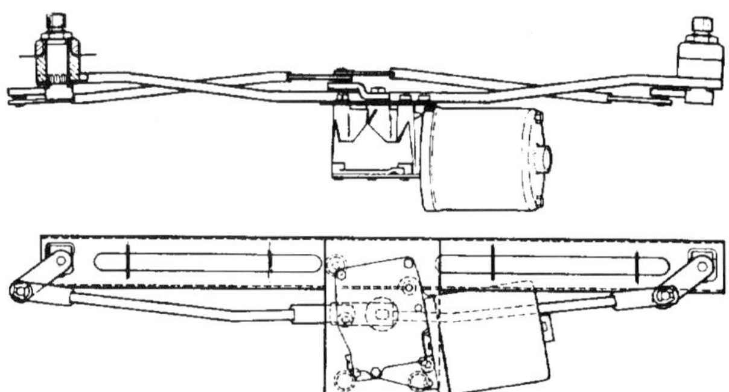

Fig 73: Typical wiper mechanism

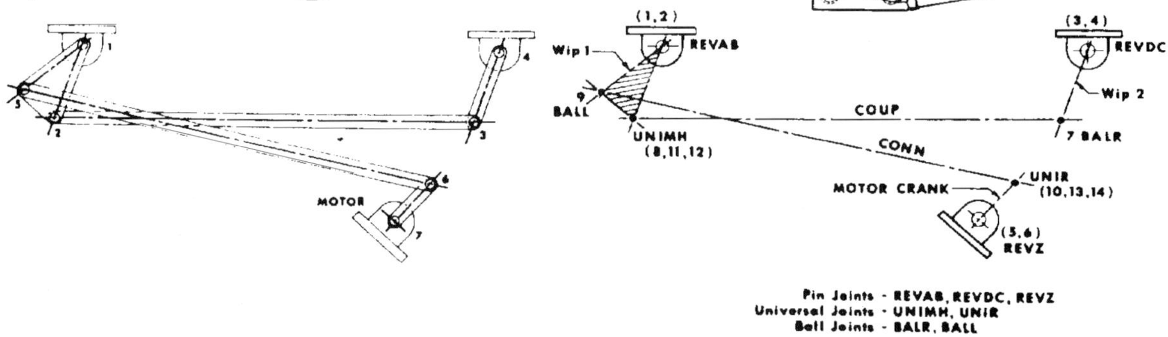

Fig 74: Mechanism modelling

Equivalent KAM model

by pin-joints, as at (c), unpredictable motion takes place. To build a more complex kinematic chain from that shown in (b), two more links may be added provided these form pairs with two existing ones — as well as themselves forming a pair, (d). Proceeding in this way, complex linkages can be constructed, for which the following general relationship applies:

$$n = 2j - 4 \quad or \quad n = (2/3)[j + 2]$$

provided only lower pairs are involved.

Strictly the word 'mechanism' should only be used when one link of a kinematic chain is fixed and the arrangement is used for transmitting or transferring motion. If the number of links is n, the same number of mechanisms can be made by fixing each of the links in turn, each being an 'inversion' of the original chain. When the mechanism has links strong enough to do work as well as just transmit motion, it becomes a 'machine'.

Straight-line mechanisms It is often a requirement to design a mechanism to produce straight-line motion and those without sliding pairs are usually the most useful for automotive purposes. The familiar Watts linkage of Fig 78 has the smallest number of links and pivots of any with only turning pairs. It has the disadvantage of a limited range of motion and an alternative is Peaucellier's mechanism of Fig 79 which has eight links but produces greater relative output. It would also be theoretically possible to combine a Watts linkage with a pantograph mechanism, Fig 80, which has the effect of magnifying the primary output to an enlarged secondary output. Point (C) in the figure could be constrained to move in a straight line by the primary linkage. While such a system could operate satisfactorily at high speeds, wearing of the pivots, especially (A) and (B), affects accuracy quite quickly.

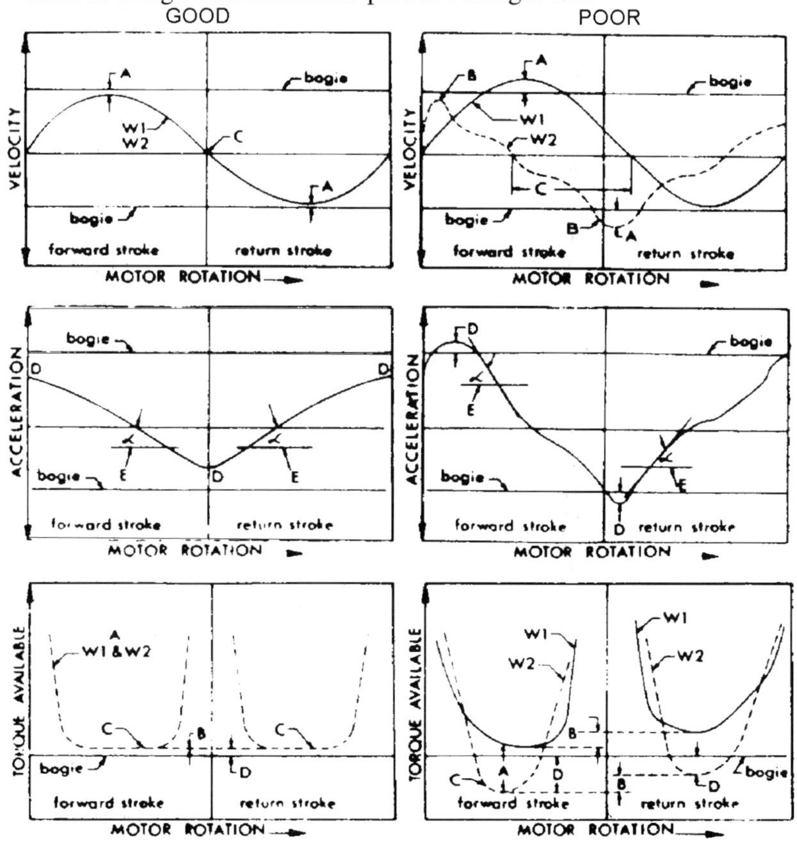

Fig 75: Simulated wiper performances

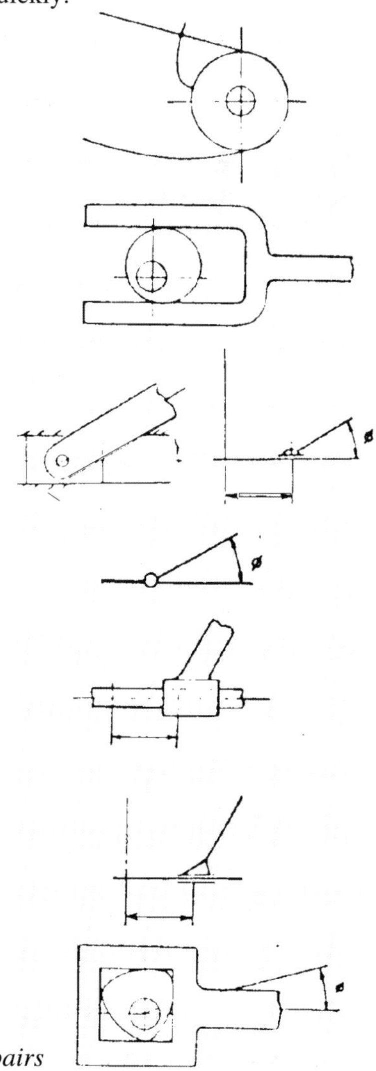

Fig 76: Kinematic pairs

Interior mechanism trends

The application of servo-assistance to body interior mechanisms is radically changing their configuration. There is also a trend for the larger suppliers to offer total system sub-assemblies, such as the door shell, which could lead, if not to integration of the mechanism, at least commonisation of the servo - power source.

The theory of mechanisms related to screen-wiper and door-latching systems was introduced in the last section. Two further important groups are seat-adjustment and window-winder systems, both of which are increasingly power-assisted as lower-cost electrical motors become more readily available. A typical manually adjustable seat mechanism for commercial vehicles uses a parallelogram linkage. At rest, this is locked by a crank which drops into slots in the frame. A tension-spring biasses the mechanism against the weight of the driver and a torsion spring holds the crank in the chosen slot. For cars, a conventional arrangement, the traditional slide adjuster, has been improved by the use of nylon-rollers and friction pads and a further refinement has the lever pivoted directly on the moving member. Now complex servo-adjust systems are employed, Fig 81.

For screen-wiping, the scissor linkage discussed in the earlier section is largely superseded by flexible rack and pinion designs, even on commercial vehicles. A commonly adopted standard for design is the SAE J 903 regulation shown in Figs 82/83, which includes requirements for 2-speed wiping. Fig 84

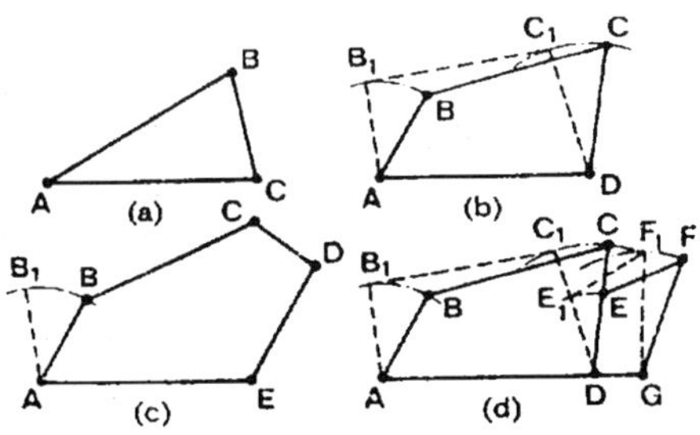

Fig 77: Kinematic chains

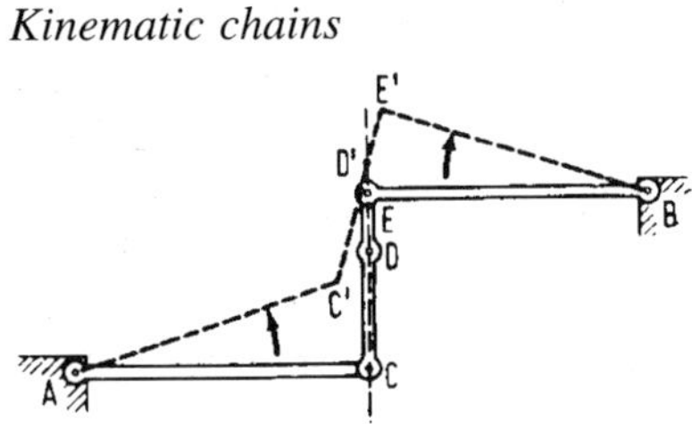

Fig 78: Watts linkage

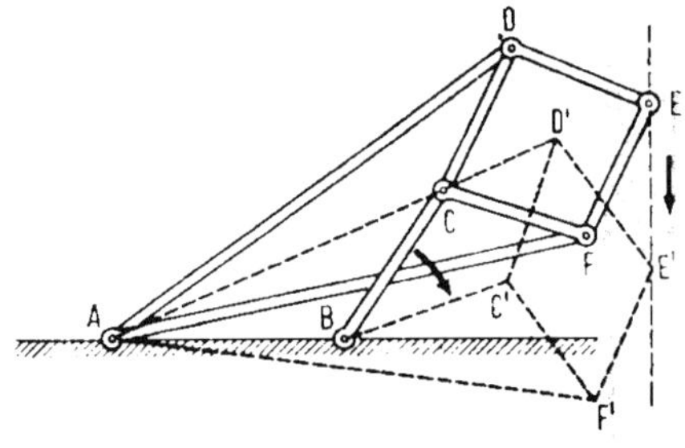

Fig 79: Peaucellier's mechanism

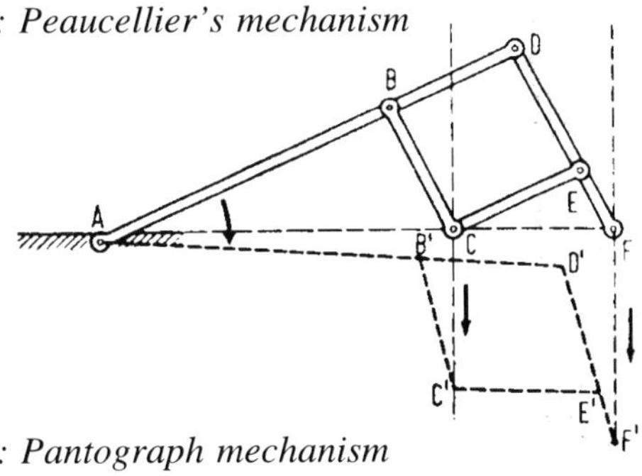

Fig 80: Pantograph mechanism

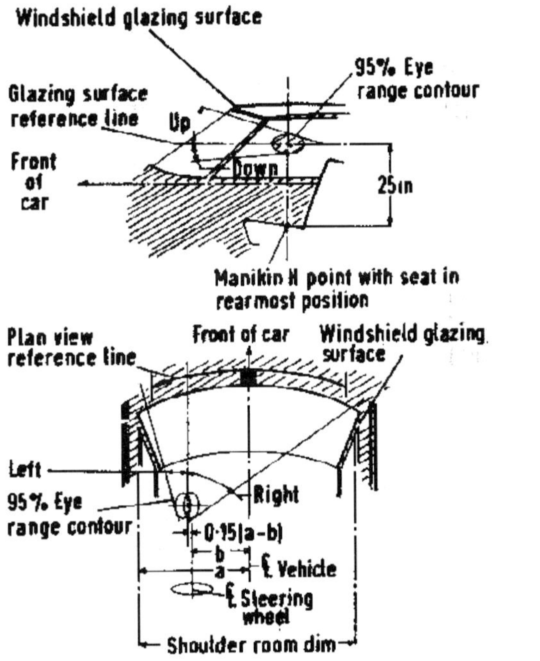

Fig 82: Notation for SAE specification

Area	Area wiped min, %	Eye range contour, %	Angles, deg			
			Left (W)	Right (X)	Up (Y)	Down (Z)
A	80	95	18	56	10	5
B	95	95	14	53	5	3
C	100	95	10	16	5	1

Fig 83: SAE specification

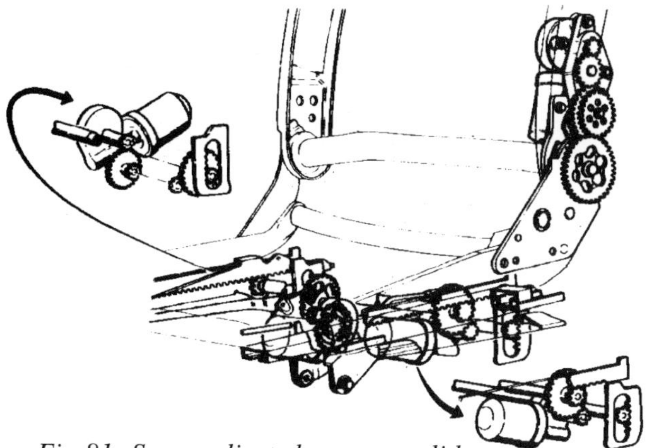

Fig 81: Servo-adjusted car seat slide

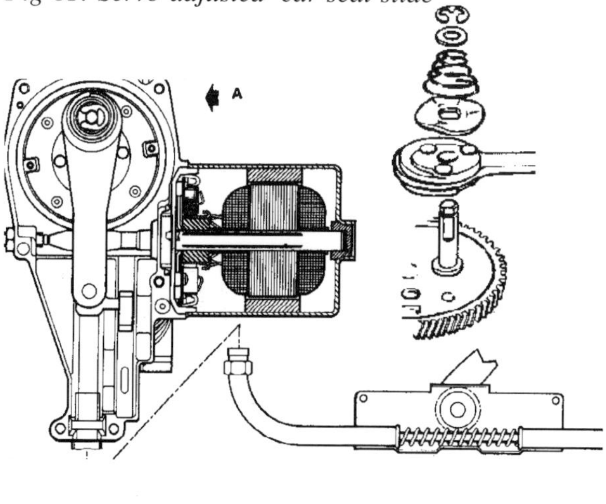

Fig 84: Flexible rack wiper drive

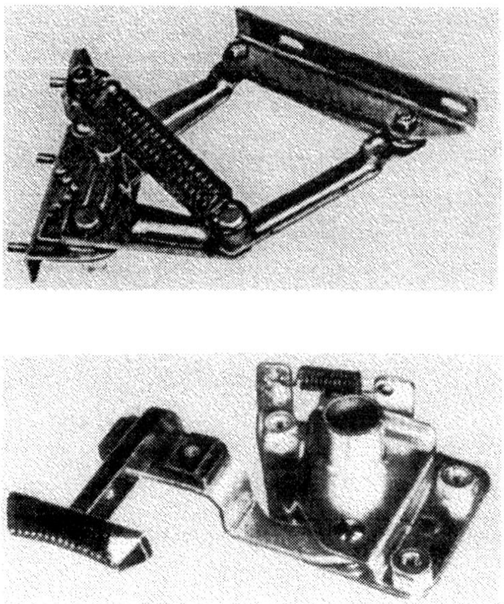

Fig 85: Bonnet hingeing and locking

shows the classic flexible-rack system which gains over the scissor-linkage in compactness. Drive from the permanent magnet motor is stepped down by about 40:1 through a plastic gear-set. The connecting rod big-end embraces a loose eccentric, frictionally connected to the crank and permitted to turn through 180 degrees; see inset. The motor is then made reversible so that in normal operation the pin travel corresponds to symmetrical blade angles. Reversal of the motor causes the eccentric to rotate and displace the travel of the pin so that the wipe is asymmetrical, and brings the blade to the edge of the screen at the limit of travel. At the same time a crosshead causes a plunger to actuate a limit switch and stop the motor, which is quickly brought to rest by the braking effect of the overrun voltage as the armature brushes are shorted out.

Bonnet hinging and locking mechanisms are typified by the systems shown in Fig 85, an assisted-lift geared hinge and a secondary latching arrangement. The former uses a 5-bar linkage to bias the bonnet into the open position, two of the links being geared together. A pawl with torsion spring is seen in the latter, access to which is usually through the radiator grille. The traditional window-regulator mechanism, exemplified by the system shown in Fig 86 for open cars is operated by a double arm geared regulator. Moulded nylon guides at A have upper and lower bearing surfaces which run in the window drop channels, B. Upper bearing surfaces C contact specially shaped upper sections of the channels, causing lateral pressure to tilt the glass inwards in the raised position. The lower bearing surfaces A raise the glass while the strip D, connecting the regulator arms, maintains the lower edge of the glass horizontal. In an electric window regulator, the mechanism consists of a lifting screw and nut with a scissor-type linkage attached to the latter. The screw is driven by a reversible electric motor through a lost-motion cou-

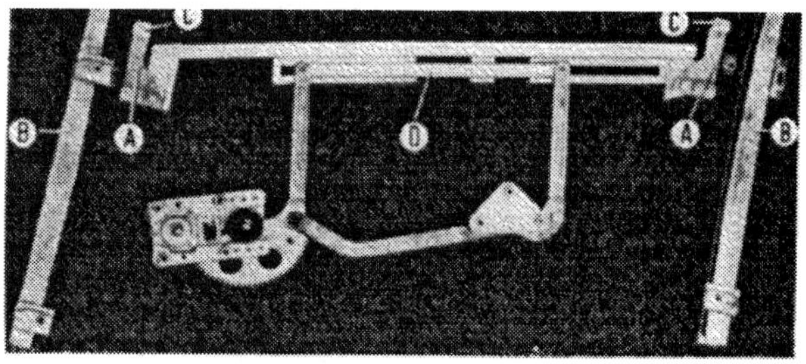

Fig 86: Manual winder

pling. A protective device is usually specified which prevents the motor windings from overheating. Fig 87 shows an alternative hydraulic drive system.

Fluid-power servo assistance can also be applied to door-hinging mechanisms, as seen by the Dunlop system shown as Fig 88. In this case pneumatic power is used in a system of inflatable sacs which act against vanes. Nylon pads shown at A, inset, prevent wear due to sliding between vane and sac. A zed linkage has also been considered which will increase door-opening from 90 to 180 degrees. Measuring 8 x 3 x 3 in (203x76x76 mm), air displacement for the powered hinge is 19 in^3 ($31.14x10^{-5} m^3$).

Seat-reclining mechanisms are the speciality of the PL Porter Company. The Sedgelok system comprises an incremental recliner, Fig 89, having an upper hinge arm which incorporates a gear segment. In the locked position this mates with a slide which prevents the rotation of the upper arm about its pivot. The lever which actuates the locking slide is spring-returned so as to have a cam action when the slide is loaded, thereby increasing locking force in response to seat back loading. Another neat way of achieving locking at different positions of angular displacement is available from the company in the form of the Mechanical Lock system, Fig 90. Here, a steel rod is gripped by two clutch coils which bear against specially configured holding bushings swaged into a surrounding tubular housing. These either have an angled coil-contact face or a flat which allows slipping action at a predictable force input. Inboard ends of the clutch coils are positioned against notches in the actuation lever so that rotation against the coil helix expands their diameter so as to allow infinite angular positioning for subsequent relocking. Another variant of it is the Electro-Hingelok of Fig 91 which can be used to power-drive a recliner mechanism, Fig 92.

Delco Products were pioneers in the development of a number of powered body-interior mechanisms for Europe following their long experience in the US market. Central locking systems based on the compact lock-actuator of Fig 93 can be used with single and multiple control points as seen in Fig 94. Direct drive actuators are available for doors, tailgate, boot and fuel-filler flap. An input at the door 'switch' causes the relay timer to be triggered. Fail-safe features built-in ensure the system does not respond to power supply interruption, open circuiting of the

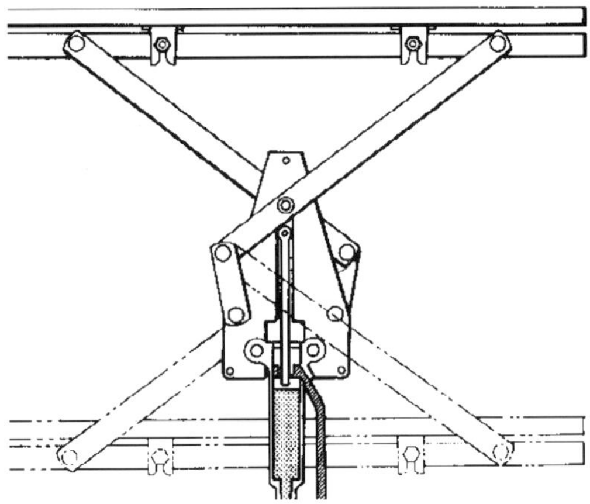

Fig 87: Powered winder

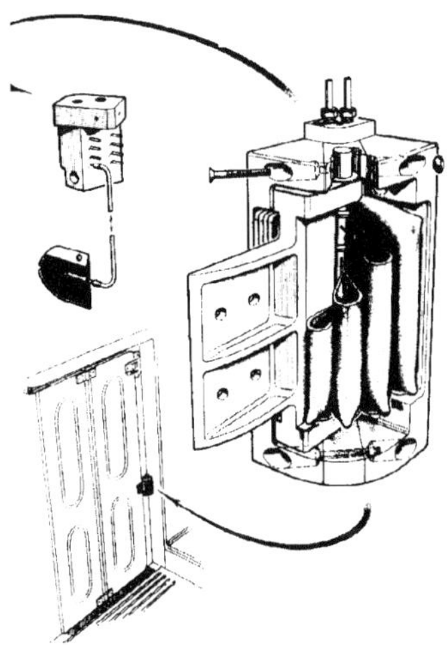

Fig 88: Powered door hinge

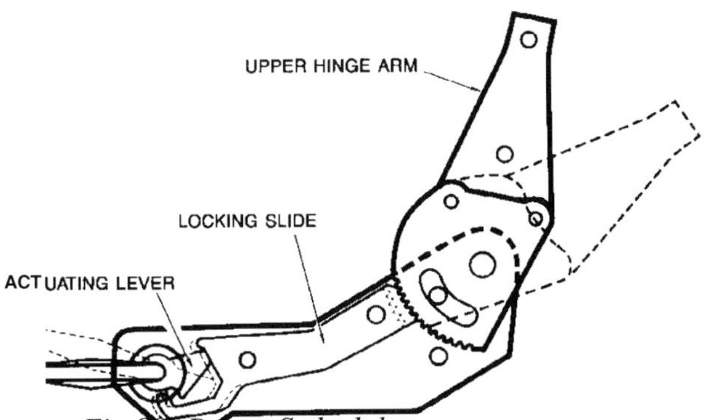

Fig 89: Porter Sedgelok

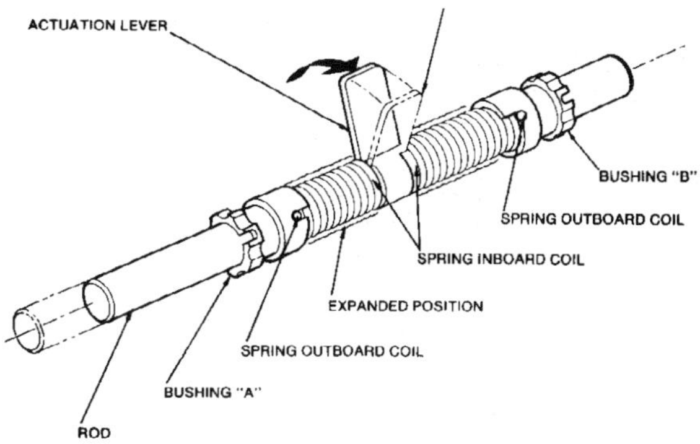

Fig 90: Mechanical lock

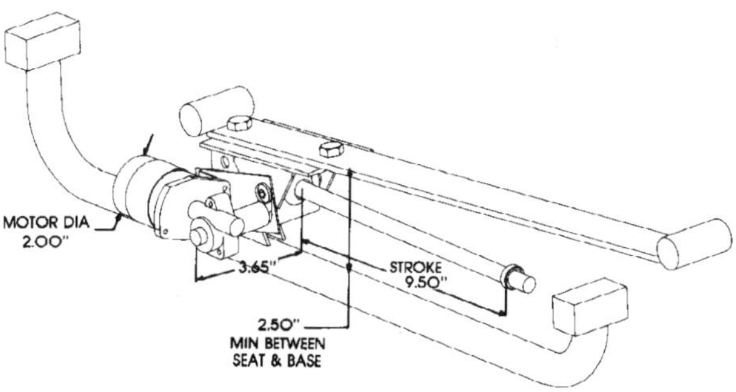

Fig 91: Electro-Hingelok

input switch or high voltage transients. It is also proof against internal operation in the event of a break-in through the windows. The company can also supply a modular door system containing the central locking facility — as well as power window lift and air-flow valve, all on the mounting panel shown as Fig 95. Even window channels, glass, seals, latch mechanisms and audio speakers can be supplied as a total pre-assembled door package.

Adjustable mirrors are themselves giving rise to some interesting mechanism designs. The one in Fig 96 by Magnetex pioneered the compact through-door configurations. Its range of adjustment is 20 degrees vertical and 35 horizontal, allowing the same mechanism to be used on either side of the vehicle. The smaller of the two adjustment knobs controls horizontal inclination, effected by rotation of a captive spigot — the internally threaded bore of which engages a corresponding thread on the acetal actuating shaft which tilts the mirror. Vertical inclination by the larger knob, held captive between a static spacer and the smaller one, rotates to move a screw-mounted plastic sleeve. This motion is further controlled by a keyway in the spacer to tilt the mirror vertically.

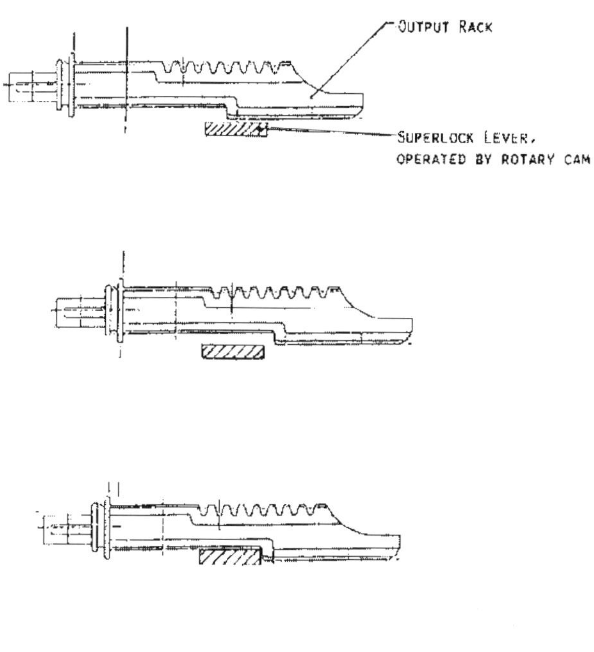

Fig 93: Lock-actuator: top, unlock; middle, lock; bottom, super-lock

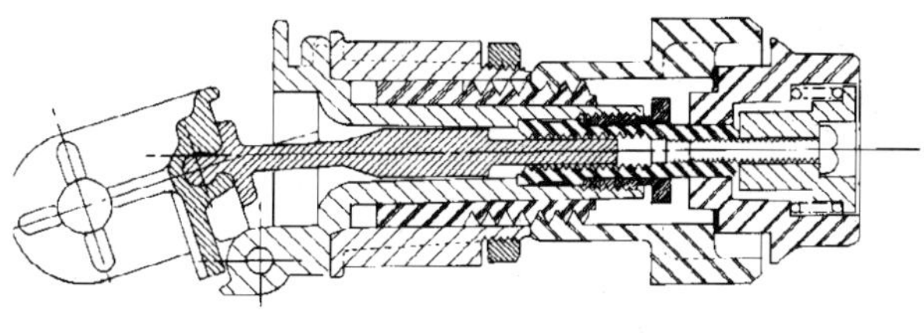

Fig 96: Mirror adjuster

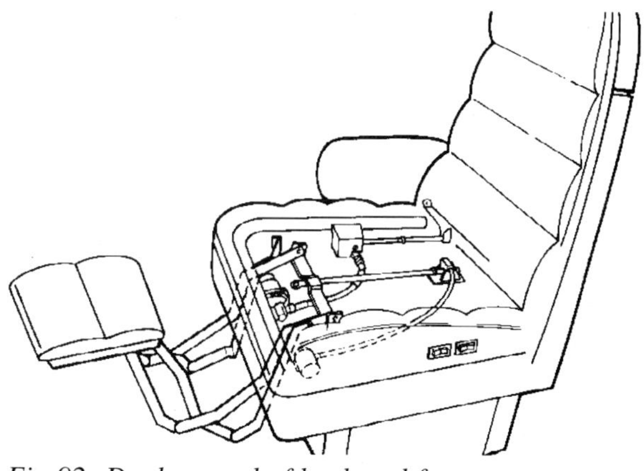

Fig 92: Dual-control of back and foot rests

Integrated door latch, lock and auto-closing system Robert Bosch recently revealed details of their electronically-controlled automatic door closing, Fig 97, and security, Fig 98, system. Only one actuator motor is involved and the functions of the device are integrated into one system. In an effort to minimise the number of mechanical parts, for closing and locking, only motor, gearbox, driving/stop plates, guide mechanism and ratchet are required; door slider and lock slider are located on one bracket. Door position is seen by the displacement of the drop latch, monitored by two proximity switches, while that of the driving plate is seen by two Hall-effect sensors.

Shown at Fig 99(a) are the door in open and the first ratchet positions; the auto-closing system is inoperative when the closing pin is introduced in the first ratchet position as this actuata the controller via a proximity switch. The ratchet then holds the drop-latch and prevents the door from springing open. The driving plate is connected to the gearbox output which turns the drop-latch into the closed position via a sprung stop-plate. The drive plate then completes one revolution and the door can then be mechanically opened by the exterior/interior handles.

A console switch signals the actuator, via the CAN bus, to enter the central locking mode. The driving plate is then motored to the door-locked

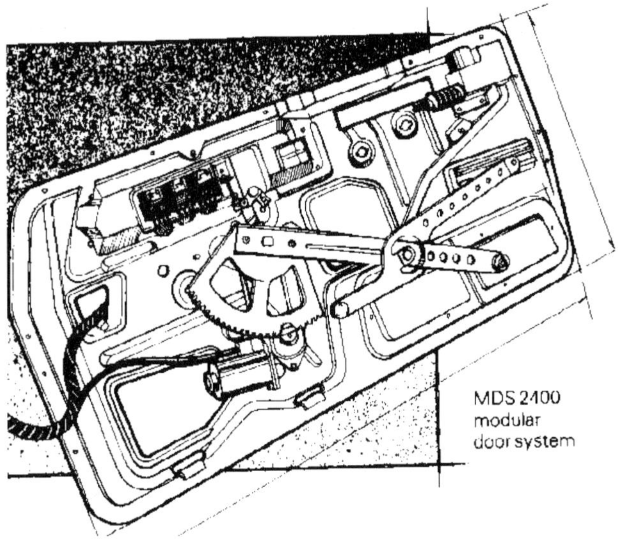

Fig 95: Delco modular door

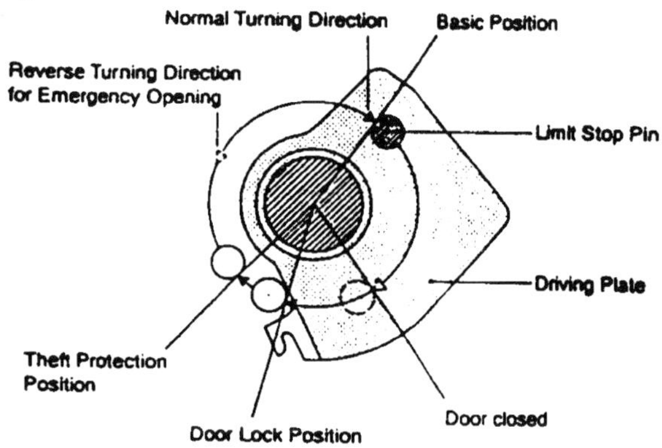

Fig 97: Door closing latch

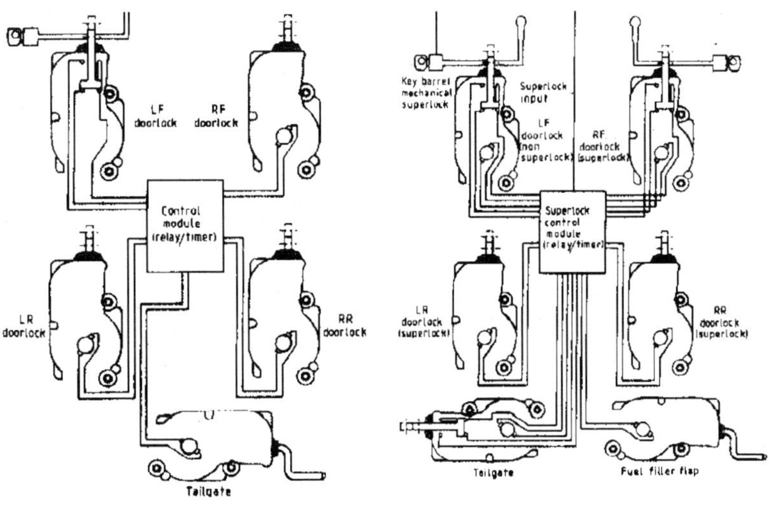

Fig 94: Delco central locking

position (b). The driving pin on the stop plate prevents the drop latch from being handle-opened from the outside - but interior handles will turn the stop plate against its spring with the slider. Turning the locking key signals the actuator to enter the anti-theft mode in which the driving plate takes up the position monitored by the the sensor; both inner and outer handles are then disconnected as seen in (c).

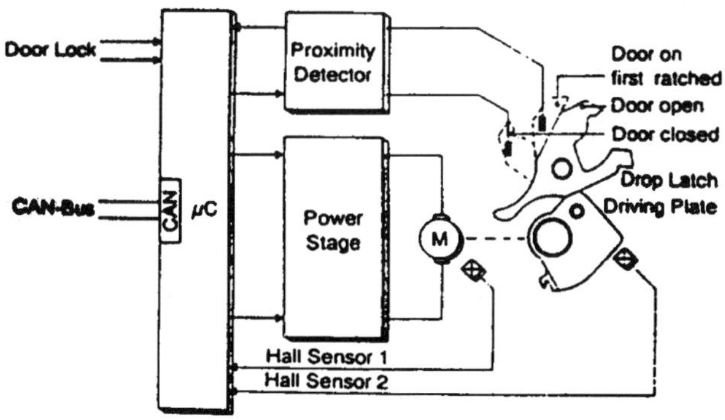

Fig 98: Security system

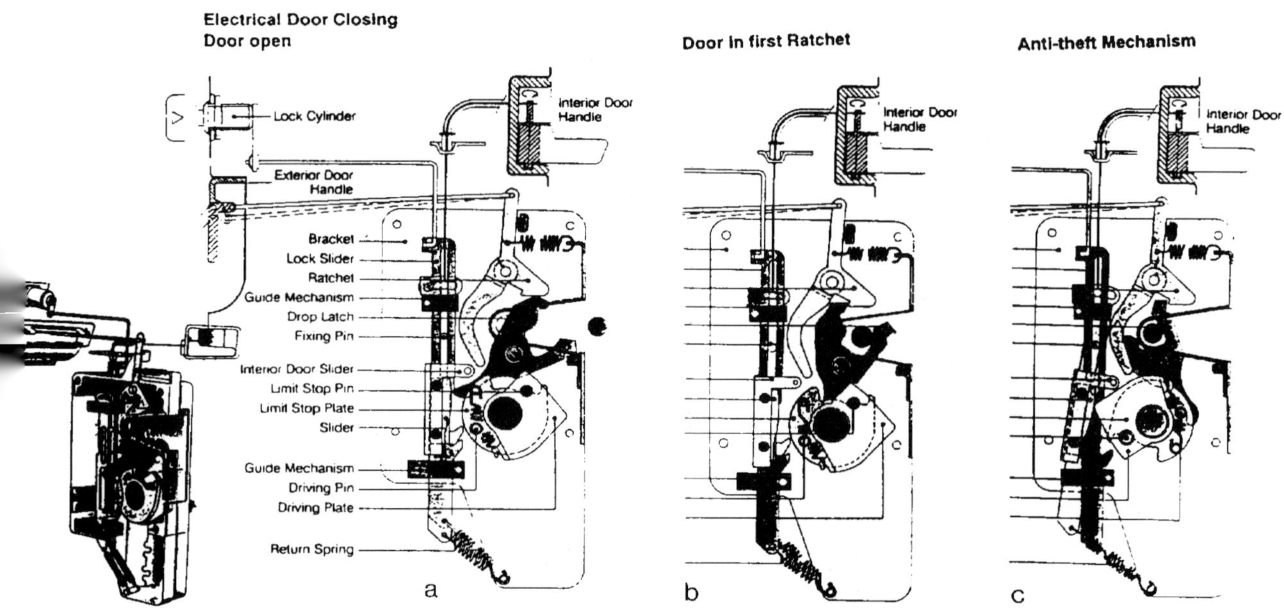

Fig 99: Bosch combined automatic door closing and security system

Chapter 5: Body electrical/electronic controls

Wiring developments; electric wiring standards; electronic control basics; engine/chassis electronic control systems; micro-electronics terminology and system developments; chassis and powertrain multiplexing; vehicle stability control; microprocessor problems and developments; multiplexing of chassis systems; Controller Area Network

The traditional mechanical engineering graduate route into professional automotive engineering has been under scrutiny in recent years with the growing quantity of vehicle electronic controls. While many courses have now been modified to take in greater modules of electrical engineering, some older engineers with a mechanical background might still find gaps in their knowledge.

Elementary texts cover the fundamentals of electrical resistance/capacitance/inductance, series/parallel circuits, heating effects of electric current and the basics of electro-magnetism. For the consideration of electronic controls, Washino[1] provides a valuable text, drawing on the considerable Japanese experience in this area and uses the table reproduced in Fig 1 to show some of the control parameters and detection variables involved. For a wider understanding of automotive elctrics and electronics the excellent company publications of Bosch and Texas Instruments[2,3] are particularly recommended.

Name	Controlled variable	Control input	Control logic	Detected variables and actuator
Fuel control	Air/Fuel ratio	Injected fuel	Optimal control (Open)	Air flow Fuel injector
EGR control	EGR rate	Valve opening	ditto	Valve position EGR valve
Spark timing control	Spark timing	Primary current	ditto	Crank angle Oscillation of cylinder pressure
Idle speed control	Idle speed	Air flow rate (current)	PI LQI	Engine speed ISC valve
Cruise control	Vehicle speed	ditto	PI control	Vehicle speed Throttle actuator
Transmission	Gear ratio	Hydraulic pressure (current)	Open loop	Vehicle speed MAP Control valve
4WD	Torque distribution ratio	ditto	ditto	Engine speed Steering wheel angle Control valve
4WS	Wheel angle	Stepping motor rev. angle	PI Open	Vehicle speed Wheel angle Stepping motor
ABS	Slip ratio	Hydraulic, pressure (current)	PI LQI control	Vehicle speed Wheel speed Control valve

Fig 1: Automotive control systems

Wiring systems

The electrical wiring harness seems, in the main, to be developing into a part-multiplexed system as many OEMs are wary of following the fully multiplexed route. New technologies of wiring and IC-construction are also revolutionising vehicle electrics/electronics. On the system side, braking and handling control are the latest to emerge.

The state-of-the-art in vehicle electrical systems is provided by Bosch[2] in the form of this useful table of installed electrical loads allowing for averaged on-times for a typical car:

Electrical loads	*Power consumption*	*Mean power*
Ignition	40 W	40 W
Radio	10 W	10 W
Side-marker lamp	8 W	17 W
Lower beam	110 W	90 W
License-plate lamp, tail lamp	30 W	25 W
Indic. lamp, instr-panel lamps	7 W	5 W
Heated rear window	80 W	34 W
Vehicle heating	120 W	50 W
Electric radiator fan	60 W	25 W
Wipers	30 W	5 W
Stop lamp	42 W	11 W
Turn signal lamp	42 W	6 W
Fog lamp	110 W	20 W
Fog warning lamp	21 W	2 W
Total		
Installed electrical loads	710 W	
Mean consumed power		330 W

With high loadings which are common on today's vehicles, dynamic system characteristic (Fig 2) is a key factor in design. A curve can be drawn which plots the interaction of battery, alternator, electrical equipment, temperature, engine speed and alternator drive ratio. This can be measured at the batteries' terminals during development and recorded on an X-Y plotter. A charge balance calculation has to be performed which allows for these parameters and Bosch has developed computer programs for determining the state of battery charge at the end of a typical driving cycle. The company point out that nature of wiring connection is also important.

If all devices are connected at the battery the charging current is a combination of the alternator current and battery charge current and charging voltage is lower owing to the higher voltage drop. If, however, all devices are connected at the alternator side then charging voltage is higher. This might have a negative effect on devices sensitive to voltage peaks so it is advisable to connect voltage-sensitive equipment, with high power consumption, to the alternator and those with low power consumption to the battery. Tabulation of a sequence of functions is a useful execise in this respect, Fig 3.

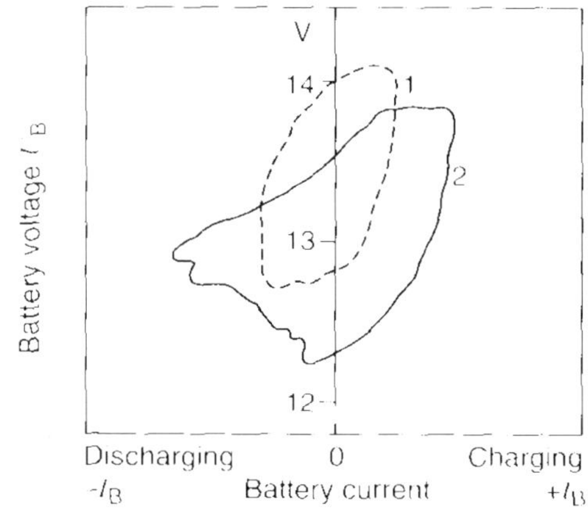

Fig 2: Dynamic system characteristic for urban driving: 1. large alternator/small battery; 2. small alternator/large battery

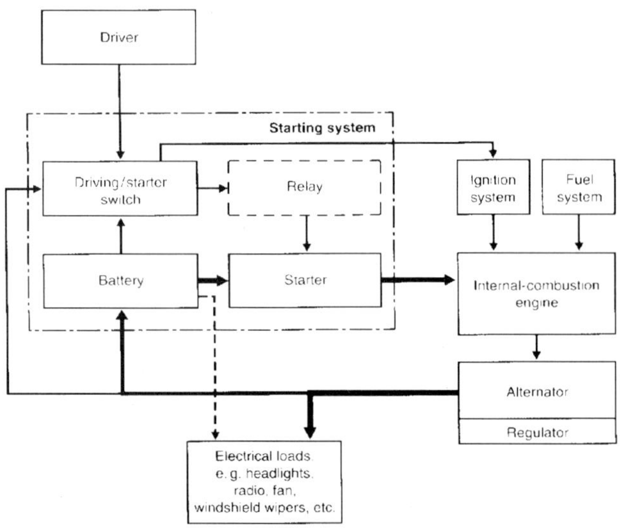

Fig 3: Sequence of functions, shown by arrows, for vehicle electrical system.

Wiring and IC advances

Several techniques for the simplification of electronic system installation have been introduced by Delphi Automotive. Flexible printed circuits, Fig 4, are now on offer in which a conductive circuit is laminated between two or more sheets of non-conductive material. Switches and other devices can be built directly on to the laminate, which can provide contact with other devices without traditional terminals, and the circuits can be integrated directly into component assemblies to simplify manufacture and reduce weight.

The GT connection system involves an innovative insertion which provides very low engage forces (40% less than its rivals) with its box-in-blade tangless terminal designed with stamped serrations to give a more stable and reliable crimp. Primary lock is provided by plastic tangs in the connector body while secondary lock is achieved by a nose-piece that also protects the female terminal. The system is particularly beneficial where access is restricted during assembly and where a high number of contacts are required per connection. Centre lines as small as 3.5 mm are allowed by the terminal design while circuit densities compare with those of 1.5 mm blade systems. So-called Integrated Switches combine electronics with elastomer-dome switch technology to give an intelligent switch, with logic capability, that is easily incorporated into high-style interiors. The units are particularly applicable to decision-making activities such as control of child-safe and one-touch electric windows.

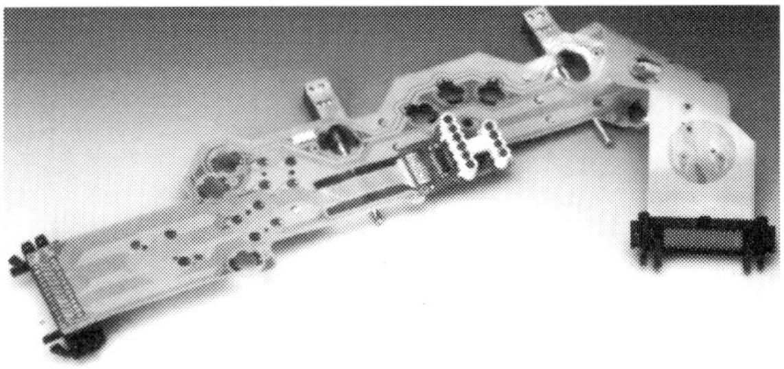

Fig 4: Flexible printed circuits for an instrument cluster by Delphi Automotive

Fig 5: Microhybrid technology used to integrate pump and ECU

Microhybrid technology, Fig 5, is now being introduced on to electronic control units by Robert Bosch, thanks to a new multi-layer ceramic technique known as LTCC (low temperature co-fired ceramic). This has led to further miniaturisation and greater temperature/vibration resistance. The Motronic control unit for a 4-cylinder engine now measures only 120 x 90 x 23 mm. Use of flash memories in the devices has also led to greater flexibility because these can be rewritten in the event of modifications being made. While the company has been involved in hybrid technology since the mid-1970s only now has the parallel processing technique overcome the rigidity of production technique associated with individual layering. It is current practice to stack the printed conductors to form multi-layer hybrids and the substrate layer now only has a quarter of the area of the original hybrids.

The company is set to introduce the first single-chip diesel engine management system, the entire electronics being combined with a 32-bit processor and flash memory. This is destined to bring greater reliability and lower cost as well as miniaturization. Already the company's VP44 has a micro-hybrid ECU directly mounted on it, enabling adjustment of control parameters while the pump is in production — so the pump can be installed as a tested module.

In making the ECU circuit, patterns are screen-printed on the unfired ceramic foils, electrical connections between foil layers being realised by punching holes called "vias" which are then filled with conductive paste. Layers are then stacked and sintered at 900 C. Resistors are screen-printed and fired on to the reverse side of the substrate, the remaining passive components being mounted on the front side using chip and wire technology.

Electric wiring standards

As vehicle electrical systems multiply there is need to simplify the wiring systems which connect them. This article introduces the reader to standardised layouts and coding conventions. Control systems are next examined and a sound introduction to on-board computer control is given with a description of multiplex networking.

A useful way to approach the design of an electrical system is to consider the balance of energy for the vehicle — from the small proportion of chemical energy burnt in the fuel to feed the electrical equipment to its division and conversion into sound, heat, light and mechanical energy. The need to supply energy when the engine is at rest in turn

Nominal cross section	Resistance per meter at 20°C	Conductor diameter Max. dimension	Cable diameter Max. dimension	Permissible continuous current (approximate values) at +30°C	at +50°C	Permissible current density, continuous operation
mm^2	m Ω/m	mm	mm	A	A	A/mm^2
0.5	37.1	1.0	2.3	11	7.8	10
0.75	24.7	1.2	2.5	15	10.6	10
1	18.5	1.4	2.7	19	13.5	10
1.5	12.7	1.6	3.0	24	17.0	10
2.5	7.6	2.1	3.7	32	22.7	10
4	4.71	2.7	4.5	42	29.8	10
6	3.14	3.4	5.2	54	38.3	6
10	1.82	4.3	6.6	73	51.8	6
16	1.16	6.0	8.1	98	69.6	6
25	0.743	7.5	10.2	129	91.6	4
35	0.527	8.8	11.5	158	112	4
50	0.368	10.3	13.2	198	140	4
70	0.259	12.0	15.5	245	174	3
95	0.196	14.7	18.0	292	207	3
120	0.153	16.5	19.8	344	244	3

Figs 6: Copper conducting cable sizes

Type of Conductor	Permissible voltage drop in positive conductor U_{vl}	Permissible voltage drop in entire circuit U_{vg}	Notes
Lighting conductors from terminal 30 of light switch to lamps < 15 W to trailer socket from trailer socket to lamps	0.1 V	0.6 V	Current at nominal voltage and nominal power
from terminal 30 of light switch to lamps > 15 W to trailer socket	0.5 V	0.9 V	
from terminal 30 of light switch to headlamps	0.3 V	0.6 V	
Charging cable from terminal B+ of alternator to battery	0.4 V at 12 V 0.8 V at 24 V	– –	Current at nominal voltage and nominal power
Control leads from alternator to regulator (terminals D+, D−, DF)	0.1 V at 12 V 0.2 V at 24 V	–	At maximum excitation current (note 1)
Main starter cable	0.5 V at 12 V 1.0 V at 24 V	– –	Starter short-circuit current at +20°C (notes 2 and 3)
Starter control lead From starter switch to terminal 50 of starter Solenoid switch with single winding Solenoid switch with pull-in and hold-in windings	 1.4 V at 12 V 2.0 V at 24 V 2.4 V at 12 V 2.8 V at 24 V	 1.7 V at 12 V 2.5 V at 24 V 2.8 V at 12 V 3.5 V at 24 V	Maximum control current (notes 4 and 5)
Other control leads from switch to relay, horn, etc.	0.5 V at 12 V 1.0 V at 24 V	1.5 V at 12 V 2.0 V at 24 V	Current at nominal voltage

Fig 7: Permissable voltage losses

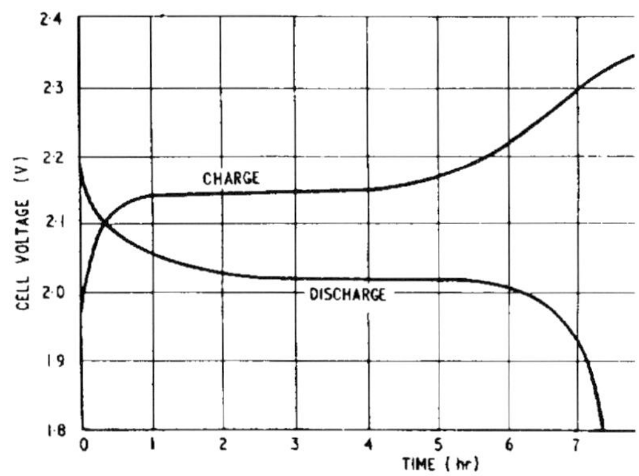

Fig 8: Charge/discharge cycles

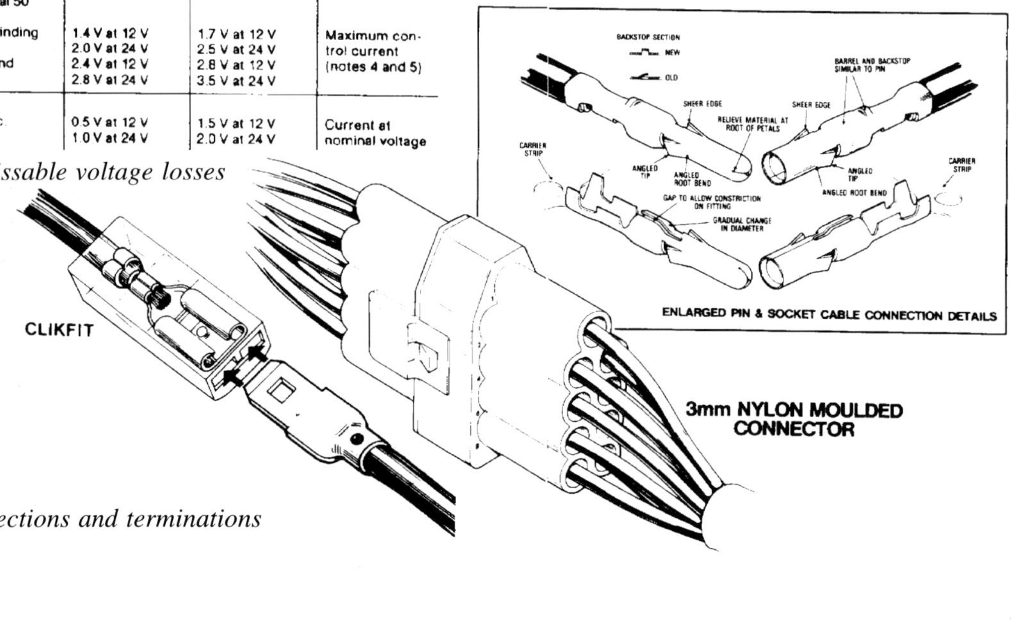

Fig 9: Connections and terminations

creates the need for a storage battery and charging system. This way of thinking leads to the consideration of electrical loads and the respective cable capacity to feed the loads.

Traditionally, the so-called 'main' feeds to control box via ammeter (or charge indicator), from the battery, used 44 strands of 0.012 in wire while from the control box to lighting switch, a cable of 28 similar strands ran to the lighting switch. Proportionally 'lighter' cables then branched out to the various auxiliaries. Where flexibility is required in the cable, to accommodate relative motion, it is usual to increase the number of strands and decrease individual diameter.

There is also a traditional terminal notation in control boxes for auxiliary equipment — with 'A' signifying a battery feed reading on the ammeter. A1 is unfused and passing around load winding of voltage regulator. A2 is fused and passing through the load winding. A3 is unfused, through the load winding and controlled by the ignition switch. A4 is the same but fused.

Standardising wiring layouts Statutory Instruments set out within Vehicle Construction And Use Regulations define the wattage and positioning of lighting equipment and cables, Figs 6/7. As alternators control output current automatically, only voltage regulation is necessary. Voltage curves for lead-acid batteries vary according to the the charge or discharge cycle as shown in Fig 8. One of the highest current consumers in the car auxiliary circuit (apart from the combined lighting load) is the windscreen wiper with consumptions from 2 to 3 amps in the case of car systems. Wiring terminations may be of the spade type for single cable interconnection or component termination — and of the pin and socket type for most multi-cable interconnections, Fig 9.

A circuit diagram for a petrol engined car using Bosch standard symbols is shown in Fig 10. Here, the traditional naming of parts has given way to a more recent standard of coding for device and terminal designation. Obviously trucks, buses and special purpose commercial/industrial vehicles are likely to have more complex layouts; use of section identification according to the DIN standard shown helps to make for relatively easy identification.

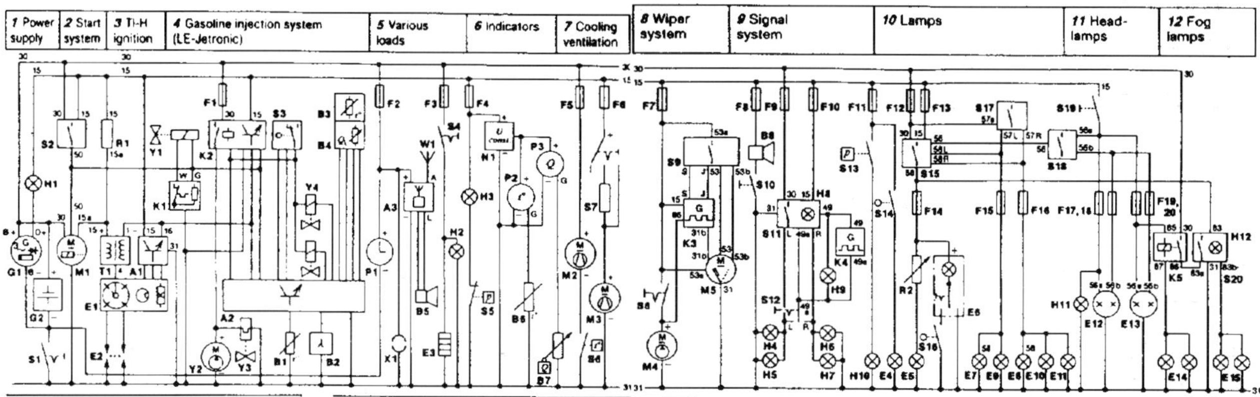

Device desig	Device	Section
A1	TI-H trigger box	3
A2	LE-Jetronic control unit (ECU)	4
A3	Car radio	5
B1	Engine temperature sensor	4
B2	Lambda sensor (only for USA)	4
B3	Air-temperature sensor	4
B4	Air-flow sensor	4
B5	Speaker	5
B6	Cooling-water temp. sensor	6
B7	Fuel-level sensor	6
B8	Horn	9
E1	Ignition distributor	3
E2	Spark plugs	3
E3	Rear-window defroster	5
E4	Back-up lamp	10
E5	Instrument-panel lamp	10
E6	Interior lamp with switch	10
E7	Side-marker lamp, FL	10
E8	Side-marker lamp, FR	10
E9	Tail lamp, BL	10
E10	Tail lamp, BR	10
E11	License-plate lamp	10

Device desig	Device	Section
E12	High/low beam headlamp, L	11
E13	High/low beam headlamp, R	11
E14	Fog lamp	12
E15	Fog-warning lamp	12
F1...F20	Fuses	
G1	Alternator with regulator	1
G2	Battery	1
H1	Alternator charge-indicator lamp	1
H2	Rear-window defroster	5
H3	Oil-pressure indicator	6
H4	Turn-signal lamp, FL	9
H5	Turn-signal lamp, BL	9
H6	Turn-signal lamp, FR	9
H7	Turn-signal lamp, BR	9
H8	Hazard-warning-signal indicator lamp	9
H9	Turn-signal indicator lamp	9
H10	Stop lamp	9
H11	High-beam indicator lamp	11
H12	Fog-warning-lamp indicator	12
K1	Thermo-time switch	4
K2	Control solenoid, Jetronic	4

Device desig	Device	Section
K3	Wiper interval relay	8
K4	Hazard-warning-signal flasher	9
K5	Fog-lamp relay	12
M1	Starter motor	2
M2	Radiator fan motor	7
M3	Fan motor	7
M4	Windshield washer motor	8
M5	Wiper motor	8
N1	Voltage stabilizer	6
P1	Clock	5
P2	Cooling-water temperature indicator	6
P3	Fuel-level indicator	6
R1	Ignition-coil dropping resistor	3
R2	Variable resistor for E5	10
S1	Battery switch	1
S2	Ignition/starting switch	2
S3	Throttle-valve switch	4
S4	Switch for rear-window defroster	5
S5	Oil-pressure warning switch	6
S6	Temperature switch (radiator fan)	7

Device desig	Device	Section
S7	2-speed fan switch	7
S8	Windshield washer switch	8
S9	Wiper switch	8
S10	Horn button	9
S11	Hazard-warning-signal switch	9
S12	Turn-signal switch	9
S13	Stop-lamp switch	9
S14	Back-up lamp switch	10
S15	Headlamp switch	10
S16	Door contact switch for E6	10
S17	Parking-lamp switch	10
S18	Low-beam switch	11
S19	Headlamp flasher button	11
S20	Fog-lamp switch	12
T1	Ignition coil	3
W1	Car antenna	5
X1	Socket	5
Y1	Cold-start valve	4
Y2	Electric fuel pump	4
Y3	Auxiliary-air device	4
Y4	Injection valve	4

Fig 10: Bosch circuit diagram for petrol-engined car

Electronic control basics

As traditional electrical equipment gives way to that involving electronic control of vehicle services the question of networking the controls together often arises. This then leads to the possibility of replacing the wiring harness by a multiplex system involving power and data 'buses'. The bus in this sense is a wire, or cable of wires, capable of simultaneous transmission of many power lines and control signals.

Control systems which are commonplace now on luxury cars are increasingly appearing on further down market vehicles. On its XJ6 model Jaguar has reduced the signal wire cross-sections from 1.0 to 0.5 mm^2 and uses multi-way connectors of just 1.5 mm diameter and a signal wire earth switching technology is employed. AB Electronics Products explain[4] that a central microprocessor is used with 30 outputs to control systems for engine management, comfort control, driver information, suspension levelling and cruise-control. The lighting is also on a logic circuit which prevents the illegal selection of certain combinations of lamps. Outputs from the circuit energise relays located in modules at the four corners of the vehicle, the relays feeding power current to adjacent lamps. The modules also contain electronic circuits which can detect bulb failures.

Electronic systems desribed It is useful to be able to describe how an automotive system works, in general, either 'qualitatively' by block diagrams or 'quantitatively' by the use of models. Subsequent systems analysis can lead to determining the function and performance of the system. The historic work of Joseph Fourier allowed any signal entering a system to be described by a combination of simple sine waves and the response of the system to be measured similarly. An equation can be written to describe the function of the system and the equation, whose plot (graphical representation) is the same as the frequency response of the system, is known as the transfer function. This can be used to show the changes in amplitude (gain) or phase (time delay) of the signal.

A control system is used to direct the operation of other systems and must be accurate, responsive, stable and have 'noise' immunity. Block diagrams for open and closed loop control systems are shown in Fig 11 for a piece of 'plant', the closed-loop system allowing the control to better meet the design criteria. The control shown is 'proportional' because the control signal is proportional to the error signal. Another familiar automotive control mode is called 'limit-cycle' in which the system output is monitored and then feedback control responds only when the output goes beyond preset limits.

Transistor technology

Solid state semiconductors are most commonly used to effect system response. The conductivity of the base silicon is altered by diffusing boron or phosphorous into it when it becomes, respectively a p-type or n-type semiconductor. Many combinations of types are used to make diodes or transistors (similar to traditional amplifying valves). A diode rectifier comprising p and n type junctions, anode and cathode,

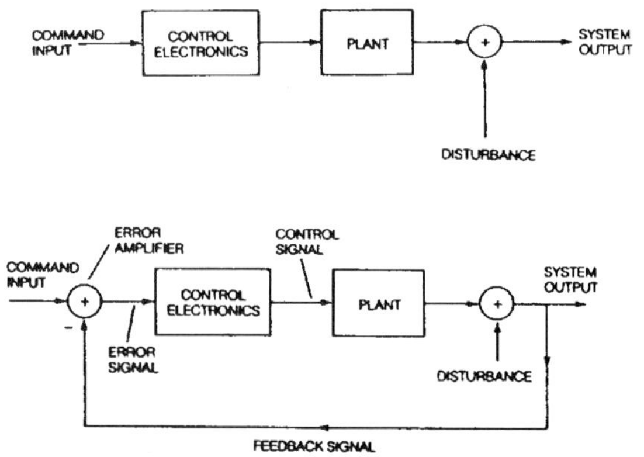

Fig 11: Open and closed loop control systems

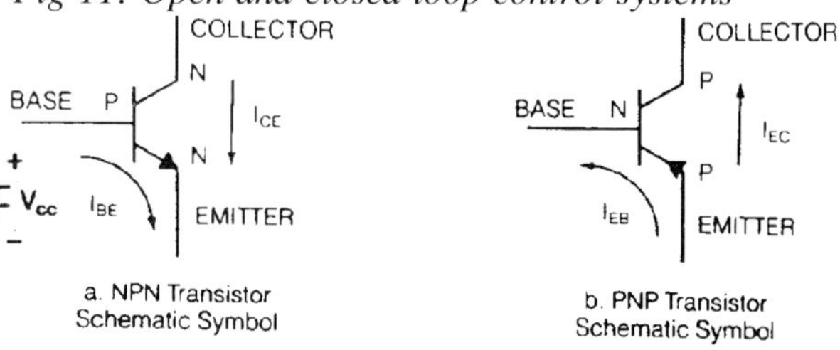

Fig 12: Common bipolar transistors

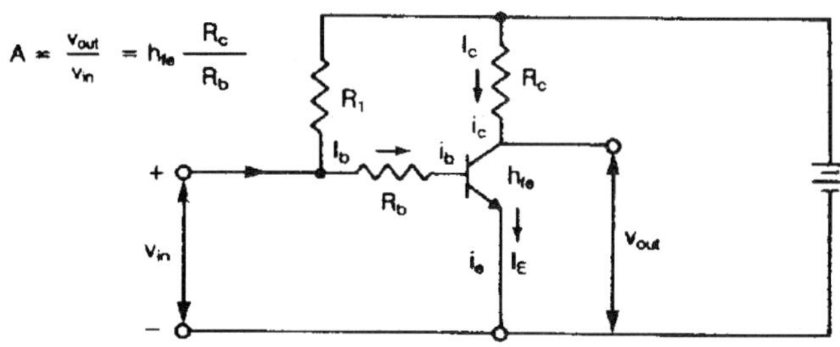

Fig 13: Transistorised amplifier

would provide a 'forward' resistance of about 510 ohms in the forward biased condition — of a positive voltage on the anode. In the opposite, reverse biased direction, the device is effectively an insulator with resistance of more than a million ohms. In contrast to diodes, transistors are active devices which either have 'gain' or store energy.

The two common bipolar types, shown in Fig 12, are the NPN and PNP transistors which differ only in respect of current flow direction and otherwise act as amplifiers or switches. The transistor is effectively a combination of base-emitter and collector-base junction diodes as illustrated for an NPN type. This would be shown as an amplifier as in Fig 13 with gain

$= V_{out}/V_{in} = h_{fe}R_c/R_b$

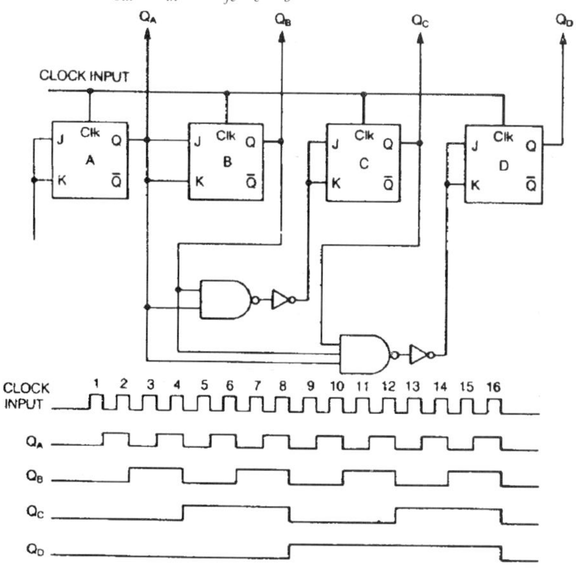

Fig 14: Four stage synchronous counter

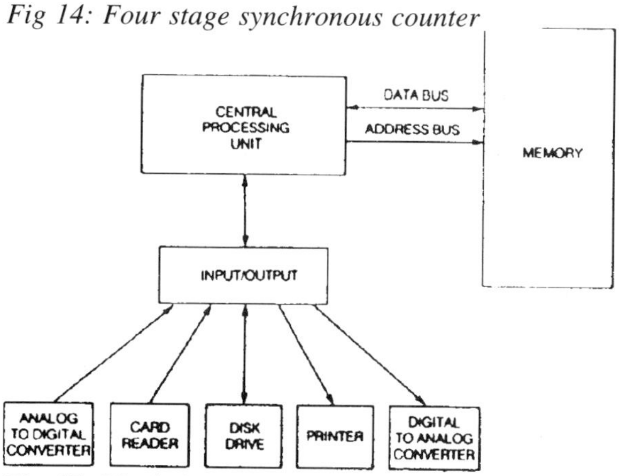

Fig 15: Block diagram for basic computer

where h_{fe} is known as the small signal current gain. With suitable choice of resistances, very high voltage gains can be obtained in the case of the operational amplifier — the basic building block of the analogue computer, used to model a variety of automotive systems.

Digital circuits

Binary digital circuits are characterised by having an output which can only be one of two different states, represented by a voltage or current level. In such circuits the transistor is used as a switch between its 'cut-off' and 'saturation' conditions. The latter is beyond the linear amplifying range of the transistor — in which a large increase in base current produces little rise in collector current and voltage drop across the collector-emitter is very small as the collector-based diode becomes forward biased.

The on-off states of the digital circuits are represented 1 or 0 in the binary, or base 2, numbering system. While the decimal system has numbers grouped from right to left, in representing ones, tens, hundreds and so on, the binary system numbers are grouped from left to right on a power base 2 instead of 10. The binary number 0010 is read as 'zero,zero,one,zero' and not 'ten' while decimal 256 is binary 100000000. Digital computers perform binary digit ('bit' for short) manipulations using three basic logic gates: the NOT, AND and OR gates. The first is an inverter for which an input of 1 becomes an output of 0 and vice versa. The second has at least two inputs and one output, the latter being high (1) when both inputs are high (1) and low (0) if both inputs are low (0). The third, like the second, has two inputs and one output which is 'high' when one OR two inputs are 'high'.

Other common gates are NAND, NOR, XOR and 'adders`. These can be combined to build simple 'combinational' circuits or more complex sequential or memory circuits in which the sequence of past input values, together with the logic state at the time, determine the present output state which may be 'latched' in one way or another. Such flip-flop circuits may be synchronised to change state at a particular time determined by a clock pulse. Fig 14 shows a four stage synchronous counter formed from such circuits which can be used to count 2 to the power 4, or 16, pulses before returning to a starting state, resulting in the timed wave-forms shown at the outputs. Such stages can be lined up to store the digits of a binary number as a temporary register or permanent memory. Because digital circuits can be made from combinations of many identical gates they are ideally made in integrated circuit form, the 1958 invention made by Jack Kilby of Texas Instruments which sparked the micro-electronics revolution.

Computer control

A block diagram for a basic computer is shown in Fig 15. When integrated circuits (ICs) are used the central processing unit (CPU) becomes a microprocessor in which all arithmetic and logical decisions are made. The memory can be 'written' into or 'read' from — and is termed a random access memory (RAM) to distinguish it from read-only-memory (ROM) circuits which can be added to introduce special facilities. This is by means of so-called 'firmware' programs — because unlike software they are unchangeable.

A typical engine control computer does arithmetic using only 8 bits, allowing decimal numbers up to 127 to be represented (plus one bit to show +ve or -ve). The machine's internal memory has several thousand data latches (flip-flops), one for each bit stored, each with a location number which can be 'addressed'. Both the instruction 'program' and data can be stored in the internal memory, once loaded from the appropriate peripheral device. The address of bits is stored as a binary number within temporary latches called registers. Fig 16 shows how the address bus is wired relative to the data bus. A 16 bit address register can access 65,536 memory locations containing 8 bits (one byte). The data bus, usually only 8 bits wide, is a two-way highway compared with the single-way highway of the address bus. A clock signal tells the memory when it can take and release control of the data bus. By combining bytes and operating on them one after another (serial operation), in time sequence, very large numbers can be handled. To start the computer, a small start-up (boot) program in memory sets all the CPU's registers to the correct values, thereby 'initialising' the system. The programmer uses an 'assembly' language of mnemonics (abbreviated command words) to compile the operating program. An assembler program is then run to convert the mnemonics into binary numbers which sets the computer into operation. The computer is connected to vehicle systems via interfaces, a parallel one implying, say, 8 lines of data being sampled at one time. Digital to analogue converters are used to control the system with an analogue voltage. A similar communication system to that inside the computer could be used to connect the service systems and their controllers throughout the whole vehicle. The pressure for such a system comes from the need to reduce the number of cables in the wiring harness as more service systems are added. Multiplex (MUX) wiring allows power and data control buses to replace many cable links. Fig 17 shows how such a system, proposed by Motorola, has the advantage of low switching currents and the ability to build in a self-diagnostic facility. Cost of intelligent power switching is the main deterrent which is likely to reduce in the course of future microelectronic development. At this stage a clock rate of 2400 Hz allows about 4 milliseconds to determine the address and store the control signal data. A serial data network uses an 8 bit word, in the system shown. Intelligent load switches being developed use unipolar Field Effect Transistors.

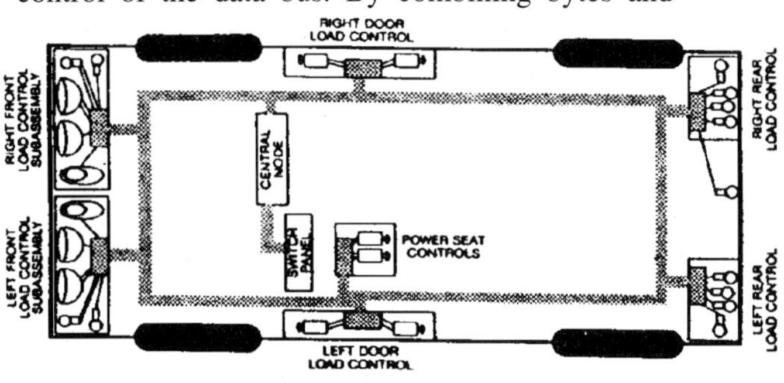

Fig 17: Vehicle multiplex system

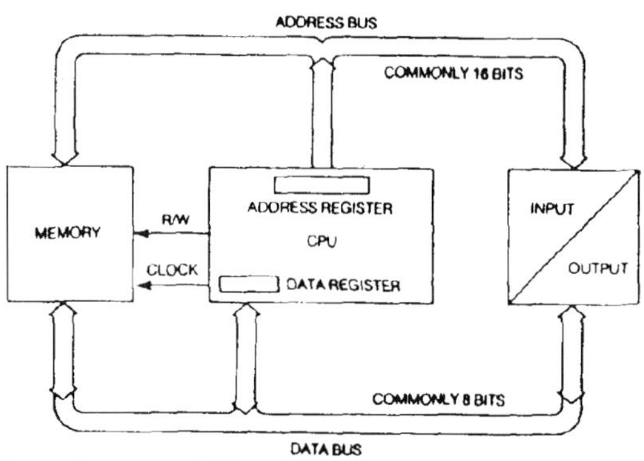

Fig 16: Address and data buses

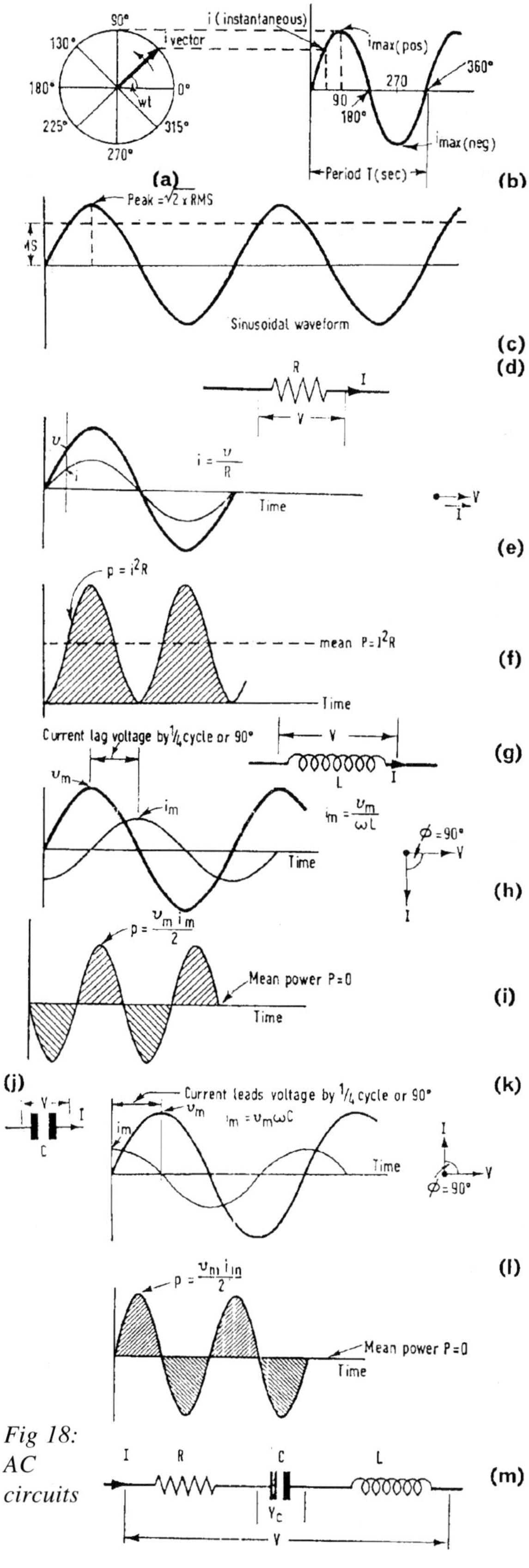

Fig 18: AC circuits

Electronic controls applied

While the previous section introduced electronic control in conceptual terms, here the solid-state electronic elements used as building blocks for electronic control circuits are examined. Some appreciation is also offered as to how these can be applied in digital logic devices for on-board microcomputer control.

Materials which can be classified as conductors, semi-conductors or insulators vary in the amount of free electrons present. The movement of the electrons can be arranged, in suitable primary circuits, to control the operation of secondary power circuits incorporating electro-mechanical actuators contained in the various systems of the vehicle. Some 6288 x 1015 electrons represents a charge of one coulomb, the passage of which each second is equivalent to a current flow of one amp. The electric force, or potential, of one volt gives each coulomb of flow an energy charge of one joule — defining the basic units for any quantitative analysis to be carried out.

Charge can be stored on a condenser, to the extent that one coulomb charge taken up when one volt is applied represents a capacity of one farad. The material's ability to store charge as the 'plate' of a condenser is expressed as a ratio to air (value unity) as its di-electric constant while the time taken for the condenser to reach 63.2% of its full charge defines its time constant.

The use of alternating rather than direct currents in control circuits is convenient in the ability of the voltage to be easily transformed into a lower or higher one — also in the absence of commutator sparking in any motors used in the associated power circuit, Fig 18. Part (a) shows how vector notation is used to express the sine-wave representation of alternating current, at (b); the root mean square quantification of voltage and current values is seen at (c) while (d,e,f), (g,h,i) and (k,l) show current and voltage relationships for purely resistive, inductive and capacitative elements of a circuit respectively.

In a resistive *R* and inductive *L* series circuit, there is a phase change such that the voltage leads the current by the 'phase angle' (between vectors) and an impedance is set up, to current flow: *Z* is voltage/current *(E/I)*— analogous to ohmic resistance in a DC circuit. With capacitance *C* substituted for *L*, the resulting *C,R* series circuit is such that the voltage lags the current by an equivalent phase angle. When all three exist together in a series circuit, (m), the capacitative and inductive impedances tend to cancel out so that when $\omega = 1/\omega C$ occurs, then in the impedance formula:

$$Z = [R^2 + (\omega L - 1/\omega C)^2]^{1/2}$$

$Z = R$ and resonance occurs so that a magnification condition can result when the voltage across the capacitance could be higher than the applied voltage. Resonant frequency of the alternating current, $w = (1/LC)^{1/2}$, the circular frequency of the rotating vector. The voltage magnification factor $= (1/R)(L/C)^{1/2}$. At the resonant frequency condition of minimum impedance, the arrangement is termed an 'acceptor circuit'. The higher the magnification factor, the better is the 'selectivity' — the ability of the circuit to select one frequency to the exclusion of all others.

In the case of parallel circuits of capacitance and inductance, some interesting characteristics are possible. In practice there is always some resistance in the circuit but in theory, without resistance, a circulating current should build up in the parallel loop. Because of the inevitable resistance present a so-called 'dynamic impedance' exists at resonance. which is the maximum condition, and defines a 'rejector circuit'.

Solid-state 'active-devices'

Voltage gain and direction control in a circuit is usually achieved by transistors, introduced in the last section, made from semiconductor materials doped with impurities to alter their characteristic. Silicon is the usual base material with boron and phophorous used respectively to achieve the p-type and n-type semi- conductors respectively. As forward-conducting rectifiers, transistor diodes have a positive voltage on the p-type material (the anode) to obtain the necessary bias. With a negative voltage on the n-type material, a reverse non-conducting bias is obtained. Fig 19a shows the symbol for the device alongside the current/voltage (I/E) transfer characteristic for a silicon diode. The rectifier circuit in Fig 19b is of the 'half-wave' type, the diode being reverse- biassed to block current flow in the reverse direction. Current flow is smoothed from half-wave to near steady value by the condenser which stores charge and opposes voltage changes by discharging to the load resistance.

Active transistors (transfer resistors) have ter-

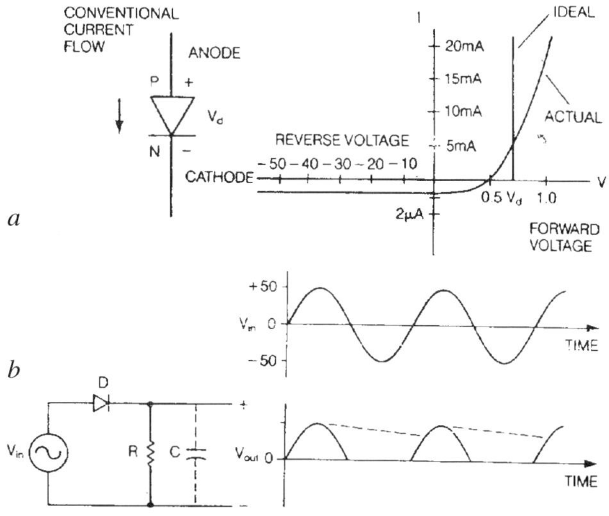

Fig 19: Transistor and arectifier circuit

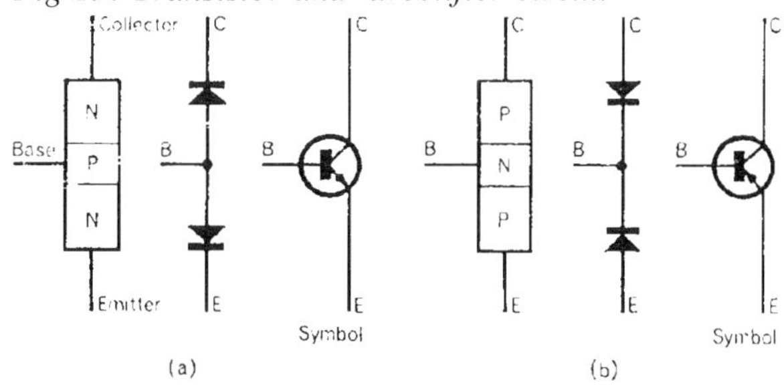

Fig 20: Back-to-back diodes

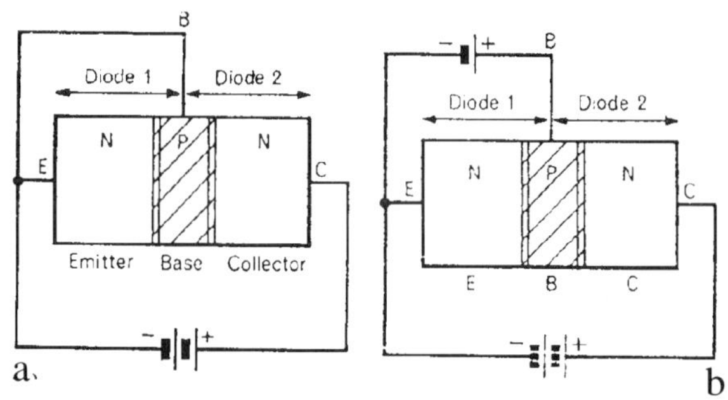

Fig 21: Transistor biassing

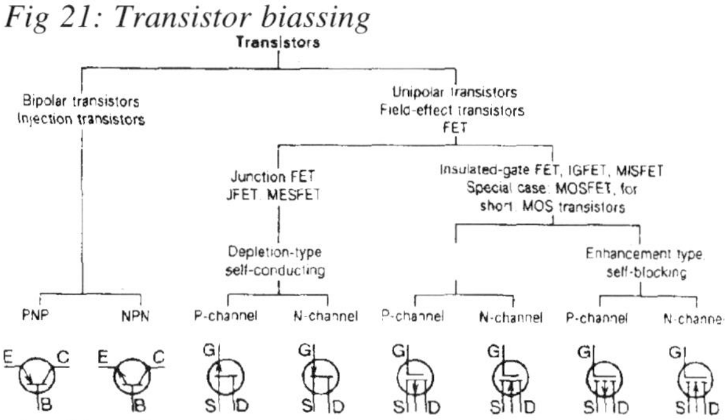

Fig 22a: Transistor types

minal connections to the circuit which can cause voltage to be amplified. They can be regarded as back-to-back diodes (Fig 20) and may be of the NPN or PNP type for so-called 'bi-polar' transistors, introduced in the earlier section, which also showed the respective current flows across the PN junction. Both can operate as amplifiers and switches, the collector/base junction becoming forward-biassed in the 'on' condition — in the switching mode. Referring to Fig 21a, the NPN transistor is shown with the base at the same potential as the emitter; no current will flow since diode 1 has the same potential at anode and cathode and diode 2 is reverse-biassed. In Fig 21b, there is a forward bias on diode 1 and when this exceeds the junction barrier potential, electrons will move from the the base (P region) towards the emitter (N region) attracted to the higher potential at the collector; small current change, base to emitter, causing large current change, collector to emitter. An amplifier circuit for a transistor was shown in the introductory section.

In uni-polar, field-effect, transistors (FET), current is controlled by an electric field generated by a voltage applied at a control electrode called the gate. The types and physical representations of different transistors are shown in Fig 22 a and b, due to Bosch, and the resistive capacative and inductive components of a series circuit in Fig 22c. The word uni-polar comes from the FET only employing one type of charge carrier, against the bi-polar's two. MOS stands for metal-oxide-semiconductor — referring to the sequence of layering of the silicon. The P

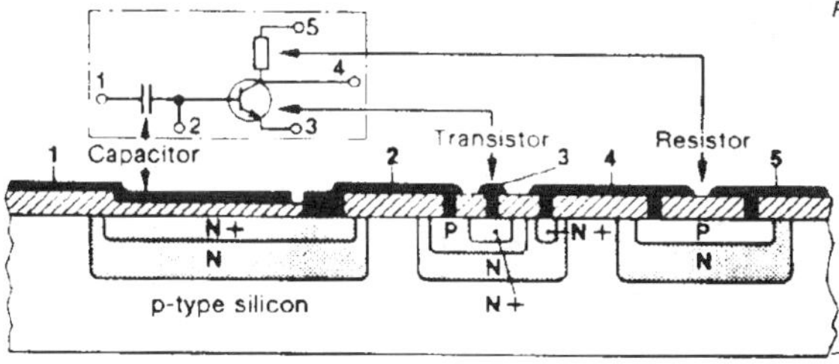

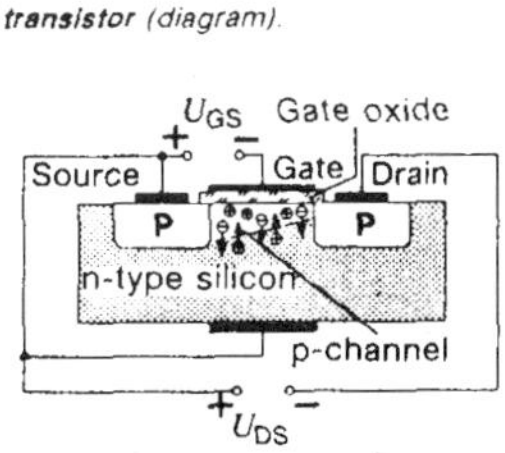

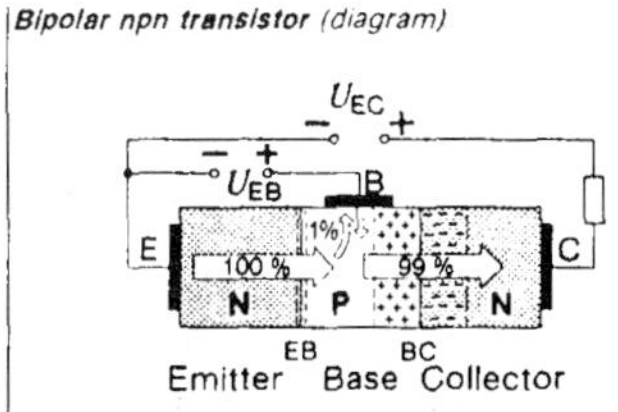

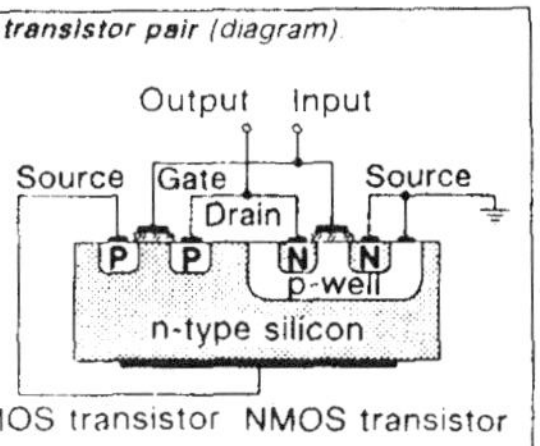

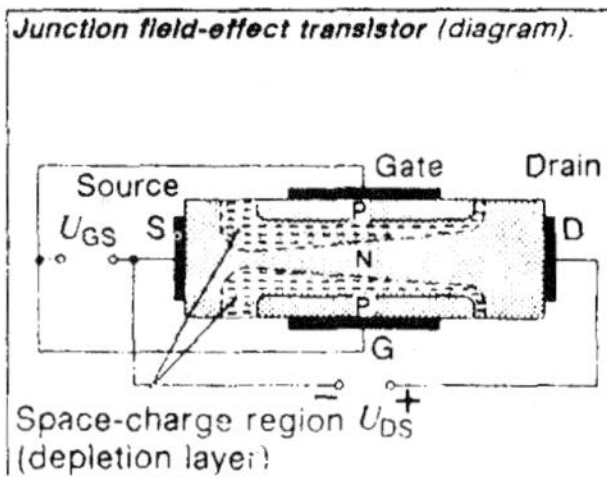

Fig 22b: Transistor types and (above) installation in an integrated circuit

	R	C	L	R & C	R & L
Circuit					
Vector diagram	V I	I $\phi = 90°$ V	V I $\phi = 90°$	ϕ I V	ϕ V I
Power $P =$	VI or I^2R	0	0	$VI\cos\phi$ or I^2R	$VI\cos\phi$ or I^2R
Impedance $Z =$	R	$\frac{1}{\omega C}$	ωL	$\sqrt{[R^2+(1/\omega C)^2]}$	$\sqrt{[R^2+(\omega L)^2]}$
Phase angle =	0	90°	90°	$\tan^{-1}(\frac{1}{\omega CR})$	$\tan^{-1}(\frac{\omega L}{R})$
Impedance triangle	R	90° $\frac{1}{\omega C}$	90° ωL	ϕ Z R $\frac{1}{\omega C}$	ϕ Z R ωL
Units for calculations	Ohms (Ω)	* Farads (F)	Henries (H)	Ohms & Farads (Ω & F)	Ohms & Henries (Ω & H)

Fig 22c: Table of R, C and L components in series circuit

and N suffixes refer to the doping by either a donor or acceptor impurity. Suffix C stands for complementary and involves the fabrication of NMOS and PMOS transistors on the same silicon chip. One further type is the thyristor, familiar to those involved in speed controllers for electric vehicles. These are semi-conductors with at least three PN junctions, switchable from the forward to the reverse-block state.

An introduction to digital circuits and computer control was given in the previous section so it is now timely to consider the basic concepts of control necessary for designing control-circuits. Fig 23, due to Bosch, compares closed and open loop systems. In the former, the directly controlled variable *x* is continuously measured and compared with a reference variable *w*. It is then changed continuously to reduce variation from the reference variable — within a closed loop. This is also called proportional control because the control signal is proportional to the error signal. In the latter case of open-loop control, one or more input variables acts on the output variable through the inherent characteristics of the system. In open loop control, a series of elements act upon one another as links in a chain. Only disturbance *z1* can be addressed as the open loop has no effect on disturbance *z2*.

Fig 24 shows a control circuit for controlling temperature within a closed space, using the closed-loop concept, This, however, uses a technique known as limit cycle control, which monitors the system's output and responds only when the output goes beyond preset limits. More sophisticated controls use logic circuits in which a certain operation takes place only when a certain set of conditions occur. Transistors in this instance are used only as switches and all data is in terms of 'signal' (1) or 'no signal' (0). According to Trevor Jones[5], the combination of logic circuits and micro-miniaturisation on integrated circuits has most transformed the prospects for vehicle control systems. Now that power devices, too, are being reduced to chip size, the way is open for new technologies to achieve control strategies without the quality problems of a multitude of soldered joints. He shows in Fig 25 the voltage and current ranges achievable with power integrated circuits. As their price reduces, he points out, the need for conventional solenoid relays for power actuation may eventually disappear.

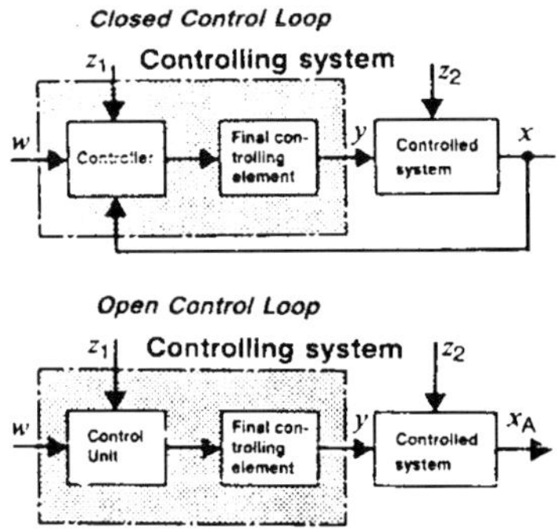

Fig 23: Closed and open loop systems

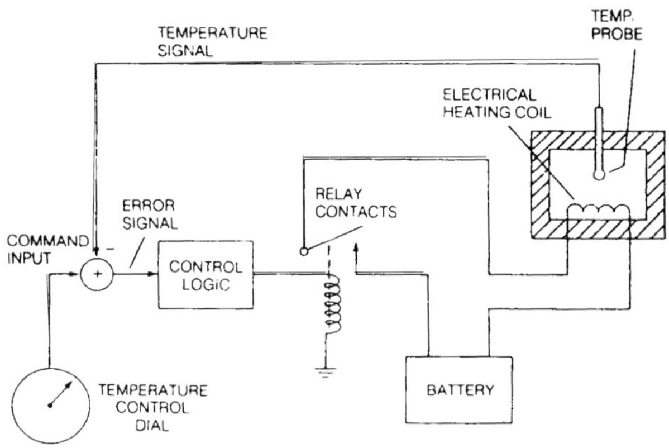

Fig 24: Temperature control circuit

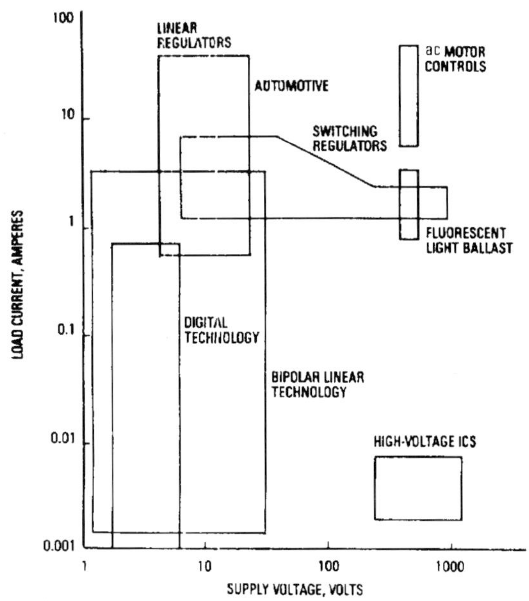

Fig 25: Achievable voltage and current ranges

Micro-electronics terminology

Before proceeding to body electronics in detail it is useful to become acquainted with micro-elctronic terms used in automotive computer technology. This valuable listing is due to Siemens Automotive, some of which company's recent technologies are shown in Fig 26:

Accumulator: In a computer, the basic work register.
Actuator: A device which performs a mechanical action in response to an input signal.
A/D: Abbreviation for analog-to-digital converter; an electronic circuit or device that converts an analog input signal into a digital signal.
ADIS: Advanced Driver Information Systems; a combination of infrastructure and in-vehicle displays that communicate information to the driver (emergency and general).
Analog Circuit: An electronic circuit which amplifies, reduces or otherwise alters a voltage signal which is a smooth or continuous copy of some physical quantity.
Assembly Language: An abbreviated computer language which humans can use to program computers. Assembly language eventually is converted to machine language so that a computer can understand it.
ATIS: Advanced Traveller Information System.
ATMS: Advanced Traffic Management Systems; a combination of highway infrastructure and in-vehicle components that yield traffic flow improvements.
AVCS: Advanced Vehicle Control Systems; vehicle systems that automate current functions, such as distance-keeping, lane-changing, ramp merging, etc.
AVI: Advanced Vehicle Identification.
Byte: Eight bits dealt with together.
Capacitor: An electronic device which stores an electrical charge.
Combinational Logic: Logic circuits whose outputs depend only on the present logic inputs.
Comparator, Analog: An electronic device which compares the voltages applied to its inputs.
CPU: Abbreviation for Central Processing Unit; the calculator portion of a computer.
Cutoff: A transistor operating mode where very little current flows between the collector and emitter.
D/A: Digital-to-analog converter; a device which produces a voltage which is proportional to the digit input number.
Bit: A binary digit; the smallest piece of data a computer can manipulate
Block Diagram: A system diagram which shows all of the major parts and their interconnections.
Damping Coefficient: A parameter which affects a system's time response by controlling the rate at which energy can be added to or removed from the system.
DEMUX: Demultiplexer, a type of electronic switch uses to select one of several output lines.
Digital Circuits: Electronic circuits whose outputs can change only at specific instances and between a limited number of different voltages.
Diode: A semiconductor device which permits current to flow in only one direction (analogous to a check valve, in fluid handling).
Dwell: In an engine, the time allowed for current to build in the primary circuit of the ignition coil for each spark generation.
ECU: Electronic Control Unit.
Electronic Feedback Carburettor: A fuel metering device in which the air/fuel ratio is controlled by continual variations of the metering rod position or air bypass modulation in response to an electronic control signal.
Electronic Throttle Control (ETC): Acceleration: peddle controls the throttle by means of a computer-controlled stepping motor in lieu of a conventional linkage from peddle to throttle.
Electronic Valve Timing (EVT)**:** A system where the opening/closing times and lift of cylinder valves are controlled by computer.
Engine Mapping: A procedure of experimentally determining the performance of an engine at selected operating points.
Frequency Response: A graph of a system's response to a different frequency input signals.
Gain: The ratio of a system's output magnitude to its input magnitude.
Inductor: A magnetic device which stores energy in a magnetic field produced by current flowing in it.
Infrared: A communication medium that employs wavelengths longer than those of visible light.
Integral Amplifier: A control system component whose voltage output changes at a rate proportional to its input voltage.
Integrated Circuit: A semiconductor device which contains many circuit functions on a single chip.
Interrupts: An efficient method of quickly requesting a computer's attention to a particular external event.
IVHS: Intelligent Vehicle Highway Systems: an umbrella term includes an advanced technology approach to traffic management.
Lead Term: A control system component which anticipates future inputs based on the current signal trend.
Limit Cycle: A mode of control system operation in which the controlled variable cycles between extreme limits with the average near the desired value.
Linear Region: A transistor operating mode where the collector current is proportional to the base current.
Logic Circuits: Digital electronic circuits which perform logical operations such as NOT, AND, OR, and combinations of these.
Lookup Table: A table in computer memory which is used

to convert an important value into a related value from the table.

Mathematical Model: A mathematical equation or equations which can be used to numerically compute a system's response to a particular input.

Microcomputer: A small computer which uses an integrated circuit which contains a central processing unit and other control electronics.

MUX: Multiplexer; a type of electronic switch used to select one of several input lines

Op Code: A number which a computer recognizes as an instruction.

Open Loop Fuel Control: A mode where engine input air/fuel ratio is controlled by measuring the mass of input air and adding the proper mass of fuel to obtain a desired ratio.

Operational Amplifier: A standard analog circuit component with two inputs, one output and a very high voltage gain.

Optimal Damping: The damping which produces the very best time response.

Peripheral: An external input-output device which is connected to a computer.

Phase Shift: A measure of the delay between the time an electrical signal enters a system and the time it shows up at the output.

Proportional Amplifier: A control system component which produces a control output proportional to its input.

RAM: Random Access Memory: read/write memory.

Random Error: A measurement error which is neither predictable nor correctable, but has some statistical nature to it.

ROM: Read only memory; permanent memory used to store permanent programs

Sample and Hold: The act of measuring a voltage at a particular time and storing that voltage until a new sample is taken.

Semiconductor: A material which is neither a good electrical conductor nor a good electric insulator.

Sensor: An energy conversion device which measures some physical quantity and concerts it to an electrical quantity.

Sequential Logic: Logic circuits whose output depends on the particular sequence of the input logic signals.

SIR Sensor: Supplemental inflatable restraint sensor.

Signal Processing: The alteration of an electrical signal by electronic circuitry; used to reduce the effects of systematic and random errors.

Software: The computer program instructions used to tell a computer what to do.

Solenoid: An electrically charged coil of insulated wire which produces a magnetic field within the coil.

Systematic Error: A measurement error in an instrumentation system which is predictable and correctable.

Transistor: An active semiconductor device which operates like a current valve.

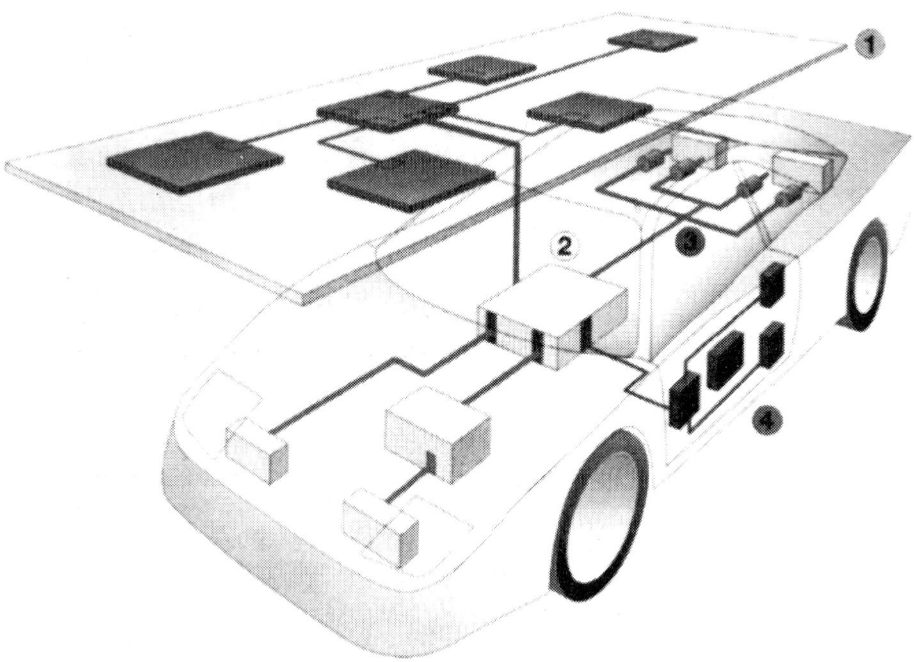

Fig 26: Electronic technologies from Siemens: 1. Information network with high speed multiplexing, tor example with implemented fibre optics 2. Junction box 3. Conventional Distribution system using wiring harnesses 4. Low speed multiplexing using copper wiring, eg subsystem on the driver's door

Engine/chassis system developments

Some 12 years of development has prefaced the introduction of WABCO's electronic braking system (EBS) which is claimed to overcome some of the limitations of conventional air braking for commercial vehicles. The increasing number of pipes and valves involved with highly developed air-braking has led to an inevitable increase in fault probability, uneven brake wear and so-called "lazy" brakes. There is also greater risk of braking instability with the prolification of circuits and devices. Performance monitoring, too, is very much a matter of visual inspection and driver opinion. With EBS, such as the system just specified for the Mercedes Benz Stratos, described in a later chapter, many of these difficulties are alleviated while higher braking torques have been accommodated and improved control of engine brakes and driveline retarders realised.

While control is electronic, actuation is electro-pneumatic on Stratos. Braking response to pedal application is speeded up and uneven wear problems due to time lags in the pneumatic transmission eliminated. Performance monitoring is improved by the sensing of many operational parameters and overall system simplification has been achieved with the additional assurance of better failure monitoring. A decentralised electronic control concept was preferred for its inherent reliability, with the multiplexing of signals to and from the different devices on a data bus. The proven CAN protocol is employed, the central module performing general functions, Fig 27, such as brake management, diagnostics and power-supply control while the individual devices, packaged with integrated electronics, use local sensor signals to control brake pressure. The EBS can also communicate with other electronic systems on the vehicle.

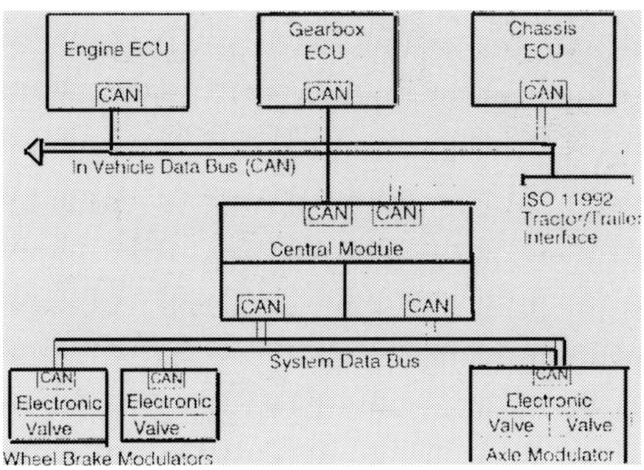

Fig 27: EBS network structure (above)

The EBS chosen by Mercedes for Stratos, Fig 28, has pneumatic back-up for the two tractor brake circuits, and for trailer control, to ensure initial market acceptance. Parts of the system are centrally controlled and other parts are decentralised, the result being a hybrid system. A brake signal transmitter at the pedal sends control messages to the front, rear and trailer axles in both electronic and pneumatic form. A proportional valve controls service pressure for the front axle and at the rear, where decentral data processing is used, a dual-channel smart pressure modulator is involved, Fig 29, processing local wheel speed and brake wear sensor signals. There is also a back-up air-control valve at the rear which can be deleted when system reliability is proven, according to Herr Wiehen from the supplier's Hannover plant. The trailer control valve is electro-pneumatic but there is a digital interface for a trailer EBS when these become more widely used. A new valve concept is used in the rear axle controller to perform both service brake and ABS/ASR control. Two solenoids control flow-rate for the valves in each channel. The so-called combi-modular provides accurate control for both small pressures and rapid brake applications as well as ABS response.

Electronic control using artificial intelligence, Fig 30, has been used by Mitsubishi to optimize drive control to each wheel of its all-wheel-drive HSR-V concept vehicle, also described in a later chapter. Four wheel steer has also been integrated to control yaw moment generated by differences in drive and brake force, on left and right hand wheels, active yaw control thus obtained maintaining optimum handling in corners. The car's INVECS intelligent control system uses sensors that monitor driver's behaviour, road conditions, vehicle movements and the traffic environment — to provide the correct control to each wheel. A suspension ride-height control has also been added to the car's traction control system, to optimize torque transmission in both low and high-ride, off-road, conditions. The system involves four wheel independent brake control, alongside 4WS and the company's torque transfer differential, to transfer drive torque between right and left sides of the vehicle. An intelligent controller picks up data from the yaw rate sensor and lateral acceleration sensor to determine cornering conditions and generates an optimum yaw moment for stable cornering. The system aids cornering firstly by transferring drive power to the outer wheels. Secondly, if a corner is entered too fast, the system applies brake pressure to the inner wheels. Should the driver inadvertently reduce throttle pedal effort during the bend, causing the car to pull to the inside, the system transfers drive force from outer to inner wheels so as to stabilise the car. In any situation likely to induce a spin the system also applies brake pressure to the outer wheels, for stability in cornering.

By adding hydraulic control, both to the suspension dampers and the adjustment of suspension link length, it is possible not just to adjust ride height according to driving conditions but also to adjust wheel alignment such that the tyre footprint is optimized for best handling response. Camber and caster angle are also optimized for cornering and high-speed driving. A computer analyses data from the vehicle speed, lateral acceleration, vertical acceleration, steering angle and vehicle height sensors as well as data from the navigation and so-called 'multi-eye' systems.

The multi-eye system is an artificial vision system designed to complement the driver's eyes. Two image and scan-type sensors at the front of the car are laser radar units which monitor fields of view. Close range optical sensors are mounted to the sides of the car and a stereo image sensor compensates for the driver's rear blind spot, as well as providing a rear field view. The system is also linked to infrastructure traffic control systems for environmental information. If the driver fails to respond to a danger warning from the system, it either applies the brakes to stop the car or, where necessary, steers the front wheels to a safe lane and decelerates the vehicle.

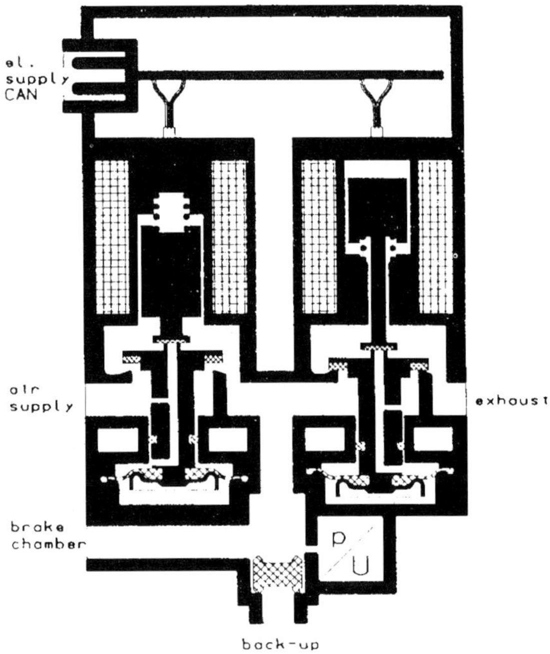

Fig 29:: Combi modulator principle

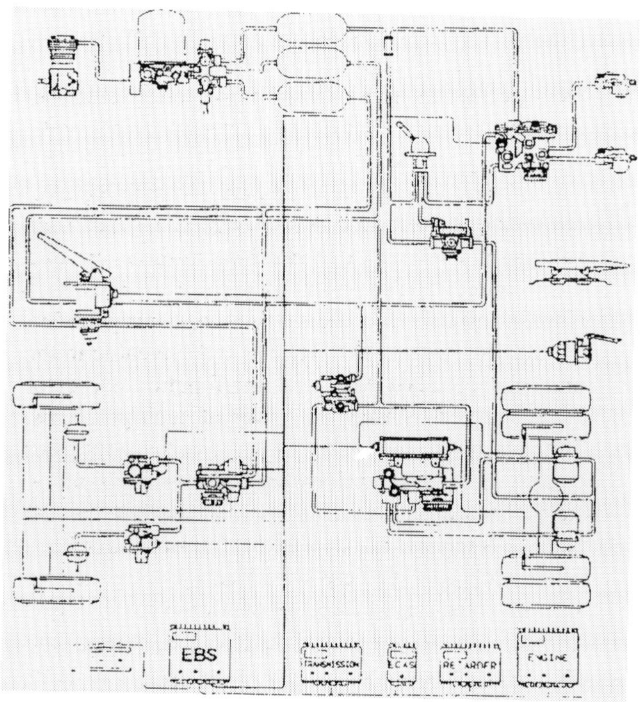

Fig 28: EBS for M-B Stratos truck

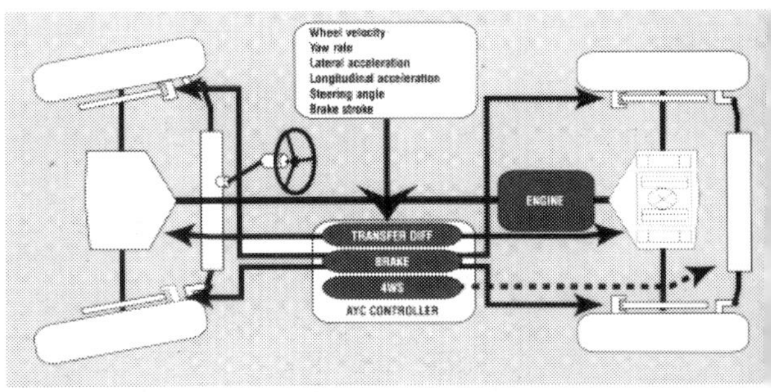

Fig 30: Mitsubishi HSR-V and its active yaw control system

Chassis/powertrain multiplexing

Lucas[6] have demonstrated CAN bus interconnection on an adapted car, Fig 31 showing the systems architecture for chassis and powertrain controls. A CAN interface was added to the control unit of the EPIC diesel fuel injection system and the controller modified to accept inputs from CAN as well as the accelerator pedal. EPIC is also programmed to transmit engine speed, engine load and accelerator position information to the master controller. A CAN and data logger interface using an additional microprocessor was added to the ABS control unit and software has been added to allow automatic brake pedal actuation using CAN inputs. Similarly, for the automatic clutch system, a CAN interface was developed in the form of an additional PCB in the control unit. The management system is programmed to receive a disengage clutch request from the master controller. Gear lever position, clutch position, and gear shift in progress signals are transmitted. A similar CAN interface has been added to the adaptive damper control to receive appropriate signals from the master controller. The master controller implements the integrated systems where sensors and actuators from multiple subsystems are used for an additional function. It receives information from each subsystem, calculates the required outputs and sends out control signals as seen in the figure. The system architecture makes all of the vehicle sensor information available in one, relatively powerful, control unit and uses the CAN to its maximum advantage. The master controller may be programmed in a high level language and is interfaced to a high speed data logger. Therefore, development of the integrated systems has been considerably simplified compared to an alternative architecture of adding the integrated system functions to the subsystems.

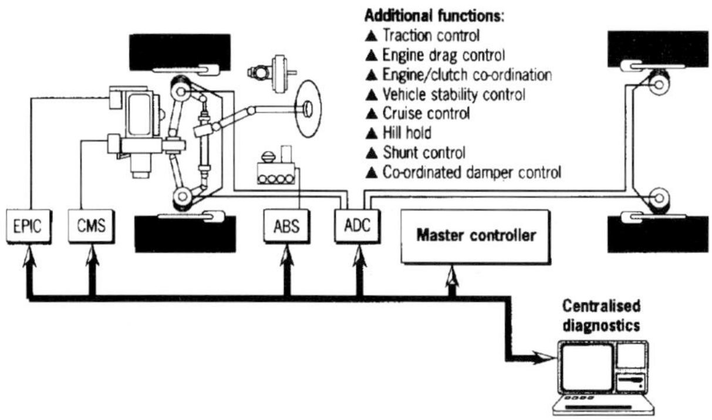

Fig 31: Lucas CAN systems architecture

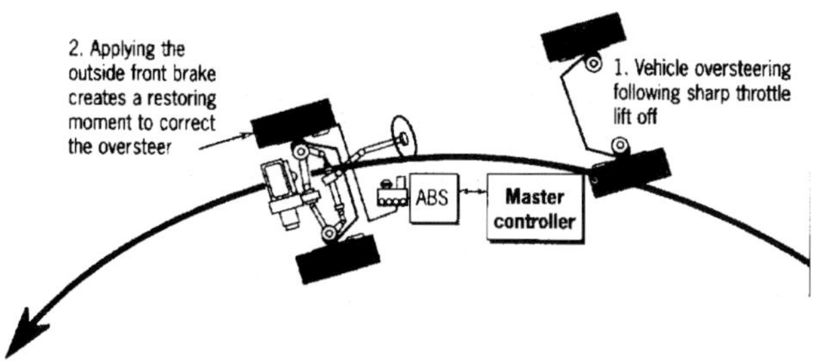

Fig 32: Lucas vehicle stability control system

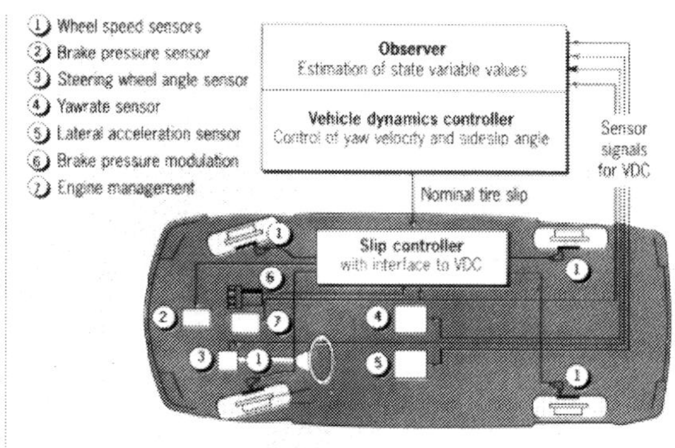

Fig 33: Bosch VDC and its control concept

Safety and reliability are further aspects of the systems architecture that have been addressed by this project. Safety has been ensured by having each subsystem check that its control signals over CAN are valid and received regularly. A second, confirming signal is sent with the most safety-critical brake control signals. Safety and reliability are also maintained by minimising the modifications to the subsystems. The system has been designed so that each subsystem can operate stand alone in the event of a CAN failure; the subsystems detect the loss of control signals and default to their normal operating modes. Hence, the vehicle remains safe and driveable in the presence of CAN faults. A malfunction indicator lamp is fitted to warn the driver of such problems. The switches interfaced to the master controller consist of a dashboard switch and steering column stalk for cruise control, and a mode switch for the damper control system. Additional sensors have been fitted including vehicle speed using a hall effect sensor to measure wheel speed, down to very low speeds, lateral and longitudinal accelerations, and yaw rate. These are used to enable integrated systems that could not be implemented with the subsystem sensors alone.

Vehicle stability control

Fig 32 shows the Lucas system integrated to the above-described demonstration car. Vehicle stability control is intended to be an enhancement to ABS that can prevent a vehicle becoming unstable when cornering even when the driver takes no corrective action. The first steps towards this goal have been taken using the integrated systems vehicle with its brake control system, vehicle dynamics sensors and efficient soft ware development system. A typical scenario for the system is when oversteer has been provoked by the driver lifting off the accelerator during a cornering manoeuvre on a low grip surface. The master controller can calculate the expected vehicle yaw rate from the vehicle speed and steering angle. If the actual yaw rate should exceed this value then the vehicle is oversteering. The master controller then sends a wheel speed demand to the ABS for the outside front wheel that is lower than the actual vehicle speed. The ABS then applies that brake using the software originally developed for brakes traction control. This brake application creates a restoring moment and a speed reduction to restore the vehicle to a stable cornering condition.

There are now several systems being readied for the market and one of the first to reach a production vehicle was the Bosch VDC system announced in 1995. At the SAE Congress, researchers at the company[7] described a simulation tool used in its development with which the effect of the system on extreme handling manoeuvres could be demonstrated. VDC actively improves the handling performance and vehicle stability in all driving situations near the limit of adhesion between tyres and the road. Using engine torque control and active control of the individual wheel brakes, the required wheel forces are established.

It is necessary for applying and controlling the individual tyre slip that the brake pressure in each wheel brake may be modulated and applied independently of the driver and as required by the control algorithm. An important difference between ABS and VDC is the following. In the ABS, the wheel is the controlled part. The wheel velocity is controlled to avoid wheel lock and to keep the slip small enough to preserve some amount of lateral force capability. In the VDC system, not only the wheel but also the vehicle is the controlled part. The vehicle motion is controlled to keep the deviation from its nominal motion as small as possible and conform with its environment. The wheel slips may be controlled to any value to get the required lateral and longitudinal forces. Typically, the bandwidth of the wheels is large as compared to the bandwidth of the vehicle. It is advantageous to choose the controller structure in accordance with the characteristics of the controlled system. For this reason, a cascade control is chosen in which the inner feedback loop controls the wheel slip and in which the outer feedback loop controls the vehicle motion. This structure is shown in Fig 33 together with the sensors 1 - 5 and the actuators 6 (hydraulic system) and 7 (engine management) of the system.

Within the outer feedback loop, corrections of the tyre's nominal slip values are derived from the difference between the nominal and actual vehicle motion and directed to the slip controller, the inner feedback loop. In the inner feedback loop, control signals are derived for the actuators to adjust the tyre slip to the nominal slip values. An observer is used to estimate quantities like the tyre normal forces. As in

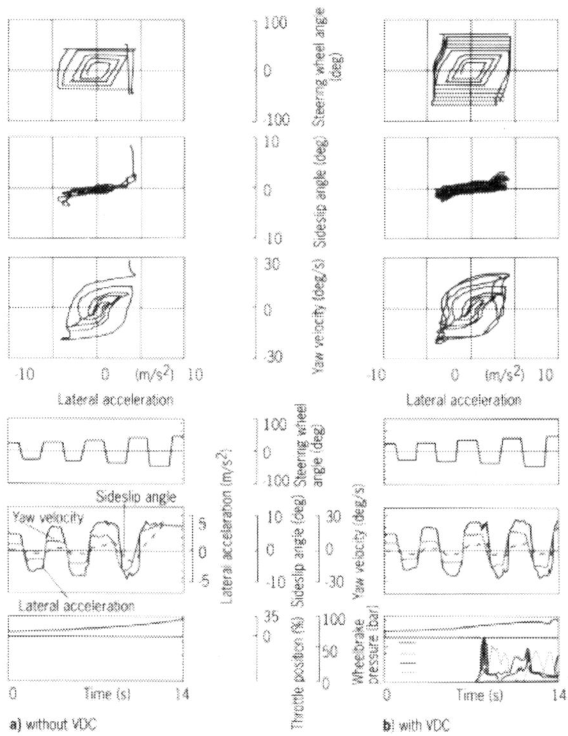

Fig 34: Vehicle behaviour in slalom with increasing steer angle

the ABS, the control algorithm is realised as a sampled data controller and it is programmed into a microcontroller. The slip controller has in case of traction slip also a cascade control structure with the engine management system as the inner feedback control loop. Nominal values passed to the engine management system are expressed in terms of engine torque and transmitted by a CAN interface.

Fig 34 shows the simulation results of a slalom manoeuvre in which the performance of a conventional vehicle without VDC is compared to that of the same vehicle with VDC. The steering wheel is rotated stepwise in opposite direction at regular time intervals and with increasing step values. The road surface is homogeneous with a medium value of the coefficient of friction (mu=0.45). During the manoeuvre the accelerator pedal position is gradually increased in order to keep an approximately constant vehicle speed. This will result in a steady increase of the traction slip at the driven rear wheels. On the left of the figure the results of the conventional vehicle are depicted while on the right the results of the vehicle with VDC are shown. The upper portion shows the results as phase plots in depending on the lateral acceleration of the vehicle. The lower portion shows the results as time histories. The conventional vehicle becomes unstable after the fourth positive step input of the steering wheel angle. Both the yaw velocity and the sideslip angle increase rapidly while the lateral acceleration stays at an almost constant value. Test results of the vehicle equipped with VDC show that the VDC system intervenes just after the third positive steering angle step input. This implicates a slight reduction of the throttle valve angle and an active braking at the front and rear wheels. The steerability of the vehicle is improved and the vehicle is kept stable. The vehicle motion also follows the sequential increase of steering wheel angle step inputs stably. The values of the sideslip angle and yaw velocity are limited by the limited coefficient of friction.

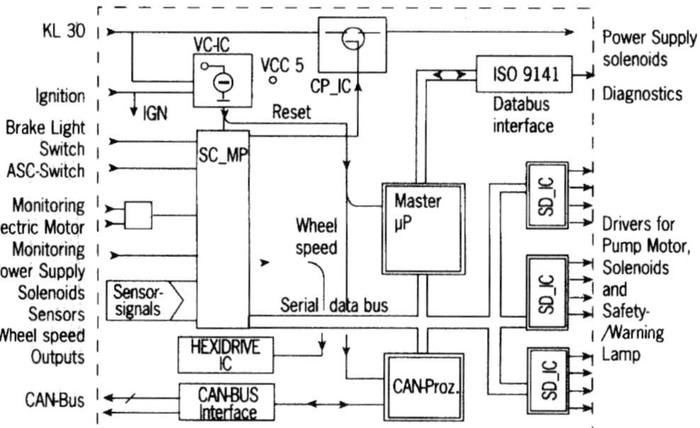

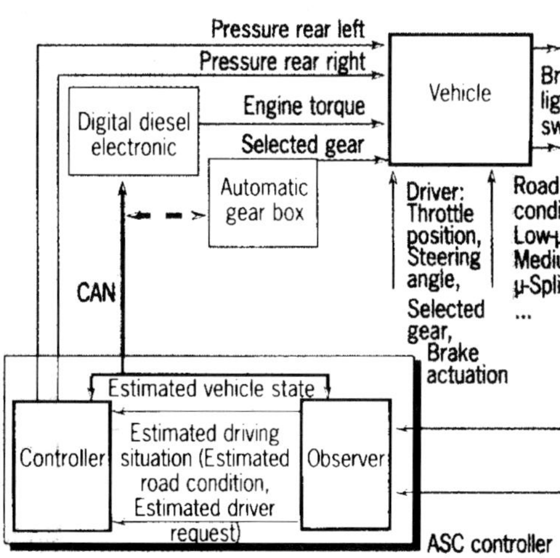

Fig 35: ASC controller as installed

BMW engineers[8] have described recent application of the ITT system to the 3-Series diesel-powered cars, including its serial data link by CAN bus to engine management system of the diesel engine. It is claimed also to be the first application of a combined stability and traction control system to a diesel passenger car.

The controller of the Diesel-ASC of the 325 tds car is designed as a separate unit, which is installed in the passenger compartment of the car. The block diagram of the ASC-ECU is shown in Fig 35. The board of the controller houses a signal conditioning microprocessor (SC MP) which serves the sensor input signals, the monitoring of the electric motor and the power supply of the solenoids. Furthermore the ASC-on/off-switch and the brake light switch are connected with the SC MP.

The wheel speed signals are processed and can be used for other sub-systems in the car. Four speed signal outputs according to the four wheels therefore are foreseen. A serial data-bus connects the SC MP with the master processor, the CAN-processor and the serial drivers (SD) which are designed as integrated circuits. The SD's serve the solenoids, the electric motor of the hydraulic pump and the various warning and information lamps of the system.

A serial diagnostic interface is foreseen to assist the failure detection of the system under service conditions. The diagnostic interface is used in the end-of-line test also, among others for the purpose of setting specific parameters in the ASC-controller. The CAN-bus is used for the communication between the ASC and the Digital Diesel Electronics DDE. This interface is used not only to reduce the engine torque in case of stability critical situations but also to overcome instabilities by reducing the engine braking moment on adhesion critical surfaces.

The block diagram shows the control circuit between vehicle and ASC-controller. The driver engages the clutch, selects the gear and presses the accelerator during different driving manoeuvres and under varying road conditions. The only signals the ASC-controller receives from the vehicle are the four wheel speed signals, the signal of the brake light switch, the status of the ASC-switch on/off and the information via CAN about the engine state such as the torque information and the engine speed (Fig 36). From this input the ASC controller has to estimate the different driving situations. The resulting control commands are either transmitted via CAN to the Digital Diesel Electronics DDE or directly drive the solenoids of the respective control valves of the ABS/ ASC-HCU. In case of an automatic gearbox an additional interface to the gear box control unit should be foreseen.

The control of the wheel slip at the driven rear axle is used to achieve the required vehicle behaviour. The ideal values of the respective wheel slip — so-called slip threshold — is calculated on the basis of the estimated driving situation. The vehicle speed, the vehicle acceleration, the wheel behaviour as well as the different left/right mu-coefficients and the driving manoeuvres are considered. During ASC-control only, two control circuits are used in case of the diesel-powered cars. The wheel-individual pressure control at the driven rear axle provides the differential brake effect, the engine control via the torque interface between ASC and DDE, provides the fast responding control circuit for the reduction of engine power.

To achieve a good control quality both control circuits provide autonomous and coordinating tasks. The autonomous tasks of the pressure control circuit are to achieve stability by the reduction of excessive wheel slip during control cycles and to improve the traction especially on split-mu roads. The coordinating tasks have to be considered in close conjunction with the respective other control circuits. The wheel-individual pressure control at one wheel must consider the dynamic of the opposite wheel, also, to prevent undesired excessive slip at the opposite wheel. The coordination of the pressure control circuits prevents phase-reversed wheel oscillations and allows the stable wheel to stay in its optimum wheel slip. Another part of the coordinating tasks is required for the transition between pressure and engine control.

Engine control depends very much on the vehicle speed, which is true for the coordinating and autonomous tasks of this control circuit, too. In the high speed range the slip control happens independently of the brake control. In this phase the control is more focused on stability. The slip thresholds for the engine control are here lower than the slip thresholds for the brake control. With reduced vehicle speed the coordinating tasks of the engine control become more dominant. Theoretical estimation of the brake pressure assist the adaptation between engine and brake control.

Information on the differences of the wheel brake pressures are considered to assist the ASC control on split-mu conditions and to achieve an optimum acceleration during such situations. By calculating individual wheel slip threshold, an optimum adaptation to stability or traction during the respective driving situation becomes possible. The different control circuits are designed such that influences by different dynamical reactions — such as response times, changing μs, changing gears, step input to the accelerator — are considered.

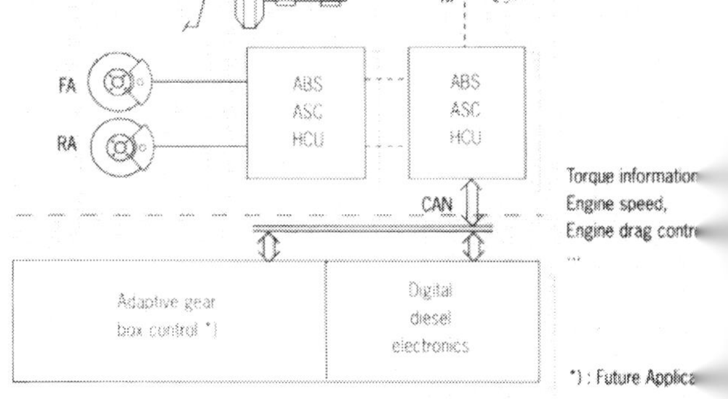

Fig 36: Interface between engine and ASC

The engine braking torque reduction uses a special engine control circuit. This control circuit is activated when the clutch is closed and at least at one of the driven wheels a suitable negative brake slip is detected. The engine drag reduction happens virtually via a simplified "inverted" ASC engine control.

The signal flow between the different control units is shown in Fig 36, where all the different signals and messages are marked. Besides torque information which consider friction, drivers demand and the indicated torque the engine speed and ASC acknowledge information is exchanged between the ASC and DDE control unit. In a similar manner the engine drag control is carried out by torque increase commands, which reduce in case of engine braking situations the disadvantageous effects of brake slip at the rear wheels. The DDE transmits information on the engine status periodically. Based on engine speed and the average speed of the driven wheels the gear ratio and thus the engaged gear is determined. This ratio allows driving torque at the wheels to be established from the engine torque, and vice versa, during phases with excessive wheel slip when prediction of the nominal engine torque becomes possible. Special check bits are used to increase the immunity against wrong information transmission. The ASC transmits the wheel speed information for other subsystems in the car. A special safety logic ensures the detection and the localization of possible failures.

Fig 37: Microhybrid electronic control unit

Electronics problems and prospects

According to Dr Martin Zechnall, President, Bosch Automotive (Semiconductors and electronic control units equipment division) electronics in a vehicle are subjected to a temperature range between -40 and +130 C, and the acceleration can reach 100 times the acceleration of gravity (100 g). In addition, dirt, dust as well as aggressive substances such as gasoline, oil, cleaning solvents, moisture or salt spray, can cause damage. Finally, voltage surges in the electrical system as well as electromagnetic fields must not affect the operation of the devices.

Bosch employs two different substrate technologies depending on the mounting location of the electronic component in the vehicle. For locations outside the engine compartment, where temperatures reach a maximum of 95 C and the vibration load does not exceed 10 g, printed circuit boards are used, Fig 38. For installation on the engine, where operating temperatures can reach 130 C or on mechanical assemblies where the vibration load can reach 100 g, the hybrid technology is particularly appropriate. Installing hybrid electronic control units on components such as engines, diesel injection pumps or ABS-hydraulic units, allows the vehicle manufac-

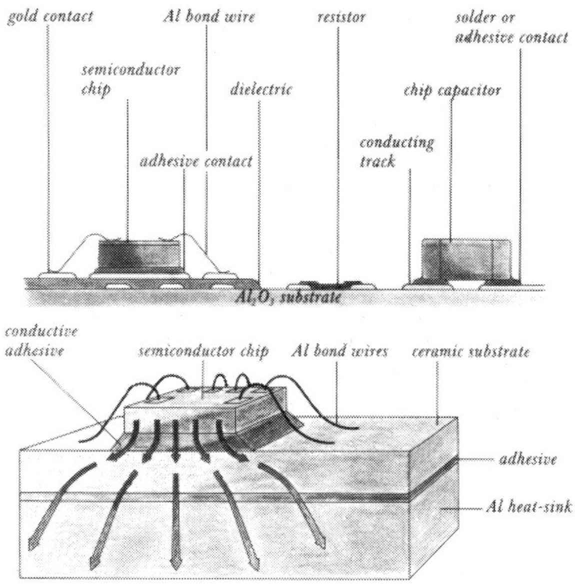

Fig 38: General design of a thick-film circuit, above, with optimized, directed heat flow from semiconductor chip to heat-sink, below

turer to simplify the vehicle's electrical harness and reduce inventory, Bosch maintain.

Printed circuit board technology The latest generation of printed circuit board control units is built in modular form. The surface mounted components are attached to both sides of the circuit board and soldered by the so-called reflow process. Heat sinks are no longer required because the dissipated heat from the power transistors is conducted via extensive copper surfaces to the metallic housing. Such standardized control units are very compact and light. This simple construction allows highly automated mass production combined with very high quality, low costs and recycling-compatibility.

Microhybrid technology Bosch claims to be the first worldwide supplier of automotive electronic control units produced with microhybrid technology. This technology makes possible a further miniaturization of the control units. Microhybrid technology is based on a special support material sintered together from several layers of flexible ceramic foil.

Circuit patterns are screen printed on the individual unfired ceramic foils. The electrical connections between the different foil layers are realized by punching holes, called "vias", which are then filled with conductive paste. The individual layers are stacked vertically and sintered at about 900 C to become a rigid multilayer substrate. Resistors are screen printed and fired on the reverse side of the substrate. The remaining passive components and the integrated circuits (IC) are then mounted on the front side of the substrate using chip and wire technology. With this technique unpackaged silicon chips are attached to the upper side of the substrate and connected to the printed circuit with wire bonds. A gel coating protects the electronics from moisture. The resulting microhybrid is very small as a result of the dense packing of the printed circuits and the small size of the unpackaged chips.

Production of Motronic control units is underway in microhybrid technology and by 2000 the company believe 30 per cent of all its control units will be produced with this technology. The unique advantages of hybrid technology are resistance to high temperature and vibration loads, combined with improved electromagnetic compatibility and reduced size. Taking advantage of the high level of integration possible with this technology costs similar to these of printed circuit boards can be achieved.

Semiconductor technology A high level of integration is required to reduce the size and weight of electronic control units and also to increase functionality. This is possible with all mainstream semiconductor technologies, including bipolar, CMOS or mixed technologies. Today, protective, control and diagnostic functions are integrated in power output drivers (Smart Power). This high level of integration of application specific circuits is key to the production of reliable and economical control units.

Sensor technology An interesting example is the intake manifold pressure sensor developed for the first mass produced electronic fuel injection system for gasoline engines. This sensor went into production for the export version of the Type 4 VW 1600, to fulfil the 1967 California exhaust emissions stand-

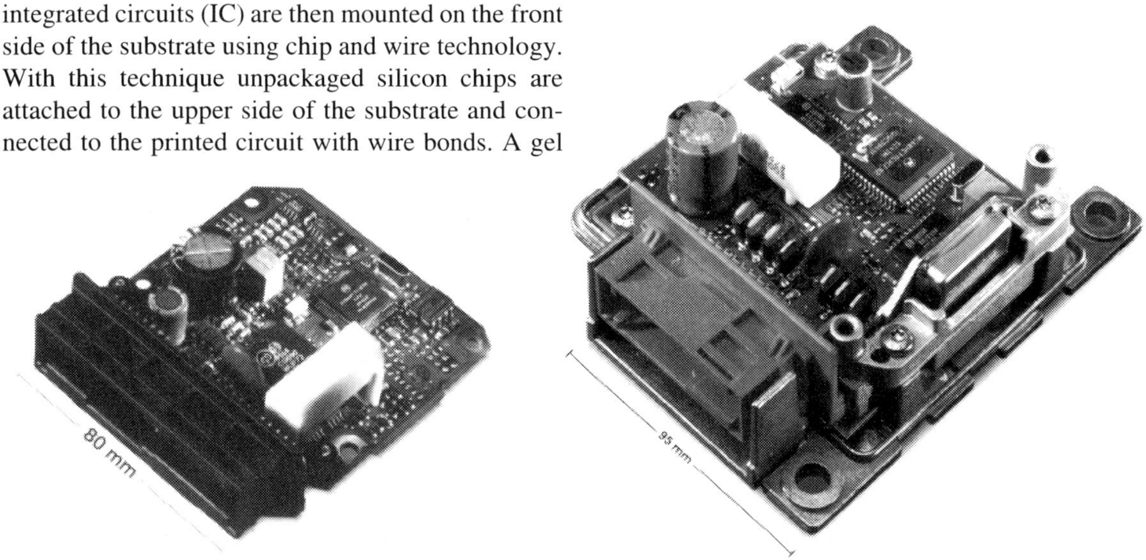

Fig 39: New and old airbag triggering devices compared, showing reduction in size

ards. Sensors for air volume, air mass, engine speed and exhaust gas oxygen concentration (lambda sensor) followed later. Bosch say the yearly growth of the world market for sensors now approaches more than eight percent, and this growth has brought a technological change in the last decade from classical electromechanical sensors to silicon-based electronic sensors. Silicon does not lend itself merely to the production of electronic circuits, but can also be used as the basis for very small mechanical devices. Its mechanical elasticity, combined with its electrical properties, are ideal for the production of micromechanical sensors. The mechanical and electrical functions of the sensor can be integrated into a single chip by adapting processes from standard semiconductor technology. The company has chosen to employ two silicon micromachining technologies, bulk micromachining and surface micromachining. In the first, the entire thickness of the silicon wafer is shaped by etching to reveal deep cavities. In the second, however, the micromechanical devices are formed only on the surface of the wafer by depositing and etching polysilicon and sacrificial layers.

An example of bulk micromachining is a sensor that measures intake manifold pressure for engine management. In this sensor the sensing element, that converts a mechanical to an electrical signal and the signal evaluation circuit are integrated on a single chip. The sensing element is etched to the thickness of a very thin membrane. The evaluation circuit is located around the membrane. The complete sensor chip is mounted on a glass base, to isolate the sensor from the thermal and mechanical stresses on the housing. The pressure to be measured is applied to the reverse side of the silicon membrane in order to protect the delicate front side of the silicon chip (with the evaluation circuit) from corrosive materials. The pressure deflects the membrane and is measured with piezoresistors. The evaluation circuit, which is integrated on the same chip, amplifies the electrical signal. Two important additional processes are involved in producing this bulk micromachined sensor: "anisotropic etching in an alkaline etching bath" to form the membrane, and "anodic wafer bonding" to connect the silicon chip to the glass base. Using the anisotropic etching process, the membrane thickness is produced to within a precision of 0.5 micrometer. Anodic wafer bonding ensures a hermetically sealed joint between the potentially corrosive gases and the delicate upper surface of the chip. An example of a surface micromachined sensor is the company's acceleration sensor, Fig 39; it triggers deployment of the airbag in a collision. The sensor consists of two silicon chips: the "heart" of the sensor, the sensing element, and the evaluation circuit. The sensing element is discernible only under a microscope. The seismic mass with its comb-shaped electrodes is suspended by fine micromechanical springs. Additional fixed comb-shaped electrodes on the chip mate to those of the electrodes on the seismic mass. If the sensor experiences an acceleration, the distance between the movable and fixed electrodes changes. This leads to a change of the electrical signal in the evaluation circuit. Surface micromachining uses only the surface of the wafer, Fig 40. Three processes are typically employed in the production of these sensors. First a so-called "sacrificial oxide layer" is deposited and structured using standard semiconductor production techniques. Second, a polysilicon layer is deposited and similarly structured. Finally, the "sacrificial oxide" is removed, leaving the shaped polysilicon. In this way moveable 11 micrometers thick polysilicon structures are realized with minimum widths as small as two micrometers.

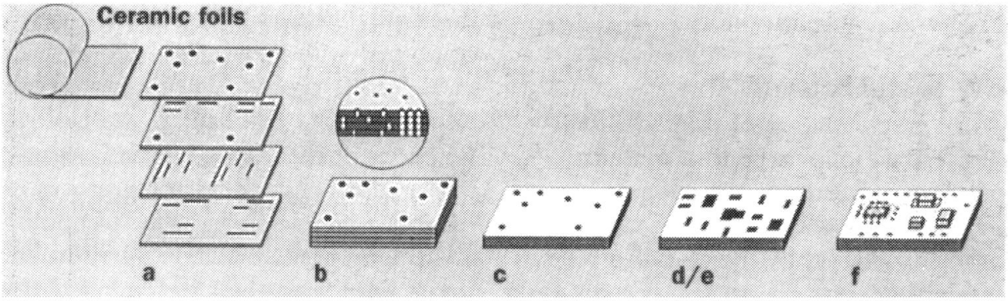

Fig 40: Circuit patterns screen-printed on individual unfired flexible ceramic foils, in microhybrid technology: a. punched-hole 'vias'; b. stacked layers: c. sintered multilayer substrate; d. screen-printed and fired resistors; e. IC mounting by chip-and-wire; f. chips attached by wire bonds

Microprocessor developments

Motorola Semiconductors explain[9] that following the progressive introduction of more complex timing functions, such as those needed to support engine diagnostics, considerable strain is being applied to the already overloaded conventional form of microprocessor. A new concept in microprocessor-based timing system, the Timing Processor Unit (TPU), has therefore been introduced.

The company explain that a processor without this hardware might capture a signal using an interrupt or by polling a bit of an input port. Once the appropriate signal is detected, software could read a timer to record the time of the event. Because of the software execution, this technique introduces a small delay between the time that the event occurred, and the time of reading the timer. While this delay can be corrected when the software execution time is known and fixed, in an interrupt driven system, various interrupts may be handled at arbitrary moments in time. Likewise, the timing of output signals are equally affected, with the error in this case manifesting itself as an unacceptable jitter in the synthesised waveform. In addition, the microprocessor's bandwidth may be rapidly diminished as software is called upon to handle every external transition. While one solution might be just to increase the microprocessor speed. Another solution, offered by some manufacturers, is to invoke some simple dedicated hardware action to replace or enhance a software interrupt handler.

An alternative solution is to provide coprocessor support to the microprocessor, the author argues. The TPU falls into the latter category, its architecture having been specifically tailored to manipulate the real time events associated with its in-built timer hardware. It can be viewed as a special purpose microprocessor geared to performing operations on time events, independently, if necessary, of the CPU. In practice, the TPU would be initialised by the CPU module, and then allowed to run without intervention except under specific predetermined conditions. In this manner very complex timing tasks can be executed by the TPU with minimum or even zero CPU overhead. The TPU can be partitioned into five main functional blocks; the host interface, the timer channel hardware, the scheduler, the microengine and the execution unit, Fig 41.

The transition from one time slot to the next occurs when the current state of a time function has been completely executed, and a pending service request is detected by the scheduler. During the time slot transition period (TST), a number of tasks must be completed before control of the microengine is given to the state which is executed in the new time slot. Most importantly, the content of the capture register is copied to a holding register (ERT) within the execution unit, and all the new channel flags are latched. The states of six of the flags — four service request flags, the pin state flag and one of the general purpose flags — are encoded and combined with the channel function number to form the address of an entry point location. The entry point can be viewed in part as a vector which contains the start address of the executable micro instructions for the new state. In addition, the entry point contains control bits to define which of the new channel's parameter RAM is

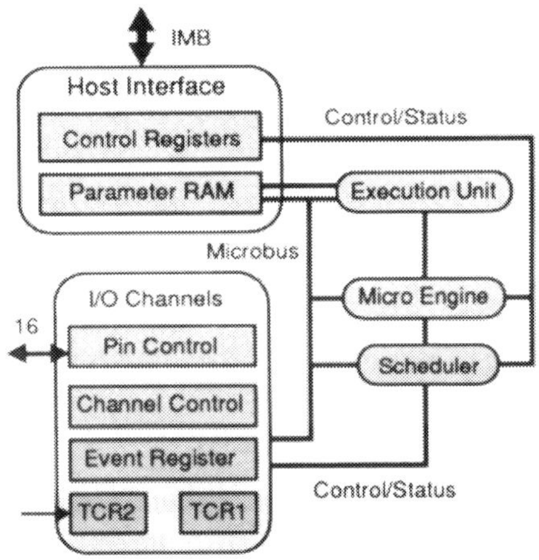

Fig 41: TPU functional block

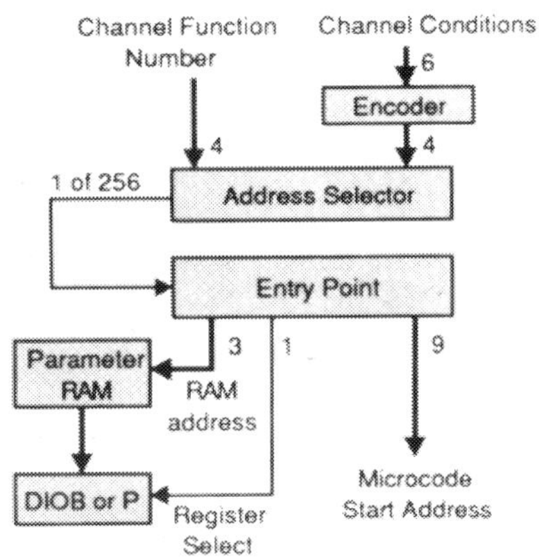

Fig 42: State start address generator

pre-loaded into a choice of two execution unit registers before execution of the new state begins. Finally, the microprogram counter is loaded with the start address of the new state. These operations complete in 600 nanoseconds with a system clock of 16.7 MHz.

Due to the partial encoding of the six channel flags, a total of 16 separate entry points may be defined for each time function. These would correspond to 16 separate states. However, by careful use of parameter RAM, and the remaining general purpose flag, a virtually unlimited number of states may be implemented for each time function, Fig 42.

ASIC developments

Texas Instruments (UK) argue[10] that while digital ASICs, and other components, are very applicable to the automotive system designer, much analogue processing of signals is still required. Traditionally, any integration of analogue functions has required a costly and time-consuming full custom design, only appropriate to high volume applications with little time constraint. To integrate analogue processing, full parametric performance must be simulated under all application conditions, and a different process technology is needed, the authors explain. They have shown how linear ASIC technology satisfies many of these requirements and use the example of an automotive oil level sensor in order to combine analogue circuitry with digital electronics.

This combination — plus cell libraries and design tools — is known collectively as LinASIC. At the time of the announcement there had been three approaches to mixed-signal ICs; arrays, full custom and cell based. Although analogue arrays have been available for more than a decade, they have yet to be widely adopted due to their limited analogue flexibility, low logic integration capability and comparatively high unit pricing.

A cell based technique reduces time and risk by using proven cells, Fig 43. As designs are executed faster, design costs are also reduced allowing the technology to be applied to a wider applications base. Die sizes will be slightly larger than full custom, but by using efficient cell libraries, this is reduced to only a minor factor. The only significant limitation is that if a suitably sized cell library is to be provided, then the number of technologies supported must be minimised. Most previous mixed-signal ASIC devices have been based on pure CMOS technologies. While this is ideal for the integration of logic and most analogue signal processing elements, its low voltage capability (usually < 10V) and poor drive capability limit its use, especially when subjected to a wide range of voltage transients. The ideal solution would be a merged CMOS-Bipolar process as this allows bipolar structures to be used for high voltage/current sections of the device, and CMOS for the analogue and digital core sections. Other features of a process which would be very welcome to automotive systems designers are non-volatile memory (such as EEPROM) for sensor calibration, on-board diagnosis and configuration; for example a fuel gauge could be programmed for a given fuel tank's characteristics.

The economical solution is to develop a truly modular technology, known as LinBiCMOS, which allows designs to use only the process modules they need for a given application, thereby offering an optimised solution, Fig 44. LinBiCMOS is in fact a

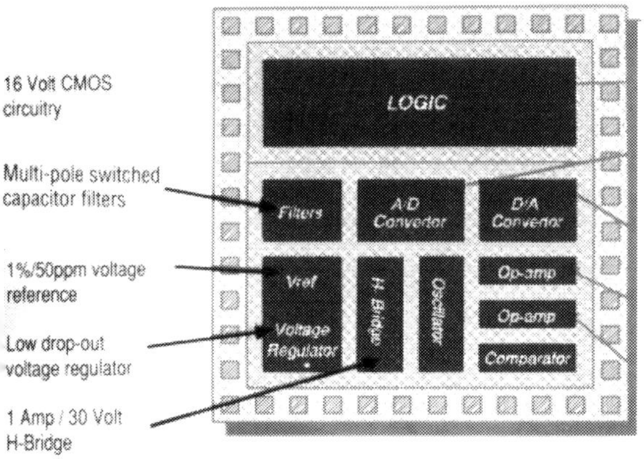

Fig 43: Block diagram of typical LinASIC

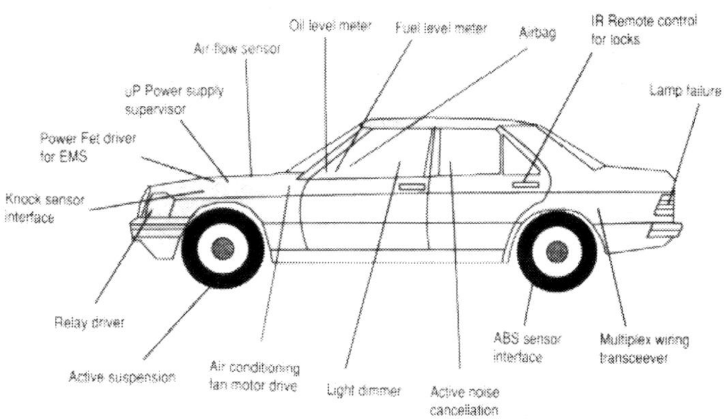

Fig 45: Application areas

pure 2 μm CMOS process (named LinCMOS) with 10V capability. Each module adds successive features, such as double level poly-silicon for filters and analogue-to-digital convertors, double level metal for high density logic and a vertical NPN device for precision circuits. LinASIC has already been used for many mixed-signal IC automotive applications including power FET drivers with on-chip diagnostics, oil and fuel level meters. Although devices are available to handle multiplex wiring protocol, they can not support the physical interface. LinASIC has been used to develop a 250kB/50mA multiplex wiring transceiver which has the capability to be included with several thousand gates of logic, and if required high current switching elements, Fig 45.

In the single chip solution to the problem of reading and displaying oil level, the principle is used of measuring the resistance of a wire passing a high current which is immersed in the oil. The resistance is measured once before the wire has warmed up, and then again once the wire is hot. As that part of the wire immersed in oil remains cold in both cases, the oil level can be found by subtracting the 'cold' value from the 'hot' one. The value is then displayed on a standard analogue display. In the solution, Fig 46, the resistance of the wire is first digitised using a 10-Bit successive approximation analogue-to-digital converter (ADC). This value is then stored in a digital latch whilst the 'hot' reading is made. The two values are then numerically subtracted from each other, giving the oil level. To generate an analog display of the value, the digital value is applied to a 10-Bit digital-to-analogue converter.

Computer-control terminology

ASICs: Application Specific Integrated Circuits
Bipolar Technology: semiconductor technology chiefly used for analog circuits.
Bulk Micromachining: technology for producing very small mechanical structures in the interior of the silicon wafer.
CAN (Controller Area Network): a digital data bus circuit developed by Bosch, designed for information exchange between the electronic systems on a vehicle.
Cartronic: a hierarchical system framework to structure the different electronic systems in a vehicle.
Chip-and-Wire Technology: unpackaged silicon chips are bonded to the upper side of a substrate and connected to the circuits with wire bonds.
CMOS Technology: semiconductor technology chiefly used for digital circuits with very low energy consumption.
DMOS Technology: semiconductor technology chiefly used for power output drivers.
EGAS: electronic accelerator pedal.
Hybrid Technology: the assembly of active and passive components on a ceramic substrate onto which resistors, conductors and electrically isolating layers are printed.
IC (Integrated Circuits): semiconductor-based components.
Microhybrid Technology: hybrid technology in which the ceramic substrate consists of several ceramic layers with printed circuits.
Microprocessor: logic semiconductor component that constitutes the central unit of a small computer.
Mixed Technology: semiconductor technology integrating CMOS, DMOS and bipolar (analog) circuits.
Surface Micromachining: technology for the production of very small mechanical structures on the surface of the silicon wafer.
Polysilicon: silicon in polycrystalline form.
Reflow Process: a soldering process in which the solder is applied as a paste and the soldered connections to the components are made by melting in an oven.
Silicon Chip: semiconductor component made of silicon.
Silicon Micromachining: collective term to indicate the technologies to produce very small mechanical structures in silicon.
Silicon Wafer: circular shaped silicon substrate from which many individual integrated circuits (silicon chips) are simultaneously produced.
Via: electrical connection between conductors through a ceramic layer.

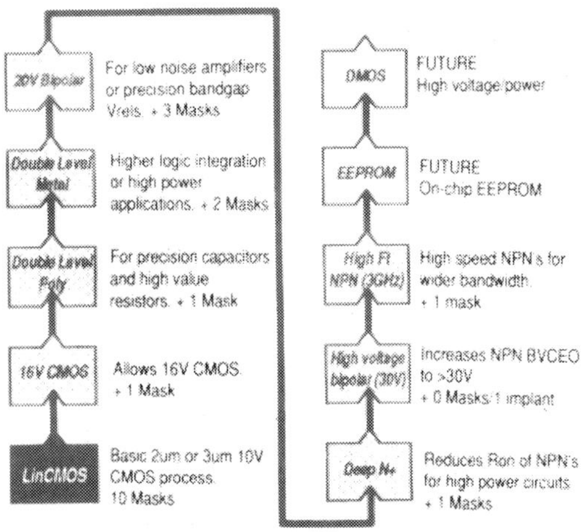

Fig 44: LinBiCMOS process modules

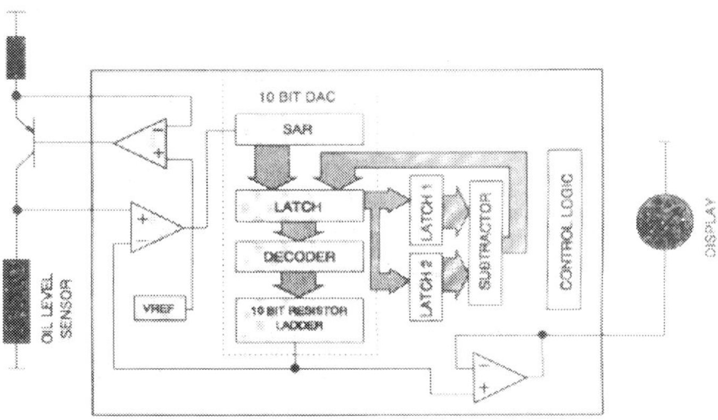

Fig 46: Oil level meter example

Economical multiplexing

Lucas Rists Wiring Systems[11] pointed out that early work on multiplexing concentrated on areas of high congestion, with the implementation of low speed multiplex links. The 'A' post link through to the driver's door switch-pack was typical where up to 60 leads could be replaced by just three or four.

Such systems had four main drawbacks: each was designed specifically to perform a certain task for a particular vehicle — as a result, they were inflexible; they only addressed problems in particular areas of the vehicle; non-industry standard protocols were often used; they could not be justified on economic grounds.

To overcome these problems, an alternative 'whole vehicle' approach was proposed, involving a multi-master architecture. This comprised a number of 'modules' distributed around the vehicle. Each one controls/monitors the actuators/switches relevant to its position within the vehicle, via a local wiring harness. They are connected together by two signal wires that form a data bus. Any of the modules may transmit/receive control information to/from others on this bus. The general principle is that the control data is sensed at the module nearest to the relevant switch/sensor. This control information is then broadcast on to the bus. All of the other modules receive this information, and those that are required to act on the data do so via their local harness. This approach has several advantages over a conventional harness: has potential to reduce total number of leads; reduces length of remaining leads; reduces current-carrying capacity of both the switches and the interconnecting leads; also saves space in critical areas — behind the dashboard, say.

Flexibility is such that a complete model range may be covered by a core of, for example, 10 common harnesses, with different local Input/Output harness from each module catering for different trim levels. Also, any new features for the vehicle can be incorporated with just the simple re-work of small I/O harnesses. To obtain the maximum benefit in terms of easier installation and handling, and better quality and reliability, a modular harness approach must be adopted, advantages being: resultant harnesses are easier to install on the vehicle; they require less complex lay-up boards; they are easier to handle and less damage prone; they are suited to automated manufacture; changes can be accommodated more

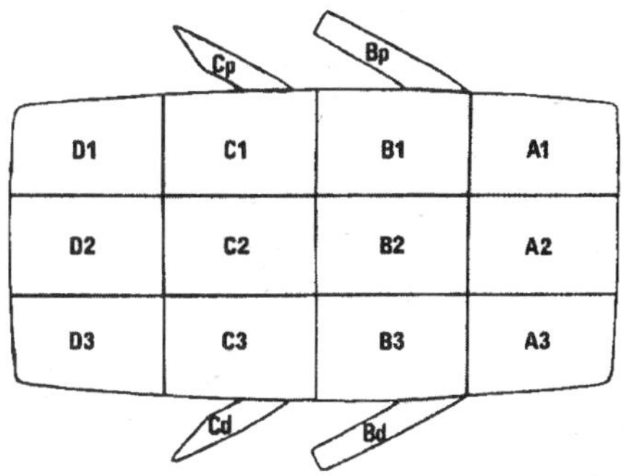

Fig 47: Functional areas of vehicle

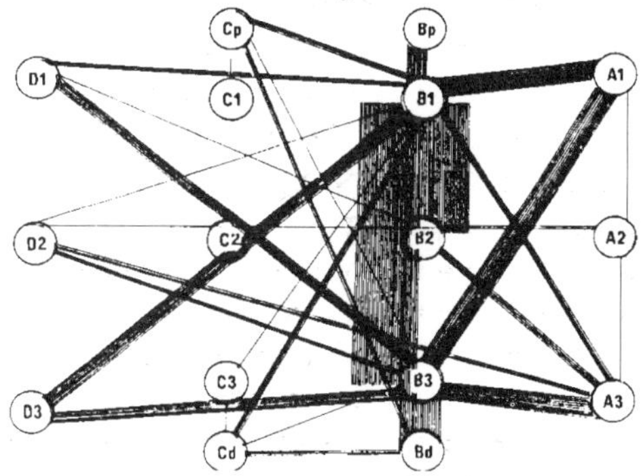

Fig 48: Source-to-destination mapping

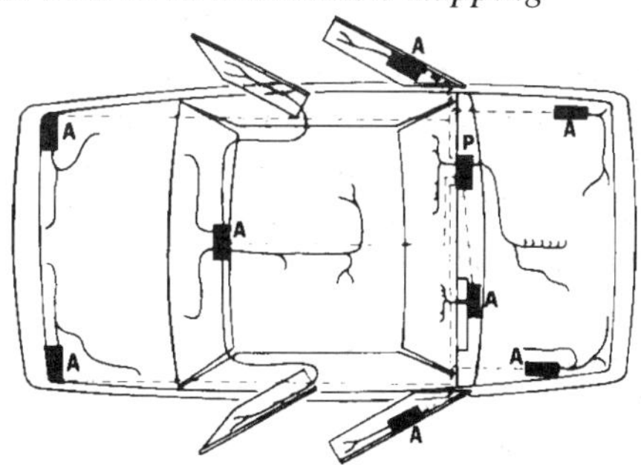

Fig4 9: AJB locations, C = conventional harness; M = modular harness; E = existing ECUs; A = AJBs; F = fusing

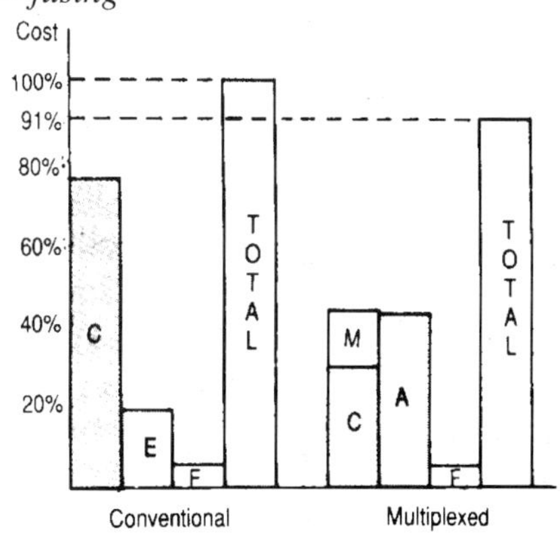

Fig 50: Multiplexing cost benefit

easily. By using junction boxes throughout the vehicle, a complex conventional harness can thus be split up into a series of smaller, more manageable harnesses, containing typically less than 40 leads.

In the project described by the authors, Active Junction Boxes were used. These integrate the databus access point and the modular harness junction box into the same housing so as to: transmit and receive data via the bus; control and monitor local actuators; accept data from local sensors; enable inter-connection of modular harnesses. It was decided to split the vehicle into 16 clearly defined areas, Fig 47, and each signal on the vehicle was mapped from source to destination (often from switch to load), Fig 48. Areas of high complexity could then be simplified with multi-master system. To provide fused supplies to the AJBs, and to enhance the modular harness structure, a passive junction box is located in B1. Thus the complete system comprised eight AJBs, one PJB, and several modular harness sets, Fig 49.

The potential cost savings of this solution are, categorised into direct and indirect savings, Direct: reduced wire count; Indirect: automated harnesses are cheaper to build; automated harnesses yield less scrap; harness sets produced to a JIT system. With these savings, the cost breakdown of this approach is compared with that of the conventional harness in Fig 50.

Connectors for multiplexing

Most of the major automotive connector suppliers are addressing this market and sample products are seen in Figs 51, 52 and 53. AMP's on-going development work with the auto-industry has resulted in several solutions. Environmentally sealed connectors were specially designed and developed for the Jaguar XJ6 vehicle electronic wiring management system, for example. For the industry as a whole, the previously labour intensive assembly of wiring harnesses has been made easy by designing connectors and cabling to be compatible with high speed termination by autoatic machine.

Type of multiplexing for initial introduction is likely to be the relatively low frequency (tens of kHz) one — for general car-body/drive controls — and of star formation. Such systems canuse filtering to eliminate interference. Limiting factor to date of such multiplexing, say the company, is the cost of power-switching. Voltage requirements in the connecting system will be 5-30 volts and currents to be switched up to 25 A. Plug-in modules are likely first to apply to headlamp/tailpiece clusters: no flying leads between module and active devices (precursor to the smart connector). Ty pes of con nectors seen by AMP as being required for multiplexing includes wire-to-module format: high current (to 25A) type; signal current (5 mA at 30 V) type; under-bonnet sealed connectors; unsealed (protected-area) connectors and connectors which suit auto-assembly. Crimp poke-home termination is likely to be accepted but some OEMs are reportedly interested in insulation-displacement terminals for high wire density, low current, applications. Making doors off-line with as many as 65 connecting wires across to the A-pillar is impossible to achieve with conventional door-to-body connections.

AB Controls & Connectors has undertaken a market survey of OEM requirements of automatic connectors and a study of some 700 connector types. One important finding, by VW,was that up to 60% of vehicle electrical faults were attributable to connector or harness malfunction. A result of this is the Autocrat connector system, Fig 54. Conventional automotive industry practice is for electrical parts

Fig51: AMP Positive Mate Mark 4

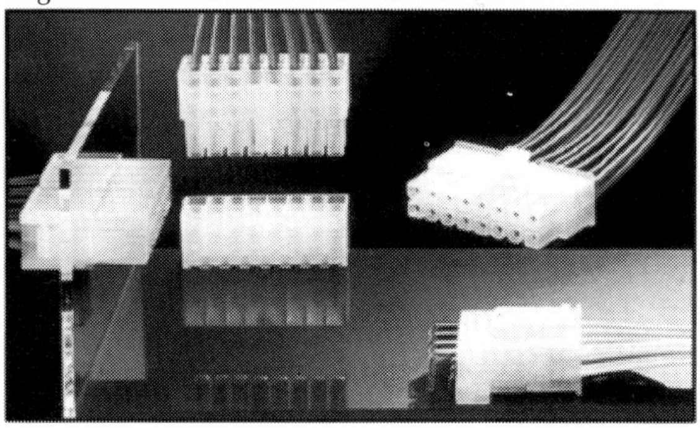

Fig5 2: Molex Minifit Junior

and components to be manufactured complete with their own unique connectors permanently attached to part leads. With AB's Autocrat system, these many different connectors can be replaced by one simple contact, crimped to every part lead. Sub-assembly of electrical parts is then simplified by inserting up to 30 such contacts into a single shell — which can then be connected into the main wiring harness. In typical vehicle harnessing applications, this results in a considerable reduction in the number of connectors required. Studies of connector systems currently being used showed potential areas of improvement of design and performance as primary contact retention, coupling forces, electrical contact characteristics, application of components used and connector latching. This led to the development of the Autocrat connector system that would, claimed AB, considerably improve reliability, reduce size, be usable in both standard and multiplexing wiring harnesses while reducing the complexity of the parts used and without increased costs.

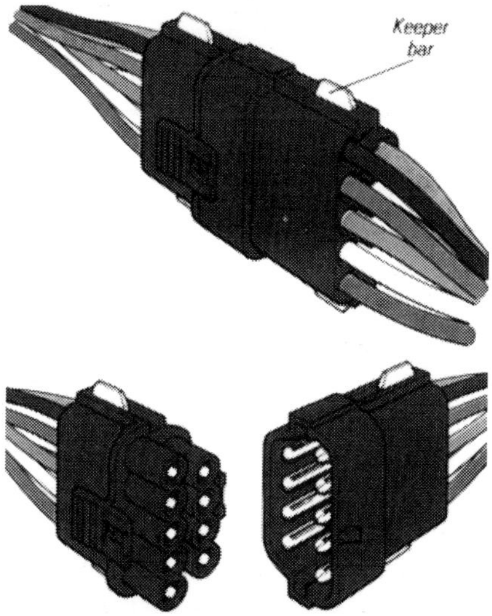

Fig 53: Rist Total terminal security

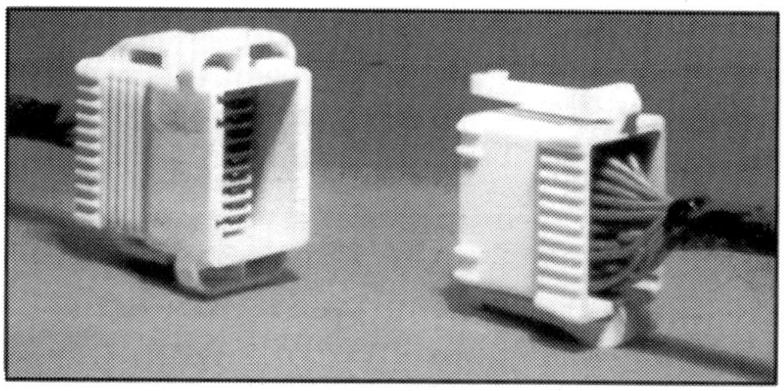

Fig 54: AB Autocrat

Connector design and the Autocrat Firstly, the problems with individual pin and socket contacts are threefold. The expansion of the socket as the pin is inserted during repeated cycling causes a degradation of electrical characteristics. In addition, the insertion of the pin into the socket produces a force which is transmitted to the primary contact locking feature. This force tends to cause the contact to back out of the housing and only the reaction of the lance prevents this happening. Weakness of the lance due to corrosion will cause connector failure. Problems can arise at the assembly stage of conventional pin and socket connectors. In Fig 55, (1) the locking lance A provides the primary locking feature into the connector housing. When the contact is inserted, the lance is deformed and should spring back into position after it has passed the latching ear. If this fails,

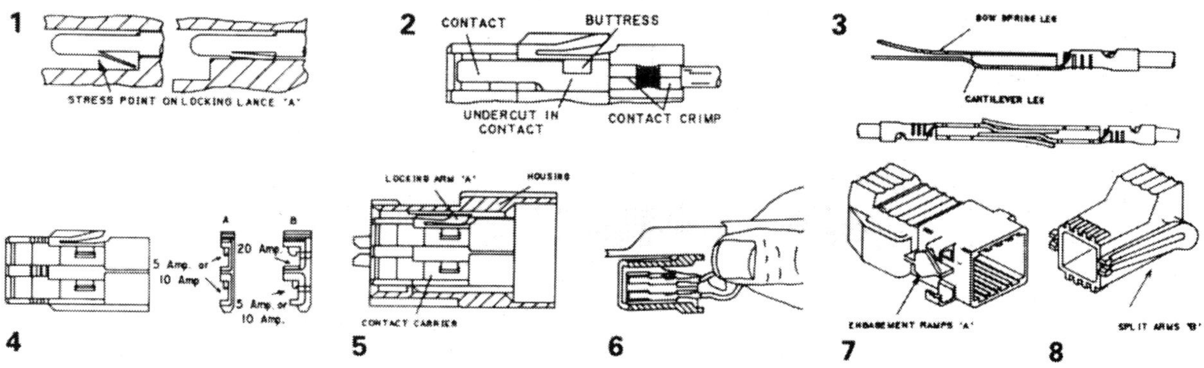

Fig 55: Detail design of AB Autocrat

then primary locking is lost. In this design the problem has been overcome by providing primary locking in the plastic carrier by means of a moulded latch . The contact has a locking slot which, during assembly, is snap-locked together and does not create deformation of either part (2). To overcome the problems of intermittency and open circuit, which can be encountered in practical use due to ovality and open seams, a new approach to contact design was adopted. The cantilever/bowspring concept was developed as an alternative to pins and sockets, and results in improved electrical characteristics. Contact pressure between mated contacts is achieved by restricting the movement of the bow-spring leg within the contact carrier and thus maintaining the force on the cantilever leg of the mating contact. This produces a spring loaded enclosed sandwich with contact pressures maintained on six faces (3). Contact pressure is distributed over a large surface area thus pressure per unit is small and contact wear is reduced. Contact forces result from low rate springs and this reduces spring fatigue. Hence there is very little change in contact resistance or voltage drop during life. This has been shown during cycling trials, say AB.

The Autocrat in detail The design concepts employed in AB's range of military connectors (these use a range of different sized shells containing a variety of inserts which mix contacts of different current carrying capacity in a range of planforms) has now been utilised in Autocrat. To satisfy the need to rationalise the connectors used, it is essential that the plug and receptacle housings can carry contacts of different current carrying capacity in a totally flexible manner. AB has achieved this by designing standard housings into which contact carriers can be securely locked. (4A) shows a contact carrier that will accommodate two contacts, 5 or 10 amp rated contact; (4B) shows a contact carrier that will hold one 5 amp and 20 amp rated contact. The carriers are stacked in the housings on a pitch which allows PCB mounting contacts to be used in the same housings and carriers. The contact carrier is common to both the plug and receptacle housings. When inserted into the housing, it is securely held in position by its locking arm A (5). By using a simple extraction tool (6) the contact can be removed from the housing allowing it to be replaced with one of a lower or higher current carrying capacity. In the unlikely event of the contact being damaged, it can be replaced with a new one. Current connector housings, AB argue, cause the following reliability and design problems: contact back out and the need to provide secondary locking features; crossmating, coupling shortcomings, lack of flexibility of use, increasing size and handling difficulties. In the Autocrat design, the position of the contact is secured by locking it into the contact carrier. The contact carrier is then locked into the housing. An audible click occurs as the lance locks into position indicating that the contact carrier is securely in place. This can also be controlled by an insertion tool at automatic assembly stage. A versatile key/keyway system is incorporated at the moulding stage, (7,8). Using this system of orientation, the keys and blanking pairs can provide up to 28 uniquely matching combinations, thus preventing an incorrect plug or receptacle being mated. The chance of incorrect mating is further reduced by the availability of a range of colour-coded housings. To ensure that the housings are fully coupled, a positive engagement

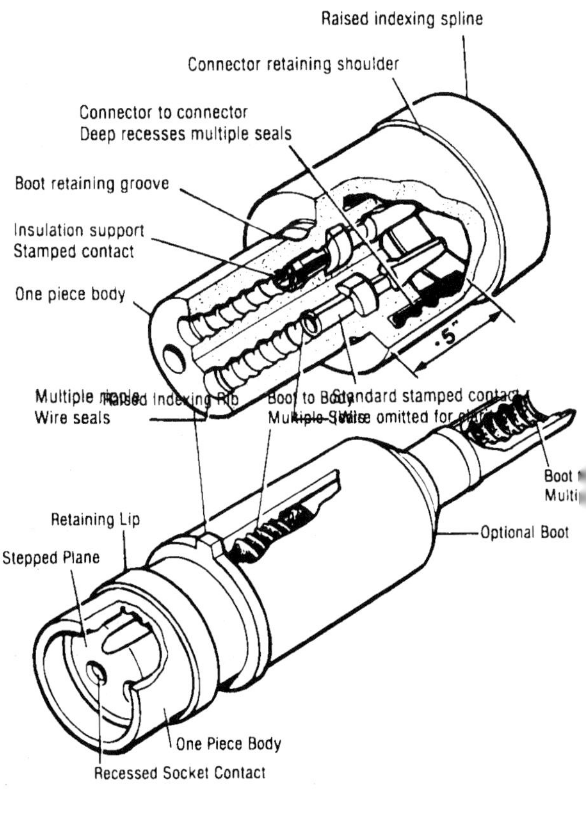

Fig 56: EITT Cannon Sure Seal

throw-apart feature forms an integral part of the moulding (7). Until the split arms B reach the apex of the engagement ramps A, there is a positive resistance to engagement that will throw the housings apart and prevent contact engagement. Once the arms pass the apex, they are drawn together ensuring that the contacts are fully engaged. The design of this feature when combined with the connector's high contact density is such that there is a significant reduction in size when compared with conventional latching pin and socket connector systems. The receptacle housing is designed to offer a wide range of mounting methods. It can be mounted to a printed circuit board, located through a panel or attached to a chassis by means of simple accessories. A cable tie can also be passed through slots on the rear of the housing, which enables it to be secured in a variety of other positions.

Most connector manufacturers offer splash-proof connectors to the automotive industry, Fig 56. These normally employ a different construction from unsealed versions and they are usually significantly larger. Harness manufacturers have to carry these as extra stock items in order to satisfy the need for splash-proof connectors in the vehicle. The Autocrat design simply requires the addition of a rubber grommet at the rear of each housing, and a small gasket for the mating surface of the housings, to produce an immersion-proof connector system.The overall sizeof the connector is only marginally increased and all parts are common to the unsealed version except for the grommets and gaskets. The grommets have a built-in membrane through which the wires of the connector are pushed. By this method, sealing is maintained around the wires and any holes in the insert which are not used remain completely sealed.

Suitability for multiplexing is inherent in the ability to separate power and data lines in one connector unit. The contact-carrier system is the essence of the connector's flexible application to any sort of circuit. Forall applications there are a total of 42 part numbers in two contact carriers, three contacts and eight shells. Under development is a low-cost fibre-optic connector system for data lines. Also being researched is a design, without bowed cantilevers, which will allow direct connection to components. Metallised plastic shells are being considered, too, for EMC protection and double cable density systems are underdevelopment involving two-sided carriers with four cable contacts.

Networking

The stage has now been reached where interconnection of systems is being used for such combination features as automation of vehicle handling, occupant comfort/entertainment and overall powertrain control.

Multiplexed body electronics The large number of controls applying to body electronics, in such applications as heating/cooling comfort control, seat adjustment, window/door systems, lighting and dashboard instruments has brought multiplexing into focus as a solution to problems of conventional wiring proliferation sooner than for chassis and powertrain electronic controls. An up-market car can now contain some 2 km of cables and 1500 electrical connections and past experience has shown the quantity to double each decade. Linking systems together by a bus network, in the multiplexing process, offers the best solution to containing this problem. Partial multiplexing solutions for body electronics have already been publicised and adopted in production vehicles. Total body system multiplexing is rarer in application but is much discussed as the next step for vehicle makers.

The CAN (Controller Area Network) protocol for multiplexed interconnection has become an ISO standard and a total body system interconnection approach has been examined by researchers at Warwick University[12]. It is argued that low-speed low-cost systems are now feasible with the availability of CAN chips developed for non-automotive industrial sectors coupled with the reduction in price of smart power FETs (Field Effect Transistors) for power switching. Additional safety features are available at no extra cost with multiplexing and costly timer units for individual control systems can be avoided. It is suggested that future vehicles will contain a two bus system, Fig 57 — one high speed ring for powertrain and chassis and a low speed body control data bus.

The authors consider the low speed bus is essential for ensuring that low priority, non-time-critical information does not cause delay of the high speed information. Most of the information on the low-speed bus is event driven and it can withstand relatively long delays (tens and even hundreds of milliseconds). Whereas the information on the high-speed bus is real-time control data that is either time critical or safety critical. This information is usually periodic in nature, and it affects a relatively high bus load. Acceptable latencies for the real-time data is typically in the hundreds of microseconds to units of milliseconds range, but rarely is it ever in the tens of milliseconds range. Therefore, non-essential information should be kept on a separate data bus to avoid any further loading of the essential data.

The authors point out that present specialised vehicles, such as the BMW 7-series, contain this network architecture, but they use different protocols for the two systerns. It is the philosophy of the authors that if there are two data busses, it would be essential to utilise the same protocol for both busses because: development of a gateway between the two data busses is much simpler; protocol development of the various controllers in the vehicle is simplified; whole vehicle diagnostics is simplified; the need for supply of different protocol controllers is eliminated and that significant cost savings on higher quantity CAN chip purchases.

Gateways will always be essential for information sharing between the two data busses. For example, the engine management system will require high load switching information (such as headlights and air conditioning) when the engine is idle. Many automotive manufactures develop their own software for several systems. The use of a standard protocol for all systems in the vehicle allows for generic software routines for the whole vehicle network. The manufacturer can also pass these generic routines on to its suppliers. As more information migrates onto the data busses, several forms of diagnostics become much simpler. End of line test of vehicle subsystems is essential before they are attached to the rest of the vehicle. For example, when a door is assembled, a simple four pin connection (+12V, -12V, CAN-high & CAN-low) is placed at the door interface, and the door is run through its motions using a hand held network emulation unit. Also, maintenance is much simpler. With information on the data bus(es), a single point connection to troubleshoot all the systems in the vehicle is possible. The following CAN chips are considered by the authors for the body control system: Philips and National Semiconductor SLIOs (used for low intelligence out stations such as lights); National Semiconductor COP/CAN (intelligent outstation such as door window anti-trap) and Philips 8-bit controller with CAN (for central control unit).

The authors point out that CAN is already well established as a high-speed control protocol. It is currently incorporated in several production vehicles, and it is under development for future production vehicles. Several CAN chips have been developed specifically for low cost body control type systems. Since the release of these chips, research is continuing at the University of Warwick on the economical incorporation of CAN in low-speed body control systems, Fig 58. Most of the outstations will incorporate the use of smart power FETs which replace conventional relays and fuses. The FETs used for this exercise are manufactured by International Rectifier.

The Central Control Unit responds to any of the driver/passenger switch inputs, as well as any other sensor inputs (cabin temperature for environment control, say), and relays this information via the two-wire CAN bus to the nodes concerned. This unit will also contain an extra CAN controller for gateway communication to the high-speed data bus which is necessary for the transfer of information between the two data buses. Most timing units are eliminated from the system since this can done in the central controller.

The smart power FETs are used at most of the outstations for switching loads such as lights and

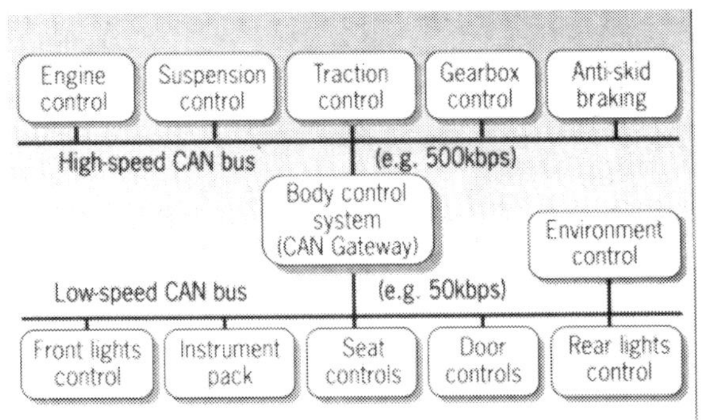

Fig 57. Dual data bus system

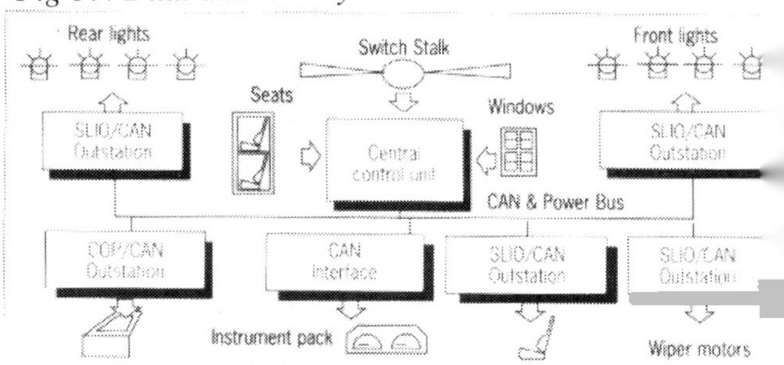

Fig 58. Example of body control system

motors. These devices are obtainable in current ratings up to 20 amps. They are used in the high side driver configuration — switching the supply side of the load. The smart FET s are available in 3-pin and 5-pin versions. The 3-pin device is connected as a standard power FET, and it has current sensing circuitry such that when it senses an over-current condition, it cuts off the load current and continues monitoring the load for recovery. The smart FET responds to a short circuit condition much quicker than a standard automotive fuse. This allows for the use of smaller gauge wire. Currently wires are specified for higher current ratings to compensate for the momentary high current surge before the circuit fuses. The use of the smart FET causes the following savings: the elimination of the circuit fuse and holder; space savings through the elimination of the fuse box; elimination of relays; reduction of circuit wire size and overall weight savings.

An earlier cost-analysis of CAN implementation by one of the authors showed that a simple reduction of 60 wires used for sharing high-speed information among the VOs of seven electronic control units (ECUs) resulted in a vehicle manufacturing cost savings of approximately £15 per vehicle. These savings did not account for several other long term savings caused by this architecture.

Controller Area Network protocol

The Bosch-Intel CAN protocol, Fig 59, for multiplexing is an important move to standardize automotive electronic controls (Intel's contribution is to market the VLSI circuit). Bosch say it is important to differentiate between the high-speed data bus (10,000 bits/second) and the low speed power bus of the multiplex system. CAN was designed with a prioritized message scheme which can offer low latency times (less than 1 mSec for transfer rates at 1 Mbaud) for high priority messages. The CAN protocol is a contention based protocol and performs non-destructive bit-wise bus 'arbitration' on up to 2032 uniquely definable messages.

The first CAN device to be produced contained dedicated circuitry to control the communication interface to and from the bus and, therefore, reduces the communication burden on the host microcontroller. The transfer of messages is initiated by the host microcontroller, but executed by CAN through a 64-byte, quasi-dualport RAM. CAN features include bus access priority by message identifier, NRZ coding/decoding, with bit stuffing, powerful error handling and data length up to eight bytes per message.

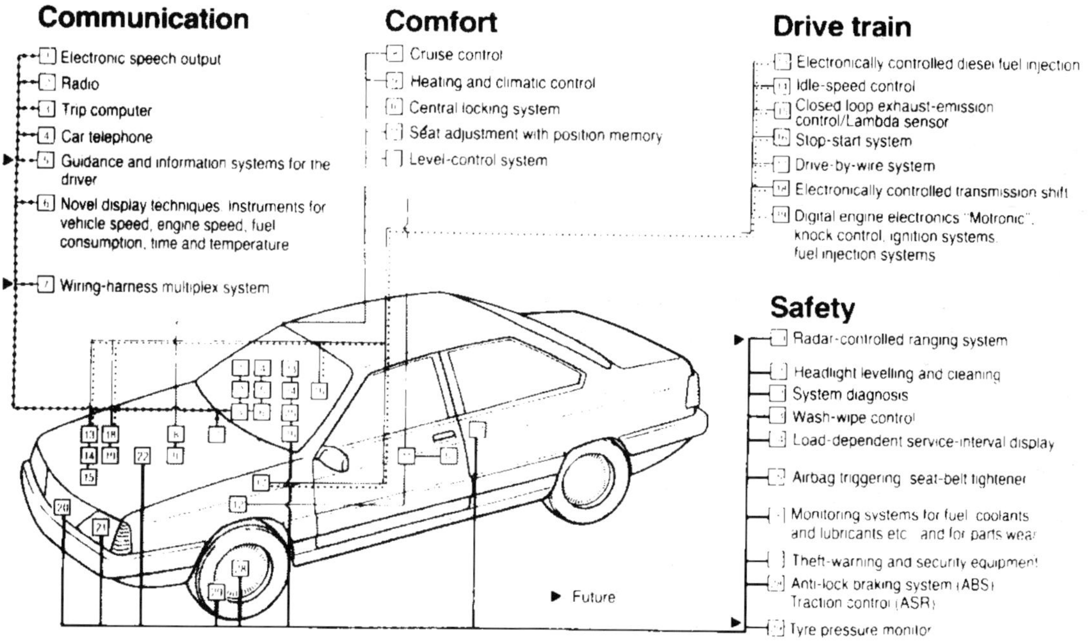

Fig 59: Controller Area Network

Fully functional CAN sample devices are available in a 44-lead plastic leaded chip carrier (PLCC) package.The samples operate over a broad temperature range and at up to 1 Mbit/sec data transfer rate. The first CAN device was designed to interface to the multiplexed eight-bit address and data bus of most microcontrollers. The CAN device features the following major functional blocks: the Interface Management Processor (IMP); a quasi-dual-port RAM (DP RAM); the Transceive Logic (TCL); a Bit Stream Processor (BSP); the Error Management Logic (EML); the Bus Timing Logic (BTL); a Processor Interface Unit (PIU); and a Clock Generator (CG). These major functional blocks are explained briefly below and are shown in Fig 60.

The IMP is the largest functional block. It executes commands from the CPU and controls data transfers to the BitStream Processor on a byte level . Two major tasks of the IMP are message acceptance and message transmission filtering. The IMP initiates the sequential scanning of the Communication Objects located in the DP RAM. Message acceptance filtering compares the incoming message identifier with the val id message identifiers located in the host microcontroller data memory. If there is a match, the message is accepted. In a similar manner, the IMP consists of a data path section with various local registers including an ALU with an accumulator and a control section for receive and transmit processes. It also computes addresses for the many communication buffer accesses to and from the DP RAM.

The PIU consists of the eight-bit multiplexed address and data bus, read/write control, address latch enable, chip select, interrupt output, external interrupt input, reset, ready output signal, two eight-bit I/O Ports — 0 and 1, and three chip select output lines for additional system peripheral devices. Because the first CAN device offers two eight-bit general purpose I/O ports, there is no system functionality loss when connecting CAN to the host microcontroller. Fig 61 shows several examples of how to connect the first CAN device to host microcontrollers.

The host CPU communicates with the CAN device through the on-chip quasi-dual-port RAM which includes a global status register and a control register. The on-chip memory serves as the communication buffer interface between the host microcontroller and the Interface Management Processor (IMP). The host microcontroller initialises the global status and control registers and creates data structures called Communication Objects within the communication buffer for reception and transmission of defined messages. The DP RAM contains 64 bytes.

TheTCL consists of bit-stuffing logic, programmable output driver logic, CRC logic and shift registers. The BSP controls the data stream between

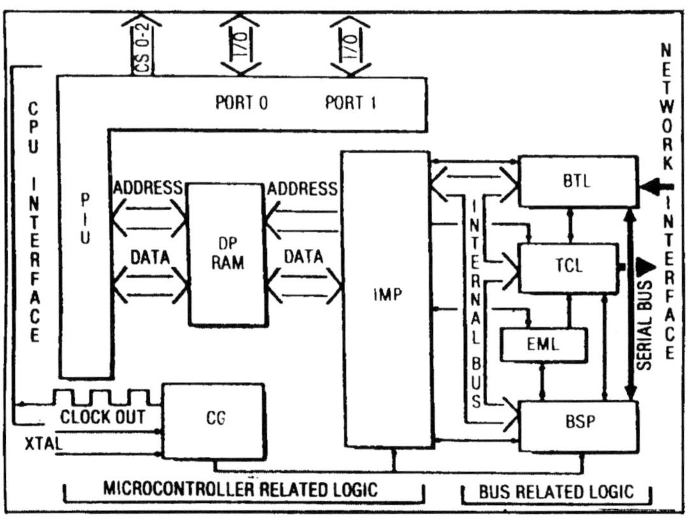

Fig 60: CAN block diagram

the IMP (parallel data) and the bus (serial data). Reception, bus arbitration, transmission, error signalling, and control of the TCL are duties of the BSP. Errors are reported to the EML by the BSP. The EML then informs the BSP, TCL and the IMP of error statistics.

The BLT block monitors the bus via an input comparator and performs the bus related bit timings. The BTL synchronizes on a bus transition at the start of a message frame and resynchronizes on subsequent transitions during the reception of a frame.The BTL also provides programmable time segments to compensate for propagation delay times and phase shifts. This feature determines the sampling point and the number of samples within a bit time slot.

The on-chip clock generator consists of an oscillator, clock divider register and driver circuit. The oscillator is driven by an external crystal or, in case of low baud rates, with a ceramic resonator. Since the first CAN device contained its own oscillator circuitry and clock output driver, the user could economise by using only one crystal in the system and allowing the CAN device to clock the host microcontroller.

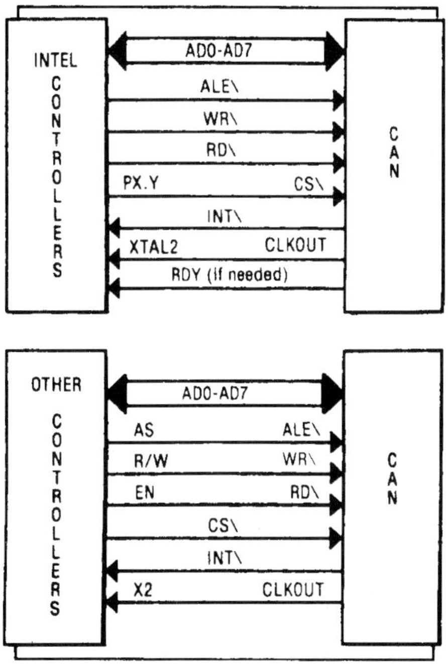

Fig 61: Connecting CAN to host microcontrollers

Chapter 6: Body system electronics and telematics

Body system electronics protocols; body electronics basics; on-board system developments; networking of body system controls; recent system advances; telematics for driver assistance; proprietary telematic systems; projected systems for volume cars; intelligent traffic systems; shape tracking for vehicle detection

Body-system electronics protocols

According to Volex Wiring Systems[1], body systems basically control functions that are operated directly by the driver or passengers. Typical body systems are lights, windscreen wipers, window lift motors, central door locking systems and mirror motors. Other systems are indirectly operated by the driver, for instance the only signal that a driver changes, out of over approximately 50 other signals in an engine management system, is the throttle setting.

It is the body electrical systems that are responsible for the majority of electrical energy used on any vehicle and consequently the wiring harness size and complexity is directly proportional to the number and positioning of loads in the body electrical system. Just as car electrical feature-content climbed to the levels now where some cars have over 1200 wires, the features of commercial vehicles are following a similar path in order to produce differential products. The increase in car feature content has resulted in such developments as modular harnesses, thin wall cable and so called intelligent wiring systems.

Protocol standards The SAE have divided multiplexed system protocols in to three areas: Class A, low transmission speed, body electrical system applications; Class B, medium speed, used for the sharing of information between ECUs; Class C, high speed, used to link ECUs that may need to exchange information in real time. To date the attempts to produce a standard have been based on Class B protocols mainly because of the need to standardise on a diagnostic protocol that can link garage test equipment to vehicle systems. The progress on these standards has been slow and it seems may not result in one Worldwide standard at all! The present contenders are an Intel/Bosch controller area network (CAN) and SAE J 1850 standard. In addition, there are the French CAN, Mazda's PALMNET, and SAE's J 1939 standard for heavy duty trucks. The number of class A protocols is huge, says the author, and is likely to remain so unless the industry enforces one standard. This lack of Class A standardisation means that the benefits possible with multiplexing are unlikely to be realised in the short term. This is due to the unwillingness of load and switch suppliers to integrate multiplex nodes specific to a vehicle manufacturer into their products. Once node integration takes place, the fully multiplexed vehicle giving maximum benefits on cost, performance, reliability, harness simplicity, weight and connector savings will be possible. Because the problems of weight and complexity of a wiring harness are mostly due to body electrical loads, these reside in the Class A application of multiplexing. Work at the company is resulting in the specific development of two separate Class A protocols for differing applications.

VMP is a synchronous time division protocol working in a master and slave configuration. This negates the requirement for the application, in appropriate outstations, of expensive complex electronics in the control of simple loads. This choice of protocol results in the most cost effective solution possible and forms the basis for most of the past and present applications. VMP+is an asynchronous full duplex protocol with separate lines for send and receive. Its application is primarily in the areas where data is being exchanged, such as feedback in control loops. Seat position control is a typical example of such an application.

The fact that a vehicle may have several protocols running at the same time should not be a cause for concern, because the company are able to make the communication fit the application. This means that the overall system performance/cost ratio is being optimised to suit the vehicle. Whether there is one or several different systems running on a vehicle the means of interfacing to the outside world via test equipment should always follow a standard. The world of test communications standards is divided into two camps, the ISO 9141 standard and the multiplex standard SAE J1850. The California Air Resources Board (CARB) specifies that all vehicles sold in California must conform to an on board diagnostic standard, SAE J 1850. However, the standard followed by most of Europe is ISO 9141, and so CARB has recently announced that a subset of ISO 9141(1SO 9141 CARB) will also be acceptable, the author explains.

Volex worked with a CV manufacturer to develop a system, Fig 1, suitable for loads such as central door locking, window lift motors, steerable mirrors, mirror wash/wipe and mirror heaters — increasingly apparent within the doors. Also these doors are physically thinner and are required to open to greater angles than the car equivalents. In order to feed these loads a 'third hinge', effectively, is created where the wires go into the door. Using the VMP protocol in this instance, has reduced the required number of wires from 15 to four, where two of those are thin signal wires. The vehicle manufacturer has defined the diagnostic standard to be ISO 9141 which will be easily accommodated within the master unit software. For all classes of vehicles, one of the most interesting future developments could be the increase in battery voltage. This would have interesting ramifications for commercial vehicle manufacturers, in that if the standard becomes the same as commercial vehicles, suppliers will have to provide loads to suit. This will expand the availability of loads to commercial vehicle manufacturers with a knock-on effect in terms of features provided; the author explains.

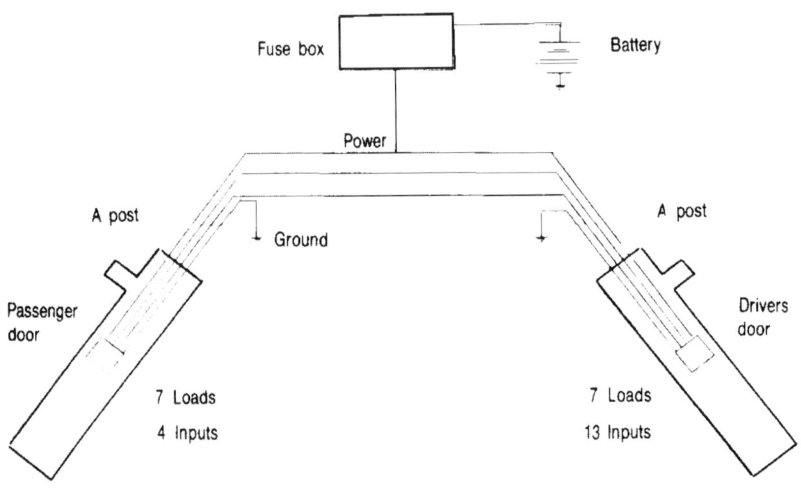

Fig 1: CV body system proposed by Volex

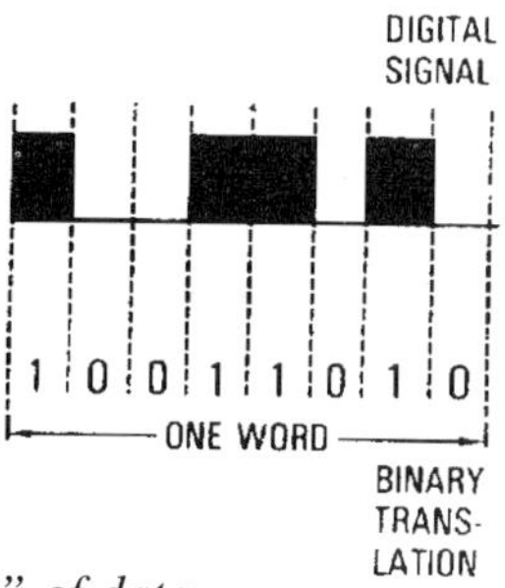

Fig 2: A "word" of data

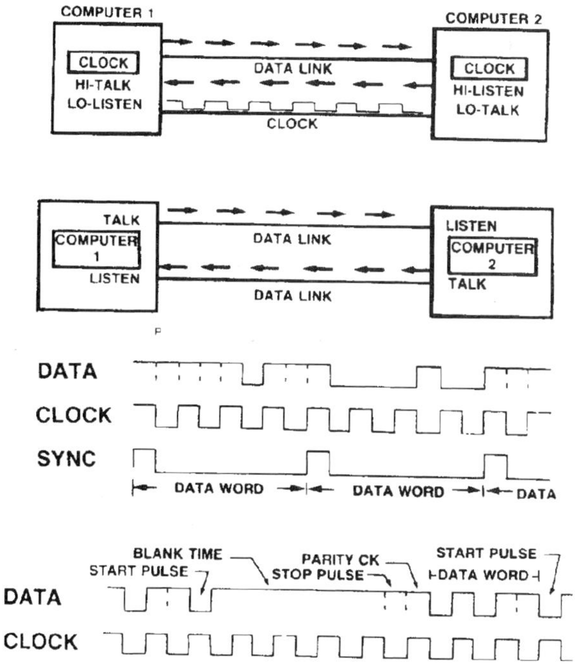

Figs 3 and 4: Parity check and clock pulse

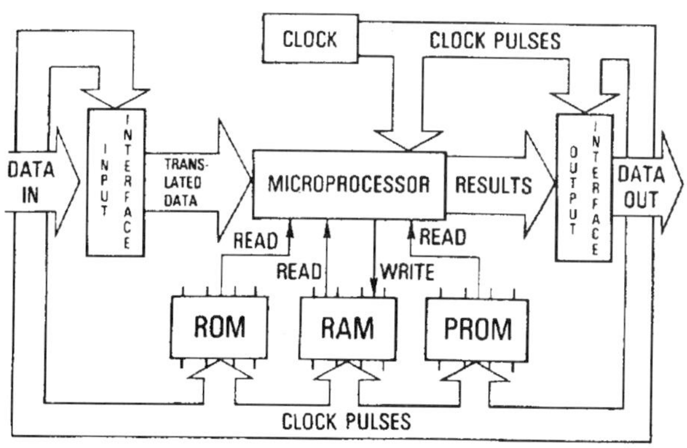

Fig 5: Parallel data transmission

Vehicle body electronics basics

Increasing data processing requirements, above those required of the electronic systems required for engine/transmission management and anti-lock braking, may lead to ways of increasing the calculation rate of microprocesssors. The three main approaches are to use high speed semi-conducting materials such as gallium arsenide, to revise the architecture of the central processing unit or increase the bit length for processing the data. The expense of providing separate processors around the vehicle for controlling the various body systems would otherwise probably preclude their use to very particular applications.

Suitable microprocessors are now being volume-produced at low price and at last multiplex wiring systems for connecting the systems together on a cable ring with a modest number of wires are appearing at the luxury end of the vehicle market. Another possibility is that so called application specific integrated circuits (ASICs or 'custom chips') will be configurable by consultants to the particular requirements of vehicle builders.

A key function of the on-board computer is the clock generator, discussed in an earlier section, used for producing single bit length pulses. The memories watch these pulses during data processing (sending and receiving information). By knowing how long each bit is supposed to last the computer can distinguish between the different binary signals, Fig 2. Important to vehicle computers are the various interfaces which have built in to the converters for turning the many analogue sensor signals to digital ones for the microprocessor.

Similarly output interfaces need circuits for converting binary code into digital output signals which suit the devices they are controlling. In cases where more than one computer is used per vehicle, data links will be needed between them. When so-called 'serial' data transmission is used, 'words' of data are sent one after the other. The computers are then arranged to be clocked together in synchronisation. Duplex serial data transmission implies a single two-way link between computers and serial data can be sent or received by either. When clock pulse is high, the first is programmed to send while the second listens and the reverse situation occurs for low clock pulse. Alternatively, in a peripheral serial bus data link, one always sends data in one direction while the other always receives it in the opposite direction.

The data link systems are said to be synchronous when the data flow is constant and the clock pulses regular. In some systems, however, data is transmitted only when it is required rather than being sent continuously — asynchronous data transmission, Fig 3. A start pulse signals the computer prior to transmission and a stop pulse signals completion while a parity check may be used to ensure that that the total words have been sent, Fig 4. Parallel data transmission, Fig 5, by contrast, involves several separate data lines with corresponding switches turned on and off simultaneously to send specific data signals.

In the case of multiplexing, a central transmitter (microprocessor) acts as a master controller for the several receivers which control the principal loads of the body electrical system. A bidirectional bus data link connects them; this comprises two wires, 'signal' and 'ground'. Switches on the transmitter each have a unique signal and the microprocessor provides bus data link control and sequencing logic. So called 'state sequencing' is used by the control software that cycles through various tables to ensure proper conditions exist to advance from one state to another. Switch input requests may be ignored, passed on as a command to the receiver or interpreted as a specific command to be issued — according to whether, say, the ignition switch is on or off, or whatever. Receivers either operate electrical loads directly or through mechanical/electronic relays. Intelligence level of the receivers is restricted to that required to signal error state (bulb-failure, say) signals back to the transmitter. The Lucas Body Systems arrangement uses the SAE J1850 (Carlink) protocol set up to standardise the datalink procedures for multiplex systems. The company have applied multiplexing to controlling electrical systems in two doors. Standard protocols open up the market so that equipment from on e supplier can be used confidently with that of another operating on the same protocol.

The following features are controlled: electric mirror, electric window lift, keyless entry, central door locking, courtesy light delay, lock illumination, lock barrel heating and rear door windowlift isolation, Fig 6a. The alternative approach (*see Chapter 5*) is that devised by Bosch and Intel which is known as CAN, Fig 6b. Philips Semiconductors have recently introduced an intelligent power switch on this protocol. It comprises a power MOSFET linked with logic circuitry and promises a solution for multiplex connection of the 30 or so motors which can now feature on top-range vehicles, Fig 7. The units are said to have lower drive power requirements than bi-polar switches. Fig 8 shows the four most common motors used in vehicles while Fig 9 shows a bridge of semiconductors used to control a sun-roof where directional change of the motor is required. Source-drain diodes in the MOSFETs eliminate the need for external switching diodes. PM brushless DC motors typically require six switches to generate the rotating field, Fig 10. Change of direction or speed are controlled by timing of the field pulses; these can be derived from a dedicated IC such as the Philips NE5570. Rotor position sensing is required to determine which windings should be energised. One switch supplied by the company is capable of driving motors up to 25A, Fig 11.

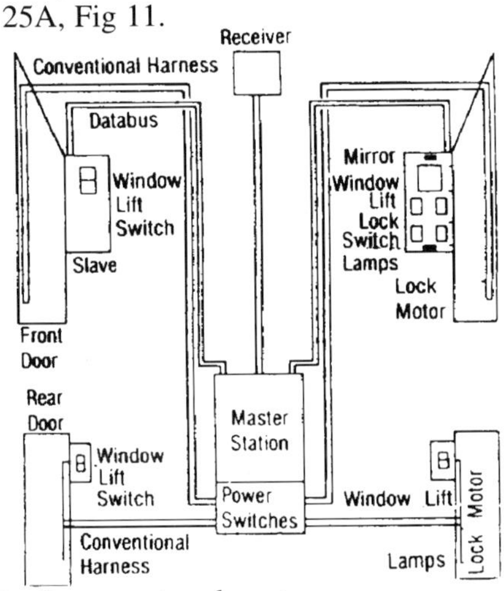

Fig 6a: Lucas network system

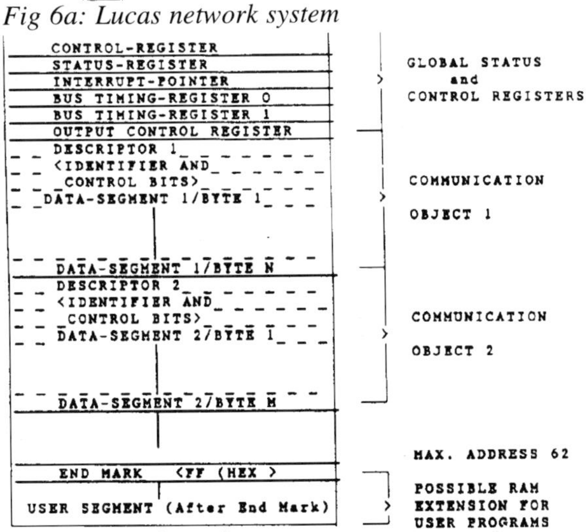

Fig 6b: Bosch/Intel system

Typical motor and switch requirements in a top of the range car

motor applications	typical power (W)	nominal current (A)	typical number of such motors	type of drive	typical number of switches per motor	proposed MOSFET: standard BUK–	L² FET BUK–	comments
air-conditioning	300	25	1	unidirectional, variable speed	1	456	556	Active suspension may also require such a high power motor
radiator fan	120–240	10–20	1	unidirectional, variable speed	1	455	555	These motors may go brushless, requiring 3 to 6 lower rated switches
fuel pump	100	8	1	unidirectional	1	453	553	
wipers:								
front			1–2			452/ 453	552/ 553	Reversing action is at present mechanical. This could be done electronically using 2 or 4 switches
rear	60–100	5–8	1	unidirectional, variable speed	1			
headlamp			2					
washers:								
front			1–2					
	30–60	2.5–5		unidirectional	1	452	552	
rear			1–2					
window lifter	25–20	2–10	2–4	reversible	4	452/ 455	552/ 555	
sun-roof	40–100	3.5–8	1	reversible	4	452/ 453	552/ 553	
seat adjustment (slide, recline, lift, lumbar)	50	4	4–16	reversible	4	453	553	
seat belt	50	4	2–4	reversible	4	453	553	
pop-up headlamp	50	4	2	reversible	4	453	553	
radio aerial	25	2	1	reversible	4	452	552	
door lock	12–36	1–3	6–9	reversible	4	451/ 452	551/ 552	
mirror adjustment	12	1	2	reversible	4	451	551	

Fig 7: Motors on top-range car

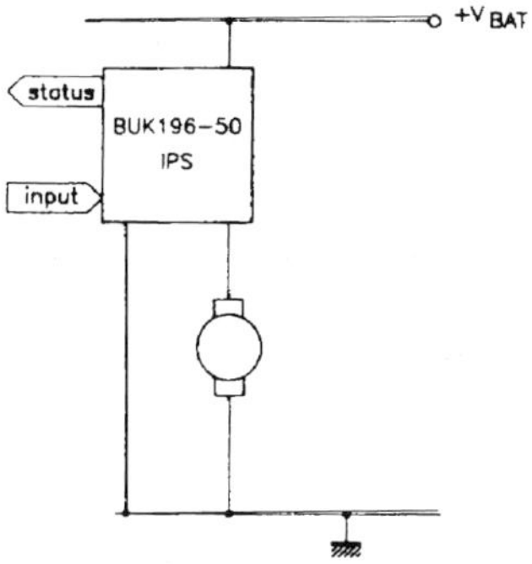

Fig 11: Driver for 25A motor

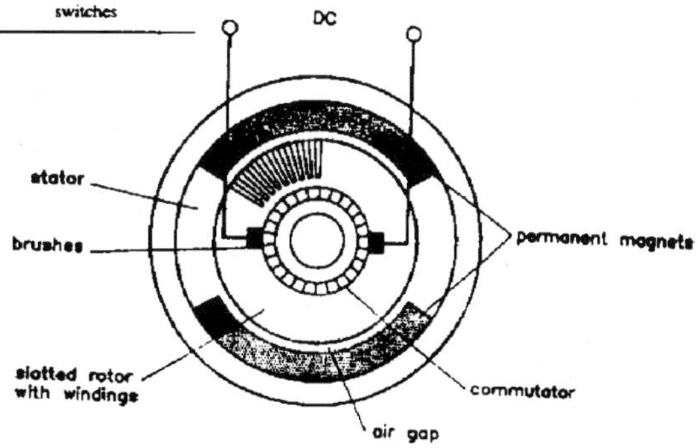

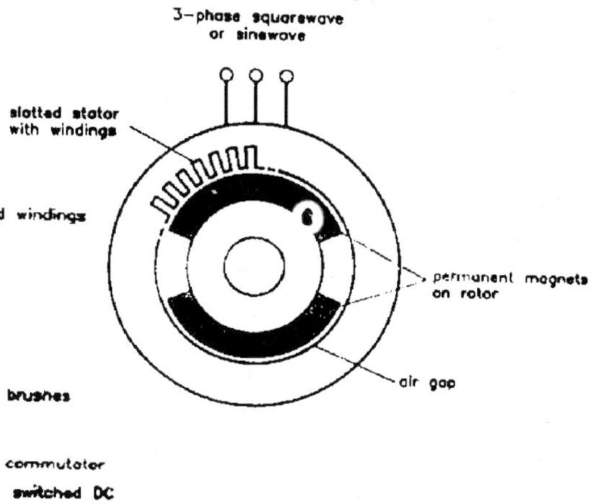

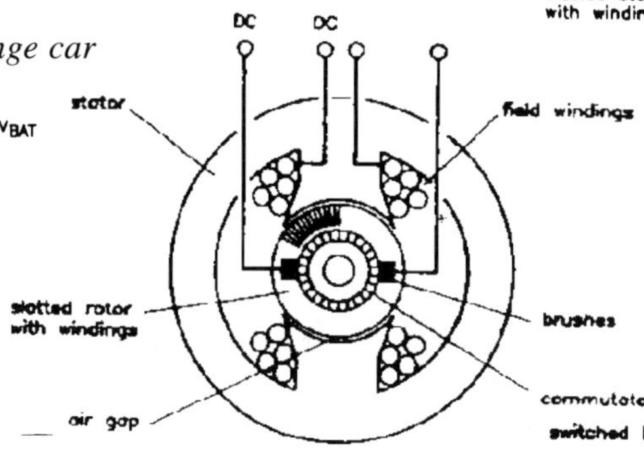

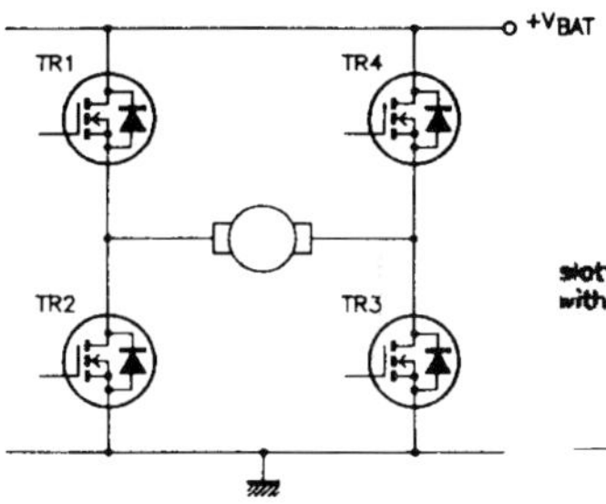

Fig 9: Bridge circuit for motor directional change

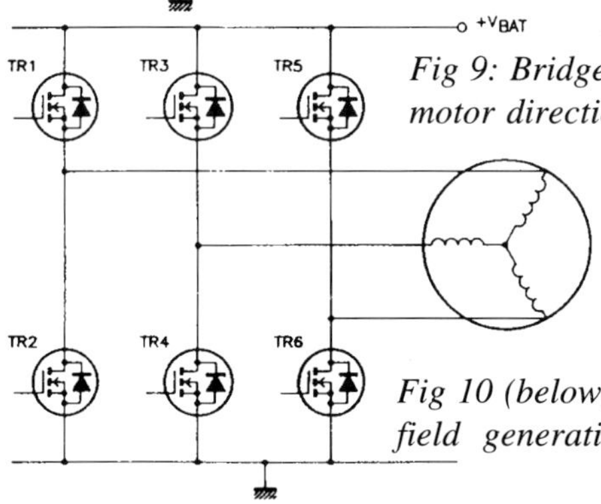

Fig 10 (below): Rotating field generation

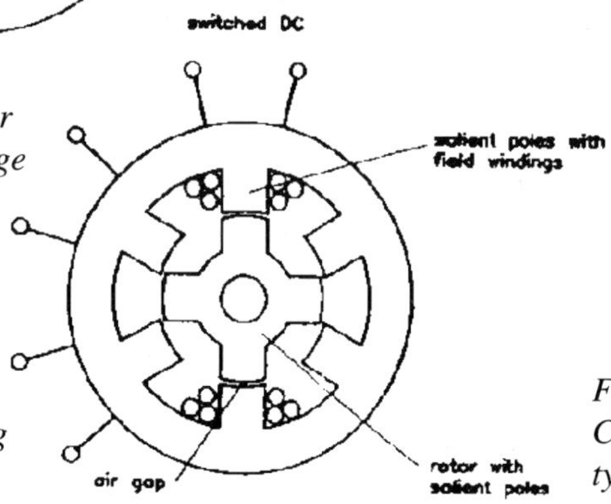

Fig 8 : Common motor types

On-board body system developments

Before multiplex wiring becomes the norm on vehicles, to handle the increasing number of on-board electronic systems, datalink connections between sub-harnesses may be the solution. This is discussed here with descriptions of some advances in the electronic systems themselves.

The practicalities of multiplexing the electrical supply to body systems seems to have put something of a brake on its application. Development of junction boxes between main and sub-sections of wiring harnesses by Lucas Rists might encourage the process when active versions are employed to establish serial links. Junction boxes are attractive, too, in conventional wiring harnesses to avoid splicing and doubling up of cables into terminals. The sub-sections of the harnesses can also involve fewer cables by automating some of the functions. Suppression components and routing diodes can also be conveniently located in the boxes — also fuses and relays when power demand is high. Making the boxes 'active' involves the addition of timing functions for electronic systems and possible inclusion of transceivers for serial datalinks. Depending on power requirements either metal stamped interconnections or printed circuit boards can be used to link through the boxes. The company describe an application to a vehicle with some 700 cables in the facia area, 70 of which are routed to the driver's door and 25 to the passenger door, Fig 12. The redesign involved breaking down the facia harness into smaller units which can have automated operation and reducing the driver's door cable number to 25. The solution was to combine active and passive junction boxes with a serial datalink. As most control signals originate within the driver's door (central-locking, mirror-adjust and window regulation) the serial data link was used between it and the A-post, Figs 13/14.

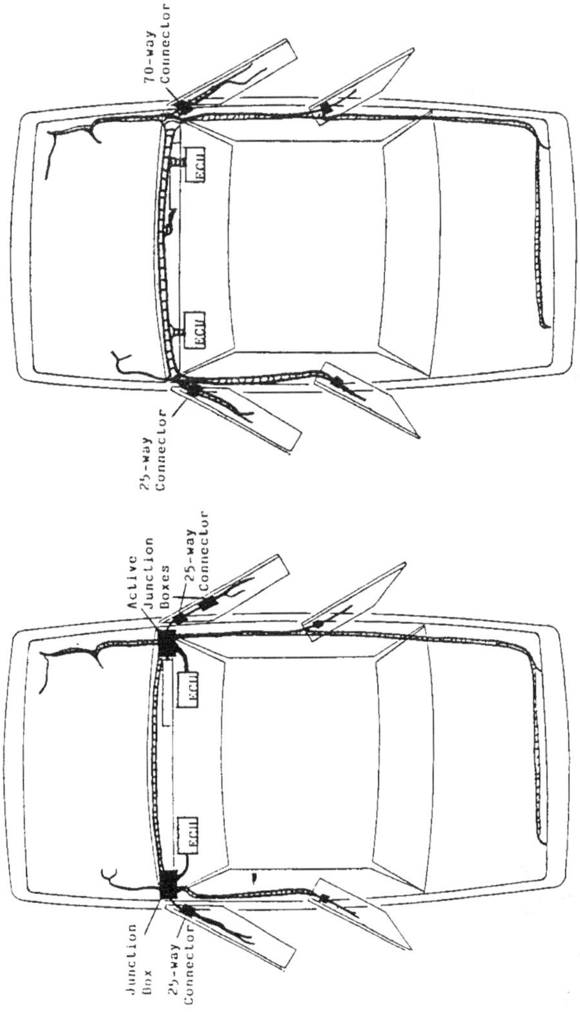

Fig 13/14: Door system application — above, serial data link; below, layout with junction boxes

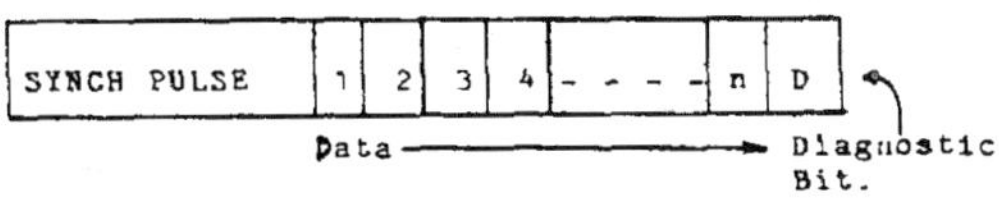

Fig 15: Bit encoding

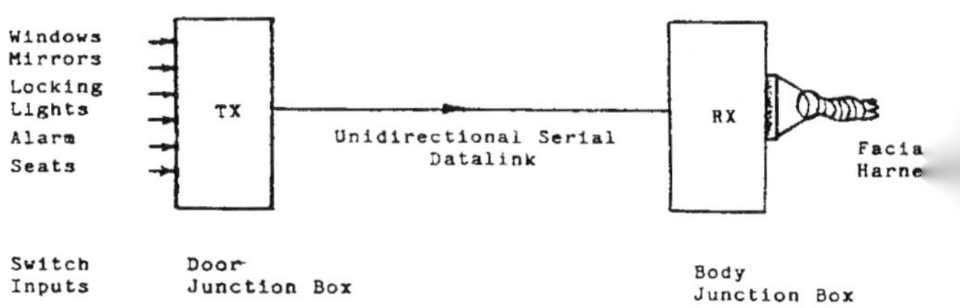

Fig 12: Serial datalink

The datalink uses pulse width modulation for bit encoding, Fig 15, and is made electrically secure in being immune to short circuits either to 0 or 12V. A diagnostic databit is transmitted in every message frame and on its receipt a light-emitting diode confirms function of the databus. As a further safety measure all systems remain operable from their local switches. In the example application the number of cable splices was reduced from 82 to 15 and the six sub-sections of the facia harness involves one quarter fewer cables.

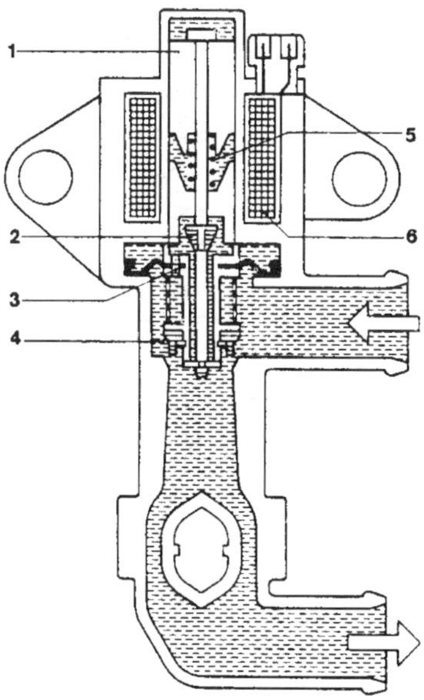

When current flows through the coil, the armature moves downward and closes both sealing cones. When no current is flowing, the valve opens.

1 Armature
2 Sealing cone
3 Bore
4 Main sealing cone
5 Spring
6 Coil

Coolant

Fig 16: Heater solenoid-control

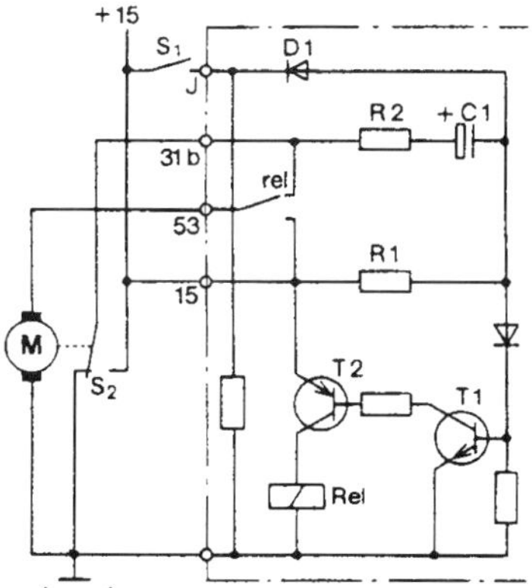

Fig 17: Wiper delay circuit

Proprietary body elctronic systems

For heating and air-conditioning, the Bosch has developed motors — of the elecronically commutated type — which dispense with brushes and mechanical contacts. Another interesting product for heaters is an automatic temperature control system based on negative-temperature-coefficient sensors and a solenoid valve for water control, Fig 16. A potentiometer is used to select the desired temperature and an electronic controller continuously adjusts the system according to signals from temperature sensors close-to and remote-from the heater. A pulse is output by the ECU to the solenoid valve at a four second interval.

For windscreen wipers, the company has introduced an optoelectronic sensor which detects rain or screen-mist and operates the wiper as necessary. Wiper operation is quieter, too, by the recent availability of variable speed motors which avoid the annoying rising and falling whir of conventional systems. A two-motor wiper system is also on offer which will make the normal wiper linkage redundant. As well as appropriately synchronising motor speeds the controllers also improve functional performance of the wipers.

Electronics also comes to the aid of automatic windscreen wash-wipe systems by Bosch. The windscreen is sprayed with washer fluid; the wipers then distribute it, clean the screen and dry it. An intermittent-wiper switch comprises a pulse-generator with a variable pulse frequency. Each pulse drives the wiper via a relay and causes one wipe cycle, the delay being set between one and five seconds, Fig 17. The unit also uses an astable multivibrator as the clock generator, Fig 18, the delay of which can be set within specified limits. Selection of the RC element in the circuit, comprising resistors R5 to R (and capacitor C1), achieves this. C1 has capacity of some 500 μF and the five resistors are connected to it in steps, their resistance varying between 30 and 300 kΩ, corresponding to the timing delays. Pulse period is set by the time constant R2.C2, at about 0.5 second. The relay switches on the motor for this period, via terminal 88a. The wash-and-wipe system overrides the intermittent wiper circuit. This circuit has an RC element which not only controls the lead time but also time for wipe-and-dry.

The system can also be purchased with a monostable, externally triggered (by the motor reverse switch), multi-vibrator. Here the RC element is as shown in Fig 19 with R1 and R2 as charge and discharge resistors, respectively determining lead and wiping times.

The company also use multivibrators as an improved alternative to hot-wire turn-signal flasher units. These have better durability and are less sensitive to prevailing temperatures. The multivibrator is the signal generator and is combined in one unit with monitoring stage, turn-signal relay with check lamp and hazard warning. Since the filament wire is a resistor with positive temperature co-efficient characteristics, high current flows initially, decreasing rapidly with increased temperature.

The multivibrator acts as controller — having two amplifier sections, one switching stage and a switch with a differential amplifier. It is an integrated circuit, R1 and C2 forming the RC element which controls the timing of the flasher-frequency, Fig 20. C1 and C3 control voltage spikes and so serve as noise suppressors. Diode D protects the IC against incorrect polarity should the battery leads become reversed. A switching transistor matches multivibrator power to the turn-signal relay. and the IC has a voltage regulator stage which maintains constant voltage despite battery condition. If a lamp fails the voltage drop across Rk changes so that the transistor conducts and the frequency of the vibrator changes as required by law.

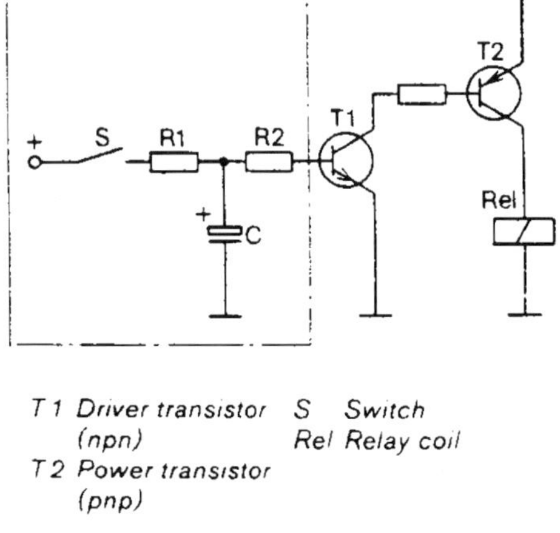

Fig 18: Clock generator

Networking of body system controls

According to Siemens, electrical connections which were part of traditional wiring harnesses are increasingly concentrated in special junction boxes. Rather than having the one central one, future vehicles are expected to have several around the vehicle as the trend to modularisation of harness sections continues. Increasingly the boxes are also expected to contain the electronic controllers for body systems to save on wiring and in the future electronic control of the connections within the boxes will make them programmable from the end of the production line.

Serial data transfer links are still uncertain, according to the company, with respect to their widespread fitment and what their data transfer rates should be. While very low rates are adequate for doors in which only power windows and central locking were involved, the eventual coming of production vehicle guidance systems will mean fast enough transfer rates to download the contents of a CD-ROM into a display system. The way in which the power supply network is separated from the

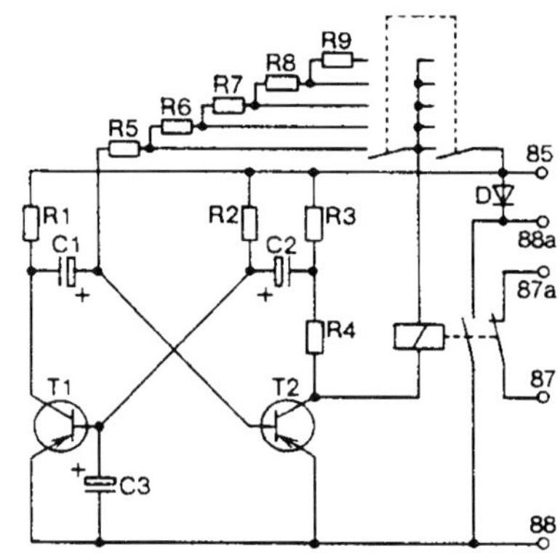

Fig 19: Externally triggered multi-vibrator

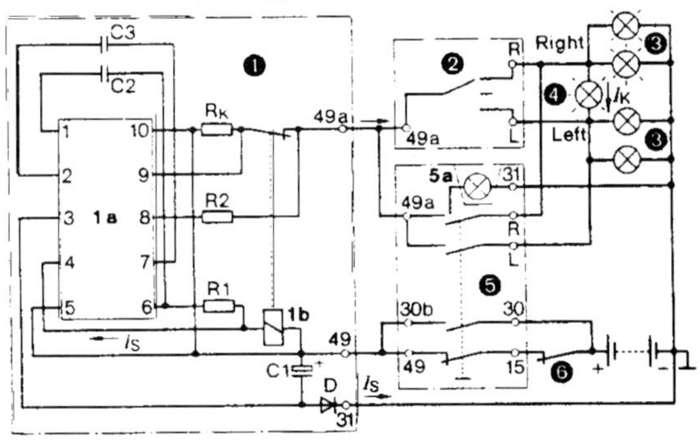

Fig 20: Flasher integrated-circuit

information network in such systems is shown in Fig 21. According to Bosch, with modular subsystems, the limited capacity of the signal and power supply lines can be taken into account. Other advantages offered by subsystems include greater system safety, decoupling development and flexibility. The partial multiplex systems communicate with one another by means of a hierarchically superior bus. This bus can, for example, relay all the subsystems' diagnostic messages. One disadvantage of segmentation is that it involves considerably more work on the wiring. A car with multiplex only requires a total of around 50 metres of four-conductor multiplex cable. A car in the top bracket with conventional wiring requires about 2.5 kilometres of cable. The CAN protocol with a data transmission speed of up to 100 k baud (= 100,000 bits per second) is especially suitable for communication between multiplex systems in the field of car body electronics. The basic CAN module is the section of the interface which is important where dealing with the protocol is concerned. This module is integrated into various microprocessors.

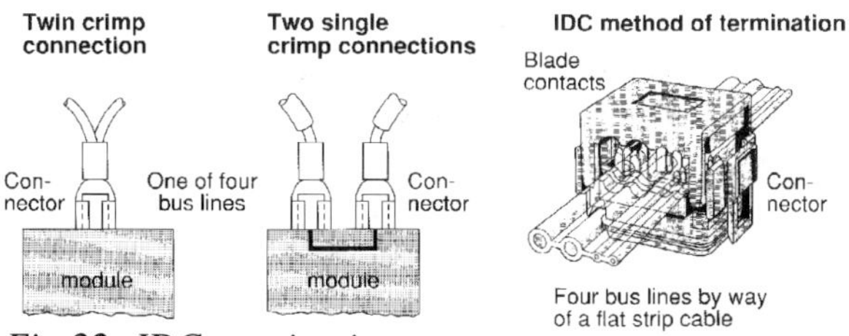

Fig 22: IDC termination system

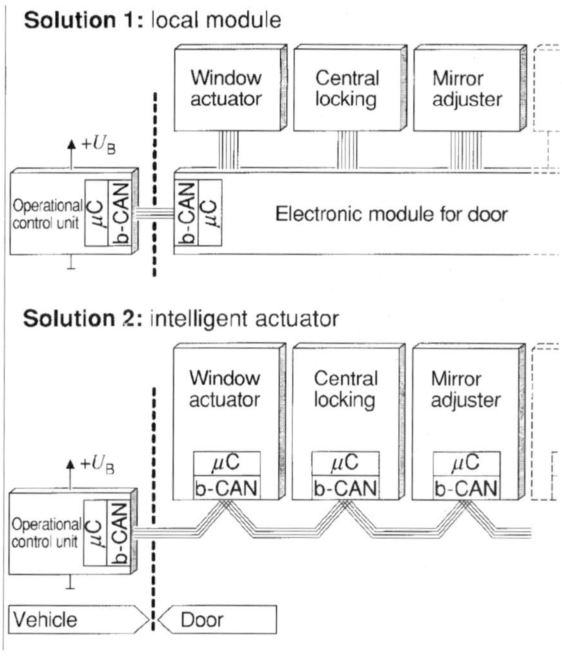

Fig 23: Local module and intelligent actuator solutions

A self-contained multiplex system also involved a change of wiring. Whereas the point-to-point connection prevails in conventional wiring, a multiplex wiring system basically consists of four connections linking the components with one another: earth line, battery voltage line and two signal lines. Traditional plug connections are only suitable to a certain extent. The so-called IDC method of termination is ideal for straight-through connections. By simply cutting through the cable insulation with blade contacts, a connection is achieved between the otherwise undamaged cables and the plug contacts. The cable can be prefabricated by way of a flat strip with four parallel lines. The plug is hermetically sealed using three gaskets. This also ensures that there will not be any corrosion at critical spots. In gold plating all the four 2.5 mm flat contacts, this provides extra reliability for low contact resistances of the signal contacts and also greatly increases the current carrying capacity of the current conducting contacts, Fig 22.

A multiplex wiring system in conjunction with 'intelligent' components can largely replace conventional control units. An example of this is a partial multiplex system in a vehicle door for which there are two solutions: in the first, a local electronic module is connected with actuators, operational elements and sensors of the subsystem in the conventional manner, in other words, by means of a point-to-point wiring system. This certainly solves the design problem of routing from the vehicle body to the door. However, a large area is required to accommodate the system which is relatively inflexible. There are more plug contacts and this reduces reliability. In the second solution, the control electronics (intelligence) system is integrated into the actuators and the operating element accommodates the electronic functions which are necessary to code instructions and relay them. The actuators with their built-in control electronics perform the operating functions (such as adjusting the position of the mirror, opening/closing the windows). The intelligence is integrated into all the basic components in the form of a microprocessor with a basic CAN interface which enables detailed self-diagnosis. A complete check at the end of the assembly line on the vehicle manufacturer's premises and rapid fault diagnosis in the workshop are both possible thanks to this intelligence, Fig 23.

Window actuation with anti-pinch The extension of a multiplex system for doors to include automatic window actuators with an anti-pinching feature is possible without using any extra cabling and control units. Extra costs are merely incurred with the connection of an intelligent actuator to the multiplex bus. Such convenience circuits are of course subject to strict safety requirements. If the driver does not have to close each window one by one it is easy to overlook the fact that something or other has got trapped in the process. For this reason the electronics system must instantly recognise any pinching and also ensure that the closing procedure is interrupted and reversed.

With this in mind the electronics system monitors the speeds of the individual motors. If the motors are slowing down but their closing power is increasing then an evaluation electronics system recognises in a fraction of a second whether something has got trapped and immediately reverses the motor's rotational direction. Of course, the drive does not switch off until the last few millimetres when the top edge of the glass is securely positioned in the ridge. It requires a lot of power to ensure excellent sealing at this point. The motors only switch off when the windows are properly closed. Bosch is working to develop intelligent actuators which can also be employed with seat adjustment motors, steering column adjustment, mirror adjustment and actuator for the parking brake.

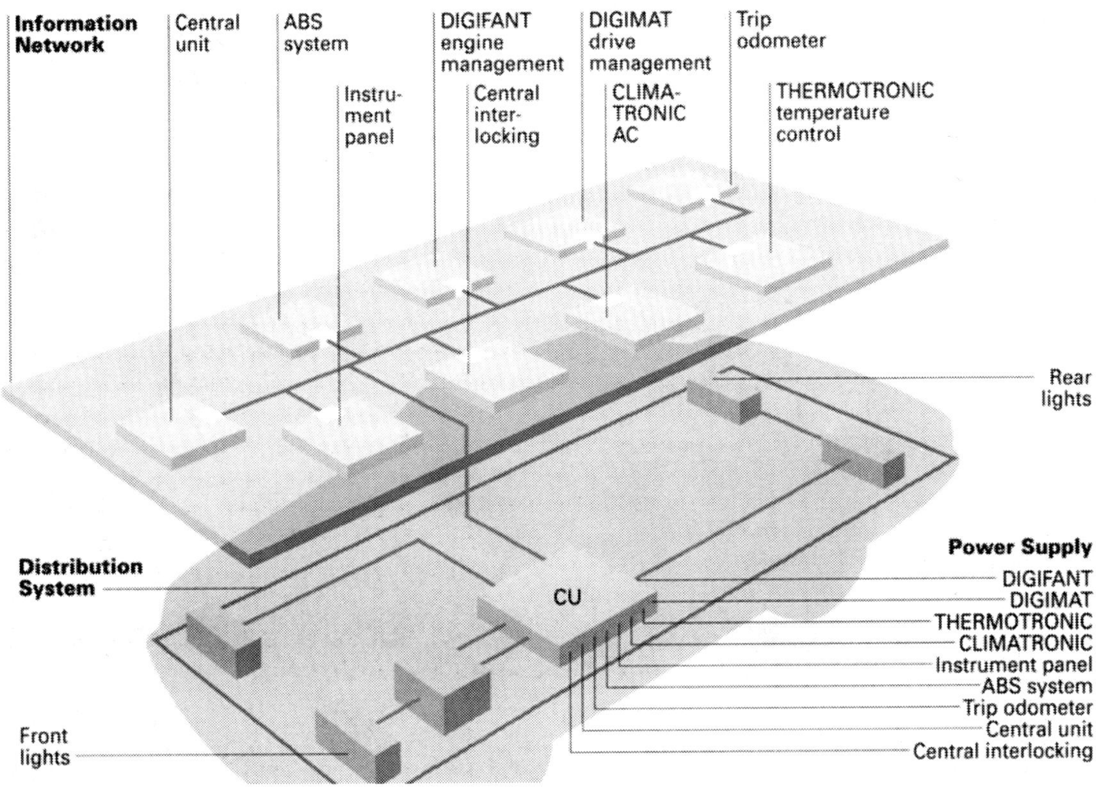

Fig 21: Separation of power supply network and data-link

Recent body system advances

Announced at a recent SAE Congress[2], the United Technologies tailgate combination unit combined wiper, lock actuator and window release solenoid in a multi-function motor employing Geneva mechanism to provide intermittent wiper motion, Fig 24, multiplexed module communicates with the body controller to fix the position of the multi-function motor and actuate heated backlight, license plate lamps and the high-mounted stop light.

The multiplexed electronic controller, communicating over a J1850 multiplex bus with a body controller, directs motor actuation in conjunction with the Geneva mechanism to control the wiper, tailgate lock/unlock, and liftglass release functions. Control of the heated backlight, centre high mount stop lamp, and the license plate lamps is included. These features are not part of the Geneva mechanism, but cost, weight, and reliability improvements are obtained by actuating them with the multiplexed electronic controller. Using the body controller and a serial data bus to the multiplexed electronic controller, significant reductions in wire circuit length are achieved.

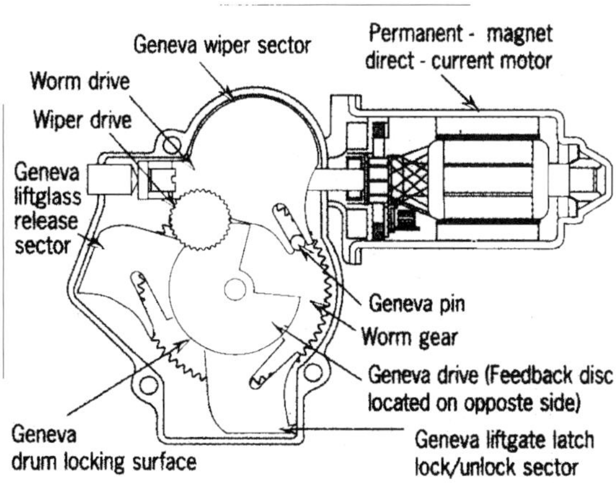

Fig 24: Integrated motor and Geneva mechanism

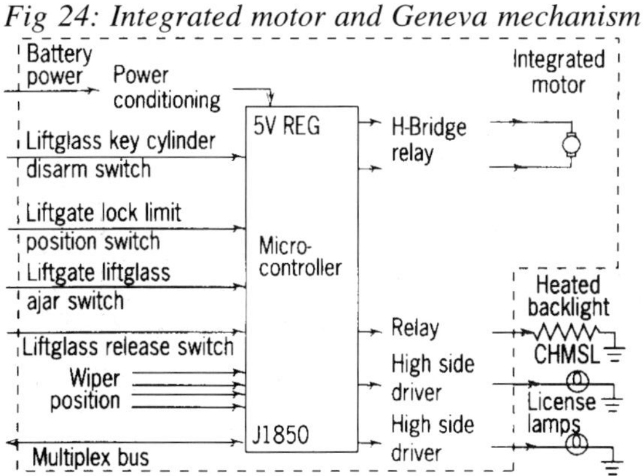

Fig 25: Multiplexed electronic controller

A worm is machined at one end of the DC PM motor's armature shaft to engage a worm gear which has an integral concentric Geneva drive and Geneva pin. Three Geneva sectors, each containing a slot, are concentrically located around the Geneva drive to actuate the wiper, tailgate lock/unlock, and liftglass release functions. Each sector is sequentially engaged and disengaged by the Geneva pin during one rotation of the Geneva drive. An output shaft at the pivot point of each Geneva sector actuates the wiper, tailgate lock/unlock, and liftglass release mechanisms. The wiper is driven through an additional pair of spur gears to provide the correct wiper speed and swept area.

The first park zone occurs after the wiper blade has completed its sweep of the tailgate glass and is in the depressed rest position and the tailgate is locked. The second park zone occurs after the tailgate is unlocked or the liftglass has been released. In both park zones, the wiper blade remains off the liftglass in the depressed rest position.

Reversing the motor moves the wiper blade clockwise to the beginning of the wipe position. The wipe cycle repeats by dynamically braking and then reversing the motor when the Geneva drive is at the beginning and end of the wipe positions. Placing the wiper switch in the off position returns the Geneva drive to the first park zone with the wiper blade in the depressed rest position. At the first park zone, the Geneva sector is locked in place by the Geneva drum. This retains the wiper blade securely in the depressed rest position. The Geneva mechanism inherently provides sinusoidal motion to the rear wiper. The benefit is that acceleration and deceleration of the wiper blade is at a minimum when the wiper blade direction is reversed at the beginning and end of the wipe positions.

The tailgate is unlocked by actuating the door unlock switch. The integrated motor causes the Geneva drive to rotate from the first part zone to the second park zone by engaging the slot in the lock/unlock Geneva sector with the Geneva pin. The shaft attached to the Geneva sector rotates to unlock the tailgate latch. To release the liftglass, its switch is actuated. The integrated motor causes the Geneva drive to rotate from the second park zone, engaging the slot in the liftglass release Geneva sector with the Geneva pin. The shaft attached to the Geneva sector rotates to release the liftglass latch. After the liftglass is released, the Geneva sector returns to the second park zone. The tailgate is locked by actuating the door lock switch. The integrated motor causes the Geneva drive to return to the first park zone by engaging the slot in the lock/unlock Geneva sector with the Geneva pin. The shaft attached to the Geneva sector rotates to lock the tailgate latch. The wipers do not operate if the tailgate is unlocked or the liftglass is open. The Geneva sectors are rigidly held in position by the proximity of their cam surfaces to the Geneva drum locking surface.

The integrated motor - Geneva mechanism is controlled using the microprocessor-based system shown in Fig 25. A Motorola MC68HC05V8 microprocessor, having an internal 5 volt regulator and J1850 hardware, is used to implement the functions. This system uses J1850 10.4K VPW hardware, 8K bytes of ROM and 256K bytes of RAM. Inputs to the multiplexed electronic controller are divided into three categories: switches within the tailgate, switches remote to the tailgate within the vehicle interior, and the integrated motor position feedback disc. The status of the remote switches is transmitted by the J1850 multiplex bus from the body controller to the multiplexed electronic controller. The position of the Geneva mechanism is continuously monitored by a feedback disc attached to the Geneva drive. Four metal contacts rest on the surface of the feedback disc which allows the position of the Geneva mechanism to be digitally encoded in eight sectors. The multiplexed electronic controller controls the integrated motor in accordance with the status of the tailgate and remote switches and the position of the Geneva mechanism feedback disc. Two relays in an "H-Bridge" configuration provide directional control and dynamic braking of the motor.

The tailgate key cylinder disarm switch disables the vehicle security system when the tailgate latch is manually unlocked using a key. In this condition, the tailgate latch moves to the unlocked position but the Geneva mechanism remains in the tailgate lock position. When a command is received to lock doors, the tailgate latch limit position switch signals the multiplexed electronic controller that the tailgate latch is unlocked. Therefore, the Geneva mechanism must first move to the tailgate unlock position to engage the tailgate latch and then return to the tailgate lock position, thus moving the tailgate latch to the lock position.

Conversely, if the tailgate latch has been unlocked by the Geneva mechanism, the tailgate latch can be manually locked with a key, but the Geneva mechanism remains in the tailgate unlocked position. When a command is received to unlock doors, the tailgate latch limit position switch signals the multiplexed electronic controller that the tailgate latch is locked. The Geneva mechanism must cycle from the tailgate unlock position to the tailgate lock position and return to the tailgate unlock position to unlock the tailgate latch. Both the tailgate ajar switch and the liftglass ajar switch inhibit rear wiper operation if the tailgate or liftglass is open. A liftglass release switch is located on the tailgate to enable opening of the liftglass only if the tailgate is unlocked for security reasons. The microprocessor controls an SPST relay which supplies battery voltage to the heated backlight circuit. The microprocessor provides a timer function to limit the amount of time the heated backlight remains switched on. In addition, it is possible to monitor the J1850 multiplex bus for load management messages making it possible to shed the heated backlight load when necessary. The heated backlight output is protected against a short circuit to ground. The microprocessor controls two separate highside FET drivers to switch power to the centre high mount stop lamp and license plate lamp circuits. Both outputs are protected against a short circuit to ground.

Fig 26: Vehicle and infrasructure systems involved in STORM

Telematics for driver-assistance

According to the European Automobile Manufacturers Association, ACEA, a safe and efficient infrastucture is now required to exploit the considerable advances already made in vehicle design on the safety and environmental front. ACEA also sees the need for a harmonised telematics strategy and efficient use of all transport modes, together with technical, tax and social harmonisation.

It is estimated that ECU 2.5 billion worth of fuel is being wasted per year in traffic congestion with the total cost of traffic inefficiency put at well over ECU 100 billion. Less congestion could reduce air pollution by up to 50%, ACEA argue, and there could be up to 35% reduction in accidents from the present level where 42,000 people die and 1.4 million are injured each year on Europe's roads. Other potential benefits include much easier navigation for motorists; the provision of car-park access information

Table 1. Telematic systems

Traffic management
Objectives:
- improvement of traffic flow and efficiency
- Lower environmental impact
- Better use of infrastructure by integrating individual and public transport - enhancement of safety

Practical applications of traffic management
- Pollution monitoring - ambient (weather) condition monitoring
- Road status monitoring - traffic monitoring - accident detection
- General and urban traffic control - public transport management
- Parking management - rescue services

Benefits for Motorists:
- Higher quality of life because of less pollution
- Less time spent commuting
- Less time spent in congestion
- Parking places easily found
- Real-time information about problems on the road ahead and alternative routes
- Easier and more convenient use of public transport

Pre-trip Information
Objectives:
- Provision of comprehensive travel planning facilities
- More attractive public transport

Systems:
- Private terminals (video, "minitel", personal portable or in-vehicle units - RDS/TMC
- Public transport information terminals at bus or train stations, for example

Benefits for Motorists:
- Saves time because less time spent in traffic jams
- Up-to-date information on public transport and parking availability

On-Trip information
Objectives:
- Real time traffic information
- Selective information in one's native language
- Dynamic route guidance
- Passenger information for public transport users
- Reduction of traffic in cities thanks to more efficient parking management, including Park + Ride

Systems:
- RDS/TMC (Radio Data System/Traffic Message Channel); communication takes place via the FM network
- Road vehicle communication via infra-red or micro-wave beacons
- Sensors installed in streets
- Cellular digital telephones (GSM)
- Short-range communication
- Route guidance systems with interfaces to RDS/TMC (mobile radio beacons)
- Screens - On board computers with electronic maps
- Head-up display

Benefits for Motorists:
- Selective information in one's native language
- Easy access to car parks
- Better safety, with current information on driving conditions and roadside warning systems
- More comfort with individual route guidance
- Tourist information

Vehicle control
Objectives:
- Increase the safety and comfort of motorists
- Road monitoring, collision avoidance and "cooperative driving"

Systems:
- Road sensors - Autonomous Intelligent Cruise Control (AICC)
- On-board radar - Laser systems
- On-board compulers for communication

Benefits for motorists:
- More safety for drivers and passengers
- Greater comfort thanks to reduced driving stress
- Reduced exhaust emissions and increased fuel efficiency because of harmonised traffc flow
- Advance warning of dangerous situations

Fleet management
Objectives:
- Reducing congestion and pollution
- Closing information gaps on, for example, delivery times, pick-up points, current position, operating problems
- Avoidance of unnecessary or empty trips and diversions
- Better monitoring of hazardous goods carriage and better intervention in case of emergency
- Increase efficiency of goods transport systems
- Increase efficiency of public transport networks
- Ensure correct functioning of complex logistic systems

Systems:
- Satellite systems for vehicle location and two-way communications
- Cellular communication systems for data and voice
- Electronic mail systems

Benefits for Hauliers:
- Fewer commercial vehicles on the road

Benefits for passengers:
- More efficient and convenient public transport

Automation of toll collection systems
Objectives:
- Free traffic flow
- Open highways
- No queues at toll booths
- No need for different currencies for toll payments

Systems:
- Automatic vehicle identification
- Automatic debiting via multi-application smart cards

Benefits for Motorists:
- Shorter travel times
- Comfortable payment by smart card or by charging to your bank account
- No currency problems

and the ability to register proximity and speed of other vehicles, for cruise control and collision avoidance strategies; improved night vision and fleet management for commercial vehicles. In the long term there would also seem to be the possibility to develop automated vehicle control for long distance motorway driving.

Already, say ACEA, there are 160 million vehicles on Europe's roads today, 140 million of which are cars, set to increase some 30% by 2010. To date around ECU 900 million have been spent on the Prometheus telematics programme and ECU 400 million on the Drive programme. Suggested reality for system becoming available to customers is up to 1998 for 'support system telematics', and 'vehicle control systems' by 2000. Advanced warning systems and some navigation devices are, of course, already available. Over all, the six main sectors of road transport telematics are seen as traffic management, pre-trip information, on-trip information, vehicle control, freight/fleet management and automated toll collection. Table 1 lists the main elements of these six groups.

An example of an integrated traffic system is operating in Europe's motor capital. Claimed to be the most comprehensive regional traffic management system in the world, it has now been through a decisive test phase. If the system proves successful, Stuttgart in southern Germany could become a model to other urban regions for intelligent traffic solutions. Daimler-Benz with its extensive know-how in the field of transportation and traffic systems is the leading industrial partner in STORM (Stuttgart Transport Operation by Regional Management), Fig 26.

It aims to better utilise and interlink existing transport systems to make public and private transport co-operate more efficiently — thus reducing people's travel time. The whole STORM project is being coordinated by means of a service computer acting as the integrated data network. This service computer not only contains static traffic data like a digital road map, but also provides an up to-the minute picture of the traffic situation in the Stuttgart area. It comprises the following six elements: travel Information; individual on-board guidance; dynamic Park-and-Ride information; connection information; fleet management; emergency call facility

STORM has made its mark on the Stuttgart area with Park-and-Ride information signs on motorways, displays over tube station platforms, and "infotheques" at various points in the city provide motorists and transport users with important information. At one station in the city, dynamic display panels inform the bus driver a bout the actual arrival time of the local trains. Through Fleet Management, comprising new techniques such as on-board computers, mobile communications and satellite-assisted vehicle location, hauliers can improve their route planning and thus become more economical.

About 100 testers have been out gathering experience in using the individual guidance system, which is supported by a network of infrared beacons. Vehicles equipped with the Dual Mode Route Guidance (DMRG) navigation device can select the most favourable route to their destination, inside and outside the city. The system receives recommendations

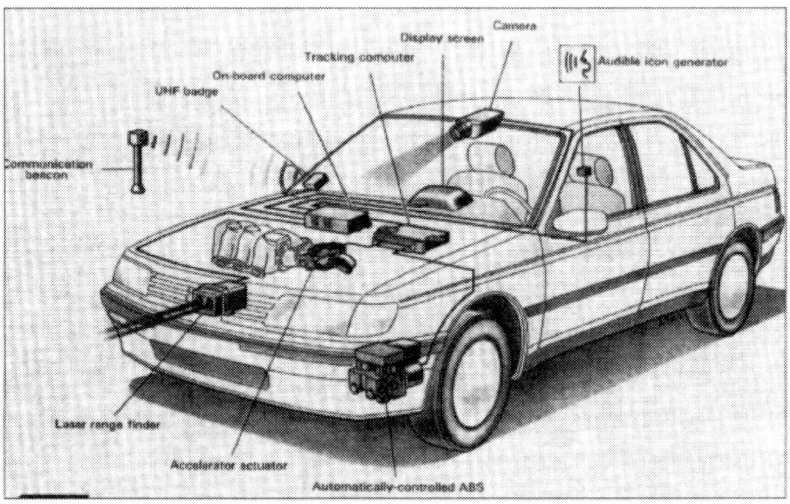

Fig 27: Intelligent cruise-control system

on the best route and transmits information on the effective travel time to the central computer in order to detect traffic hold-ups.

The automatic emergency call system can appreciably shorten the response time of rescue units to accidents. Previously it took an average of 14 minutes in the city to get assistance to accident victims. With the emergency call system it now takes only eight minutes for help to arrive. Outside town, rescue time can be cut from 21 to twelve minutes.

The cost of the STORM project amounts to over DM 60 million, 15 percent or which is covered by industry in the form of support from companies like Daimler-Benz, Bosch and Siemens. The project is promoted and funded by the European Community, the federal state of Baden-Wurtenberg, and the City of Stuttgart. STORM was prompted by the European research programme PROMETHEUS. During the testing phase, the parties of the STORM system have established that all elements interact smoothly and that the objectives set can be met.

Vehicle systems

ALTO (Advanced Leading Technology On-board) is said by engineers[3] of PSA Peugeot Citroen to be the first vehicle in Europe to incorporate all the functions of the Prometheus programme: vision enhancement, lane keeping support, co-operative driving, autonomous intelligent cruise control and dual mode route guidance. Based on a Citroen XM, these functions are integrated on a multiplex network of fibre-optics and a multifunction interface display for the driver. The enhanced vision system involves a combination of direct drivervision and projected images on the display. The vehicle is equipped with additional infrared headlamps, a camera sensitive to IR and a head-level display screen for this purpose. It displays real-time images identical to those viewed in high-beam. Additional headlamps light up the near-IR, thus illuminating the field in front of the vehicle without dazzling oncoming drivers.

For lane-keeping support, whose objective is to keep the vehicle under control if the driver falls asleep, the vehicle is equipped with a camera, image-processing computer for optimal path calculation, a warning system and stepper-motor controlled steering mechanism. The camera records the oncoming scene and the computer processes the data to extract white-line position and road-profile for determination of the path upon which the steering correction will be based.

For co-operative driving, the objectives are improving safety by information exchange between vehicles and between vehicle and road infrastructure, so that the driver can adjust his driving behaviour accordingly. Vehicles are equipped with 57 GHz UHF radio, a longitudinal repositioning system, an on-board computer and display unit, for communication with road-side beacons UHF communication between similarly equipped vehicles relays position, speed and direction-changing information while that from beacons advises on external conditions, such as traffic build up and weather-change, as well as providing positioning signals to help vehicles maintain relative spacing; the beacons also receive signals from vehicle for rescue requests which are service through a network of support services.

PSA have developed the second-phase ele-

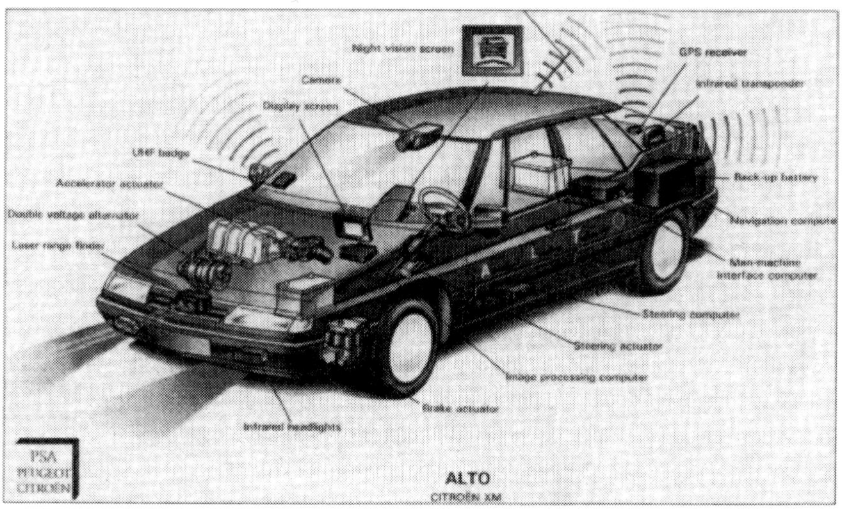

Fig 28: ALTO total-systems vehicle

ments of Prometheus on different vehicles. Fig 27 shows intelligent cruise control fitted to a Peugeot 605 which was developed to achieve automatic spacing with the vehicle ahead without the aid of roadside beacons. Equipment includes a laser range finder, a control computer for acceleration and braking, including anti-skid control, also a head-level display. The vehicle has lane-keeping support warning, too, so warning-free lane-changing is only allowed with trafficators operational; also vehicle-to-infrastructure communications are provided to ensure prevailing speed limits are kept.

Fig 28 shows how all these systems are combined, together with dual-mode route guidance comprising compound navigation system, on-board road map database and RDS/UTI decoding, on the XM-based ALTO concept car. The navigation system works in a way which lets the driver indicate his/her destination, whereupon he/she is guided along a digital road map, the compound navigation system plotting the vehicle's course on the map and giving the driver directional instructions. Dual-mode signifies the ability to automatically integrating traffic information using RDS-UTI. If the driver fails to follow the prescribed route, the system recalculates vehicle position and a new route is provided to the programmed destination.

Proprietary telematic systems

Siemens have developed autonomous vehicle control using electronic throttle control, automatic transmission and brakes; the system was described by engineers[4] at a recent conference. The authors point out that different stages of automatic cruise control (ACC) have been realised over time, involving combinations of the three vehicle systems mentioned above, starting with throttle only. When all three are combined a continuous fuzzy-logic control is involved. With such a combination, the company equipped a vehicle with a 3-beam radar sensor, the lateral beam characteristics being shown in Fig 29, so that cars could be detected at up to 170 metres, trucks up to 200 metres and pedestrians up to 75 metres. After evaluation it was decided to use a 5-beam sensor which allowed pedestrians to be detected up to 100 metres. Fig 30 shows a block diagram of the hardware involved.

The headway sensor provides information on distance and relative speed of potential obstacles. The radius of curvature of the road also needs to be detected and this can be done by wheel speed sensors, steering wheel position sensor, gyrosensor or video camera. The first three can only provide real-time detection while the fourth can foresee curvature. Currently this will probably be exceeded on cost grounds, the authors explain.

Digital inputs acquired include cruise command switches, brake switch, driving style and road conditions. In the throttle only access system ACC functions are executed via the conventional cruise control of the engine management system (EMS). When access to the automatic transmission is enabled, the gear ratio is down-shifted while decelerating. With additional brake access, very modest braking torques have to be used otherwise a virtual obstacle a short distance ahead could easily result in emergency braking, endangering following vehicles. Continuous acceleration control is available using electronic throttle control (ETC). For deceleration the booster of the braking system was appropriately modified.

When the driver himself makes a braking or accelerative input all ACC functions are automatically disengaged to ensure his total control. These re-

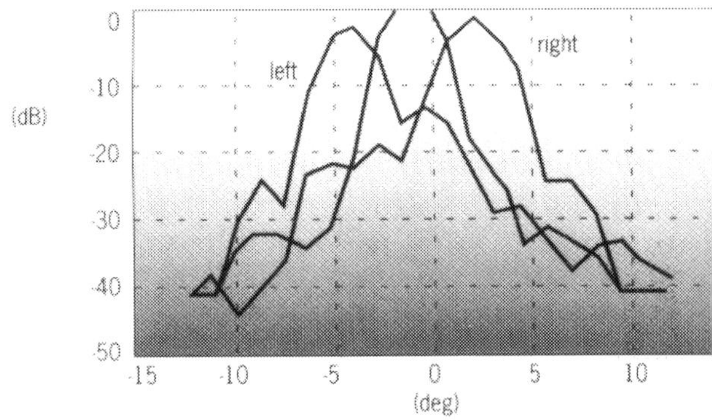

Fig 29: Lateral beam characteristics

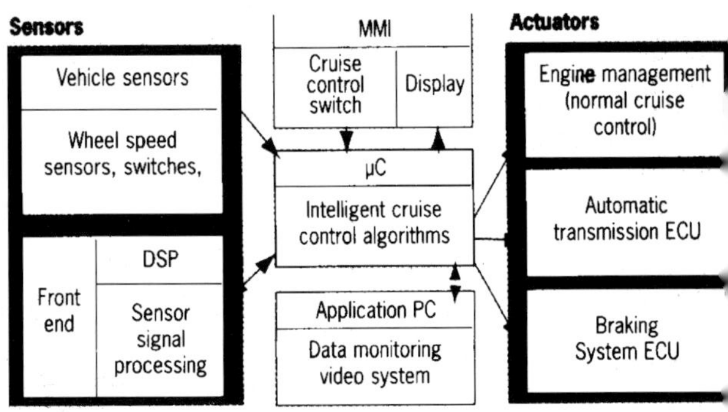

Fig 30: Hardware set-up for Siemens ACC

engage after an input such as decelerating or after an overtaking manoeuvre. When the ACC functions are on, primary objective is to control spacing distance when a vehicle ahead is detected; when no vehicle is detected the system acts like conventional cruise control, Fig 31. Vehicle speed, target distance and relative speed are used as inputs for ACC algorithms and safe spacing distance computed.

Bosch experience with adaptive cruise control involves a unit[5] built as a distributed system using common ECUs with other vehicle systems but including one more containing range sensor and ACC controller, mounted in front of the vehicle, Fig 32. The range sensor is the key component of ACC. It has to supply the ACC's controller with data about the position and speed of preceding vehicles, relative to the ACC-equipped vehicle. The considered ACC system contains a mm-wave radar sensor in the 76.5 GHz frequency range. With the FMCW modulation (frequency modulated continuous wave), resolutions of better than 1 metre distance, and 0.5 m/s in relative speed, are achieved. Not only the values for distance and relative speed directly used in the ACC controller are of interest; the lateral position or the angle relative to the vehicle's longitudinal axis is important to select the relevant preceding vehicle. An angular resolution of better than 1 deg within a range of +/- 5 deg, can be achieved using multi-beam techniques. This results in sufficient detection quality for straight and curved roads. Multiple target values are passed to the ACC controller after signal processing, preselection, tracking and filtering for further selection and computation, the authors explain. A semi-planar technology is used for the transceiver to meet the requirements of high integration, high reliability and low cost. The average output microwave power is very low (<1 mW) due to simultaneous multi-beam measurements. Although the unit contains the 76.5 GHz transceiver, the signal processing and the ACC controller, the entire unit is reasonably small for integration in the front part of the vehicle.

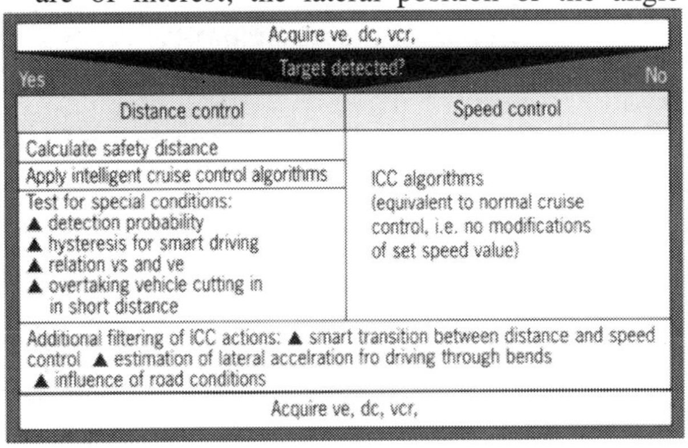

Fig 31: System logic

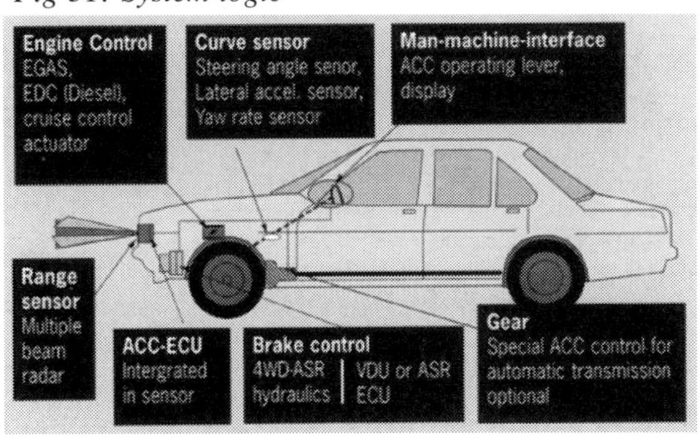

Fig 32: Bosch system

In comparison to other sensor concepts only mm-wave radar sensors are said to be sufficiently robust against bad weather influences. There is only a small degradation by fog and rain. Detection range will always remain higher than human visibility for the specified detection range (150 m). It should be noted that, for bad weather robustness, it is necessary to mount the radar antenna so that it cannot be covered by snow.

Curve sensors To detect and select the "relevant" preceding vehicle it is essential to estimate the future course of the ACC car. Fig 33 shows two typical examples where a wrong "target" will be selected without a proper course prediction. In the simplest case, course prediction assumes movement in a straight line. Under this assumption, no dedicated curve sensor is needed. It is obvious that this way of predicting the course does not work in curves or during lane changes. In a more complex case, a constant course is assumed, meaning constant speed and curvature. This works well most of the time, since estimation errors usually only occur at the beginning and end of curves. Several ways of sensing the actual curvature are commonly used today, the authors explain; evaluation of the relative difference of the wheel revolutions, using a steering angle sensor or measuring the lateral acceleration. A yaw rate sensor which is needed for Vehicle Dynamic Control (VDC) systems can also be used. All methods possess characteristic errors. Hence a combination is most robust.

A higher level of curve prediction can be

provided by using navigation systems (digital maps and course calculations). The limits of this technique depend on the agreement of the maps with the actual road and on the system's capability in determining the car's actual position. The prediction is faulty in road-construction areas and for new roads. Moreover, today's systems do not have a position resolution which is sufficient for ACC course prediction. Furtherrnore, the price has to be significantly reduced to guarantee a broader distribution.

Several high-tech radar development laboratories promise a road and lane prediction purely based on the radar information, say the authors. Stationary objects like reflector posts and crash-barriers detected by the signal processor are used to reconstruct the road boundaries, Fig 34. Little is known, however, about the quality and robustness of these new techniques, so far.

Extracting road markers from video images is another way to predict the road course. Lane classification of the detected objects is possible by combining a video system with an angle resolving distance sensor. Lane prediction using video seems to be the most promising, from a technical point of view, in spite of current limitations. Firstly, ambiguous lane markers as in construction areas can not be distinguished. Secondly, visibility in bad weather and darkness is limited. In general, the prediction range will be limited to about 80 metres.

The infornation shown on the display, Fig 35, includes the set speed, since the driver might forget it after long periods in slow traffic; the set time gap is also displayed, as well as the prevailing ACC mode (following control or speed control). There is also, usually, an audible warning to request the driver takes over control as necessary.

Near-production products

According to Bosch, the most basic navigation systems use GPS, with a flashing marker showing the location of the vehicle on a map of the road network. The road network itself is stored as a digital map on a CD-ROM which is then displayed on a colour monitor. The accuracy of this design does rely on good GPS reception, but can be made more accurate using additional sensor signals from the car's ABS and compass systems. It was such a principle that was used in the Bosch TravelPilot IDS launched in 1989.

If these locating systems are expanded to include a route finding function then it is possible to indicate the ideal route on the road map. However, it is very difficult for a driver to identify both the driving route and his current position on the map with just a quick glance at the monitor while the vehicle is in motion. Systems need to be able to provide clear spoken instructions backed up by unmistakable direction symbols on screen. As complex maps are no longer required, the expensive colour monitor can be replaced by a cheaper monochrome matrix display, significantly reducing the overall cost of the system.

Surveys have identified the following as primary customer demands: quick and reliable route guidance; easily understandable driving instructions which are delivered early to reduce driving stress and

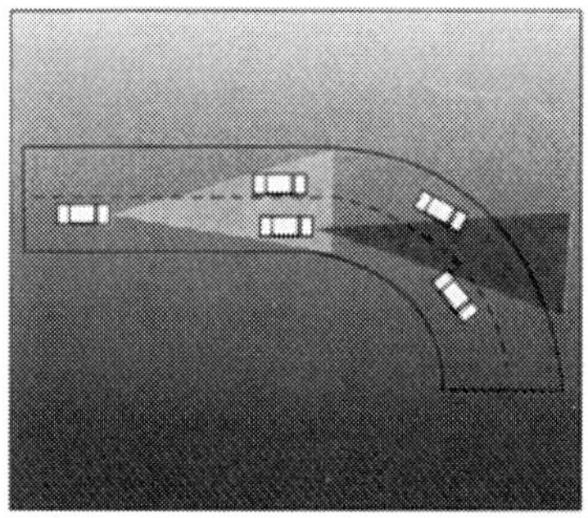

Fig 33: Curve sensors

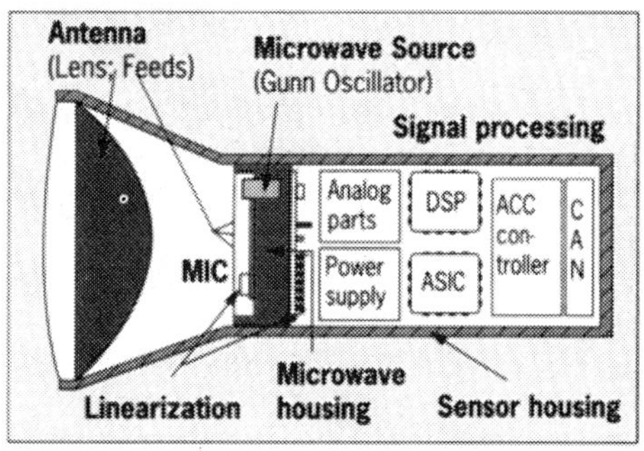

Fig 34: Radar sensor

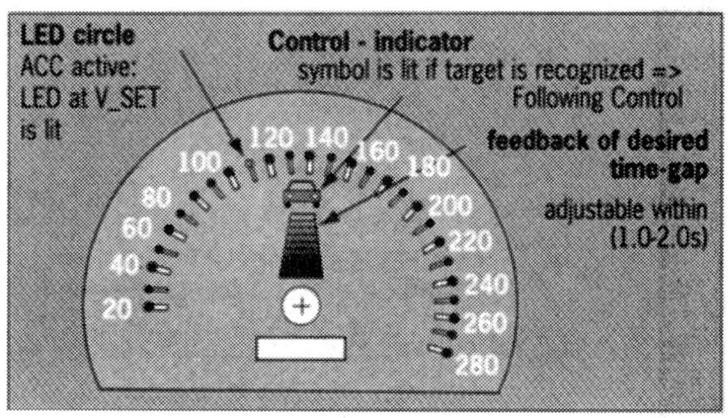

Fig 35: Bosch ACC display

increase road safety; easy to use; an ability to provide alternative routes to bypass congestion; direct selection of important destinations such as airports; reasonable price

For the foreseeable future a keyboard will remain the most effective method of entry. As far as software is concerned, CD-ROM's are an ideal medium to store the data for the digital road map and other additional information. While it is possible to store a complete road map of Europe on just one super-density disk, it does not make much sense as motorists would only ever use a fraction of the information contained. It is more likely that one CD would contain more detailed information on just one country or region. There is no reason why these CDs should be restricted to pure route guidance. A travel guide CD, for instance, could locate and direct motorists to selected restaurants hotels or other places of interest.

The top-of-the-range Blaupunkt Berlin RCM 303-A in-car audio/video system already on the market has an open systems architecture which allows integration of navigation products. A navigation system, described later, will give spoken instructions as well as visual guidance in the form of direction arrows on the display screen. Instead of marking position and destination on a road map, it calculates the route between any two points and provides recommendations for every possible deviation.

Fig 36: Blaupunkt Berlin system

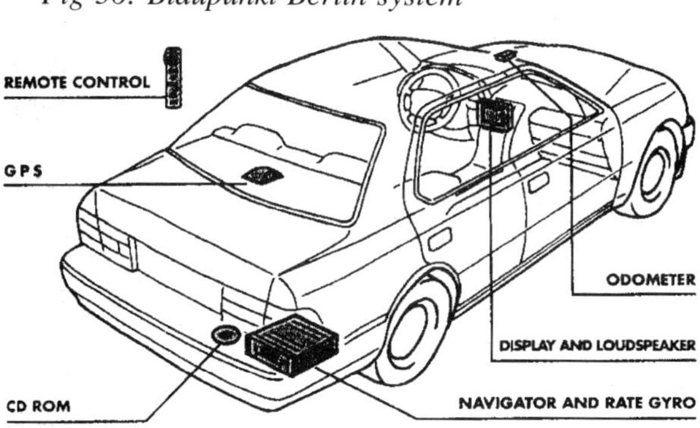

Fig 37: Bosch electronic braking

It provides visual and spoken route-guidance information at each road junction. A CD-ROM holds digitized voice commands as 100 standard phrases. Location of the vehicle is fixed by monitoring wheel location together with a magnetic compass for fixing orientation. A GPS receiver processes satellite signals for inter-city travel. An improved version of RDS (Radio Data System) switches to the strongest FM station for program continuity while driving. The Berlin stores the last seven RDS announcements in a digital voice memory for play-back after short-term parking and journey restart.A TV interface can also show weather information for rear seat passengers.

The company's current ACC system involves a 3-beam radar system for detecting upcoming traffic. Two beams evaluate distance and relative speed while a third identifies angular position for lane position determination. The ACC controller adjusts speed through intervention in engine, auto-transmission and brakes. It is also interfaceable with the company's traction control, vehicle dynamics control systems and a newly developed smart electronic braking system, Fig 37.

Driver information

According to engineers[6] of VDO, traffic telematics in the Mannesmann group is based on linking bi-directional mobile data telecoms (GSM) and satellite position determination (GPS). The latter is, of course, available without any need for additional infrastructure like beacons or induction loops. The actual position is calculated by processing the GPS signals The accuracy that can be achieved is in the range of +/- 100 m. For special applica tions the accuracy can be improved, to about 10 to 20 metres, by using a GPS reference station (the base station of the mobile telephone network) which trans mits correction data. This method is known as DGPS, Fig 38, the author explain.

In the case of bad or missing reception of the

GPS signals (between high buildings or in forests) the position will be determined by dead reckoning methods. The direction of the path and the distance travelled allow an incremental continuation of the path vector. The change of the direction is detected by the use of a gyroscope (or if available by processing the ABS signals) and the distance travelled is recovered from the speedometer signal. VDO found it was necessary to combine GPS positioning and dead reckoning in an optimal way for satisfactory results.

GSM mobile telephone networks are available in many European countries with nearly complete coverage and connection across borders. The services provided by the GSM net are continuously extended and thus facilitate their use for traffic telematics applications. For data transmission there exists one cost effective solution in the Short Message Service (SMS) which allows package oriented data transmission with a capacity of 140 bytes per package. SMS uses free capacity on th control signal channel which is transmitted independently and parallel to the speech. The overall system of course requires efficient performance by service provider, mobile data communication network provider and the in-vehicle unit.

The in-vehicle unit has basic functions as follows: 1 The user selects a service through the man-machine interface of the in-vehicle unit (keyboard, touchscreen, speech through a speech recognition system or the operator in the service centre). 2 Transmission of the service request via mobile data communication to the service computer. 3 The service computer and/or the operator processes the data transmitted according to the requested service. 4 Transmission of the response data and/or speech to the vehicle. 5 The data are processed in the in-vehicle unit and presented to the user through the man-machine interface.

The car driver does not need a telephone number to transmit a service, breakdown or emergency call using the in-vehicle unit. All he needs to do is press a button to discuss his current situation with the service centre. The position of the vehicle is determined automatically and transmitted, an inestimable advantage in remote areas and irreplaceable when every minute is a matter of life and death. The dnver can call up a display of all traffic information in a particular direction. Here an information segment is selected from a pie chart and can be updated as it is "taken on the journey". Traffic information which is "behind" the driver, and of no further interest, is suppressed. The highlight of the system is that the

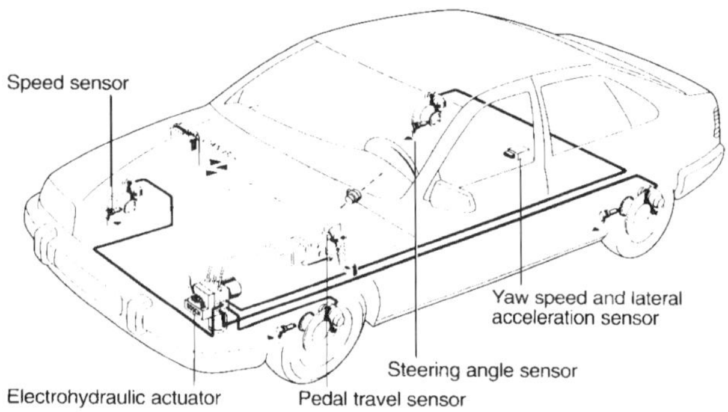

Fig 39: Magneti Marelli RoutePlanner

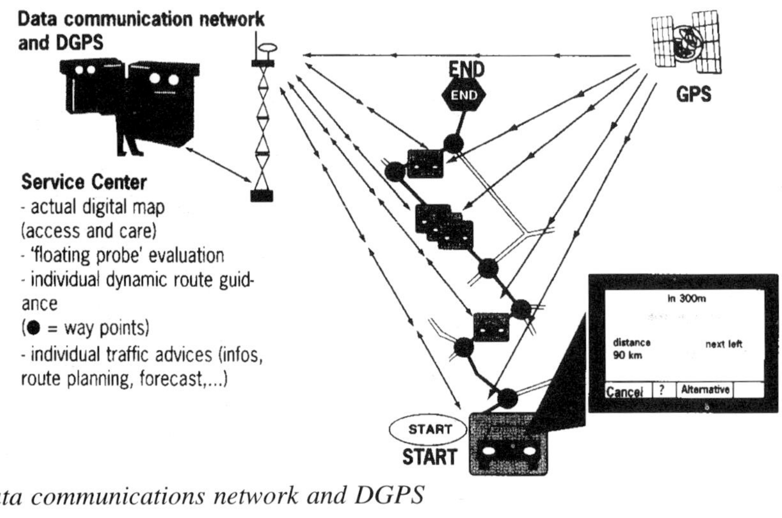

Fig 38: Data communications network and DGPS

service centre receives the current traffic situation via the same terminal equipment in the vehicle. The on board unit transmits the traffic conditions (speed, stop-go traffic, time links between highway points) to the service centre. The system is based on many of these mobile measuring devices which "float" with the traffic flow. This principle is called "floating probe" evaluation.

The transmission of anonymous vehicle data (position, speed) to the service centre for recording the location of traffic is an important function of the system, giving the community of users an indirect advantage when interrogating the traffic information. This function makes an additional contribution to the traffic information available and is the first step in the direction of an optimized "traffic feedback" system. On the basis of histoncal data and the current traffic situation, the service centre is then able to make predictions regarding likely traffic volume developments.

Mobile navigation system Magneti Marelli's recently announced RoutePlanner system (Fig 39) is of the type which continuously displays the position of the vehicle on a map, as well as providing the driver with the usual back-up infrastructure information. The driver has a IR data-entry remote-control as well as colour monitor and five-language integral speaker, which are removable to avoid theft. Vehicle location is by both dead reckoning and map matching, whereby position is first estimated by information from sensors and than corrected by algorithms designed to relate estimated position with the corresponding point on the road system.

Only two sensors are used for dead reckoning operation: an odometer, already available, to measure distance travelled, and a ceramic gyroscope to measure angular velocity. The response curves for each of the two sensors, and in particular the effect on them caused by variations in environmental parameters, and by the presence of interfering noise sources, both have been thoroughly researched in order to improve estimation accuracy. An automatic initial resetting operation is also provided.

The map matching process involves a section-by-section correlation carried out between the estimated route and the digitised road system available on board. The location of the vehicle is always identified by a region of probability (CEP) that arises from the indeterminate nature of the estimated position, which in practice, represents a probability distribution of the position. This region contains a number of road segments which have different probabilities at each step of the process. The system identifies the segment with the highest probability and positions the dead-reckoning-estimated position of the vehicle at that point. A GPS is used for the acquisition of vehicle position and direction information.

The cartographic database for the digital map contains the urban and out-of-town road system represented in vector format, which means that each road is stored as a set of oriented segments called branches that are joined at points called nodes. As well as the road system, many other geographical elements are included to provide a host of visual references. Besides the geographical co-ordinates of the branches and nodes of the roads valuable additional information such as house numbering, road class. One-way systems and pedestrian zones are indicated.

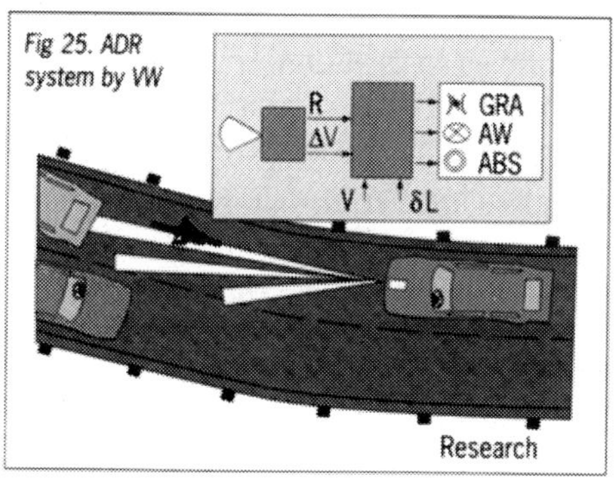

Fig 40: ADR system by VW

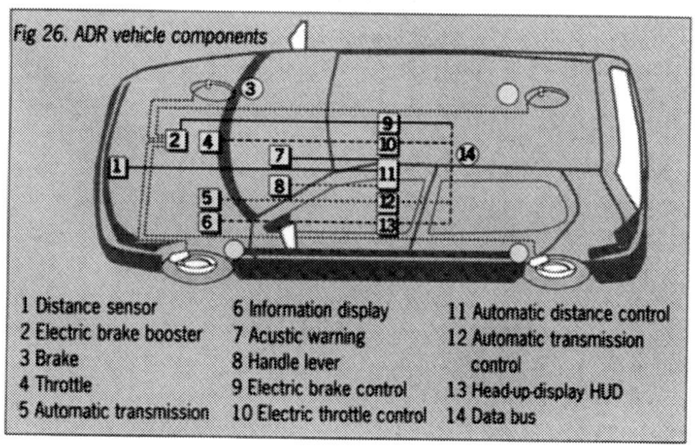

Fig 41: ADR vehicle components

Projected system for volume cars

Some insight has been given[7] as to how a volume vehicle builder might actually apply intelligent highway systems. An embodiment of VW's Automatic Distance Regulation is seen in Fig 40. A three-beam distance sensor monitors the area in front of the sensor-mounted vehicle. Automatic docking-on to the vehicle in front is achieved by closing the throttle or, if necessary, by moderate application of the brake. If the vehicle ahead accelerates, the sensor-mounted vehicle follows suit up to a maximum speed set by the driver, and then carries on in cruise control mode. The driver receives information on the system status by means of a head-up display, so that he is able to track the system response at all times. The system only docks on to vehicles travelling at more than 25 km/h, but braking is applied down to standstill. When pulling away again, the driver must consciously press the accelerator. The braking deceleration is limited to 2.5 m/s^2, sufficient for comfort braking but not for emergency braking. If the deceleration calculations show that the braking power is not sufficient, the driver receives an acoustic signal telling him to brake more forcefully himself. ADR increases driving comfort by relieving the driver of tasks; it increases traffic safety, reduces fuel consumption and helps protect the environment.

When the system is active the accelerator or brake pedal are not needed almost for the entire journey, representing a considerable relief for the driver. The intended application is primarily on motorways and well-structured trunk roads. The longitudinal guidance of the vehicle, especially in nose-to-tail and stop-and-go traffic, places a great strain on the driver, which can be substantially reduced by the ADR system. The more vehicles are run with ADR, or similar systems, the more convoy-style traffic will be created, with the vehicles travelling all at the same speed and with the same distances between them. At current actual traffic densities, and with frequent speed limits, nose-to-tail traffic is common, but with high-risk deviations in speed and distance to the next vehicle. Simulation results show, ADR harmonises the traffic flow, thus contributing to increased traffic safety. Consumption will inevitably be reduced if unnecessary acceleration and braking are eliminated in a smooth-running traffic flow. The reduction in traffic jams resulting from accidents will also reduce consumption.

Fig 41 shows a schematic of the ADR components, key ones being the distance sensor, electric brake booster and the distance control algorithms. The image sensor is integrated in the right-side headlight; the advantages of this fitting location are that installation does not affect styling; it can be cleaned with standard headlight washer system; de-icing by heat of headlight bulbs

As vehicle deceleration by means of the engine drag torque alone is not enough for the ADR function to dock on in normal traffic, the system requires an electric actuator to build up brake pressure. The requirement is for gentle, smooth pressure build-up, especially in the event of little braking deceleration, in order to meet comfort demands. And the existing

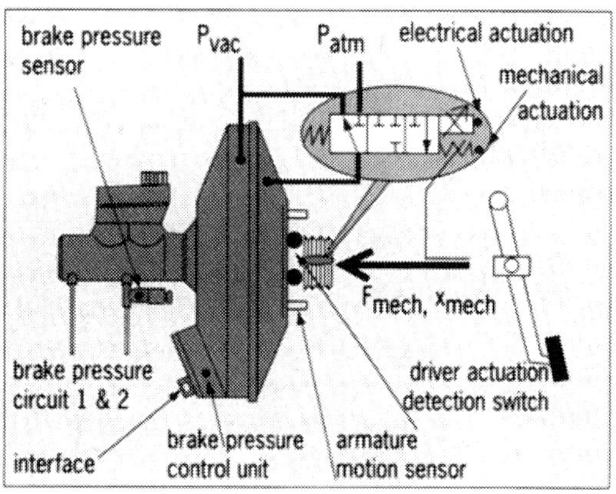

Fig 42: Smart brake booster

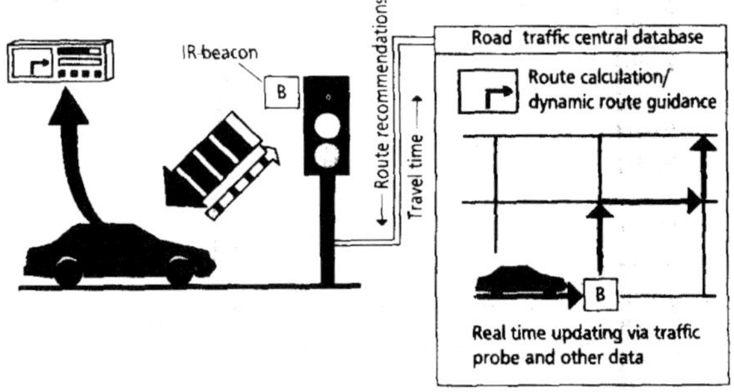

Fig 43: Dual Mode Route Guidance

brake system, including the ABS, is to be modified as little as possible. The problem is solved by expanding the brake booster to a "Smart-Booster", Fig 42. It includes an additional pneumatic valve which can apply more vacuum to the booster diaphragm, dependent of the electrical actuation signal. To be able to adjust the valve gap continuously by means of a simple magnet, the assembly includes a position control circuit incorporating a motion sensor. A brake pressure control circuit is superimposed on the position control circuit, so that a brake pressure is set as the electrical input variable for the booster. The quality of the distance signal and the relative velocity signal derived from it are decisive for the functioning of the system. There is typically serious disturbance of the raw sensor data, especially resulting from characteristic signal failures and superimposed measured values of immobile objects on and next to the road. Intelligent tracking algorithms generate the relevant distance and velocity signal.

The Dual Mode Route Guidance system, Fig 43, being developed as part of the PROMETHEUS project is planned to offer a choice of the shortest as well as the fastest route, on the basis of a digital road map and traffic information. An on-board computer determines optimal routes and chooses information systems that are more localised, accurate and up-to-date. For tracking vehicles, worldwide, BT's Globetrak constantly transmits a vehicle's position back to its base which can also issue instructions to vehicles in the field.

With the Euro-SCOUT mentioned earlier in this section, Fig 44, vehicles are anonymously linked by road-side infra-red beacons to a central guidance computer and database. Guidance updates are transmitted to vehicles at five minute updates. A variety of fixed detectors can transmit this data but the key source is traffic probe data in the form of Euro-SCOUT equipped vehicles which transmit travel times as they negotiate the hold-ups. These provide 'historic' travel time patterns which are revised by incoming traffic probe data which allows prediction of journey times. The in-vehicle unit has a keyboard for inputting a new destination

Working with Ford and Jaguar engineers, Lucas have used optical, infrared and radar detection devices to develop the independent systems seen in Fig 45 for intelligent cruise control, collision avoidance and vision enhancement. The first controls acceleration and braking at rates up to +0.15 and -0.3 9, using radar. The second uses a video camera, the system processing 7.5 million characters per second via 11 transputers and a graphics processor. The third uses a GEC Marconi infra-red camera with lOOX100 array pyro-electric sensor and has an important input from Pilkington Technology.

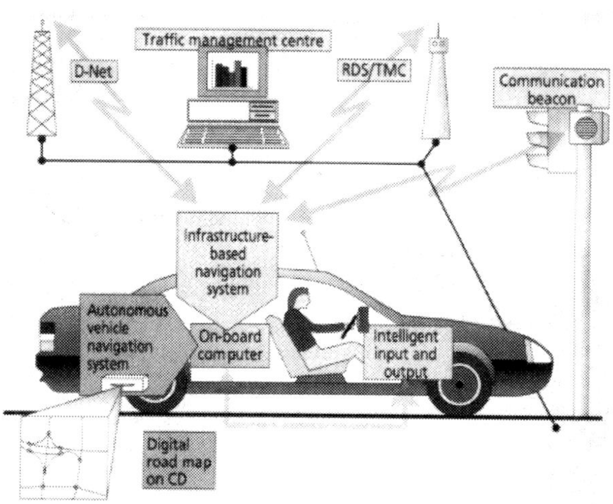

Fig 44: Euro-SCOUT dynamic route guidance

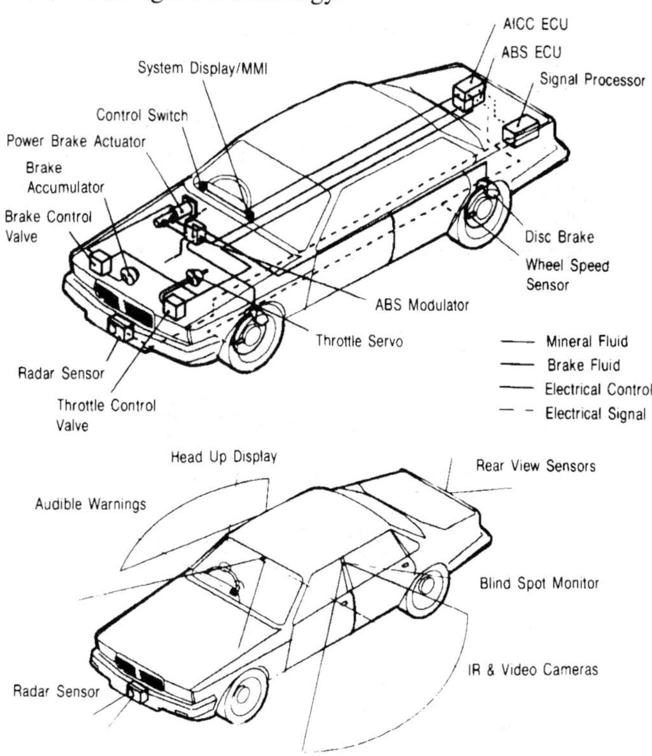

Fig 45: Lucas collision-avoidance systems

Towards intelligent traffic systems

As traffic densities will soon rise to a point where 'co-operative driving' intelligent-systems are necessary on many routes, the French company GSI Tecsi have proposed[8] a path towards improved in-vehicle driver support systems, termed Vehicle Distributed Executive.

Its proponents argue that the current embedded electronic system of a car is composed of heterogeneous, distributed units (ECUs), interacting with a non-cooperative environment under strict time constraints. There are many similarities with the control systems of aircraft, but aircraft manaufacturers are not submitted to the same level of economic constraints and can use highly powerful and expensive hardware architectures. The car manufacturers must cope with the same kind of problems, but by using cheap hardware at the peak of its performance.

A real-time system required on a vehicle is characterised by its ability to respond under deterministic time constraints to asynchronous external events. According to GSI Tecsi, a task-based real time system requires a dedicated execution support, responsible for the allocation of the processor to tasks. The RealTimeCraft line of products developed by them is based upon the SCEPTRE normalisation scheme for execution support. These provide support for the local processing of each ECU, but do not include support fo distributed applications. By request of Renault, an extension layer providing communication features over a network was developed in association with GS Tecsi, the layer being known as VDX (Vehicle Distributed eXecutive).

It includes features for: sharing data and peripherals between tasks (critical regions, mutual exclusion semaphores); synchronising the execution of correlated tasks, or tasks and interrupt routines (events, counting semaphores); communicating between tasks (mailboxes); time management (delayed execution, timed waits). The VDX layer is also message-oriented and supports this set of communication objectives: datagrams (unstructured messages with a bounded size); signals (logical conditions, for reactivating a remote task); groups of parameters (each parameter being an individual value); unbound messages (with a rendezvous protocol). Communication objects types are paired, with a Transmitter and a Receive subtype for each kind of message. Their connection scheme is very flexible, and allows for the

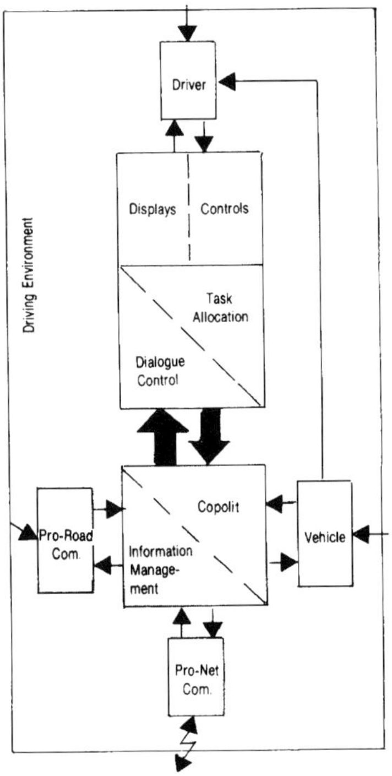

Fig 46: Saab Information and Assistance system

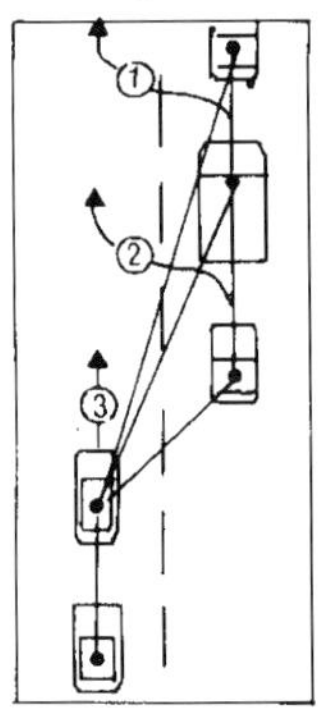

Fig 47: IMC system

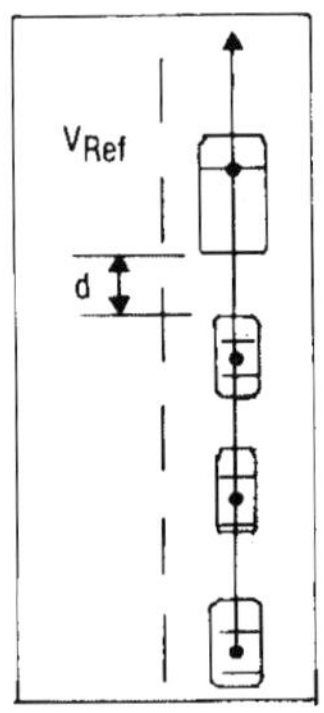

Fig 48: ICC system

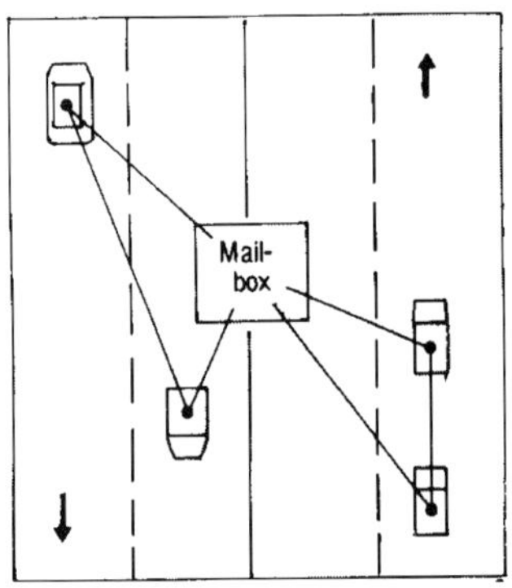

Fig 49: MRP system

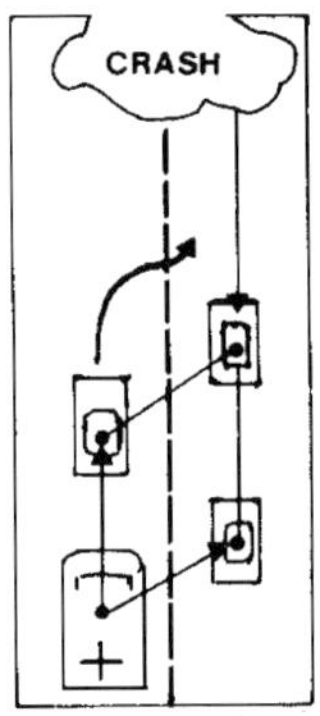

Fig 50: EWS

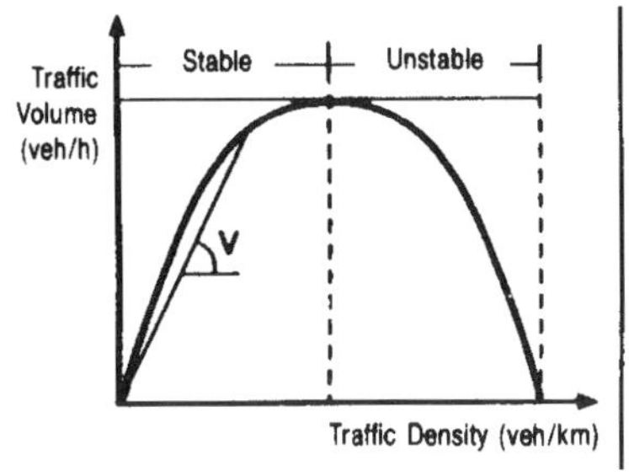

Fig 51: Fundamental diagram of traffic flow

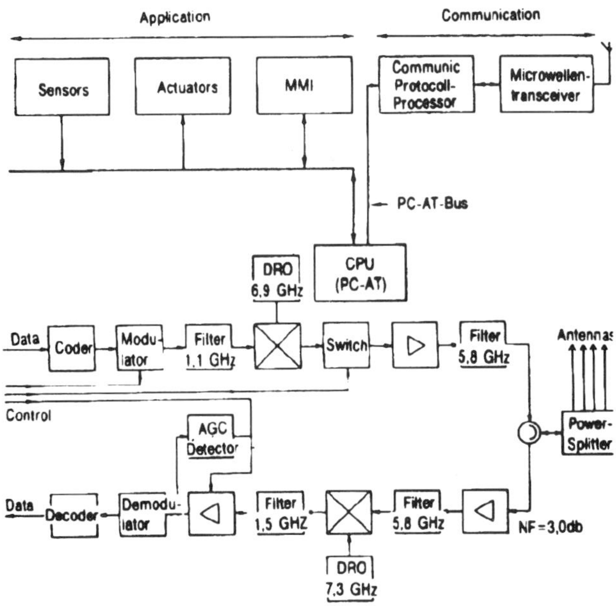

Fig 52: Prometheus demonstrator communication equipment

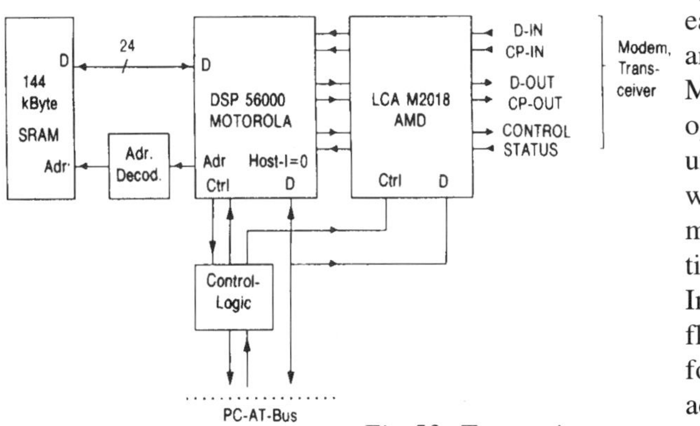

Fig 53: Transceiver

connection of one transmitter to several receivers, the result being the broadcasting of the messages. This can be of particular interest for safety critical parameters, that can be broadcasted periodically to all nodes in the system, thus providing a primitive, but efficient scheme for safe sharing of data.

Information/assistance system Acording to engineers[9] of Saab-Scania driver/vehicle (man/machine) interfaces (MMI) can in principle be realised as the Information and Assistance System shown in Fig 46. Such a system will act as a link between all subsystems and elements in the road traffic system. He emphasises that the MMI design cannot be reduced to display and control optimisation only — neglecting their use in real traffic situations, their role for different tasks and their interpretability and acceptance by different users. The ideal system design will make use of the capabilities of the user in an adaptive way. At the same time the system will support the driver in situations where the human senses and the human motor system have limitations. A systems approach in the design includes both top-down analysis and bottom-up synthesis performed in parallel. The interactive links between the two activities can be described as continuous evaluation of different steps in the design work. This process gives the possibility of approaching MMI design problems from both a theoretical and a practical point of view.

Co-operative driving According researchers[10] at VW the Prometheus sub-programme PRO-NET provides for co-operative driving, offering prospects for medium-term introduction into tomorrow's traffic systems. This is via the concept of Intelligent Cruise Control (ICC) based on inter vehicle and vehicle-roadside communication. Self-organised co-operation between vehicles which are influencing each other should significantly improve the safety and efficiency of road traffic, it is argued. Intelligent Manoeuvring and Control, for safeguarding and optimisation of lane changing and overtaking in unidirectional traffic flow (Fig 47) is one aspect while Intelligent Cruise Control is seen as just harmonising speed and vehicle-spacing in unidirectional traffic flow (Fig 48). Other aspects include Intelligent Intersection Control for optimising traffic flow through intersections; Medium Range Pre-information for providing advance information for adaption to changes in traffic situation ahead (Fig 49) and Emergency Warning Systems for local short-range warning of emergencies and manoeuvring for supporting rescue services (Fig 50).

Intelligent cruise control (ICC) The main objectives of ICC are: harmonising speed and distance; keeping in longitudinal traffic flow; performing basic safety functions for the individual vehicle; providing dedicated recommendations to the individual driver. The merits of these objectives can be clarified by considering them in the context of the 'Fundamental Diagram' of traffic flow (Fig 51). This diagram describes the volume-density relationship in longitudinal traffic flow which primarily depends on the type of road and not on traffic as such.

Each point in the curve marks a traffic situation characterised by the volume (number of vehicles per unit time) and the density (number of vehicles per unit distance). The mean speed for such a point is given by the gradient of the radius vector to this point. In the left hand part of the curve traffic flows freely- — vehicles do not influence each other very much. While traffic volume increases average speed, and speed-dependent distances between vehicles drop, this entails increasing traffic density. There is a maximum of traffic which a given road can handle in a stable condition — 'road capacity'. Beyond this maximum, in the right hand part of the diagram, traffic flow is unstable. When traffic density increases average speed drops more and more and traffic volume drops as well. This is known as the 'stop and go' phenomenon. Traffic density reaches its maximum when vehicles stand bumper to bumper. In this case traffic volume is zero!

Microwave data transmission According to researchers[11]at Bosch, the company have provided UHF radio units (470 MHz range), and some of the required data interfaces, for vehicle-to-vehicle data transmissions. The transmission rate was 2.4 kBit/s (gross), about 1 kBit/s (net). In order to suit the considerably higher communication requirements (more participants and shorter response time), Bosch proposed eventually to use the frequency range of about 5.8 GHz as this range serves especially for industrial/scientific/medical purposes (ISM band) so no problems in terms of test admission would arise. Here, a transmission rate of 2.5 MBit/s (gross), about 1 MBit/s (net), is feasible, which is thus increased by a factor of 1000 compared to UHF. The communication equipment (Fig 52) will be based on a microwave transciever and communication protocol processor (CPP) on a PC plug-in card. A PC-AT is also used as the linking element between the 'communication unit' and the 'application'. The transceiver (Fig 53) represents the link between the on-board PC and the mobile radio channel. Broadband FSK (Frequency Sift Keying) is used as modulation method. The carrier frequency is about 5.8 GHz.

The data stream to be transmitted is first encoded and band limited. The Bosch modulator consists of a frequency synthesizer for the transmitter IF of 1.1 GHz. The transmission data are modulated on to this carrier. An FSK signal having a bandwidth of 120 MHz and a power of about 10 mW is available at the output. The signal is then filtered to suppress unwanted waves. In a mixer the transmitter IF is modulated to the output frequency of 5.8 GHz by means of an oscillator signal (6.9 GHz). The following output stage amplifies the output signal to a power of about 500 mW. This transmitter power can be adjusted with an accuracy of five bits by a damping element. The blanking of the transmitter is done with a PIN diode switch employed before the output stage. The switch-on time for the transmitter is about 2-3 microseconds. Another filtering process ensures that postal regulations are fulfilled.

The received signal is amplified by means of a low-noise amplifier such that the mixing for an IF of 1.5 GHz can be realised without any sensitivity loss. The band-path filter employed before the mixer provides for the necessary reliability of the image frequency. The adjacent channel selection is determined mainly by the following filter. The output of the demodulator offers the receivable baseband data with an amplitude of 1 volt at 50 ohms. An AGC circuit (Automatic Gain Control) is provided in the IF module. The signal is then fed to the decoder which determines whether the signal level becomes Low or High. In addition, the clock pulse is re-gained by means of a fast-locking circuit. Data and clock pulses are then fed to the communication protocol processor of the personal computer for further processing. A selector switches the antenna over between the transmitter and the receiver. Also, protective circuits are employed to protect the receiver against damage during transmission. The transmitter power is distributed to three or four antennas by means of a power splitter.

In order to obtain the maximum transmission range in both directions, 4four-element patch antennas are used; each aperture amounts to about 40 degrees. In case of a three-antenna solution a waveguide antenna is attached below the vehicle, radiating in a lateral direction of the vehicle so that lateral transmission is improved. The transmitter power is fed to the antennas either through flexible microwave cables or by low-cost semirigid cables. The Communication Protocol Processor (CPP) represents the linking element between the μW transceiver and the in-car personal computer.

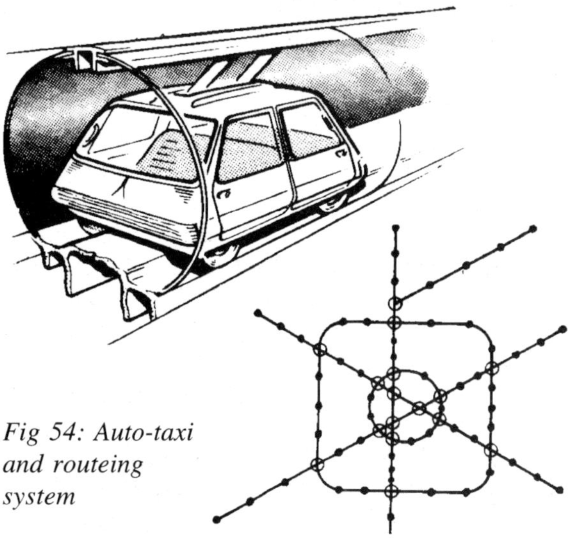

Fig 54: Auto-taxi and routeing system

Fully guided vehicles

The idea of cars electrically propelled through transparent tubes was conceived by Hawker Siddeley whose Dr. Blake[12] suggests the Auto-taxi as a method for the carriage of persons or light freight in cities, Fig 54. A person enters the station, obtains a punched card or plastic token, which issue is prevented by an attached computer if any part of the system is overloaded during the time he is travelling on it, climbs into the first car on the platform, inserts his ticket in the slot and sits down. The doors close, the car moves onto the acceleration section, accelerates at 0.24g up to system speed (35 mph) at which it slips into a space in the main stream and continues until just before the chosen destination where it slips out of the main stream into a deceleration path coming to rest in another station where the doors open and the client alights.

To operate the system, a network of stations is required, each with its acceleration and deceleration path, used also for vehicles changing lines. Stations would be in convenient places like stores, sports grounds or airport buildings and they would be dotted about in residential areas. The car, electrically driven and electronically controlled, would be in continuous communication with the computer sending and receiving messages of speed and car position. If one car slows down on the main track, the section will be de-energised and all cars within it halted at about one g, the section by-passed and the stationary cars taken off to be examined.

Boyd, Plotkin and Tang[13] describe a system based on existing vehicles, which, for the interurban stage of their journey, are loaded into electrically propelled rail vehicles, Fig. 55. A special 6 ft gauge track would be used for the 17 ft wheelbase vehicles which would travel at 150 mile/h. Vehicles are assigned to moving spaces on the guideway, called slots, and control is such that there is only one vehicle at a time in a slot, there being acceleration and deceleration tracks at joining and leaving points.

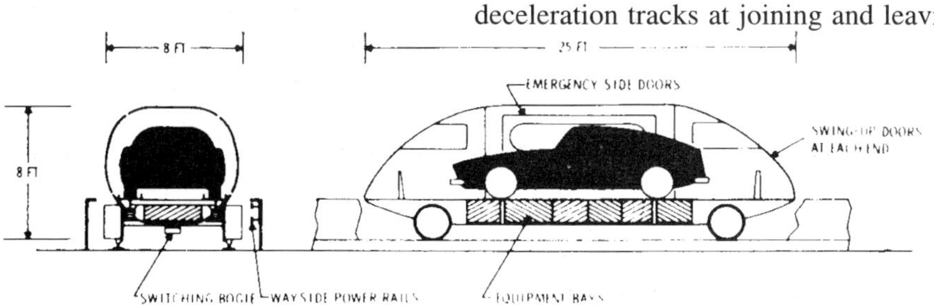

Fig 55: Throughway transit system

Close-to-market driver-assistance

According to BMW research engineer G Reichart[14] a pure technological approach will not lead to (driver assistance) systems that are accepted and optimally used by drivers. Systems, Fig 56, must address a real support-need of the driver and not overburden him with devices that could very easily lead to driver's tasks being made more complicated. What is needed is to provide the driver with appropriate feedback on vehicle performance quality, based on measurements and system modelling, which facilitates stabilising the required performance level. Besides the obvious requirements of availability and reliability there must be the ability of any semi-automated system to hand the task back to the driver and he/she must be informed when and what to take over. There must be complete transparency as behaviour of the system must be predictable by the driver at all times.

In assistance systems for lateral control of the vehicle, so as to ensure lane-keeping, BMW's Heading Control System, Fig 57, is one of the first successful prototypes. It involves a video-camera behind the windscreen scanning the road in front of the vehicle. From the digitized image a computer continuously calculates optimum driving line in a particular lane. With deviation of driver steering behaviour the system generates additional forces in the steering column by means of an electrical servo drive. The driver can sense these forces and follow them by correcting the steering behaviour or they can be easily overridden, Fig 58.

Dynamic route guidance systems are destined for the near future and already in Germany the broadcasting services operate RDS-TMC . On-board monitoring systems are already a reality on BMW 7-Series cars although the navigation facility can thus far only be used in Germany. The company claims to be the first to have introduced acceptable route planning and voice guidance to such systems[15] and point out the self-explanatory operation of the systems. The radio has a 5.4 inch colour screen, Fig 59, which is also used for message displays and user transparency is ensured by the ability to throw up full versions of the abbreviated messages, on screen, as required. Trip-computer functions are available in the system and a seven-band equaliser can be thrown up on screen for the sophisticated ICE sound system. Additional temperature control functions are also

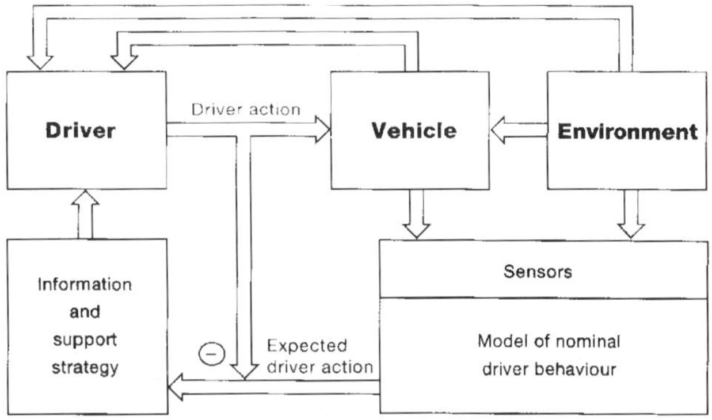

Fig 56: Basic structure of BMW driver assisting system

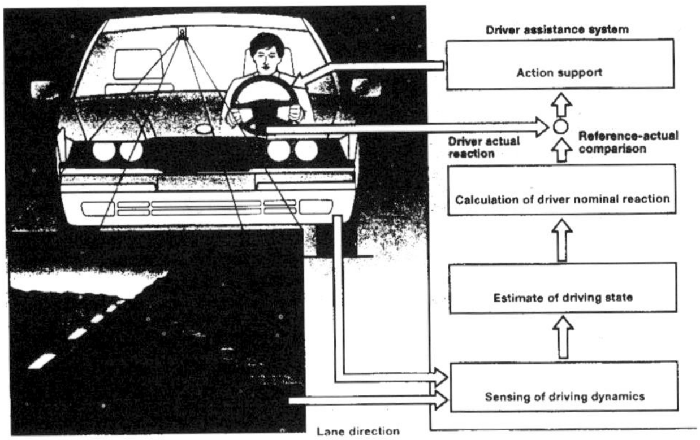

Fig 57: Heading Control driver assistance for lane keeping

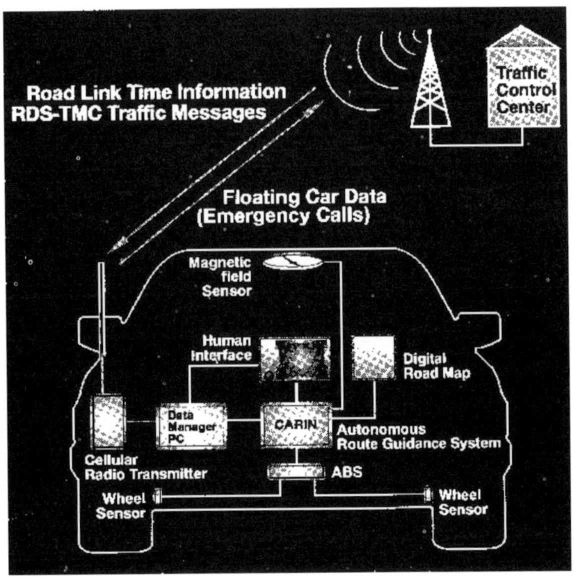

Fig 58: BMW dual-mode route guidance system

incorporated, and displayed on-screen, for the car's comfort-control system. The in-car telephone is also planned into the unit and a touch-screen dialling device for most commonly dialled numbers is incorporated. A single digit is entered on a dial display to obtain the desired connection. Navigation system used is the Carin and the on-board monitor displays the position of the car at all times. The usual route planning facilities are provided; the whole of Europe should be digitized by 1998. The incorporation of television into the system is not just for entertainment but in order to use Teletext for checking such information as flight arrival times. TV is switched off as the vehicle moves away but an additional set for the rear passengers can be plugged into the system and a VCR can be used with it if required.

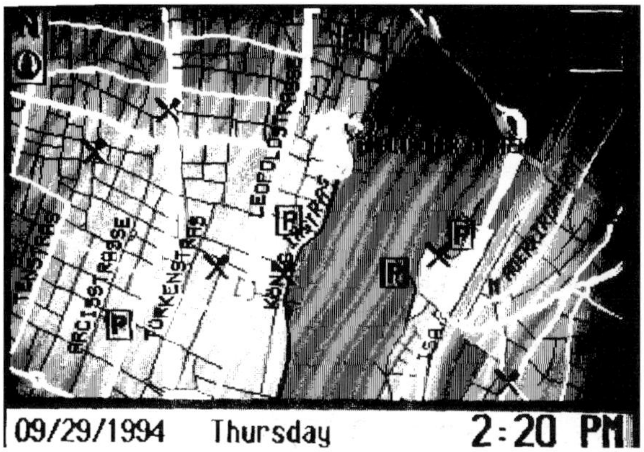

Fig 59: Screen display for BMW route map

Automated driving

As part of a broad move to bring automation to the driving experience, Honda is developing an array of "intelligent technologies" towards evolving Automated Highway Systems (AHS), Fig 60. They will assist drivers in getting to their destinations more quickly and safely until cars virtually become capable of driving themselves. Ultimate AHS automation will take a building block approach and include: anti-spin control; blind spot warning systems; brake support; and cornering speed control which reads the road ahead and slows the vehicle down for curves. Full-range-following allows the car to "lock on" to the car in front of it and automatically respond to its speed and braking patterns.

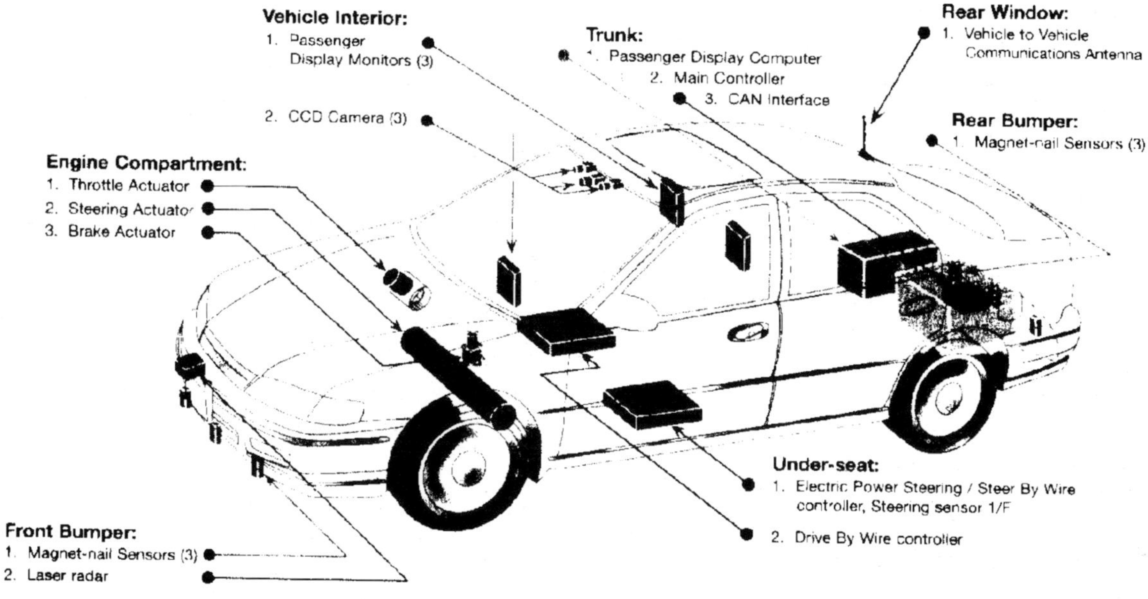

Fig 60: Honda Automated Highway System

Shape tracking: a solution for vehicle detection?

The UK Defence Research Agency have adapted a shape tracking and image processing system from one used to identify enemy tanks and aircraft, into one with potential for intelligent 'reading' of other road vehicles in an automated guidance system. This is the interpretation of the agency's SM Smith in his report, on the system, given to the editor by DRA.

Image sequences taken by a moving video camera may be processed, to detect and track moving objects against a moving background in real time, in a motion segmentation and shape tracking system. The system requires no camera calibration or motion registration, and performs all calculations in the image plane. Information gained from the motion of two dimensional features is integrated with one dimensional boundaries. It achieves automatic instantiation of tracking for each new object detected, and any number of objects can be tracked. Occlusion of one object by another is correctly handled. Work is now under way on extending the system to incorporate knowledge of the world so that simple estimates of the three dimensional positions and motions of objects in its field of view can be made; this has already been done for the simple case of motorway traffic monitoring.

The system is built around feature-based image motion estimation. The features used primarily are two dimensional (often referred to as "corners"), and edges can be used to refine the results obtained by using two dimensional feature motion. The main part of the system has solely two dimensional calculations and not three. This means that feature matching is performed in the image plane, as is object segmentation and tracking. The only area in the main body of the system where three dimensional behaviour is acknowledged is in predicting which one of two touching objects will occlude. The reason for staying in the image plane is that no more information about the world processes can be obtained by working in three dimensions than by staying in the original two. Unless interaction with outside processes and measurements is needed at an early stage in a vision system, there is no need to move to three dimensions until the final interpretation of the processed data is made, DRA explain.

A brief summary is seen in Fig 61, the system being fed a stream of digitized video pictures, usually taken from a moving platform. An example sequence, Fig 62, is taken from ROVA (ROad Vehicle Autonomous, the autonomous vehicle at DRA Chertsey). In this sequence ROVA is travelling forwards and turning slowly to the left with the curve of the road. The ambulance is overtaking ROVA and the Land Rover is travelling with almost identical speed to ROVA. At about frame 103 the ambulance passes behind the Land Rover, and becomes occluded. Each frame taken by the video camera is initially processed to find two dimensional features and edges. The two dimensional feature list is passed to a feature tracker which uses a two dimensional motion model to match and track features over as many frames as possible. A two dimensional vector field can then be created by taking either feature velocities or displacements over a fixed number of frames. The resulting vector field is passed to a flow segmenter which splits the list of flow vectors into clusters which have similar flow within them and are different to each other, using the assumption of first order flow variation.

DRA point out that this cluster list is compared with a temporally filtered list of clusters, and the filtered list is updated using the newly found clusters. This gives good results as it stands, but the boundaries of the clusters in the filtered list can be further refined by using image edges, if available. This list of clusters, which are spatially and temporally significant, is finally used to provide information about the motions of objects viewed by the video camera. The three dimensional positions and motions of these objects can be estimated by making simple assumptions about the world. In the case of an autonomous vehicle, this information is of practical value; for example it can enable the avoidance or the following of other vehicles.

The decision to use image features to find optic flow was made for several reasons. Firstly, optic flow found using two dimensional features should contain as much information about the scene motion as is available, at the places in the image where the process of flow recovery is most well conditioned and where the information is most relevant. Directly related to this, flow discontinuities are not smoothed in the proposed method, as they are with most others, and no constraints are necessary in order to find full flow. The use of features to find optic flow naturally leads on to a sensible and simple representation of object shape. Finally, the process of finding image flow (to the accuracy required here) is relatively computationally expensive using other methods of motion estimation.

The two dimensional features are found using a corner detector. Feature tracking is performed using either constant velocity or constant acceleration models, depending on the application (few applications benefit from using the higher order constant acceleration model). The first stage in tracking features is to instantiate a motion model for each new feature, by matching features from the first two frames. If image motion is smaller than the distance between image features then there is no difficulty in finding a list of correct matches, and the problem is trivial. However, even in the case of stereo matching, where epipolar geometry constrains the possible disparity vectors, ambiguous matching possibilities usually arise. In monocular motion estimation this occurs even more frequently. An intelligent algorithm to disambiguate different possible matches is therefore necessary.

The matching done is based on local operations. Each feature in the first image of an image pair is looked at in turn, and an attempt is made to find for each a matching feature in the second image. The criteria which have been used to determine the quality of a match vary greatly in their complexity. A method of finding feature matches was chosen because it is much faster than a correlation-based method but is not much less reliable. This method uses a smaller amount of information describing the structure of the image at the feature to perform matching. Possible functions suitable for matching include image brightness and local image spatial derivatives. The image functions chosen are typically combined into a vector which is compared with the corresponding vector in the second image. The vector created from n functions will be n dimensional, and these functions need to be weighted so that they affect the vector to an extent roughly equivalent to their reliability as matching components.

It was found that the best attributes to use were the image brightness (not smoothed) and the x and y components of the position of the centre of gravity, when using the corner detector. Obviously, when a corner has been matched once, the velocity estimate thus obtained allows a large reduction in the necessary search space for future matches. The motion model update filter used is a simplified two dimensional Kalman filter in which the search space is reduced over time and the model estimates are given increasing importance over time. In Fig 62 at frame 40 of the sequence described, the Land Rover has very little motion, the background is generally moving to the right, and the ambulance is moving to the left and away from the observer. Note the negative divergence in the flow of the ambulance. Fig 63 shows the set of velocity vectors found by the feature tracking stage of the system

The next step after flow estimation is to segment the flow list into clusters which represent different objects in the world. The flow within each cluster is fit to a linear space dependent (affine) model, giving a total of six parameters for each cluster, using a least-squares fit. The vector values used for the segmentation process may either be the current velocity estimate for each feature, or the feature's displacement over the past N frames, where N is constant, and typically set to 10.

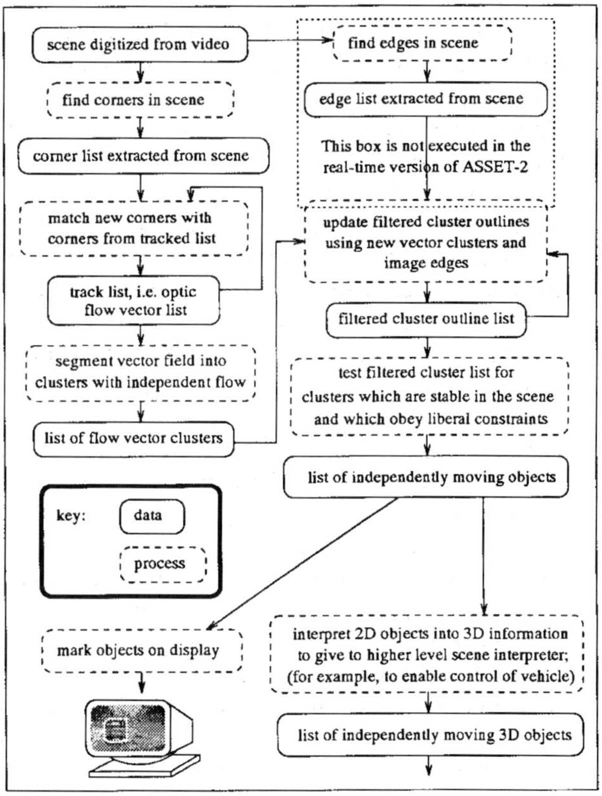

Fig 61: System overview

The list of flow vectors is processed to find separate clusters using simple segmentation similar to the concept of the minimum spanning tree. Each new candidate vector is compared with the value expected by the running flow model. When a new cluster is started, the initial flow model is that of constant flow, two parameters, this is replaced with the affine model once enough vectors have been included into the cluster.

Finally, once the list of independent clusters of flow vectors has been found, each cluster has its bounding box, centroid and motion model calculated, and added to the cluster description. (The centroid calculated is the mean position of the flow vectors which make up the cluster. This has more meaning than some "centre of shape" centroid, as outlying vectors — vectors incorrectly joined to the cluster — will not have so great an effect on the centroid.) This information is then passed on to the part of the system which tracks clusters over time.

Fig 64 shows a typical set of clusters found by the flow segmentation stage. No cluster has been formed where a flow vector has no neighbours to form one. Note the small errors in the two important clusters, particularly in the Land Rover cluster, which should be filtered out over time. The cluster in the foreground will not appear in the filtered cluster list as a "good" cluster because it will not be consistent over time. The background cluster is ignored here by the display part of the system.

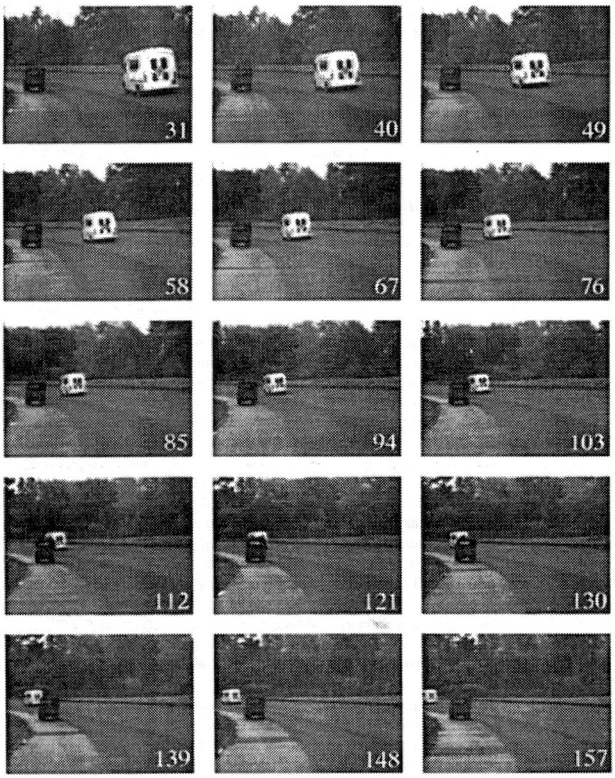

Fig 62: Example video sequence

Newly found clusters are now matched with the clusters in the filtered list, using time symmetric matching; both elements of a potential matched pair must prefer their proposed partner more than any other in the partner's list for the matched pair to be valid. The attributes used for matching are the cluster's motion model and the shape of the cluster. The reverse matching check is applied as before, except that the best backwards match (within a maximum search space and within a maximum allowed difference in matching attributes) is chosen to be the oldest cluster. This is so that a cluster will be consistently tracked over many frames even if it is occasionally incorrectly split into two clusters at the segmentation stage.

Each cluster's motion model (including modelling of acceleration of the whole cluster if appropriate) is updated in the same way that individual features' models are updated. To update the shape of each cluster requires more complicated processing, as it is clearly not possible to take an "average" of two shapes which are defined by lists of two dimensional points. The method used is based around a radial map shape representation. This is an array of distances from the centroid of the cluster to the "boundary", calculated at equal angle increments; each radius is initially set to the smallest value which just includes all of the features found between the neighbouring radial divisions, Fig 65. The map is then forced to be convex, a heuristic which is appropriate for most applications and which is desirable given the sparse nature of the feature-based optic flow field. The radial map clearly does allow for non-convexities, and these may develop when image edges are used to aid boundary estimation. Notice that the radial map approximation to the convex hull gets worse at vertices as the number of radii is decreased. Typically, 32 radii are used, which is easily enough to match the accuracy of the rest of the system.

The radial map is now a useful representation of cluster shape. It allows two shapes to be combined very easily, using for each updated radial estimate a similar method to that described for the cluster centroid and motion, a simplified Kalman filter, taking the previous best estimate and the new estimate as input. Apart from cluster tracking and filtering, this stage also performs a test for cluster occlusion. The test is simply for two dimensional overlap calculated using the bounding boxes of the clusters. If overlap is found, the cluster which will be occluded is predicted using the safe assumption that this cluster will have a higher image position for its lowest point than will the other cluster. This will hold for a ground-based situation, and could be replaced by more detailed analysis of cluster shape change in a more general situation.

Once a cluster has been identified as being occluded its shape is kept constant as it moves behind the object in front of it. If it is at all visible the part which can be seen is used to update the position and velocity of the cluster; if the object is completely hidden, estimates of its position are made. Once the object appears from behind the object in front then it is tracked as before.

The filtered list of clusters can be passed to a higher level stage for scene interpretation. Not all clusters in the filtered list are output as reliable objects. All the clusters found in each image are passed to the filtered list, and until they can be discarded as being spurious they remain in the list. Therefore only clusters which have been seen for longer than a minimum lifetime and which have been seen for more than a minimum percentage of this lifetime are counted as reliable objects. Also, in order to reject spurious objects which have some kind of consistency but which are physically untenable (for example, on the ground immediately in front of the camera), the change in position of the centre of gravity of each cluster over some time interval is compared with an integrated measure of its velocity. This determines whether the object is self-consistent.

Boundaries of objects in the real world almost always give rise to image edges. Therefore a section has been built into the system which uses image edges found by the edge detector to "fine-tune" the cluster boundaries so that nearby edges can pull the boundaries into them. This will of course not always be successful, as background edges may confuse the issue. However, this part of the system has usually been a valuable addition to the core processes. In a situation where edge information is not relevant, it can run without using the "edge enhancement" stage.

Edge correction of a cluster's boundary only takes place after it has been tracked for a certain number of frames. This is partly to prevent the wastage of computational effort on spurious clusters, but, more importantly, because, until the estimate of the cluster boundary is close to the actual object boundary, edge correction could degrade the object boundary instead of improving it.

Next the radial map describing the cluster outline is allowed to "follow the forces" which the vector field defines in the following way. The map

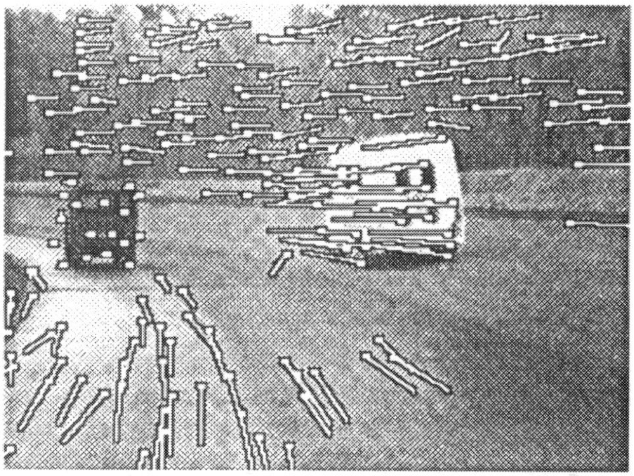

Fig 63: Set of velocity vectors found by the feature tracking stage

Fig 64: Typical set of clusters found by the flow-segmentation stage

boundary is split up into sections, centred at the end of one radius (say radius *i*) and running halfway towards the adjacent radii ends (radii *i - 1* and *i + 1*). Thus each section will be made up of two line segments.

Finally, the corrected cluster shape is used to improve the original filtered radial map. To provide good temporal stability a weighted mean of the input to the edge correction algorithm, and its output, is taken.

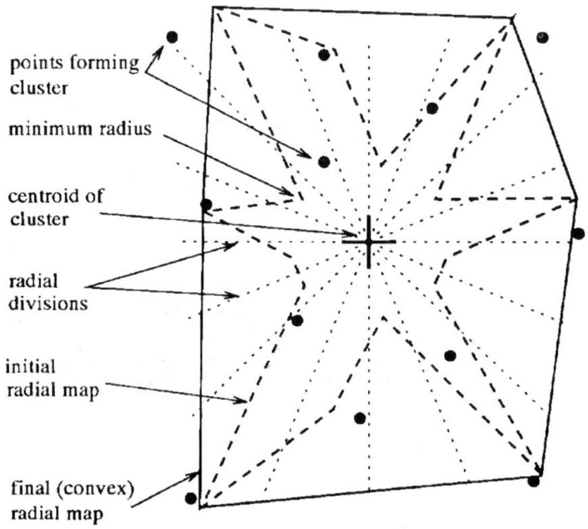

Fig 65: Converting image point list into a convex radial map

Chapter 7: Volume roadcars

Style and package trends; super light-weight construction;advanced styling trendsetter; totem of product engineering development; packaging design for sub-compacts; structural/suspension innovation; mid-engine configuration; power unit and driver-control innovation

The conformity in design of contemporary volume-produced cars has been attributed to the increasing influence of aerodynamics. But this may be more of a marketing 'smokescreen' than a physical airstream effect. The American practice of testing political opinions with forums of subjects registering approval or disapproval of floated policy moves is now spreading to Europe via the UK and some broadsheet newspapers are beginning to say that inter-party sameness in policy is attributable to that practice. It seems quite likely that the clinic testing of new car designs, and floated changes in trim and specification, may be the real culprit of conformity in vehicle design.

It has been said that the American yearning to confirm is born out of a combination of the insecurity found in the world's first serious democracy and the desire of the huge influx of immigrants to find a common norm within it. Only the boardrooms of the huge international motor corporations will be privy to the actual thinking and outside observers can only make guesses based on visible signs. The requirements of European markets, and the effects of European and Pacific-rim imports on the US market, may be perhaps behind Ford's remarkabe New Edge styling initiative to break the conformity inflicted by the marketing men.

Ford's MC2 concept car, Fig 1, predates an all-new Mondeo based Cougar sports coupe package to be introduced by the company into Europe during 1998. The latest interpretation of New Edge styling will complement the similar Puma model and involves smooth, sculpted surfaces with crisp intersections designed to sharpen the silhouette. Roof comprises five louvred glass panels which appear "one-piece" when closed. Lights and grilles at the front-end form distinctive bands over the face of the car. The initiative has produced a beauty in form missing from some of the international companies offerings aimed at the fleet market, for example. Its interpretation in the mini-car sector is seen in the Ford Ka, examined in detail in a later section.

Fig 1: Beauty of line in New Edge styling

In the budget and mini-car sectors, compact packages put tougher constraints on stylists. It is therefore interesting to see how the leading Pacific-rim maker might design for the future, Fig 2. Toyota's trio of new concepts are variations on a common platform, shorter at 3.61 metres than the company's Starlet car. The Funtime, a, uses the styling trick of a downward sloping dividing line between rear quarter and front-wing to reduce visual bulk by dividing the body in two, the high waistline, so-formed, reducing the impression of height and providing a protective look. The shoulder shape below the divider also lowers the car's visual stance. Interior styling, d, provides an uncluttered feeling of space, with centre-console instruments allowing a small dash area. The Funcoupe, c, has removable targa hood and rear hatch and colour scheme follows the dayglo shades of wetsuits and surf-boards. Highly embossed trimcloths provide a stronger texture than

c d b a

Fig 2: Tototа style concepts for minicars

traditional simulated leather. Fun cargo, b, is a mobile base for outdoor pursuits; with body styled around the image of a helicopter it features an unusual stepped roof. Rear fender flares are emphasised to give a feeling of movement not usually found in slab-sided vans.

Packaging initiatives

European car makers have a history of highly innovative package designs typified by VW's rear-engined 'Beetle', France's Citroen 'Traction Avant', Italy's super-compact Fiat Topolino and the UK's east-west engine configuration Mini. One of the reasons offered for the UK not staying in the volume manufacturing sector was its failure to percevere with advanced packaging concepts for the burgeoning fleet sector, being seduced by American marketing thinking into trying to combat Ford's successful Cortina with a somewhat poorer imitation, the Marina.

In fact the UK volume maker BMC was researching an advanced package for a medium sixed car at about the time the Mini was under development, but with a consultancy rather than an in-house job. The archives of the British Motor Heritage Museum have thrown up an interesting design case

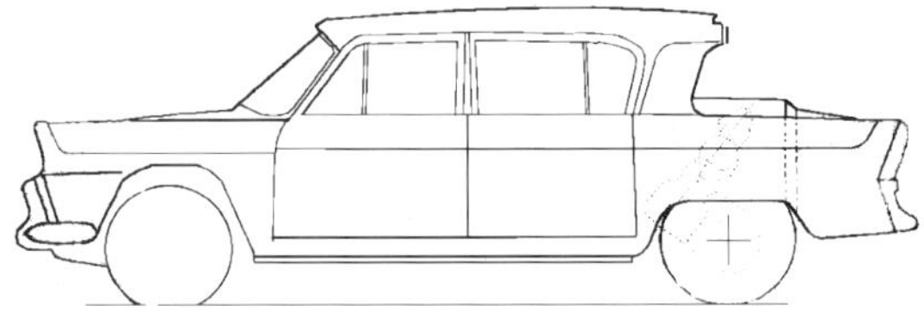

Fig 4a: Engine/transaxle location in ERA 378

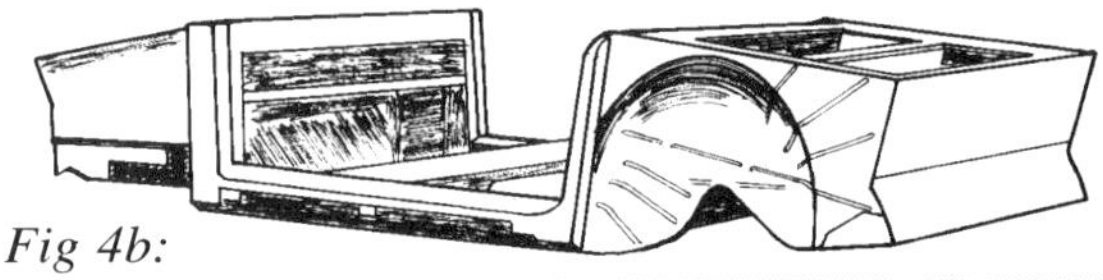

Fig 4b: Main structure

Fig 5a: Mobile tyre-tester

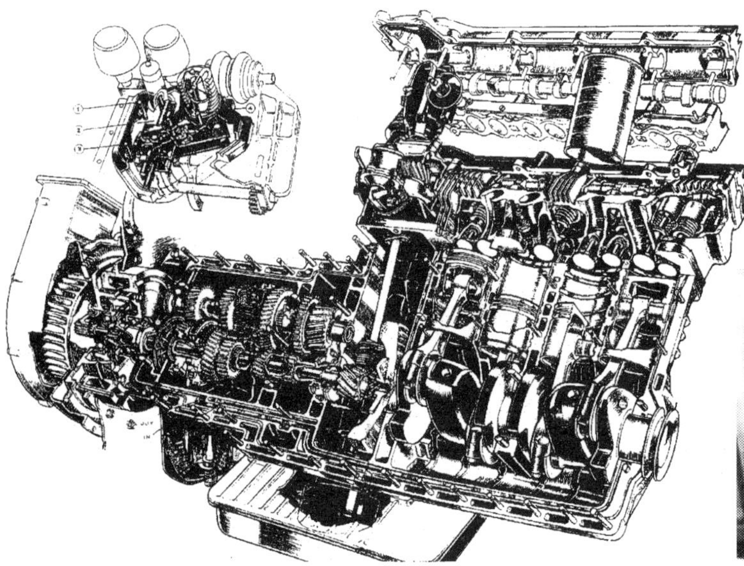

Fig 3: Centreline-split engine/transaxle

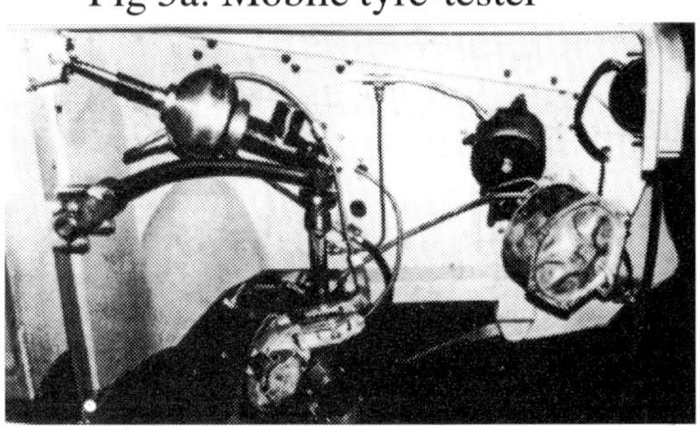

Fig 6: Front suspension and braking

study which reached prototype build but not production. At a time when Citroen had launched its advanced DS 19 car it was felt that BMC should prepare itself with an equally radical design and, through the initiative of Laurence Pomeroy junior, ('senior' was the chief engineer of the British Vauxhall company in the 1920s at the time of its acquisition by America's General Motors), Consultants ERA were contracted to engineer a state-of-the art vehicle incorporating the very latest technologies at the time. Design work on the project started in May 1956 and two prototypes were running by July 1959.

The engine itself had much innovative in engineering design, Fig 3, and included a split-down-liner-centreline construction in magnesium alloy. Thus two halves of the air-cooled integral cylinder block and crankcase assembly divided on the mid-plane of the engine so that liners could be accessed and the oil-galleries on the slit-plane of the two halves, revealed. There was a common lubrication system for engine, gearbox and final-drive.

Generally the objectives were to provide a light compact vehicle which maximised the use of the space for six adults within its overall dimensions. High cruising speed with relatively low engine speed was a further objective. Good packaging of the mechanicals was achieved by placing the in-line four-cylinder engine and gearbox unit transversely in the vehicle within the triangular space between the rear seat back bulkhead and the one ahead of the luggage boot, Fig 4a. Best use of space within the vehicle was also obtained by designing the structure as a braced floor with four vertical pillars at each corner braced by box sectioned crossmembers, Fig 4b.

At the time of the project ERA had a unique knowledge of dynamic tyre characteristics through its mobile tyre tester, Fig 5, which was able to apply steer and camber angles up to 30 degrees and measure realistic road load responses. In this way confident predictions of vehicle handling response could be made. In the prototype vehicles, despite the rear-engine configuration, weight distribution was held at 40:60 and front rear spring rates were arranged so that roll transfer occurred mainly at the front. For a rear-engine rear-drive car, applying roll transfer more heavily at the front avoided loss of tractive power caused by lifting of the inner wheel on cornering. Air-jet damping was by orifices in the pipes of the air suspension. Fig 6 shows the front suspension system

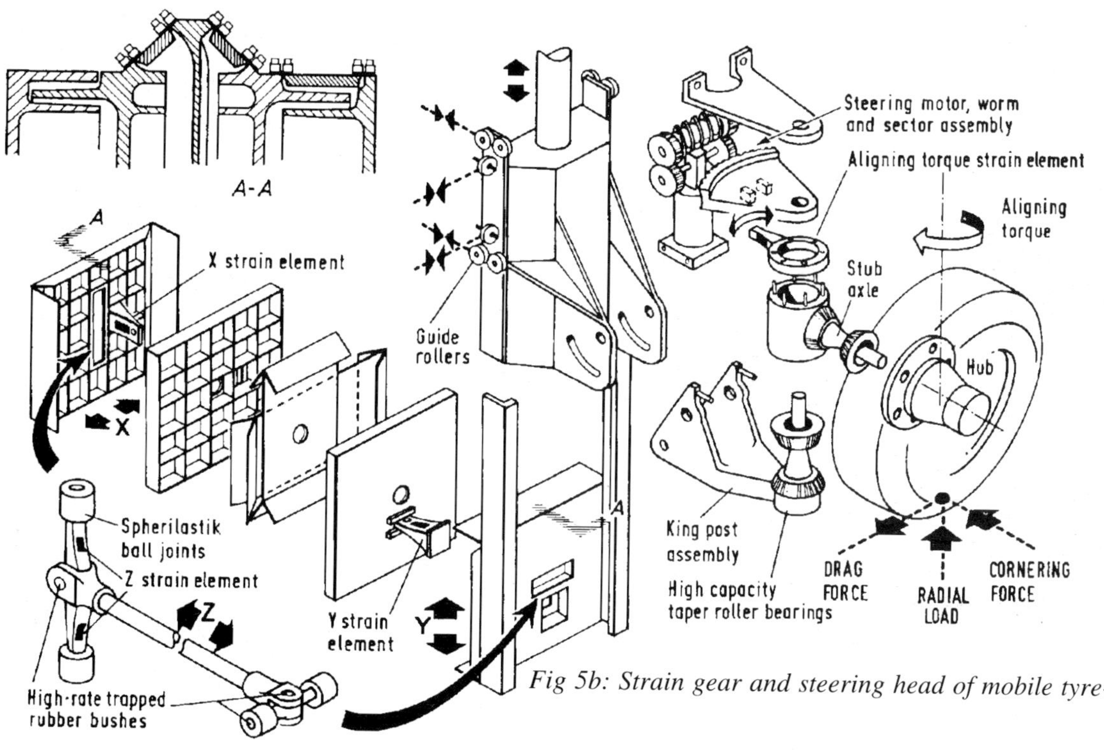

Fig 5b: Strain gear and steering head of mobile tyre-tester

set up on a rig for steering geometry check. There was relatively low levels of steer fight between full bump and full rebound. Between 10 and 35 degree steer angle left and right fight was below 0.5 degrees, with a maximum of one degree at 40 degree steer angle and full bump. Camber and caster angle change with bump travel showed an inherent steering stability under braking, Fig 7. The rig also served to verify the handbrake and speedometer cable runs, as well as tyre to body clearances throughout the full range of steer and bump travel. The mechanically operated front handbrake also caused concern with respect to lost motion caused by deflection of the callipers under application loads. Substantial stiffening of the pad carrier arms was a necessary outcome, Fig 8.

The front pillars held the overhead lever type front suspension, a patented design later to be used on the Rover 2000. The rear suspension was built on to the engine transmission unit. The lower wishbone pivoted from below the differential unit while the upper one was formed by the universally jointed drive shafts. To allow for the substantial change in weight between laden and unladen condition for a lightweight 6-seater, self levelling suspension was required and this was based on air-springs, in a system which allowed full bump travel to be available, laden or unladen. Light steering, without power-assistance, led to the use of zero caster angle and king-pin inclination but large, 2.5 inches of steering offset. Disc brakes were found to be more easily installed than the conventional drum brakes of the time and handbrake callipers were fitted to the front wheels to minimise the cable runs. The engine structure was designed for cold-chamber pressure die-casting with 3/16th inch wall-thicknesses used where possible. By splitting the engine vertically, the die-casting could be carried out with a minimum of coring and cross-slides. A cork-rope in a groove cored into one face sealed the two halves when assembled. The only machining required on the two halves, after lightly skimming the faces, were bolt holes, main-bearing and cylinder liner housings. Each half had just one small draw angle vertical to the split line while one of the halves had an additional draw at 30 degrees to the split-line for producing the sump-joint face and cast-in filter head.

In the integral gearbox, helix angles were minimised to reduce axial loading and gear widths were increased beyond that required for strength and wear, to provide high overlap ratio and reduce noise. Crankshaft and flywheel were integral and cast in SG iron. The worm timing gear was mounted outside the flywheel. Weight of the 1500 cc engine was reduced by the use of air-cooling, for which a cenrtrifugal blower was provided. Gearshift was by servo-shift by a finger-tip switch on the steering column, for up or down changes, which were clutchless. This required synchronizers with 100% baulk. There was also a system of automatic safety changes built in to prevent engine over-revving. A hydraulic governor was used for this purpose and provision of a further switch provided for the option of fully automated shifting if required. The engine oil-cooler served as the matrix for the interior heater and was positioned in the front boot on the vehicle.

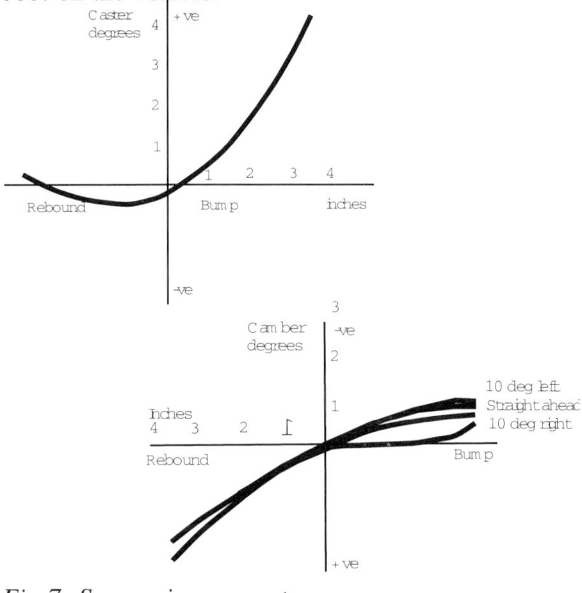

Fig 7: Suspension geometry

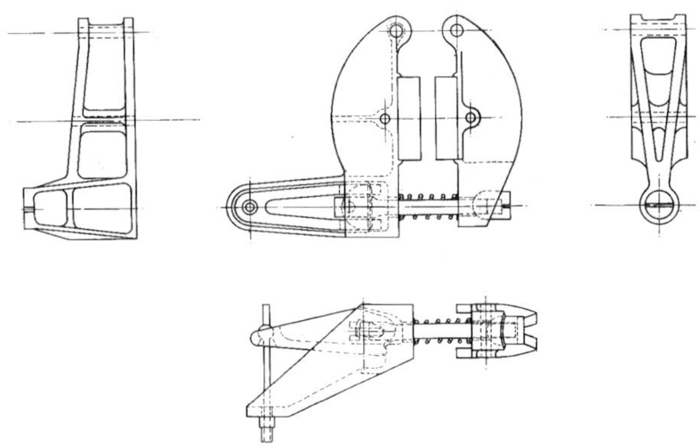

Fig 8: Handbrake calliper

Super light-weight construction

The world's largest car producer, GM, seems to have concentrated much of its future design initiatives into viable alternative production cars such as EV1 and its Ultralite concept, Fig 9, has shown the massive step forward possible for achieving light weight construction that would make such propulsion systems viable.

Just-in-time local manufacture, and direct-delivery to customers in the surrounding area, were among the forecasts in an historic 1993 paper[1] Despite some notable lightweighting initiatives, there is little sign yet of affordable super-light production cars. While vehicle makers continue to spend fortunes on refinement of vehicles that will need revolutionary redesign if they are to meet supercar standards, it is suggested that the move is akin to refining the typewriter after the desktop computer had been invented. The real fuel economy gains are to come from extensive weight and aero-drag reductions rather than further engine development on systems that might soon succumb to hybrid drives. There is also substantial potential in reducing tyre rolling resistance; when best experimental-car fitment figures are compared with those for production cars, it is clear that substantial savings are available.

The problem for steel is not necessarily its high density, because specific strength can be high with appropriate alloying. The problem is that its low raw-material cost blinds people to the massive tooling and processing costs required for its successful volume exploitation. In pressworking alone, thousands of engineers are employed over a two-year period in producing expensive dies for new model introductions. It is interesting that the GM Ultralite concept car involved around 50 technologists for a 100 day introduction lead-time for this polymer-monocoque structured vehicle. By achieving drag co-efficient of 0.192, forward-projected cross section area of 1.71 m^2, mass of 635 kg and road-speed rolling resistance of 0.007 at 4.4 bar tyre pressure, it is still a standard against which other new cars can be measured.

Fig 9: GM Ultralite concept car

The six-piece 191 kg monocoque body involved foam sandwiched between laminates of carbon-fibre reinforced epoxy resin. The CF cloth was interwoven with single-strand roving at high-stress areas and press critics foolishly dismissed the car for its carbon fibre composite material cost, per lb, being greatly higher than steel without consideration of the enormous tooling/process cost of steel fabrication or the fact that much less weight of material was used per car. It should also be noted that the company's press-release also hid this point!

Ultralite in detail The tapered look of the car conceals a front track which is five inches wider than the rear. It was designed to package four large adults comfortably with an overall length of 165.6 inches and a 110 inch wheelbase. The use of advanced materials — thin Duofix seats — allowed designers to create a spacious passenger compartment; the front-seat passengers have 37.2 inches of head room, 57.1 inches shoulder room, and 42.8 inches leg room. Rear-seat passengers have 36.3 inches of head room and 33 inches of leg room.

Besides the *CdA* of just 0.329, another aerodynamic challenge was to provide proper air flow through the vehicle as the structural requirements prevented any openings in the front or rear of the car which has an interchangeable, modular "power pod" located in the rear of the structure. This allowed designers to minimize the car's frontal area, but posed a challenge to get sufficient air flow to the engine, radiator and heat exchangers. Air flow channels, designed as NACA ducts, were incorporated into the vehicle's underbody structure to deliver air to these components.

Before outside air reaches the occupants, it passes through an electrostatic pollen filter. Cool and warm air for passenger comfort is then moved through the centre tunnel to vents placed throughout the passenger compartment, Fig 10. The driver position has air ducts around the speedometer, and a pivoting vent that functions as windshield defogger, can also be used to direct air to the front seat passenger. Rear seat passengers benefit from vents mounted in the centre tunnel and vents positioned under their seats. The air conditioning unit is packaged inside the rear of the centre tunnel to conserve space and utilize the thermal characteristics of the carbon fibre composite material. The energy-efficient lighting system consists of several different technologies: light-emitting diodes (LED), fibre optics, electroluminescence, fluorescent tubes and incandescent bulbs; it consumes 60% less energy than a conventional system.

The modular power pod has an 111 bhp, two-stroke engine with a four-speed automatic transaxle. Engine and transaxle, as well as exhaust and suspension systems, are all housed in a rear power pod. The tubular space frame unit is pre-assembled and pre-aligned to make it interchangeable, as future powertrain technologies become available. For now, the 1.5 litre two-stroke has three-cylinder engine generates maximum power at 4500 rpm and moves the 1,400 pound car from a standing start to 60 mph in 7.8 seconds, the engine weighing 173 pounds. A small panel on the power pod's carbon fibre skin

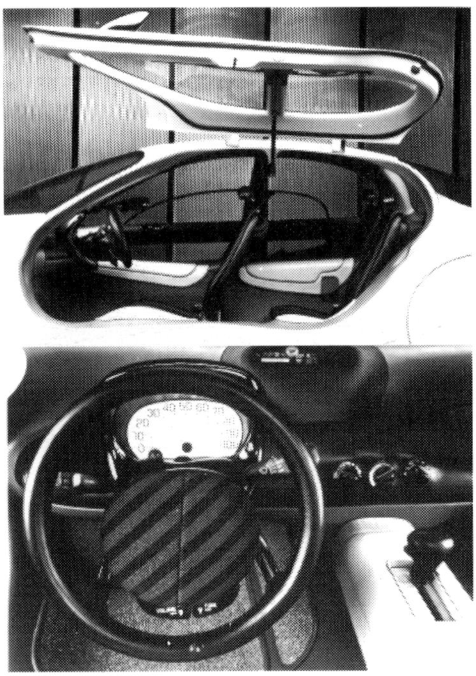

Fig 10: Cockpit layout

makes spark plugs accessible. The two-stroke engine's lubrication system eliminates the need for oil changes and disposal of used oil. The Saturn four speed automatic transaxle is of the clutch-to-clutch type and designed for low spin losses and minimal internal friction. The mostly stainless steel exhaust system incorporates an aluminium muffler that serves a dual purpose. The muffler is configured in the lower bumper area as an aerodynamic wing for efficient airflow under the car.

A difficult challenge was engineering good ride and handling into such a lightweight vehicle. An advanced computer-controlled, automatic levelling suspension system was essential; depending on vehicle speed, the nose will dip up to 25 millimetres to lessen aerodynamic drag. Its tail can be lowered up to 10 millimetres to further improve aerodynamics. The automatic levelling is also necessary to compensate for load differences — crucial for good ride and handling. Passengers can account for more than 40 percent of the gross vehicle weight.

Further weight savings are engineered into the brakes with the use of aluminium silicon carbide rotors. The aluminium composite contains 20 per cent silicon carbide particulates which provide an excellent friction surface. The material offers a 50 per cent weight reduction compared to an iron rotor. Lightweight aluminium callipers decrease brake drag. While the entire system is designed so that power assist is not needed and GM's ABS VI system is specified. Self-sealing tyres contribute weight savings by eliminating the need for a spare tyre. Goodyear decreased the rolling resistance of the tyres to about 2.5 to three pounds per tyre at highway speeds; production cars have about nine pounds of rolling resistance per tyre. The 18-inch tyres are large for the car's weight class but are optimized for low rolling resistance rather than smaller size and weight.

Table 1: Ultralite specification

Powertrain details, dimensions and performance		Wheel size	18 x 4.5 in.
Bore x Stroke	86 mm x 86 mm	Wheel type	Forged aluminium
Torque	127 lb-ft @ 4000 rpm	Tyres	Goodyear 175/65R18
Final-drive ratio	4.00:1	Inflation	65 psi
Gear ratios	2.238, 1.266, 0.811 and 0.480:1	Zero to 60 mph	7.8 seconds
Overall dimensions		Standing 1/4 mile at 90 mph	16.0 seconds
Wheelbase	110 in.		
Length	165.6 in.	Top speed	135 mph
Width	64 in.	EPA city economy	45 mpg
Height	50.8 in.	EPA highway	81 mpg
Ground clearance	5.7 in.		
Curb weight	1400 lb.	Constant 55 mph	96 mpg
Fuel capacity	5 gal.	Speed at 100 mpg	50 mph
Turning circle	38 ft.		

Ford Ka: advanced styling on an established platform

The revolutionary approach to small car design taken by Ford engineers has lessons for many involved on this sector of vehicle manufacture. Bold styling and ingenious modification of the Fiesta platform has produced a trailblazing vehicle.

Ka, Fig 11, features elements of Ford's 'New Edge' design concept which integrates planes and angles with complex structural forms, creating a different approach to surface design. The body of the car features sheer, simple metal surfaces that flow together to create sweeping feature-lines that work with the bumper and bonnet cut-lines to define the shape of the head and tail lamps. The result is edge highlights where the curved planes intersect.

The vehicle platform, Fig 12, has the longest wheelbase in its class (2,448 mm against the average of 2,260 mm) but is only 3,620 mm long — and exceptionally low drag coefficient of only 0.36 is achieved. By combining long wheelbase, with extremely short front and rear overhangs, Ka provides compact overall dimensions (3,620 mm) and interior space for four passengers. Ka is 210 mm shorter than the new Fiesta, and, with a turning circle of under 10 metres (kerb to kerb). The glass area has been maximised throughout the car to improve visibility, with a deep direct-glazed tailgate glass and one-piece side door windows. Tinted glass is standard.

The unique, curving shape of the instrument binnacle has easy-to-read instruments, linked to an integrated centre console that is inclined slightly towards the driver, Fig 13. This brings the controls of the uniquely styled audio bezel and heater into hand's reach. One of the most striking design features of the interior is the oval-shaped analogue clock which features uniquely shaped grey hands on a white face. A total of eight separate stowage locations are provided, including a facia bin, two door pockets, three cup-shaped depressions and a lidded compartment beside the rear seat.

The Ka platform features MacPherson struts at the front, a twist beam rear axle, Fig 14, and gas dampers. Cars fitted with the optional power assisted steering have a different wheel geometry, lower profile tyres (165/65 versus 155/70) and modified gearbox ratios to provide even more responsive performance characteristics. The 44 kW (60PS) engine provides fuel economy of 47.9 mpg (93/116/EC) The Endura-E engine develops 44 kW (60PS) and has class-leading levels of low end torque (105 Nm at only 2.500/min). Engine features electronic multipoint fuel injection controlled by the latest Ford EEC-V engine management system.

A completely new three-point engine installation system that distributes loads evenly within the front end structure, provides good insulation from powertrain noise, vibration and harshness (NVH). The five-speed transmission has a floating gearshift

Fig 11: Ford Ka

mechanism which minimises powertrain NVH transmitted through the gearshift lever.

The car's front structure is specifically optimised for energy absorption, utilising widely-spaced longitudinal side members that also support the engine mounting system. High-strength door beams and a cross-car beam behind the instrument panel contribute to the crash performance. It has been tested according to the proposed EC-98 test procedures involving a 40 per cent offset impact at 56 kph with a deformable honeycomb barrier on the driver's side. This is intended to replicate the relative stiffness of the front of another car, providing closer simulation to a real-world vehicle-to-vehicle impact. Ka has also been tested for the proposed EC-98 side impact, which represents a typical T-bone impact that accounts for almost half of all European road traffic accidents. The test involves a 950 kg deformable barrier striking the side of a vehicle in the area of the B-pillar at 90 degrees with an impact speed of 50 kph at a height of 300 mm.

New-Edge styling The unique style is accomplished through high, linear belt lines and a 'shape-upon-shape' technique that results in edge highlights where curved planes intersect. The exterior features sheer, simple sheet metal surfaces that meet to form lines that give the vehicle an unusual blend of flowing, rounded panels and dramatic edged detail. The simplicity of the body panel elements is strongly emphasised by moulded polypropylene front and rear bumpers that wrap round to integrate with the wheel-arches and by bold design graphics. The bumpers and mesh radiator grille are self-coloured in a mid-shade of blue to contrast with the nine paint colours (two solid and seven metallic) and reinforce the design statement.

The headlamp shape has a striking effect on the exterior appearance. The design intent for Ka has been to integrate large, fixed lens mouldings between the bumper and the bonnet top, to further emphasise the wrap-round corners and provide a sharply focused break line. The headlamps also make a major contribution to the futuristic look. The smooth bodyside profiles are relieved by two feature lines each side with the overall appearance further enhanced by black door and tailgate handles, a black tailgate wiper arm and black aerodynamic door mirrors.

Distinctive interior design results from the instrument panel having a diagonal centre stack that flows smoothly from the instrument cluster to the central tunnel. This original facia treatment leaves a low panel with a softly-rolled edge in front of the passenger to provide an unusually roomy effect for such a small car. The passenger side airbag (when specified) is installed in the upper instrument panel, leaving the lower section clear for stowage. A large diameter centreline speedometer is flanked by a fuel level gauge on the right with a group of 8 warning lights above. Front doors are part covered with trim pads that incorporate integral armrests with door pull handles and lower bins. Reinforced door structures are designed to operate in conjunction with moulded foam blocks behind the trim panel, for maximum side

Fig 12: Ford Ka platform

impact protection. A deep stowage bin on the passenger side of the facia is covered by a removable storage container when a passenger airbag is not fitted and a rotating drum-shaped storage container with the airbag option. The rotating drum container has stowage for tape cassettes in the first position and for small personal items in the second. Further storage is provided by cup-shaped carpet depressions ahead of the gearshift lever and each side of the rear seat cushion. A divided depression is centrally located in the rear between the separately shaped rear seat cushions. There is also a useful stowage compartment on the right-hand side of the rear seat covered by a lid. Door bins are incorporated in the lower door trim panels.

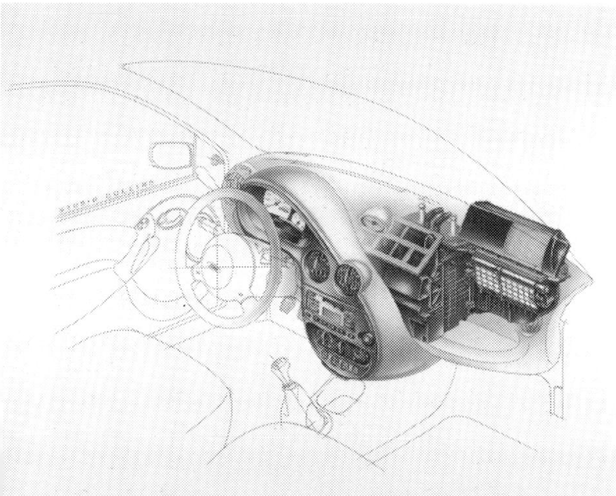

Fig 13: Ka cockpit

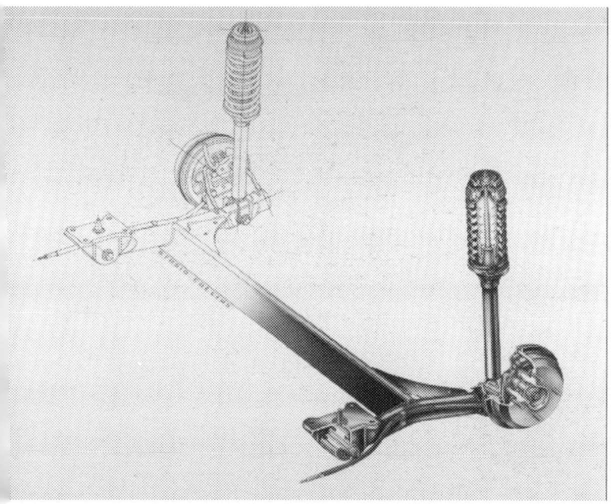

Fig 14: Twist-beam rear suspension

Body systems A new headlamp design uses integrated double can reflectors with separate bulbs for high and dipped beams behind a single moulded polycarbonate lens. H7 halogen bulbs are used for the dipped beams to increase light output and H1 halogen bulbs for high beams. Each bulb has an individually optimised third generation Complex Surface reflector, developed by computer modelling to correct any variations in the pattern of the filament's incandescence. The system uses the high beams to fill in the low beam cut-off pattern, all four operating together for long-range use. Electrically operated headlamp level adjustment is provided, operated by a thumbwheel on the facia.

Instead of a conventional air blending temperature control, output of the new heating system is governed by a solenoid-operated double-spool shuttle valve situated near the entry point to the hot water matrix. This provides a pulse-width modulated flow from the engine block to vary the air discharge temperature which is selected by the facia-mounted rotary control. The optional air conditioning system is available with power-assisted steering because its compressor is direct mounted to the structural oil pan specified on these models. The condenser unit is direct mounted to the cross-flow radiator tanks and the evaporator unit is installed in series with the heater matrix. The EEC-V engine management module delays compressor clutch engagement if necessary while it adjusts the operating variables, to maintain smooth idling and smooth power delivery.

Safety-related features of the structural design include specific reinforcement of the cowl structure and A-pillar connections, upgraded upper A-pillars, B-pillars, C-pillars and roof rails, reinforced pressings between rocker panels and floor and between floorpan pressing and central tunnel. A cross-car beam behind the instrument panel, braced to the central tunnel, is designed to absorb and distribute side impact energy more uniformly and braces the steering column mount.

Other structural features that contribute to the crash performance include high strength steel door beam pressings intended to improve side impact resistance and connect front hinge posts with the centre pillars, reinforced waistline pressings, centre tunnel, door apertures and door hinges, and B-pillars designed to resist roof crush forces.

Product-engineering development: Honda Civic

For buyers wanting a good deal more than mere virility symbols, Honda seems to offer products which are engineering-development led. The Civic was considered quite a dramatic new departure when the car was first launched yet as each year sees quite substantial development of the marque.

The sixth generation Honda Civic, Fig 15, had involved extension of the five-door car range by the addition of reworked three and four door variants, an extended series of VTEC engines and the specification of CVT auto-transmission. Body shell torsional stiffness was 10% up on earlier models and crashworthiness had been improved by an increased front overhang and the fitment of a bumper beam extension; front side and floor panels were also redesigned for better energy absorption. The engine now had an EMS with knock sensor which adjusted the timing so as to tolerate a wide range of fuel qualities. Five engine mounts were now used, Fig 16, the two on the inertia axis no longer fitting direct to engine and body but had stiff brackets and were of the hydraulic type, to damp noise in the 250-500 Hz frequency range. CVT fitment had followed a 10 year development programme and its adoption by such an

Fig 15: Sixth generation Civic

engineering-conscious company was significant.

In structural design, Fig 17, the latest model has achieved compliance with 30-degree oblique and 50% offset impact situations as well as meeting all relevant US and EC frontal, rearward, rollover and side impact tests. This mainly is involved in the design of the new front-end structure. Rearward of the bulkhead three separate structural designs, for 2-door, 3-door and 4-door Japanese-built variants as well as the UK-built 5-door. All have a common platform, and front-end, of 2620 mm wheelbase. Some 10 % increase in body torsional stiffness has been obtained in the new design despite the narrower and more steeply raked C-pillar, mostly by means of gusset plates and a rear parcel shelf stiffener, and larger rear hatch opening, by means a stiffener around the complete hatch opening and gusset plates around the rear damper housings. Bending stiffness is in-

Fig 16: The Civic package

creased by 65 %.

Front-end redesign has involved increasing front overhang by 50 mm and providing a bumper beam extension. Side frames were stiffened and correct stiffness distribution between its constituent components was arranged to obtain the desired controlled-collapse properties. A new doorframe and floor-pan were, in turn, designed to correctly spread residual loads efficiently over the main structure, the latter being stiffened by increased A-pillar inner panels and sill reinforcements. An additional crossmember has been introduced at the A-pillars for better safety inside-impacts; this member is also the mounting for the steering gear and has resulted in reduced steering vibration.

Structural design changes have also been made to improve interior refinement. Idling vibration has been countered by increasing the thickness of the front roof rail and the stiffness of the steering column. General engine noise transmission has been countered by re-insulating the upper dashboard area and providing thicker insulation behind the dashboard as a whole. The insulation has also been made more efficient by arranging for the wiring harness to be installed after assembly of the insulating material. Thicker insulation has also been provided on the cabin and boot floors, and over the rear wheel arches, to reduce the transmission of road noise. Some 88 % of body panels, including the large outer side panels, are now made from double-sided hot-dip galvanised steel. Panel dimensional tolerance and weld-fit have also been improved over the high standards already achieved in the previous model.

VTEC power units Among the technical developments common to all the new engines, Fig 18, are lighter pistons with shorter but stiffer skirts, and lighter, computer-optimised cylinder blocks; a die-cast aluminium sump with internally stiffened walls replaces the previous pressed-steel component, to help reduce air-transmitted engine noise. The engine used in the three-door base Civic is a single overhead camshaft 1.4-litre unit with a bore of 75mm and stroke of 79mm, for an actual capacity of 1,396cc. For UK variants, this engine produces 90PS at 6,200rpm, and 124 Nm of torque at 4,500rpm. The 1.4i unit is equipped with a new cylinder head design in which specially shaped inlet ports tumble the air

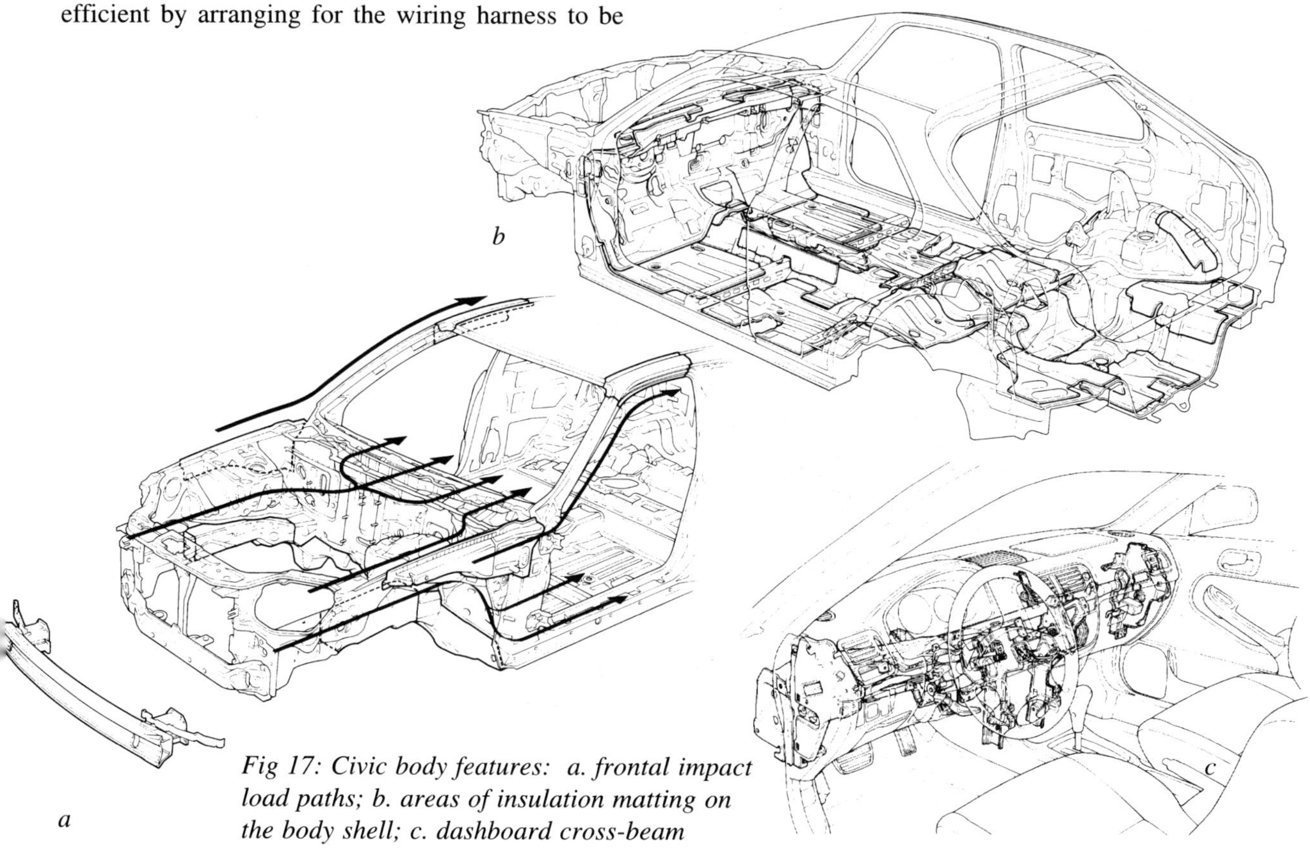

Fig 17: Civic body features: a. frontal impact load paths; b. areas of insulation matting on the body shell; c. dashboard cross-beam

into the combustion chamber, so that the air and fuel remain evenly mixed and that after ignition, the flame front spreads quickly to all parts of the chamber. This engine is also a lean burn unit, capable of operating on air/fuel mixtures as weak as 22.0:1 (compared to the theoretical ideal of 14.7:1). Combining tumble with lean burn technology not only saves fuel, but also reduces exhaust emissions, especially during the warm-up phase.

The 1.4i engine is equipped with a new Honda-developed fuel injection system which feeds four injectors as two pairs. This gives improved performance when compared with the simultaneous injection systems more often used in engines of this size, while not being as complex as full sequential injection. The ECU developed for the new system is especially effective in computing the amount of fuel needed during transient conditions . The engine is also equipped with a new down-flow intake system which incorporates a multi-chamber resonator whose acoustic characteristics are tuned to reduce engine noise, especially during acceleration. Both external (drive-by) and cabin interior noise benefit from this feature. Cabin noise levels are also reduced by a new cast aluminium stiffener which reduces flexing of the engine/transmission face.

Compared with the existing VTEC-E engine, the new 1.5-litre SOHC VTEC has larger (30mm instead of 27mm diameter) inlet valves, increased valve lift, higher compression ratio (9.6 instead of 9.3:1), and an ECU whose control loop works on a new principle, using a sensor which directly detects incipient misfire by measuring variations in crankshaft speed. This provides a more accurate and faster-reacting way of controlling the engine at the most efficient operating point, compared with the linear air-fuel ratio (LAF) sensor previously employed. The ECU also controls the operation of the VTEC actuator, at engine speeds varying from 2,700rpm at full throttle, to 3,300rpm under light load. The engine is fitted with a "dual-mode" crankshaft damper unit which controls both torsional and bending vibrations.

An enlarged version of the new SOHC VTEC has been developed specifically for use with the new continuously variable automatic transmission. Compared with the 1.5 litre unit, the stroke is increased to 90mm, giving a capacity of 1590cc. The engine develops the same maximum power as the 1.5 litre VTEC, but has slightly higher maximum torque

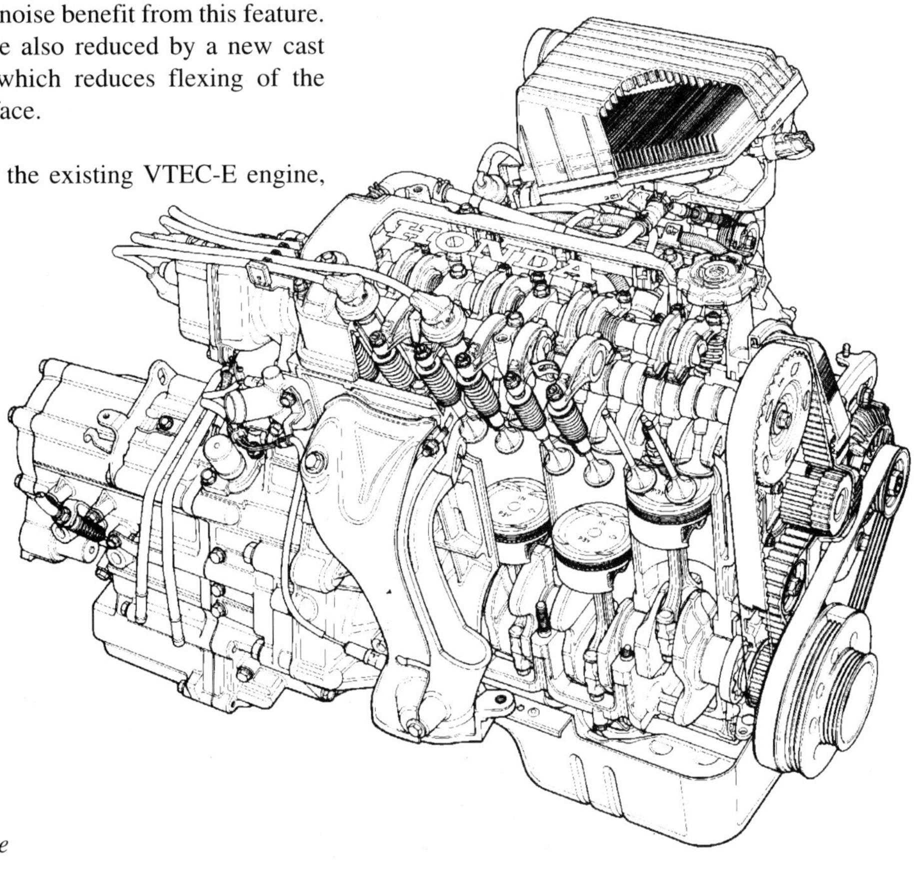

Fig 18: VTEC engine

output and a broader torque curve better suited to the needs of the CVT. As before, the most powerful engine in the Civic range is a 1.6-litre unit (81mm bore x 77.4mm stroke for 1,595cc capacity) with twin overhead camshafts, again using VTEC technology which overcomes the engineering compromise implicit in all units with fixed valve timing and lift, and allows the timing and lift to be optimised for both top-end performance and for mid-range flexibility and pulling power. This DOHC VTEC engine is equipped with a new engine management system, used for the first time in an engine in this class, based on the use of a knock sensor. This system works according to a new control algorithm, in which the control unit deduces the effective octane number (RON) of whatever fuel is in use, and adjusts the ignition timing accordingly.

All new Civics are equipped as standard with a five-speed manual transmission embodying several improvements over the previous model, to make the gearshift lighter and faster, and reduce noise. Two versions of the same basic gearbox are used, with different gear ratios: one for the 1 .6-litre DOHC VTEC engine (Civic VTi) and one for all other versions. The synchromesh mechanisms have been reprofiled and the material of which they are made has been improved. A new type of one-piece, shaft-slide selector fork has been introduced; its greater stiffness makes for more accurate selection, while it also reduces the shift effort. The gearbox internal lubrication has also been improved. The result of these changes is 30 percent lower effort in downshifting, and 20 percent faster synchromesh operation. Another new feature, previously used in the Civic VTi but now extended to all versions, is the use of the fifth gear synchromesh mechanism to bring the gearbox mainshaft to a halt when selecting reverse gear. This reduces the gear crunch which could formerly occur if reverse gear was selected too quickly. In this case, the synchromesh is engaged by means of a one-way cam mechanism.

All versions of the new Civic in EC markets may be optionally equipped with Honda's Electronic Logic Controlled Automatic Transmission (ELCAT) 4-speed automatic transmission with lock-up. The ELCAT resembles the EAT transmission offered in other markets, but is equipped with a revised selector mechanism with a seven position quadrant (P-R-N-D4-D3-2-1). Thus ELCAT provides a first-gear hold position in which full engine braking is also available. In common with the EAT, the ELCAT transmission also embodies a new and more efficient torque converter (already fitted to the Civic 5-door) with a more "solid" driving feel and reduced creep at idle. The transmission also uses revised hydraulic valving which substantially reduces accelerator pedal effort by 20 percent. The shift mechanism itself has been modified to reduce operating noise. Finally, a CVT is offered as outlined in our first report.

Refined suspension Among the changes made to the front suspension of the new Civic, the castor angle has been increased from 1.14deg to 1.67deg, to improve straight-line stability and steering feel. Minor changes to the suspension geometry have lowered the front (static) roll centre from 77mm to 59mm, while camber change over bumps has been

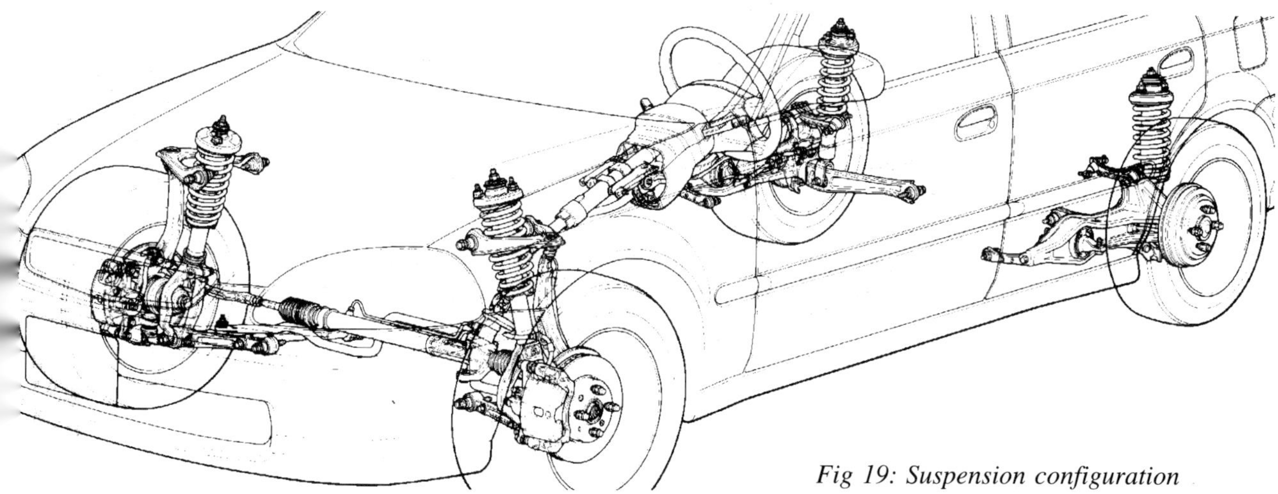

Fig 19: Suspension configuration

reduced, in order to balance the revised camber characteristics of the rear suspension. Detail changes have also sharpened the turn-in response and reduced the speed of roll when entering a corner, Fig 19..

The rear suspension shows rather more physical changes. The lower wishbone arms are 65mm longer than in the previous model, and the upper and lower wishbones are 12.5mm farther apart. The static negative camber has been increased from 0.5deg to 1.0deg. Together, these adjustments reduce the change in wheel toe-out during rebound, which improves high-speed braking stability (during which weight transfer tends to unload the rear wheels) and also the amount of toe-in during bump, which improves cornering stability. The method of attachment of the anti-roll bars has also been changed. In the previous Civic, the bars were connected directly to the suspension arms via rubber bushes; in the new Civic, the bar connections adopt racing car technology, being attached via ball-jointed droplinks. These have a much more linear response, especially towards extremes of cornering performance, and result in more predictable handling. In addition, body roll angles are reduced by up to 30 per cent when cornering with 0.3g sideways force. The dampers have also been respecified for the entire Civic range. The new units have improved valve configurations which change the way in which the damping force varies with the speed of vertical wheel movement. The force has been reduced for slow vertical movements, thus making the ride feel softer when the car is being driven gently. The damping force now increases more steeply as vertical wheel movement and speed increases, leading to a feeling of great stability when the Civic is being driven quickly.

Accuracy and predictability of handling was also the reason for designing a much stronger and stiffer front crossmember, to which the front wishbones are attached. By eliminating as much deformation as possible from the suspension attachment points, the behaviour of the suspension can be far more accurately calculated and adjusted. In addition, an intensive development programme was carried out to improve levels of refinement by preventing or cutting down on the transmission of suspension and tyre-induced noise at source. The main results of this work were the introduction of new suspension arm mounting bushes, using rubber with better vibration damping characteristics, and the addition of sound-absorbent pads to several more areas of the body interior. These measures, plus the many noise-reducing alterations to the driveline as already described, have resulted in a cabin interior noise level of 65dBA when cruising at 88 mph, compared with 68dBA at 62mph in the previous model, Fig 20.

Physically, the most obvious change in the new Civic front suspension is a change from forged steel to hollow, fabricated pressed-steel lower wishbone arms, achieving a weight saving of 1.4kg per car and benefiting ride comfort by reducing the unsprung weight. Further unsprung weight has been saved by the use of a new heat treatment process for the outboard drive shaft constant-velocity joints, allowing each pair of joints to be made 1.9kg lighter.

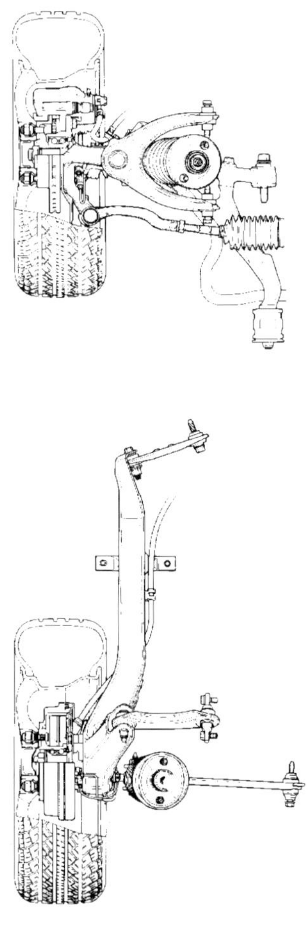

Fig 20: Civic wheel suspension

Fig 21: Peugeot, abve, and Fiat packages compared

Packaging design for sub-compacts

Peugeot's new 106 and the Fiat Cinquecento are interesting approaches to mechanical and occupant packaging, Fig 21, within compact exterior dimensions. While at 3.56 metres long, the 106 is not as sub-compact as the Fiat, its interior dimensions are generous with its more bullbous body. The aluminium alloy vertical four-cylinder-in-line front transverse engine gearbox unit of the 106 contrasts with the twin horizontal power-pack of the Fiat which lies 'north-south' in the vehicle, Fig 22.

Front Macphcrson struts are favoured for both vehicles, with relatively high-mounted rack-and-pinion steering. Rear suspension is similar on each car but disposition of the springs and dampers is very different, the Peugeot's damper being subject to a very high mechanical advantage and springing based on transverse torsion bars. The Fiat's 3.23 metre length is less than most Japanese 'minicars' yet provides comfortable accomodation for four. One reason is the hanging of the 704 cc two-cylinder engine ahead of the 'front' axle; the larger 903 cc variant uses the now more common practice of 'east-west' layout in line with the gearbox, Fig 23.

It is also interesting to see the approaches taken

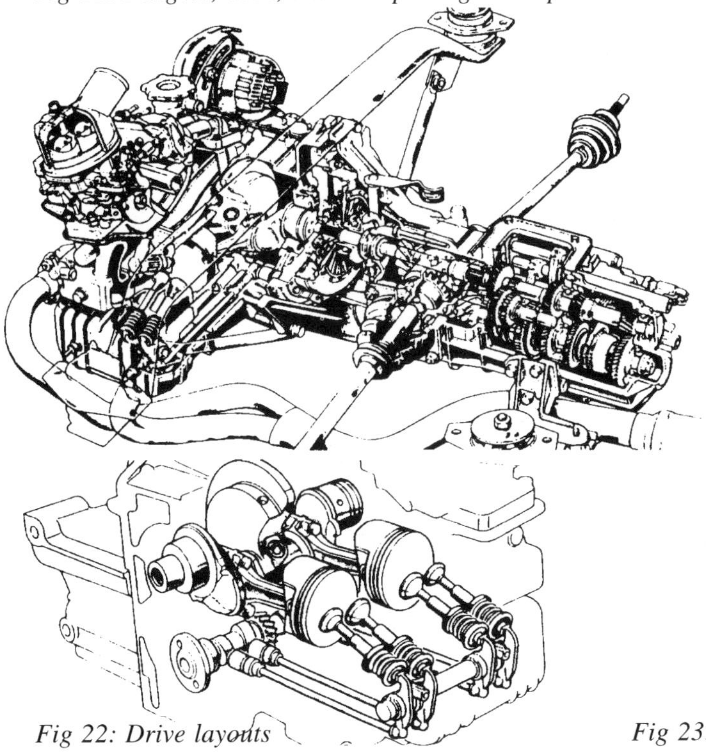

Fig 22: Drive layouts

Fig 23: Suspension layout

by each company to packaging electric-propulsion verions of the cars, Fig 24. Based on the 106 saloon, PSA's EVs have Leroy-Somer motors, Sagem electronic controllers and Saft nickelcadmium batteries. At the current stage in development, top-speed of 90 km/h has been achieved with 0-50 km/h acceleration time of nine seconds and a range of 90-160 km with eight hour recharging cycle. There is a 40 km range on the 20 minute quick-recharge cycle. Lead-acid batteries are being developed in parallel by Sonnenshein. Initial sales were in the city of La Rochelle where the local council built the necessary infrastructure of charging stations. A self-service rental scheme was introduced to operate with the user's standard bank credit card.

The Elettra version of the Fiat Cinquecento has a kerb weight of 1020 kg when fitted with nickel-cadmium batteries, the alternative lead gel-acid ones weighing 90 kg more. In EV form it is just a two-seater but notable indeed for being released in productionised form. It uses a chopper-controlled DC motor and drives through the same five speed gearbox as the petrol-engined versions. A top speed of 85 km/h is claimed. On NiCd batteries the range is 180 km. Twelve 6-volt battery packs are involved which are series-connected; they weigh 270 kg and can provide up to 2000 discharge cycles.

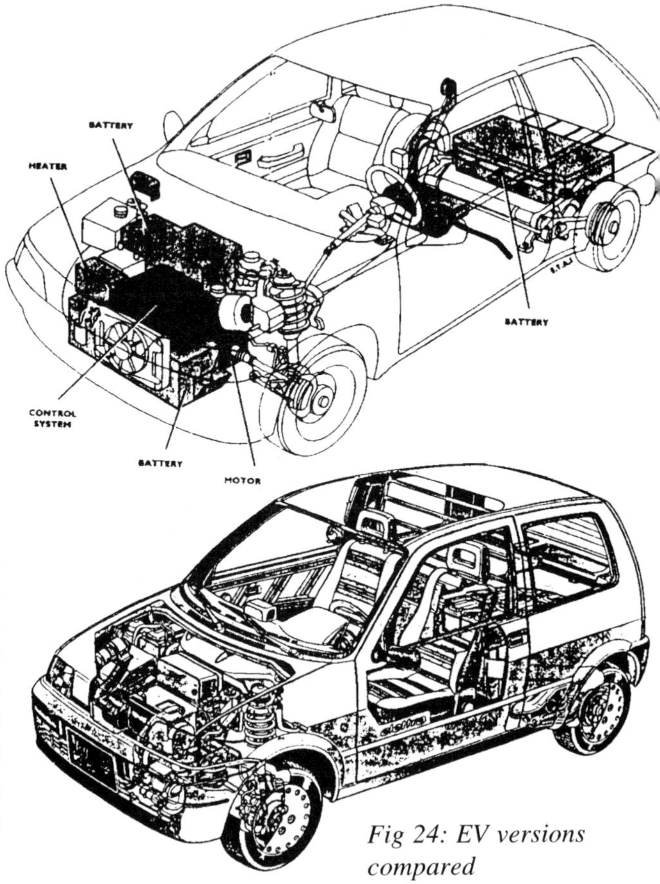

Fig 24: EV versions compared

Feature design in recent car introductions

The following pages examine the development of passenger car features which perhaps reflect particular expertises of some of the companies concerned.

Volvo's 850, Fig 25, has a so-called Delta-link suspension for its rear non-drive axle which is a nice compromise between a live-axle and independent rear suspension, Fig 26. As well as trailing links which carry the wheel-hubs, tapered half-axles join these links transversely. While acting as a single unit, nevertheless their narrow ends are solt-mounted to provide lateral motion. In straight-ahead driving they serve as a de Dion tube while in cornering individual vertical wheel motions are permitted in the manner of IRS. Seats have built-in members for side impact protection, Fig 27, and belts are adjustable to shoulder height by means of a reel which is wider than the webbing. Transverse installation of the five-cylinder

Fig 25: Volvo 850

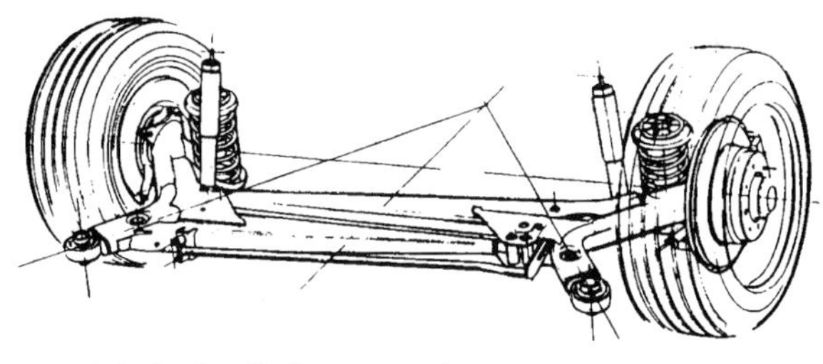

Fig 26: Delta-link suspension

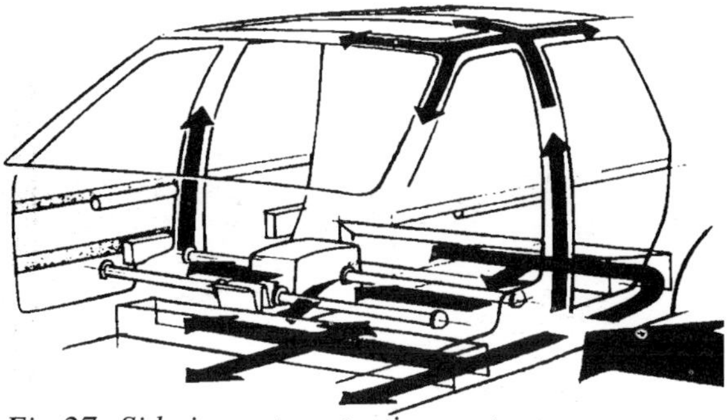

Fig 27: Side impact protection system

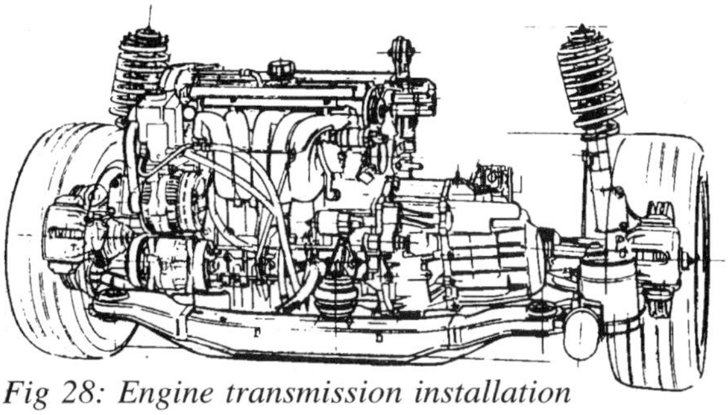

Fig 28: Engine transmission installation

engine/transmission unit is also a notable achievement, Fig 28.

Volvo question the safety, long-trip comfort and height restriction with multi-story parks which afflict most conventional purpose-built 4x4s, in advancing their four wheel drive 850 AWD saloon version, Fig 29. Lower aero drag and more stable cornering are also suggested while offering higher ground clearance than a number of 4WD saloons. The permanent AWD system is set to divide torque 95 % to the front and 5 % to the rear for motorway speeds but automatically transfers it to the wheel needing most grip in off-road and hill-climbing conditions. Viscodrive couplings are employed to distribute the torque from the 193 bhp 2.5 litre five-cylinder turbo engine. There is a self locking differential at the rear and TRACS electronic traction control at the front. A freewheel is combined with the viscous coupling, immediately ahead of the rear axle, which produces zero drive to the rear axle when braking. The freewheel has an automatic locking function which allows the car to be always in 4WD when reversing. Final drive is incorporated into an aluminium alloy subframe containing the multilink (aluminium alloy) rear suspension; final drive ratio is 3.31:1. There has been a 110 kg weight penalty but no loss at all of volume capacity.

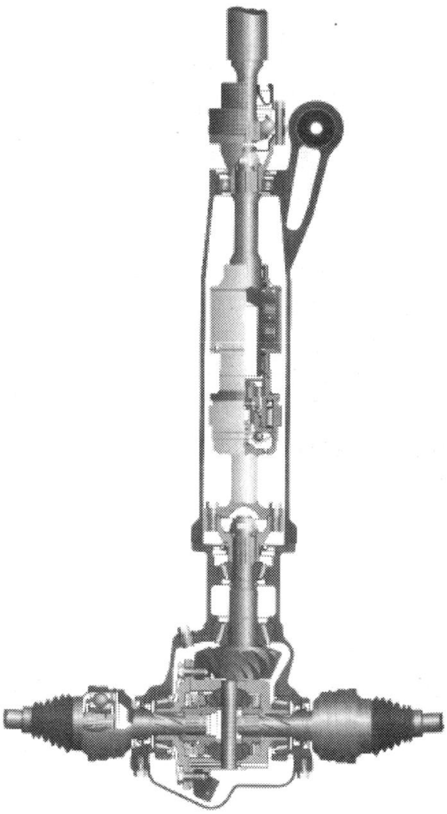

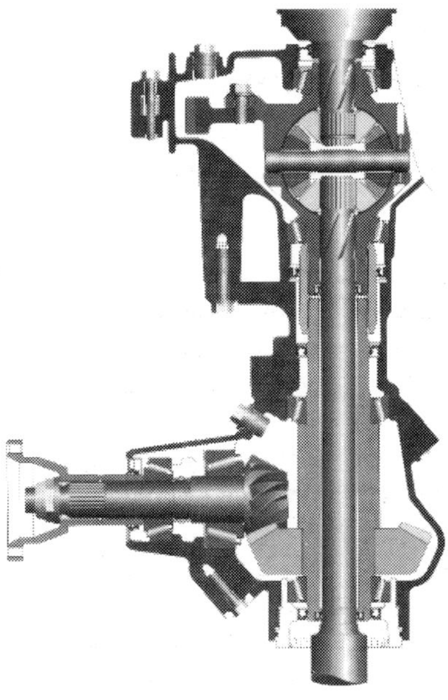

Fig 29: Four-wheel drive system

European market leader: VW Golf

The recently announced fourth generation of Europe's longtime best selling family car, Fig 30, is the most spacious to date; the length increases to 4131 mm and width to 1735 mm, Fig 31. The body structure, Fig 32, is fully galvanised and permits VW to provide the new Golf with a 12 year warranty against rust protection. A new era of body quality has also resulted in very narrow shut lines between closure panels and unusually high torsional and bending rigidity

The new Golf employs 19 tailored blanks on the body-in-white accounting for around 21 per cent of its mass. Additionally, 27 per cent of the mass of the doors are also made from tailored blanks. The blanks cut weight and costs by joining thick and thin sections of sheet steel into single components, which eliminates the need for reinforcement. High strength steel is used for 35 parts of the body-in-white accounting for around 10 per cent of the its mass. Moreover, high strength steel also accounts for around 36 per cent of the mass of the doors.

Despite increased dimensions, Volkswagen has successfully pursued a policy of lightweight construction — resulting in a weight for the body-in-white of 242 kg (two-door). The bodyshell's exceptional rigidity is achieved by manufacturing techniques such as laser welding and special adhesive bonding. Laser welding is used on the two-door model to connect the roof to the outer side panel at the C-pillar and on the four-door model between the B and C-pillar. Laser welding saves weight by eliminating the flanges required for spot welding, and improves torsional rigidity by creating an uninterrupted join. Using these innovative technologies, Volkswagen has succeeded in achieving a high level of dynamic rigidity. In technical terms, this is expressed by the figure for the first dynamic torsional characteristic frequency, which amounts to 40 Hz in the new Golf (compared with around 30 Hz for the Mk 1 Golf). The torsional stiffness (16,500 Nm/degree) is almost twice that of the original Mkl Golf (8,620 Nm/degree).

On the interior the dash panel is redesigned, behind a more steeply raked windscreen, and built by slush-moulding, with a compact instrument panel featuring blue backlighting and including a navigation system, a first for this class of vehicle for which the Golf enjoys over 30 % market share (over 17.6 million units have been sold since the first introduction in 1974). A new 16-valve engine for the entry level version displaces 1.4 litres and can develop 75 bhp. It is an all-aluminium alloy unit. Its torque curve allows particularly flexible performance, some 90 % of maximum torque being available over a speed range from 2000 to 5000 rpm. Other engine options, up to the V6, provide powers up to 150 bhp.

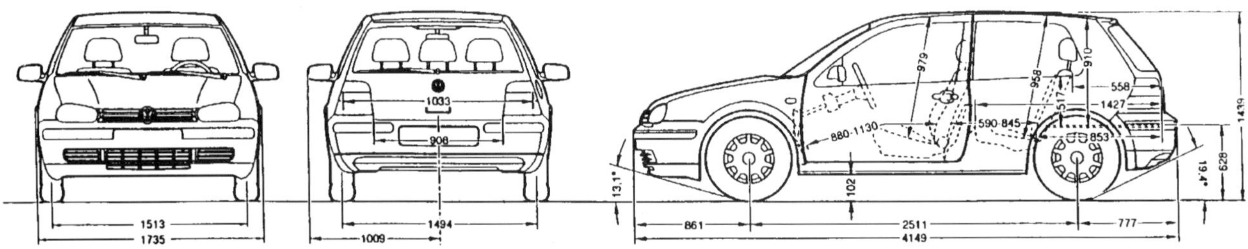

Fig 31: Key dimensions

Fig 30: VW Golf

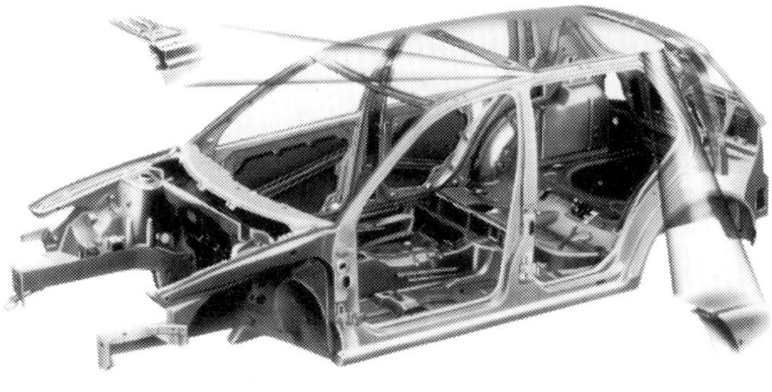

Fig 32: Golf structure

Structural/suspension innovations: Lotus Elan

Torsional stiffness of 6600 lbs ft /degree was achieved for the open body structure, Fig 33, of the Lotus Elan with a steel and polymercomposites combination structure. Body panels are produced by the VARI process, a constant curing temperature and newly modified resin systems having improved the production cure time; also a new preform system is used to produce the panels which enables sharper corners to be achieved. Body shape gives 0.34 drag coefficient and front/rear lift coefficients of 0.30 and 0.065 respectively.

The chassis structure departed from the traditional Lotus pure backbone with the integration of the composite platform which is a single-piece 3 mm thick moulding, including inner sill, toe-board, heel-board, 'A' and 'B' posts. The resin is an isopthalic polyester reinforced by continuous filament glass fibre with additional reinforcement at the mating points with the backbone outriggers. The latter are made from 18 gauge steel which is electro-coated and wax injected. Body panels are joined by elastomeric polyurethane adhesives. There is also a steel subframe reinforcing 'A' and 'B' posts, sills and rear suspension anchorages. The relatively large number of separate 2 mm thick skin panels makes for quite easy future face-lifting, according to the company. Bumpers are of RRIM and doors, unusually, swing in an arc inside rather than outside the front A-panel.

A patented 'interactive wishbone' suspension assembly, Fig 34, is mounted on a separate raft of heat-treated aluminium alloy. The raft allows very stiff fulcrum bushes to be employed but without passing road noise; it also permits abnormally low caster angle and accurately controls suspension geometry under all conditions. Torque steer effects are also reduced while allowing longitudinal wheel compliance without changes in handling. There is not the normal excessive loss of caster during braking, caused

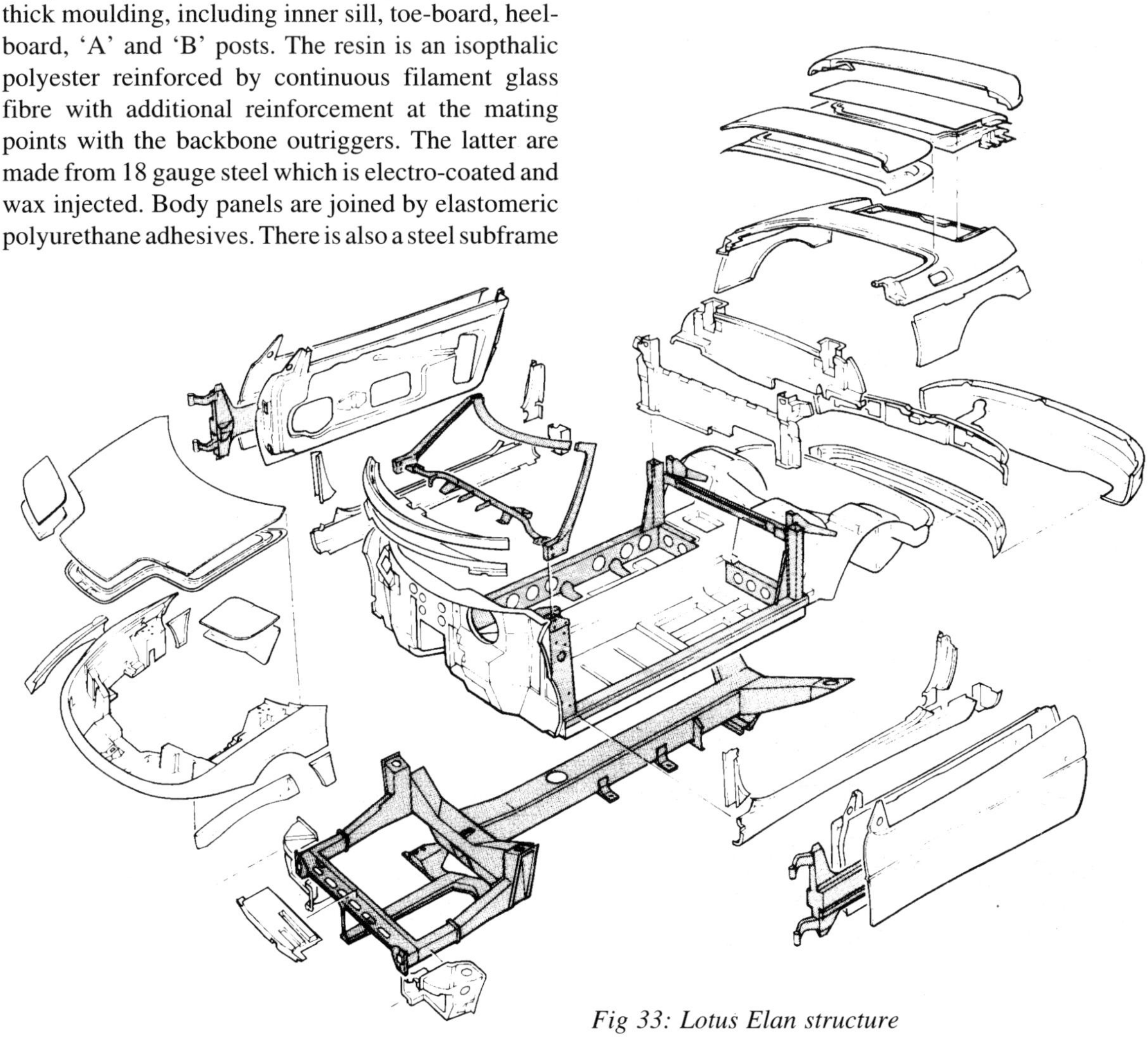

Fig 33: Lotus Elan structure

by deflection of the fulcrum bushes and braking stability has actually been enhanced by offsetting the hub spindle rearwards relative to the steering axis. Tracking stability is also optimised by commonising camber change geometry, with bump travel, on both front and rear suspensions. Roll centre heights (30 mm, front, 60 mm, rear) remain virtually constant for normal roll angles and steering geometry of 60 per cent Ackermann has been chosen.

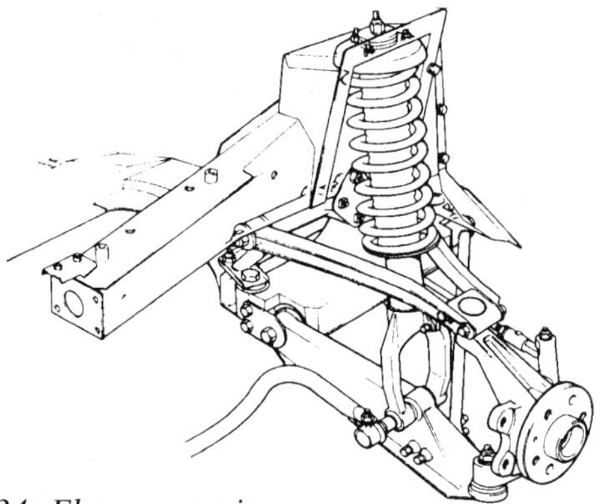

Fig 34: Elan suspension

Radical approach to occupant-protection: M-B A-class

Mercedes-Benz's A-Class, Fig 35, claims to have overcome the problem of limited crush distance available to protect the occupants of a small car. The front-mounted engine gearbox unit are sandwiched between the interior floor and exterior under-panels in a way that provides controlled collapse as the power-unit moves rearwards on front-impact, without penetrating the passenger survival space.

Fig 35: M-B A-class

Advanced suspension linkage configuration: BMW 850i

Active rear axle kinematics is the name given by BMW for its rear suspension design, Fig 36, on the 850i coupe. Besides providing anti-squat and anti-dive, wheel-camber is adapted to reduce tyre. Strain is built-in at high speed and roll-steer which achieves load reversal reactions.

As well as four wheel control links, a fifth joins two of the others to modify their relative motions. Two lower and one upper transverse link form the 'wishbones' while a trailing link provides fore-aft location. This link has a secondary connection via an intermediate link to the upper transverse link. Coil-spring and damper are mounted adjacently on the rear wishbone. The wheel-carrier pivot is above and ahead of the wheel-axis, to aehieve anti-dive/squat. The intermediate (fifth) link is a strut which transfers braking torque on the wheel carrier, through the upper 'wishbone' to the trailing link. Because the strut lower pivot has less velocity than the wheel-carrier pivot on the trailing arm, during vertical travel of the wheel, this velocity is also seen at the upper wishbone. This causes the wheel carrier to rotate more slowly than the trailing link about the wheel-axis and twisting of the wheel carrier on braking is suppressed without losing the horizontal compliance in the anchorage bushes of the diff-carrier at the root of the transverse links.

The linkage compensates for unequal horizontal forces on the wheel either side of the vehicle so no swivelling of the whole suspension, with the diff-carrier, takes place. The elasto-kinetic function arises from the interaction of the anchorage bushes with the linkage geometry. In braking, for example, the two links forming the lower wishbone are subjeet to differential tension and compression. At the same time the anchorage bush of the trailing arm yields. Transverse link motions are such as to ensure that resulting motion of the wheel carrier is purely horizontal, without causing steer efrects as the wheels more vertically.

With the short wheelbase of the coupe, compared with the saloon, and the comparatively high driving torques, smooth behaviour under load reversal conditions is ensured. This is helped again by the design of the trailing arm anchorage. At high lateral accelerations, wishbone links are distinctly skew to the body of the tilting vchicle. With load reversal caused by sudden removal of driving torque, the diff-carricr swivels due to the resilience of its mounts.

Simultaneously it pulls the pivot point of the upper wishbone link with it, horizontally, and the front/rear lower wishbone links are displaced by (fv) and (fh) respectively. Tilt of the wishbonc also causes lateral displacements or the Pivot poinls (fv') and (fh'), causing understeer in both wheels. This effect is only achieved because the suspcnsion of the trailing link is independent of that of the diff-carrier from which the transverse links pivot.

Full active suspension: Ford US compact

The distinction is drawn between 'narrow and broad bandwidth 'active suspension by engineers at Ford (USA) in describing a recent system installation. A number of existing narrow-bandwidth varieties have just dealt with active control of lower suspension frequencies, leaving the higher ones to the 'passive' suspension of the vehicle. In this more broad-bandwidth system, fitted to one of the company's concept cars, the actuators are fitted in parallel rather than in series with the passive suspension, Fig 37.

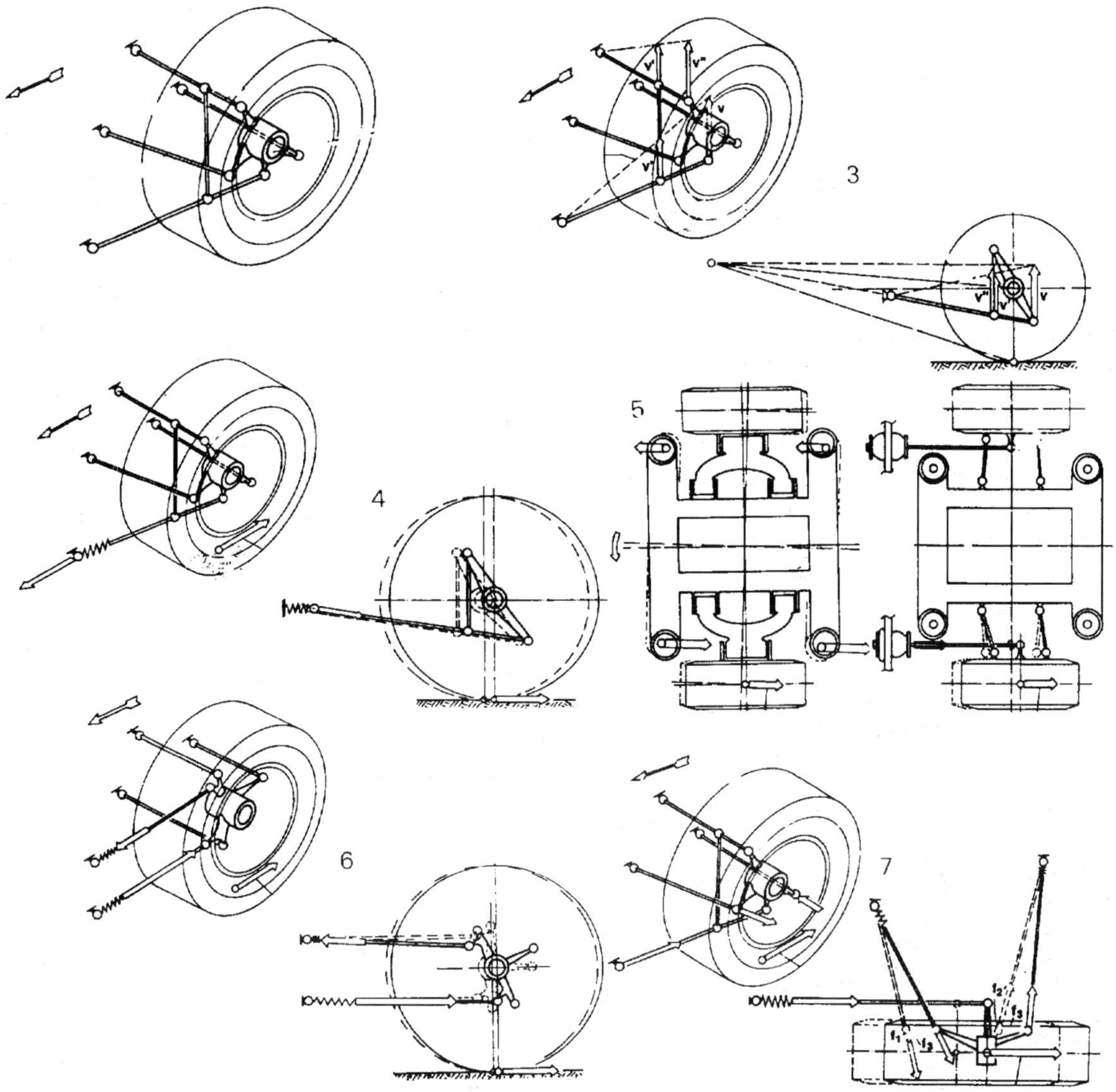

Fig 36: BMW 850i suspension linkage

Ride performance and power usage of the narrow bandwidth type deteriorate when a soft passive suspension is used and the systems become oversensitive to suspension-friction. Broad bandwidth systems were found by the authors to use less power in systems which improved ride and handling performance beyond the expectations of conventional passive systems. On this installation, the computer control system is based on a distributed network — with a processor at each wheel and a central processor monitoring the remaining vehicle systems as well as communicating with the wheel stations.

The wheel computers take the commanded body forces sent to them and make the actuators generate the correct force, in response. The body force command is updated every 20 milliseconds while the wheel processors update in a one millisecond loop. This loop involves measurement of the actuator-generated force, its displacement, the wheel acceleration and the pressure in the air springs. The last two derive the force being generated by the air springs; damper forces are similarly computed and both are used to modify the required body response force that the actuator produces. In the 20 ms loop of the central processor, there are two steps: pitch and roll are controlled using longitudinal and lateral accelerations of the body and actuator force to level the body is calculated; low-frequency vertical body motion is controlled by estimating its velocity relative to an inertial plane (skyhook) from the accelerations recorded above each wheel; this then determines the necessary correcting force using a hypothetical damper.

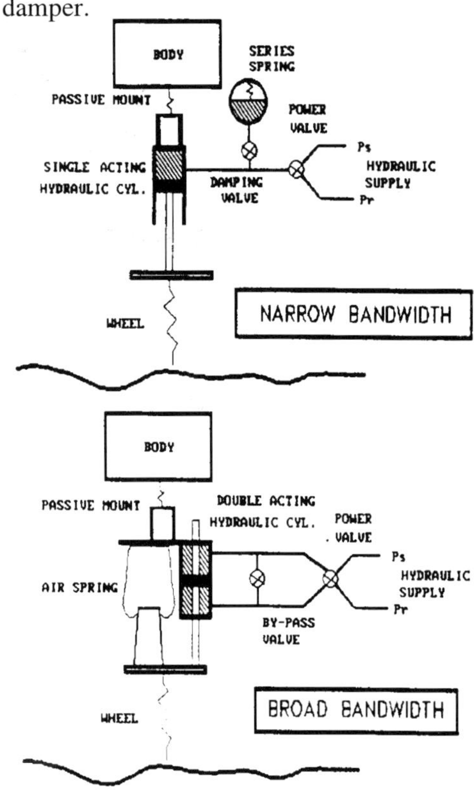

Fig 37: Active suspensions compared

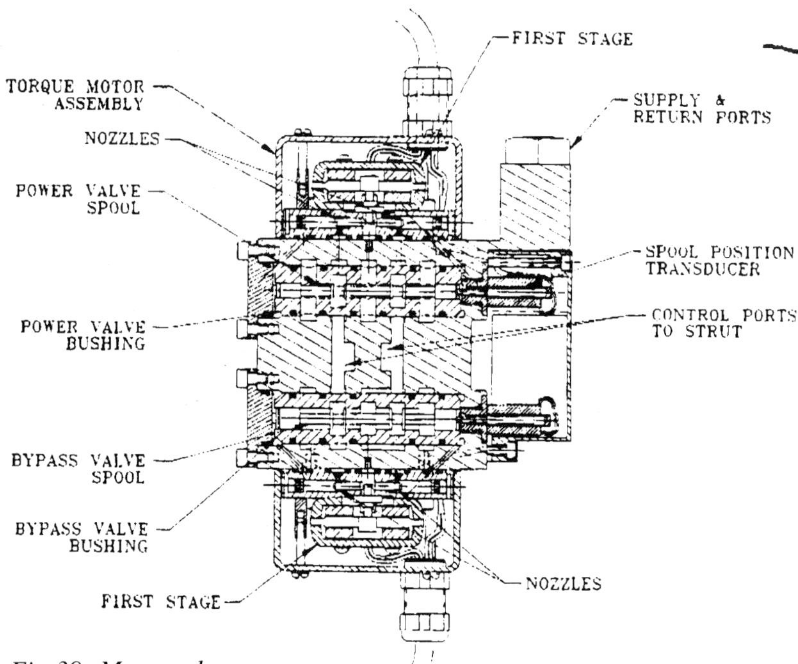

Fig 39: Moog valve

The system has a number of operating modes representing vehicle start-up, ride or handling manoeuvres. The actuators replace the dampers of the passive suspension — and comprise double-acting hydraulic cylinders, servo-valves and load/displacement sensors, Fig 38. Because of limited packaging space at the rear, the rear-unit has a cylinder folded in on itself, having concentric retraction and extending chambers. Fig 39 shows a section through the two-stage, electronic feedback type, front-valves developed from Moog units. These are mounted in the engine compartment above the corner unit while the rear-valves are U-shape and wrap around the hydraulic cylinders. Fig 40 is a diagram of the hydraulic power supply alongside an installation diagram for the whole system.

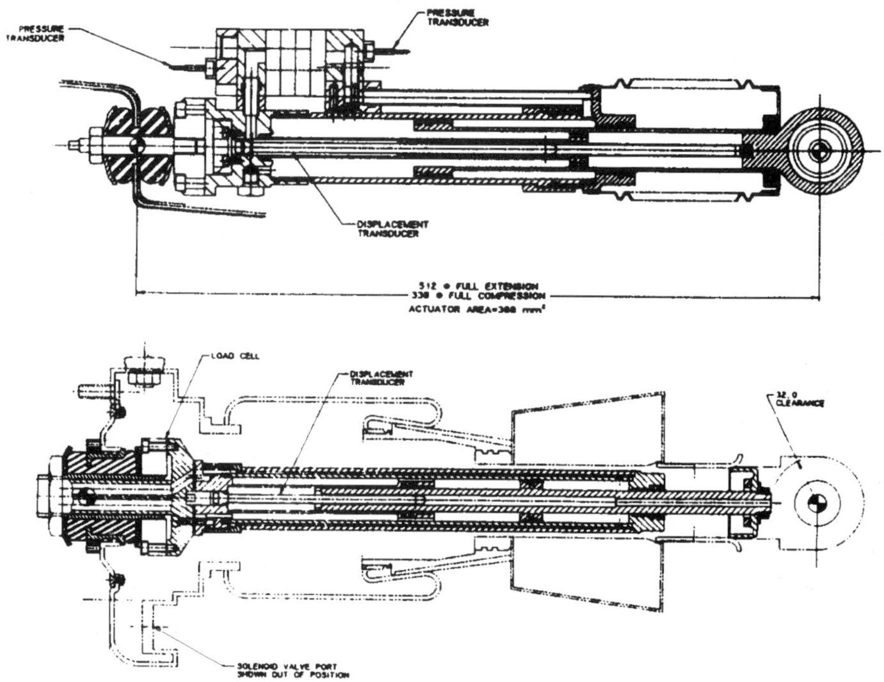

Fig 38: Section through front valves

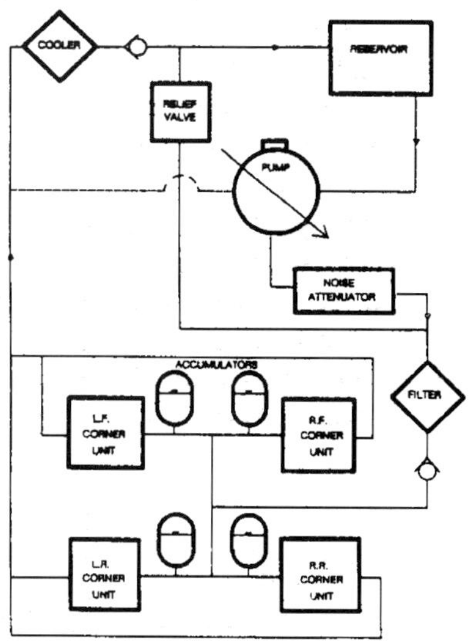

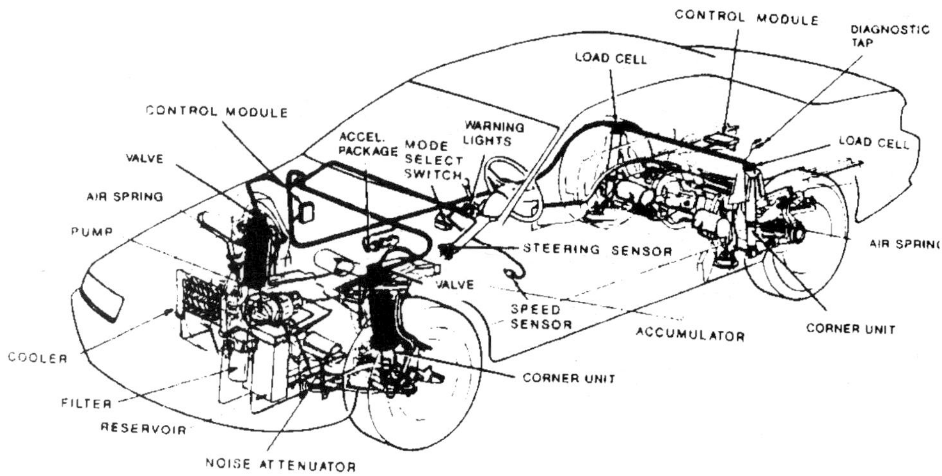

Fig 40: Total active suspension system alongside hydraulics diagram

Mid-engine configuration for volume sports car, Rover MG*F*

The mid-engine, rear-drive, layout of the car (Fig 41) was partially chosen to free the front steer axle of driving torque and avoid unduly high front axle weight for a high performance car. Nose-crush space is also maximised, Fig 42. Good ride/handling balance arises from the use of interconnected Hydragas springs alongside the favourable front/rear weight distribution, Fig 43. Electric speed/load sensitive steering is used on the 1.8i VVC. A torque sensor in the steering column measures the required driver effort and feeds data to an ECU which also monitors vehicle and engine speed; it calculates the ideal level of power assistance and signals the column-mounted assist motor. The 1.8i VVC develops 145 PS at 550 rpm and maximum torque of 174 Nm at 4500 rpm. Use of Rover's large-car five-speed manual transmission is the first application of this gearbox with the K-series engine.

A stretched version of the K-series engine features variable valve timing. So-called 'dampliner' technology has been used involving bore walls just three mm thick and a water cooling jacket of just 0.65 mm. An extra long crank throw then brings swept volume to 1.8 litres. The VVC mechanism, Fig 44, continually varies the inlet cam period to optimize power. An eccentric disc controlled by the EMS alters the phasing of camshaft to crankshaft so as to

Dimensions and Weight		
Length	3913 mm	
Width	1628 mm	
Height (hood up)	1264 mm	
Wheelbase	2375 mm	
Front/rear track	1394 mm	
Weight	1060kg (1070kg - 1.8i VVC)	
Luggage capacity	7.4 cu.ft.	
Cd/Cda	0.36/0.62	
Performance (provisional)	**1.8i**	**1.8i VVC**
Max speed	120 mph	130 mph
0 - 60mph	8.5 secs	7.0 secs
Fuel consumption mpg (provisional)		
Euromix l/100km (mpg)	7.1 (39.8)	7.3 (38.7)

Fig 41: MGF side-view and specification

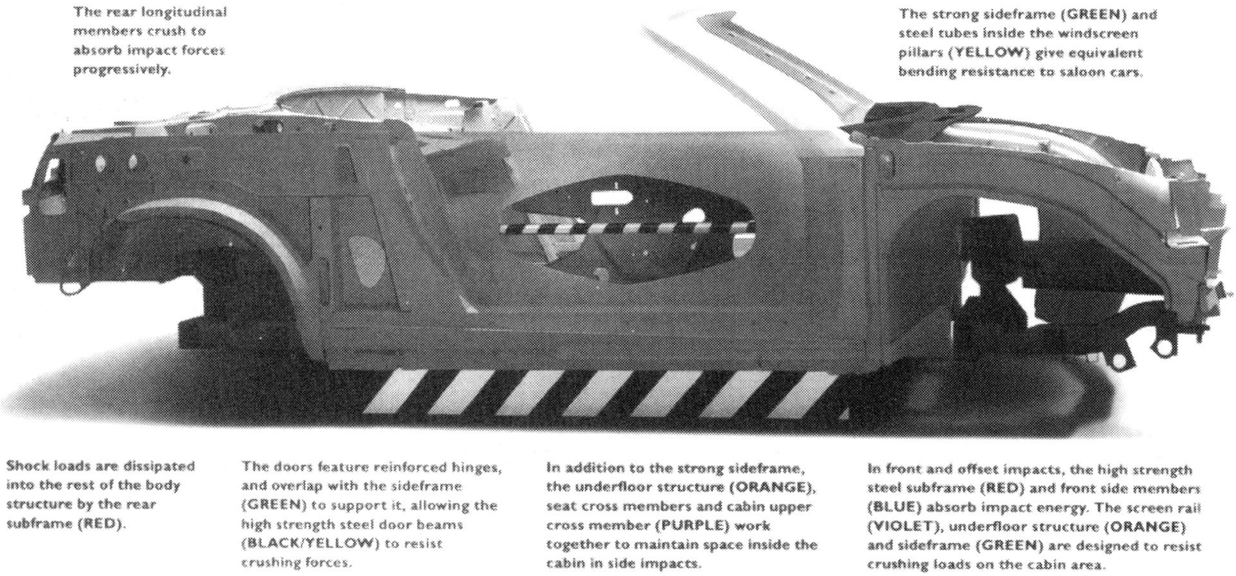

Fig 45: Body features

Fig 43: Front view and engine configuration

lengthen the inlet valve opening period above 400 rpm. A maximum power of 143 bhp at 7000 rpm propels the 2359 Ib vehicle at its maximum speed of 130 mph.

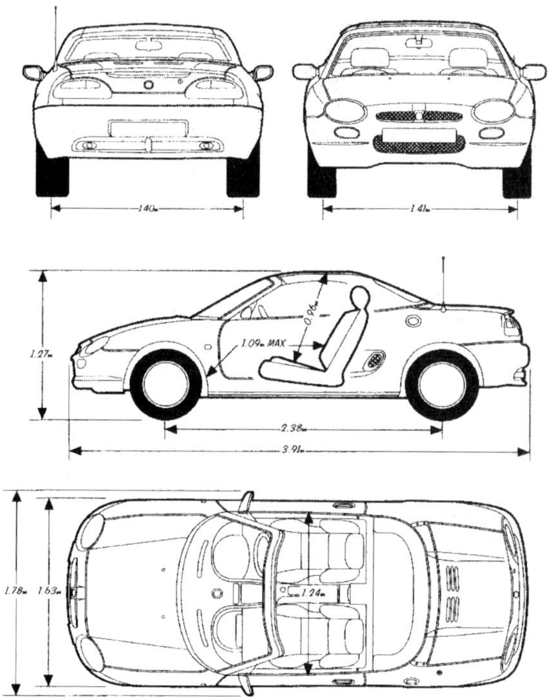

Fig 42: Key dimensions

Special points of body design/construction, for all MGF derivatives, include a woven acrylic fabric hood engineered by Pininfarina and a body shell designed by Rover in combination with Mayflower's Motor Panels Coventry. Trimming on the 1.8i VVC model includes a leather covered steering wheel and "half-leathered" seats. The shell is engineered to withstand the 55 km/h offset barrier impact and has torsional stiffness of 8500 Nm/degree (Fig 45).

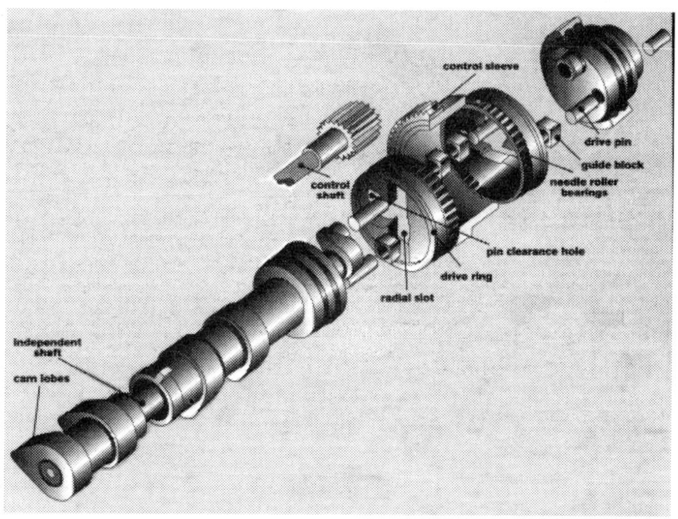

Fig 44: VVC mechanism

Mazda Zedos: power unit innovation

With its Xedos 9 luxury car, Fig 46, Mazda has brought the Miller cycle engine, Fig 47, into series production with the claim that they have resolved the traditional conflict of high output while reducing CO_2 emissions. Belt-driven exhaust camshafts, in turn, drive the intake camshafts by phasing spur gears, resulting in a 27 degree narrow angle between the valves. Each exhaust manifold connects via catalytic converter and collector into a single exhaust system. Shown inset is the Lysholm twin-screw blower which is an intercooled unit which turns its 2.3 litre engine into a performance package comparable with a 3.3 litre unit, the company claim. This is achieved by the cycle's ability to provide high expansion ratio (10:1) while maintaining relatively low compression ratio (7.6:1), as seen in Fig 48. Additionally, an electronic traction control system modulates engine power by fuel cut-off to two through six cylinders by control of the air bypass valve. The car's multi-link IFS is also the first such design to appear on a Mazda model and body structural stiffness is said to be among the world's highest for the vehicle category.

Fig 46: Mazda Xedos

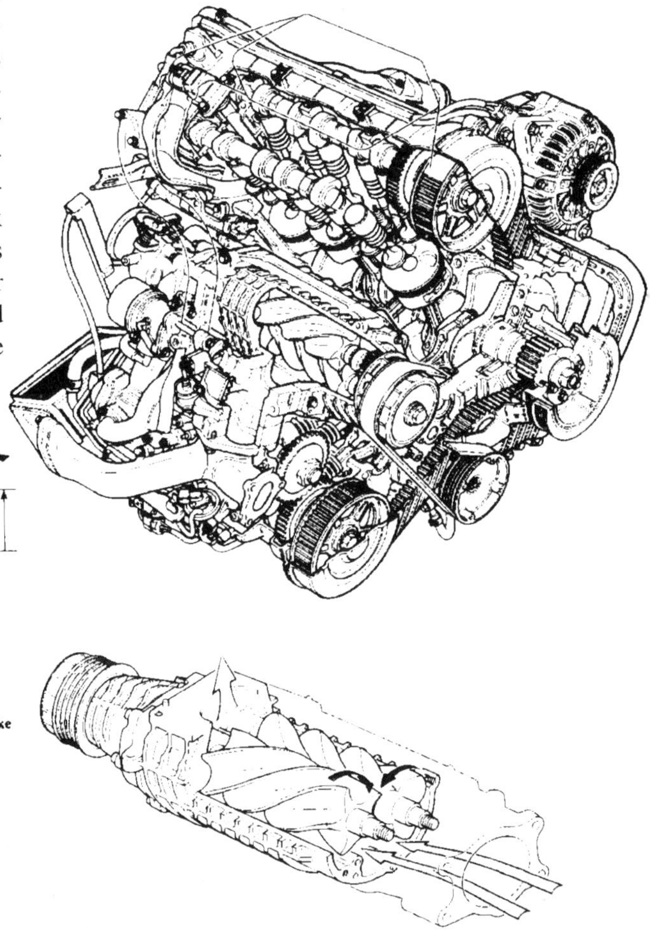

Fig 47 : Miller cycle engine and supercharger

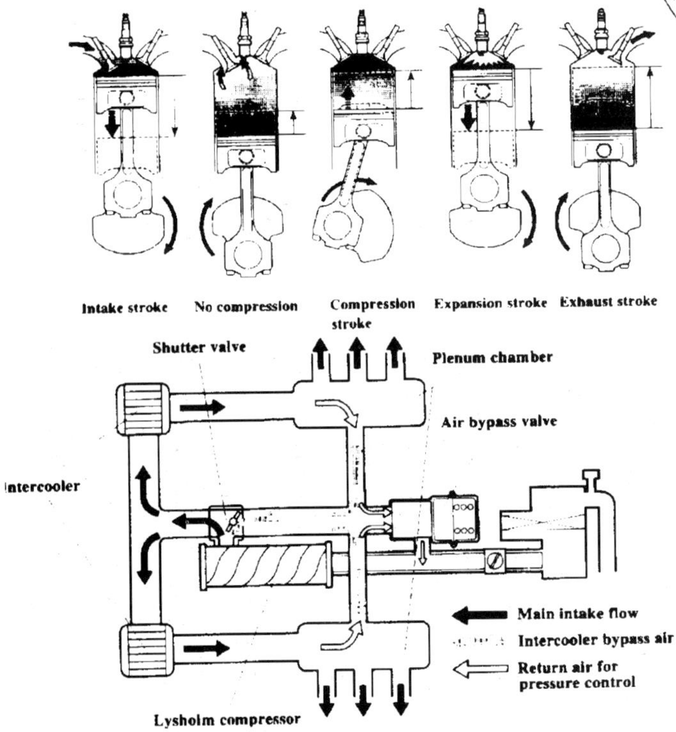

Fig 48: Cycle stages and exhaust system

Mitsubishi HSR-V, car as "man-machine" vehicle system

Creating a feeling of unity between car and driver is the objective of Mitsubishi's HSR-V concept car, Fig 49, notable for its forward driving position and multi-mode open-top which can be switched between coupe and targa modes, Fig 50. Suspension allows adjustment of vehicle height, and gives on and off-road performance. All wheel control technology, developed in the previous generations of HSR, has been further refined by incorporating alignment control into the suspension, yaw control into the brakes and implementing 4WD.

Auto-drive systems that assist the driver include the Accident Avoidance system and the Autonomous Headway Control system for traffic jams. These systems combine with the intelligent alarm that warns the driver if the HSR-V is too close to another car in any direction, and the Driver's View Detector system that makes HSR-V capable of providing the optimum field of view and of perceiving any irregularities in the actions of the driver.

Sufficient collision protection, with forward-control driving, is ensured by using CFRP (carbon fibre-reinforced plastic) inside the internal aluminium frame for superior collision absorption. To reduce the moment of inertia on the z-axis, the engine is mounted forward of the rear axle in a mid-ship layout.

In keeping with the cab-forward design that optimizes visibility, the nose continues the sharp slope of the windshield down to the front of the car. The overall effect is a sharp front mask that strongly expresses forward motion. When the open-door switch is pressed, the height adjusters lift the car, the steering column tilts up, the seat slides to the rearmost position, and the door opens automatically. Once seated, the driver presses the close-door switch to effect the reverse procedure. A so-called vision-optimisation system adjusts the driver's seat to the optimal position regardless of the driver's size. At high speed, the system lowers the driver's seat to ensure the driver can concentrate on looking farther ahead.

An on-demand information display comprises three 4-inch liquid crystal displays. Under normal

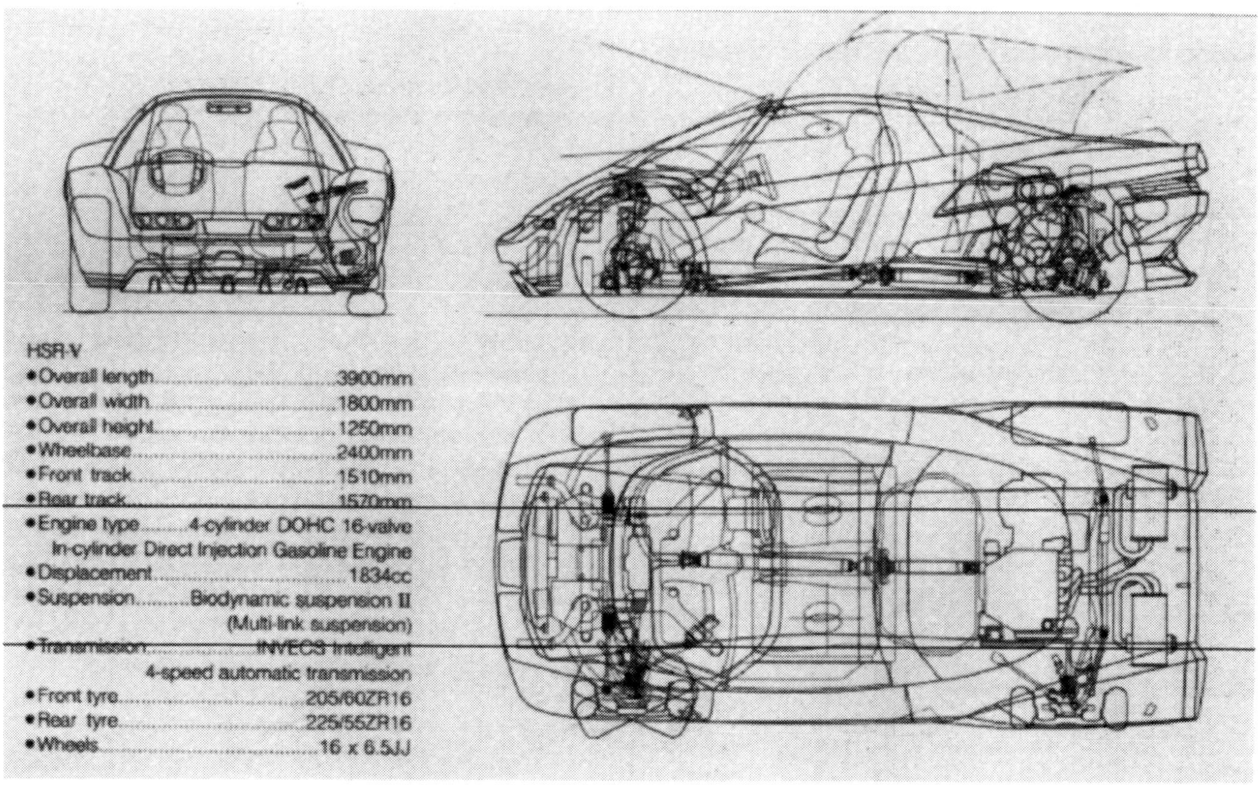

Fig 50: Package and specification

driving conditions, the display shows engine speed in bar graphs and digitally displays the car's speed. However, when another car follows from behind, the system determines how much danger that car presents. When necessary, the system displays video images of the rear on the left and right monitors as captured by the multi-eye system. This ensures safe lane changing. Also, when the multi-eye system detects a large obstacle in the vicinity of the car, a voice-alarm is sounded, a `caution' sign flashes in the centre display, and the system recommends immediate evasive action. When the navigation function of the multi-communication system receives road direction instructions, a direction indicator is displayed on the centre display screen. At these moments, the left and right display screens project images of the rear of the car, facilitating safety in turning corners.

A large display screen is mounted on the passenger's side dashboard, and is designed to pop-up and rotate in place. When the navigation system is in use, the screen displays a map. The screen can also display cabin information, such as the air conditioning settings. Since this may be distracting to the driver while the car is in motion, the system rotates the display screen away from the driver and toward the passenger's side when the car begins to move.

Microphones mounted in the headrests detect road noise and engine noise of 500 Hz or less that is generated while the car is in motion. The active noise-cancelling system then emits a noise-cancelling signal from a speaker also mounted in the headrest.

Fig 49: Mitsubishi HSR-V

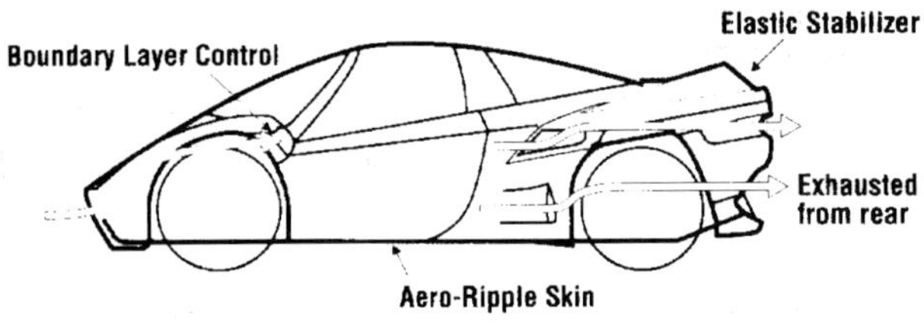

Fig 51: Aerodynamic features

Boundary layer control has been further refined for HSR-V. Air is forced in at the nose and exits at the rear of the front fenders. In addition to cooling the radiator and other components, this air deters the air running along the side from peeling away from the body, thus improving aerodynamics. Also, air brought in from the two air ducts in the front of the rear fenders cools the engine, then is forced out of the rear of the body and recovers air pressure at the car's tail for reduced drag. The car employs a compact tapered rear shape that generates effective down-force. The upper portions of the rear wheel arches form horizontal stabilizing fins. The air flow coming down off the body passes through the fin and generates negative lift, increasing rear wheel traction. Key to the rear fin design is a thermoplastic elastomer skin-like material with non-linear elastic properties which causes the surface of the tail section to rise in the centre and changes shape to form an aerodynamic pyramid that reduced lift without increasing drag. Twin rear fins located on the right and left sides rise 100 mm from the surface to function as a vertical stabilizer. While the basic wing-shaped body of HSR-V creates minimal drag, the company have introduced a revolutionary new way of further reducing drag. Aero-ripple skin is a boundary layer control system that takes its lead from research on the soft skin of dolphins that helps them move through water at high speeds. Specifically, with a specially designed oscillating panel and an under-cover, vibration is created to control air turbulence in the boundary layer and reduce drag from friction. Therefore, the flow of air sent to the rear of the car is streamlined, resulting in Bernoulli's effect at the rear under-cover. This reduces lift on the rear wheels, Fig 51.

Mitsubishi's in-cylinder direct injection gasoline engine, Fig 52, combines better engine performance than conventional engines with fuel efficiency that exceeds that of diesel engines. The 1.8 litre DOHC 16-valve engine injects fuel directly into the cylinder in the same manner as a diesel engine. Also, the improved volumetric efficiency of the upright straight intake ports, combined with in-cylinder injection, allows the use of compression ratios comparable to those of racing engines, and supplies greater output and more torque than conventional MPI engines of equal displacement. The ultra-lean combustion system greatly improves fuel economy and achieves stable combustion even at an air-fuel mixture ratio of 40:1. The upright ports control airflow within the cylinder for optimal fuel injection; the curved-top pistons control combustion and the electromagnetic high-pressure swirl injectors control the atomization and stratification of the fuel spray. The engine achieves fuel economy 40% better than conventional engines. In Japanese 10-15 mode tests, considered representative of typical urban driving, the engine achieves a 25% fuel economy improvement compared to conventional gasoline engines, and exhibits even better fuel economy than a 2-litre supercharged diesel engine. In addition, the engine is capable of supporting rapid combustion even when EGR is introduced while running an ultra-lean mixture. This results in a 90% reduction in the NOx emitted from the engine's exhaust ports.

Features of the all-wheel-control include the integration of 4WS with yaw moment generated by differences in drive and brake force on the left and right wheels, and active yaw control to help maintain control when cornering. Artificial intelligence allows the system to apply the appropriate amount of control to each wheel. A form of intelligent traction control uses sensors to pick up information from the navigation system on the type of corners ahead, the grade and road conditions, then uses this in combina-

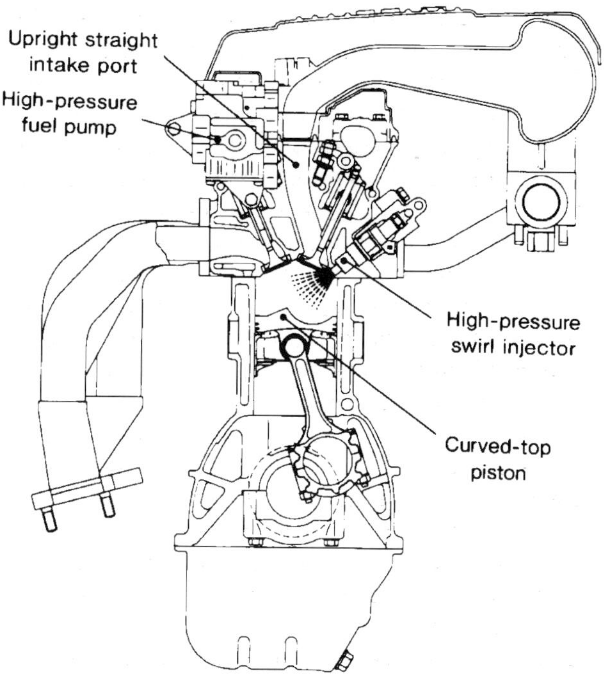

Fig 52: DI gasoline engine

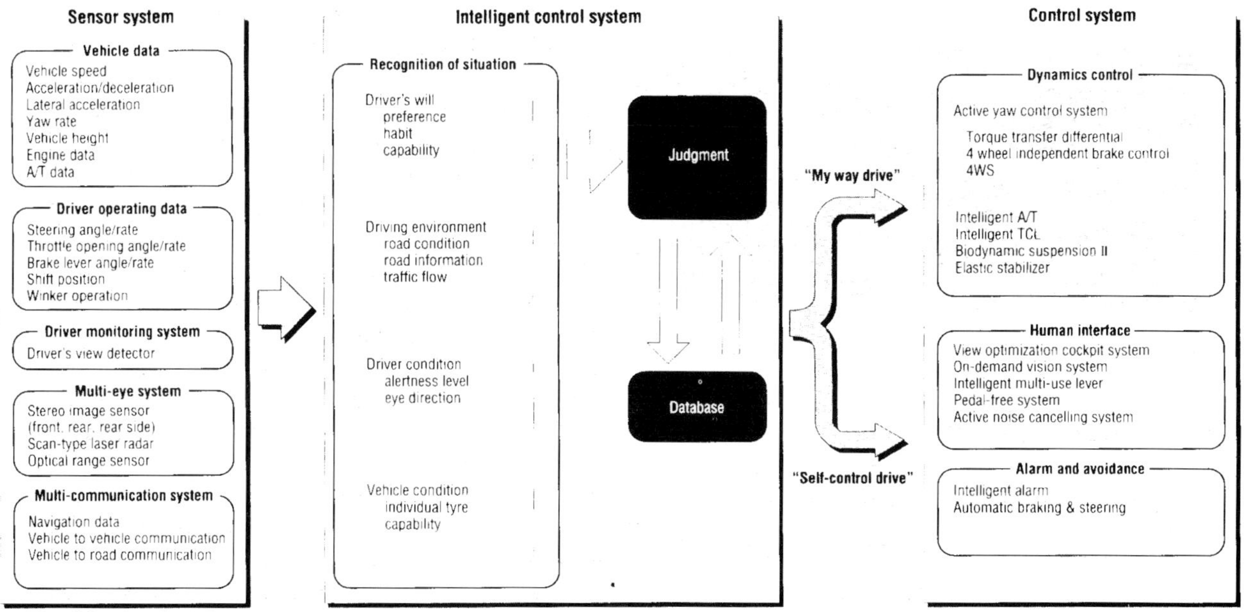

Fig 53: Artificial intelligence system

tion with image sensor data from the multi-eye system to predict what the road conditions ahead will be. Of particular note here is an artificial intelligence system, Fig 53, that updates navigation data with information about the condition of roads the car has previously travelled. The transmission then sets a shift pattern that suits the road being travelled and regulates the output to supply the right amount of power for the driver.

With what is claimed to be the world's first active yaw control system, HSR-V provides a high handling stability. The system is comprised of 4WS, 4-wheel independent brake control and a new torque transfer differential that transfers drive force freely between the right and left wheels. The intelligent controller that governs the system picks up detailed information from the yaw rate sensor and lateral *g* sensor to determine conditions as the car corners. It uses this data to generate the correct yawing moment for easier cornering, and also to improve safety by suppressing spins.

Hydraulic control of the so-called Biodynamic Suspension II is administered by a computer that analyzes data from the vehicle speed, lateral acceleration, vertical acceleration, steering angle, and vehicle height sensors, as well as data from the navigation system and image sensors of the multi-eye system.

Chapter 8: Specialist cars

Specialist road cars; track-racing cars; race-car structural and aerodynamic trends; sports/racing cars; design of a supercar; systems design for kit cars; CG determination

Specialist road cars

The increasing disenchantment shown by the general public for the stereotyped designs from the volume producers is in turn leading to improved prospects for specialist builders who produce individualised designs.

The particular specialist car function will often affect the design and packaging of mechanical systems. The earliest road vehicles, with modestly powered engines, usually had their power units tucked beneath the rear seat. As powers increased, power-units moved forward to the position once occupied by the horse — in the days before the horseless carriage. Here it has remained despite the almost universal adoption of rear-engine/rear-drive by racing cars. The breakthrough for specialist road cars came with the so-called mid-engine position pioneered by the Lamborghini Miura, Fig 1, in which the dictates of mechanical efficiency and body aerodynamic function began to coincide. The layout allowed the freedom to lower the front end while accommodating a large transverse engine towards the rear where the body upswept into a 'wedge' profile.

The use of integral structures on many of the latest specialist sports cars also allows body function and user comfort to dictate the vehicle design — without having to compromise wheel position, say, to suit a particular chassis layout. An early stage in design for good vehicle handling is the determination of roll centre heights given by the wheel suspension systems, front and rear. These are one of the first things to be entered on a general arrangement drawing of the complete vehicle, having sketched in the packaging of occupants and major mechanical units on the basis of preliminary performance-prediction and ergonomic-dimensional calculations. The suspension linkages can next be laid out and the vertical wheel travel determined — wheel clearance envelopes being a major constraint on the overall vehicle

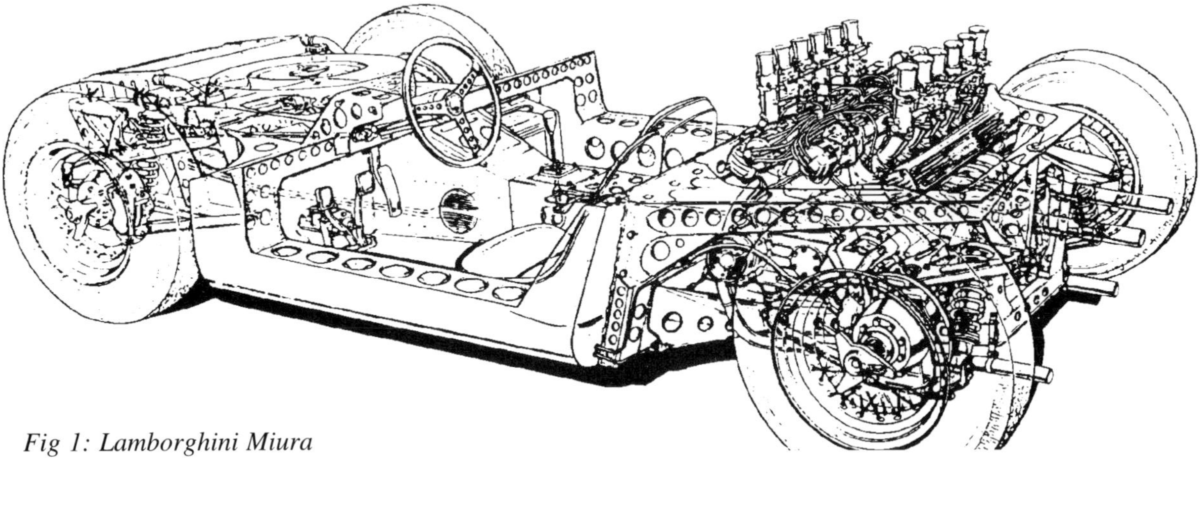

Fig 1: Lamborghini Miura

package. The undertray line can next be drawn in to determine ground clearance at full bump motion of the suspension. The specialist car builder can at this stage benefit considerably from ergonomic techniques for positioning instruments and controls (Chapter 8). This helps to position the major bulkheads of the vehicle structure at seat-back and dash-panel locations. A most important adjunct to this is the accurate prediction of weights for each element of the vehicle structure, power-train and running-units. For this purpose, alignment charts as shown in Fig 2 can provide a valuable guide to the use of different materials in sheet and rail form. This will lead to an early appreciation of front-to-rear weight distribution upon which ride and handling performance is dependent — as well as the selection of tyre size and construction. Fig 3 shows a tabular approach to gravimetric analysis and the use of moment equations to determine axle loadings.

Trendsetting specialist cars

Aside from the Miura design mentioned above a number of recent introductions have marked substan-

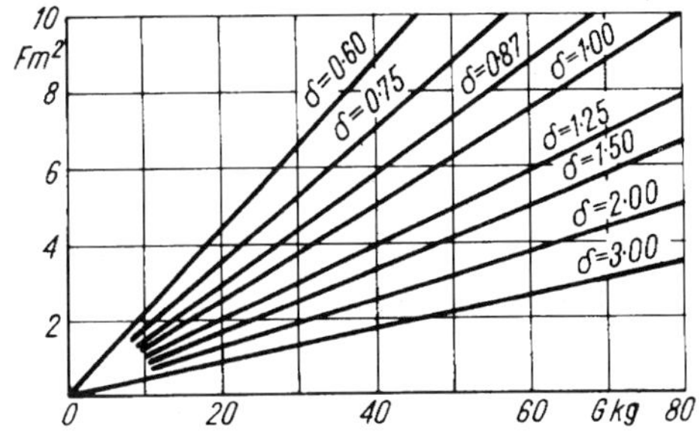

Alignment chart for the weight of steel sheets of different thicknesses. For small areas, 0–3 m², a larger scale graph is used

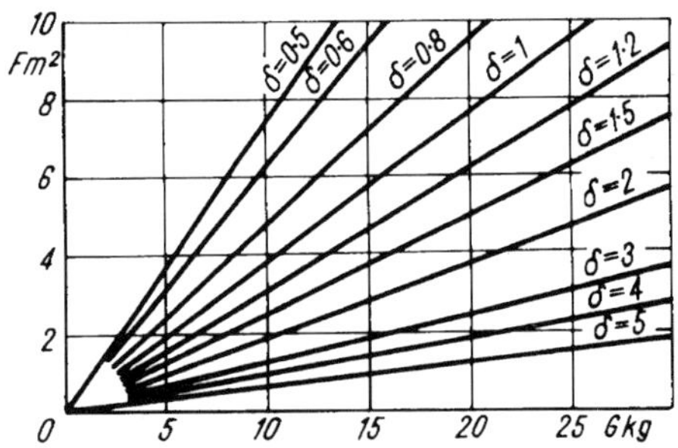

Alignment chart for the weight of aluminium sheets of different thicknesses

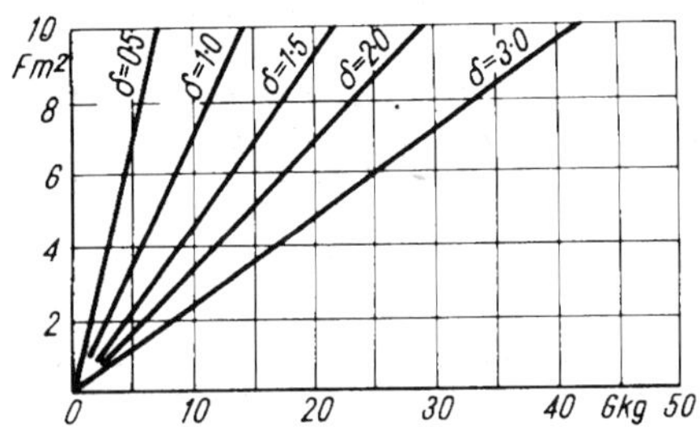

Alignment chart for the weight of trim material of different thicknesses

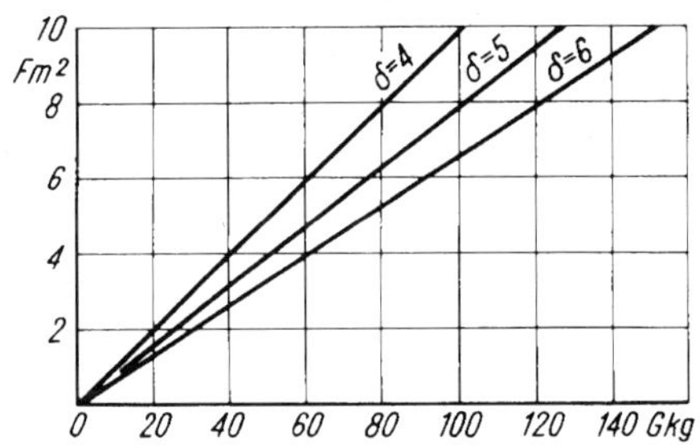

Alignment chart for the weight of glass sheets of different thicknesses

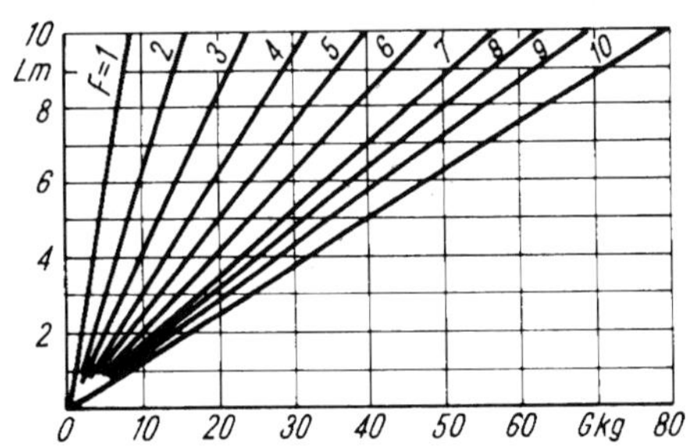

Alignment chart for the weight of various sections

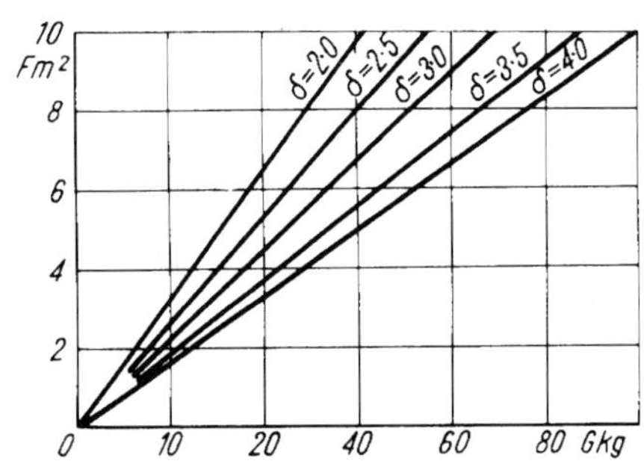

Alignment chart for the weight of rubber mats of different thicknesses

Fig 2: Alignment charts

tial new directions in specialised road cars. Jaguar's XJ220, Fig 4, is a road car which is almost on the boundary between road and track design — being originally conceived as a vehicle which could compete successfully in Group B racing, had regulations allowed. Crucial to the design was the means used to achieve 4-wheel drive — a quill shaft between cylinder banks of the vee engine and linking the engine via torque tube to the front drive, Fig 5. The car also had to achieve very high speeds without generating aerodynamic lift and without resort to 'inelegant' wings and spoilers. A 48-valve variant of the Jaguar V12 engine was developed to power future variants of the vehicle giving an eventual potential of 700 bhp. This required purpose-designed rear transaxle and front final drive units to be produced in conjunction with FF Developments. Viscous control units are used in the centre, rear and, optionally, front differentials. The centre differential is incorporated within the rear transaxle, forward and concentric with the gearbox output shaft. Some 70% of the torque passes further forward to the rear final drive pinion while the remainder passes to a transfer gear set into a shaft passing above the rear differential and into the vee of the engine. The mid-engined layout has the gearbox overhanging the rear final drive. By designing the front final drive with a negative hypoid offset, and by slightly inclining the drive-line downwards towards the front, roughly equal heights are achieved for front and rear final drive shafts. So, unlike many 4WD vehicles, both power unit and drive to the front lie on the centre line of the car. This allows well forward positioning of occupant's feet within the wheel arches. The body was designed with completely flat floor and

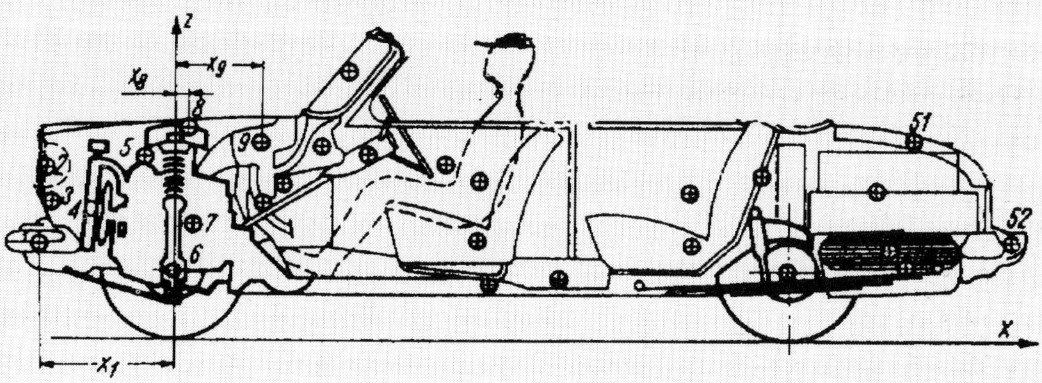

1—front bumper; 2—headlamps; 3—radiator grille; 4—radiator with water; 5—front mudguard; 6—front suspension complete; 7—engine, etc., up to 52—rear bumper

Lp.	Element	Q (kg)	X (m)	ΔM (kg.m) −m	ΔM (kg.m) +m	Correction	Remarks
1	Front bumper	4	−0·62	2·5			
2	Headlamps	6	−0·58	3·5			
3	Grille	5	−0·57	3·0			
4	Radiator with water	15	−0·50	7·5			
51	Boot lid	9	+3·10		27·8		
52	Rear bumper	4	+3·50		14·0		
		$\Sigma Q = G_c$		$\Sigma \Delta M = M$			

Fig 3: Results of gravimetric analysis

fully enclosed transmission tunnel, Fig 6. Box member bulkheads join the large sill sections and bonded aluminium alloy was employed. Double curvature glass forms a structural part of the vehicle's upper skin.

Moving down the power scale, Volvo's LCP 2000 concept car, Fig 7, also broke new ground in design. This is an extension of the sports estate type vehicle pioneered by Ogle for their Reliant Scimitar. LCP stood for Light Component Project and stressed the overwhelming importance of light weight in achieving low fuel consumption and exhaust emissions. Dry weight of 645 kg was achieved against an industry standard of 830 kg for a similar specification vehicle — resulting in a diesel fuel consumption of 100 mpg from a three-cylinder 1300 cc engine designed by Ricardo for multi-fuel capability. The vehicle base structure was made from aluminium alloy using the weld bonding technique to reduce spot welds from 4000 to 500. Skin panels were in thermoplastics including side windows in clear polycarbonate. Most of the major power train castings were in magnesium alloy and carbon fibre was used to reinforce the slender door arches and other parts of the plastic skin. A key departure was the use of a 'platform' structure, to which other major assemblies of the vehicle were attached as modular units.

In the 'mini' car class, Fiat's experimental CNR project car, the X1/75, broke new ground for that end of the specialist road car spectrum. The metal

Fig 4: Jaguar XJ220

Fig 5: Drive layout of XJ220

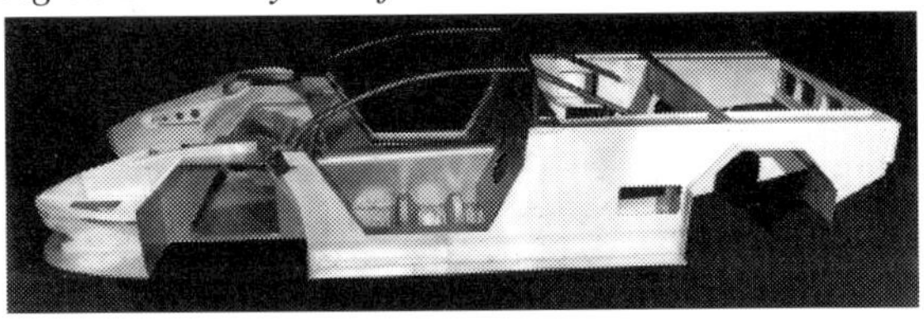

Fig 6: XJ220 structure

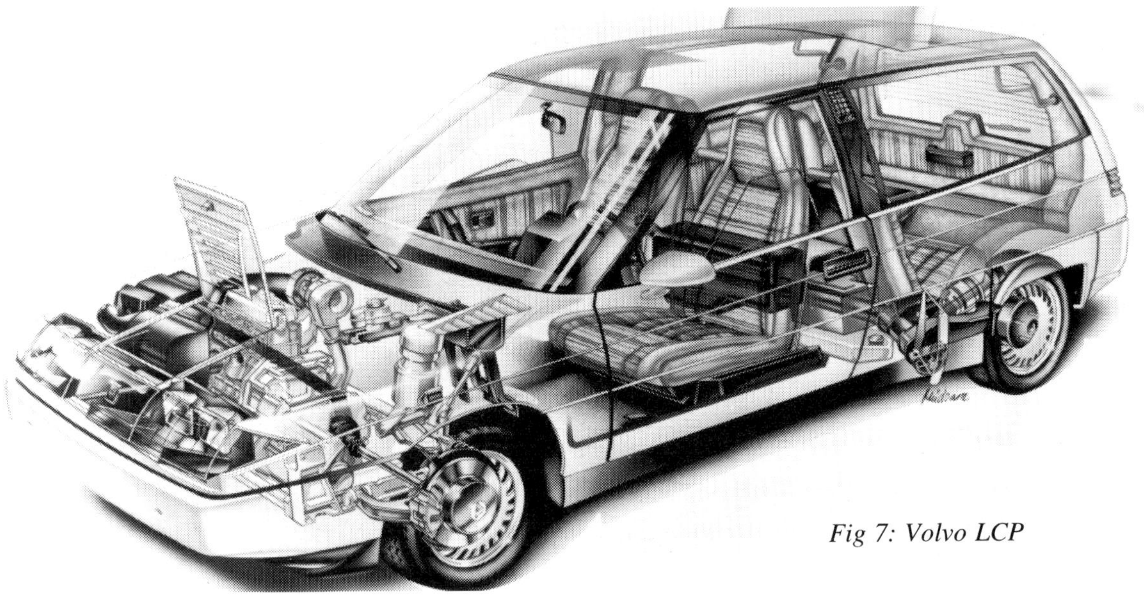

Fig 7: Volvo LCP

base structure is of interest for its stiff-jointed space frame construction used to attach removable plastic panels, Fig 8. Because of the wide load variation on the rear axles of mini-cars, according to passenger load, the designers chose a rigid axle with levelling compensation using air springs. The powertrain comprised a two-cylinder engine coupled to a CVT gearbox arranged for front-wheel-drive. Key dimensions for the vehicle were 3,43 m long x 1.525 m wide x 1.41 m high with 2.27 m wheelbase.

In the small sports car class, Suzuki's Cappuccino convertible, Fig 9, is going squarely for the market sector once served by BL's MG Midget and Austin-Healey Sprite cars. It is a T-bar roadster with optional targa top. Using traditional front-engine rear-drive layout will give a nostalgic feel to the small sports car, but improved by a close-ratio five-speed gearbox and 12-valve DOHC three-cylinder engine. This provides 63.1 bhp from 657 cc while allowing 52.3 mpg at 56 mph. Electronic spark-advance and multi-point fuel-injection are two of its advanced features alongside sophisticated electronic engine-management. Kerb weight is 1599 lbs and turning radius 14.4 feet — almost as good as the Triumph Herald!

Fig 9: Suzuki Cappuccino

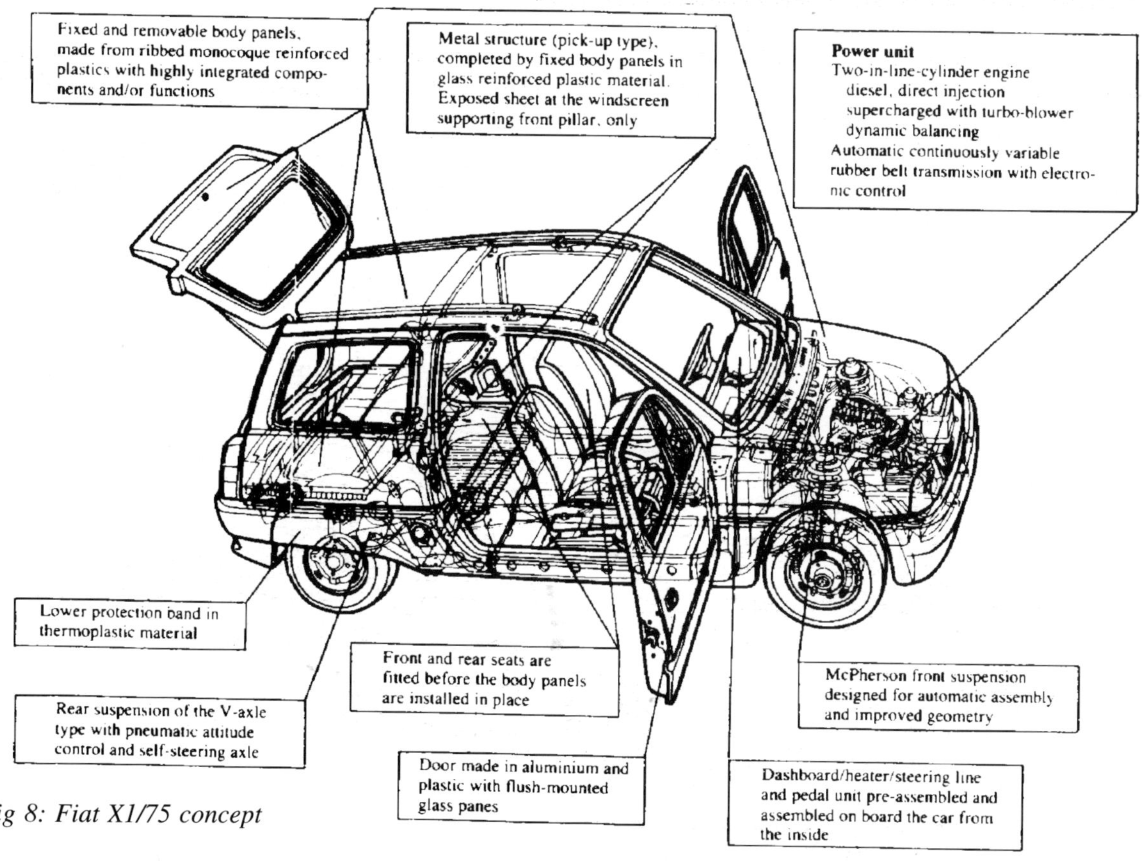

Fig 8: Fiat X1/75 concept

Racing car engineering

Changes in the method of financing racing cars in recent years, with the now widespread adoption of advertising sponsorship, has also had a marked effect on design in the respect of monies spent on product engineering to obtain competitive advantage. Nonetheless there are still developments which have their roots in the days when manufacturers financed their own cars — but more widespread is the use of advanced aerospace techniques which bring performance at almost any price.

The first move towards the current rear-engine layout in racing cars can be traced to the success of pioneering Cooper and Lotus cars in the early 1960s. Minimisation of power loss and reduced frontal area were the principal stimuli — coupled with low centre of gravity. In 1963, the Lotus 25 also pioneered the use of monocoque construction and the near reclining posture of the driver, from which the Lotus 33 in Fig 10 was derived with its 'monocoque-chassis' structure, shown inset, and wide-based double-wishbone IRS with familiar long trailing radius-rods, Fig 11. The Coventry Climax engine became an integral part of the structure as a rearward cantilever supporting rear suspension, gearbox and drive. The width of the engine governed the width of the car. The 1960s also saw the emergence of aerofoils, the negative lift device appearing on Formula One cars of 1968. However, 1969 saw the establishment of regulations which prohibited them being mounted on the unsprung elements of the car and restricted their size. But it was not until 1977 that Lotus introduced the first ground-effect car, aided by skirts, in which the full plan area of the car contributed to the development of negative lift — raising it to twice the weight of the car — at 250 km/h.

The main structure of the car amounts to an inverted four square metre plan area wing, Fig 12,

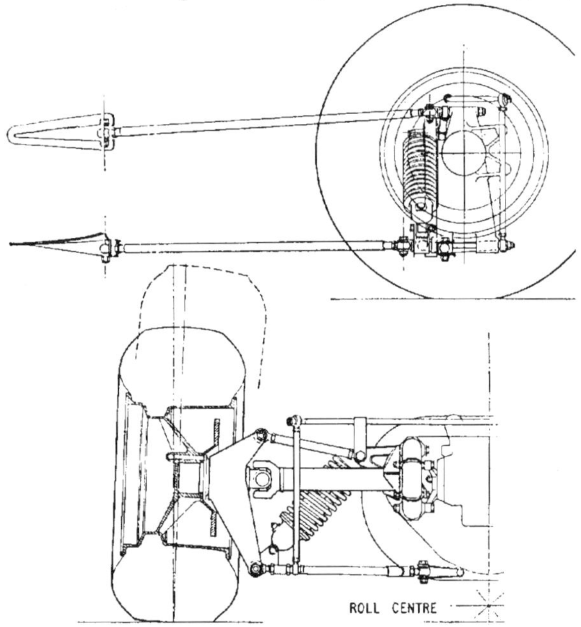

Fig 11: Lotus 33 suspension

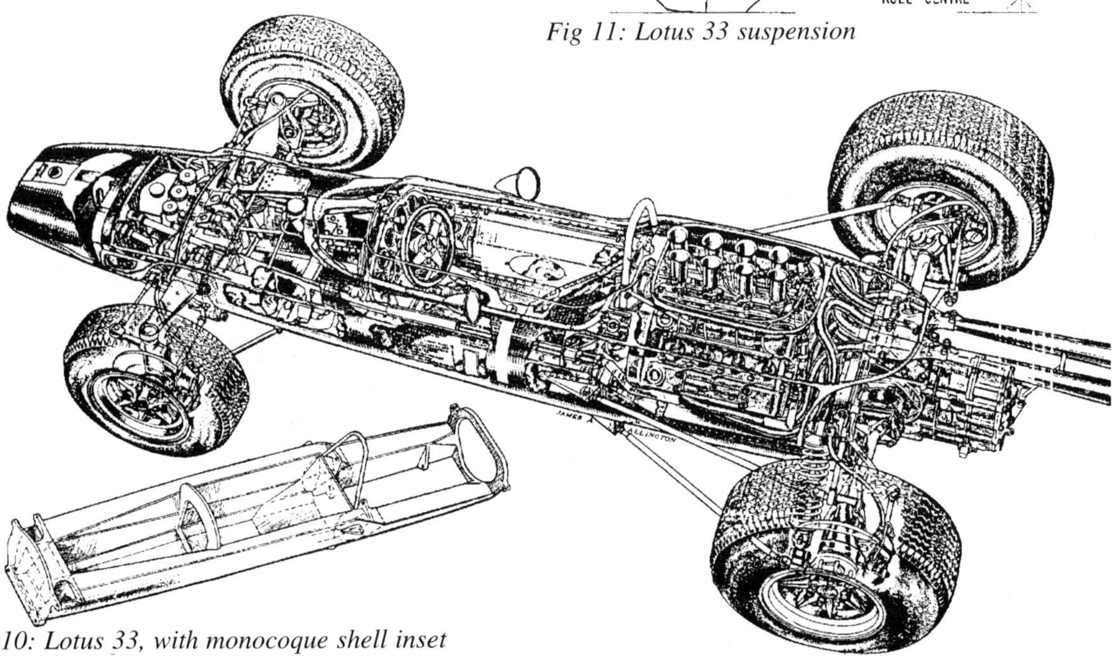

Fig 10: Lotus 33, with monocoque shell inset

with Fig 13 showing the pitching and heaving characteristics. The aerodynamic loading was a spur towards the very stiff monocoque structures of composite construction which are commonplace today. Aerodynamics were also an important factor behind the development of active suspension systems in the need to maintain the critical ground clearance on ground-effect cars. Another development was towards two separate suspension systems, one supporting the 'chassis' structure, the other supporting the body, Fig 14. The aerodynamic part maintained a constant attitude relative to the ground, while the inner part moved separately with respect to the body.

Structural trends in racing cars

Important contributors to high-strength composite laminates for modern racing car structures are carbon and para-aramid fibres. Sandwich panels of Nomex honeycomb core materials are also a key ingredient. This material is introduced in the section on GRP and is typically bonded to skins of high strength laminates by an adhesive film such as the Ciba Geigy Fibredux. For curved surfaces, the core is heat-formed at 170-200 C. Key property of this high performance sandwich panel, in this application, is very high resistance to buckling for very low weight, Fig 15. Typical edging methods, joints, corners and facings are shown inset.

Builders favour carbon-fibre reinforced laminate skins for ultimate strength and para-aramids such as DuPont's Kevlar for toughness. The latter is available in a variety of cloth weaves in weights ranging from 60 to 460 grammes per square metre. Typically an aramid/epoxy-resin laminate of 1.33 specific gravity has an elastic modulus of 42 N/mm^2 x 10^3 and tensile strength of 760 N/mm^2. A carbon-fibre/epoxy-resin laminate using a material such as Courtauld's Grafil has modulus up to 190 N/mm^2 x 10^3 and tensile strength of up to 1900 N/mm^2 with individual fibre strengths up to 1.9 N/mm^2 x 10^3. Weave of the cloth is important to laminated strength: plain weave, alternate warp and weft, is stable and easily handled while satin weave (weft passing over three yarns and under one) is suitable for draping over deep-form surfaces. For high strength applications, twill weave, in which groups of warp and weft yarns are woven alternately, is particularly suitable.

Sandwich laminates are formed either in a vacuum bag or autoclave, an intermediate perforated membrane and blotting paper being used to absorb surplus resin squeezed out during the compression process. For aerodynamic wings, foam cores rather than honeycomb cores are often used. In the case of the Tyrrell 011 Formula One car, shown in Fig 16 as

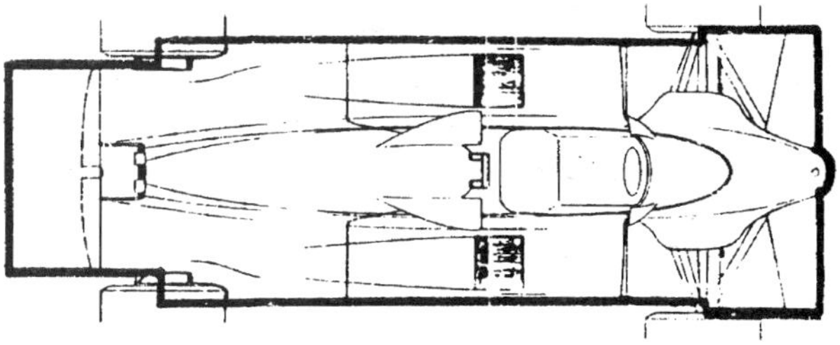

Fig 12: Wing plan area

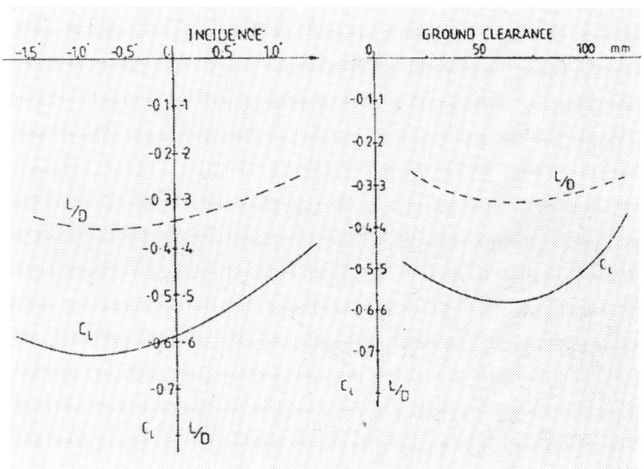

Fig 13: Pitching and heaving characteristics (C_l = lift co-efficient; L/D = lift/drag)

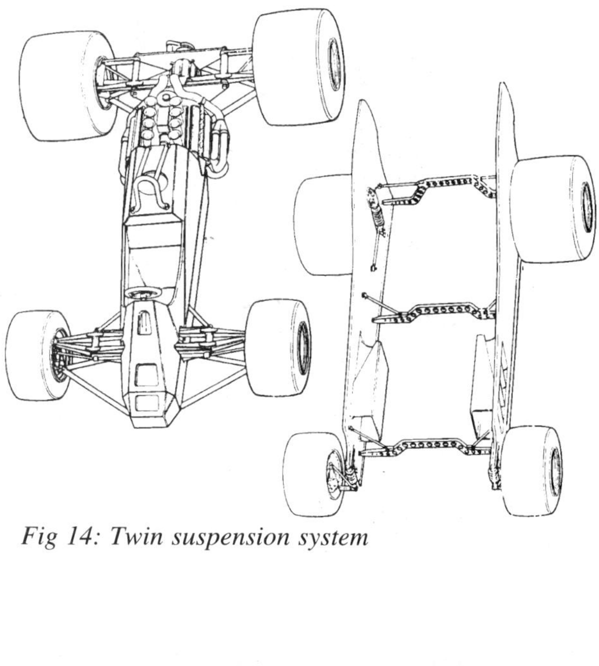

Fig 14: Twin suspension system

a finite-element mesh of its structure, the main body has two full length side panels forming the outer skin and joined by a Nomex honeycomb core to a Kevlar reinforced inner. Bulkheads down the body are machined from aluminium alloy. Weight of the car is 585 kg, with oil and water, for an overall length of 6.65 metres. Another example is the Williams Grand Prix car whose main structure is shown in Fig 17. This follows the now classic construction of a central semi-monocoque accommodating driver and front suspension; the rear-attached engine supports the gearbox casing on which, in turn, is mounted the rear suspension. The purpose of the bulkheads, Fig 18, is to feed attachment loads (and those due to driver and fuel inertia) into the main structure. Roll-over protection structures in front of and above the driver must withstand 7.5 g downward, 5.5 g rearward and 1.5 g sideways without collapsing. By regulation, for this formula, the car must not exceed 500 kg.

Fig 15: Composite sandwich panel

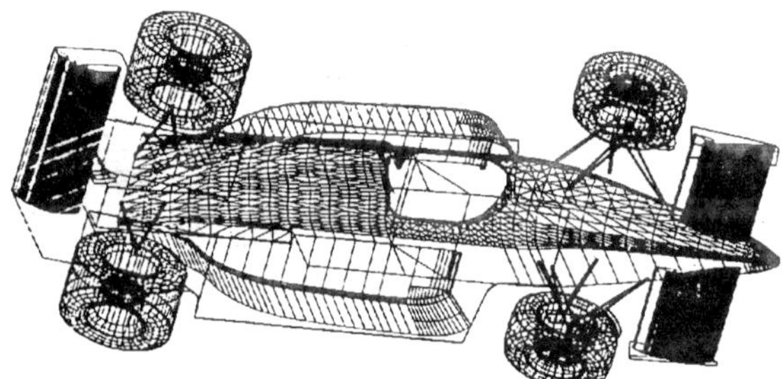

Fig 16: FE mesh for Tyrell F1 car

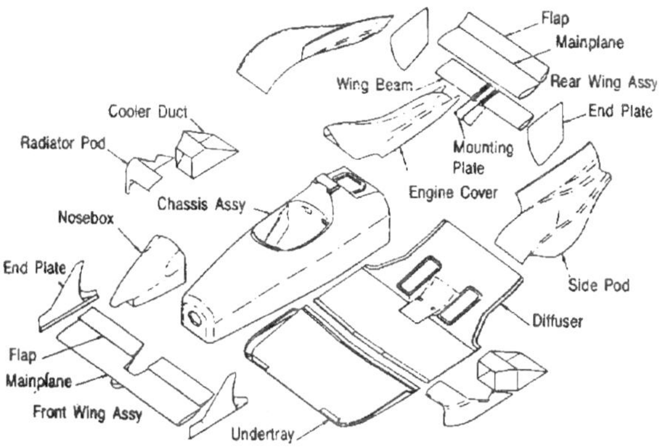

Fig 17: Williams structure

At Reynard, innovative techniques have been used to produce substantial batches within an economic framework — annual production now running at some 240 cars. A recently installed autoclave, Fig 19, operates at 7 bar and can be programmed to a temperature cycle which optimises cure of the laminate monocoques. As can be seen, the part is first enclosed in a vacuum bag to ensure no air permeates the laminate. Layup, Fig 20, is carried out in clean-room conditions at 21 C and 50 % relative humidity, materials having been previously stored in a cold room at 18 C for a maximum of six months. Cyanamid 77 carbon fibre is the currently favoured material for the body shells which must withstand a downforce in excess of 750 kg under track conditions. The latest cars have asymmetric nose cones which have U-shaped honeycomb reinforcements on their lower, extended, regions to withstand 740 kg test pendulum load impacting at 20 metres/second. Recent additional legislation also requires that a 2000 kg sideforce, applied over a 20 in^2 (130 cm^2) area, must also be withstood.

Race-car mechanical developments An important milestone was reached in the 1960s, for establishment of future racing car design trends, by Len Terry's work for BRM. The designer had a back-

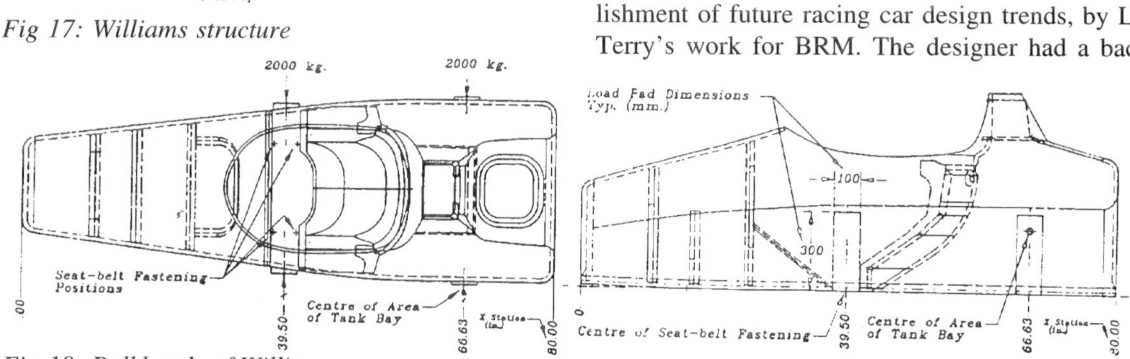

Fig 18: Bulkheads of Williams car

ground in Indianapolis car design which no doubt affected the Formula One contender for the 1968 racing season, shown here alongside a contemporary Lola, with which it bears considerable resemblance in profile and driver posture, Fig 21. The structure was a full monocoque with spring-damper units inside it and actuated by rocker type top wishbones. At the rear, for the first time, a bottom trailing link of the parallel arm type was used to overcome the disadvantages of previous designs of the period — with rearward inclined wishbone axes to minimise bump steer. The wishbone became a four-bar linkage which ensured the wheel plane remained parallel on bump.

Following the banning of the twin chassis Lotus 88F1 car, referred to earlier, came the need for another major change in suspension design, this time to maintain a constant ride height for skirted ground effect bodywork of recent years. This led to the Lotus active suspension system using double-acting hydraulic actuators in place of conventional dampers. Fig 22 is a schematic of the control system, the heart of which are the Moog servo valves.

The body-to-ground clearance-space beneath a car can be likened to a half-venturi having an expansion ratio which increases without limit until the duct stalls and causes a sudden drop in suction load as the distance to ground changes. At speeds in the order of 240 km/h, with the previously mentioned high downthrusts, any cornering can also cause massive additions in load on the suspension. Cornering accelerations in excess of 3.5g are possible and drivers can get transient loading of 4g/second. Stiffnesses are thus inevitably high to tolerate 3 to 1 changes in vertical loading.

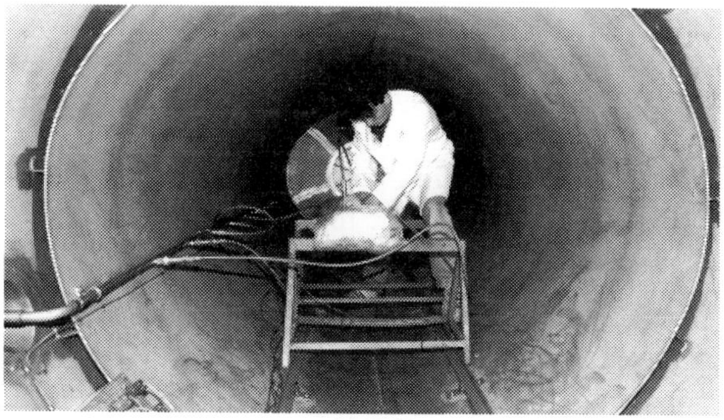

Fig 19: Autoclave at Reynard

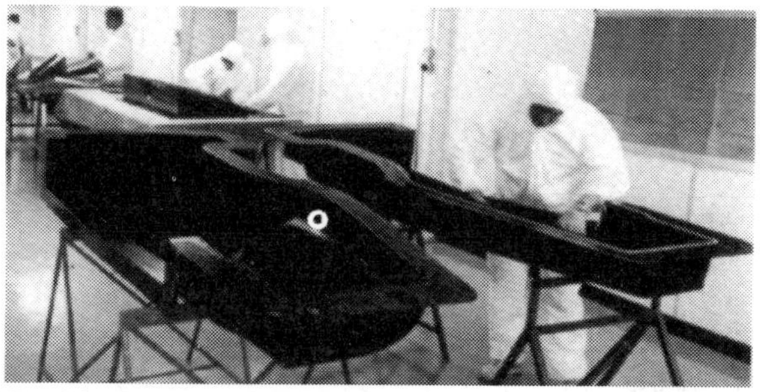

Fig 20: Lay-up of the laminate

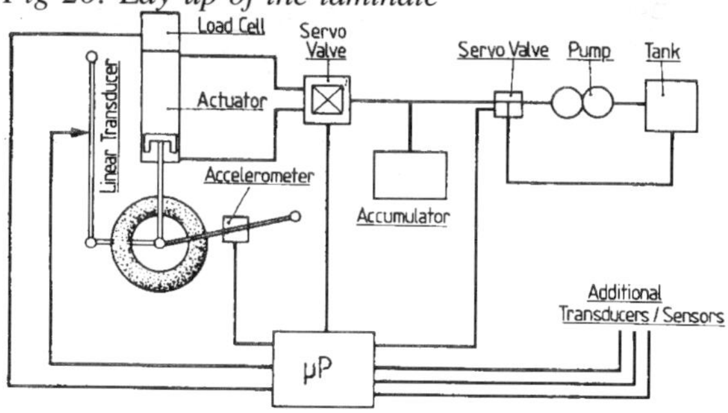

Fig 22: Control for Lotus active suspension

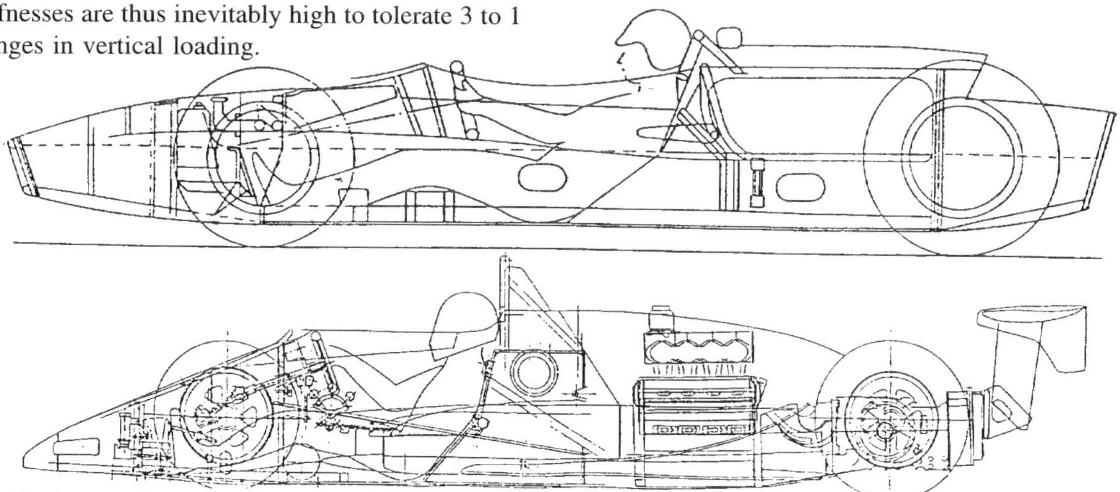

Fig 21: BRM of 1968 (above) compared with recent Lola design

Building to F1 race specification

The 024 F1 car, Fig 23, is the creation of Tyrrell's rejuvenated 13-man and 2-woman design team headed by Harvey Postlethwaite. He has found that the most common motor racing corner is a 2nd gear, 75 mph (120 km/h) corner. As a result he went for a high-nose design because he needed more downforce. The side pods on the car are quite short and he has changed the shape for aerodynamic reasons.

At the time of writing, the 650+ bhp Yamaha OX11A engine was supplied exclusively to the Tyrrell team which has created a totally new 6-speed transmission, a narrow longitudinal gearbox that is torsionally very stiff. The unique air-powered pneumatic gear shift has been retained, mated to an even faster up-shift. Suspension is unusual in that both the front and rear systems use a three-spring - rather than two-spring - layout. The third spring, which controls pitch but not roll, is an interim solution until the re-introduction of Hydrolink. Both the three - spring and Hydrolink suspension systems use tailor-made components from the team's technical partner Koni

Yamaha's OX11A has been developed from scratch as a V10 engine designed specifically for the 3-litre class. Result is a high potential speed range of over 16,000 rpm, delivering a max. power output of over 650 bhp. It is 50.5 mm shorter in overall length, 44.0 mm shorter in height, 56.0 mm narrower in overall width, and 20 kg lighter in weight than the OX10C. An important specification change is the adoption of a barrel type throttle valve that improves response and fuel economy while its outstanding compatibility with the fly-by-wire actuating system also give a big boost to the engine's drivability. The development of the new OX11A has been conducted in cooperation with Engine Developments Ltd.

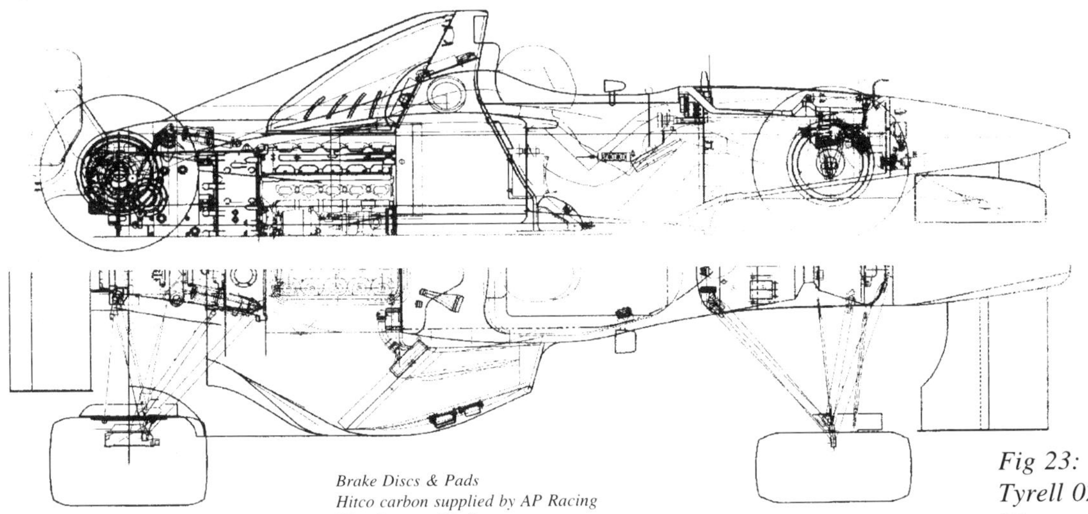

Fig 23: Tyrell 024 F1 car

Technical Specification
Chassis
Moulded carbon fibre and honeycomb composite structure to fully meet the FIA 1996 strength and impact regulations
Front Suspension
Combined spring and damper units operated by push rods and rockers, third spring, mechanical anti-roll bar
Rear Suspension
Combined spring and damper units operated by push rods and rockers, third spring, mechanical anti-roll bar
Dampers Koni
Wheels Fondmetal cast magnesium:
Front 4-spoke, 11.5 in wide, 13 in dia
Rear 5-spoke, 13.7 in wide, 13 in dia
Tyres: Goodyear
Brakes
AP Racing: Front 2 x 6 piston callipers
Rear 2 x 4 piston callipers
Brake Discs & Pads
Hitco carbon supplied by AP Racing
Pipe Systems: Titeflex
Steering: Tyrrell rack and pinion
Steering Wheel: Personal
Driver's Seat
Anatomically formed carbon composite
Seat Belts: Willans 6 point harness (75 mm wide shoulder straps)
Extinguisher: Tyrrell carbon fibre moulded tank to FIA 1996 specification, pressure filled with Spraylance AFFF extinguishant.
Fuel Tank : ATL Kevlar bladder to FIA FT5
Battery: Tyrrell 12v 5Ah lead acid
Instrumentation: Pi Research System V, integrated digital data display and telemetry.
Engine: Yamaha OX11A V-10 3litre
Injection/ignition: Zytek
Spark Plugs NGK
Transmission Gearbox
Tyrrell longitudinal three-shaft 6-speed
Gear selection: Pneumatic, sequential
Drive Shafts: Tyrrell
Clutch: AP Racing carbon plate
Differential: Tyrrell / X-Trac viscous coupling
Dimensions
Overall length 4430 mm
Wheelbase 2940 mm
Front track 1700 mm
Rear track 1610 mm
Overall width 2000 mm
Height 950 mm
Engine
Cylinder format
72 degree V format 10 cylinder
Displacement 2996 cc
Max. power output 650 bhp
Valve mechanism
DOHC, cam driven direct lifter type
Fuel injection Zytec
Ignition Zytec
Dry weight 105 kg
Overall length 575 mm
Overall width 499 mm
Overall height 373 mm
Spark plugs NGK

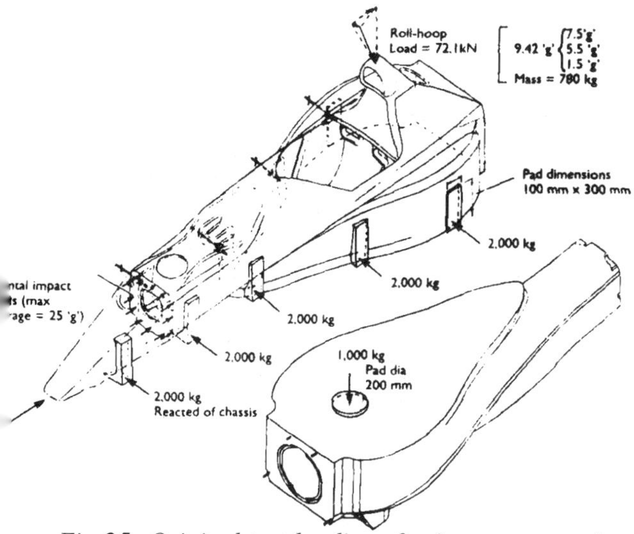

Fig 25: Original test loadings for impact protection

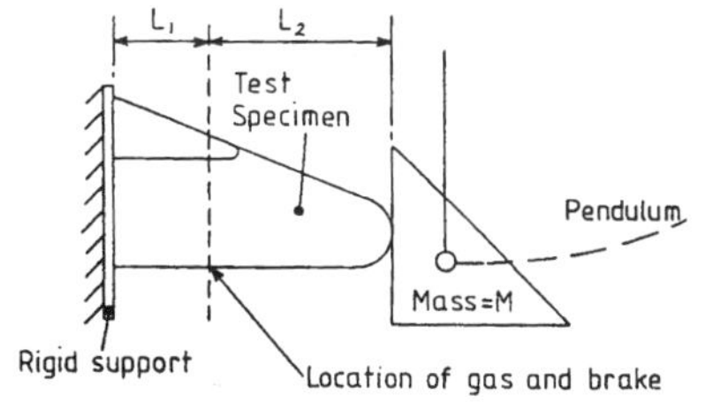

Fig 26: Original test requirements

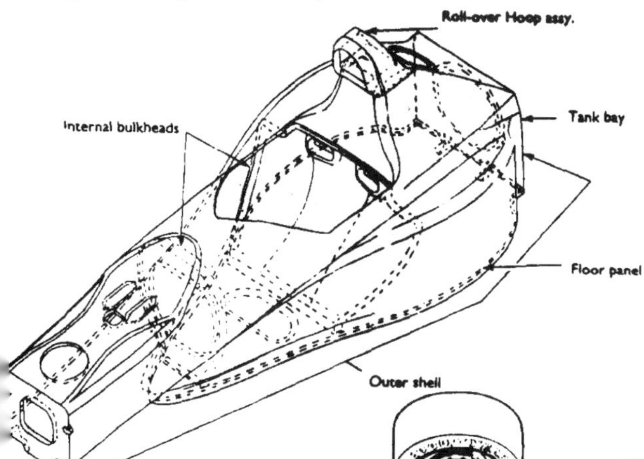

Fig 27: Typical construction

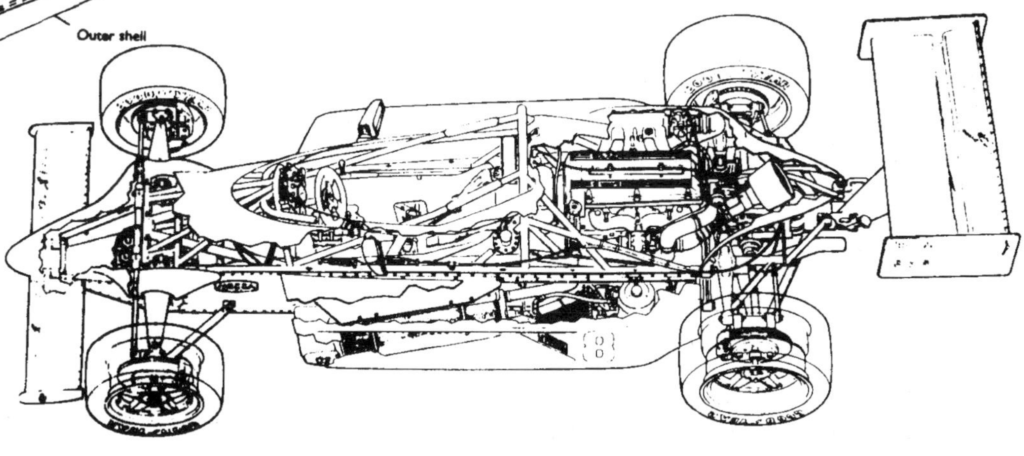

Fig 24: Pro-series race-car

Pro-series space-frame race-car This vehicle, Fig 24, is powered by a turbocharged 16-valve two-litre twin cam Saab engine as fitted to the company's current road cars. Purpose-built by Mondiale, who also design and fabricate the spaceframe, it uses the Hewland five-speed gearbox with limited slip diff. Engines have been converted to dry-sump lubrication and tuned to 225 bhp for a top speed of 165 mph.

Structural impact in race-cars

Racing car structures, subject to testing within FISA regulations governing motor sport, are being subject to stiffer demands being made on them in both the legal and the structural sense. Now an additional 2000 kg lateral test load is required on the nosebox. This checks the integrity of fastening between nosecone and the main monocoque shell. There is also a proof load test for the main roll-over hoop structure involving 7.5g downwards, 5.5g rearwards and 1.5g sideways loads, based on a mass of 780kg. A further load of 9.42g (721 kg) must not deflect the rearward part of the structure more than 50mm. Changes to the earlier side impact test requirements, Fig 25, with the 1000 kg load at the bottom of the fuel tank bay being increased, have now been made.

A Grand Prix car carries some 2000 gallons of fuel which occasionally sustains lateral loading up to 4.5g even in the non-impact situation. Design load cases of 10g asymmetric bump loads on individual wheels stations are now quoted, and braking loads up to 3g. Aerodynamic downthrusts on the front wings are put at 5kN maximum. These show how design operational loads relate to test loads simulating impact.

Test requirements involve firstly the well-known frontal impact pendulum test on the nose cone for which the energy absorption requirement is 47 kJ; see Fig 26 based on the report sheet used at the Cranfield Impact Centre. Impact speed is 11 metres/second and trolley mass representative of the maximum weight car, is 780 kg. The loads of the fuel (the bladder tank must be filled with water for the test) and driver are also simulated for checking integrity of bulkheads and belt anchorages. A test such as an inclined sled track set-up is thus required, involving the main 'survival cell' as well as the nose-cone, Fig 27 and any impact damage must be contained within an area forward of the driver's feet and peak deceleration recorded in the test must be below 25g.

In the static load test for side impact integrity, load pads of 300x100 mm are stipulated and the load must be held for 30 seconds without damage to the inner skins. Side loads of 2000 kg must now be sustained and 1000kg at the fuel tank location, Fig 25. A further side load in the cockpit area is required, in a vertical plane passing through the centre of the seat belt lap strap fixing, also on a vertical plane passing half way between the front wheel axis and the centre of the dashboard hoop, as described for the rollover test, as follows. The general requirements set down by FISA for the safety structure of the car is that two rollover protection systems be provided, the first in front of the steering wheel not more than 25 cm ahead of the rim. the second to be not less than 50 cm behind the first and high enough for a line joining at the top to be 5 cm above the driver's head/helmet.

The facilities at MIRA for impact testing of racing cars, Fig 28 include a frontal impact rig which in this case does not involve an inclined plane. Fig 29 shows examples of accelerations and durations applying to frontal impact. The 25 g limit for F1 technical regulations refers to the average acceleration level throughout the test. The 20g/30ms pulse is an idealised one for car seat testing while the 60g/3ms one represents the limit adopted for chest acceleration. This is because whole vehicle decelerations of 20g are typical but those of components and occupants can be much higher; high peaks can also occur due to the slack in seat belts.

Aerodynamic factors in race-cars

The basic considerations for aerodynamic design of racing cars seen by a Durham University researcher[1] suggests that the stability of regulations has led to a convergence of design which means that competitive circuit performance now depends on fine tuning of aerodynamic design. He shows the major aero-D features and the corresponding forces and moments as Fig 30, of which he singles out the front wing as the most efficient device. As ground clearance beneath it is reduced flow can no longer tolerate the pressure gradient and the aerofoil stalls in the limiting condition. The limited permitted span of the aerofoils makes it necessary to control the associated tip effects. The endplate solution aimed at maintaining approximate two-dimensional flow increases loading at the wing tips but does not eliminate strong ti vortices — for which semi-tubular guides along the lower edges of the end-plates has been one of the ways of containing them.

To overcome the compromising effect of the front-wing on engine cooling performance, the use of flapped aerofoils has been introduced — with chord and camber at the outer end of the flap greater than at

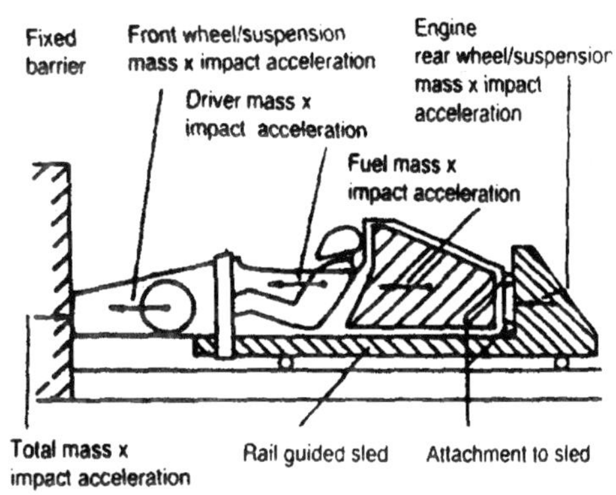

Fig 28: MIRA frontal impact rig

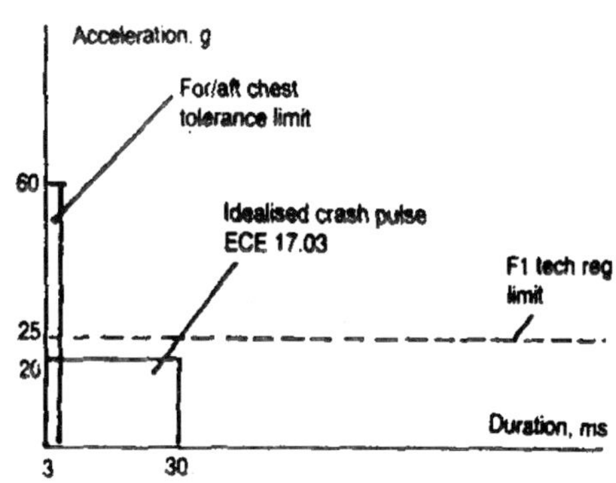

Fig 29: Typical accelerations and durations

the chassis end. Combining this with taller and narrower engine air-intakes ensures that only the wake from the inner section of the wing impinges on the intake. With rear wing aerofoils the requirement for high downforce to be generated from already disturbed airflow has necessitated the use of highly cambered multi-element configurations which are inefficient in terms of drag. The trailing edge of the wing is governed by regulations while leading edge position is chosen as a compromise between a high one better exploiting air-flow and a low one allowing a larger aerofoil to be used. The undertray of the car is the other important factor, Fig 31, air flowing beneath the car leading to a high-pressure stagnation point on the upper surface and a separation bubble below the entry lip. Addition of a diffuser creates a further zone of low pressure and its design is crucial to obtaining marginal advantage.

Design of sports-racers

The design and construction of sports-racers is very much geared to economic factors and varies widely between budget kit-cars and expensive supercars. The latter category increasingly draw on the techniques of the volume sector with respect to CADCAM and the use of sophisticated shell structures while the former relies on workshop ingenuity

Specialist sports racers are often based on the fitment of proprietary running units into tubular space-frames on a try-and-test basis. Space frames, too, can be purchased as ready built units from such suppliers as Spydersport, a company which established itself with substitute frames for such cars as Lotus Europa and Elan, Fig 32, for which it also manufactured proprietary suspension assemblies and designed structures which would accommodate pro-

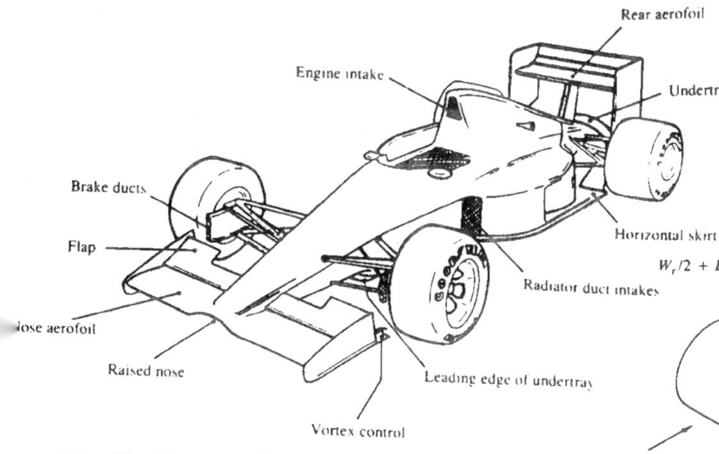

Fig 30: Key aerodynamic elements

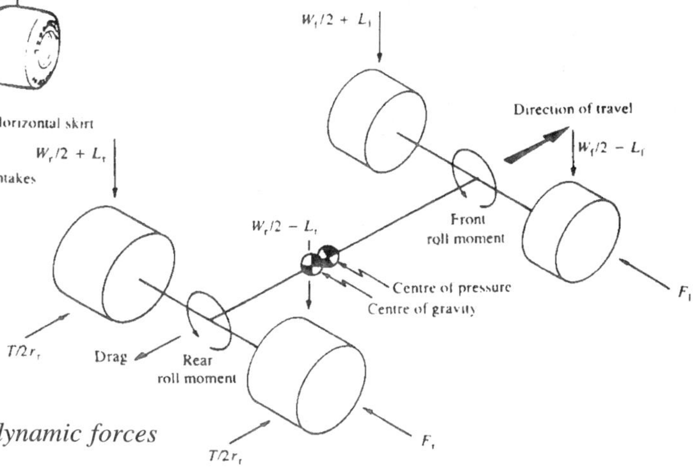

Fig 31: Aerodynamic forces

Fig 32: Spydersport chassis

Fig 33: Spydersport race-car rolling-chassis/structure

prietary transmissions. Now the company has extended into rolling sports-racer chassis assemblies, Fig 33, either of pure space-frame or semi-monocoque construction. These provide well-engineered bases for the small-series builder who need only concentrate on body design and construction.

Other small-series builders have taken the root of devising replica designs based on notable past successes such as Ford's GT40 and GT70 projects. The GTD 40 is a case in point, Fig 34a, where the best in the earlier technology is retained and updated to be fully competitive with contemporary designs. The original GT 40 car, Fig 34b, was built for Ford by JW Automotive which was an early pioneer of monocoque construction for this class of vehicle. The later GT 70, Fig 35a, was to have a simpler structure based on a perimeter rail, Fig 35b. This design was due to Len Bailey who went on to create the Panther Solo, the structure of which, Fig 35c, was a composite sandwich construction of woven glass-fibre and carbon-fibre reinforced epoxy resin skins with an aluminium-alloy honeycomb core. The car was also of interest for its four-wheel-drive configuration.

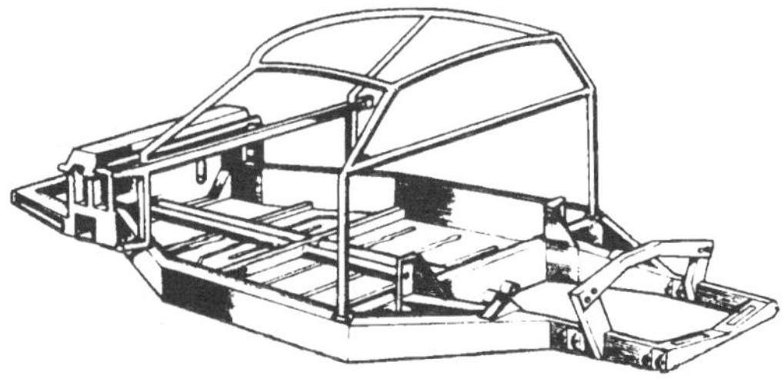

Fig 35b: GT70 structure

Fig 35c: Panther Solo structure

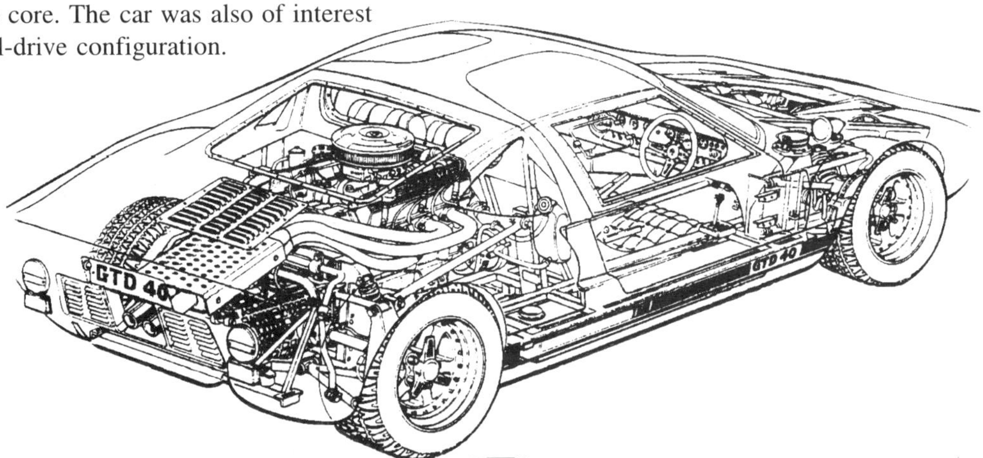

Fig 34a: GTD40

Fig 35a: Ford GT70

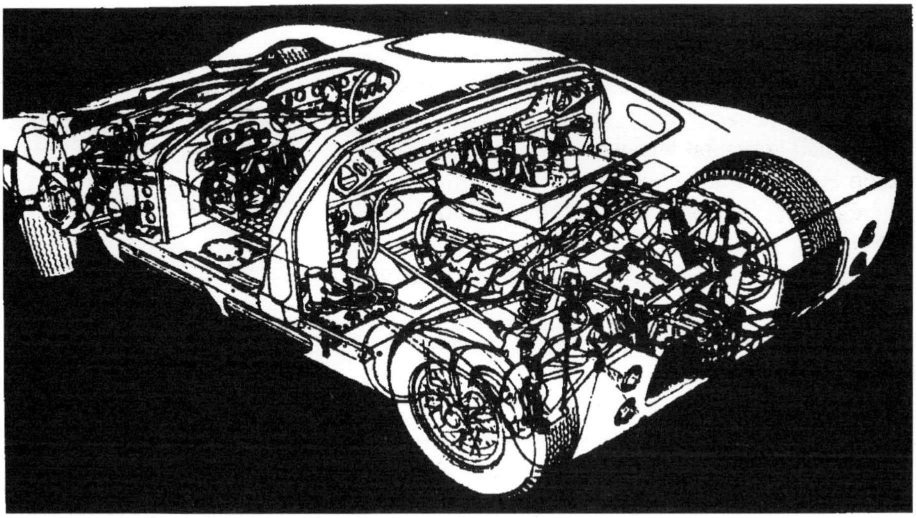

Fig 34b: Ford GT40

The special edition Mercedes-Benz 300 SL, Fig 36a, is a reborn version of the early post-war classic with some innovative technology added. Three and five litre power unit variants produce powers up to 326 bhp and a top speed limited to 156 mph. The safety shell body structure, Fig 36b, has energy absorbing front and rear sections surrounding the tubular base structure. There is a pop-up roll-over bar and magnesium alloy seat frames to give extra protection in lateral impact. Seats have 10 way electric adjustment with memory pre-set and a structure, Fig 37, which includes anchorage for the shoulder belt inertia reel. It was designed with the aid of the material supplier, Norsk Hydro. Possibility to adjust the seat-belt harness to almost any occupant size is an important contributor to safety in an open car which otherwise has to rely on safety-reel anchorage by the half-height c-pillar. Since the crash loads on the seat from the safety harness are very considerable the integrity of the structure and its slide mechanism is crucial and required a 5800 element FE analysis in the SL design.

Fig 36a: M-B 300SL

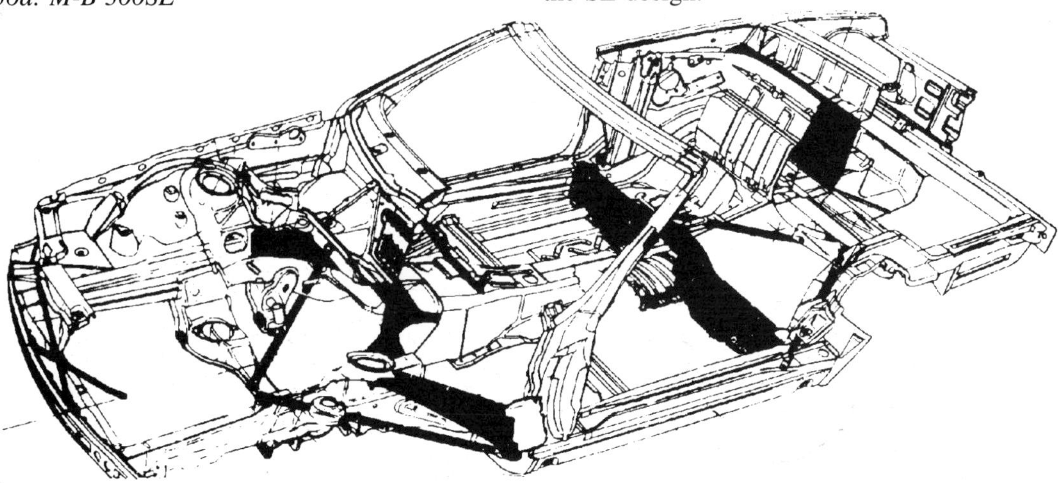

Fig 36b: 300 SL structure

A revamped version of the classic Porsche 911 is expected by the makers to revise the popularity of rear-engine/rear-drive in this exceptionally well balanced design. Latest version of the 911 Carrera, Fig 38, has a cross-flow water-cooled boxer engine replacing the long-running air-cooled unit but still in the familiar rear-end position. Swept volume is down 0.2 litre to 3.4 litres but output is up substantially to 300 bhp and fuel consumption down by 10%. The six cylinder unit has 96 mm stroke and bore reduced down to 78 mm, the resulting overall package is 70 mm shorter and 120 mm lower than the previous unit. Water cooling was forced by stricter exhaust-pollutant and noise emission regulations. Cylinder block is pressure-cast aluminium alloy, split along the crankshaft centre-line and with separate bearing bridge; dry-sump lubrication is retained.

The Vario cam system switches to a torque-enhancing timing mode at 1500 rpm, then at medium speeds keeps valve overlap to a minimum; at 5820 rpm the system switches to a high-overlap condition. Forged rather than cast pistons are employed and the company claim a world-first in featuring a mechanically-split forged connecting rod, rather than the lower stress tolerant sintered one of its rivals. The engine is cooled by two radiators, located on each side of the car's front valance and for air-intake, each bank of cylinders has an individual air-collector, doubling as a resonance generator comprising two chambers linked by a pipe switched by a butterfly valve open until 2700 rpm and staying closed until 5100 rpm and thereafter until opening again finally at maximum engine speed, to optimize engine air charge. Sequential fuel-injection is employed and the single ignition-coil for each cylinder is wound around the ignition lead.

The rear suspension arrangement involves a sophisticated subframe structure, Fig 39, which is elastically mounted to the body. The LSA lightweight strut axle comprises a five-link wheel location system, the links being arranged in two planes, two forming the upper 'wishbone' and three, including a tie-rod forming the lower one. As the wishbones are effectively each a pair of tension/compression members, these are largely relieved of twisting or bending moments. Combined with its brakes and drive-shafts the whole is a pre-assembled module, Fig 40. Each suspension strut is pivot-mounted at either end, to wheel-carrier and sub-frame respectively. The suspension 'struts' are die-cast by the Acural process and are of C-section, with webs reinforced by diagonal ribs. These are in aluminium alloy AlSi7Mg with 240 M/mm yield strength. The sub-frame members are of similar construction but use an H- rather than C-section. Total weight of the assembly is 17 kg.

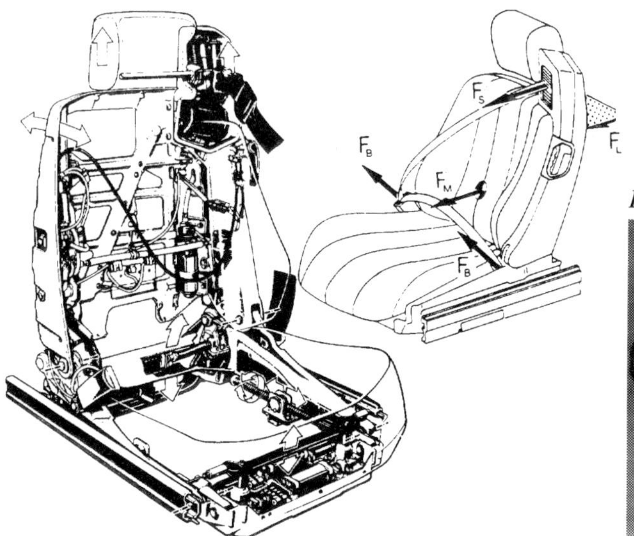

Fig 37: 300 SL seat structure

Fig 38: Porshe 911 Carrera

Fig 39: Rear suspension structure

Nissan's AQ-X, Fig 41, was developed as a next-generation vehicle that would both reduce fuel consumption, and associated emissions while enlarging the package size to fall between Primera and President. The low flat floor-pan has been designed to adapt for the carriage of batteries in the event of a future electric drive version being offered and the high waist body gives a remarkable drag co-efficient of 0.28. Body structure is half steel and half aluminium, a newly developed clad material being used as a sandwich between welded joints. The hydraulic engine mounting involves an electro-magnetic actuator which actively absorbs vibrations. This is said to give the 4four-cylinder engine the feel of a six-cylinder one. The concept version of the car, floated in advance of its launch, used an interesting automatic inching system designed to keep the car at a fixed distance from the vehicle ahead in dense traffic conditions. The system acts on engine and brakes to maintain a speed, below 15 km/h, and distance from the car ahead to relieve the driver of throttle and brake pedal application in slow moving traffic streams.

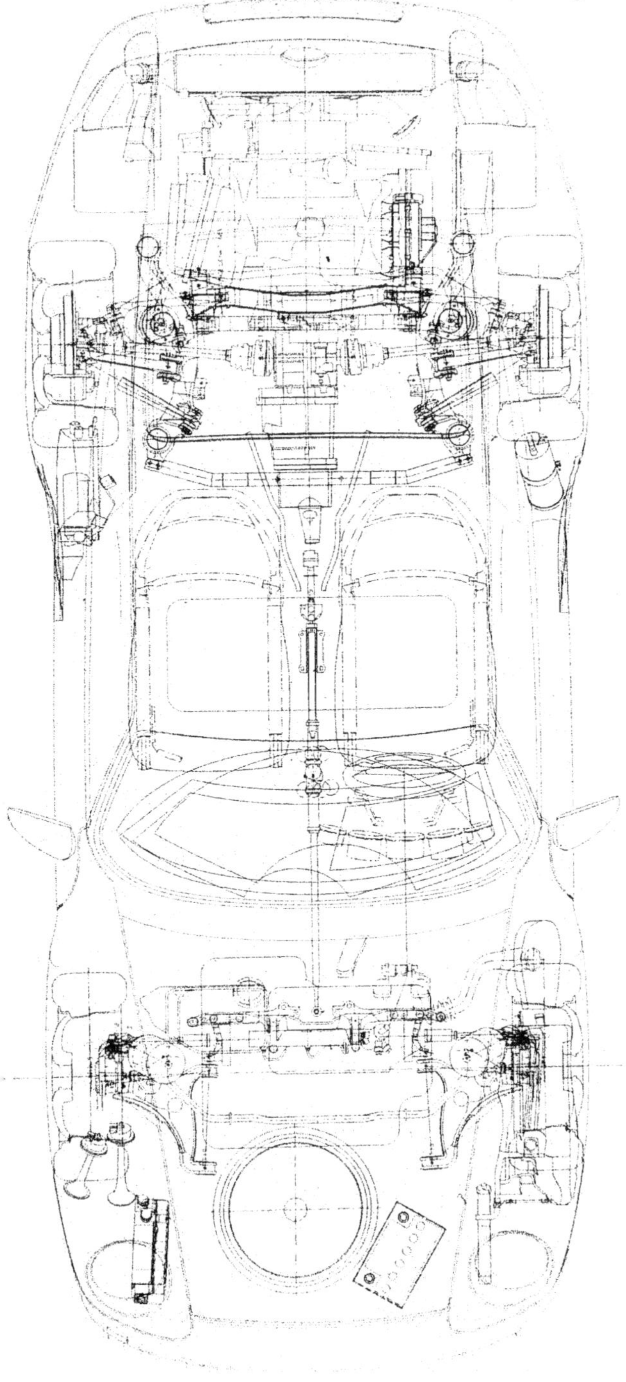

Fig 40: Porsche 911 mechanical package

Fig 41: Nissan AQ-X

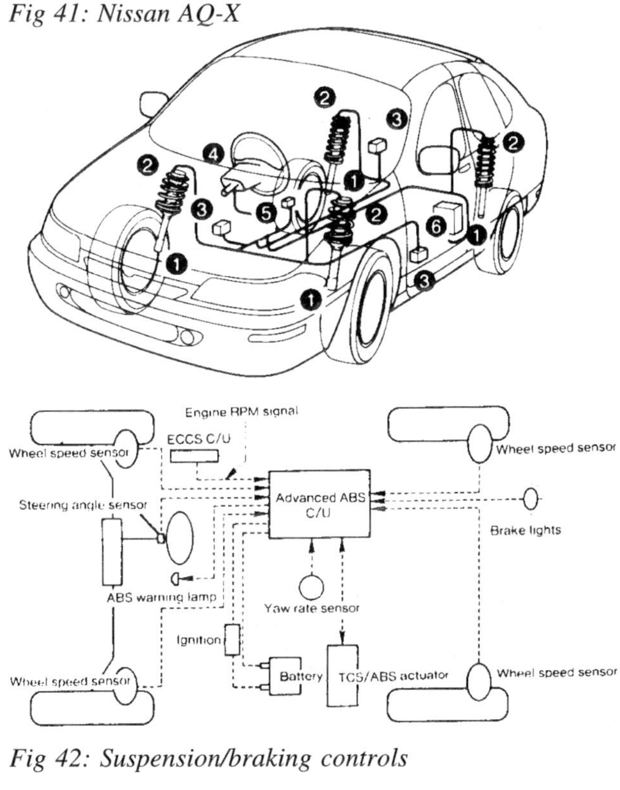

Fig 42: Suspension/braking controls

Rear suspension involves the company's interesting multi-link beam layout with an active damping system superimposed, Fig 42. The beam design, Fig 43, will be applied to all the company's front-drive models. Its objective is to provide camber control characteristics which always keep the tyres perpendicular to the ground while ensuring proper steer alignment. Low friction and low inertia for vertical wheel motion was another requirement. A control-rod and lateral link are mounted on to the torsion-beam axle which is configured in such a way as to virtually eliminate scuff changes which cause uncomfortable lateral body motions. At the same time, difference in roll characteristics between left and right hand bends is minimised. The active damping control uses the 'sky-hook' damper control principle which damps body motion regardless of wheel inputs. Constant ride height is maintained, with accelerometers sensing vertical wheel travel. Signals from them are modulated by road speed and steer-angle inputs to change damper settings on a continuous basis.

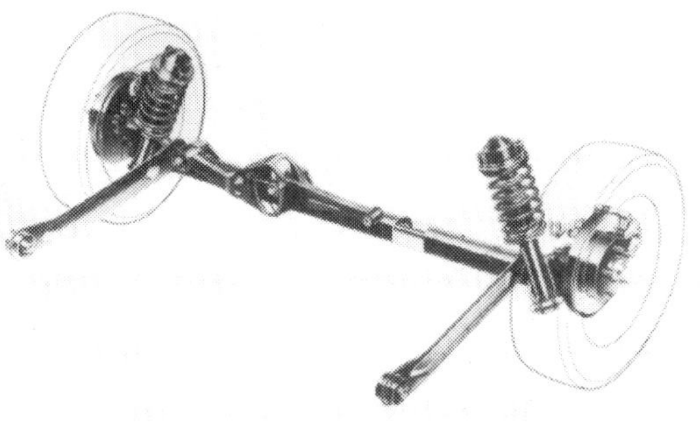

Fig 43: Twist-beam rear suspension

Fig 44: F1 elevation

Designing a supercar

The ultimate automobile has been the challenge of specialist car designers over the years and many see the McLaren F1, Fig 44, as fulfilling that dream. The A-Z stages in that fulfilment are discussed here alongside the latest development stages in a rival contender, the Lamborgini Diablo and that of two supercars in the luxury class, the Mercedes S-class and Bentley Continental R.

One of the early sketches for the McLaren F1, Fig 45a, showed the first steps towards packaging evolving around the idea of a central driving position. Another sketch set out the basic requirements for weight distribution and low polar moment of inertia, Fig 45b. Weight target was 1000 kg and width had to be kept within 1.8 metres. A stable centre of pressure was deemed another essential for avoiding high speed instability problems. The vehicle was pack-

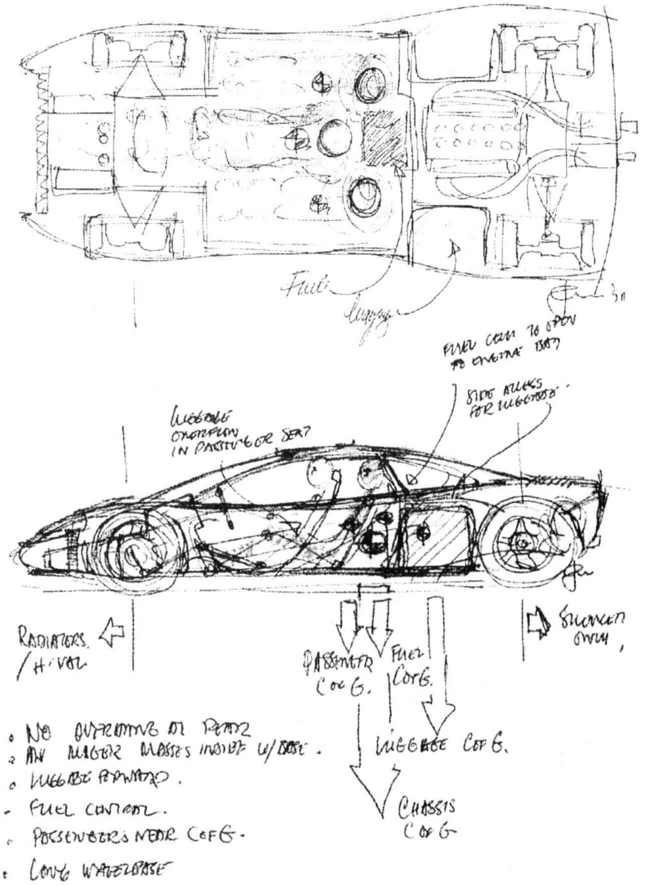

Fig 45b: Designer's concepts sketches for F1

aged around a five litre V-12 engine tuned to output some 100 bhp/litre — about half that of a racing engine — and fitted with a transverse gearbox to keep the main masses within the wheelbase. BMW Motorsport suggested going to six litres to guarantee the 550 bhp target but McLaren insisted it had to be no more than 600 mm in length. An offset final drive gear alongside the flywheel made the powertrain particularly compact.

Aesthetic design of the car was undertaken by Peter Stevens and interior packaging by Barry Lett who used a seating buck to establish the body height of 1140 mm. No styling sketches were undertaken until the package had been finalised so that no unfortunate compromises could result. At the stage when cabin area and windscreen rake were known, as well as height and width, wind tunnel of model variants commenced. Key parameters were rear glass slope, tail height, nose shape and wing design. A modest 0.35 Cd was required with the centre of pressure almost coincident with the centre-of-gravity; 160 lb of downforce at 150 mph was needed and very little change to the centre of downforce as pitch or attitude changed. Pop-up headlights were quickly discarded as causing high increase in downforce ahead of the lights as the centre of pressure would be moved ahead of the car at high speeds.

Main constraints on the style were an air-intake on the roof, for clean air and easy water separation, committing the designers to a spine frame. Twin radiators required a high nose line. While the door arrangement, scissor style, pioneered by the Countach/ Diablo was seriously considered, this was not possible for reasons of access and egress, and something akin to that used by the Porsche 962 was adopted.

Suspension design was helped by McLaren's insistence that weight distribution (42/58, front/rear) would change by less than one per cent from a full to empty fuel load. Longitudinal weight transfer in cornering was to be minimised, too, by equalising the distances between suspension roll centre and body mass centroid at front and rear. Since roll centres had to be low, to avoid jacking effects, the engine too had to be as low as possible in the body, a situation helped by dry-sump lubrication.

Progressive rate springing was avoided on the grounds that in excess it can suddenly increase weight transfer when a wheel hits a mid-corner bump, making handling unpredictable. Adaptive damping and ride height control were also avoided, this time on grounds of weight. Finally wheel travel front and rear was set at 3.5 inches bump and 3.1 rebound while target unladen bounce frequencies were 1.43 Hz front and 1.80 rear. Such relatively soft springing had to be carefully matched to available downforce since high speed handling would otherwise have been dangerously unpredictable.

The double-wishbone suspension had longitudinal compliance built in to allow rearward wheel movement, on striking an obstacle, to improve ride quality. Compliance varied between front and rear wheel stations; at the front, priority was to prevent castor wind-off under braking (limited on the car to about one degree per one *g* of braking deceleration), which otherwise compromises stability. The company adopted the technique of ground-plane shear centre in which the wishbones were mounted on subframes having suspension bushes arranged to constrain arcuate motion so that it centred on the tyre ground contact patch.

At the rear, the inclined shear axis technique was adopted which accounts for traction reactive forces acting through the wheel hub. Engine and suspension mounts were designed as an integrated system aimed at minimising toe-in or toe-out under traction and braking. These were limited to 0.08 and 0.04 degrees per g of acceleration/deceleration respectively. Front castor angle of 4.6 and king-pin inclination of eight degrees were accompanied by a larger than usual wheel offset of 25 mm.

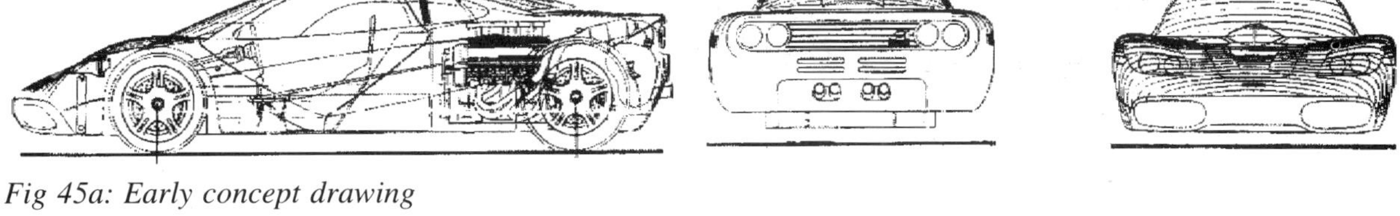

Fig 45a: Early concept drawing

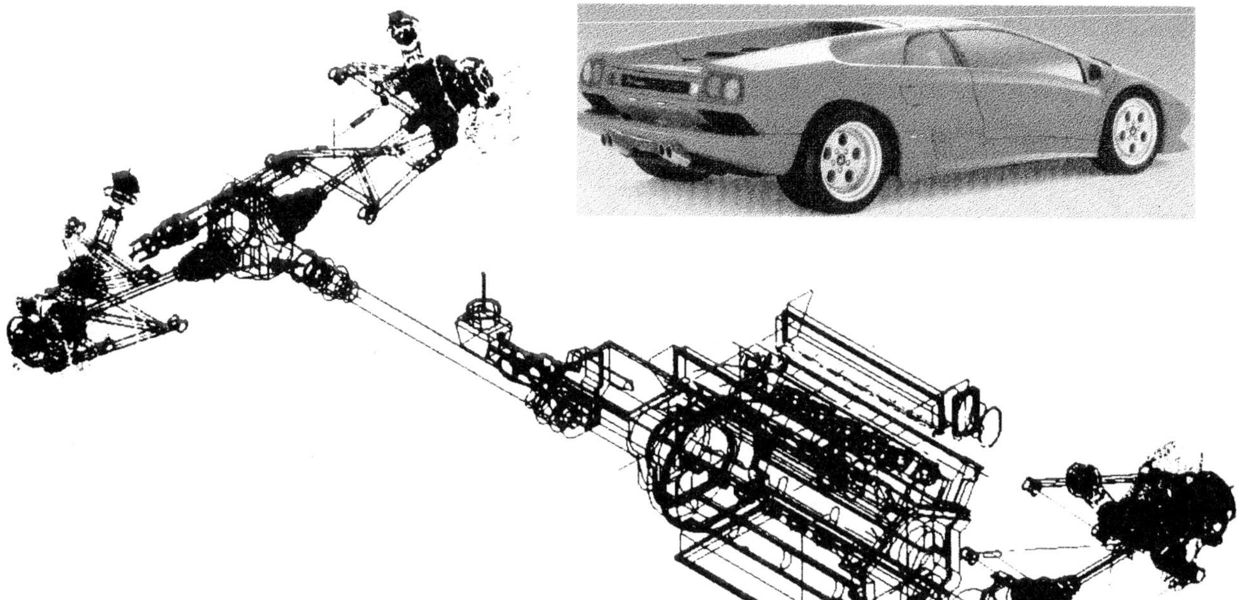

Evolution of a super-roadster

The beginnings of the Lamborghini Diablo date back to 1971 when the company's Countach, conceived by Bertone, was launched. Marcello Gandini was to interpret Bertone's concepts into a classic Italian supercar design. The Diablo was the heir to this classic, to be unveiled in June 1986 shortly before the Chrysler acquisition of the company. Subsequent changes resulted in the production version with power of 492 bhp, top speed of 325 km/h and 0-100 km/h acceleration in 4.1 seconds, Fig 46. 1991 saw the announcement of the Roadster version whose evolution from the basic car was a triumph of CADCAM.

Fig 46: Diablo with its mechanical package

The Roadster's evolvement owed much to Control Data's Integrated Computer Aided Engineering (ICEM) system since the essential surface information of the base Diablo was already held on this system. The package was able to create surface design faster and more accurately than had previously been possible. With the associated SURF software basic surface data from concept renderings could be made to produce a computer-design suitable for producing machinable surfaces in just one step, without the time-consuming iterative steps of earlier systems. Surface smoothness on generated curves is said to be within 0.25 mm of measured inputs from the styling physical model. Once the data is organised by the package, surfaces can be directly manipulated by control-points, lifting-points and boundary conditions, Fig 47.

The closed version of Diablo has a skeleton and roof of steel, front and rear ends in aluminium alloy and closure panels in carbon-fibre composites. For the Roadster a significant number of these panels were changed. The drag co-efficient of the base car was a relatively high at 0.31 but the design is purposely arranged to exert maximum downforce. The space frame chassis is made of tubes with cross-shaped sections, welded to a centre tunnel, to which the front and rear wishbone suspensions are mounted.

The car's V-12 engine has a separate electronic controller for each bank of cylinders. Combustion chambers are trapezoidal with narrow valve angle and central spark plug; piston crowns are lenticular with squish lips.

Camshaft grinding is carried out on a CNC machine with which cam profiles can be computer-generated by the same software that generates lift and acceleration diagrams for the cams.

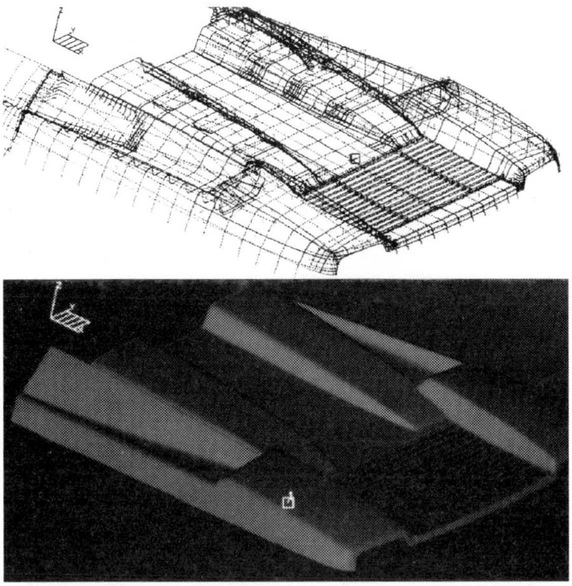

Fig 47: Stages in surface development of tail panel

Fig 48: Mercedes S-class

Fig 49: Bentley Continental R

Fig 50: Mercedes V12 of over 400 bhp

Luxury closed-coupes compared

The latest Mercedes Benz S-class coupe variant, the S420, completes a trio started by the S500 and S600 derivatives of the main-line saloons, Fig 48. The latter are smaller than equivalent Rolls-Royce models but the V-12 engine introduced in 1991 develops 400 bhp at 5200 rev/min from six litres making it one of the world's most powerful passenger car engines, Fig 50. This uses variable intake camshaft timing and Bosch LH Jetronic fuel injection. As in BMW's V-12 the two cylinder banks are independently managed with each cylinder having its own program. The S models use a front subframe system, for engine and suspension, and the front suspension is unusual in the aluminium alloy upper arms pivoting around an oblique axis such that those of the left and right hand arms converge towards the front. The five-link rear suspension is effectively a double wishbone system but incorporates an extra link for toe-in/out correction. This is also subframe-mounted in common with the final drive.The Bentley Continental R, Fig 49, is the first uniquely styled version of the marque since 1952 — with distinctive coupe body. Integral bumpers, air-dam and rear skirt emphasise the low side profile of the vehicle which now has reduced frontal area and flush glass to aid aerodynamics. The interior transformation is marked by a through centre console which extends through to the rear seats which are individual contoured units. Powered by the Turbo R engine coupled to a 4-speed auto-transmission, top-speed is now 145 mph and 0-60 mph acceleration a creditable 6.6 seconds for this 2.42 tonne kerb weight vehicle. There is a two-piece prop-shaft and final-drive ratio is 3.07:1; wheels too are new and sized at 16 inch diameter and 7.5 inch rim width. The 6.75 litre 90 degree V-8 has a compression ratio of just 8:1 and is fuelled by the Bosch K-Motronic system. Mid-range power of the engine has been increased by 20 per cent despite reductions in fuel consumption and exhaust emissions. Larger inlet ports and exhaust valves have been introduced to improve volumetric and combustion efficiency, to effect the 18 per cent fuel consumption reduction, as well as multi-point air injection and transient boost control for the turbo-charger. Mechanical layout of R-R models dates back to the company's switch to integral construction in 1965 commencing with the Shadow model and is particularly notable for very widely spaced anchorages for the front suspension wishbones, Fig 51a/b.

Specialist supercar The Alpine A610, Fig 52, from the Renault associate company is a stylish new supercar powered by a 3-litre V6 aluminium alloy engine developing 250 bhp. Top speed of 165 mph is recorded for this successor to the GTA V6 Turbo which has had 80 % of its parts re-worked. The rounded nose now incorporates retractable head-lamps and new air intakes ahead of the rear wheels distinguish the side view. Variable rate power steering and a more even weight distribution are further attributes. Drag factor has been lowered from 0.33 to 0.30 and increased torsional strength is provided by a Heuliez-designed central backbone chassis. The forward chassis is now constructed from steel pressings rather than square-section tubing as is the floorpan which was a GRP moulding. The main body shell is in laminated polyester made by the RTM process while exposed panels, front and rear, are of impact-absorbing PUR. Front wings, by contrast, are of an un-named thermoplastic co-polymer.

MPV-Limo The Italdesign Columbus, Fig 53, is a research study for a seven-seater MPV aimed to displace the US-style limousine in the long term. Its somewhat stange appearance belies the fact that much greater interior comfort is provided than by a stretched-limo. It is just over 2 metres high and just below 6 metres in length, with a track of about 1.9 metres. Driver and two front passengers share the 'cockpit' with the main elements of the vehicle drivetrain. The four rear passengers are on a lower-level deck in a more isolated situation. The transverse mid-engine position is quite novel but admirably suits an all-wheel drive-line.

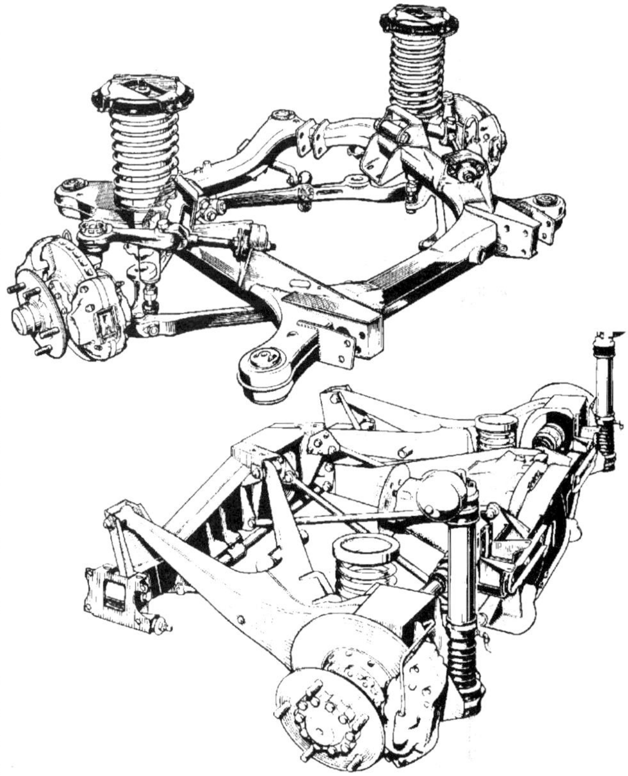

Fig 51 a: Bentley front and rear suspension

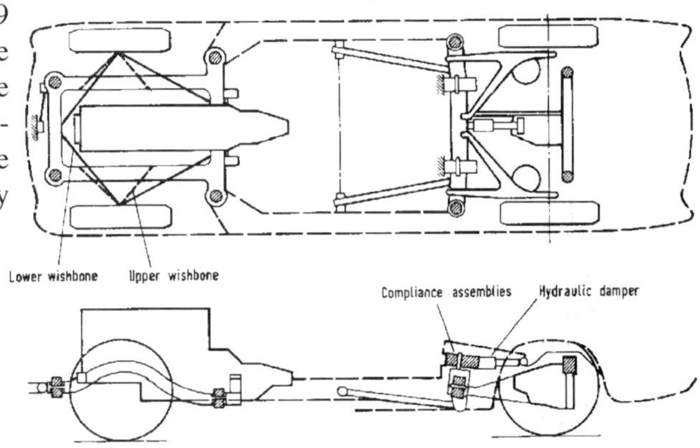

Fig 51b: Wide anchorage pivots for the R-R wishbones

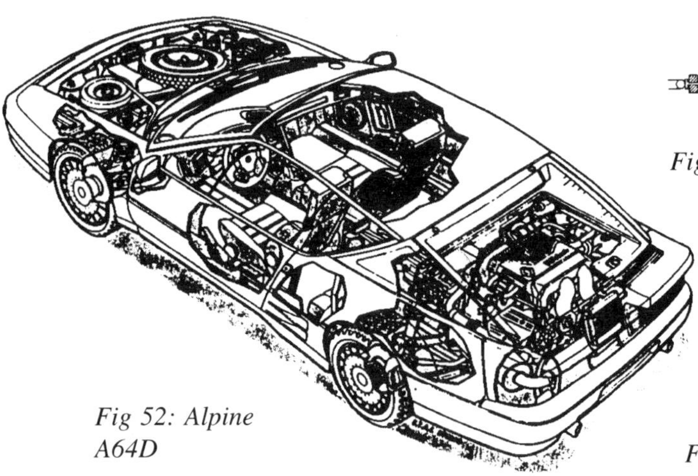

Fig 52: Alpine A64D

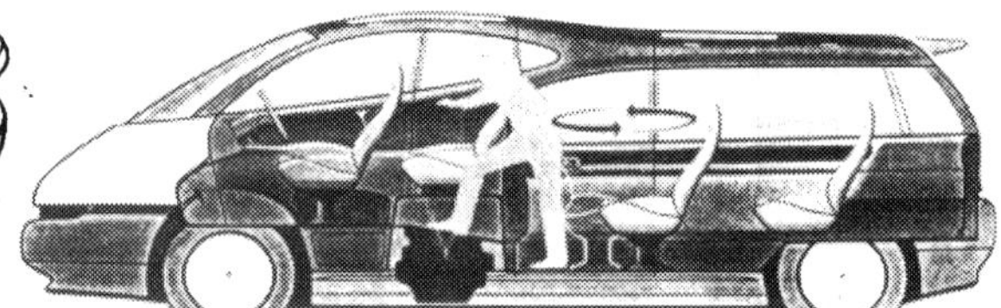

Fig 53: Italdesign Colombus

Fig 54: Audi A8

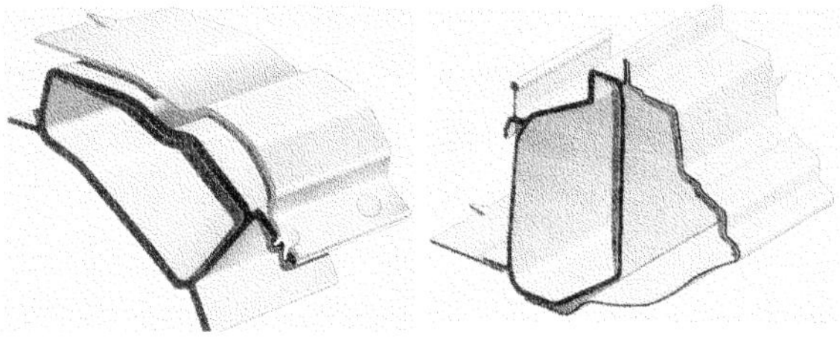

Fig 56: Cant rail and sill sections

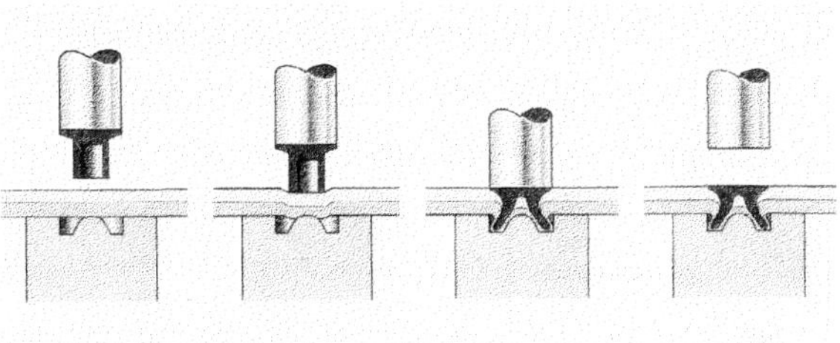

Fig 57: Punch-riveting process

Light-weight supercar Audi's aluminium-structured A8 car, Fig 54, has brought this production technique into medium-volume production. The frame structure of extruded sections, Fig 55, has sheet aluminium panels integrated with it to share in load-bearing. The frame elements are connected together by vacuum-pressure die-castings at the nodal points, to create a lattice-tube space frame.

The elements are mostly of closed sections, Fig 56, and are pressed into their final lengthwise curvatures prior to insertion into the node castings. The latter are highly ribbed for maximum stiffness; they are cast in a way which allows air to be evacuated from the dies prior to injection of the molten aluminium. Skin sheets are flush-joined to the space lattice using the Henrob punch-riveting system, Fig 57. Each completed body-shell has to be heat-treated for 30 minutes at 210 C to develop its full strength, a factor which will probably prevent the method into full volume production.

The resulting body shell is 40 % lighter and yet 40 % stiffer than a steel equivalent, according to Audi. While the current vehicle gives buyers the choice of 4.2 litre V8 or 2.8 litre V6 engines it is rumoured that the company might make the car their flagship model and fit it with their new 4.8 litre 3-bank W-12 gasoline engine which can develop 260 kW. Another possibility, seen in ASF prototype version is the 3.4 litre V8 direct-injected turbo-diesel.

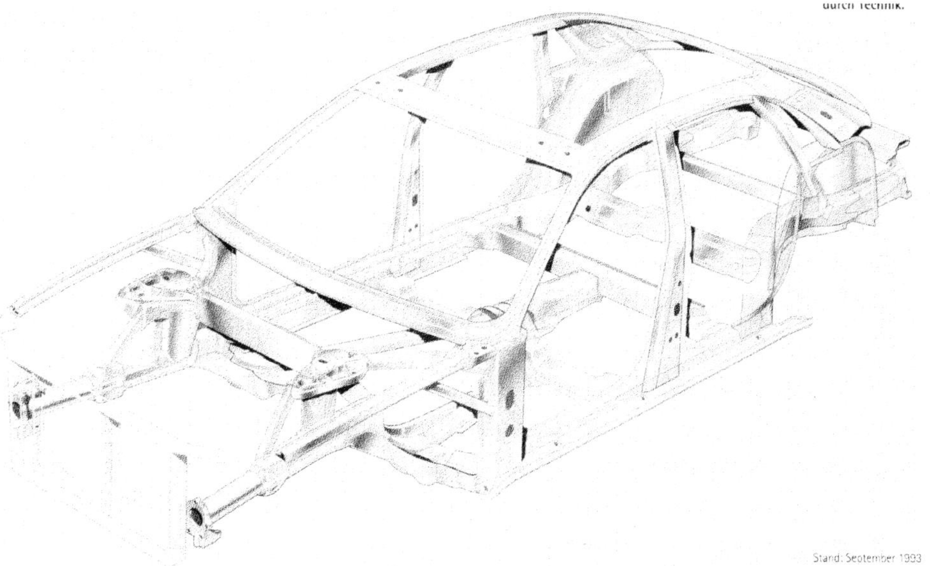

Fig 55: Aluminium alloy space-frame

System design for kit-cars

In stark contrast to the purpose-built supercar, the kit-car provides a valuable contender in that niche of the market which requires individuality of style at a readily affordable price. By applying a systematic approach to design, best use can be made of the systems which come from the donor vehicle(s) and the use of GRP body construction techniques can achieve high-style economical bodies.

The ever increasing popularity of GRP as a body material, which allows widespread choice of shape, has led to the materials-supplying companies providing a valuable back-up of design and construction data. New materials entering the market are also helping to achieve better engineered bodies. Combinations of woven roving and chopped-strand material are now available, for example. Such bi-directional materials reduce crimp tendency and allow the full tensile properties of the reinforcement to be realised. It also allows more consistent resin to glass ratio over the surface of the body and the moulded panel's strength as a laminate is enhanced.

Fig 58 compares bi-directional woven roving with conventional woven roving. Fig 59 compares the physical properties of different reinforcements and Fig 60 is a nomograph giving glass-weight per square metre against laminate thickness. The two examples illustrated concern a glass weight of 450 grms/metre^2 (1.5 oz/sq ft) and laminate thickness of 1.5 mm, giving glass content of 21.5 %; the second shows that, for a glass content requirement of 30 %, a 1 mm laminate corresponds to 425 grms/m^2.

Design safety factors range from 2 for static to short-term loads, 5 for repetitive loads, and 10 for impact loads. Joining mouldings together is crucial to overall performance of the overall body shell — thickening of the laminate being involved, Fig 61; sometimes mouldings can be butted together at a

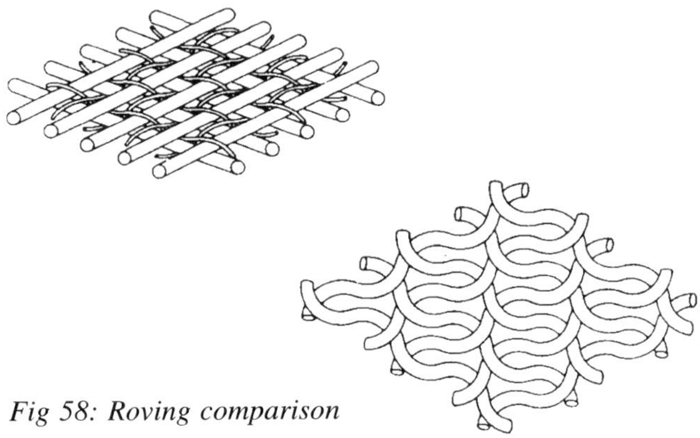

Fig 58: Roving comparison

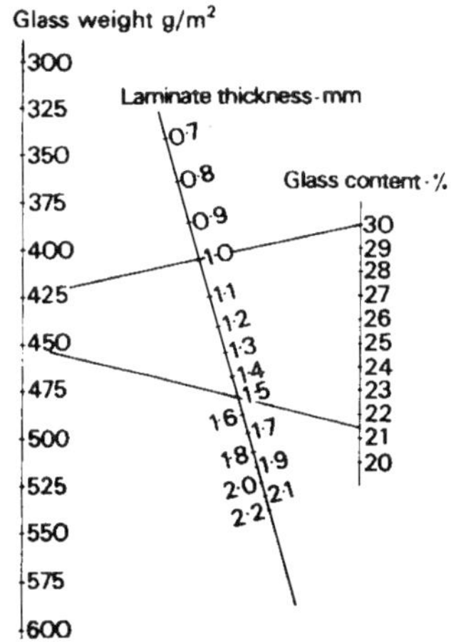

Fig 60: Nomograph for glass weight

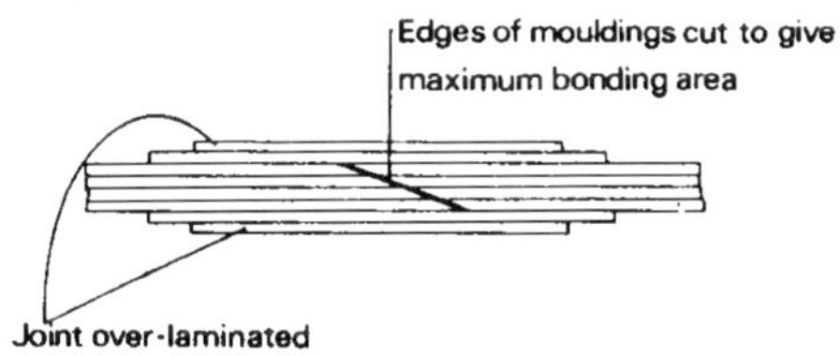

Fig 61: Joint design

Properties	*Unit*	*Chopped strand mat*	*Woven rovings*	*Satin weave cloth*	*Continuous rovings*
Glass content	% weight	30	45	55	70
	% volume	18	29	38	54
Specific gravity		1.4	1.6	1.7	1.9
Tensile strength	MPa	100	250	300	800
Tensile modulus	GPa	8	15	15	40
Compressive strength	MPa	150	150	250	350
Bend strength	MPa	150	250	400	1000
Modulus in bend	GPa	7	15	15	40
Impact strength, Izod, unnotched	kJ/m^2	75	125	150	250
Coefficient of linear expansion	$x10^{-6}/°C$	30	15	12	10
Thermal conductivity	W/m K	0.20	0.24	0.28	0.29

Fig 59: Reinforcement properties

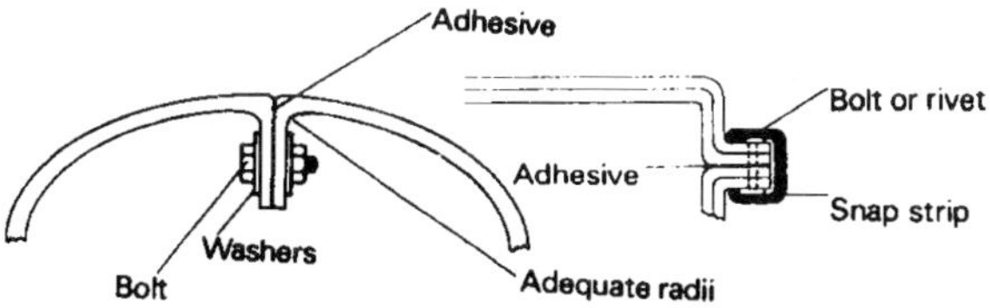

Fig 62: Bolting on a style line

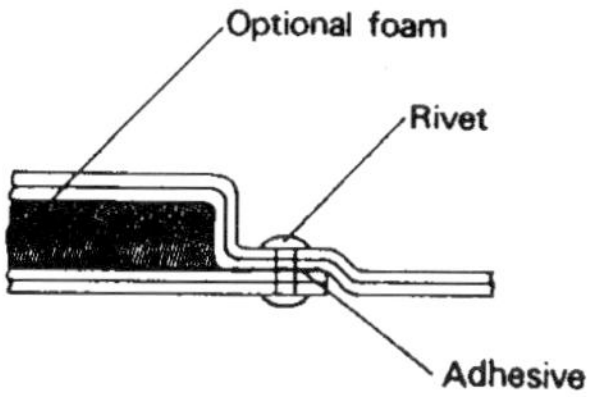

Fig 63: Foamed core floor joint

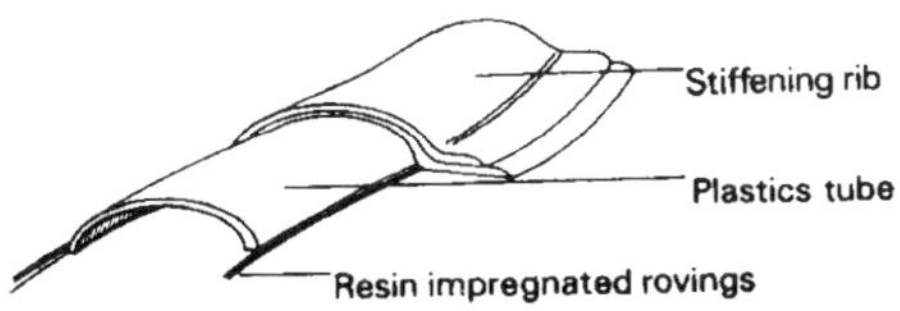

Fig 64: Stiffening ribs

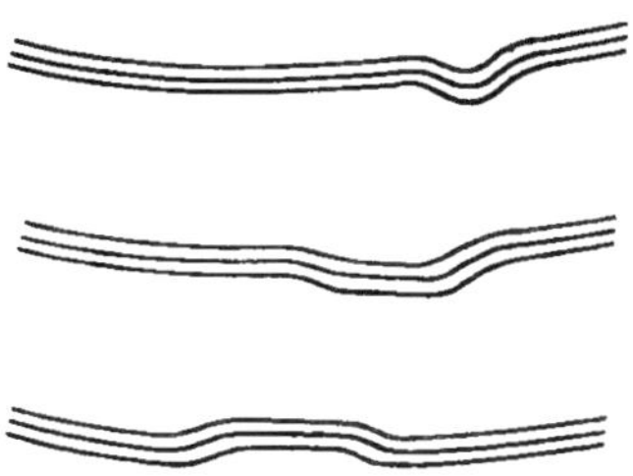
Fig 65: Rigidity from style lines

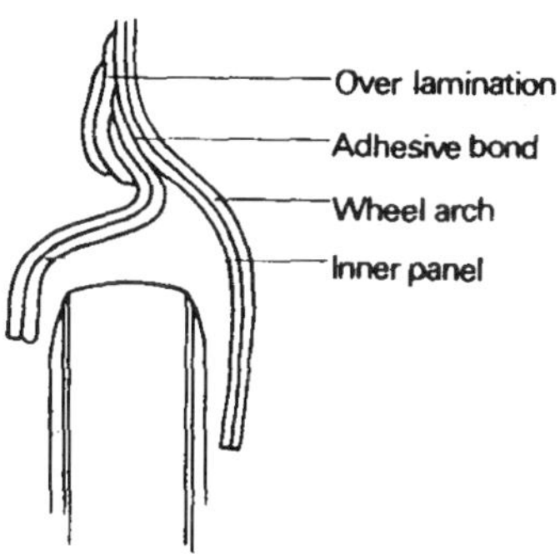

Fig 66: Double-skinned wheel-arch

style line of the vehicle, Fig 62, where they can have returned edges and be joined by adhesives or bolts. When mouldings join at right angles, a snap-on channel can be used to cover the join. In floor sections requiring high rigidity, double skinning may be involved, with a foam core, in which case the joint of Fig 63 is suitable.

When moulding-in stiffening ribs, Fig 64, the latter should be bedded down against the moulding, in the mould just prior to gelling. The over-laminating operation should ensure there is adequate area of the laminate in contact with the panel at each side of the stiffener. General technique in three-dimensional design is to build in maximum contour and where necessary supplement the stiffness with double-skin construction. Exterior style lines also serve to give rigidity, Fig 65. Double-skinning of wheel arches, as in Fig 66, can also be used to enhance rigidity.

Initial styling can be performed on a Plasticine model to a scale of typically 1:12 and a plaster mould cast from it — which is split to allow the undercuts to be withdrawn. After smoothing, this mould can be waxed, re-assembled and a GRP moulding laid up using surface tissue followed by layers of the lightest chopped-strand mat. After curing, the GRP model is removed and lightly sanded — then scribed-off at equal intervals using the front-axle centre-line as datum. The model is then cut into sections at each station interval and the outlines traced on to squared paper for scaling up to obtain the full-sized plywood templates, Fig 67. Fine-mesh wire netting between templates serves as an armature for subsequent plastering, to form a master pattern. When hard, the plaster surface is given several coats of shellac and a smooth coat of PVA release agent.

To produce a monocoque structure, two major moulds would be made, one for floor and bulkheads, the other for the outer shell. A scale drawing taken from the master model would then be used to construct full-size mock-ups of the floor and shell, Fig 68. A 3 x 2 in (75 x 50 mm) timber frame with conti-board fill panels has been suggested. The body-shell 'plug' uses formers over bulkheads and hardboard is used to support the clay. Female templates made from the master model are then offfered up to the clay to achieve the design shape. A mould can then be taken from it for subsequent body production. Panels may incorporate foam infills such as that shown in Fig 69 for a door-sill.

Kit-car calculations

Some simple calculations and graphical constructions help to rationalise the choice of mechanical systems. Tractive effort P in lbs = $\varepsilon mt/R$ where ε is overall mechanical efficiency, m the overall gear ratio, t the engine torque (lb-in) and R the tyre-rolling radius (inches). Thus $P = k_1 t$ where $k_1 = mt/R$. The road speed of the vehicle can be expressed as:

$$V = (N/m)(2\pi R/12)(60/5280) = k_2 N$$

where $k_2 = Rm/6723$; N is engine rpm.

An engine torque speed curve would then represent a tractive-effort x road-speed curve — if appropriately scaled — and *T-E* vs road-speed curves can readily be plotted for each gear ratio as in Fig 70. The motion resistance forces to the vehicle can be calculated using performance prediction techniques (approximated as in Fig 71) — and the gear ratios selected by conventional methods.

Curve AB is the top gear tractive effort and CD etc are those for the lower gears. JK is the total resistance curve and others for several gradients are shown. Where AB intersects JK is the point L where maximum level-road top speed is attainable. Any intercept such as MN gives net tractive force available for acceleration. PQRS is the tractive effort that would be available if the gearbox ratio could be continuously adjusted as in a CVT. Time required for a given change of speed can be obtained from:

$$T_2 - t_1 = (W/g)\int_{v2}^{v1} [1/f(v)]dv$$

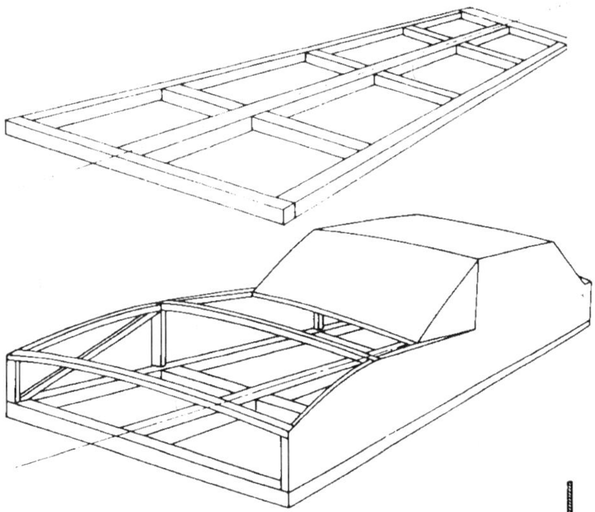

Fig 68: Floor and shell mock-ups

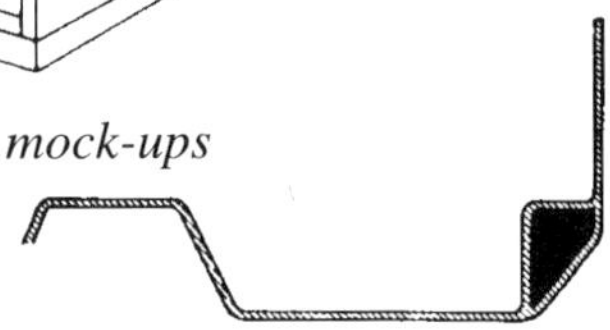

Fig 69: Foam infilling

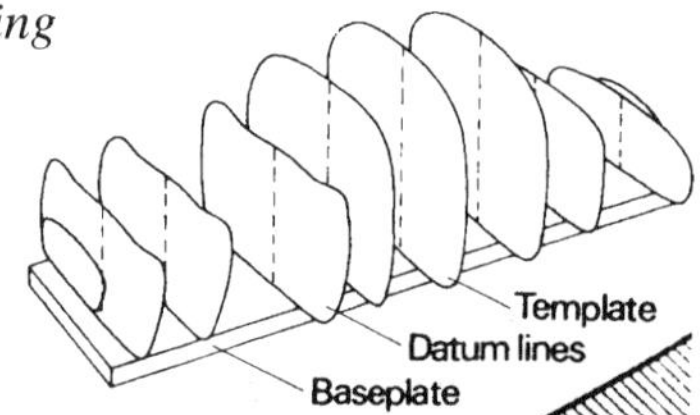

Fig 67: Body section templates

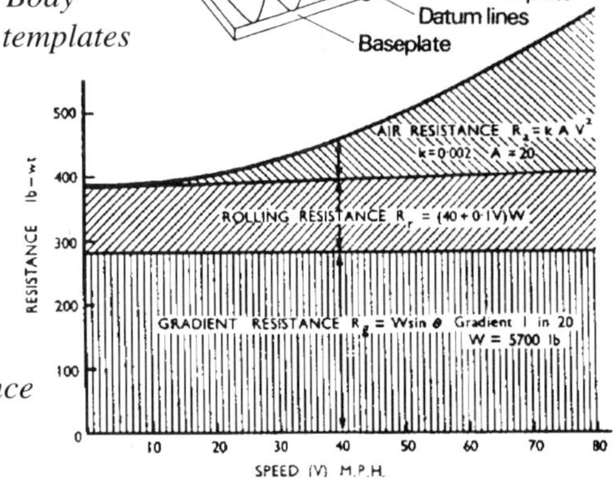

Fig 71: Motion resistance forces

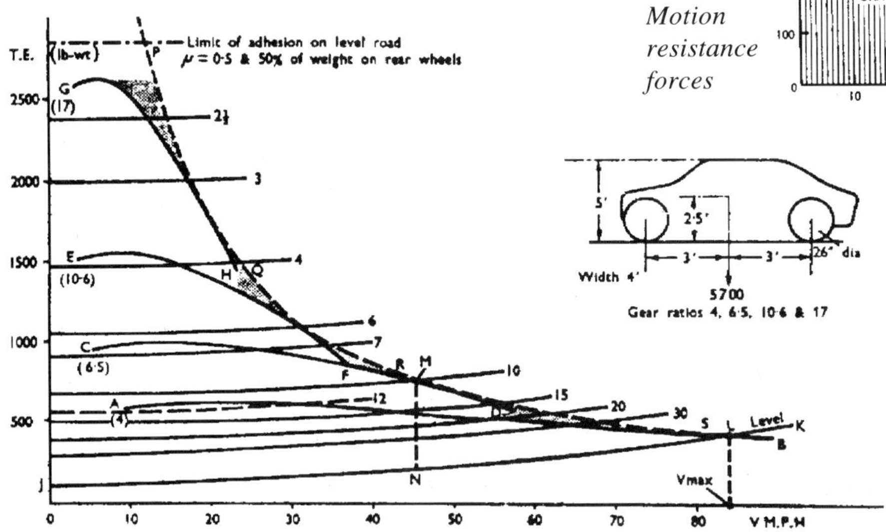

Fig 70: Tractive effort curve

where $f(v)$ is the function giving tractive effort with respect to speed. This can be obtained from a graph made by plotting the intercepts corresponding to MN against the corresponding speed as in Fig 72a, curve AB. The integration can be performed graphically by drawing the curve of $1/f(v)$ on a speed basis as in CD, the ordinate EF being made equal to 1/EH to a convenient scale. The shaded area between speed limits of interest can lead to a corresponding time interval — to provide the time/speed graph of Fig 72b with ordinate JL made equal to the corresponding area in Fig 72a. Thus:

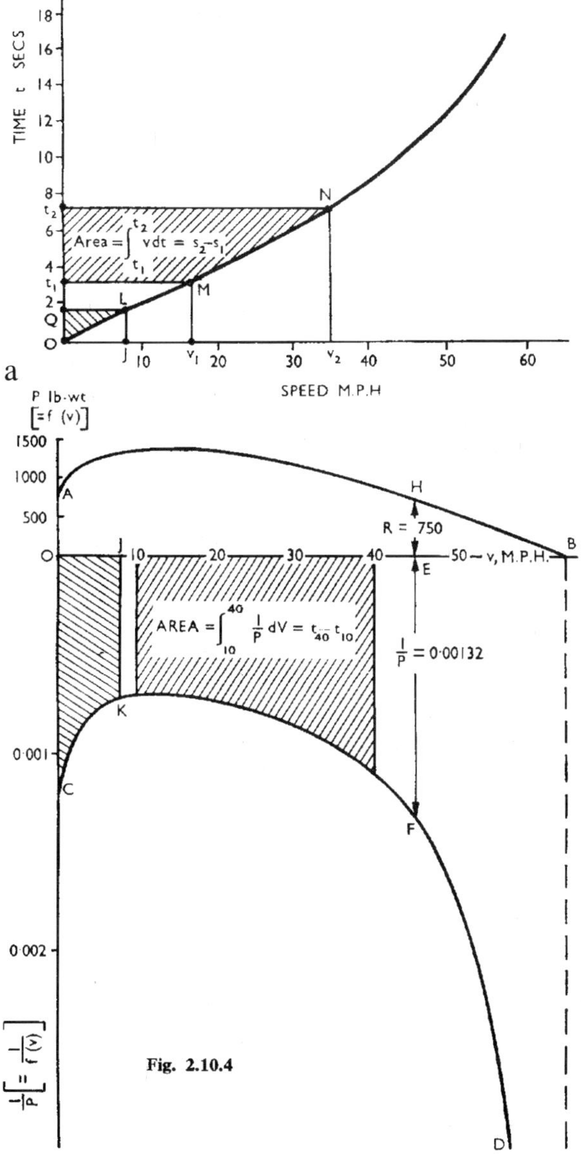

Fig 72: Speed/time curves

$$v=(ds/dt)\int_{v2}^{v1} v\,dt = \int_{v2}^{v1} ds = s_2 - s_1$$

This integral is the area between the time-speed graph and the time axis. The shaded area then gives the time taken $t_2 - t_1$ to accelerate between v_1 and v_2.

Case study in kit-car design

A feel for how the series-production specialist approaches the design of a new model can be gauged by case-study. The celebrated designer Frank Costin recently advanced the design of a single-seater commuter car whose road-performance targets were more than a match for many kit-cars. Looking for a 90 mph top speed and 0-60 mph acceleration in less than 10 seconds he chose the Reliant 850 cc engine supplied by Beans Engineering to power the vehicle shown as a concept layout in Fig 73a. That company also supply the 'matching' Reliant gearbox and drive-axle. His other targets were 50 mph fuel consumption and 0.8g cornering force. Engine, gearbox and axle weigh 128, 31 and 72 lbs (58.06, 14.06 and 32.66 kg) respectively. Price of the structure is kept down by

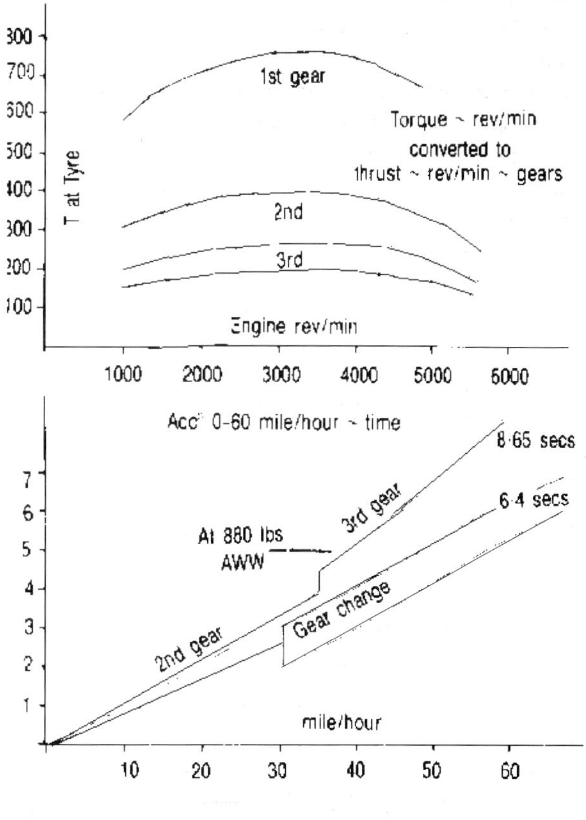

Fig 73b: Performance predictions

the use of standard rectangular steel tube of constant section throughout the frame.

Maximisation of interior space dictated the use of radius rods in fore-aft wheel location and lateral control by panhard rods. At this stage of the design an unladen weight could be reliably forecast and the graphs in Fig 70 were used in performance prediction — resulting in 0-60 mph acceleration time of 9.4 seconds with the engine tuned to 55-60 bhp. Body cost is minimised by limiting the unit to five principal parts. Low-cost pre-preg GRP sheets are used, involving a vacuum-bag lay-up comprising two prepreg laminates followed by Nomex honeycomb — finally enclosed by a single prepreg laminate. This proved to be less costly than a 3-4 oz hand lay-up but very much stiffer as a torsional shell.

Fig 74: Caterham Seven

Kit car now in series production

The Caterham Seven kit car is now to be offered in fully-built form as the K-series with fuel-injected 1.4 litre Rover K-series engine. This will coincide with the introduction to the UK of Low Volume Type Approval. Another variant is powered by the Swindon Engines prepared Vauxhall H PC unit and is claimed to be the fastest accelerating car in the world with 0-60 mph figure being achieved in 3.44 seconds. This has carbon-fibre and Kevlar body panels and seats in an ultra-light body. Aluminium alloy is used for radiator, steering box and 4-pot race-developed ventilated disc brakes. The two litre 16-valve engine from the Calibra is used and a race series is to be developed. Twin 45DCOE Weber carburettors give the engine an output of 175 bhp — in a car kerb weight of 1300 lb. Spring rates and anti-roll bar sizes have been changed to accommodate the increased power and an aluminium bell-housing has been designed in house to connect the engine to the Sierra gearbox. Track version of the car has a dry-sump system with integral oil-tank and bell-housing and a Formula One style fuel cell has been built in behind the rear axle. In Fig 74 is the 1990 version of the road car which was recently updated to include De Dion rear suspension and a differential unit based on the Sierra 4 x 4 front drive, Fig 75.

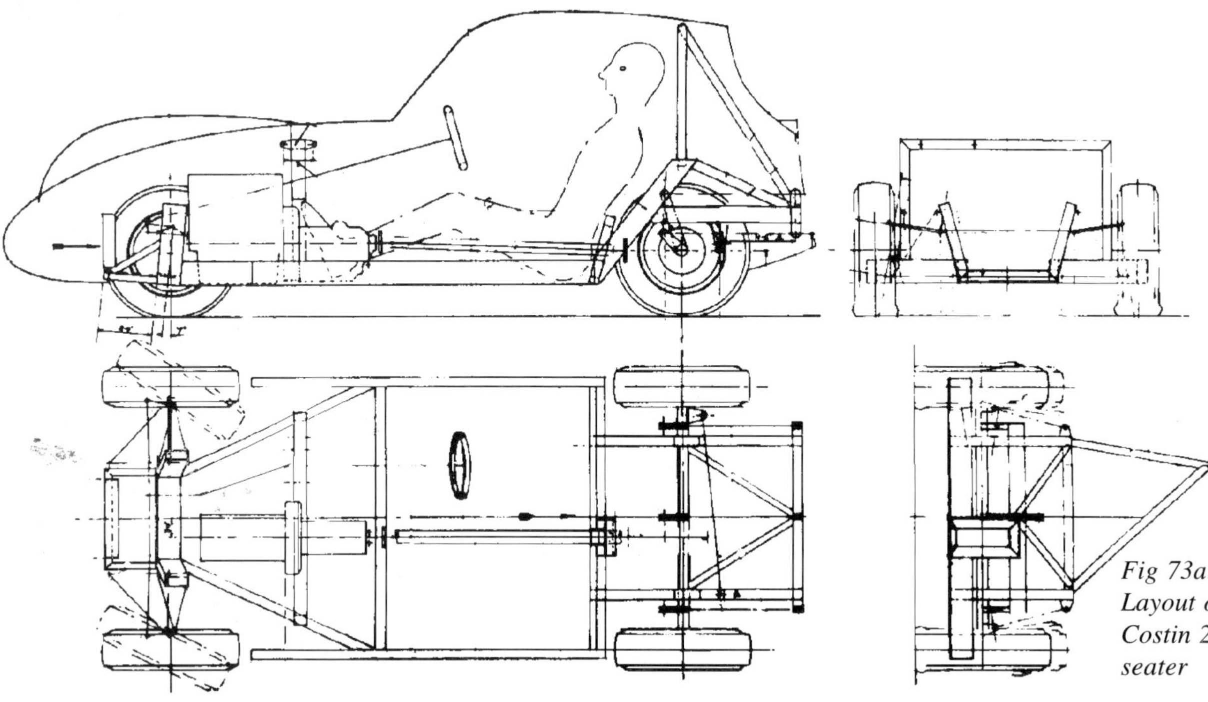

Fig 73a: Layout of Costin 2-seater

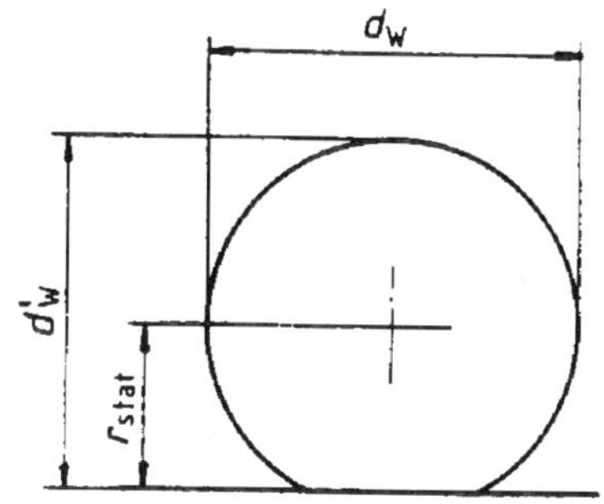

$$r_{stat} = d'_W - \frac{d_W}{2}$$

Fig 78: Laden radius in centre of gravity (CG) determination

Technique for kit-car CG determination

As much specialist vehicle design work involves adaptation of existing vehicles or body-building on to an existing rolling chassis, the need to find centre-of-gravity position of the base vehicle is paramount. BS Al 246 is a useful standard (ISO 10392) from BSI, Linford Wood Milton Keynes, MK14 6LE, which overcomes the difficulty, establishing CG height without the need for elaborate equipment such as tilt tables. The procedure is to lift one axle in steps (three or more are recommended) and record the axle load of the other axle, and the lifting angle, for each position, then repeat the procedure in the unload direction to take hysteresis into account. Mean axle loads are then plotted against the corresponding lifting angles and the procedure repeated in lifting the opposite end axle. Location of CG longitudinally,

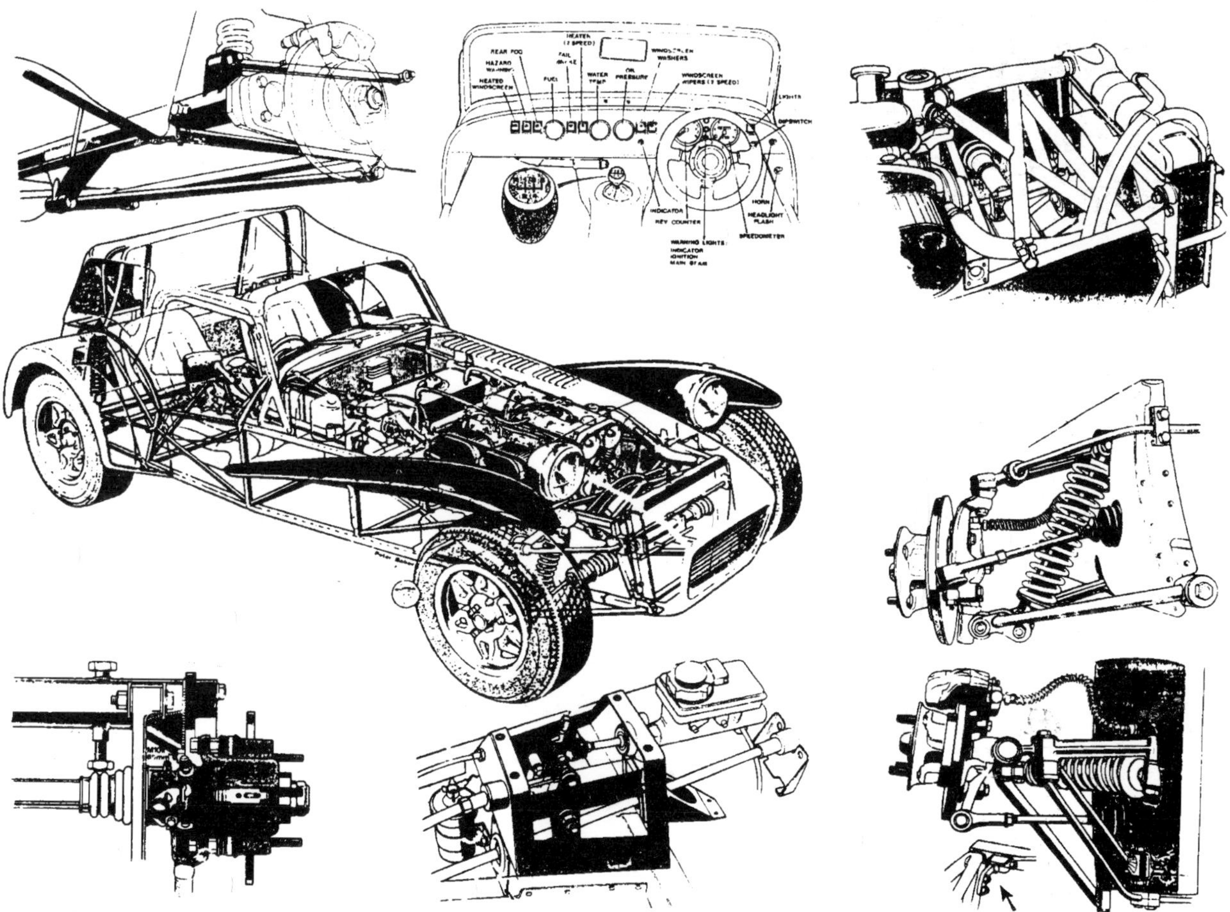

Fig 75: Key features of the Seven

laterally and above ground are obtained for the expressions in Fig 76 using the notation of Fig 77. Determination of static loaded radius of the tyres is shown in Fig 78 and the required degree of accuracy is shown as Fig 79.

$$m_r = m_3 + m_4$$

$$m_v = m_1 + m_2 + m_3 + m_4$$

$$l = 0{,}5(l_{left} + l_{right})$$

$$x_{CG} = \frac{m_r}{m_v} \times l$$

$$y_{CG} = \frac{b_f(m_1 - m_2) + b_r(m_3 - m_4)}{2m_v}$$

$$z_{CG} = \frac{l(m'_r - m_r)}{m_v \times \tan \theta} + r_{stat,r}$$

Fig 76: Calculation formulae

Absolute axle load value:		± 0.2 %
Change in axle loads due to lifting (applies to scales, which do not measure absolute loads, but changes in loads):		± 2.5 %
Dimensions:	< 2 000 mm :	± 1 mm
	> 2 000 mm :	± 0.05 %
Angles:		± 1 %

Fig 79: Required degree of accuracy

$m_f = m_1 + m_2$ = front axle load, in kilograms;

$m_r = m_3 + m_4$ = rear axle load, in kilograms;

$r_{stat,f} = 0{,}5(r_{stat,1} + r_{stat,2})$ = static loaded radius, front, in millimetres;

$r_{stat,r} = 0{,}5(r_{stat,3} + r_{stat,4})$ = static loaded radius, rear, in millimetres;

m_1, the wheel load, front left, in kilograms;

m_2, the wheel load, front right, in kilograms;

m_3, the wheel load, rear left, in kilograms;

m_4, the wheel load, rear right, in kilograms;

$r_{stat,1}$, the static loaded radius—front left, in millimetres;

$r_{stat,2}$, the static loaded radius—front right, in millimetres;

$r_{stat,3}$, the static loaded radius—rear left, in millimetres;

$r_{stat,4}$, the static loaded radius—rear right, in millimetres.

l_{left}, the wheelbase, left, in millimetres;

l_{right}, the wheelbase, right, in millimetres;

b_f, the track, front, in millimetres;

b_r, the track, rear, in millimetres;

Fig 77: Notation

Chapter 9: Truck engineering

Production diversification; market leading HGV; one truck for supply/distribution; CV body technology advances; specialist goods vehicles; heavy haulage vehicles; road tankers; pressurised tankers; truck technology: 1. structures; 2. braking in a bend; 3. cab/body; 4. HS steel; 5. aluminium alloy; 6. total system

Diversification in truck production

An important truck industry development in recent years has been the virtual disappearance of light-medium truck chassis-cab making by divisions of the large volume car producers. With it has gone the division between such vehicles and the so-called heavier "premium" trucks from the specialist builders which often had to be bought direct from the factory rather than through a distributor network. The following pages will show how the large truck makers are diversifying into a wide range of CV products and considerable commonality in design is spread over products designed for supply on the one hand and distribution on the other.

Whilst specialist vehicle builders continue to make products for particular segments it is not always recognised to what an extent the volume builders are increasing in flexibility. Here the wide variety of niche models currently under development by the world's largest CV builder, Mercedes-Benz is examined. Its recently launched Sprinter range of panel vans is even challenging similar products from the volume car makers while the Actros HGV series, reviewed in the next section, is showing signs of moving down the gross vehicle weight ranges. A number of technologies such as crash impact simulation, Fig 1, are being read across from car engineering

EV Developments A hybrid truck project based on the model 1117 chassis, Fig 2, combines a Euro 2 diesel and electric motor in its drive train, for operation in city centre 'emission free' zones. This involves a parallel hybrid configuration and uses a three-phase AC motor which is electronically controlled, the motor being permanently connected to

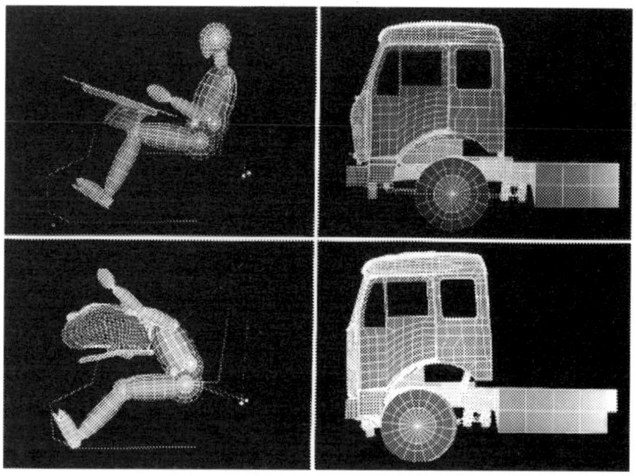

Fig 1: Simulated crash testing for HGVs is a new departure at Mercedes-Benz. Finite element models are used in both structural deformation and occupant restraint investigations. The two programs can be linked to study dynamic interaction

the six-speed gearbox, without a clutch. It is positioned behind the gearbox and inputs its torque through the PTO hence avoiding serious drivetrain modification. The 11 tonne prototype has 125 kW diesel and 46kW induction motor, delivering 171 kW in combined mode. Lead gel batteries weigh two tonnes and fit in front of the rear wheels as shown.

Fig 2: M-B 1117 chassis converted to an EV

Fig 3: Production electric panel vans

Fig 4: U90 Funmog

They give 50 km range in the urban cycle of 400 metre cycle length and maximum speed of 30 km/h. Ni-Cd batteries give the same range but that option means battery weight is down to 1.2 tonnes. The company's two electric vans (Fig 3), unlike the hybrid prototype, are already in production for municipal service. Both have 36kW motors and a maximum speed of about 70 km/h; operating range is 50-80 km. An electric retarder, with energy recovery, supplements the service brakes.

Container handlers Another city vehicle is based on the 1524 truck and has a swivelling subframe to mount twin, 2.5 x 2.5 metre, cargo containers. So that these can be unloaded selectively the subframe can be rotated hydraulically; two hydraulic struts brace the chassis against the ground during the operation which takes about 200 seconds. A further city vehicle is a central spine bottle float the body of which can be hydraulically lowered 400 mm, to the ground, for loading/unloading. A special demountable truck for handling containers for a lower than normal level is another innovation. This can apply to containers with 2.9 metre interior height requiring set-down platform of 1020 metres rather than the standard 1.320. Based on a 2538 3-axle truck with all-round air suspension, having shortened wheelbase for 7450 mm containers, the trailer coupling is mounted beneath the frame and 60 % aspect ratio tyres are fitted. Two pneumatic lifting arms raise the containers 200 mm while the air suspension provides a further 150 mm of movement. The low-mounted coupling allows long drawbar trailers to be transported allowing the carriage of both 7150 and 7450 mm containers.

Off-roaders The company's Gaggenau division has built a special version of its Unimog vehicle for the emergency services. This 155bhp vehicle has all-terrain mobility and can carry 2400 litres of water which is dischargeable at a rate of 1600 litres/min. On the U1650 model Unimog a new facility is available of interchangeable left/right drive giving the driver an opportunity to change en-route during snow-clearing or similar operations. The model also has a hydrostatic drive system operated by a foot-pedal which allows equipment working speeds to be adapted continuously. A further variant, the U90, is called Funmog and is fully equipped for the leisure market as well as being able to haul a trailed load of 24 tonnes, Fig 4.

Market leading HGV

Top consideration has been given to reliability and durability during the development programme, resulting in a class-leading truck with exceptional projected life-span and high mileage intervals between servicing

Mercedes-Benz new generation Actros range of heavy-duty trucks has been developed with the aim of minimising the life-cycle costs; increased driver comfort, Fig 5, and substantially better safety for other road users have also been key aims. The world's largest truck maker will produce Actros at a rate of 53,000 units per year by 1998. These heavy-duty trucks with permissible gross vehicle weight ratings between 16 and 32 tonnes, and gross combination weight ratings up to 40 tonnes, are fitted with the new 500 Series engines, with output ratings ranging from 230 kW (313 bhp) to 420 kW (571 bhp).

This is a modular truck family incorporating engineering from many sectors of vehicle development, including unit injection pumps, full electronic engine management, electronic braking, intelligent semi-automatic transmission and the first ever multiplexed wiring system to be used on a truck. The so-called Telligent maintenance system provides for servicing intervals between 100,000 and approx. 120,000 km, which means that a long-distance truck, for instance, has to come into the workshop for servicing only once a year. Engine, transmission and axles are designed for a service life of at least one million kilometres. The aerodynamic efficiency of the newly designed cabs and the economy of the new engines combine to reduce fuel consumption by a considerable margin.

Actros is the first truck range to be fitted with disc brakes on all wheels as standard. A seat belt with belt tensioner is integrated into the driver's seat and a driver airbag is optionally available. With its new electronically controlled compressed-air brake system, the truck or combination comes to a standstill some 25 % earlier than with conventional brakes, depending on road and traffic conditions. At the speeds customary on long-distance roads, this shortens the stopping distance by up to 20 metres compared with conventionally braked trucks. Protection for other road users is enhanced by improved and reinforced underride barriers at the sides and rear.

The automated gearshift system is based on a 16 speed (top three are overdrive ratios) synchro-

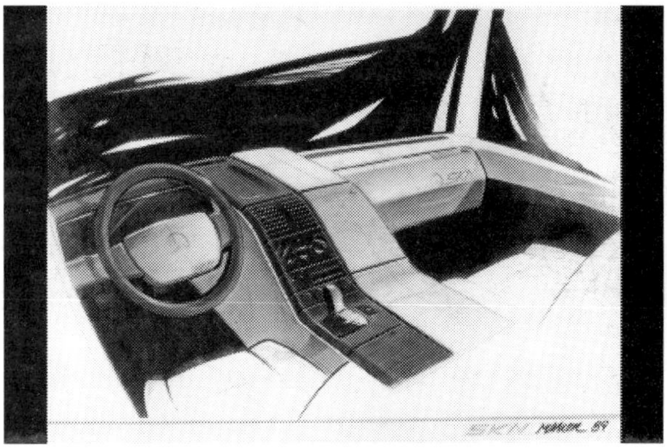

Fig 5: Stylist's renderings, above, compared with the real thing, below

mesh gearbox and a pre-select facility is incorporated. The shift lever moves only in a fore-aft direction while buttons on the side of the lever select additional functions. A fully-automated variant of the transmission can calculate the weight of the laden vehicle, together with the current road gradient, and choose the appropriate shift pattern.

Actros cabs come in three sizes: standard (S), medium (M) and long (L). For the German market, the L cab for long distance transport will have a standing height of 1.92 metres as standard, Fig 6. A completely new version, based on the L, is the 340 mm higher mounted "Mega-space long-distance cab" with a level floor and host of driver-comfort additions, Fig 7. The S cab has 140 mm greater interior length than that in the current truck with SK day cab.

Some 5 % reduction in fuel economy, even while meeting Euro 2 emissions standards, has been achieved by the adoption of unit pump injectors. Further improvements have been made by adopting a 12 degree windscreen rake angle. Actros will offer the lightest 2.5 metre wide cab on the market and some 500 kg has been pared from the chassis weight of its replacement. Adoption of high strength steel and more slender chassis frame sections has been an important contributor as has standardisation on air suspension for the drive axles. Engine and driveline weight reductions account for a total of 390 kg and a chassis cab weight of 7035 kg has been achieved.

Engines in detail The 500 Series engines, Fig 8, come in 12 litre V-6 and 16 litre V-8 forms. The unit-pump fuel system, Fig 9, can be regarded as a unit-injector minus its nozzle section and it is driven by the main engine camshaft to derive much higher injection pressures (up to 1800 bar for Actros) than typical of systems with separate pump camshafts. Injectors are conventional hydraulically opened Bosch types but fuel lines between them and the one-per-cylinder unit-pumps are reduced to a minimum length. Injection occurs over a shorter crank angle than usual so that fuel economy trade-off is reduced to cut NOx and PMs to a minimum. M-B also consider system stiffness to be greater than that available with conventional unit injectors because of pushrod/rocker clearances otherwise involved. Positioning of the unit-pumps inside the vee space, Fig 10, of the cylinder banks helps to provide noise shielding in meeting the new EU drive-by noise limits for heavy trucks which has just come down from 84 to 80 dB(A). The top of the 'vee-space' is closed by turbocharger-to-intercooler air-ducting.

The engine management system, Fig 11, which

Fig 7:The Megaspace high-cab version

Fig 8: 500-series engine cross-section

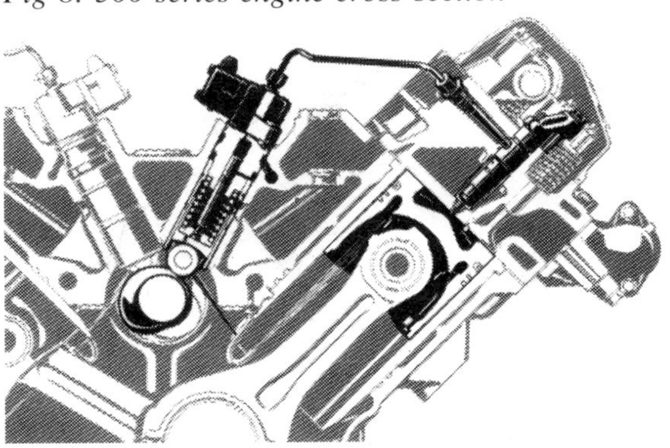

Fig 9: Unit-pump injection system

Fig 6: Service access with cab front lifted

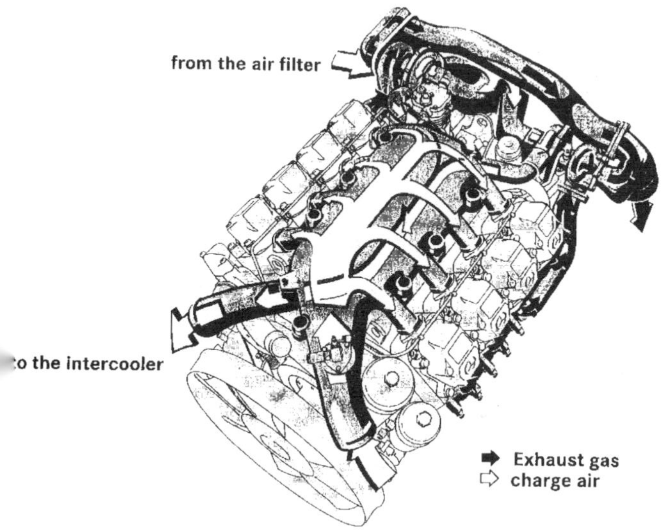

Fig 10: Exhaust gas and charge air flows

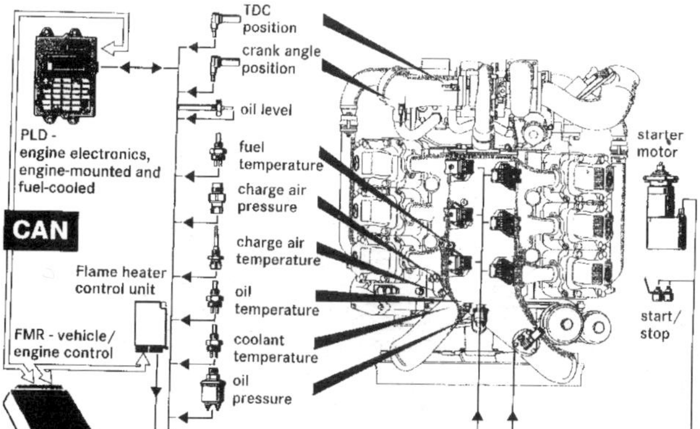

Fig 11: Engine management system

features two control units linked by a CAN interface. The first is the Vehicle Engine Control Unit responsible for vehicle functions such as registering the accelerator pedal position, actuating the engine brake, speed regulation, increasing idling speed, setting working speeds, reduced torque etc. The second is the engine control unit, its main task being precise electric actuation of the solenoid valves of the unit pumps. For this purpose, the optimal start of injection and the optimal injection quantity are computed on the basis of control maps using engine and periphery data registered by sensors (optimum torque, engine speed, atmospheric pressure, boost pressure). Further functions include error recognition, engine protection, diagnosis, connections with other electronic control systems (special functions such as the flame start system) and the immobiliser.

The crankcase was the central element in the development of the new 500 engine series. The main design parameters were a high power-to-weight ratio, with the associated high peak pressures and incorporation of the new unit pump injection system. A compact engine layout was obtained through integration of oil cooler, coolant pump coil, unit pumps and the coolant, oil and fuel lines. A central, internally located coolant line is situated in the V of the crankcase, with separate upper and lower water chambers around the cylinders. A rigid crankcase cover has low positioned threads for cylinder head attachment to ensure that bolting and engine firing forces are absorbed as far as possible without warp via the rigid collar of the wet cylinder liner.

A further improvement in the engine braking effect was achieved by means of the pneumatically operated fifth valve, the decompression valve. When the engine brake is actuated, the decompression valves open a precisely defined cross-section between the combustion chamber and the exhaust system. The intake air is compressed but at top dead centre it flows into the exhaust system and thus does not assist the downward travel of the piston. In conjunction with the engine brake valve, this achieves an increase in the engine braking power. The specific engine brake power has been increased by around 30% compared with the present 400 series engines. This was achieved by optimally matching the decompression valve cross-sections to the new swept volume of two litres per cylinder and by a new engine brake valve which is attached on just one side.

One truck for supply/distribution

The world's third largest truck maker, after Mercedes Benz and Volvo, Renault VI has launched long-distance and distribution models in one range known as Premium which replaces the Major and Manager ranges of heavy and medium weight trucks.

Contemporary styling lines, Fig 12, that blend into a soft-looking shape devoid of aggressiveness are intended to reconcile road transport with the environment. Good accessibility comes, on the Distribution models, with a floor 1,100 mm, and the first step 500 mm above the ground — a revolutionary driver-side access concept, Fig 13, features programmed steering wheel release and quick lowering driver's seat, with good side visibility thanks to the Vision door — an exclusive forward and rear vision system for street/road maintenance vehicles.

The chassis can have the new Airtronic air suspension (standard on Long Distance) and front disc brakes. A new 11-litre, 380 hp engine; two further new engines: 6.2-litre, 250-hp, and 9.8-litre, 255-hp; a new J engine brake; an electronic I.T.C. injection system anda new manual gearbox with no clutch pedal, EASY Shift, also add to the list of innovations. Low weight is the range's key parameter; for example, the Long Distance Privilege 385-19 t 4x2 tractor is the lightest tractor in its category, with a 6,370 kg GVW. It offers 600 kg more payload than the Major, and up to 900 kg more than similar vehicles available in the market, say Renault VI.

The Long Distance cab features four-point air suspension, a high driving position with three access steps, a floor height of 1,350 mm, the first step 500 mm above the ground; low engine cover for better mobility inside and choice of day, sleeper or so-called Privilege cab. High-stress areas are made from IFHR high tensile steel and the cab meets impact and crushing resistance standards (ECE 29; VVFS 1993; 18 for Sweden). The chassis front cross member, Fig 14, is at the same height as the zone of maximum deformation in cars. This design will cut the risks to car drivers in a front-end collision. The rigidity of the cab's front section is enhanced by using a bonded windscreen. Aerodynamic design features include flush mounted windscreen, angled at 15 degrees; smooth front face with integrated handles and windscreen wipers integrated roof lights. There are wide-radius cab corners; profiled sideview mirrors and

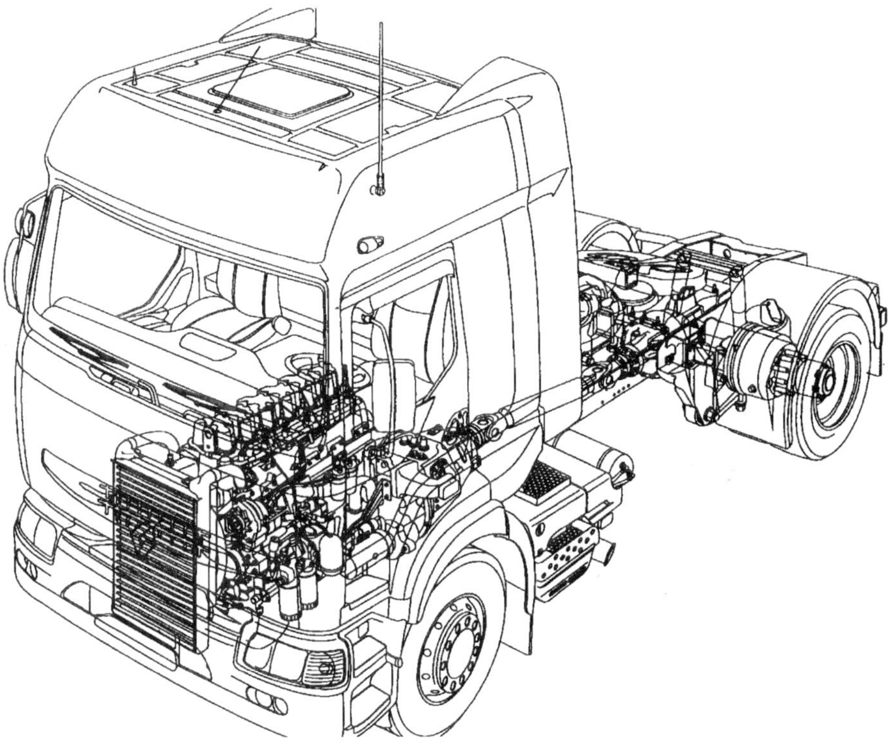

Fig 12: Styling concept, above; final versions of the two variants, below; mechanical package, right

arms; wrap-around front bumper; spoiler integrated to the front of the high roof cab, curved doors and sides of cab flared at the back. Cabs come with a five-year anti-corrosion guarantee and the structure is made entirely of electro-galvanized sheet metal. After cataphoretic pre-treatment, the cab gets a water-soluble coat of primer and a top coat, for a total protective layer 100 microns thick. Many parts exposed to weather, splashes and spattering, such as the grille, the fenders or the bumpers, are made of SMC (Sheet Moulding Compound).

For both cabs, the front overhang of 1,395 mm provides ample tread area in the axis of climb. Doors, which open to 90 degrees, are fitted with offset hinges for a wider opening width at shoulder level (2,219 mm). Two continuous access handles are arranged on either side of the door aperture to assist access and a lamp located in the lower rebate of the door automatically lights up the step area. Programmed steering wheel release is by exerting heel pressure on the steering wheel adjustment control, the wheel moves forward towards the windscreen; the fast lowering control vents the air suspension, which quickly drops the seat down to its lowest position, Fig 15.

Fundamental work has been done on the entire drive line with a view to reducing noise at the source. Filtration of engine vibration is first achieved via reinforced engine suspension. The aerodynamic shapes of the cab ensures optimal stream guidance and thus reduces slipstream noise. A set of acoustic screens has been developed around the engine compartment in order to reduce soundwave propagation. The floor is covered with heavy gauge tar-based compounds (IFF) and laid with a very thick rubber floor mat for even better absorption. The steering wheel is adjustable for 90 mm in height and 30 deg in rake; wheel is 460 mm in diameter, with hub having a fold-away A4 writing panel.

Cab width (2.5 m or 2.36 m depending on model) offers the best compromise between compactness and roominess, say the makers. Under the bunk located behind the seats on the Long Distance model are three compartments to store tools, clothes and shoes. There are two large storage areas above the windscreen: a centre storage area can accommodate a CB radio; to the right, an area is set aside for personal belongings.

Powertrain The new 11-litre MIDR 06.23.56 (A41 ACE Electronic) engine has peak power: 280 kW (380 hp) at 2000 rpm; peak torque: 1750 Nm (180 mkg) at 1200 rpm and I.T.C. electronic injection system. This engine is from 250 to 300 kg lighter than the bigger engines commonly used to deliver the same power, and thus makes a positive impact on the profitability of the Premium 385. On 300 to 380 hp engines, the company offers as an option the new J engine brake which delivers braking power of nearly 230 kW at 2300 rpm. The operating principle is based on an hydraulic valve closing system and compression of the air inside the cylinders. Control of the J engine brake is coupled with the FOWA exhaust brake control.

Fig 13: Interior styling and key cab dimensions

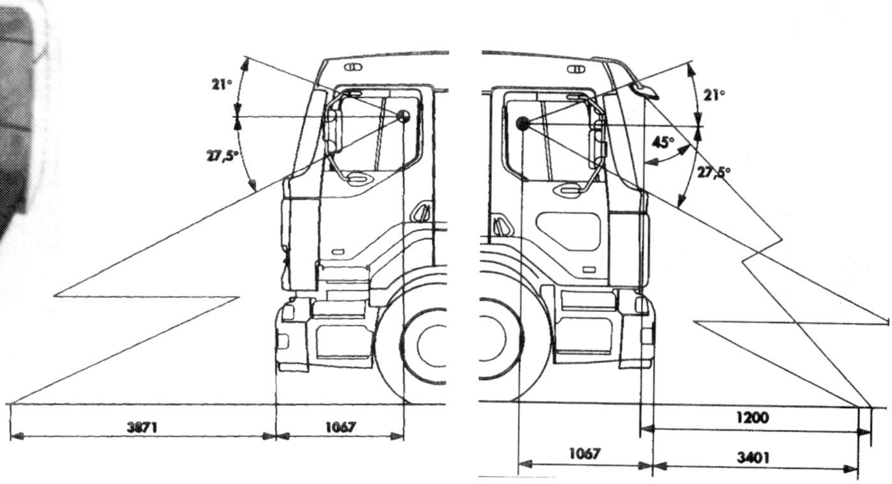

Visibility angle for Long Distance (left) and Distribution (right)

The EASY Shift automatic clutch system, fitted on conventional manual gearboxes, does away with the clutch pedal which is replaced by a sensor integrated to the gearshift lever. The computer then operates the clutch through a hydraulic system. A set of sensors enables the computer to take into account all the vehicle's operating parameters. The system is best fitted to applications making heavy use of the clutch, such as refuse collection and pick-up and delivery in cities and suburban areas.

Chassis A modular, lighter chassis, introduces Airtronic air suspension and makes front disc brakes a standard on trucks over 16 tonnes GVW. There is a new approach to the way servo systems are sited and the new two-bag air suspension is an integral configuration. From the back of the vehicle to the cab, the frame is 800 mm wide. At the front end, the frame is 900mm wide to fit the engine and gearbox. Sidemembers come in two different heights: low, for lower chassis height, at 258 mm with thickness: 7 mm (260 mm for a thickness of 8 mm); high, for better loading capabilities of 290 mm, thickness: 7 mm.

Suspension, Fig 16, air bag inflation pressure (12.5 bars) is regulated by means of an electronic levelling valve while a remote control unit in the cab which is used to change chassis height. The difference between the high and low positions is 240 mm on both rigid and tractor models. The remote control unit can also be used to memorize the height of different loading platforms. Suspension weight is about the same as a conventional suspension on rigids, is 90 kg lighter on tractors. 4x2 and 6x2 rigids can be fitted with integral air suspension. In this case, the front suspension offers the same remote levelling adjustment.

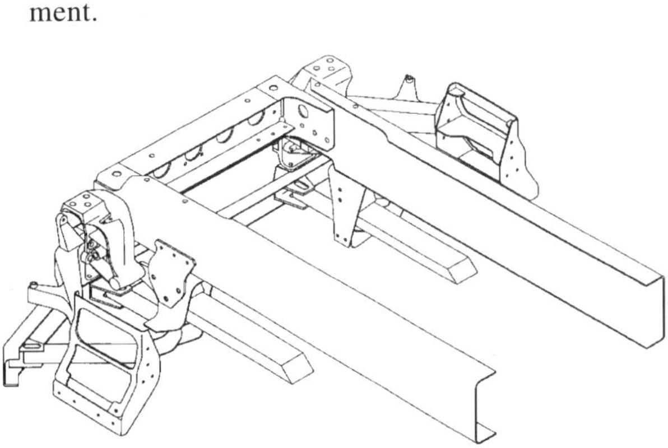

Fig 14: Chassis front-end structure right

Fig 15: Cab appointments, clockwise from top-left: distribution steps; adjustable steering wheel; rear-view mirrors; writing ledge

Fig 16: 6x2 Airtronic suspension, above; front air suspension, below

Fig 17: Volumaker sandwich-panel box-van body

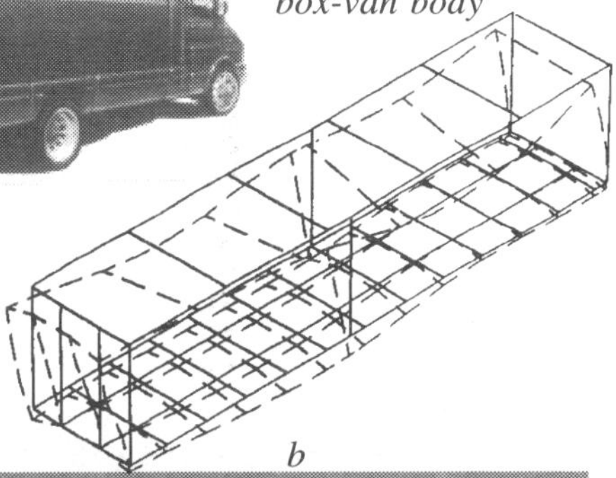

b

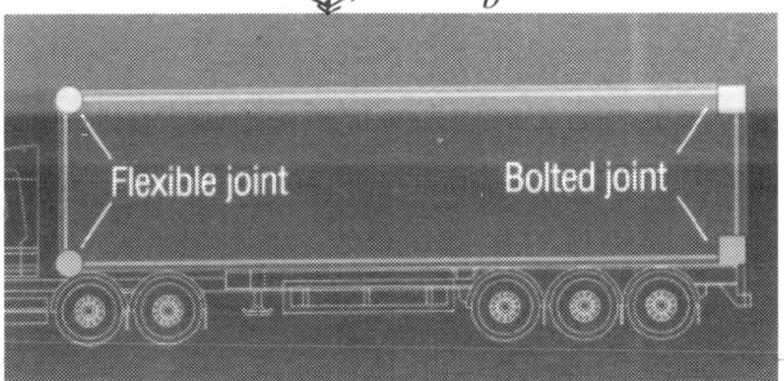

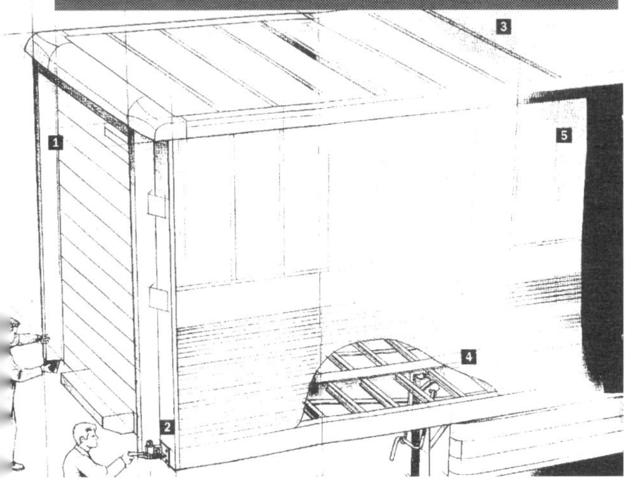

Fig 18: Flex-jointed curtain-sider

CV body technology advances

A significant advance in van body construction has been registered by Volumaker, Fig 17, with a new resin injection process for GRP moulding. It involves pre-fabrication of roof, floor, end-wall and sidewalls and bolting them up to form a monocoque construction with conventional framing. Floor is completely flat and smooth interior wall are ensured by the sandwich construction into which resin is drawn into the spaces between skins and core by machine. The core is high-rigidity expanded PVC hitherto used principally in aerospace applications. Side-walls are 14 mm thick, slimming down to 8 mm at the flanged edges which provide attachment joints. Unidirectional glass-fibre reinforcement strands are used in the skins and are configured to lie in horizontal vertical and diagonal directions to provide good shear strength. Interior finish can thus be as smooth as the exterior and depend on the quality of moulds. Normally these give a fine-textured high-gloss finish. A slight curvature on the panels adds to the attractive appearance and contributes a measure of structural stability. Panels are coloured during the moulding process and the complete body is mounted on a sub-frame of galvanised steel. The front panel is pleasingly contoured to the shape of the particular chassis-cab to be used while the rear panel houses a roof slide shutter of hollow extruded aluminium alloy sections to cover a rear aperture of 1950 x 1700 mm.

*Flex-jointed curtain-sider*An important beakthrough in CV body construction, Fig 18a, has been achieved by Lawrence David in conjunction with the Cranfield Impact Centre. Hitherto, no historical data has existed on the stresses in corner connections and pillars of conventional welded curtainsiders. Eight different body types were therefore subject to finite-element analysis which revealed very high corner stresses even under static laden conditions. To counteract unacceptably high stresses under dynamic loading Lawrence David's solution was to flexibility mount the body to the chassis, using a bolted rather than welded construction. This resulted in a reduction in stress levels from 124 to 15 N/mm^2 obtained from the stress plot of Fig 18b. Dynamic testing of two prototype vehicles at MIRA's rough terrain track were then used to record vibration mode and corresponding body deflections over a two-year period. This provided the necessary data for the company to asses a whole range of flexibly-mounted curtainsiders and calculate an expected life cycle under a variety of loads for each model. This also led to more structurally efficient bodies with lower peak stresses more evenly spread throughout the structure. The company was also able to prepaint the constituent components prior to bolted assembly and obtain better corrosion resistance. Product consistency was also improved by the need to have 100 % accurate parts and fixing points prior to assembly and any weaknesses hidden by welding removed.

The introduction of galvanised rear frames is made possible by the removal of welded joints which otherwise destroy the zinc-coating locally. The steel sections are hot-rolled and punched prior to galvanising by the hotdip process. The front-bulkhead, 1, comprises a steel framework surrounding aluminium snap-lock sections faced by plywood on their internal surface. The flexible joints involve substantial slabs of neoprene rubber between two rows of bolts. A ratchet tensioner, 2, is singlehanded in operation. The roof, 3, has a one-piece aluminium roof-sheet which is bonded to the roof bows. Trailer construction, 4, is a key to the weight reduction from a standard 7300 kg down to 6300 kg for a tri-axle air-suspended vehicle. Armoursheet curtains,5, are made from 2 x 2 heavyweight polyester cloth, pvc-coated on both sides.

New approach to demounts Maximum time of four minutes is claimed for the direct-to-floor Body Swap demountable system, Fig 19. It is designed to suit chassiscabs up to 3500 kg GVW and reduces operator handling to a minimum. The load/unload procedure involves connecting the male adaptor to the unit, at A, and drawing it up to the vehicle and lifting, B. It is then pulled up on the body frame, C, and then into the Autolock position, at D. A reverse procedure is adopted to unload.

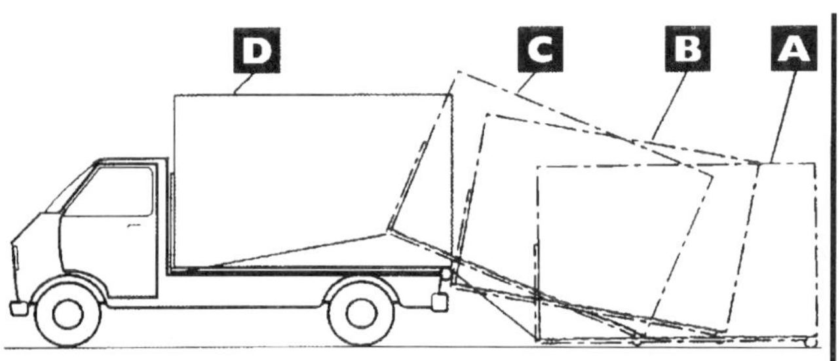

Fig 19: Body-swap demount system

Box-vans based on kit-bodies

CV kit bodies are an attraction to operators for their ability to be assembled, and altered to suit later requirements by add-on modules, by skilled fitters at the operator's own premises, and because of the perceived quality and proven design of the kits.

Because kit-bodies have usually been the subject of joint engineering expertise by body-builder and supplier of panels and sections there is good expectancy of optimum exploitation of component properties. Body-builders, too, stand to benefit from purpose-designed systems which give some confidence of structural integrity. The past two decades have seen a considerable growth in CV kit-body and systems producers with the widening appreciation of the versatility of aluminium alloy extrusions and their adaptability into relative simple-to-assemble structures. Many options become availabe to builders from systems suppliers of made-to-measure bodywork such as Alusuisse; a typical kit from this company is seen in Fig 20.

Dawnpress Ltd supply their Eurokit system, which is available in aluminium alloy or GRP forms as either dry-freight (Glasply) or insulated (Glasfoam) bodywork for refrigerated operation. The former are plywood cored sandwich panels with polyester resin bonded GRP skins, lamination taking place under vacuum following wet lay-up, curing taking place at 120 C. Seamless panels in lengths up to 13.42 metres can be supplied, with widths up to 3.66 mm, and

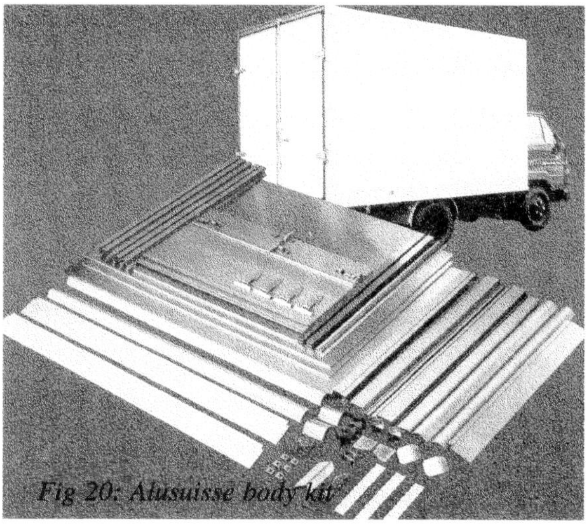

Fig 20: Alusuisse body kit

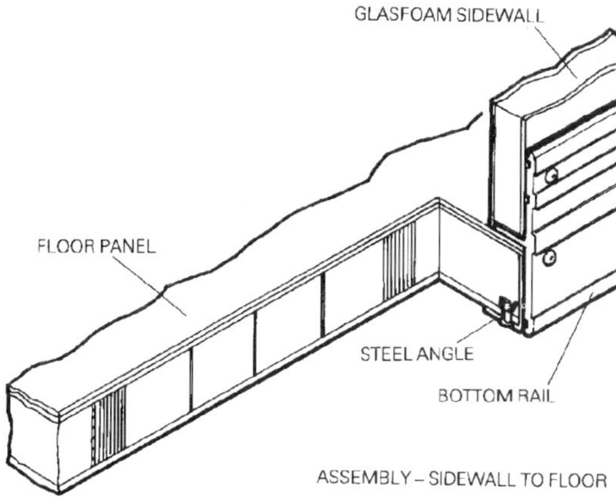

Fig 22: Glasfoam corner arrangement

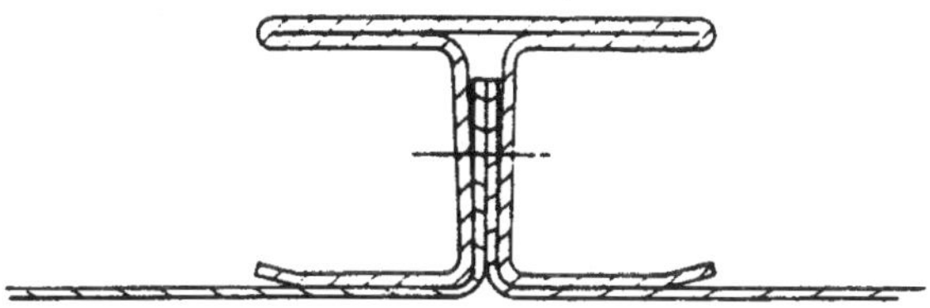

Fig 23: Aluvan panel clinching system

thicknesses ranging from 6 to 24 mm in seven steps. Steel inserts are provided within the core to form anchorages for cargo restraint systems. Glasfoam panels come in the same lengths and widths but with thicknesses up to 150 mm and containing the slabstock polyurethane foam core which is interspersed with GRP vertical webs at 100 mm centres to give high beam strength. Variants include 2LF construction where duplex layers of foam are separated by a GRP membrane and the webs are spaced at 600 mm. For the floor-panel variant timber supports are provided at 600 mm spacing between the ply-backed GRP skins. Fig 21 shows the Aluminium Eurokit in which section details of the members and joints are shown. Fig 22 shows how Glasfoam floor and sidewall panels are assembled.

Aluvan recently announced the Klemmbau kit which has the distinction of offering a smooth exterior walled body. This is achieved by joining panels with a unique clinching system, Fig 23. The clinching member also acts as a stiffener member which in the

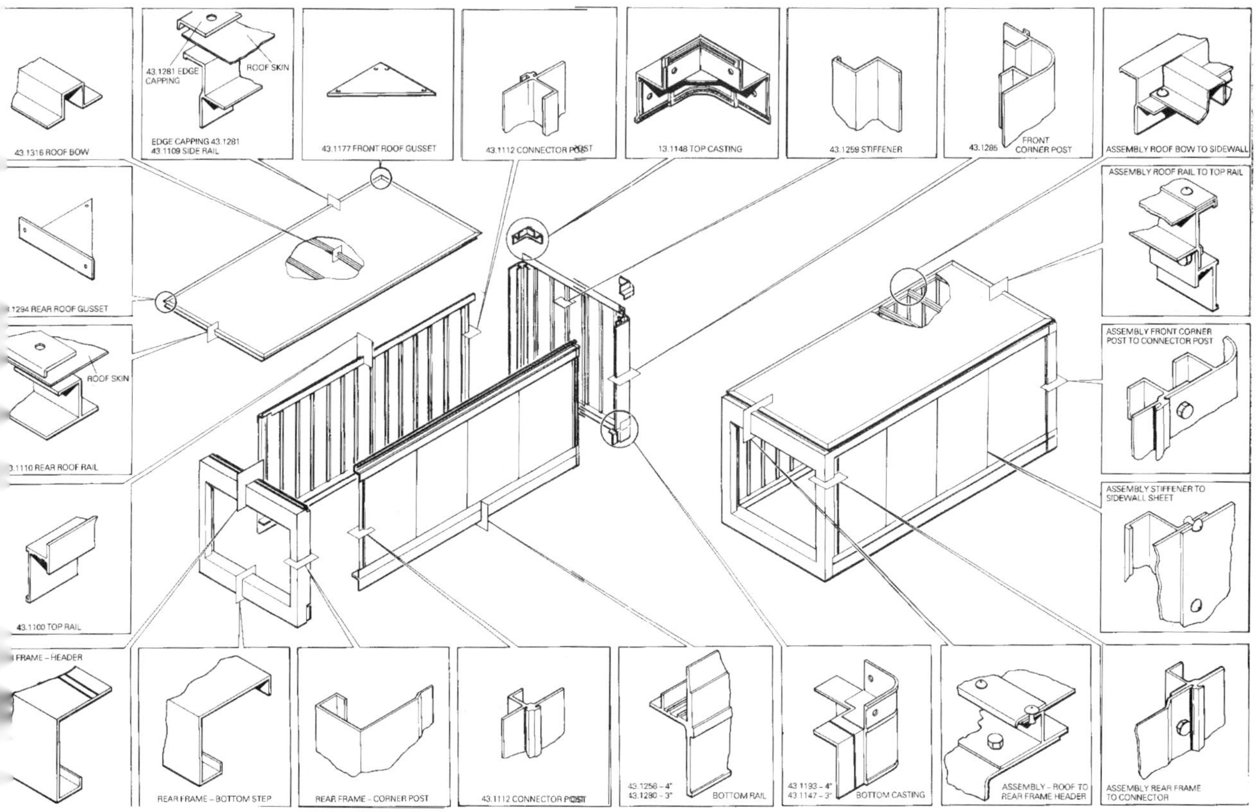

Fig 21: Eurokit by Dawnpress

latest version is just 25 mm in section depth. These are at a pitch of 406 mm on the S-type body, to give the construction shown in Fig 24. Options in the kit include single and double side doors, a selection of rear frames, transparent roof-panels, sliding side-windows and lashing rails. Reinforced models are also available which have stiffeners spaced at 305 mm pitch.

A standard procedure has been established for assembly, comprising the following steps: first set up the front-end frame vertically and apply sealant along the length of the connection profile post. Then offer up a side-wall panel to it using positioning pins through a selection of the mating holes and a small number of securing bolts through some others. This is repeated for the second side-wall and the rear-frame is also applied with sealant and offered up to the sub-assembly. The resulting structure is then accurately levelled and made square and sealant applied to the upper edge prior to mounting the roof panel. Bolting should start from the front, one side at a time and an alignment check made prior to final bolting all round. Finally the corner post castings are finish-drilled to align with the body mounts on the vehicle chassis.

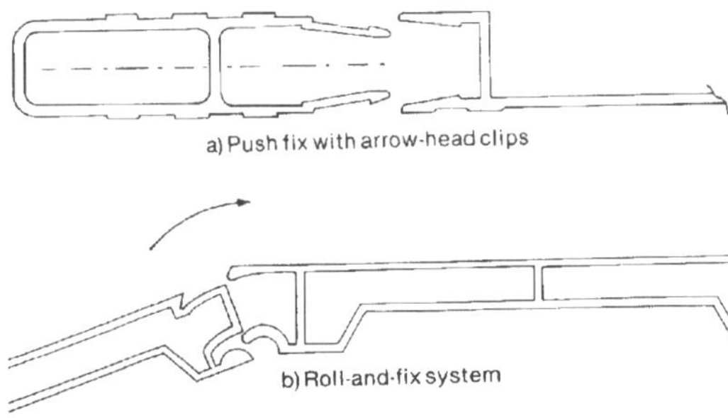

Fig 25: Sections for drop-siders

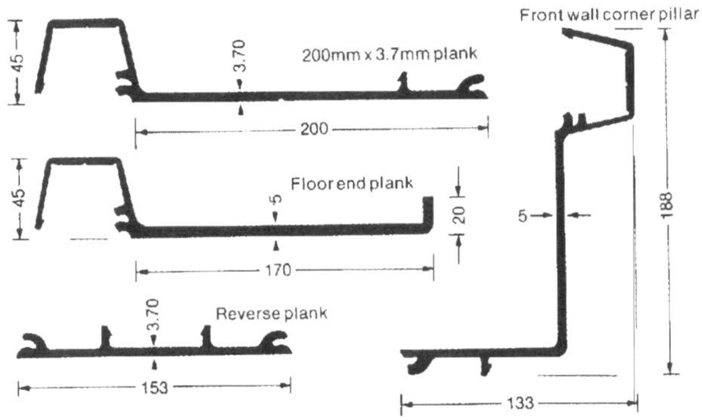

Fig 26: Sections for tippers

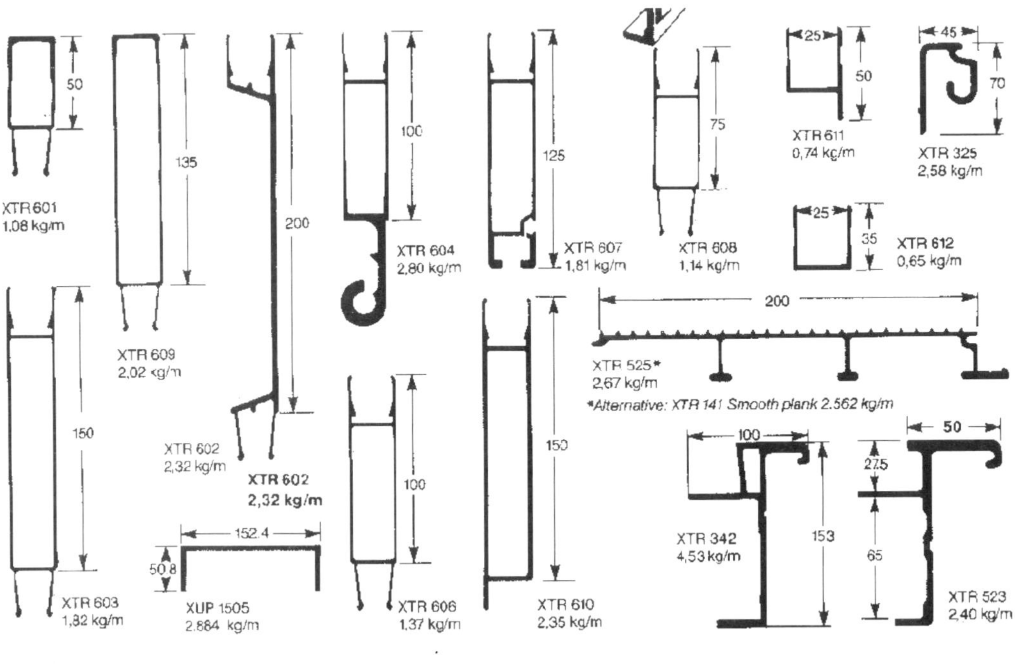

Fig 28: Sections for floors and walls

Tailor-made bodies from proprietary profiles

Body-builders wanting greater flexibility than is available with kits can still take advantage of extruded sections which have been specifically designed for CV bodies. The wide variety of sections available is somewhat daunting but the Aluminium Federation co-ordinates the activities of the various extruders in the UK, for example, through its member organisation, the Shapemakers. For a drop-side truck body two interesting sections are shown in Fig 25; such sections usually employ 6063 alloy in T6 temper. The first fixing method shown, for joining the sections, is a permanent one and requires only end-stiffeners. Continous hinges can also be built into the lower section.

For sidewalls and end-walls, a range of sections up to 250 mm width is typically available in 6082 alloy at T6 temper, particularly suiting heavy-duty applications such as tippers, Fig 26. These are clipped together by a roll-and-fit system, with sufficient holding power to allow jigless assembly before welding, the channelled portion of which doubles as a panel stiffener. Fig 27 shows a typical set of extrusions for a floor and wall assembly from Hydro Aluminium which illustrates a dropside hinge design. Another major supplier of profiles is Reynolds Transport Systems.

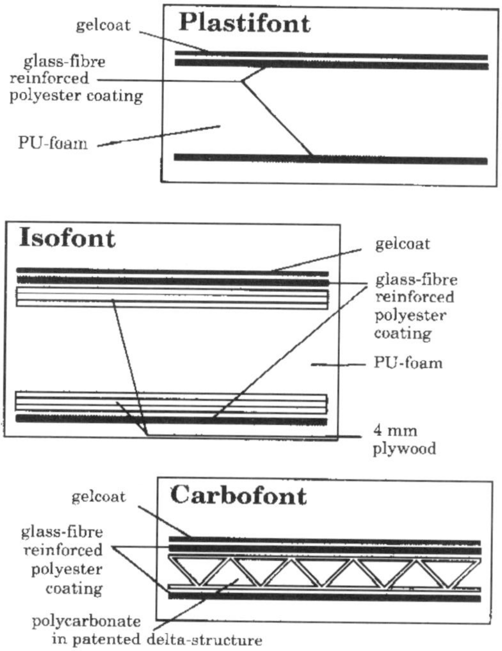

Fig 30: Insulated panel alternatives

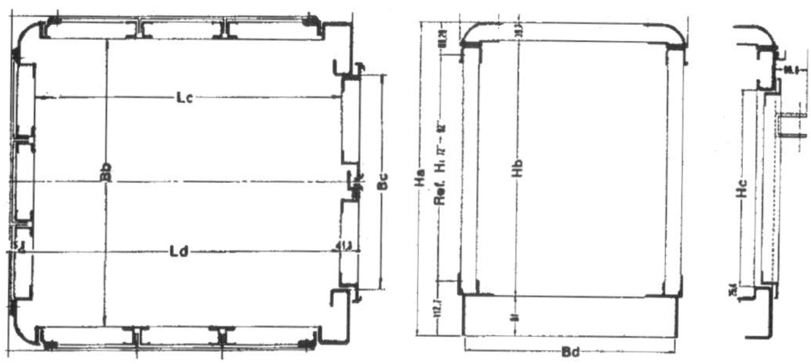

Fig 24: Aluvan body assembly

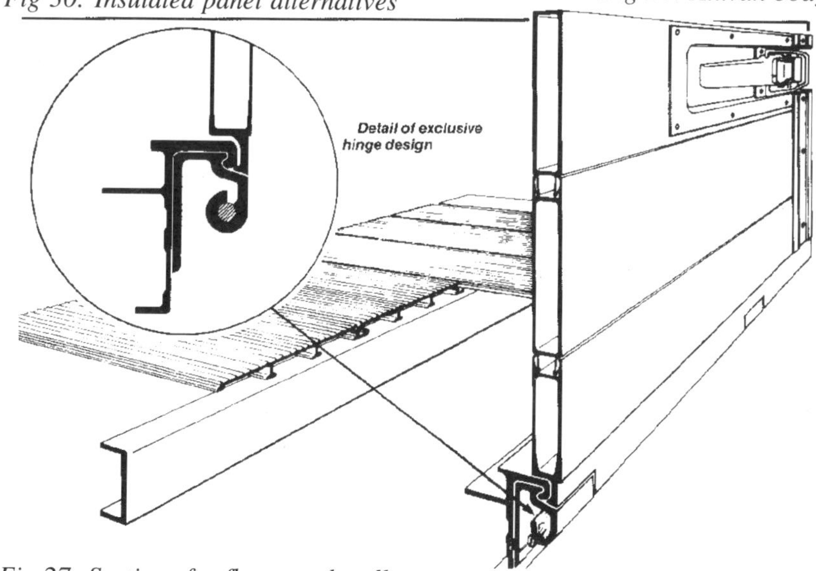

Fig 27: Sections for floors and walls

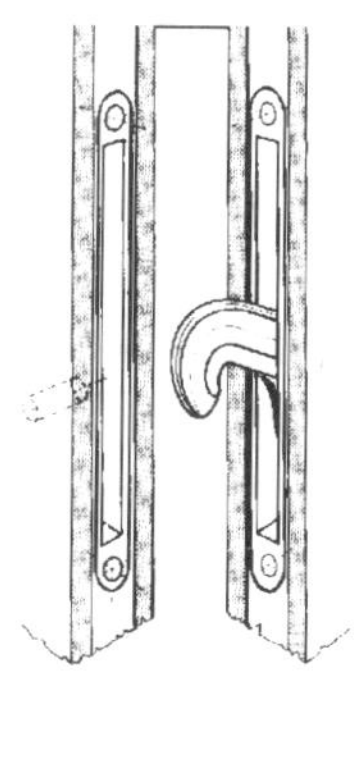

Fig 29: Isofix system

Box-van body systems are based on snap-lock extruded "planks" to form either single or double-walled bodies. For insulated bodywork, the company have developed the Isofix system with which, again, panel members are coupled together without the need for pop-rivets or coupling strips, Fig 29. A synthetic hook-fastener avoids any thermal bridging at the panel intersections. For insulated bodies, Plastifont panels are used, while for fully-refrigerated bodies, Isofont panels, Fig 30, are available; for special-purpose light-weight applications, Carbofont is a third alternative.

The Isofix system involves a sandwich of 0.7 mm galvanised steel sheet over a PUR core of thickness tailored to thermal requirements. The bottoms of the panels have a hardwood edge and sizes of panels range up to 1200 mm wide x 2700 mm high, each containing 4-5 hook-fasteners on the vertical edges, which a secured by a socket-head wrench after PUR adhesive has been applied.

Specialist goods vehicles

The need to meet the requirements of a wide range of special vehicle operations has led most builders of goods vehicles to establish SVO departments which are effectively special vehicle units within the volume chassis-cab builders. From them, strict regulations are issued to CV body-builders for producing specialist variants; a selection of the rules are seen here.

Because the chassis-cab is essentially a compromise design to meet a multiplicity of operations, considerable engineering effort has to be devoted to adaptation for special purpose. With volume-build constraints on the number of available wheelbase and chassis length options, the most frequent adaptation is change in frame length. Some manufacturers, while forbidding the welding of attachments to chassis frames, do allow welding to be used in the cut-and-shut process of frame shortening — albeit under carefully controlled conditions. Requirements normally include use of dry welding-rods and the use of visual inspection methods after fabrication. The weld hump has usually to be ground flat and any strike marks or spatter ground off to avoid stress raisers. When holes are closer than normally allowed following the 'shut', selective ones must be filled by bevelling the hole to two-thirds metal-thickness then fitting a copper plate to the inside of the hole — and

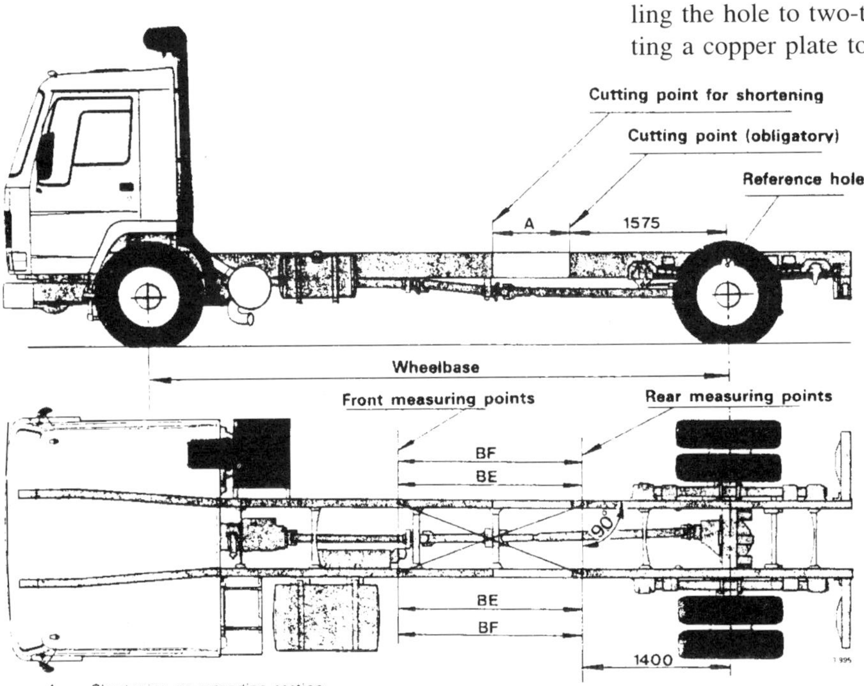

A = Shortening or extending section
B = Distance between measuring points before wheelbase alteration
BE = Distance between measuring points after wheelbase alteration

Fig 31: Recommendations for chassis cutting

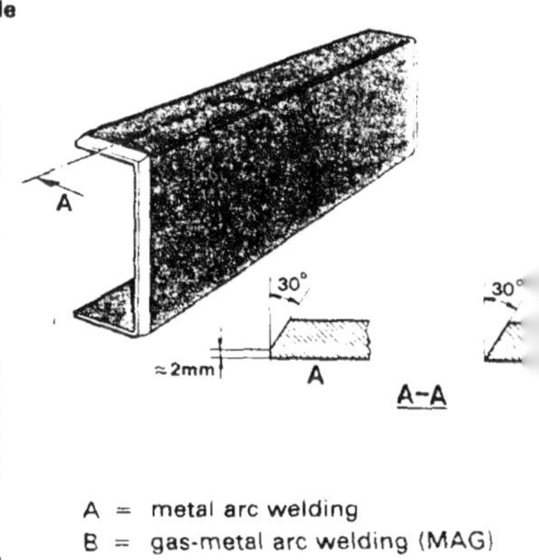

Fig 32: Bevelling of the joint to be welded

when the hole is larger than 20 mm diameter there may be a requirement for using a bevelled washer to block the hole prior to plug welding, which must be finished from the inside. Generally there is the requirement to strengthen the frame joint with a reinforcing profile to cater for any loss of temper in the steel as it is welded. This will typically extend 400 mm either side of the joint and in the case of a wheelbase extension a single reinforcing profile should cover both joints. Where the lengthening operation increases the allowable distance between crossmembers, additional members must be fitted. Where the crossmember position coincides with the reinforcing plate inside the chassis sidemember channel profile, the part should be cut at its centre span, reduced by the appropriate length, and re-welded.

Chassis-makers will normally recommend cutting points, those applying to the Volvo FL7/10 being shown, for example, in Fig 31. That company's procedure is to stand the truck on a flat floor and remove prop-shaft, rake-lines and electrics in the vicinity of the joint; then to mark out the position of the rear measuring points, making sure they are perpendicular to the web of the channel section; the distance BF to the front measuring point is measured

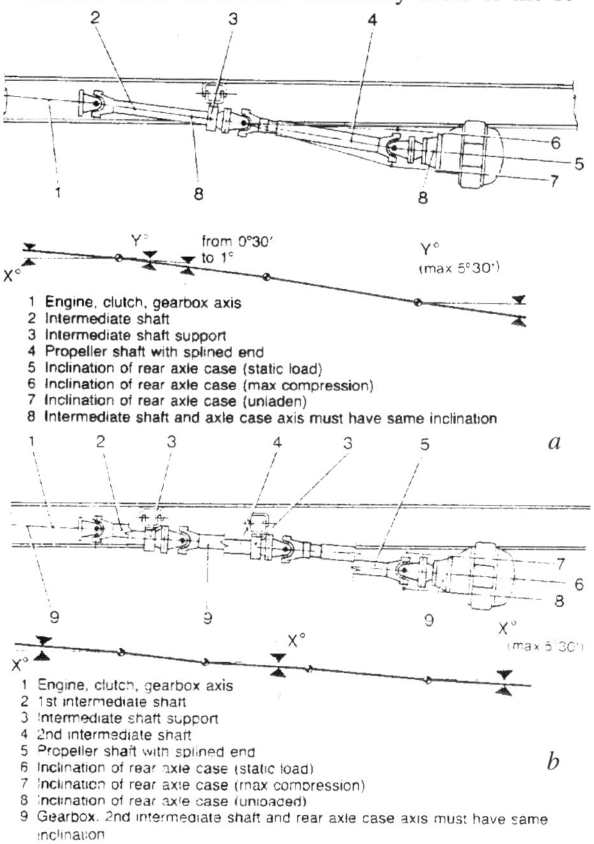

Fig 34: Single and double intermediate shafts

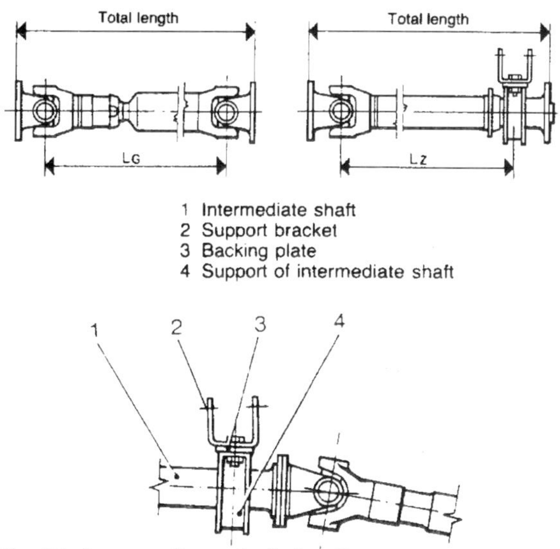

Fig 33: Intermediate shaft detail

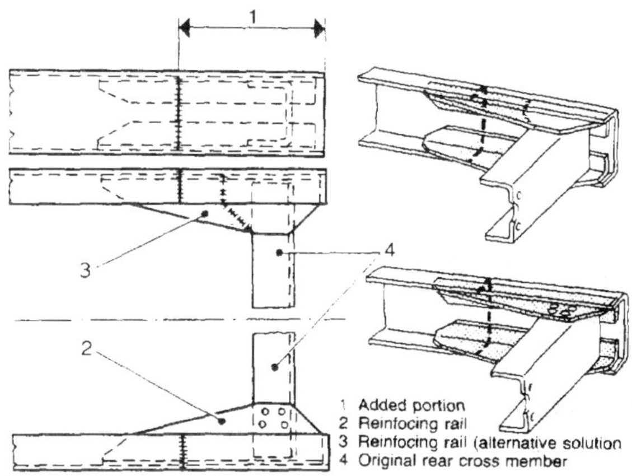

Fig 35: Welded crossmembers

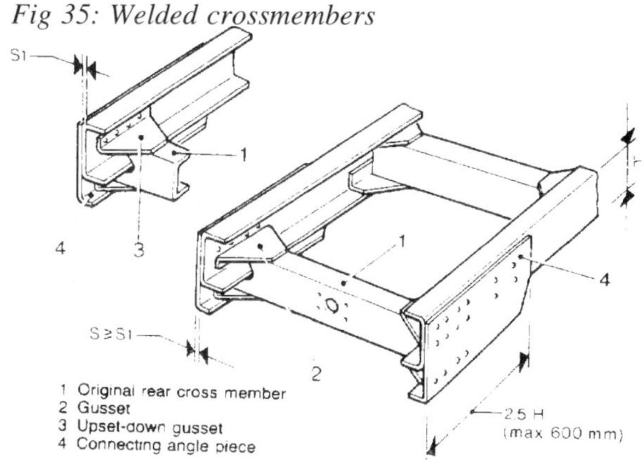

Fig 36: Riveted crossmembers

so that the checking measurement BE, after alteration, is at least 1200 mm. Trestles are then placed under the chassis to keep the frame truly horizontal; the member is then cut to a true right angle and its weld surface bevelled, together with that of the extension profile, as in Fig 32. The frame is then repositioned to obtain the correct BE measurement on both sides and that diagonals are equal to within 2 mm. This measurement procedure is repeated after tack welding — before the full weld is applied.

HGV drive-line

Prop-shaft dimensioning is critical on the new-length vehicle and manufacturers will define strict dimensional limits between flanges and between bearing supports, according to maximum rotational speed of the installed shaft — Fig 33, from Iveco. It is a company rule that individual shafts must be approximately the same length in multi-segment systems, and the difference in length between a driven and splined shaft must not exceed 600 mm (400 mm for similar duty shafts), Fig 34a. A margin of 25 mm minimum should be allowed for travel when the splined shaft is totally retracted. When extended the sliding sleeve should cover the splined stub for a length at least twice the spline diameter. Difference inclination between axle pinion axis and that of the gearbox output should not exceed limits of 0.5 - 1.0 degree. This can be adjusted by wedging the spring support pad on the rear axle — or by adjusting the torque rods of an air-suspension — but maximum axle pinion tilt is 5.5 degrees. When the vehicle is laden, these inclinations must also be checked: according to whether the rear-axle flange is lower or higher than that of the gearbox, inclination of the axle pinion and driven shaft must greater or less than, respectively, the inclination of the gearbox output axis. For substantial wheelbase lengthening an intermediate shaft, Fig 34b, may be necessary — requiring the same inclination as these axes.

Towing provision for heavy vehicles

Increase in frame overhang, rather than wheelbase, brings particular recommendations with respect to rear crossmember. In the case of Iveco, views in Fig 35 show ways of arranging a welded crossmember and Fig 36 a riveted one. Fig 35a covers length increases of not more than 300-400 mm — the reinforcing L-gussets being the same thickness as the original plate. Connections are made with bolts, to class 8.8, having the same diameter as the original rivets; an alternative welded connection is also shown. For length increases over 350 mm, the configuration in Fig 35b is rcommended. A lowered rear crossmember may be required for certain towing applications and alternative configurations in Fig 37 are suggested by Iveco. Thickness of the outer reinforcing angles must be not less than that of the sidemembers and they must extend over an area at least 2.5 times the height of the sidemember and bolt sizes must be such as to account for the greater moment that is transmitted by the downward displacement of the towing pintle.

The towing of trailers with rigid drawbars imposes additional stress on the towing vehicle chassis for which the rear crossmember must be further strengthened. The alternative proposal by Iveco in Fig 37, uses a C-section reinforcement — the company insisting that the connection between

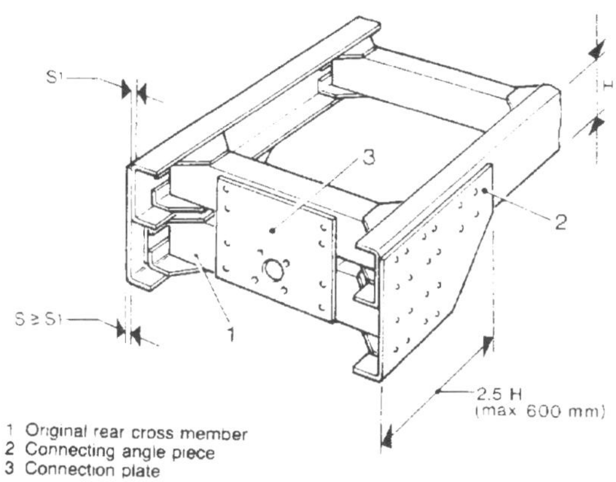

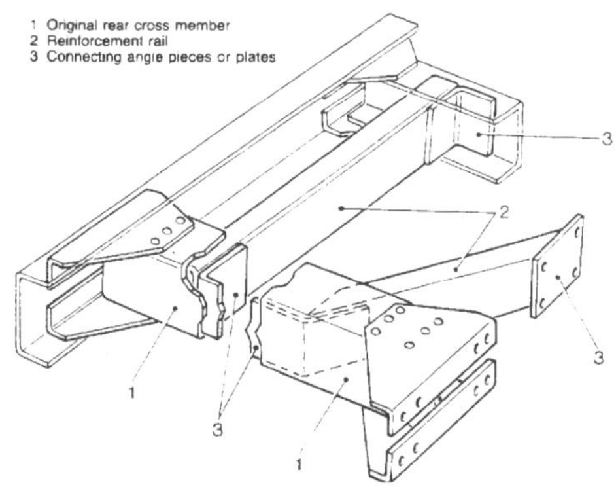

Fig 37: Alternative reinforcement schemes for drawbar mounting

sidemember also be reinforced. This shows the reinforcement on the inside of the crossmember, joining it to the vertical web of the sidemember — or to the following crossmember of the frame. The alternative is to mount a box-section beneath the crossmember — anchored at its ends to the same web and joined at its centre to the crossmember.

Additional axles for HGVs

The fitting of auxiliary axles requires modification to chassis frames such as to upgrade their loading capacity according to the additional carrying capacity of the axles. Fig 38 shows some possible alternatives, each of which must extend over the full span of the sidemember. Body mounting points should be used as a means of connecting the reinforcement to the sidemember in these Iveco proposals and a shear resistant connection should be used between midspan and the rear of the frame. Figs 39 a and b compares installations forward and aft of the drive axle, the former requiring reduction and the latter, extension.

Power-take-offs (PTOs) for HGVs

A later chapter deals with gearbox PTOs but other types and other locations on the chassis are available. Fig 40, due to Leyland DAF, shows some of the preferred ones for different SPV adaptations. The company recommend that for a Z or W drives, the configuration should satisfy the criteria shown as Figs 41 a and b. Angles described by the drive-shafts should be equal and not exceed eight degrees (six for front-end PTOs): inset is shown how to obtain proper alignment for ensuring constant-velocity operation of the driven equipment. The sliding joint should provide at least 8 mm, forward, and 6 mm, rearward, of movement. In PTOs from the front of the engine the company gives recommendations as in Fig 42 for vee-belt drive from a pulley co-axial with the crankshaft for two typical model series. For flywheel PTOs, Fig 43 applies.

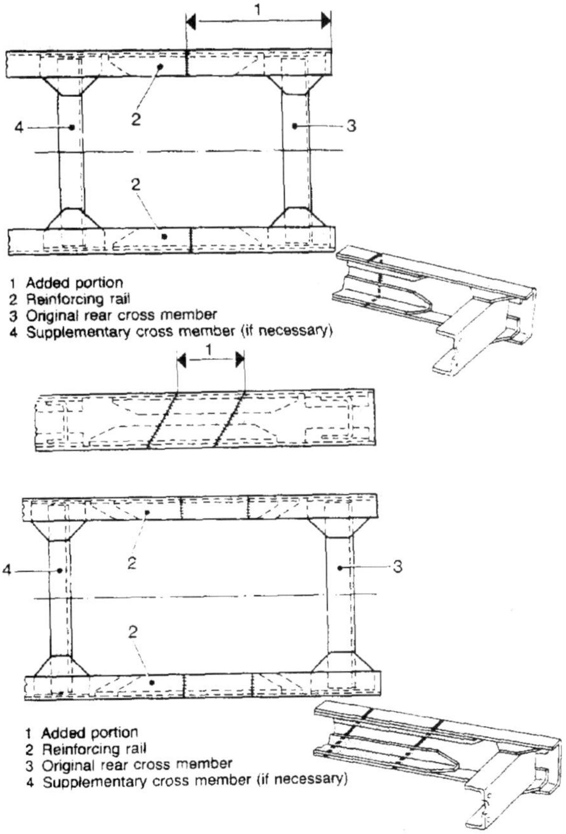

Fig 38: Frame bracing for extra axles

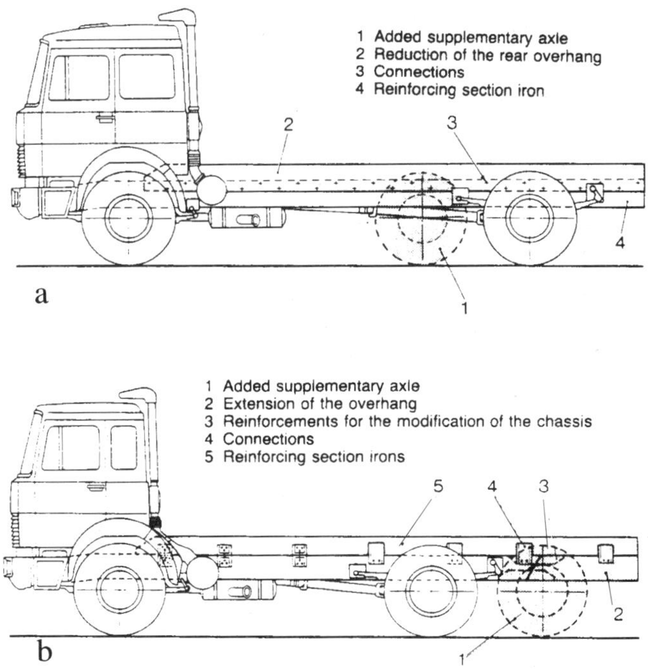

Fig 39: Fore and aft added axles

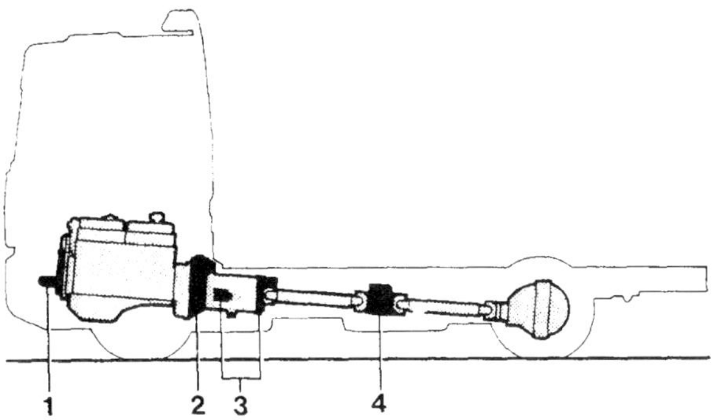

Application	P.T.O. 1	2	3	4
Refrigerated transport	x			
Air-conditioning	x			
Fibre transport			x	x
Tankers			x	
Bulk transport			x	
Aircraft refuellers		x		
Refuse collectors	x		x	
Road sweepers	x		x	
Gully emptiers	x	x	x	
Sewer cleaners		x	x	x
Tippers			x	
Demountables			x	
Concrete mixers		x		
Concrete pumps		x	x	x
Vehicle loading cranes			x	
Recovery vehicles			x	
Winches			x	
Hydraulic platforms			x	
Turntable vehicles			x	
Fire engines	x	x	x	

Fig 40: Preferred PTO locations

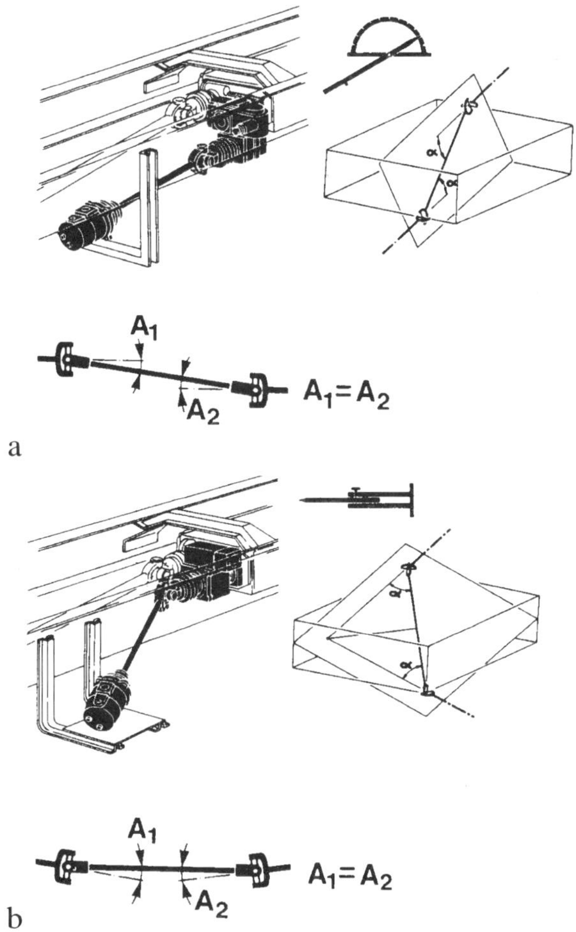

Fig 41a and b: Z and W drives

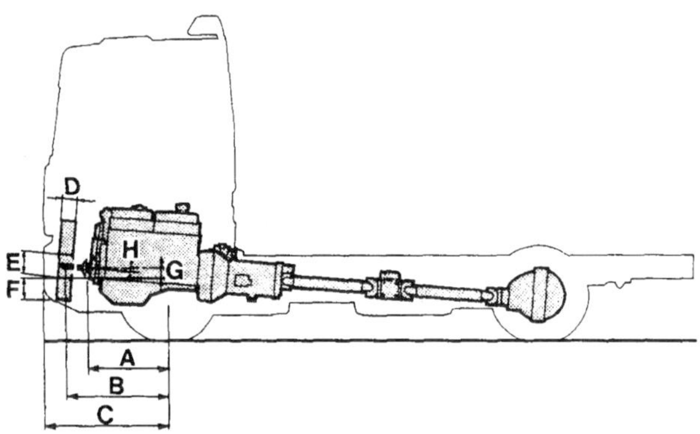

Fig 42: Engine-front PTO

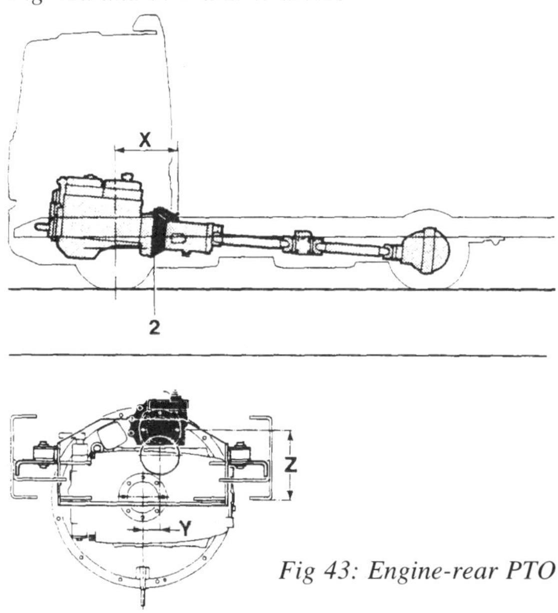

Fig 43: Engine-rear PTO

Heavy haulage vehicles

This vehicle category is required to operate outside the normal limits of the UK Construction and Use regulations and is subject to Special Types Orders covering the carriage of abnormally large and/or heavy indivisible loads. An understanding of the legal requirements is crucial to the design of such vehicles, which are usually tractor/trailer combinations. Trailer frame design for abnormal loads has some commonality with bridge design as seen in the following analysis.

The UK Special Types orders sub-divide heavy vehicles into three groups according to gross vehicle weights — 46, 80 and 150 tonnes.The lightest category vehicles have to comply with C & U regulations with respect to axle weights and spacings, as well as braking requirements, between 38 and 46 tonne GVW; see Fig 44. It also allows vehicles to be fitted with four closely spaced axles — beyond the normal remit of C & U. In the next category, upweight, axle weight increases up to 12.5 tonnes are permissible with the provision of 1.35 metres minimum axle spacing. Down to a 1.1 metre spacing maximum axle weight is just 12 tonnes. In all cases, vehicle tractor-trailer combinations must have at least five axles and and each 'bogie' is restricted to a total imposed load of 50 tonnes.

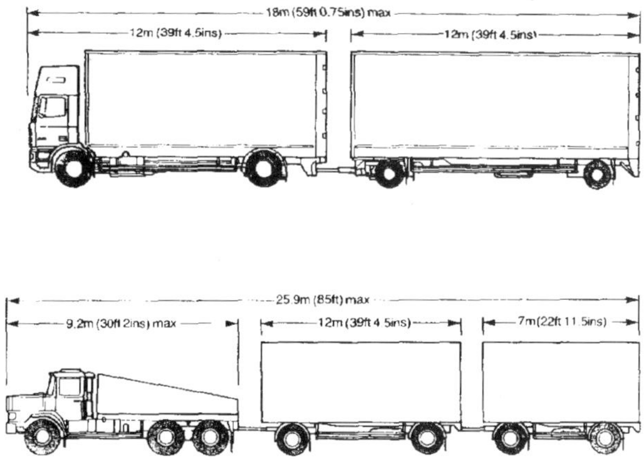

Fig 44: Length limits

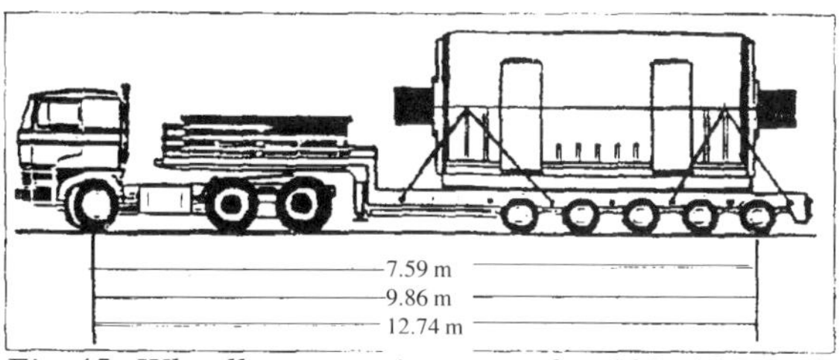

Fig 45: Wheelbase requirements for 100, 125 and 200 tonnes GVW

	Motorways mph	Dual carriageways mph	Other roads mph
Category 1	60	50	40
Category 2	40	35	30
Category 3	30	25	20
Vehicles carrying loads exceeding 4.3m but not exceeding 6.1m	30	25	20

Fig 46: Speed limits

There are also minimum wheelbase requirements as set out in Fig 45. In the heaviest category, the same axle spacing dimension limits are imposed but the respective axle weights rise to 16.5 and 15 tonnes. The vehicle combination is required to have at least six axles and bogie limits of 100 tonnes apply, reducing to 90 tonnes where inter-axle spacing falls below 1.35 metres. Full details are given in section 44 of the UK Road Traffic Act and general Order SI 1198/79 of the Motor Vehicles (Authorization of Special Types) and careful reading is required before committing expensive special vehicles for this purpose to metal. Overall widths must not exceed 2.9 metres unless specially required to support extra wide loads in safety and then an absolute maximum of 6.1 metres applies — when a police escort will usually be required for its movement. Overall lengths must not exceed 27.4 metres unless by special permission of the Transport Secretary.

Differential speed limits are imposed for the three categories, Fig 46, to regulate the possible incidence of road damage. Special dispensation is given for the movement of engineering plant giving GVW increase to 152.4 tonnes provided individual wheel loads do not exceed 11.43 tonnes. Overall axle spreads for the three main categories are shown in Fig 47.

Since the imposition of these special types orders in 1988 weight restrictions applying to existing heavy haulage trailers have led to the development of multi-wheeler tractors such as Leyland DAF's FTD which uses a Ginaf second steering axle added in front of the rear bogie of a three-axle tractive unit. This 8 x 4 vehicle entitled the FTD 95.380, Fig 48, can operate at up to 200 tonnes GTW. The hydro-pneumatic suspension for the tri-axle bogie uses the spring/reservoir units shown in Fig 49. Steering for the first and second axles is interconnected while the second third and fourth axles share a four-spring fully-compensating suspension.

The vehicle has an imposed load capacity of

47.5 % as a tractor. Extra traction is obtainable on the third and fourth drive axles by using the interconnected oleo system to reduce the load on the second, trailing, axle by some 25 %. In unladen running the axle can be lifted clear of the ground. The bogie suspension as a whole is self-levelling and proximity switches sense the piston positions in the six oleo springs.

The Unipower Contractor 6 x 4, Fig 50, is an on/off road tractor with gross train weight capability of up to 224 tonnes under the earlier regulations (on-road). This would be as a general service slow-speed drawbar combination, or as a special purpose one, up to 488 tonne GTW, in double formation road trains. The vehicle is also familiar as a high-speed tank-transporter in articulated form as illustrated, with 152.5 tonne capability on-road. Special 6 x 6 versions were designed for off-road duty. Each axle of the tandem through drive bogie has a torque capacity of 1.35 kNm and an inter-axle differential is provided. Gear ratio reductions allow a tractive effort potential of 26.75 tonne, gradient-ability up to 1 in 7.26 and road speeds up to 60 mph.

More recent designs include Volvo's NL 10/12 normal control tractor and the Faun Goliath, Fig 51. whose speeds, gradabilities and tractive efforts are listed in Fig 52. Trailers for heavy haulage are also built in a very specific way and are usually affected in a major way by the means used to on- and off-load indivisible payloads. Low-loading semi-trailers may be of the detachable back-axle variety or of the type using a detachable goose-neck, Fig 53. An interesting variation was the folding goose-neck by Dyson in Fig 54; for very heavy loads multi-wheeled trailers are often built up as shown in Fig 55. Here two

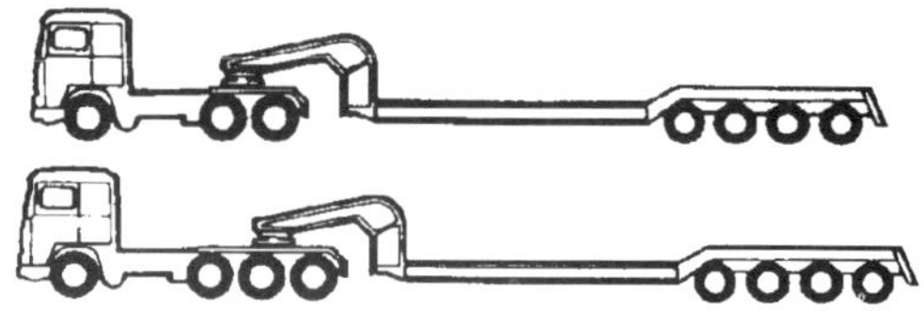

Relevant axle spacing in metres*	Weight (kg)
at least 6.5	40,000
at least 7.0	42,000
at least 7.5	44,000
at least 8.0	46,000

*Meaning distance between rearmost axle of semi-trailer and rearmost axle of tractive unit.

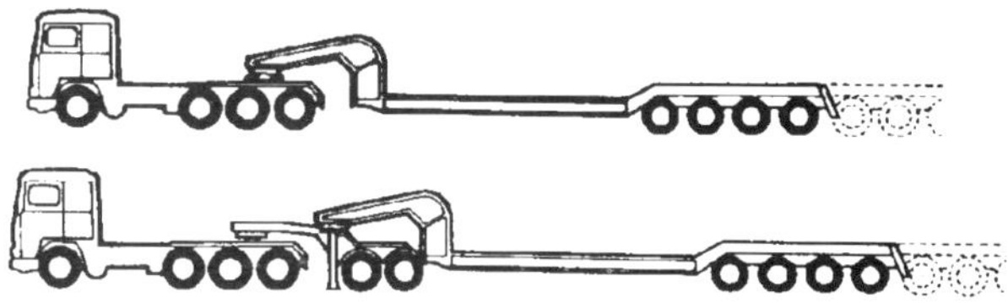

Distance between foremost and rearmost axles (metres)*	Weight (kg)
5.07	38,000
5.33	40,000
6.0	45,000
6.67	50,000
7.33	55,000
8.0	60,000
8.67	65,000
9.33	70,000
10.0	75,000
10.67	80,000

Distance between foremost and rearmost axles (metres)	Weight (kg)
5.77	80,000
6.23	85,000
6.68	90,000
7.14	95,000
7.59	100,000
8.05	105,000
8.50	110,000
8.95	115,000
9.41	120,000
9.86	125,000
10.32	130,000
10.77	135,000
11.23	140,000
11.68	145,000
12.14	150,000

Fig 47: Axle spreads

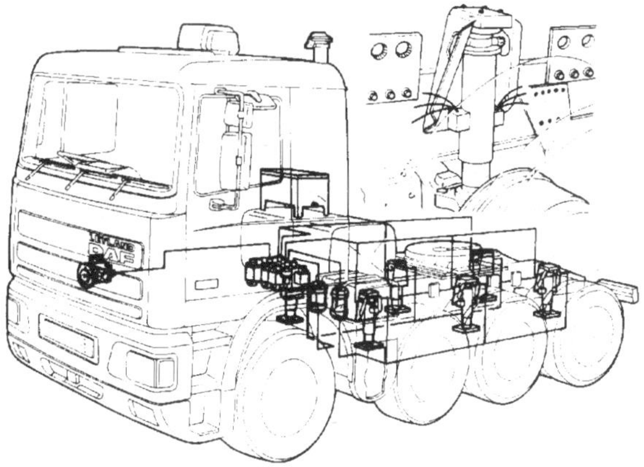

Fig 48: DAF FTD 95.380

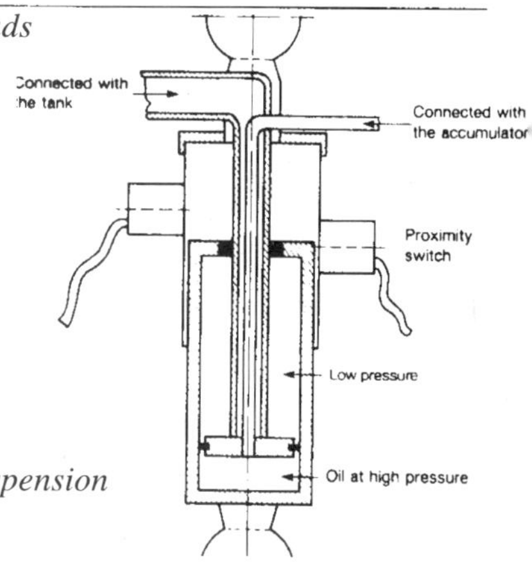

Fig 49: Oil/air suspension

twelve-wheeled bogies are attached to a common load carrying frame; in this German design wheels are evenly spaced rather than divided into bogies. It is possible by altering the direction of the steering motions to couple up to four of these units together in order to transport such items as heavy transformers while still retaining the wheel tracking facility. A 'standard' range of step-frame semi-trailers up to 60 tonne payload capacity are shown in Fig 56.

The design of heavy haulage trailer chassis inevitably leans heavily on beam theory developed for girder bridges. When the span or the loading are too great to permit the employment of rolled steel beams, singly or in combination, plate girders are used. Flanges are built up of plates or flats and attached to a web plate by means of flange angles. The web plate is stiffened by a series of vertical members, Fig 57. Whereas the most economical form is that with parallel flanges in which the area of the flanges is everywhere proportional to the bending moment, sometimes plate girders of varying web depth are feasible, Fig 58.

Occasionally more than one web can be used, in the form of a box girder, Fig 59. Built up beams of this type are recommended to have mid-span depth not less than 1/20 th of span. Ratio of breadth to

Gearbox step	Speeds in km/h at max. engine r.p.m.	Speeds in km/h at 70% converter range	Speeds in km/h at max. engine torque	Gradabilities in (%) at a total train weight of 100 t	200 t	250 t	300 t	Tractive effort in N
Off-Road								
1st gear	6.0	2.1	4.3	41.9	18.9	14.7	12.0	393786***
				30.6	14.0	10.9	8.8	301634**
				19.5	8.9	6.8	5.4	202632*
2nd gear	8.9	3.2	6.4	26.4	12.1	9.3	7.5	264723***
				19.5	8.9	6.8	5.4	202771**
				12.5	5.5	4.1	3.1	136210*
3rd gear	11.9	4.2	8.5	19.0	8.6	6.6	5.2	197878***
				14.1	6.2	4.7	3.7	151566**
				8.9	3.7	2.7	2.0	101802*
4th gear	17.8	6.3	12.7	12.1	5.3	3.9	3.0	132903***
				8.9	3.7	2.7	2.0	101789**
				5.5	2.0	1.3	0.8	68338*
5th gear	24.0	8.5	17.1	8.6	3.5	2.5	1.8	98447***
				6.2	2.3	1.6	1.1	75386**
				3.7	1.1	0.6	0.2	50565*
6th gear	35.8	12.7	25.6	5.2	1.9	1.2	0.7	65959***
				3.6	1.1	0.6	0.2	50472**
				1.9	0.2	0.0	0.0	33731*
rev. gear	4.7	1.7	3.3	57.6	24.9	19.4	15.8	504046***
				40.9	18.5	14.4	11.7	386093**
				25.8	11.8	9.1	7.3	259373*
On-Road								
1st gear	10.3	3.6	7.3	22.5	10.3	7.9	6.3	230284***
				16.7	7.5	5.7	4.5	176391**
				10.6	4.5	3.3	2.5	118484*
2nd gear	15.3	5.4	10.9	14.4	6.4	4.8	3.8	154809***
				10.6	4.5	3.3	2.5	118572**
				6.6	2.6	1.7	1.2	79625*
3rd gear	20.4	7.2	14.6	10.4	4.4	3.2	2.4	115718***
				7.6	3.0	2.1	1.5	88622**
				4.6	1.5	0.9	0.5	59479*
4th gear	30.4	10.8	21.7	6.4	2.5	1.7	1.1	77721***
				4.6	1.5	0.9	0.5	59496**
				2.6	0.5	0.1	0.0	39843*
5th gear	41.0	14.6	29.3	4.4	1.4	0.8	0.5	57571***
				3.0	0.7	0.3	0.0	44030**
				1.5	0.0	0.0	0.0	29349*
6th gear	61.2	21.7	43.7	2.4	0.5	0.1	0.0	38573***
				1.5	0.0	0.0	0.0	29394**
				0.5	0.0	0.0	0.0	19233*
rev. gear	8.0	2.8	5.7	29.8	13.6	10.6	8.5	294764***
				22.0	10.1	7.7	6.2	225783**
				14.1	6.2	4.7	3.7	151672*

Fig 52: Goliath performance based on 0.015 rolling resistance co-efficient, 0.7 static friction co-efficient, 12-764 axle reduction ratio and 1.71:1 transfer box reduction

Fig 51: Faun Goliath

Fig 50: Unipower 224 tonne tractor

unsupported length should not exceed 1/40. Where the ratio l/B exceeds 20, the compressive stress should not exceed:

9.4 - 0.13l/B

Joints in web plates are made with butted joints, usually with butted joints having double cover plates, Fig 60. Such joints must be of sufficient strength to carry the shear force at the joint as well as bending moment and a web stiffener may be used to form a joint cover. As well as stiffening the web against buckling by the shear force, the stiffeners must also prevent the girder racking sideways and distribute any concentrated load over the depth of the web plate. Girder bridge makers recommend safe shearing force per unit length of web depth as:

5.5t - d/20

where t is web thickness and d is the stiffener spacing.

The curve of the goose-neck can be accounted for in design analysis by using the Unit Load thereom. In the example shown in Fig 61, the downward deflection of the point B due to the bending deformation of AB and BC is found using the analysis alongside:

$$\delta_B = \frac{1}{EI}\int M_0 m_1 \, ds \text{ then:}$$

	M_0		m_1
AB	$R_{0A}.s$	$\frac{ws^2}{2}$	$R_{1a}.s$
BC	$R_{0C}.r\sin\theta$		$R_{1C}.r\sin\theta$

(θ is measured from OC and s from A)
and

$$\delta_B = \frac{1}{EI}\left[\int_0^l \left(R_{0a}.s - \frac{ws^2}{2}\right)\left(R_{1a}.s\right) ds + 2\int_0^{\frac{\pi}{2}} R_{1C}.R_{0C}.r^2.\sin^2\theta.r.d\theta\right]$$

$$R_{1a} = \frac{1}{3};\ R_{1C} = \frac{2}{3};\ R_{0C} = \frac{20 \times 4 \times 4/2}{2+4} = 26{\cdot}7;$$

$$R_{0a} = 80 \quad 26{\cdot}7 = 53{\cdot}3\ kN$$

Given the 2nd moment of area of the sidemember

$= 0{\cdot}25 \times 10^{4}\ m^4$ and $E_{steel} = 210\ GN/m^2$;

$$\delta_B = \frac{10^3 \times 10^3}{210 \times 10^9 \times 0{\cdot}25 \times 10^{-4}}\left[\frac{64}{3}\left(\frac{53{\cdot}3}{3} - \frac{80}{8}\right) + 26{\cdot}7 \times \frac{2}{3} \times 8 \times \frac{1}{2} \times \frac{\pi}{2}\right]$$

$$\therefore \delta_B = 53 \text{ mm}$$

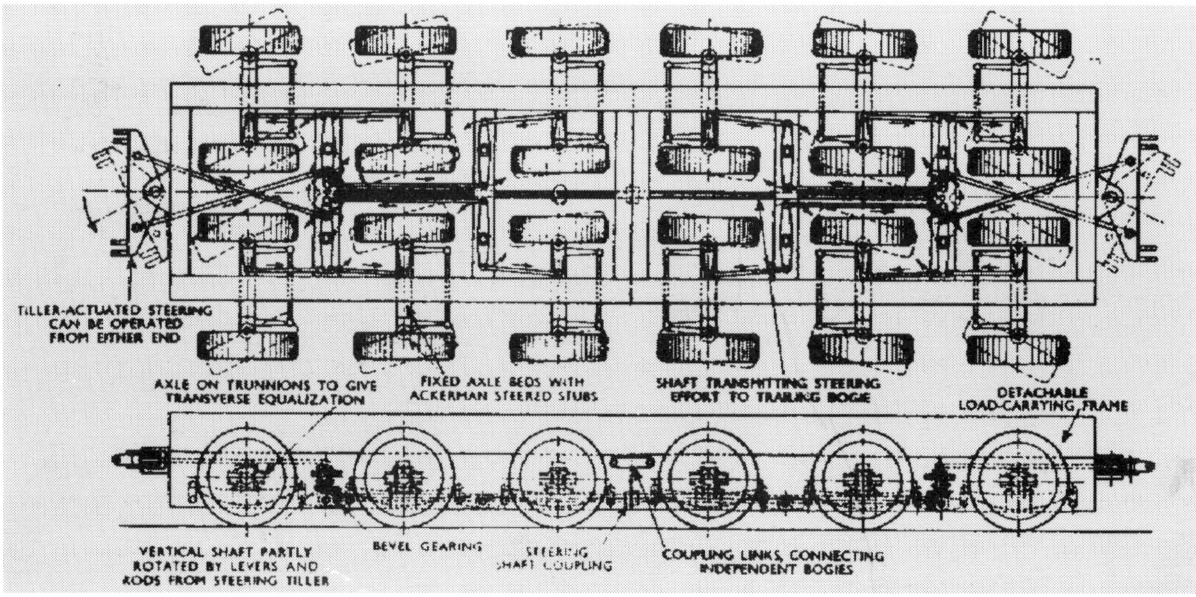

Fig 55: Multi-wheel trailer

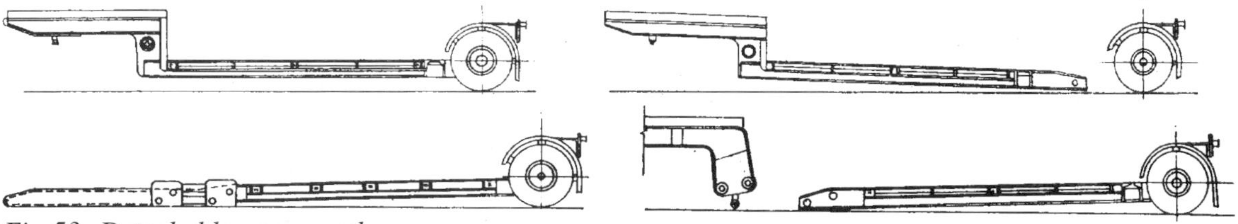

Fig 53: Detachable goose-neck

A NECK LENGTH	B PLATFORM LENGTH	C UNLADEN HEIGHT	D OVERALL LENGTH	E SWING CLEARANCE	NOMINAL PAYLOAD KG	KING PIN LOAD KG	BOGIE LOAD KG	F FIFTH WHEEL HT	W OVERALL WIDTH	TYRES	SUSPENSION
2.9 m	7 m 9 m	1.14 m	9.9 m 11.9 m	1.98 m	15,000	10,000	10,000	1.32 m	2.5 m	10.00 x 20-16 (4 off)	Single TST 1/7
2.9 m	7 m 9 m	1.14 m	9.9 m 11.9 m	1.98 m	25,000	10,000	20,000	1.32 m	2.5 m	10.00 x 20-16 (8 off)	Tandem TST2/7
2.9 m	8 m 10 m	1.22 m	10.9 m 12.9 m	1.98 m	30,000	15,000	22,000	1.32 m	2.5 m	10.00 x 20-16 (8 off)	Tandem TST2/8
3.5 m	8 m 10 m	1.30 m	11.5 m 13.5 m	2.58 m	40,000	20,000	30,000	1.38 m	2.9 m	11.00 x 20-16 (8 off)	Tandem DN7
3.5 m	8 m 10 m 12 m	1.32 m	11.5 m 13.5 m 15.5 m	2.58 m	45,000	20,000	35,000	1.38 m	2.9 m	12.00 x 20-16 (8 off)	Tandem DN8
3.5 m	8 m 10 m 12 m	1.25 m	11.5 m 13.5 m 15.5 m	2.58 m	50,000	25,000	39,000	1.38 m	2.9 m	10.00 x 20-16 (12 off)	Tri-Axle TST3/8HD
3.5 m	8 m 10 m 12 m	1.30 m	11.5 m 13.5 m 15.5 m	2.58 m	60,000	30,000	45,000	1.38 m	3.0 m	11.00 x 20-16 (12 off)	Tri-Axle TST3/8HD

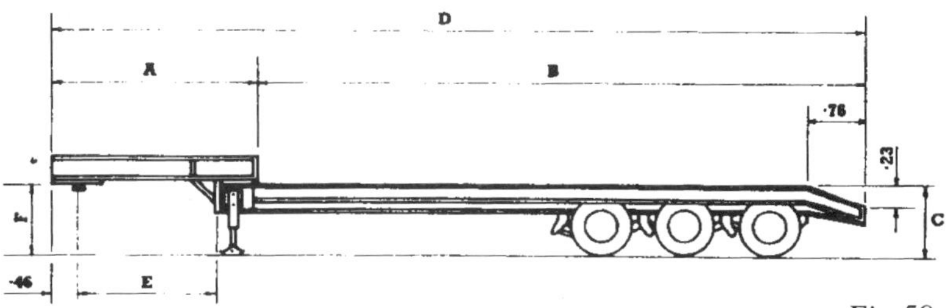

Fig 56: Typical trailer range for heavy haulage

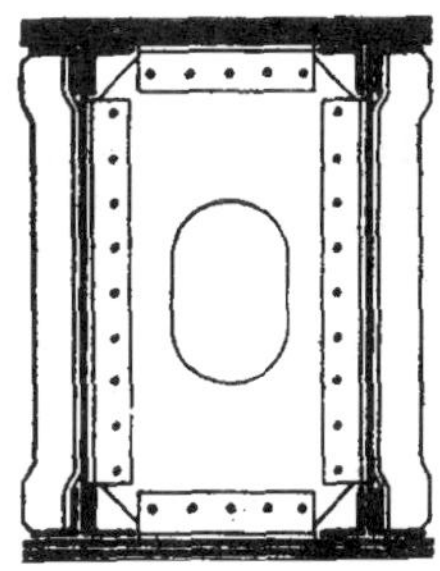

Fig 59: Box girder

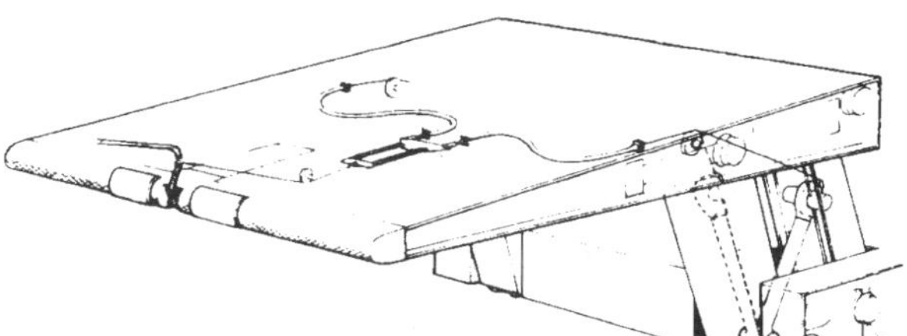

Fig 54: Folding goose-neck

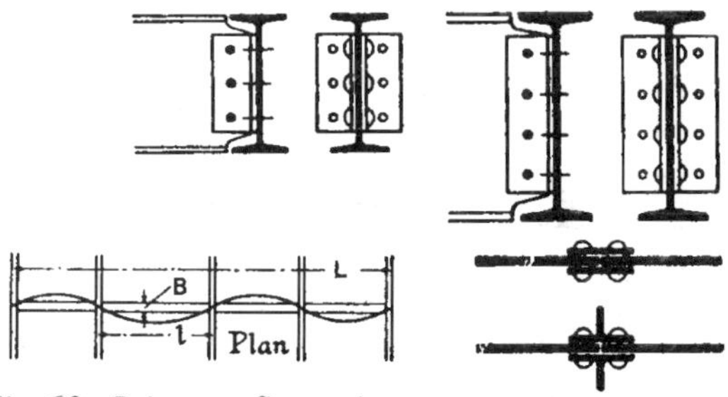

Fig 60: Joint configurations

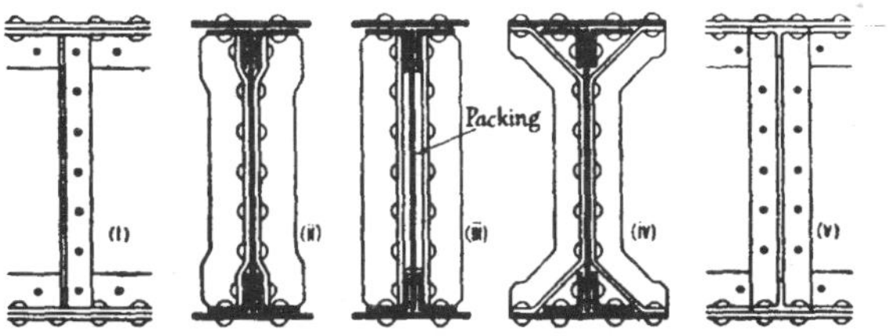

Fig 57: Webstiffeners

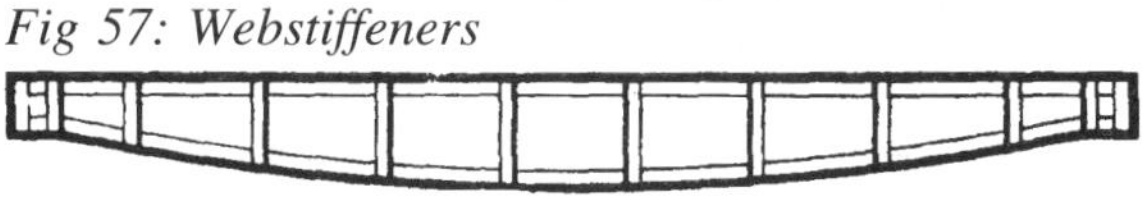

Fig 58: Tapered beam span

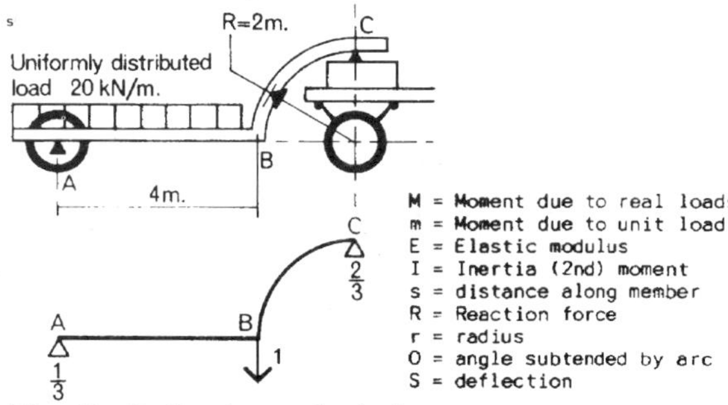

Fig 61: Deflection calculation parameters

Truck technology: structures

Mercedez Benz engineers have explained[1] how far FEA techniques had established themselves in commercial vehicle development. Simultaneous working of design, analysis and test departments at M-B has been used to accelerate the development of new vehicles, the analysis department contributing substantially to the decision-making process, as shown in Fig 62. Three dimensional computer-aided-design of cab components is applied to those items shown in Fig 63. The geometrical data is also used for establishing panel fits and designing connection joints. As well as optimizing the strength and stiffness of the structure, the techniques are used to design out structure-borne noise and to improve acoustics of the cab interior. First step is to ensure resonant frequencies of bonnet cover, cab and chassis-frame are outside the vibrational range of the axles. Whereas shell elements are used to represent the panels in FEA work, the connecting joints are modelled by bending elements. In acoustic analysis, the vibrating air in the interior cavity is represented by solid elements having reduced numbers of degrees of freedom. Mesh generators are used to build the models.

Low frequency vibration characteristics of the vehicle as a whole is obtained with a global model, Fig 64, with relatively coarse mesh density. The model contains single vibrating elements to model road inputs and engine vibration. Measured road irregularities of a pave track are also input to the model — for checking response co-relation with later prototypes of the vehicle, fitted with the necessary recording equipment. At the subsequent fine-mesh analysis stage for the cab, some 15,000 degrees of freedom are involved. Load inputs represent a whole series of body and wheelbase variants collected from pave-testing of similar models. Detail areas are even more closely investigated (with some 45,000 degrees of freedom per detail) as shown in Fig

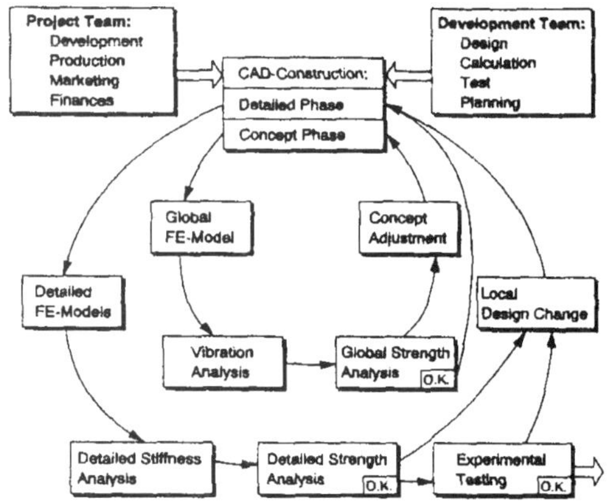

Fig 62: Structural analysis within system development

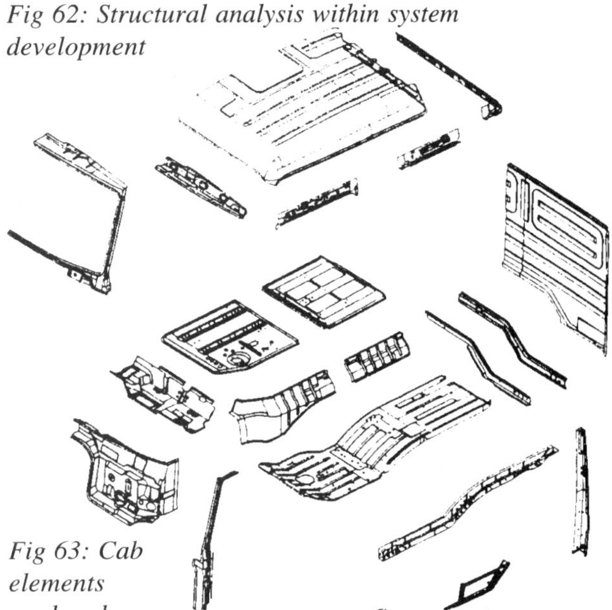

Fig 63: Cab elements analyzed

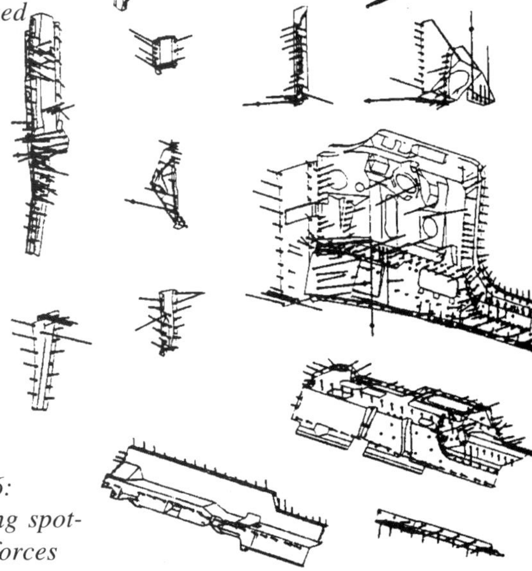

Fig 66: Plotting spot-weld forces

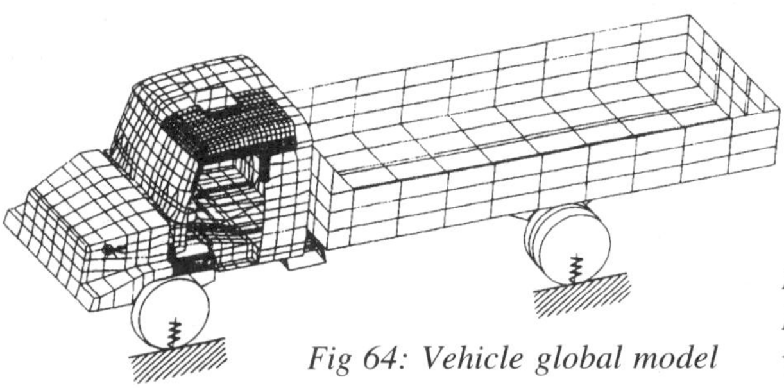

Fig 64: Vehicle global model

65. Spot-weld forces are plotted as in Fig 66 which allows both shear and tensile forces to be visualized. Chassis frames are analyzed under additional load cases to the those of the pave inputs. Steady-state inputs representing severe twisting on site operation and lozenging forces representing split-mhu braking are simulated. Stress concentration plots in Fig 67 are typical, from which individual parts of the structure are separately examined. Bar charts record critical stress levels at sidemember/crossmember junctions as percentages of allowable stress. Torsional and flexural responses are also evaluated, Fig 68.

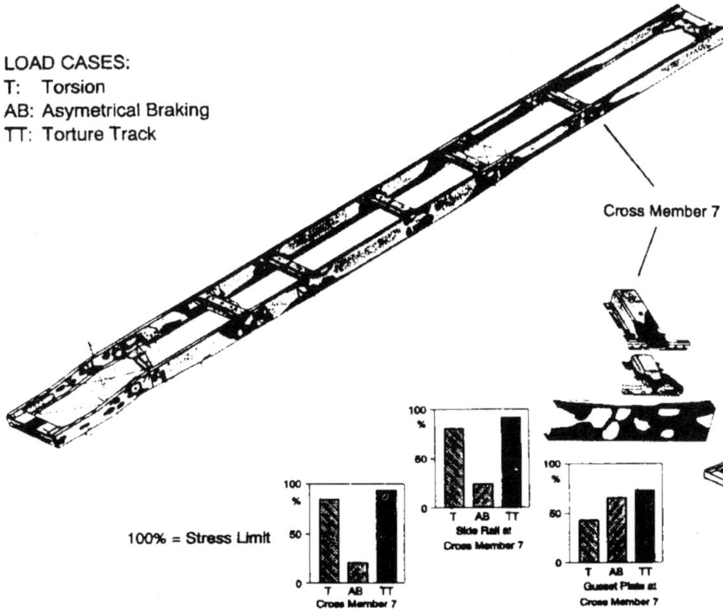

Fig 67: Stress concentration plots

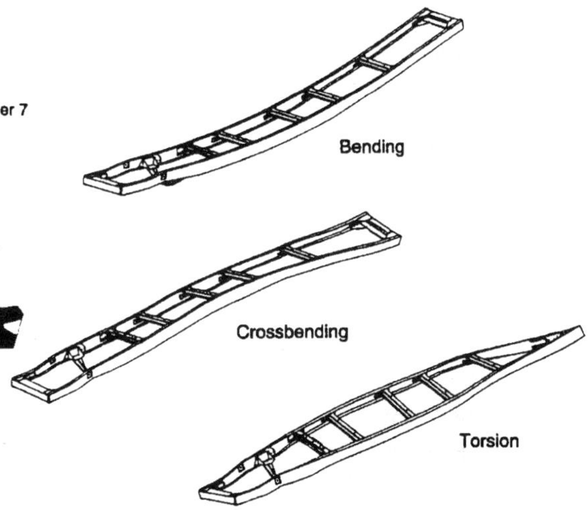

Fig 68: Torsional/flexural responses

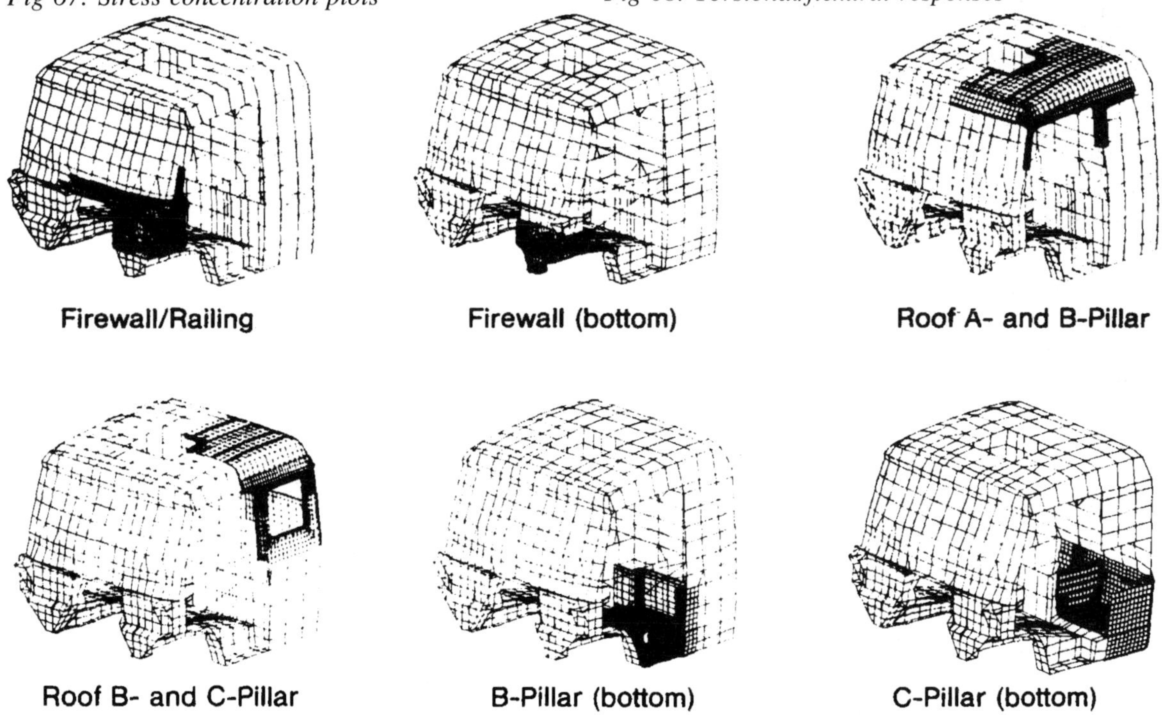

Fig 65: Areas for detailed FEA

Truck technology: braking in a bend

Mercedes Benz engineers have examined[2] truck braking while cornering according to the test procedure established by the ISO/TC 22/SDC working group. This involves driving on a skid pan at 0.5g constant lateral acceleration. Changes in the yaw velocity while braking is shown in Fig 69 comparing vehicles of contrasting weights. The assessment is based on the ratio of the instantaneous yaw velocity in braking, one second after application, with that applying to steady-state cornering — dubbed specific yaw velocity. The figure shows, however, that the critical point can vary from the required 1 second according to vehicle and brake system. A Mercedes truck was modelled for analysis as shown in Fig 70. The subsequent simulation showed that when a vehicle is decelerated while cornering there is a change in distribution of the lateral forces at the front and rear wheels. This change produces a yawing moment and a yaw velocity increasing with deceleration. Vehicle stability reduces as rear axle lateral forces reduce. Below 0.3g the yaw develops as in Fig 71a and above this value as in Fig 71b. The crucial effect of the instantaneous braking force distribution is revealed in Fig 71c for which choice of brake lining material is a major decider.

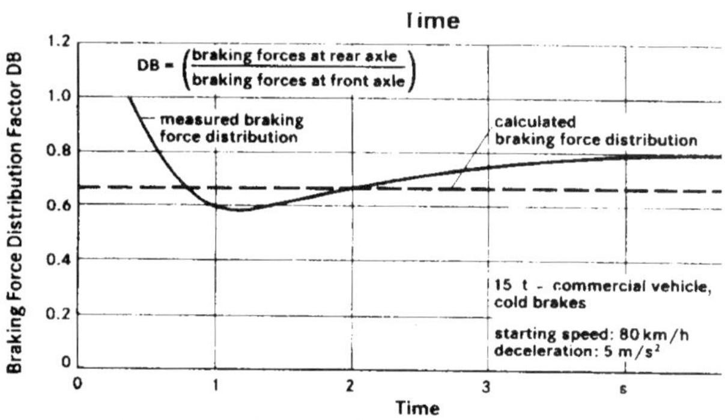

Fig 69: Yaw velocity when braking

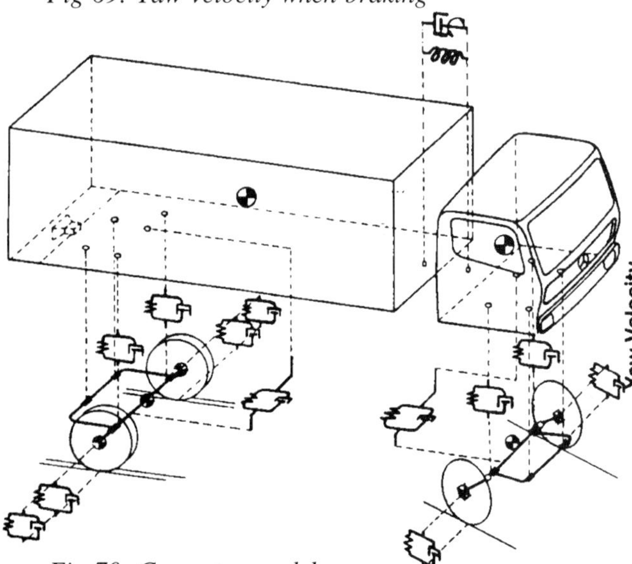

Fig 70: Computer model

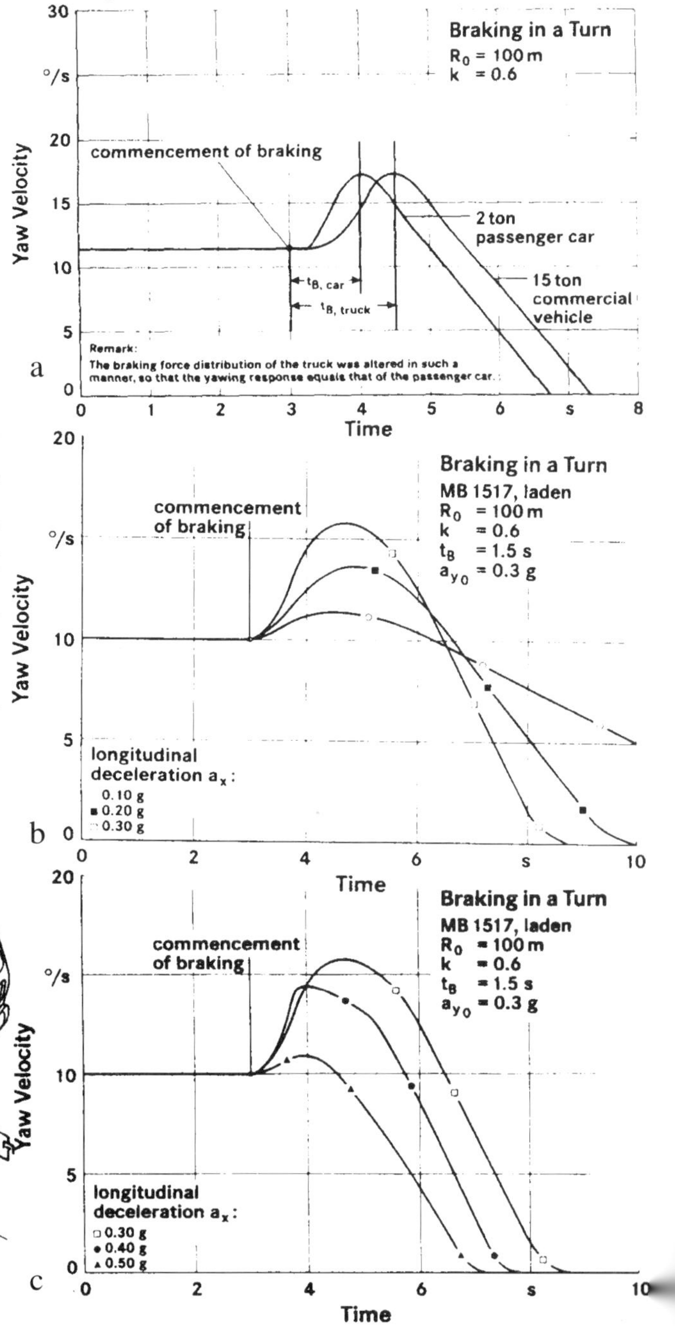

Fig 71: Brake force development with yaw

Truck technology: cab/body

Geoff Matthews of the Hawtal Whiting company Styling International has challenged the traditional layout of the commercial vehicle cab. Why, he argues, should the motor coach driver/operator have accepted lower and further forward driving positions than was generally the case in CVs — while the trucker is perched high in the air and subject to some uncomfortable ride motions from cab-nod and related phenomena. Ducting technology which was now better understood could be used to re-site the radiator from its frontal position, normally requiring high-drag grille and a disproportionate control over packaging the whole cab layout. He suggests the four alternatives in Fig 72. By placing the radiator on the roof of the cab, it could be cooled through a ducted fairing doubling as the aerodynamic deflector for the load-container body. Alternatively, why cannot it be mounted to the rear of the cab and cooled by ducting from the top-sides or underneath. It could even be horizontally under the engine to double as a noise shield. Resiting would allow the driver and passenger to move forwards and downwards — as well as allowing the cab to be shaped for greater environmental friendliness, allowing easier entry and egress with a rearwards shift of the front axle.

Fig 73: AS&A aerodynamic proposal

Aerodynamic container van A determined attempt to break the conservative tradition in container van design has been made by an enterprising joint venture between Besco Bodies, Leyland DAF, Ricardo and consultants AS + A. The project,Fig 73, aimed to tackle head-on the problem that large box vans conventionally have to overcome a wind drag at motorway speeds that is twice that due to rolling resistance and gearing losses. One interesting approach is the rear end treatment, avoiding the trap of going for a deep skirt below the rear doors — every effort being made to ensure there were no restrictions to exit of the underside air flow into the low pressure area behind the vehicle. Drag co-efficient of just 0.325 for the vehicle was achieved (for an equivalent untreated box van body behind the same cab, the figure is 0.685!). Interesting aspect of the front-end design is the collar fitted to the box-body leading edge — having a distinct gap between it and the cab and its air-deflector. In motion the space enclosed by

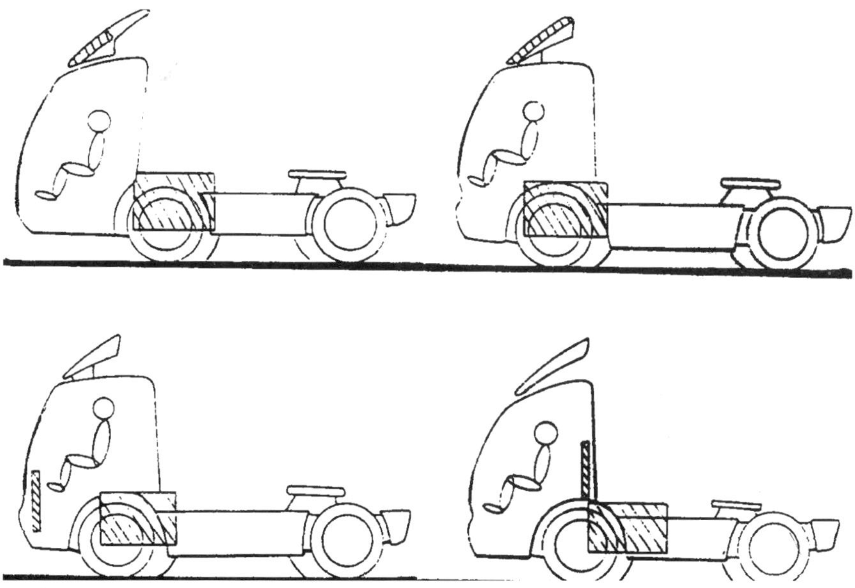
Fig 72: Rethinking engine packaging

the collar develops an air-pressure; this has the effect of creating an air barrier at the gap which performs the valuable function of smoothing oncoming air past the sides and top of the cab, Fig 74. Further back along the top of the box body a downward slope gives the vehicle a form of Camm tail. The aerodynamic package adds around 250 kg to vehicle weight. Power demand at 60 mph is reduced from 137 to 82 as a result — with a consequent reduction in fuel consumption of 40 %. Even at 40 mph there is 22 % improvement. There are substantial side benefits, too, in reduction of aerodynamic noise, improved crosswind stability and reduced side spray from the tyres. Final drive ratio of the rear axle was also reduced from 5.4 to 4.8:1.

Optimized for deliveries Seddon Atkinson's Leader urban-delivery vehicle, Fig 75, successfully incorporates a low-entry cab layout with central seating position for the driver and access ahead of the front wheels, to provide one-step walk-through facility. It is based on the company's Pacer chassis and a Jack Allen developed Britannia cab. The normal cab has three-person capacity but an optional five-person crew-cab is available. Cab construction is GRP over a steel support frame, with carbon-fire reinforcement in impact-vulnerable areas. Doors are pneumatically operated and an electrically-operated full-depth sunblind is provided for the deep windscreen. The back of the driver's seat drops to form a table for the two rearward offset passenger seats. Drop-down safety rails are fitted ahead of the crew seats. Crew access is available from both sides of the cab. An Allison five-speed automatic transmission is featured and either 4 x 2 or 6 x 2 axle configurations are available. The 4 x 2 version is rated 17 tonne GVW.

M.A.N. have built a working prototype delivery truck - for the future - in the form of the SI,W 2000 7.5 tonner, Fig 76. It has a stepped chassis frame to give low deck height for the box-body which is

Fig 75: Leader easy-access cab

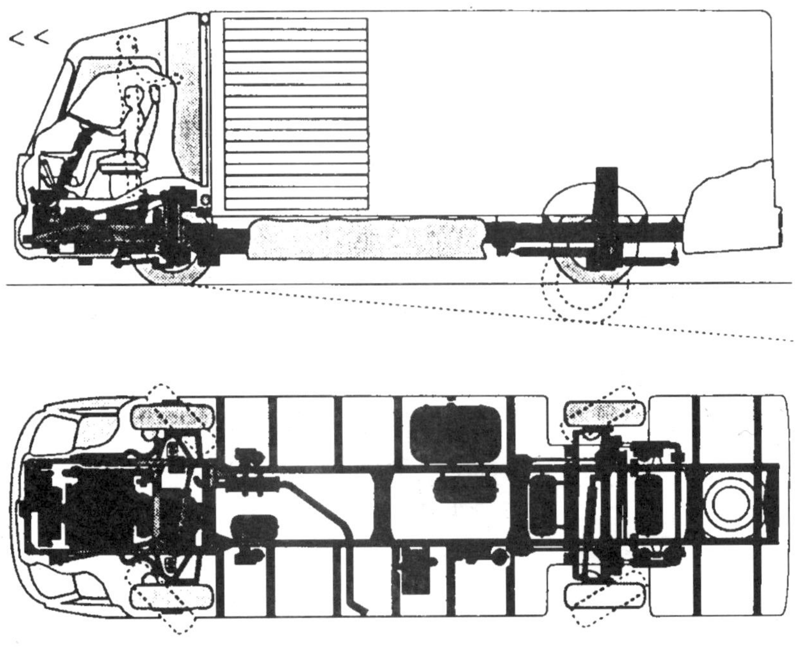

Fig 76: M.A.N. concept delivery tuck

accessible through the back of the cab as well as via its rear loading-doors. The four-cylinder engine is mounted horizontally beneath the floor and drives through a ZF automatic transmission. There is steering on all four wheels, rear steering being made possible by the front wheel drive layout. The 4.48 litre engine is centrally mounted beneath the cab.

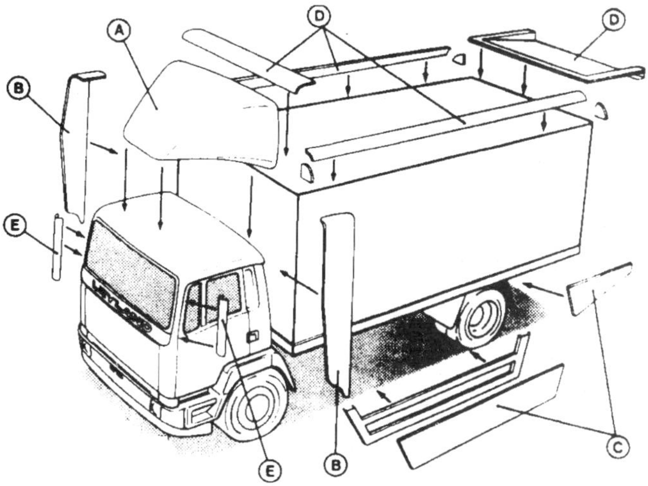

Fig 74: A cab roof deflector; B collar between cab and body; C side and rear guards; D corner fillets; E windscreen pillar blinkers

Suspension is hydro-pneumatic all-round, the rear being height adjustable by remote-control to match the vehicle to different loading banks. Plug-type cab doors move outwards and forwards to ease dismounting in cramped conditions.

Design for ecological advantage A gas turbine/ electric hybrid truck of 32 tonne GVW could maintain 60 mph up motorway gradients, Fig 77, is proposed by AS&A consultants. A diesel engine of equivalent power to maintain this performance would make the truck uneconomically overpowered for much of the truck's other operating service. Under this extreme performance requirement the gas turbine prime mover would be supplemented by electric power from some two tonnes of lead acid batteries. The 250 kW gas-turbine output would be directed through electric generators which power wheel motors and charge the battery pack. Fuel is natural gas compressed to 400 bar. The drive-by-wire vehicle would be particularly quiet in moving off from rest in sharp contrast to the noise emitted from heavy duty diesels driving through multi-speed gearboxes. A cruise control system would select the optimum speed on motorways for the gradient conditions and computer satellite link to base control would give advance warning of traffic conditions.

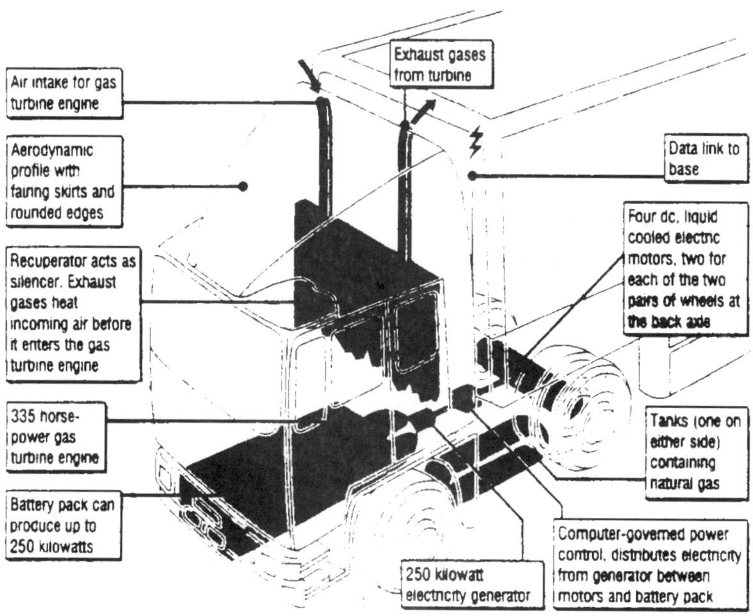

Fig 77: Eco-truck proposal

Truck technology: HS steel

Swedish Steel[4] cite the structures of refuse collection vehicles and freight containers, Fig 78, as typical of those which have benefitted from HSS sheet, 20% weight saving being achieved in each case; details of a tipping-frame for a container truck and a trailer draw-bar are also shown, Fig 79, where over 30% weight saving was achieved.

In a merit rating for comparable materials the point is made that any assessment of materials should be made on what the user is prepared to pay for a given weight saving. In the automotive industry the figure of $1-4 per kg compares with $200 and more in aerospace. The types of HSS are categorised in Fig 80 and the design capacity and stiffness of box-sections is nicely shown, on a comparative basis, in the figure. Loss of stiffness can in many cases be compensated by alterations in section shape while plate elements can be stiffened by grooves and other forming techniques.

It is a mistake, say the company, to link buckling wholly with elestic modulus. For the same sheet thickness, design capacity can often be raised in proportion to the square root of yield strength-increase. Dent resistance, too, can be improved by HSS and the manual quotes a typical door outer panel in 270 N/mm^2 yield steel which allows a 30 % decrease in thickness for the same dent resistance compared with mild steel, for a 30 % price increase.

Initial calculation in relative design can be based on the assumption that the ratio of thicknesses between new and old material is inversely proportional to the square root of the ration of yield strengths of existing and replacement materal. The above relationship leads to the chart of Fig 81 and graphs of Fig 82. The relationship is based on the general assumption:

$$E = AR^m t^n$$

where E is the property concerned; A is a constant; R is yield stress R_e os R_m; t is material thickness; m,n are constants defined by the table in Fig 83.

Design practice The first premise of thin-walled stuctures must be obeyed: that loading must where possible be directed into the plane of the sheet, Fig 84. Stiffeners should be used to raise the buckling load. Change in stiffness due to local buckling must be considered and so too must flange curling effects

Fig 78: HSS freight container

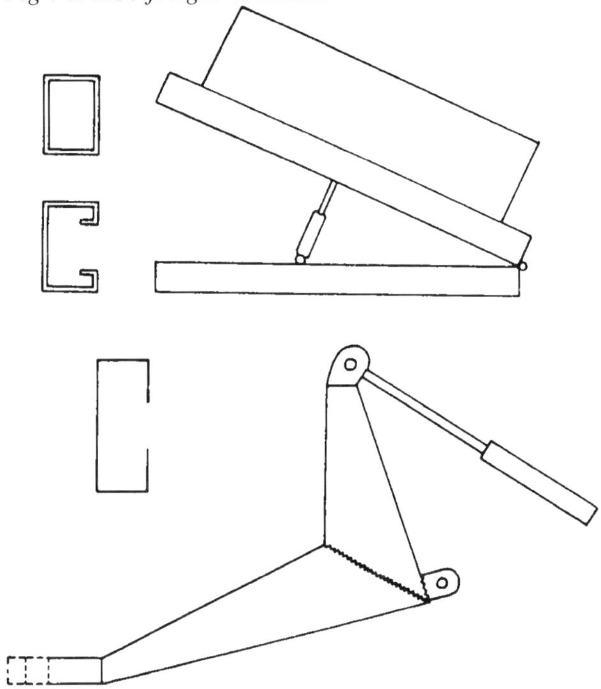

Fig 79: Structural detail in HSS

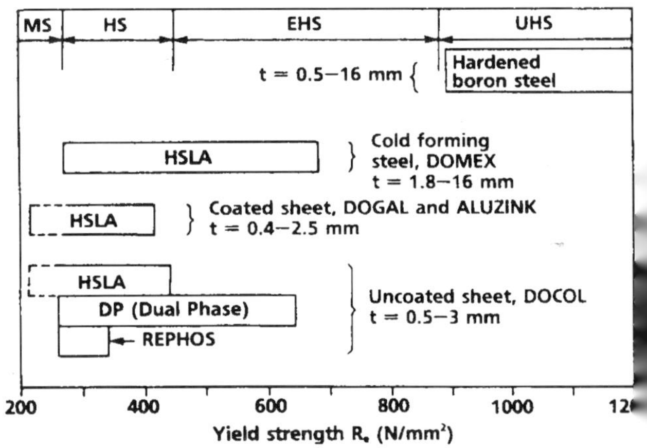

Fig 80: Types of HS steel

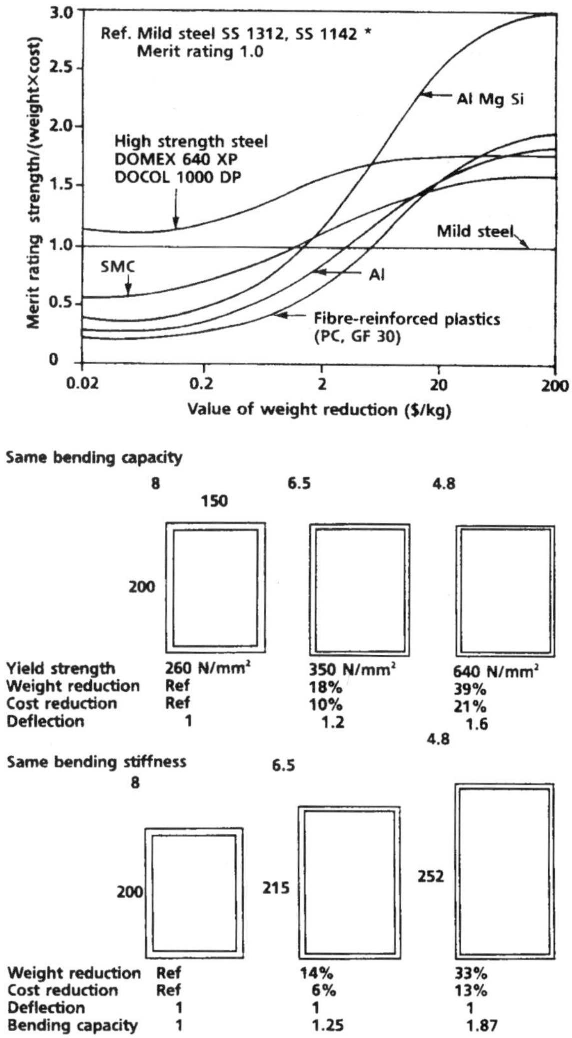

Fig 81: Material efficiency comparisons

in open-sections where buckling deformation can lead to changes in section properties. In the case of dump-truck bodies, a measure of permanent deformation is allowable in the handling of harsh payloads. This is in relation to the spacing of stiffeners, as shown in Fig 85.

Design of stiffeners in the general case where anti buckling duty is coupled to providing support against lateral loading is given in Fig 86. Use of the load-carrying capacity of the sheet in its own plane also applies to intersections of sidemembers and crossmembers, Fig 87. Bending moment due to the force P is resisted by a reversed pair of forces in the flanges and the shear force is resisted by shear forces in the web.

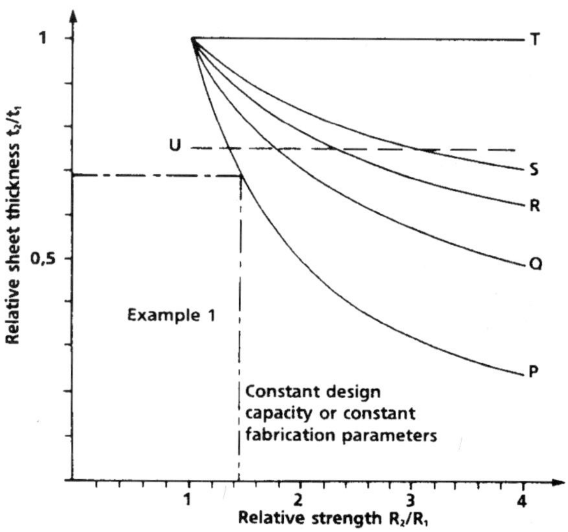

Fig 82: Steel characteristics, graphical

Property	Relationship	m	n	Remark
Design capacity, static load				
– tension, member	$\sim R_e\ t$	1	1	
– bending, sheet	$\sim R_e\ t^2$	1	2	
– bending, beam	$\sim R_e\ t$	1	1	Flanges
Local buckling	$\sim R_e^{0.5}\ t^2$	0.5	2	Worst case
Energy absorption				
– maximum load	$\sim R_e^{0.4}\ t^{0.9}$	0.4	0.9	
– energy absorption (14 m/s)	$\sim R_m^{0.6}\ t^{1.8}$	0.6	1.8	
Dent resistance	$\sim R_e\ t^2$	1	2	
Fatigue strength				
– parent material	$\sim R_e\ t$	1	1	
– welded joint				
small N	$\sim R_e\ t$	1	1	
large N	independent of strength	0	1	
Punch clearance	$\sim R_m^{0.35}\ t$	0.35	1	
Punch force	$\sim R_m\ t$	1	1	
Bending force	$\sim R_m\ t^2/w$	1	2	w - die opening

Fig 84: Calculation constants

DESIGN CAPACITY \ Curve	Constant sheet thickness						Constant design capacity						R
	A	B	C	D	E	F	P	Q	R	S	T	U	
Static load: tension, member	×						×						R_e
Static load: bending, sheet	×							×					R_e
Static load: bending, beam	×						×						R_e
Local buckling	o	×					o	×	o	o			R_e
Energy absorpt.: maximum load		×	o					×	o				R_e
Energy absorpt.: energy absorption (14 m/s)	o	×							×				R_m
Dent resistance	×							×					R_e
Fatigue: parent material	×	o					×	o					R_e
Fatigue: Welded joints: Small N ($\sim 10^3$)	×						×						R_e
Fatigue: Welded joints: Large N 10^5, 10^6: Without TIG dress.					×						×		R_e
Fatigue: Welded joints: Large N 10^5, 10^6: With TIG dress.				o		×						×	R_e

Fig 83: Steel data, tabular

Truck technology: aluminium-alloy

A recent announcement of the new British Standard code-of-practice for the Structural Use of Aluminium puts the confident prediction of structural behaviour in this material on a new footing. The availability of both parts of BS 8118, from BSI Linford Wood, Milton Keynes, MK14 6LE, now offers a quite new approch to structural design in the material. While compiled with a number of industries in mind. particularly civil engineering, there is a number of aspects which are of value to the CV end of the automotive structural spectrum.

The 'limit-state' of static strength is an interesting new concept which is based on the premise that the factored resistance of a member should not be less than the magnitude of that action-effect arising from factored loading. Rules for obtaining resistance to different actions are given for different structural members. Limiting stresses p are defined for bending and overall (o) yielding; for direct (a), shear (v) and buckling stability (s) as in Fig 88 for the heat-treatable alloys.

For beams a moment check is carried out such that at any section, moment M under factored loading should not exceed the factored moment resistance M_{RS} as derived below. Similarly the shear load at the cross-section is checked such that factored shear foce V does not exceed factored SF — resistance V_{RS}. First the section is classified as being either compact, semi-compact or slender, based on the least favourable of its component elements. The latter are defined as 'flat outstand', 'flat internal' or 'curved internal'. The section is classified on its susceptibility to local buckling and it may be stiffened or not as in Fig 89.

A parameter β describes element buckling tendency and depeneds on b/t or d/t — also breadth and depth b,d. For an element under uniform compression, $\beta = b/t$ while under a stress gradient, resulting in neutral axis at the centre, $\beta = 0.35d/t$ or $0.35b/t$; for any other stress gradient, $\beta = g$ times the above where g is read from Fig 90, y_c and y_o being distances from the neutral axis of the gross section to the more heavily compressed edge and other edge respectively of the element.

When the flat elements are reinforced two separate β values are found for each of the following modes: when the reinforced element buckles as a unhit, taking the reinforcement with it — and, when the sub-elements themselves buckle and the junc-

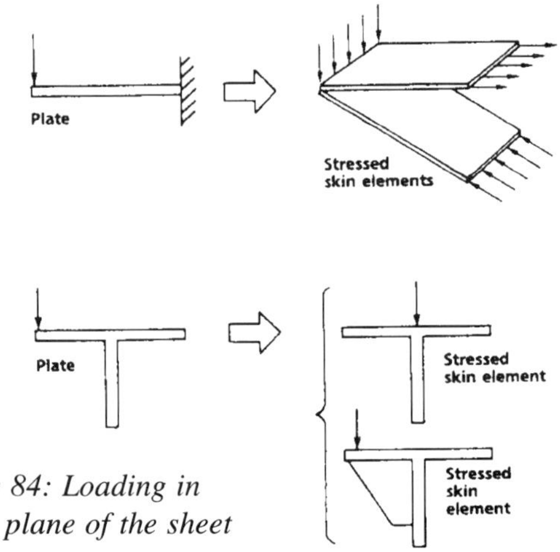

Fig 84: Loading in the plane of the sheet

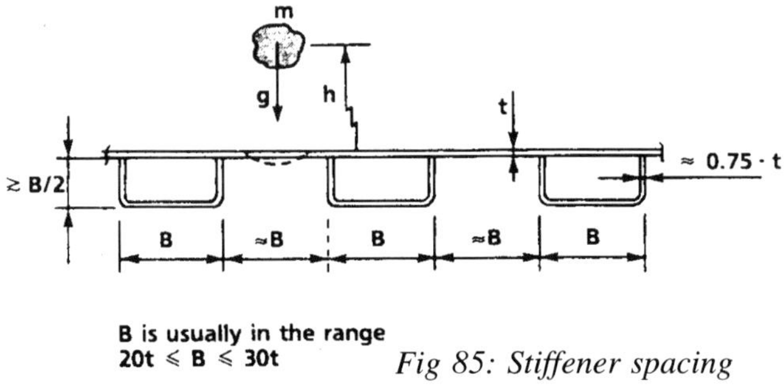

Fig 85: Stiffener spacing

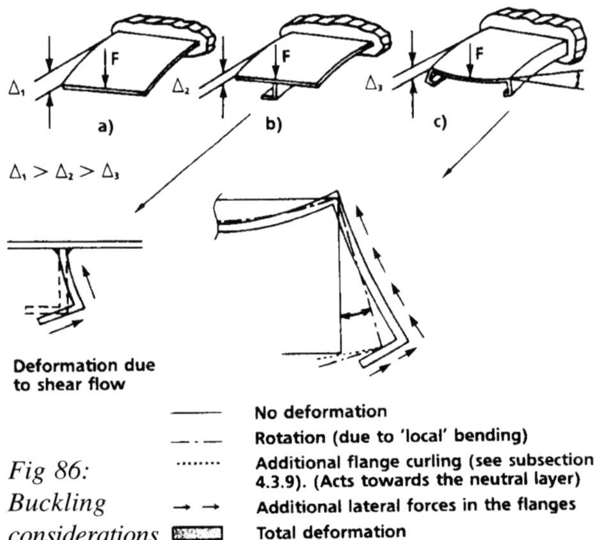

Fig 86: Buckling considerations

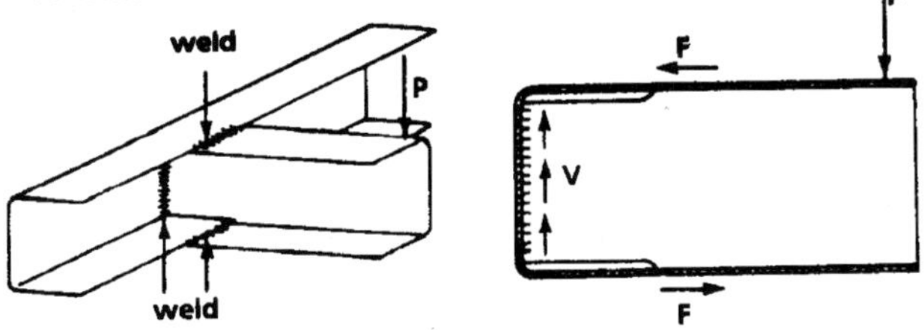

Fig 87: Sidemember/crossmember intersection

Alloy	Condition	Product	Thickness		Limiting stress		
			Over	Up to and including	p_o	p_a	p_v
			mm	mm	N/mm²	N/mm²	N/mm²
6061	T6	Extrusion	—	150	240	260	145
	T6	Drawn tube	—	6	240	265	145
			6	10	225	260	135
6063	T4	Extrusion	—	150	65	85	40
	T4	Drawn tube	—	10	95	120	60
	T4	Forgings	—	150	80	100	50
	T5	Extrusion	—	25	110	130	65
	T6	Extrusion	—	150	160	175	95
	T6	Drawn tube	—	10	180	190	110
	T6	Forgings	—	150	160	170	95
6082	T4	Extrusion	—	150	115	145	70
	T4	Sheet	0.2	3	115	145	70
	T4	Plate	3	25	105	140	65
	T4	Drawn tube	—	10	105	140	65
	T4	Forgings	—	150	115	145	70
	T6	Extrusion	—	20	255	275	155
			20	150	270	290	160
	T6	Sheet	0.2	3	255	275	155
	T6	Plate	3	25	240	265	145
	T6	Drawn tube	—	6	255	280	155
			6	10	240	275	145
	T6	Forgings	—	120	255	275	155
7020	T4	Extrusion	—	25	185	230	110
	T4	Sheet, plate	0.2	25	160	205	95
	T6	Extrusion	—	25	280	310	170
	T6	Sheet, plate	0.2	25	270	295	160

Limiting stresses, non-heat-treatable alloys

Alloy	Condition	Product	Thickness		Limiting stress		
			Over	Up to and including	p_o	p_a	p_v
			mm	mm	N/mm²	N/mm²	N/mm²
1200	H14	Sheet	0.2	12.5	90	95	55
3103	H14	Sheet	0.2	12.5	110	120	65
	H18	Sheet	0.2	3	150	150	90
3105	H14	Sheet	0.2	3	145	150	85
	H16	Sheet	0.2	3	170	175	100
	H18	Sheet	0.2	3	190	200	115
5083	O, F	Extrusion	—	150	105	150	65
	O	Sheet, plate	0.2	80	105	150	65
	O	Drawn tube	—	10	105	150	65
	F	Sheet, plate	3	25	130	170	75
	H22	Sheet, plate	0.2	6	235	270	140
	H22	Drawn tube	—	10	235	270	140
5154A	O, F	Extrusion	—	150	65	100	40
	O	Sheet, plate	0.2	6	65	100	40
	O	Drawn tube	—	10	65	100	40
	H22	Sheet, plate	0.2	6	160	200	95
	H24	Sheet, plate	0.2	6	225	250	135
	H24	Drawn tube	—	10	200	220	180
5251	F	Welded tube	0.8	2.0	220	230	130
	H22	Sheet, plate	0.2	6	125	155	75
	H24	Sheet, plate	0.2	6	175	200	105
5454	O, F	Extrusion	—	150	65	100	40
	O	Sheet, plate	0.2	6	60	95	35
	H22	Sheet	0.2	3	180	215	110
	H24	Sheet	0.2	3	200	235	120

Fig 88: Properties of heat- and non-heat-treatable alloys

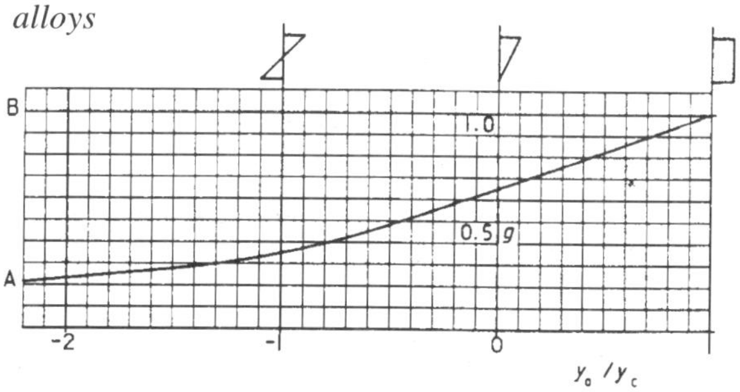

Fig 90: g-values

tions between them stay straight. In this second case β is found separately for each sub-element while in the first, the following procedure is adopted. For the kind of reinforcemenet shown in Fig 91, $\beta = hb/t$ and h is read from the figure. For a different kind of reinforcement the technique is to assume an equivalent size reinforcement of the standard configuration, c being chosen to give equivalent section second moment of area.

For shallow curved elements under uniform compression:

$$\beta = (b/t)/[1+(0.006b^4/R^2 t^2)]^{1/2}$$

where R is mid-thickness radius of curvature and b is the developed width of the element. R/b must be not less than $0.1b/t$ to define the degree of shallownees for which the method is applicable. In the case of round tubes, $\beta = 3(D/t)^{1/2}$ where D is the mid-thickness diameter. For struts in general classifcations are made according to slenderness ratio as gauged by the β values for the elements — such that $\beta \leq \beta_l$ is fully compact; $\beta_l < \beta \leq \beta_o$ is semi-compact and $\beta > \beta_0$ is slender in terms of moment resistance.

Local buuckling can be ignored for fully compact struts; for semi-compact struts the section can develop a moment of p_0 times the section modulus; for slender struts, moment resistance is reduced by premature local buckling at an extreme-fibre stress below p_0. Limiting values of β are given in the table of Fig 92. The manual also gives criteria for axial

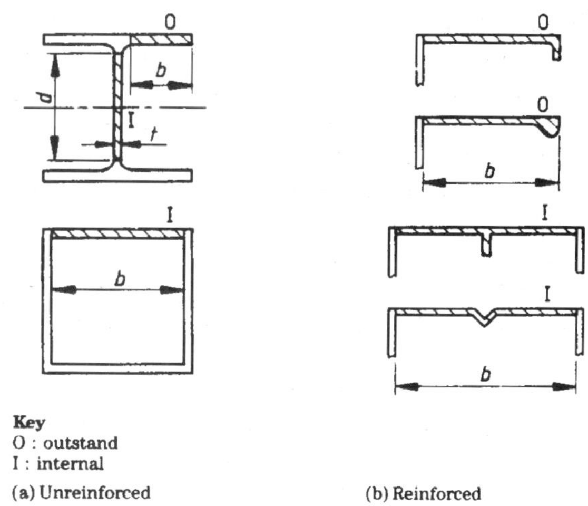

Fig 89: Stiffening of section

compression resistance as well as those described here for moment resistance.

For beams, the factored moment resistance M_{RS} is as follows:

* unwelded, fully compact: $p_0 S_n/\gamma m$
* unwelded, semi-compact: $p_0 Z_n/\gamma m$
* welded, fully compact: $p_0 S_{ne}/\gamma m$
* unwelded, slender: $p_0 Z_e/\gamma m$
* welded, slender: $p_0 Z_e/\gamma m$

where S_n and Z_n are plastic and elastic section moduli; S_{ne} and Z_{ne} are plastic and elastic net effective section moduli: Z_e is elastic modulus of the effective section; p_0 is the limiting stress and *gm* is the material factor from the table in Fig 93. Net section means including deduction for holes only while net effective section includes reduced thickness taken in the vicinity of welds to allow for the heat-affected zone; effective section includes the same thickness reduction allowance plus local buckling effect, but not holes.

Many other aspects are considered in the mannual, the above only demonstrating the technique and thoroughness of the method which would be ideal for setting up as a computerised design routine using one of the excellent maths programmes now available for evaluating expressions. Despite bearing a BS number it should be understood that the work is a Code of Practice and not just a set of standard procedures — and represents excellent value as a textbook of structural design.

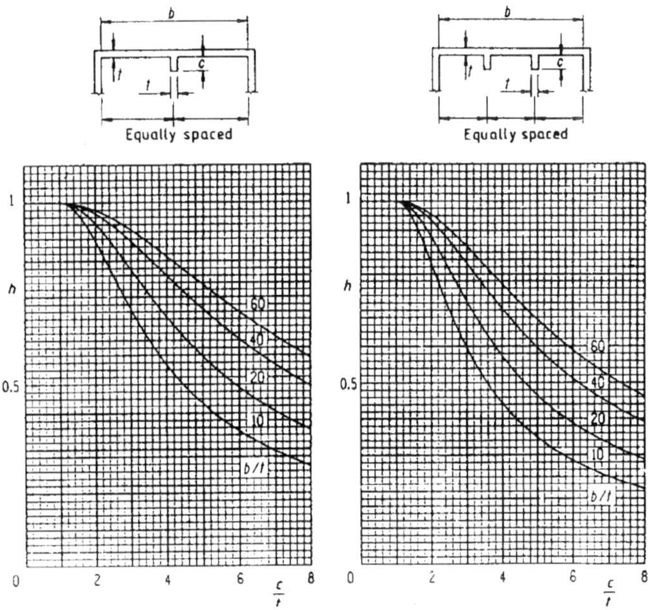

Fig 91: Reinforcement types

Truck technology: total system

DAF's future concept vehicle (FCV), Fig 94, is a fully driveable working unit whose features all have prospects, individually, for eventual production fitment. The chassis-frame is a fabricated rectangular section aluminium alloy box beam; the box girder design gives high torsional rigidity necessary to mount the independent suspension assemblies. Outboard of it are also hung fuel tanks and air reservoirs, while inside houses the the major elements of the drivetrain. The apertures in the box tube are for access to various parts of the drive train, the front rectangular one giving engine access. A flat-bottom sump is fitted to the engine which is installed as low as possible within the box-tube, forward of the front wheels centre-line.

While engine, semi-automatic transmission and prop-shaft are otherwise conventional, the rear wheel drive is novel, in the form of a 'double drive differential system', Fig 95. A torque apportioning device transmits two-thirds of it through a straight-through shaft to the second axle, the other third passing, via an annular gear drive, to the input pinion of the lead axle. Being a hypoid gear, its pinion offset allows the through-shaft to pass beneath the cross-shaft of the front differential. A short propshaft then connects the

Elements	β_0		β_1	
	Unwelded	**Welded**	**Unwelded**	**Welded**
Outstand elements	7ϵ	6ϵ	6ϵ	5ϵ
Internal elements	22ϵ	18ϵ	18ϵ	$15\ \epsilon$

NOTE 1. The quantity ϵ should generally be taken as follows (except for certain flange elements in beams, see 4.3.3.5):

$\epsilon = (250/p_0)^{1/2}$

where

p_0 is the limiting stress (in N/mm²)

Type of construction	γ_m	
	Members	**Joints**
Riveted and bolted	1.2	1.2
Welded	1.2	1.3 1)
Bonded	1.2	3.0

Figs 92 and 93: Beta (above) and gamma values

drive to the second axle.

Independent suspension, Fig 96, is fitted to both front and second 'axles', the latter both driving and steering. Special low-profile air-springs are employed which bear on saddles on the stub-axles. Lower links of the wishbone suspension, whose roll-centre is at ground level, are cranked downwards to clear the inboard disc brakes. In the second axle, drive extends outwards to the wheel hubs via a ring-type kingpin housing. Suspension travel is 80 mm to bump and 160 mm to rebound stops. For the third axle, twinned wheels on special two-flange hubs are supported by stub axles on a rectangular frame. It's slender, deep sidemembers have brackets extending outwards, beyond the inner road-wheels, to give as wide a spring-base as possible.

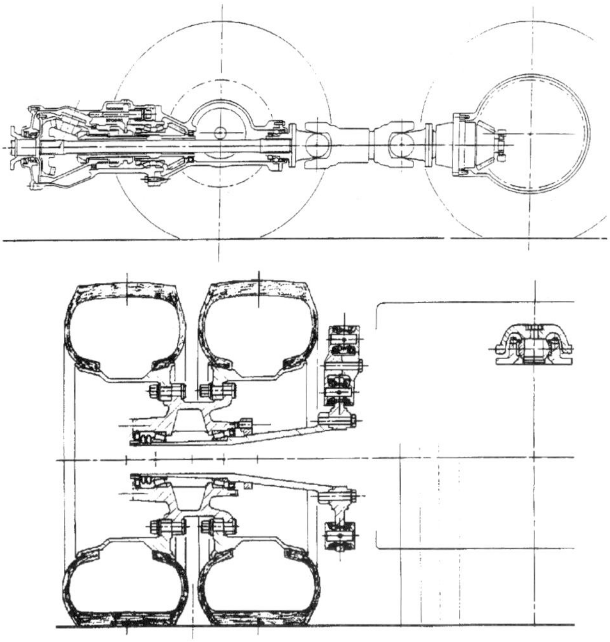

Fig 95: Double drive differential system

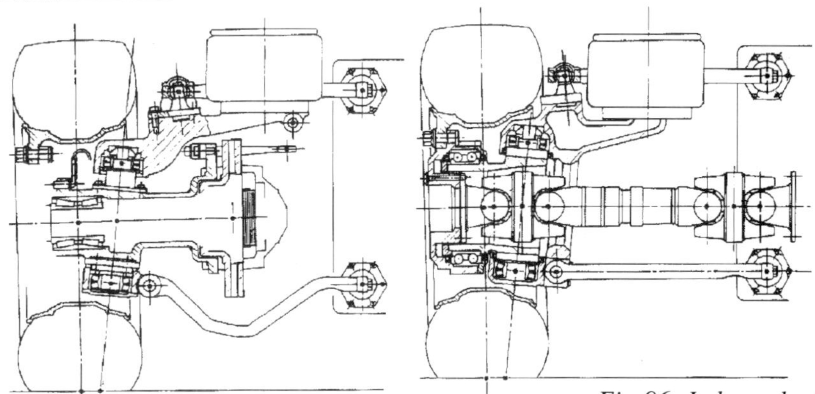

Fig 96: Independent suspension

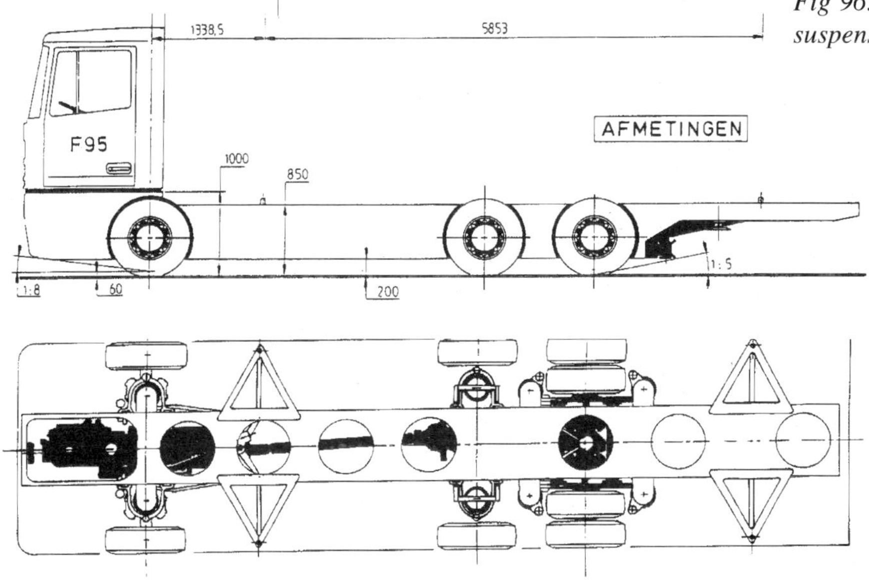

Fig 94: DAF FCV

Chapter 10: Bulk-carrier vehicles

Road tankers; tanker stressing; pressurised tankers; bulk powder vehicles;hopper design; open top bulkers/tippers; bulker charge/discharge; calculating tipper stability

Technology leads to constant change in the bulk carriage of liquids and powders and even the fundamental operating problem, of such vehicles, the inability to carry a return load has been addressed. Dual purpose tanker/platform trailers are seen in recent announcements by Wilson Double Deck Trailers of and the Finnish company Saalasti. The Wilson Decktank, Fig 1, allows 25,000 litre outward journeys for liquids and an equivalent weight of dry freight for the return load. The tank shell is made of 2.5 mm thick 316 stainless steel and has 100 % drainage to a single outlet point; the shell is mounted in an outer container and the 70 mm air gap between them filled with expanded polyurethane to form an insulated and impact-resisting structure. Covering the load deck, of height ranging from 1.53 to 1.69 metres according to model, is a 26 mm hardwood floor mounted above RSJ cross-bearers which will support 22,000 kg payload. A patented triaxle suspension system allows the floor to be just 0.5 metres from the ground and the vehicle rides on 265/70 R 19.5 low-profile tyres. The Saalasti vehicle, Fig 2, with a higher deck height of 2 metres, has a capacity of 27,500 litre or 27,500 kg of dry freight in a 44 tonne articulated rig. However, the company also make a drawbar combination with both towing vehicle and trailer having load-bearing tanks. That one can carry 40,000 litres or 39,000 kg of freight.

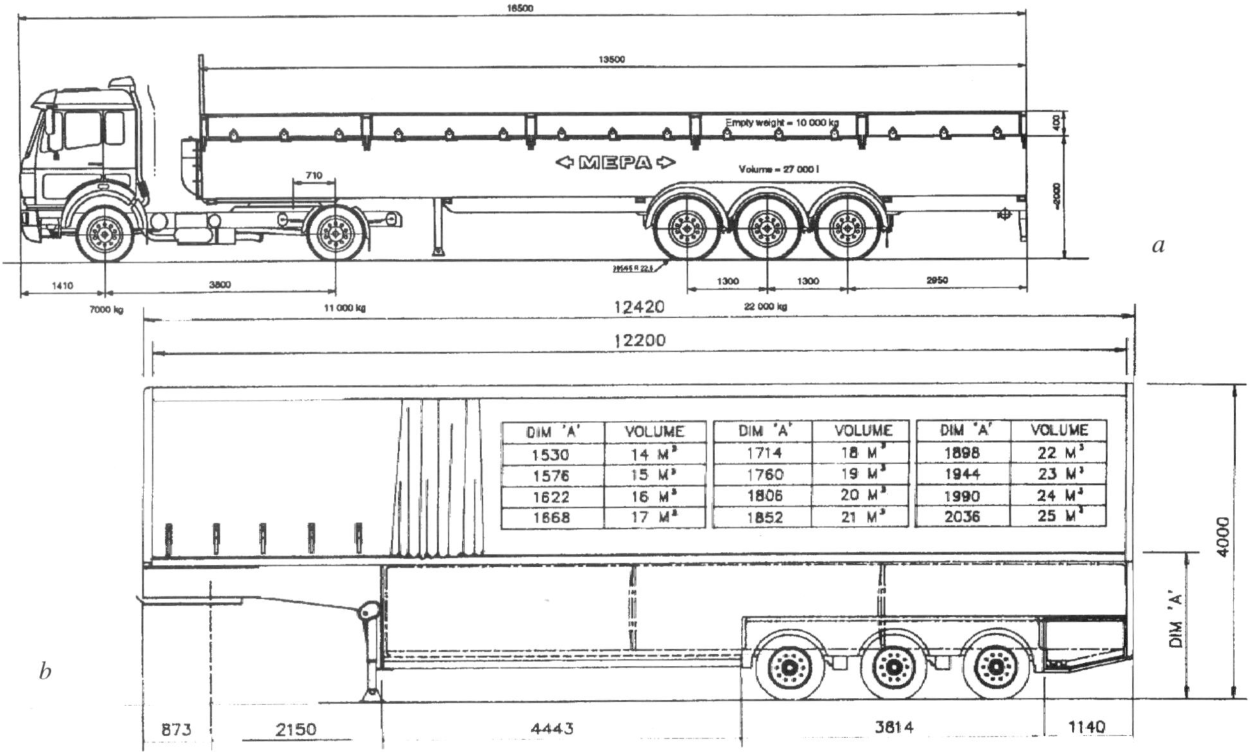

Fig 1: a. Wilson Decktank, top, and b. Saalasti dual-function vehicle

Road tankers

As one of the first truly chassisless vehicles, the monocoque road tanker was also one of the first vehicle categories to exploit the use of structural theory in design. According to the commodity being carried, the design can also be considerably influenced by Construction and Use regulations as well as the chemical and physical properties of the product.

The original 5000 litre/compartment limit for road tankers was one of the restrictions which forced the use of bulkheads along the length of a tanker, in the case of road-going vehicles. For airfield-only operation, a later regulation allowed refuelling tankers to have single compartment shells as large as 20,000 litres and, of course, very different design constraints apply to that particular category. In practice, however, these large airfield tankers do normally have baffles every 1.5 metres or so which help stabilise the shell structures.

In an effort to improve the operating safety of road tankers, the West German authorities recently constructed the TOPAS experimental vehicle, Fig 2, which could well affect future designs. This 39,000 litre unit has a centre of gravity some 300 mm lower than conventional designs and uses 70% aspect-ratio tyres at the maximum legal width dimension of 2.5 metres. The Haller-made tank has built-in roll-over bars and foam-filled double-skin front and rear bulkheads. Like most modern petrol tankers it is built for bottom loading and vapour recovery avoiding the risk of fuel spillage in roll-over.

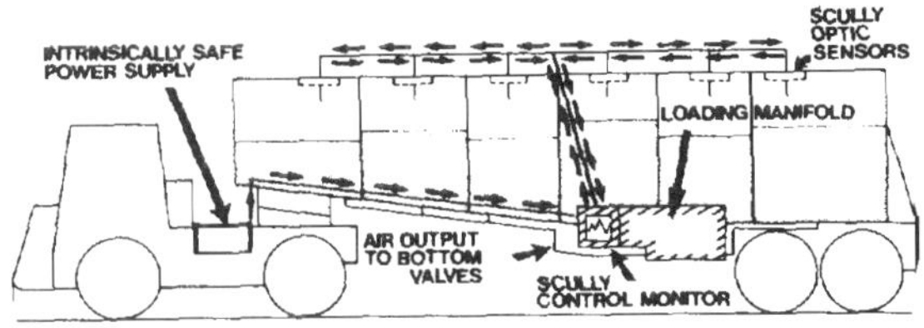

Fig 3: Skully overfill prevention system

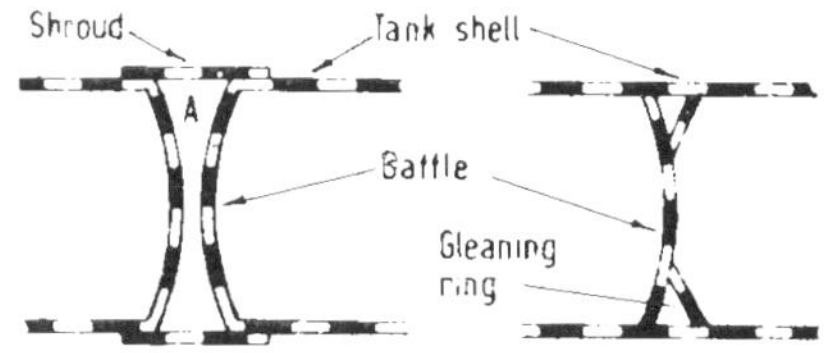

Fig 4: Small tanks joined by shrouds

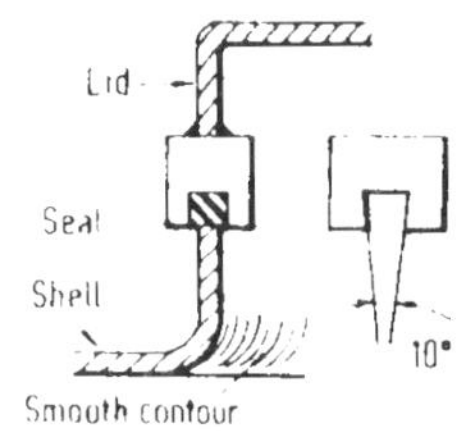

Fig 5: Man-lid details

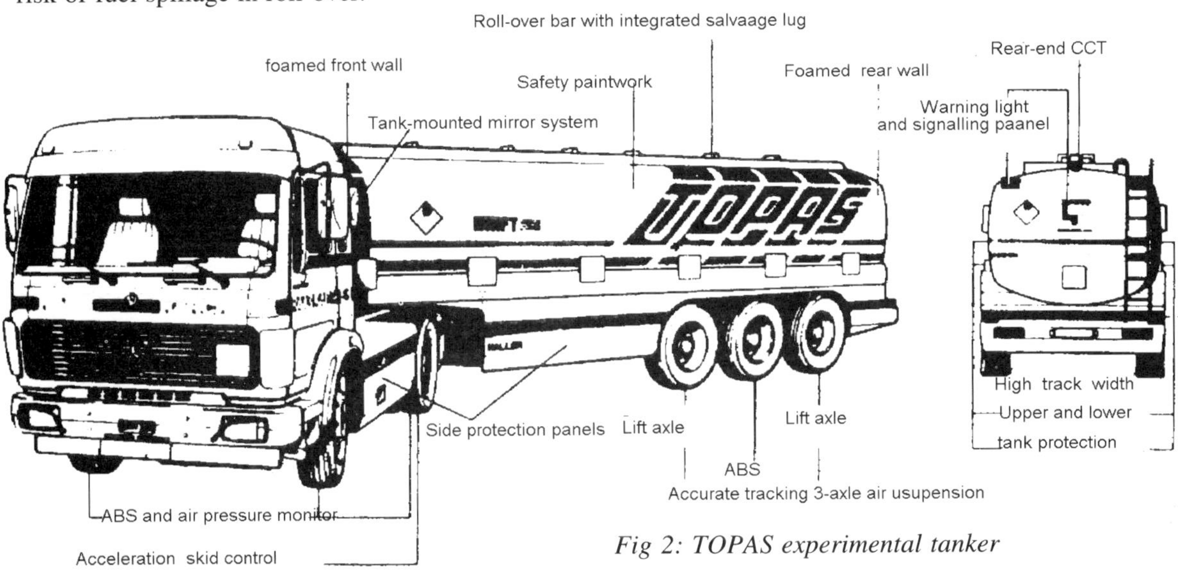

Fig 2: TOPAS experimental tanker

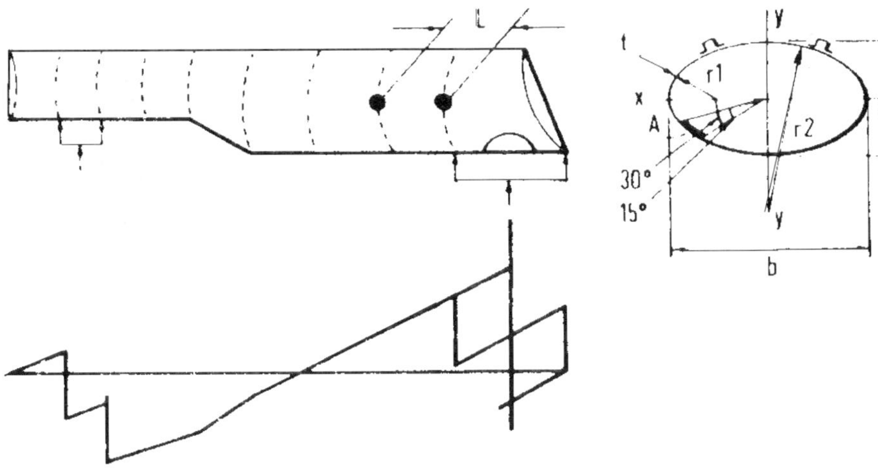

Fig 6: Tank dimensions and shear force distribution

Under the Dangerous Substances (Conveyance by Road in Road Tankers and Tank Containers) Regulations 1981, minimum tank specifications were prescribed for different commodities. Health and Safety Regulations broadly followed those applying to HM Goverment Petrol Regulations. A vessel was required to have a minimum wall thickness of 5 mm and to withstand a design pressure of 2.7 bar. In conjunction with the Institute of Petroleum, the HSE also supported a design code calling for centre of gravity of the load to be lower than 90 % of the wheel track.

With tanks being filled under pressure, from beneath, through the same hoses used for discharge, an overfill protection device is often used. One such is the Scully Dynacheck, Fig 3, which uses electronic pulsing detected by optic sensors which can shut off the air valves on the delivery hoses. Some tanks are required to resist vacuum which puts different structural demands on the shell. The 3500 gallon KiloWhale by Whale Tankers is an example, used to lift waste through heights of up to 10 metres. Others, for special commodities like liquid sugar, use gravity discharge involving special shaped tanks to ease flow.

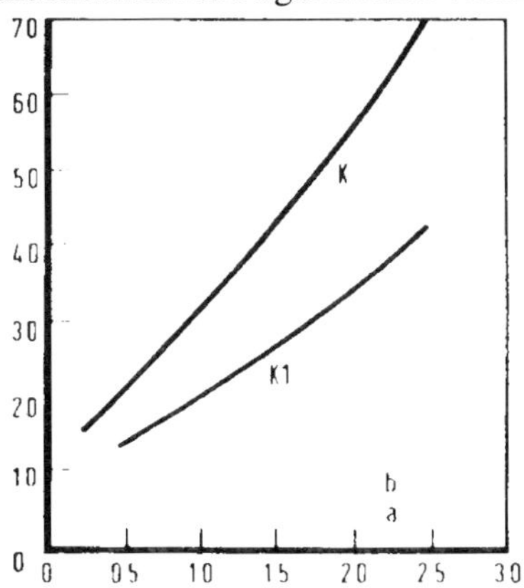

Fig 7: Shell shear stress factors

Shell construction techniques

Aluminium alloy of NS6 grade has been commonly used for conventional tankers of chassis-less construction. The aluminium companies have been a valuable source for structural design procedures used in tanker shell design. Useful codes have also been drawn up by the major oil and chemical companies with large tanker fleets. One such draws the distinction between conventional and pressure discharge tanks; in the latter case, individual small tanks may be joined up as in Fig 4, by shrouds, to make up the overall shell. Pressure entry points must be made to the enclosed spaces (A) as well as the compartments. An alternative approach is to use dished bulkheads braced by gleaning rings (also shown). Man-lids of 20 in (508 mm) diameter are typically provided, arranged as in Fig 5 to avoid contamination traps in the neck-ring to shell joint. Where ring stiffeners rather than bulkheads are employed, on elliptical section tanks the arc of the bottom surface is recommended for the suspension subframe mounting.

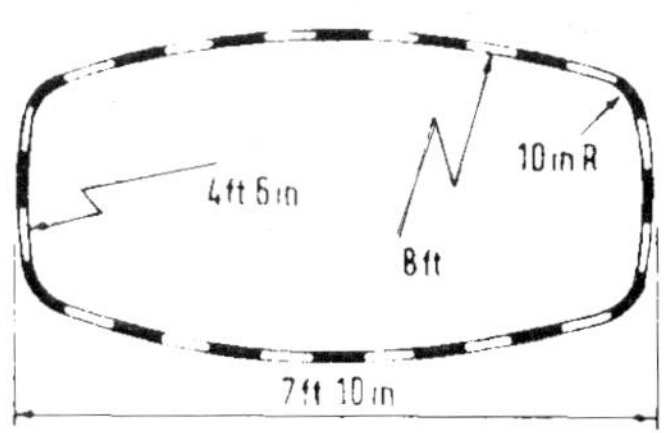

Fig 8: Max-section

Simple stressing of road tankers

A stressing procedure given in an ALCOA publication[5] gives the critical shear stress at which buckling might take place in the side walls of an elliptical tank under transverse shear as:

$$0.313EB(r_f/t)^{3/2}$$

where E is elastic modulus and:

$$B = 1.2[(2r_f/t)^{1/2}/(L/2r_1)]^{1/2}$$

The lower diagram Fig 6 shows the vertical shearing force V distribution along the length of the tanker. This gives rise to a corresponding shear stress $VQ/2tI$ for section moment of inertia I about the neutral axis and Q about the point on the section farthest from it. These values can be found from Fig 7 using $Q = K_1 a_2 t/4$ and $I = Ka_3t/8$, the safety factor derived on the vertical axis, relating maximum stress to critical stress.

Higher volume capacity of the tank is obtained by going from an elliptical to a so-called 'max-section' cross-section as shown in Fig 8. For the dimensions given in this example, a radius to thickness ratio of under 500 is recommended to avoid panel 'panting'. Circumferential welds around the shell should be arranged to avoid the pitfalls shown in Fig 9 and tank ends should be flanged so as to lap into the main shell by 3/4 to 1 in (19-25.5mm), keeping the weld away from the flange radius.

A worked example for the stressing procedure is provided by the stainless steel tank design shown in Fig 10, devised some years ago by International Nickel. The tank has weights and dimensions as shown in Fig 11. Rear shell section is straight-sided and the front section is conical in shape. Ring stiffeners are 12 gauge top-hat section welded to the shell. One of them is directly adjacent to the girth weld where the two sections of the tank are joined. Load reactions are first determined at A and B as R_a = *16,200* lb (7.3 tonnes) and R_b = *30,515* lb (13.75

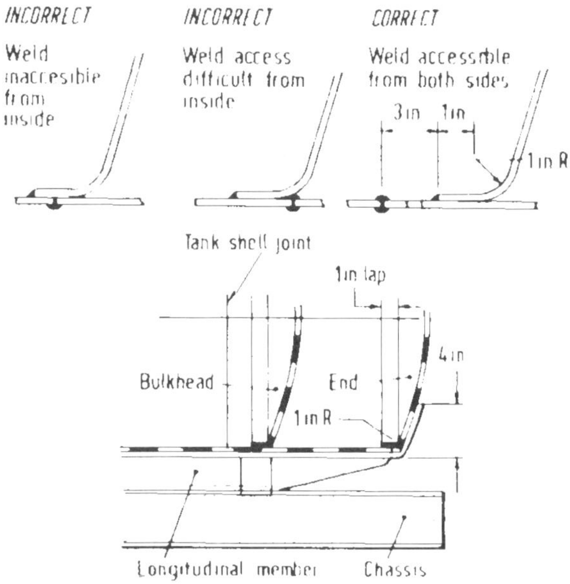

Fig 9: Shell welding recommendations

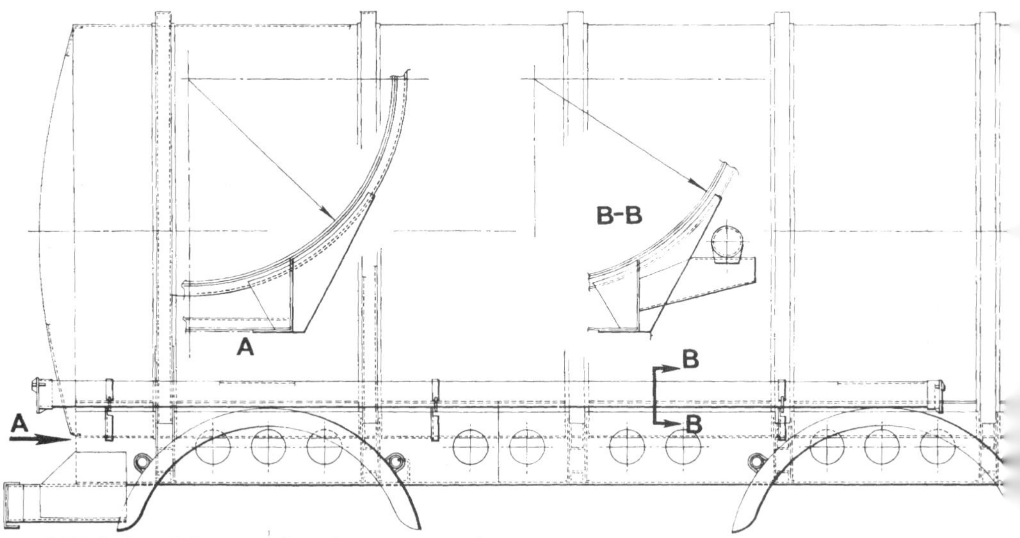

Fig 10: International Nickel stainless steel tanker construction

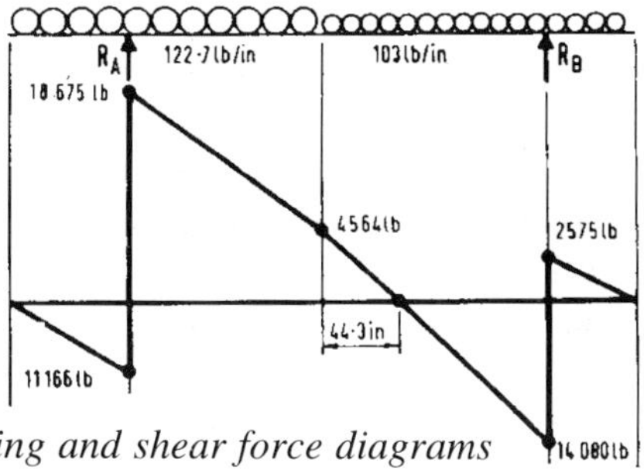

Fig 12: Loading and shear force diagrams

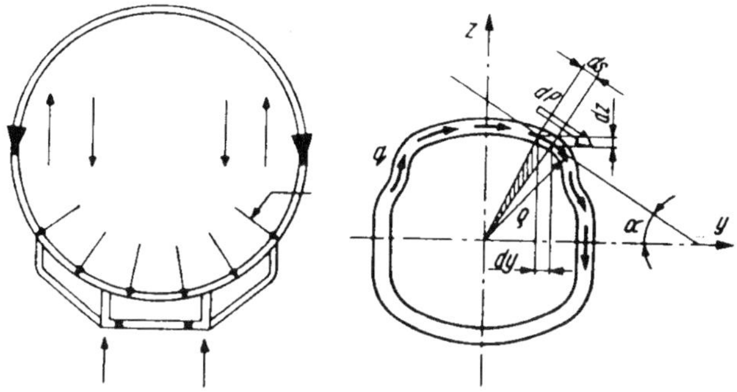

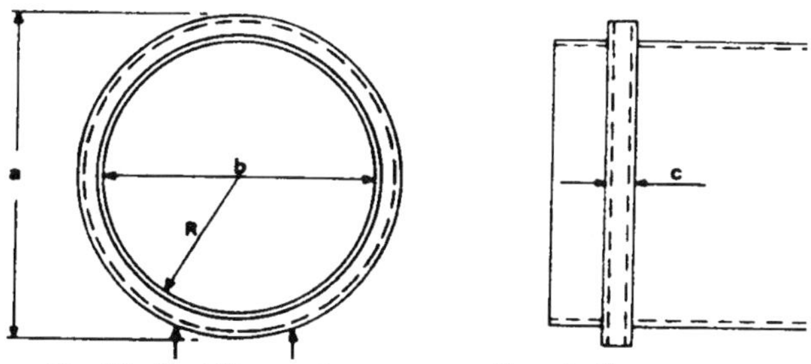

Fig 15: Saddle or ring support for shell

tonnes), Fig 12 shows the laden weight distribution over the rear suspension and front king-pin reaction points — also the shear force diagram. Critical shear stress at the point of maximum shear is 9660 lb/in^2 (6.66 MN/m^2), from the expression above, while the actual shear stress, similarly derived, was 4800 (3.33) — to show a safety factor of approximately 2.

Detail design of road tankers

Further details of tank construction recommendations are given in an Alcan Industries publication[6]. It points out that the shell of an aluminium alloy tank is usually composed of sheets, curved to the desired radius and welded together. Sheets are butt-welded to form a complete cylinder and the bulkheads inserted and welded in place. The lower sheets can be wrapped around the bulkheads with the upper sheet and then lap-welded. Longitudinal extrusions often form the upper and lower parts of the tank; see Fig 13 for these alternatives. Instead of simple dishes, bulkheads may be stiffened or swaged flat sheet as shown in Fig 14a and have end-heads the same or simple curved sheet welded to the shell as shown in Fig 14b.

In designing supports, it should be noted that the weight of the cargo is carried first by circumferential tension in the shell. These tension stresses create shear stresses in the shell which are maximum at the supports where load feeds down from the shell walls via the bulkheads. The support reaction can be

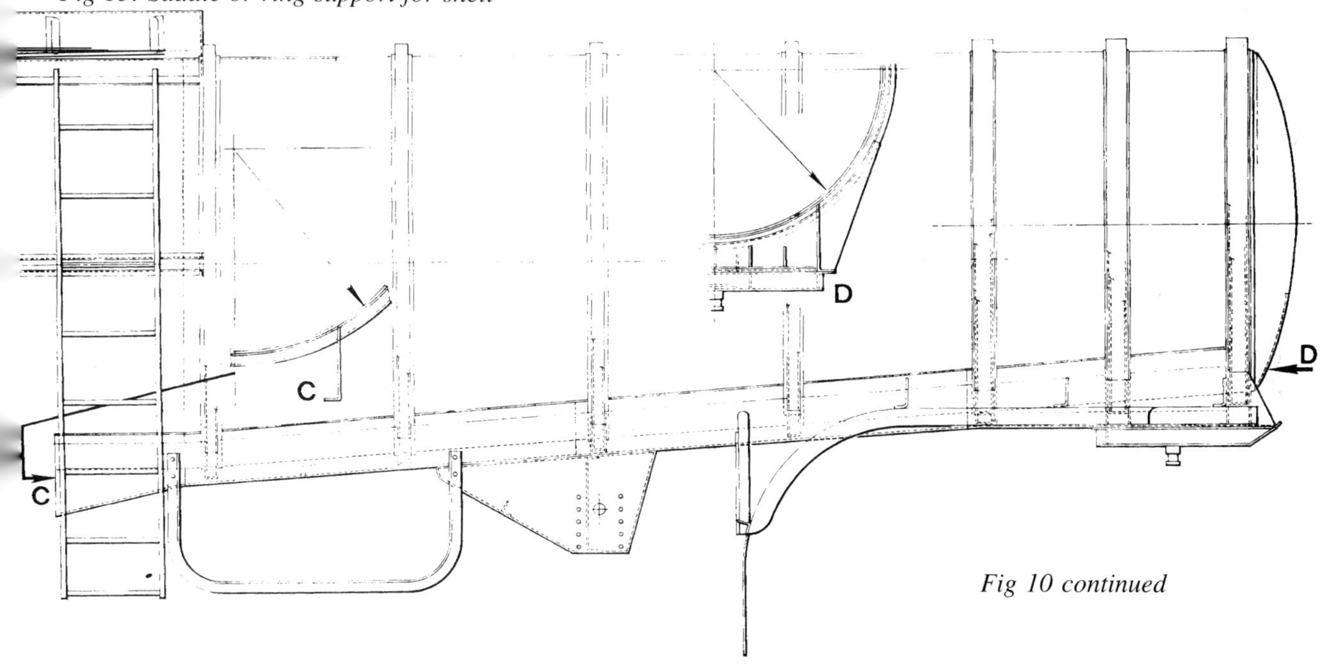

Fig 10 continued

provided by a saddle or by total exterior rings, Figs 15 a and b. The collapse load for the ring shown, not welded to tank, is:

$$2t[c + 2(Rt)^{1/2}]\sigma_y$$

where t is shell thickness, c ring width, $R = b^2/2a$ and s_y the alloy yield strength. In calculating transfer of load downward to the vehicle suspension, stress calculations should be based on the thickness of fillet weld either side of the saddle webs. Dos and donts of alternative bulkhead and support arrangements are shown in Fig 16.

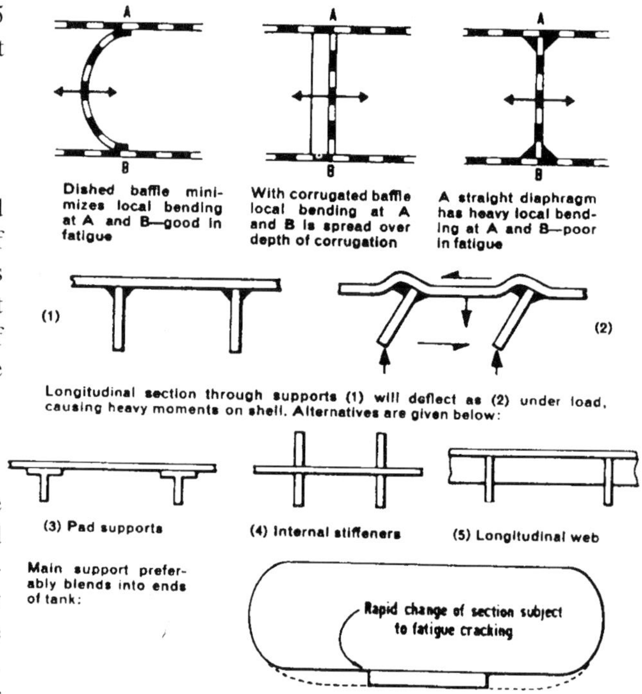

Fig 16: Dos and donts of bulkhead welding

Torsional loading of tankers

Because of the very large diameters of tanks relative to the twisting moments applied to them, by road wheels striking assymetrical bumps at speed, torsional stresses are comparatively low. However they can combine with the shear stresses computed above to lower the safety factor from the critical stress. Theory of structures texts give expressions for shear stress arising in twisting a relatively thick walled cylinder. Thin-walled closed tubes, by contrast, resist torsion by the area of the cross-section bounded by the mean thickness of the wall so that shear stress $\tau = T/2A$ for torque T and bounded section area A while the twist/unit length is $(T/4GA^2)\backslash\Sigma ds/t$ for shear modulus G, wall thickness t.

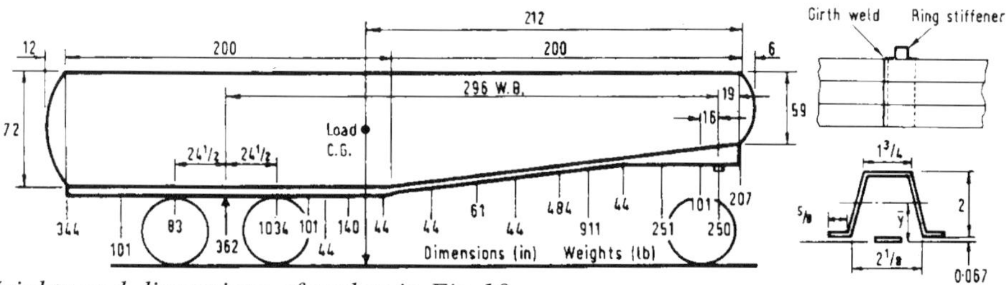

Fig 11: Weights and dimensions of tanker in Fig 10

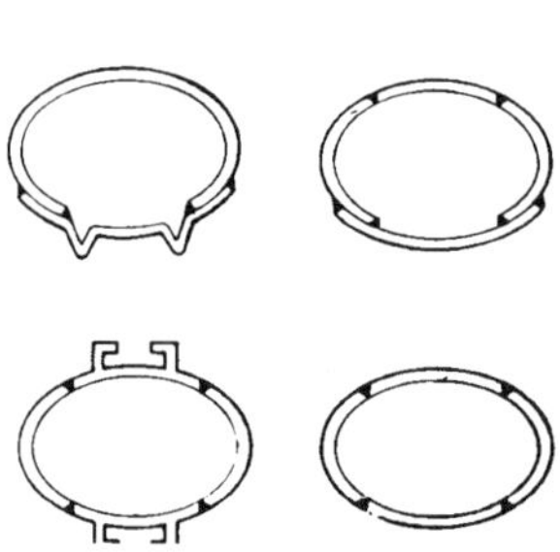

Fig 13: Alternative welding configurations

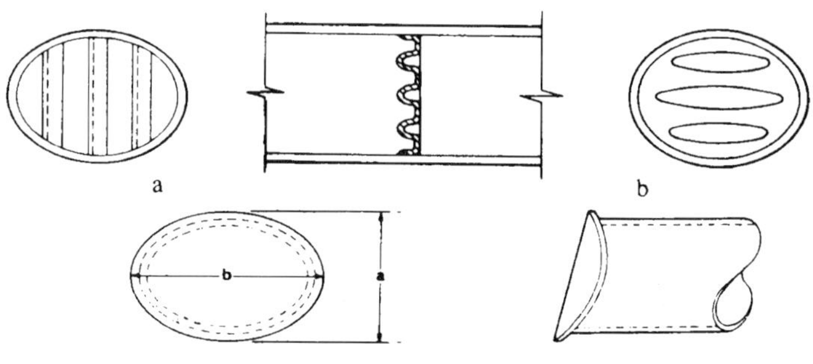

Fig 14: Bulkhead and end alternatives

Pressurised tankers

The design of low-pressure bulk powder tankers is generally carried out to BS 5500. When higher pressures are required for special purpose, use can be made of analytical work catalogued by ESDU (Engineering Science Data Unit) whose pressure vessel studies have now been written into computer programs. In all cases an understanding of the basic theories which underline these practices is useful in appreciating the design limitations.

An introduction to semi-empirical stressing methods (simple Engineers theory of bending factored by values obtained from the results of experimental measurement) for road tankers was given in the last section. Pressure tank design is also discussed in the next section, on bulk powder vehicles, based on the methods of BS 1500, Part 3. However, a better understanding of the factors involved in pressurised road tankers for bulk liquids comes from the basic theory of thin shells subject to internal pressure, Fig 17.

The force system shown arises from a half full tank of liquid, in this example. For internal pressure p and tank length l, $P = pld$ and the transverse skin stress $\sigma_1 = pd/2t$. Pressure on the ends of the tank also produce a complementary longitudinal stress of $\sigma_2 = pd/4t$. In a cylindrical shell with hemispherical ends of a different thickness, Fig 18, the hoop strain in the main tank will be:

$$(pd/4\tau_1 E)(2 - \nu)$$

where E is elastic modulus and ν is Poisson's ratio; τ is shear stress. The corresponding value for the dished ends has t_2 in place of t_1. These two strains should be equal, for compatibility, at the junction of tank and ends. so that:

$$t_2/t_1 = (1 - \nu)/(2 - \nu)$$

This ratio is 7/17 where ν is typically 0.3; for equality of stress between tank and ends the ratio should be reduced to 0.5. In the case of a rigid-vehicle tanker for which torsional loading due to road irregularities is combined with pressure, the shear stress on transverse planes needs to be added to direct stresses to obtain a principal stress:

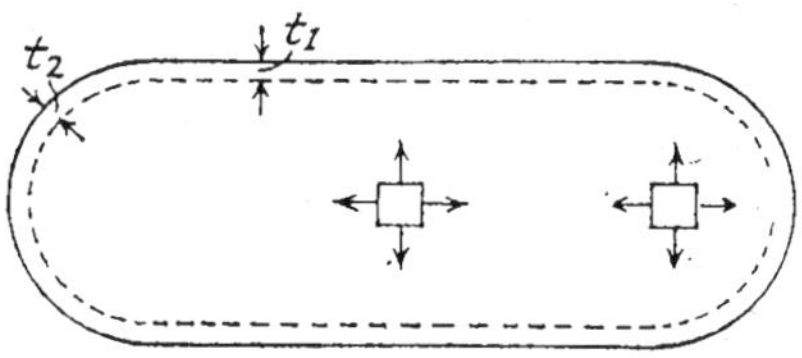
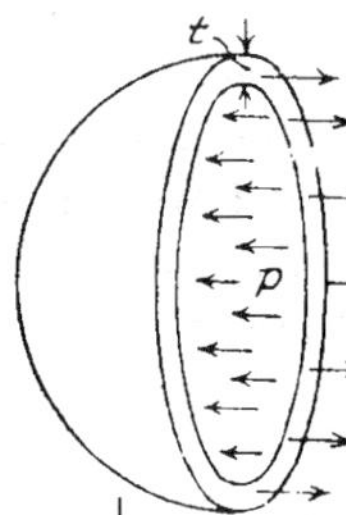

Fig 18: Cylindrical shell

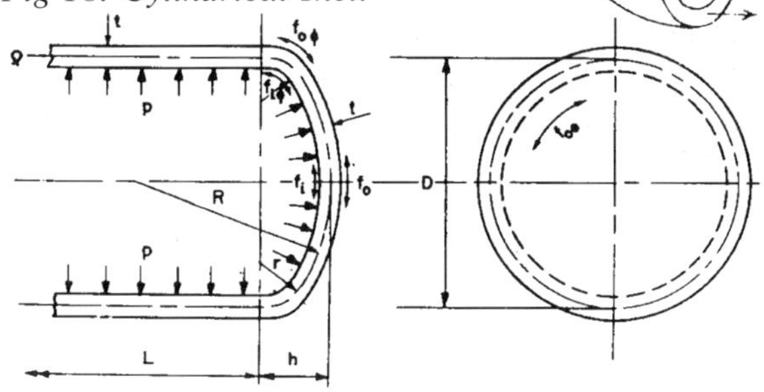

Fig 19: Dimensional notation: Dished-end cylinder (stress σ notated f)

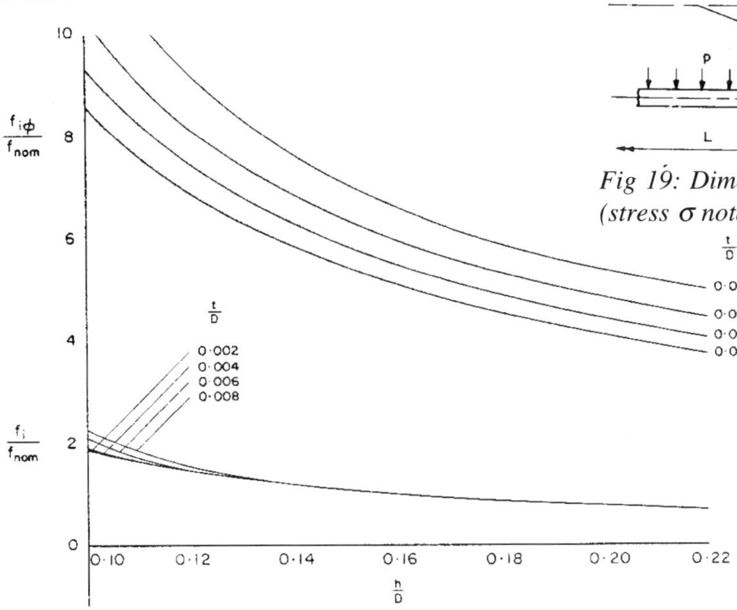

Fig 20: Stress characteristics: dished-end tank

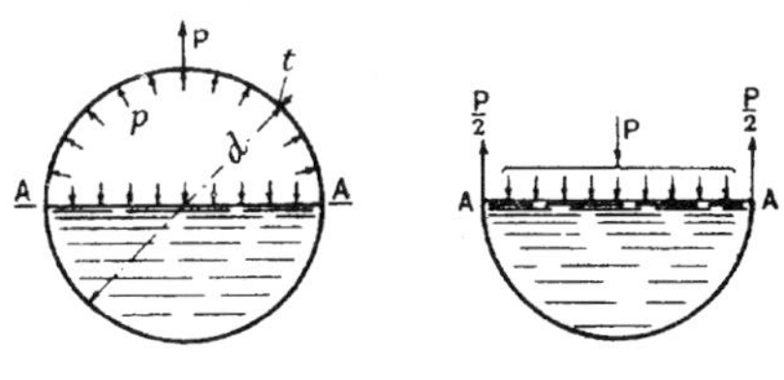

Fig 17: Shell under internal pressure

$$(1/2)(\sigma_1 + \sigma_2)+/-[(\sigma_1 - \sigma_2)^2 + 4\tau^2]$$

More exact analyses of pressure vessels of many different shapes and loading conditions can be found in 'Roark's classic work[7]. Many bulk powder tankers use conical ends where tipping is used to assist discharge. To obtain even stresses in tank and ends for the same material thickness usage the ends are made with an angle of 60 degrees to the axis of the cylinder. When the cone is truncated the ideal diameter of the flat portion, from the standpoint of pressure retention is:

$$[16t^3\sigma/3p]^{1/2}$$

The computerisation of pressure vessel design has been eased by the recent release by the Engineering Science Data Unit of PV Designer-BS, a program geared to BS 5500. Its calculation procedures deal with cylindrical shells, stiffening rings, heads and bases; flanges, nozzle openings, flat plates and cone-cylinder reinforcing.

ESDU also publish three useful conventional data sheets of value in pressure vessel design. Cylindrical tanks with flat heads are dealt with in 66010. Stresses and deflections in the flat ends are dealt with in 65002/3 and torispherical ends in 67017. Fig 19 shows the force system and dimensional notation for the cylindrical tank and Fig 20 is typical of the curves derived from thin-shell theory. The shell stresses are nominal ones at the junction of tank and end resulting from a combination of bending stresses due to the effect of the end on the tank as well as direct stresses due to pressure thrust of the end and circumferential membrane stresses. Fig 21 shows the load case and notation for the torispherical dished end and Fig 22 a sample curve. Usually the maximum tensile stress occurs in or near the knuckle.

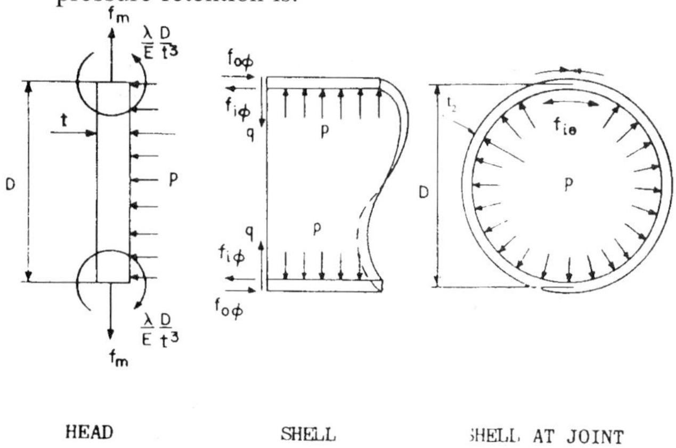

Fig 21: Dimensional notation: torispherical end tank (stress σ notated f)

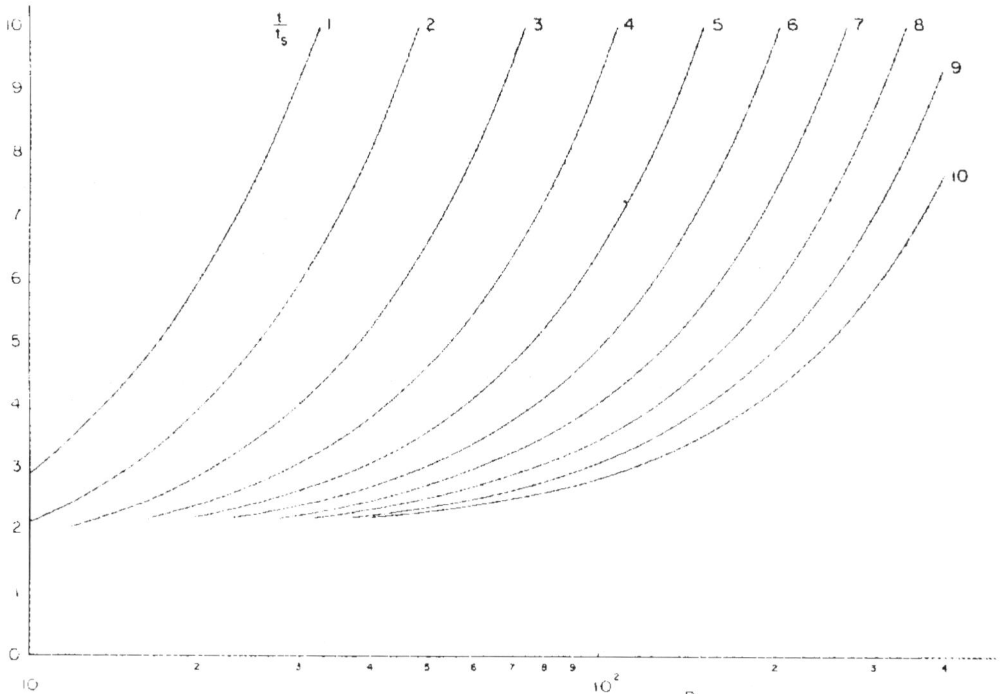

Fig 22: Stress characteristics: Torispherical end tank

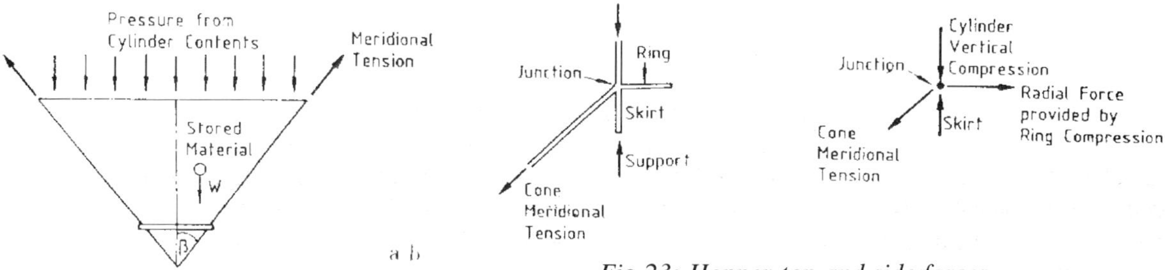

Fig 23: Hopper-top and side forces

Vertical cylinders for bulkers

A number of road-going low-pressure vessels for powders and viscous fluids are conical-bottomed vertical cylinders using pressure-assisted gravity discharge. To some extent stressing methods used for elevated silos can be used in these cases. The walls of such vessels are subject to both normal pressures and the frictional shears from the retention and discharge of the product. The former cause hoop stresses and the latter vertical compression stresses. Modes of failure include bursting, buckling under vertical or circumferential compressive stress or buckling under membrane shear stress. Work at Edinburgh University has found that the tension at the hopper top, Fig 23, may be from global equilibrium (a), the vertical component of this tension being resisted by the support (b) and the radial component by a ring. The ring thus carries a high circumferential compressive stress and if it is not provided by addition the hopper, cylinder and skirt will deform to form an effective ring. Failure modes of the ring may be either plastic collapse or buckling.

Metalair's tipping and non-tipping bulkers are each built to BS 5500; the company's tank-container design of up to 50 cubic metres capacity which can withstand pressures up to 2 bar is seen in Fig 24.

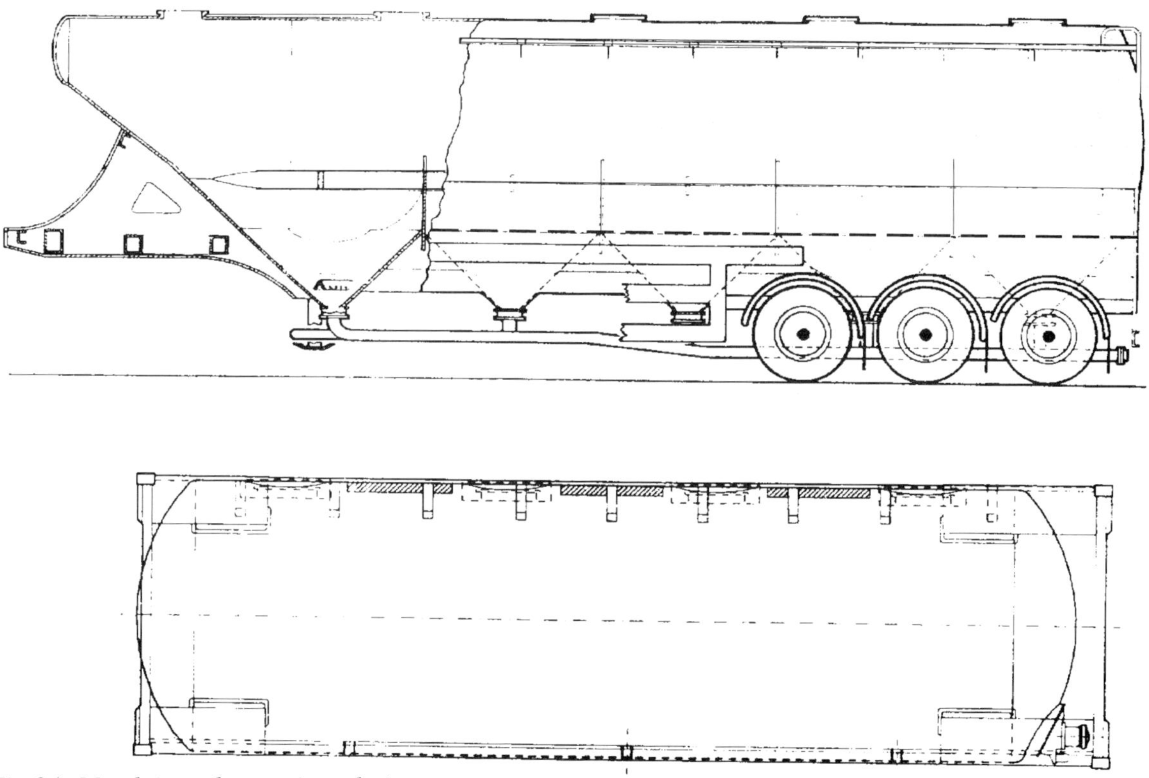

Fig 24: Metalair tank-container design

Bulk powder vehicles

The transport of bulk powders by so-called bulker vehicles involves careful design and selection of body type. If the material being carried is only required to be dumped at ground level then obviously a high-sided tipper could well suffice. If access to the dumping point is restricted then the use of a screw- or extending flat belt- conveyor would also be valuable — and a non-tipping body might suffice. For below-ground material 'dumps', or those accessible by overhead tracks, then gravity-discharge through the bottom of a hopper body might prove to be the solution.

Where both the horizontal and vertical distance between the point of vehicle access and the material 'dump', or store, is particularly large then pneumatic conveying is probably going to be required. This could involve a special-purpose pressurised body. Two common options are a fully pressurised body or a non-pressurised body having a rotary valve feeding a small pressurised compartment. The valve acts like a revolving door — it is a mechanical seal allowing material to be fed from an area of atmospheric pressure into one that is connected to a pipe used to convey the material at some 2/3 bar above atmospheric.

The economics of the operation often depend on the most intensive utilisation of the vehicle. Thus a quick turn-around of the vehicle, offered by a high pressure body operating at say 40 tonne/hour discharge rate may be preferred to one that might take twice the time — by a low-pressure, rotary valve, method. The relative potential payloads of the body types must also be considered and even the possibility of employing highly specialised rectangular section bulker tanks with the prospect of carrying 're-turn-loads' on top. Another factor is the weight of the thicker walls required for a high pressure bulker — which might adversely affect the maximum payload of the vehicle. In some cases a site air-pressure supply might be available for obtaining pneumatic discharge, avoiding the need for a vehicle-mounted compressor.

The ability for a vehicle to suck up material from bulk bay storage can also be an advantage, where overhead hopper feeds do not exist. Maintenance costs can also be crucial to overall economics and an operator may decide to have different demountable bodies for different materials — that

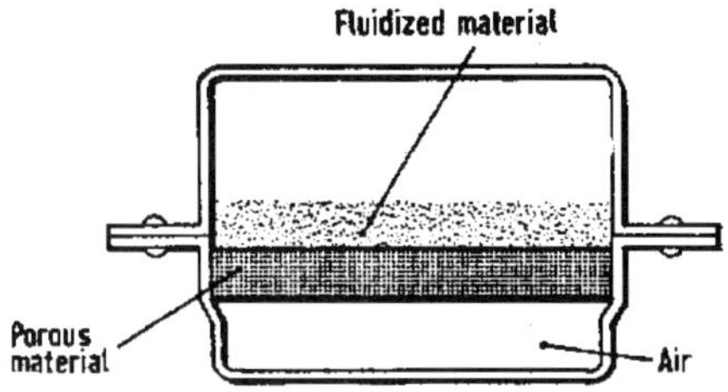

Fig 25: Fluidised bed

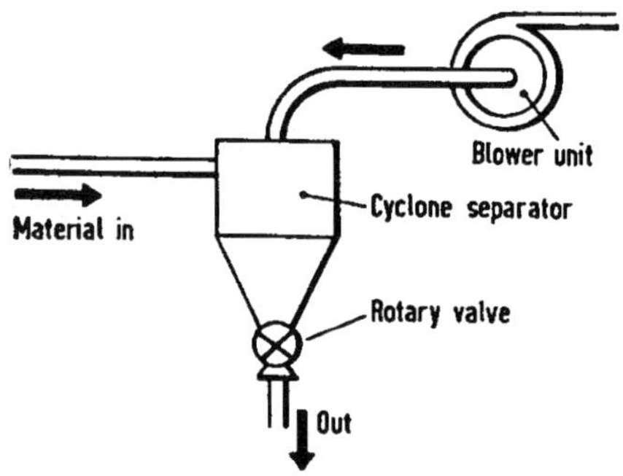

Fig 26: Suction handling system

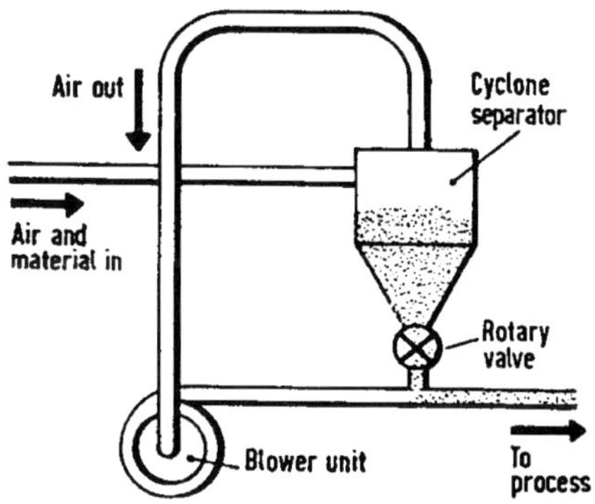

Fig 27: Combined suction and blowing system

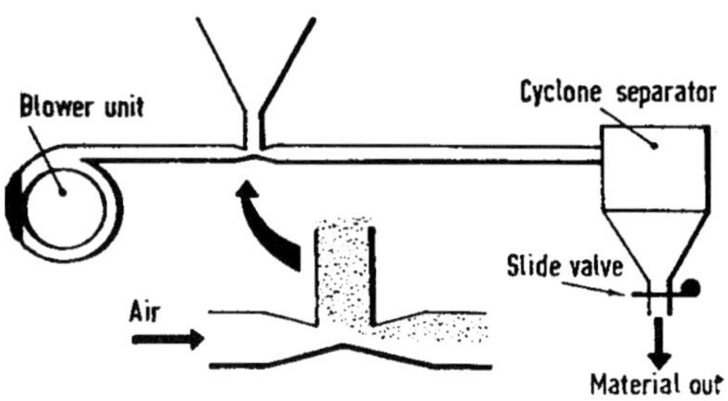

Fig 28: Low-pressure blowing system

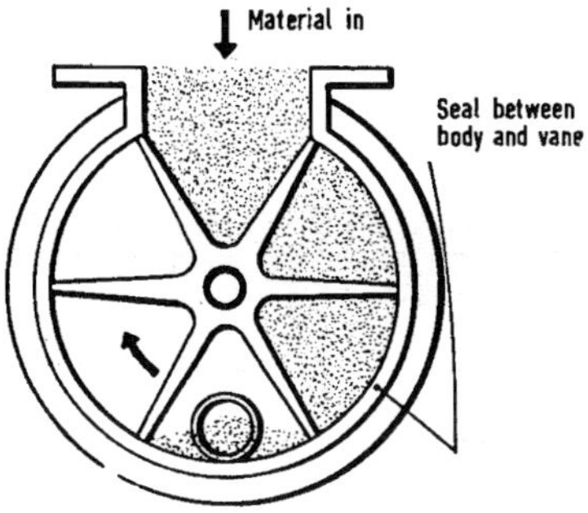

Fig 29: Medium pressure blowing slide-valve

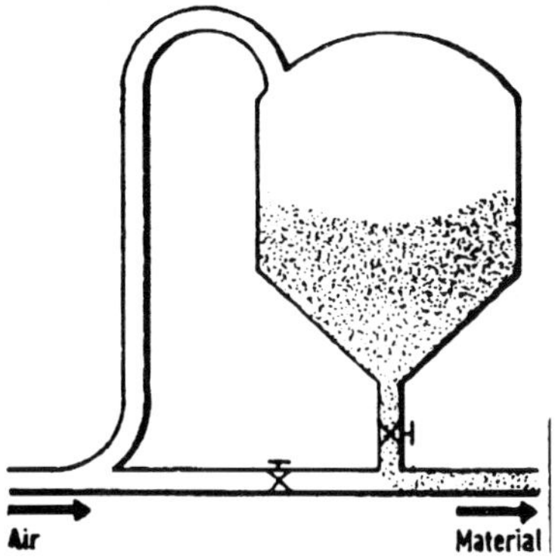

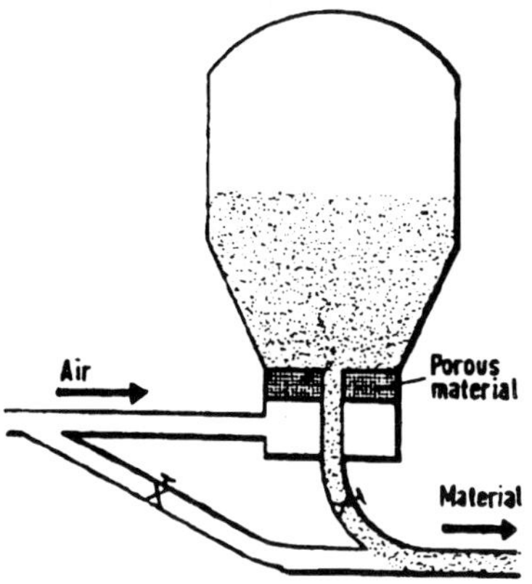

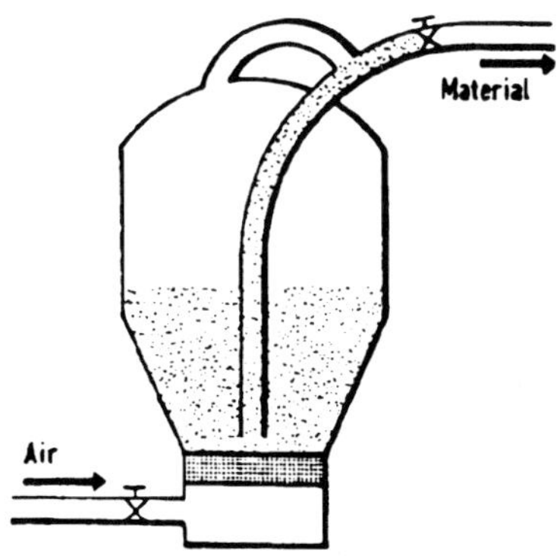

Fig 30: Fluidisation systems

might otherwise cause undue wear of the rotary valve, for example. Air conveying itself can vary in kind — between that in which the material is moved due to the energy contained within the air stream and a fluidised gravity flow arrangement. In the latter case the bulk-powder is converted effectively to a 'liquid' which will flow down a slight incline to convey horizontally without added kinetic energy. Fig 25 shows a fluidisation bed in which a porous membrane is subject to an air pressure beneath it, to disperse the powder above it.

Figs 26 - 30 show alternative blowing systems. A variety of factors might affect their selection not least of which are the properties of the individual materials to be carried. Abrasiveness affects any moving parts, for example; particle size governs the quantity of compressed-air required — while variation in size could lead to particle segregation problems with some equipment. Friability could affect the generation of unwanted dust — which can become explosive with certain air mixtures. Corrosive and heat-sensitive materials need special consideration and the materials' proneness to compaction can cause difficulties in fluidisation.

Fig 26 shows a suction system, limitation of which is that the inlet to outlet pressure differential is limited to less than one atmosphere — so that large bore pipes and big quantities of air are necessary. However, there is little difficulty entraining the particles in the air stream and it is easy to manoeuvre the inlet nozzle so that manual trimming of the product in the receipt-container is minimised. A combined suction and blowing system as shown in Fig 27 may be the best solution. Careful control is needed, however, over separating air from dust — any not trapped in the cyclone could pass through the blower.

Low-pressure blowing, Fig 28, can have problems of entraining particles in the airstream, leading to the requirement for a venturi constriction for assisting direct flow of the material. Throughputs are thus usually restricted to 3 tonnes/hour — however, only a limited power requirement can lead to a moderate cost installation. Medium pressure blowing, Fig 29, involves the use of a mechanised seal to feed the material into the compressed air pipe — while high pressure blowing involves pressurising the vehicle container tank itself and the material entraining in the airstream by gravity/fluidisation. Greater efficiency is obtained because less energy is required to entrain the powder and more material can be moved per kilo of air employed. Fig 30 shows different means of fluidising the powder solids.

Hopper design theory

For gravity-discharge vehicles, care is needed with hopper design to obviate against 'bridging' or 'rat-holing' disturbing the free-flow of bulk solids. Work by AW Jenike[8] is valuable in helping to solve such problems. He showed that 'doming' occurring in gravity feed of the product will collapse for slot-width less than or equal to σ_c/ρ (unconfined yield-strength/bulk density) or $\sigma_c/2\rho$ where the slot is not narrow but tends more towards a square or circular hole of diameter equivalent to the width dimension. σ_c strength depends on the hopper shape according to the flow factor *ff*, a high value of which signifies arching of the product. See lower curve of Fig 31; when the *ff* curve intersects the strength curve, this indicates critical consolidation pressure.

Stresses at the surface of the material on a 'dome' or 'pipe' divided into this pressure yield the critical flow factor. Plots show material characteristics from which average effective angle of friction is obtained. This can then be used in conjunction with Fig 32 for a particular hopper-side angle (30 deg in the example shown) to determine *ff* for constructing $x = ff$ lines on Fig 31. Intersection of these lines with the material property curves then gives the consolidating pressure P and σ_c.

A more accurate friction angle can next be obtained by inserting this pressure value as the upper curve in Fig 31 and a new value used in Fig 32 to obtain a more correct *ff* — leading to the more accurate blobbed curve in Fig 31. This in turn leads to a new consolidating pressure and a value of bulk density obtained from the figure. The minimum outlet size to prevent 'piping' is obtained in a similar manner but using Fig 33 and the expression $4(\sigma_c/\rho)A$ for controlling diagonal (or diameter) where A is the piping factor.

In the case of conical hoppers, Jenike suggests the minimum hopper angles shown in Fig 34 applying to hoppers having smooth transition radii to the vertical walls of the vessel. It is also assumed there are no impact conditions that might cause consolidation of the material during filling. A shelf should be provided in the path of the filling trajectory to avoid 'direct hits'. For a container with a long rectangular

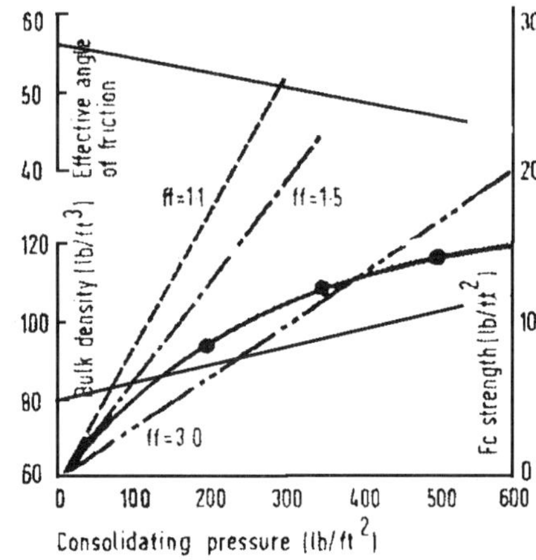

Fig 31: Critical consolidating pressure determination

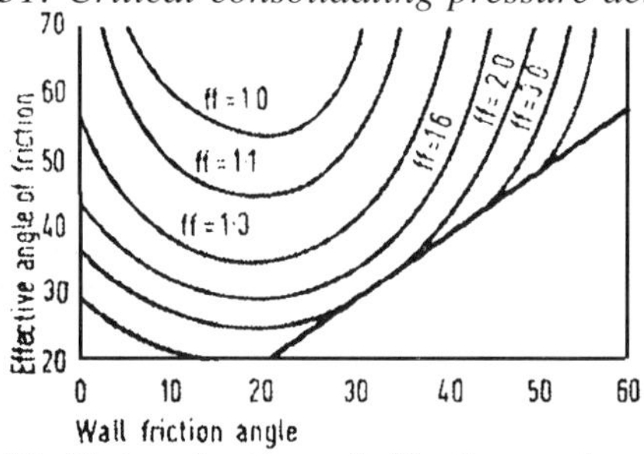

Fig 32: Determination of effective angle of friction

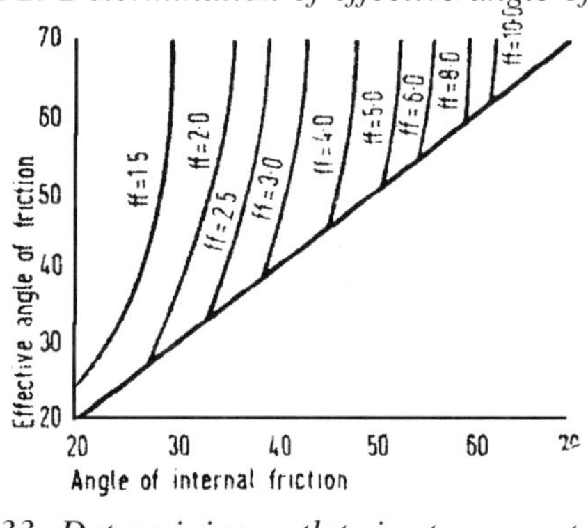

Fig 33: Determining outlet size to prevent piping

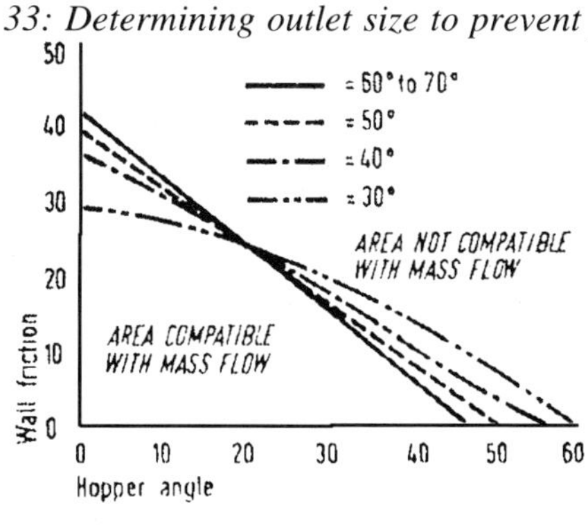

Fig 34: Minimum angles for conical hoppers

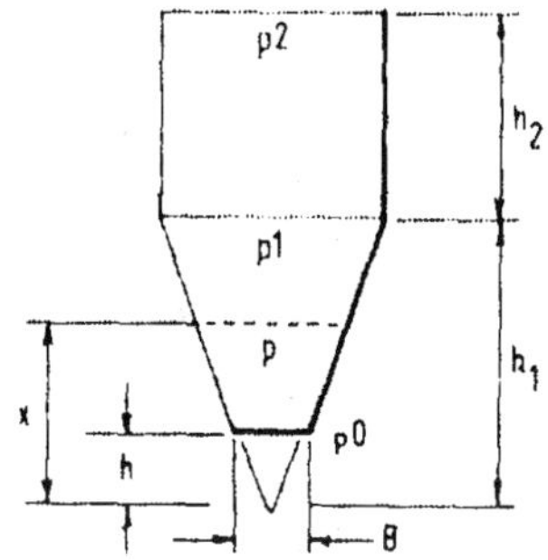

Fig 35: Typical hopper configuration

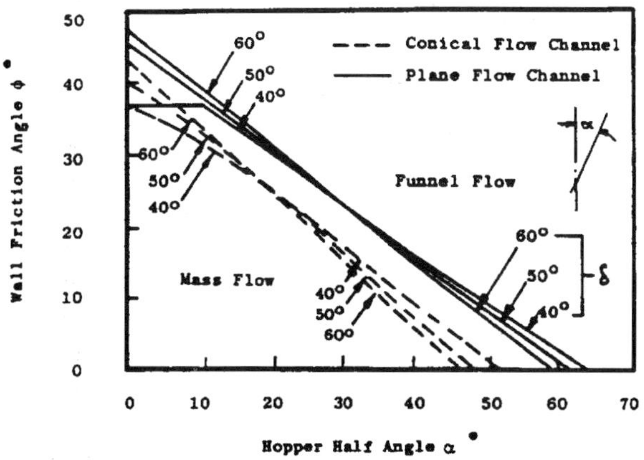

Fig 36: Limits for mass flow in conical and plane flow channels

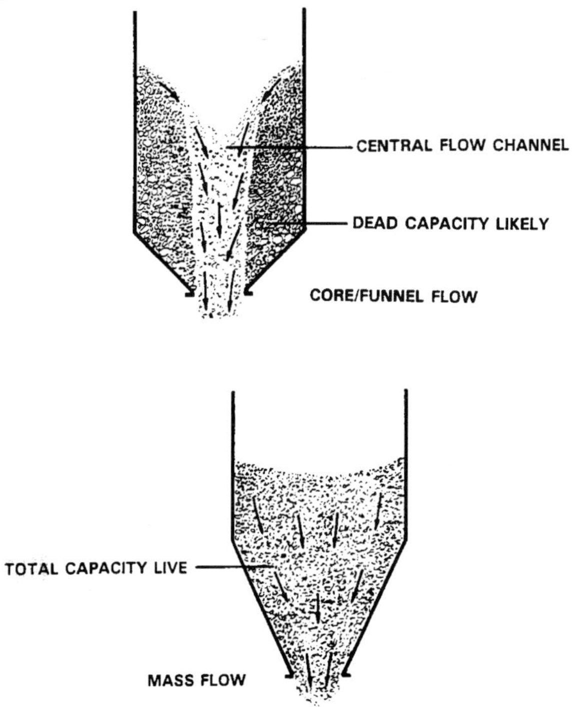

Fig 37: Basic patterns of discharge from storage containers

plan section, plane-flow conditions are said to exist. In a configuration such as that in Fig 35, top-pressurisation causes a pressure gradient to build up from atmospheric p_0 at the bottom to p_2 at the top. The difference between them equals:

$$agh[1 + h(h_2 - h_1)/h_1^2]$$

and in the design of the hopper, critical flow factor is divided by $(1 + a)$. For all values of h within the hopper, a must be sufficiently large for $ff/(1 + a)$ to be less than or equal to Ps_c. The required rate of pressurising air flow is given by:

$$(A/R)ag(h/h_1)^2$$

where A is the section area of the vessel and R is the resistance to air flow of the material.

Hopper design practice

Valuable insight into the advantages and disadvantages of the principal discharge patterns of material flowing in hoppers, silos, bins and bunkers was provided recently that should have useful design indications for bulk powder vehicles. Researchers at the University of Greenwich[9] gave important data on the shapes of such facilities that lead to these flow patterns.

In general, he maintains, for free flowing products where particle segregation is of no concern, a funnel/core flow mode of discharge is an acceptable and economic approach. Determination of which flow type to adopt requires undertaking a simple bench scale test to determine the frictional characteristics between samples of the product to be carried and the material from which the internal surfaces of the converging section of the vehicle discharge hoppers are to be constructed. This test will yield an 'angle of wall friction', which, by consulting graphs of the form shown in Fig 36, will, depending on the basic shape of the converging section under consideration, suggest the angle of the converging section that is necessary to bring about the required discharge pattern.

The two basic flow/discharge patterns, mass flow and funnel flow (sometimes called core flow), are seen in Fig 37. With funnel or core flow the material sloughs off the top surface and discharges through a vertical channel which forms inside the bin. This channel extends from the outlet upward through the material, with the material outside of this channel being stationary. With free flowing products, that material in the stationary region will then flow into the discharging channel once the level of the surface moves down to permit it to do so. Funnel flow is often referred to as a 'first in, last out' mode of flow, since with this pattern of flow much of the material that is first to fill the container, particularly around the periphery, will be the last to flow out.

Funnel flow usually occurs in flat bottom bins or facilities where the angle of the converging section, measured from the horizontal, is relatively shallow. For obvious reasons, such a configuration serves to maximise the stored volume for a given headroom. Prior to developing an understanding of how to design for mass flow, a common, but very often unsatisfactory approach to hopper design was to place an unconstrained sample of product on a piece of material from which the hopper was to be constructed and then by tilting the plate noting the angle at which the produce flowed off the plate. Since the majority of materials will flow in such a situation at angles less than 60° to the horizontal, this figure together with 45° in the case of very free flowing materials, were chosen by many designers of silos as 'convenient' angles of manufacture of the converging section. Subsequent work has show that even with the most free flowing materials it is likely that 60°, and certainly 45°, will result in a funnel flow discharge pattern. However, the fact that many silos are operating perfectly satisfactorily under these conditions demonstrates that this is an acceptable discharge pattern with many products.

With mass flow, the bulk solid is in motion at substantially every point in the bin whenever any material is allowed to flow through the outlet. As such, the flowing channel of material coincides with the walls of the bin, which is brought about when the walls of the converging section are sufficiently steep and smooth. This flow pattern is often referred to as 'first in, first out', in that the first material to be charged into the bin will be first discharged, thereby preserving the loading sequence.

Just how steep the angle of converging section should be to bring about this mode of flow will depend on the frictional characteristics exhibited between the stored product and the material from which the internal surfaces of the converging section are to be constructed. However, to put the geometry of a bin operating in a mass flow pattern of discharge in perspective with funnel flow, the former will necessitate the angle of the converging section typically being in excess of 70° (measured from the horizontal) for a conical configuration of hopper. Obviously, such a geometry can result in a significant reduction in stored volume for a given headroom

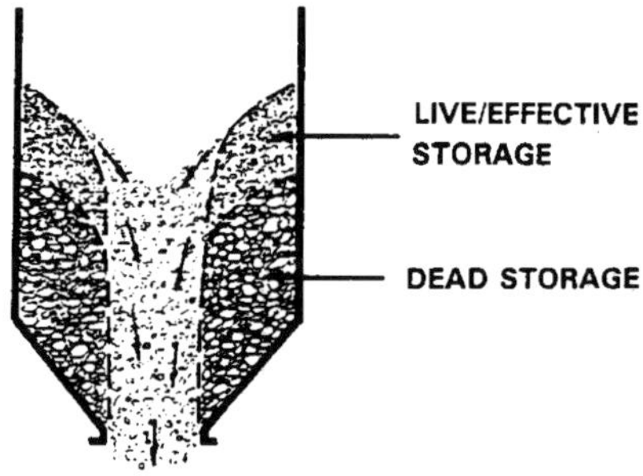

Fig 38: Reduction in the effective storage capacity of a funnel flow hopper as a result of 'topping-up'

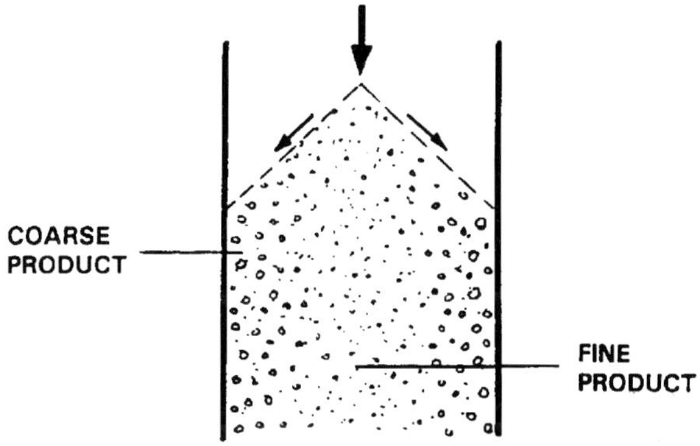

Fig 39: Particle segregation on charging a container

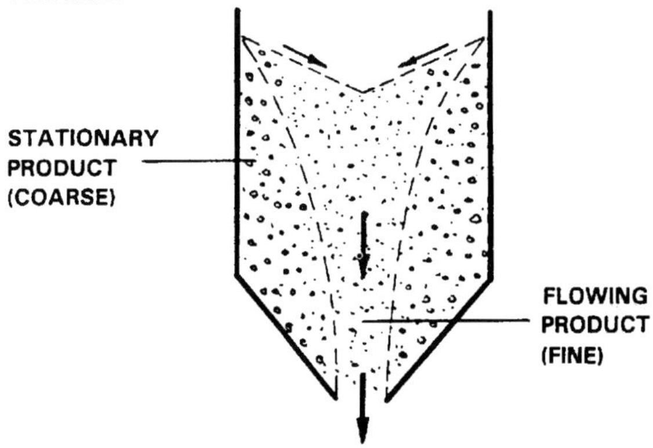

Fig 40: Particle segregation on discharge in funnel flow

when compared with operating in a funnel flow mode on discharge. Other requirements to achieve a mass flow pattern of discharge are that the container has no abrupt transitions, the converging section has no inflowing valleys and the feeder is active across the whole area of the outlet.

Clearly, the first in, last out' characteristics of funnel flow can, with some products, lead to potential product contamination and or reduction in live capacity if the bin is continually topped up before being permitted to empty completely, Fig 38. With some products, the extended periods of storage of material in the dead zone, can, for a variety of reasons, lead to the material setting hard or 'caking', such that when product in this region is given the opportunity to discharge, it will not do so under the action of gravity alone. The main reason for this effect is often a difference in temperature between the stored product and the walls of the bin, which, in turn, leads to the migration of moisture between the particles comprising the bulk. Sugar and salt are examples of materials that can exhibit this behaviour.

Materials that are essentially free flowing, and have a wide particle size distribution, can be readily segregated in terms of particle size as a result of bulk handling in funnel flow mode. On filling the vessel, larger, and hence heavier particles will roll towards the periphery of the bin, with the finer elements remaining in a central core, Fig 39. As a consequence, a funnel mode will, in the first instance, discharge material that is rich in the finer elements, Fig 40. The proportion of fines will gradually decrease through the discharge cycle, until, at the end of the cycle, the material will be rich in the coarser elements. A similar phenomenon occurs with materials where the constituent particles are all of a similar size, but vary in density. Clearly, these segregation effects can result in quality control problems, and may have a significant impact on the operation of downstream processes. Whilst it is important to realise that this basic problem results from the segregation occurring on filling, which in itself is largely independent of hopper shape, the funnel flow mode of discharge resulting from the hopper shape serves to accentuate this effect.

Another potential drawback associated with funnel flow containers is that of product 'flooding' or 'flushing'. This is a particular problem with material of small particle size, i.e. typically less than 100 - 150 microns. It is known that such materials when aerated can take on liquid-like characteristics and flow uncontrollably; an effect commonly know as 'flooding' or 'flushing'. Such an effect can occur when such materials are discharged from containers that operate in a funnel flow mode. This is a particular problem when there is no feeder at the bottom to regulate or control discharge. The effect is caused by the material flowing off the top surface of the stored product into the flowing channel. Depending on the required discharge rate of product from the container, this might result in the material within the flowing channel acquiring a significant downward velocity. As the particles move through the interstitial air in the flow channel, this effect might be sufficient to aerate the product. Clearly, this situation can lead to wide variations in product density, which can result in difficulties in proportioning such materials for subsequent operations.

In general, many of the advantages and disadvantages associated with a facility operating in a funnel flow mode of discharge are reversed with mass flow. For example, it is clear that the steep angled converging section that is often necessary to bring about mass flow, will, when compared with the

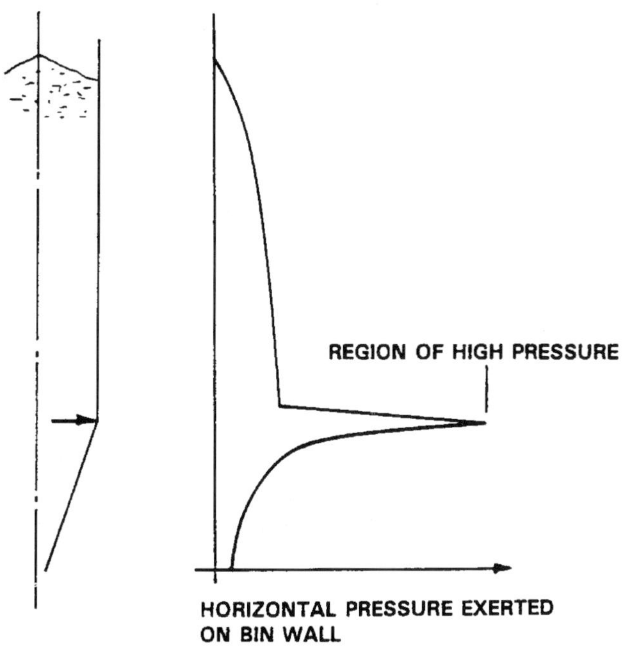

Fig 41: High radial pressure in region of transition from vertical to converging section of a mass flow container

squatter proportions of a funnel flow container, reduce the potential storage capacity corresponding to a given headroom. Given that mass flow is essentially a 'first in, first out' mode of operation, with no dead regions in the stored product, it is clear that the time for which material will be in storage will be minimised. This, in turn, will minimise or eliminate problems such as product spoilage/degradation or caking, which are often time dependent. The 'first in, first out' pattern associated with mass flow means that the loading order is preserved and segregated material is re-mixed in the converging section. Therefore, mass flow is commonly recommended for products that readily segregate on handling. In a somewhat similar vein, the more controlled discharge associated with mass flow serves to lessen or eliminate potential flooding or flushing, thereby resulting in material of a more uniform density flowing through the outlet. This then makes it easier to proportion or meter the material into subsequent processes.

Structural design of hoppers

Although not necessarily a drawback, the necessity for significant structural support for a mass flow container in the region of the transition from the vertical to converging section is seen by some as a disadvantage. The necessity for such support stems from the fact that, when discharging in a mass flow mode where a significant vertical surcharge is present, the pressure exerted is a horizontal direction on the walls of the bin in the region of the transition from the vertical and converging sections can, compared with the pressures present in funnel flow, be considerable, Fig 41. Therefore, to account for this high pressure considerable reinforcing might be required in this region so as to ensure the integrity of the resulting structure. Failure to recognise this important point has led to the structural failure of some containers.

Hopper detail design

It is important to recognise that attention to detail is essential if mass flow is to be achieved. As well as ensuring smooth transitions, consideration should also be given to the wear of the walls of the container that will occur as a result of the abrading/sliding action when handling abrasive materials. This is likely to be more noticeable in the lower regions of the converging section where the velocity of the discharging particles will be at their greatest. With regard to the feeder being active across the whole area of outlet, failure to observe this requirement may destroy the required flow pattern in an otherwise properly designed mass flow container, Fig 42.

It is possible to blend some of the advantages of both funnel flow and mass flow into a single facility, by way of a composite/expanded flow hopper, Fig 43. This is achieved by way of a funnel flow container interfacing with a mass flow hopper section. The

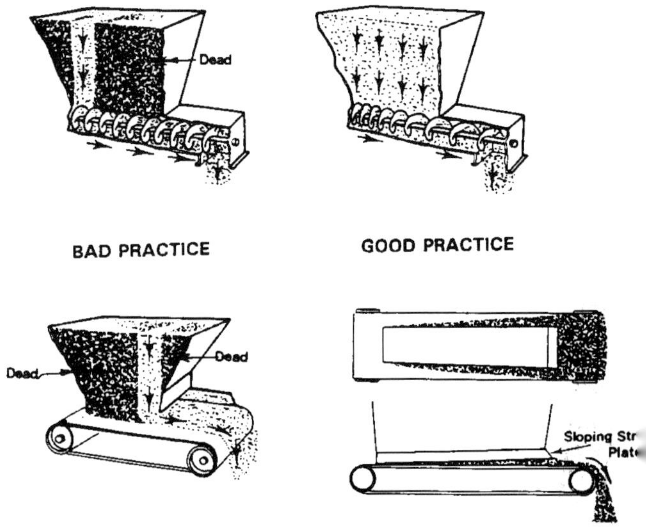

Fig 42: Importance of container/feeder interface to effect mass flow

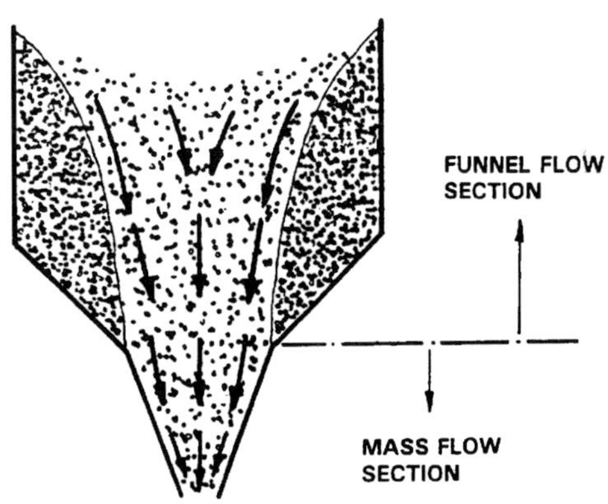

Fig 43: Composite flow

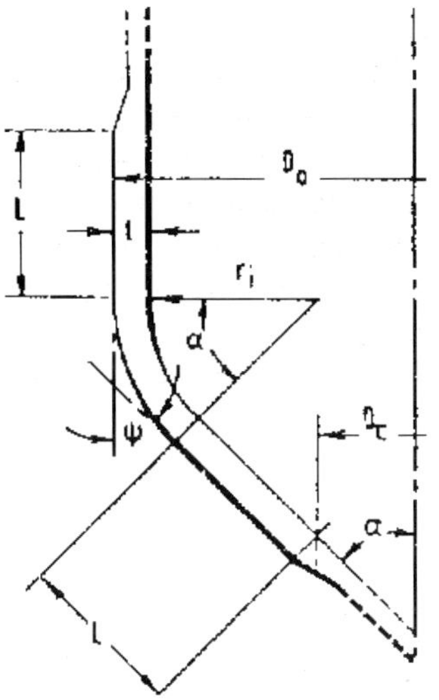

Fig 47: Analysis of tank top corner

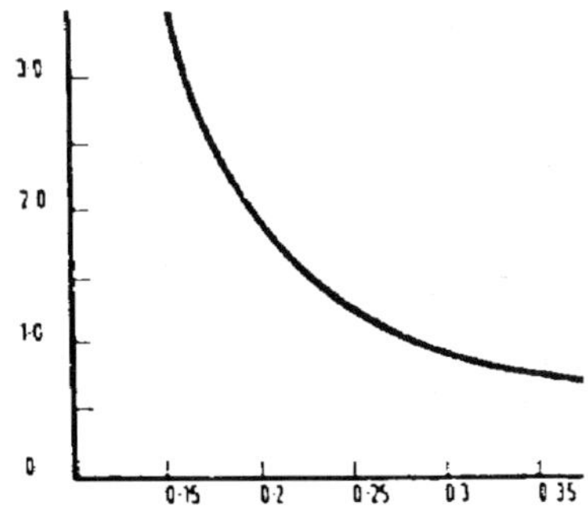

Fig 44: Tank-end design factor, k (x-axis) against h_o/d_0 (y-axis)

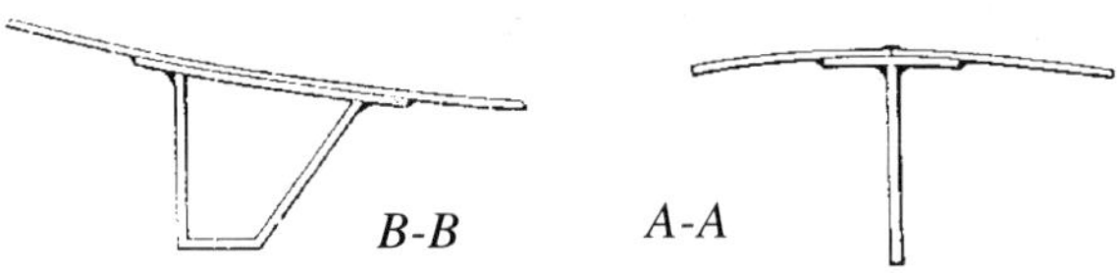

Fig 46 : Key sections

major advantage of such an approach is that if affords a potentially significant increase in stored volume when compared with a pure mass flow approach, whilst, at the same time, with cohesive products, it achieves reliable flow through an outlet that is smaller than would be necessary for a container operating solely in a funnel flow mode. Additionally, the density of feedstock flowing through the outlet will be more uniform than might otherwise be the case with a bin operating in a funnel flow mode. Although the application of such an approach is usually limited, when carefully applied it can be a very effective means of combining these advantages.

Pressure tank design

Aluminium construction for pressure tank design is covered in British Standard BS 1500 Part 3 for static units — but generous safety factors (2-5) have to be allowed for in vehicular duty. While tanks rarely actually travel under pressure, wracking and surging loads have to be designed for. With cylindrical tanks, shell thickness is determined from:

$$t = (f.D_i)/(2PJ - f)$$

where f is design stress, D_i is internal diameter and P is design pressure (joint factor J). The expression assumes support conditions along the length of the tank.

With gravity-discharge bodies, the first consideration is to ensure the hopper sides are steeper than the angle of repose of the material, after which dimensions can be fine-tuned, as above, to avoid compaction effects. Undue bulging of the hopper walls can be avoided by designing within deflection limits prescribed by established plate-deflection (under-pressure) formulae, the pressure on the plate being given by $k\rho h$ where:

$$k = (1 - \sin\gamma)/(1 + \sin\gamma)$$

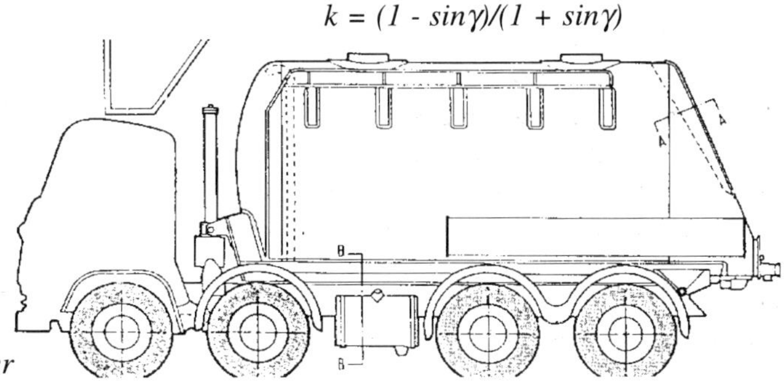

Fig 45: Inclined-axis conical bulker

and where g is the angle of repose, r the density of the material and h the depth of the plate below its surface.

Design of pressure-tank ends is dealt with in BS 1966; these ideally should be hemispherical — and thickness is given by:

$$fD_i/(4PJ - F)$$

Should the semi-ellipsoid design be needed, to save on vehicle length, thickness is obtained from $fD_0k_1/2PJ$ where k_1 is obtained from Fig 44.

Conical ends of symmetrical pattern are covered in BS 1500; however, as these are almost universally of the inclined axis type (Figs 45/46) the worst-case condition of the top corner of the tank (Fig 47) needs to be considered. Thickness at the junction is given by:

$$[(fD_0)/(2PJ - f)][1/\cos a]$$

Detail design of both gravity and pressure discharge tanks is important and a typical layout is shown in Fig 48. The use of doubler panels to avoid stress concentration at critical areas of pressure tanks is particularly important.

It is also helpful to consider the valley angles between the hopper walls when evaluating flow inside a gravity-discharge vessel. Valley angles are always less than the limiting angle of repose that causes product to hang up in the corners. Fig 49, from a DATA design guide[10], enables valley angles to be read off when the hopper side angles are known.

An expression for the deflection of a uniform plate constrained at its edges is given by Rees[11]. Depending on the values predicted, the option of using tie bars could be considered with the possibility, in turn, of reducing wall thickness.

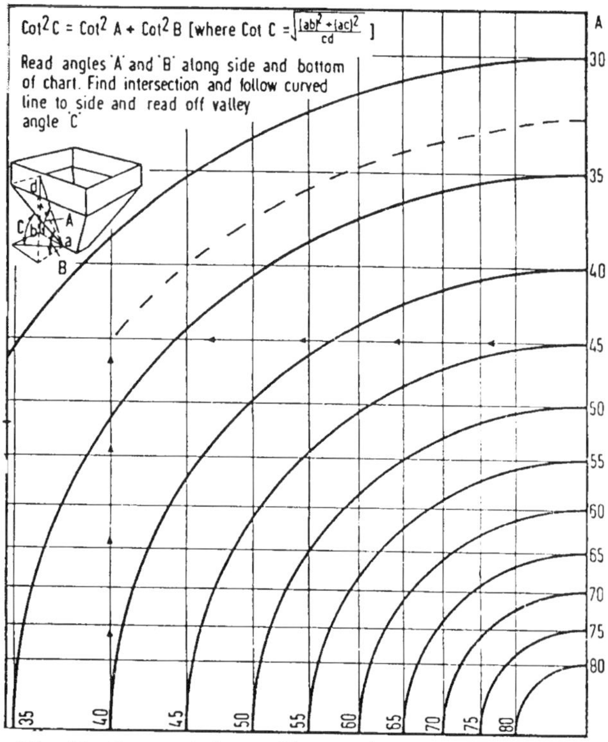

Fig 49: Determining valley angles

Fig 50: Bulk tipper

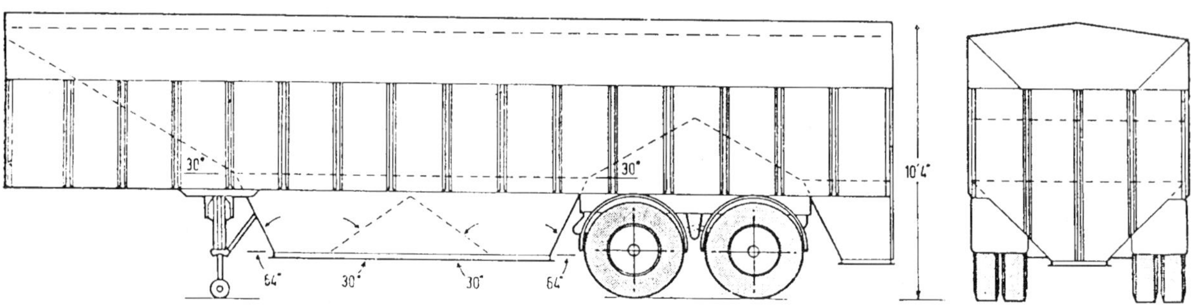

Fig 48: Gravity-discharge bulker

Breadth of plate, in.	1/16	⅛	3/16	¼	5/16	⅜
	Thickness of plate in in.					
6	3.32	13.3	30.0	53.2	83.4	120
7	2.45	9.80	22.0	39.2	61.2	88.0
8	1.88	7.50	16.9	30.0	46.8	67.6
9	1.49	5.93	13.3	23.72	37.0	53.2
10	1.20	4.80	10.8	19.20	30.0	43.2
11	0.99	3.96	9.93	15.84	24.8	39.72
12	0.84	3.34	7.50	13.36	20.8	30.00
13	0.71	2.84	6.40	11.36	17.7	25.60
14	0.61	2.45	5.51	9.80	15.3	22.04
15	0.53	2.13	5.07	8.52	13.3	20.28
16	0.47	1.88	4.22	7.50	11.7	16.88
17	0.41	1.66	3.74	6.64	10.4	14.96
18	0.37	1.48	3.33	5.72	9.26	13.32
19	0.33	1.33	2.99	5.32	8.30	11.96
20	0.30	1.20	2.70	4.80	7.50	10.80
21	0.27	1.09	2.45	4.36	6.80	9.80
22	0.25	0.99	2.23	3.96	6.20	8.92
23		0.908	2.04	3.632	5.67	8.16
24		0.834	1.87	3.472	5.21	7.50
25		0.769	1.73	3.076	4.80	6.92
26		0.71	1.60	2.84	4.44	6.40
27		0.66	1.47	2.63	4.11	5.92
28		0.61	1.38	2.45	3.83	5.52
29		0.57	1.28	2.28	3.56	5.12
30		0.53	1.20	2.13	3.33	4.80
31		0.50	1.12	2.00	3.12	4.48
32		0.47	1.06	1.87	2.93	4.22

Correction factors for use with the above Table for other than square plates. Pressures given above should be multiplied by the appropriate factor.

L/B	1	1.1	1.2	1.3	1.4	1.5	1.6	1.7
Factor	1	0.895	0.83	0.783	0.752	0.73	0.713	0.7

L/B	1.8	1.9	2.0	2.25	2.5	2.75	3.0
Factor	0.69	0.682	0.677	0.665	0.66	0.656	0.65

Fig 52: Pressure in lb/in² to give stress of 10,000 lb/in² in flat square mild steel plates with fixed edges

Breadth of plate, in.	1/16	1/8	3/16	1/4	5/16	3/8
	Thickness of plate in in.					
6	**20.1**	**80.4**	**273**	**643**		
7	**10.9**	**43.4**	**147**	**347**		
8	**7.0**	**28.0**	**86.5**	**204**		
9	**4.0**	**16.0**	**54.0**	**127**		
10	**2.7**	**10.8**	**35.4**	**83.3**		
11	**1.78**	**7.11**	**24.2**	**56.9**		
12	**1.26**	**5.03**	**17.1**	**40.2**	**78.8**	
13	**0.91**	**3.65**	**12.4**	**29.2**	**57.3**	
14	**0.68**	**2.71**	**9.2**	**21.7**	**42.5**	**73.6**
15	0.51	2.05	**7.0**	**16.4**	**32.3**	**56.0**
16	0.40	1.61	**5.4**	**12.7**	**24.9**	**43.2**
17	0.31	1.25	**4.25**	**10.0**	**19.6**	**34.0**
18	0.25	1.00	**3.38**	**7.95**	**15.6**	**27.0**
19	0.20	0.80	2.72	**6.39**	**12.6**	**21.76**
20	0.163	0.65	2.21	**5.21**	**10.2**	**17.68**
21	0.135	0.54	1.82	4.29	**8.41**	**14.56**
22	0.123	0.49	1.51	3.56	**6.98**	**12.08**
23	0.099	0.39	1.27	2.98	**5.85**	**10.16**
24		0.31	1.07	2.51	4.94	**8.56**
25		0.269	0.91	2.13	4.19	**7.28**
26		0.228	0.77	1.82	3.58	6.20
27		0.196	0.66	1.57	3.08	5.33
28			0.58	1.36	2.66	4.61
29			0.50	1.18	2.31	4.00
30			0.44	1.03	2.02	3.50
31			0.38	0.90	1.77	3.07
32			0.34	0.80	1.56	2.70

Pressures shown in bold figures result in a stress of more than 10,000 lbf/in² and should not be used for this reason in normal tanks.

Correction factors for use with the above Table for other than square plates. Pressures given above should be multiplied by the appropriate factor.

L/B	1	1.1	1.2	1.3	1.4	1.5	1.6	1.7
Factor	1	0.885	0.812	0.756	0.722	0.695	0.676	0.664

L/B	1.8	1.9	2.0	2.25	2.5	2.75	3.0
Factor	0.65	0.645	0.638	0.637	0.621	0.617	0.614

Fig 53: Pressure in lbf/in² to give central deflection of 1/32 inch of square steel plate

Open-top bulk vehicles

The last two sections covered closed bulk vehicles for carriage of solids and liquids respectively. Here, open bulk-carriers are considered with particular reference to their design to withstand structural loading. These vehicles are, of course, characterised by more arduous operating conditions, particularly during loading and unloading, and involve considerably more rugged construction — especially in the case of off-road operation.

Open-top bulktippers (Fig 50), for on-road, and dump-trucks (Fig 51), for off-road operation, have been constructed in steel and aluminium alloys and to a limited extent in GRP. The ability to withstand considerable drop loads in the case of dump-trucks, which might be loaded by tractor shovels pitching in massive rock boulders to their bodies, is matched by the need to withstand large side-wall pressures, for the case of low-viscosity payloads carried in high-sided vehicles. By far the most important input for the vehicle designer is operating history of earlier designs. Only when a very radical departure in material usage, or operating condition takes place will any recourse be made to other than comparative techniques in structural design.

The prospective use of GRP focuses again on the use of predictive techniques to calculate the internal loads in the structural elements. Finite-element techniques of dividing the structure into small inter-connecting parts, prior to the analysis, would be the normal way to prove the final design before commitment to hardware. At the earlier concept stage, a return to fundamental engineering principles is a good way to assess the effects of loading extremes. How much stress, for example, could be induced from a boulder falling on the floor of a dump-truck. Individual rocks, each weighing 3 tonnes are not unknown in the spoil being removed during blasting to provide the core of an 'earth' dam.

Gravitational acceleration will result in the boulder striking the floor and giving up its kinetic energy into the structure to an extent which depends on the local elasticity of the floor panels as well as the overall compliance of the vehicle suspension. The impact velocity can be found by equating the loss of potential energy in falling height h with the gain in kinetic energy as $2h^{1/2}$. The transfer of energy to floor also depends on the respective co-efficients of restitution of the two impacting materials and this effect will typically factor the relative velocity ratio before and after impact by about a half. Thus in the worst case of the floor having virtually no structural compliance there is a gain in strain energy across the area of impact far in excess of that caused by gradually resting the rock on the surface. A factor of 50 times increase in resultant stress would not be untypical.

Most materials carried in open tippers have an angle of internal friction which governs the angle which the body has to be tipped in order for the material to flow from it. Depending on the degree of moisture present, the angle reduces until a virtually liquid-like behaviour applies to wet soils — which can apply substantial forces to body sidewalls as well as to floors. How much bulging of the wall plates can be tolerated is debatable. Some say bulging to the extent of up to one hundredth of the supported plate-width (up to a 1/32nd of an inch limit) is a 'visually' acceptable maximum. A formula due to Grashoff is used to determine the deflection of a flat rectangular panel subject to lateral pressure as:

$$l^4pb^4/(l^4 + b^4)32Et^3$$

where length l, breadth b and thickness t are in inches while pressure p and elastic modulus E are in lb/in^2.

The table in Fig 52 due to C W Rees provides a quick reference for steel plates up to their visually tolerable bulge limit. This sort of approach can be used to check the effect of rail-pitch for giving edge support to the side panels. The table in Fig 53 can be used to further check whether panel stresses are likely to be kept within safe working limits. Lateral pressure is, of course, the product of 'head' (height to the surface) times density for a liquid material (with zero cohesion). This results in a triangular distribution of pressure intensity toward the bottom of the tank — but for design purposes it would be reasonably safe to assume uniform distribution at the maximum level. Fig 54 shows the pressure at the base of the body sides for a range of depths and material densities, assuming a 30 degree internal friction angle (of repose) and the table of Fig 55 the proposed spacing of ribs in the case of an aluminium-alloy body, with recommended panel loadings in the table of Fig 56. The stress can be also calculated in absolute terms as:

$$b^2pl^4/2t^2(l^4 + b^4)$$

Fig 51: Dump truck

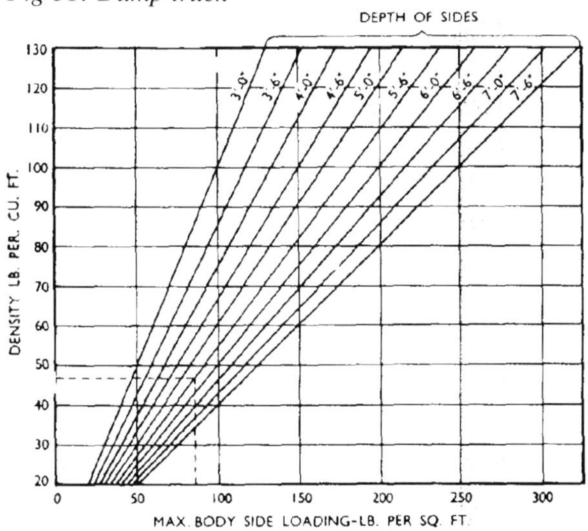

Fig 54: Pressure on body sides

Height of side (ft.)	4	5	6	7	8
Maximum body side loading (lb./sq. ft.)	67	83	100	117	133
Maximum pillar spacing (in.)					
2¼" Top-hat section 12873	117	60	33½	22	15
1¼" Top-hat section 12674	48½	25	14	9	6

Fig 55: Spacing of side pillars for product density of 50 lb/ft^3

Side Panels (BA.25 WP)*	Side Pillar Spacing 12"	15"	18"	21"	24"	30"
SHEET						
12 s.w.g. (0·104")	460	300	205	150	115	75
14 s.w.g. (0·080")	275	175	120	90	70	45
16 s.w.g. (0·064")	175	115	80	55	45	
18 s.w.g. (0·048")	100	65	45			

Fig 56: Side panelling — recommended maximum loading (lb/ft^2)

The design of ribs to support the side plates will be affected both by the need for support against the above pressure and by a robustness necessary to survive the hazards of operation on construction sites and similar conditions. The ribs must also stabilise the sidewalls against shear deformation in the overall bending and twisting of the body as a stiffened box tube. Top and bottom rails of the body must also withstand direct stresses due to overall bending and torsion.

One approach to assessing the pressure-retention capacity of stiffening ribs is to assume each rib supports a strip of plating extending half way to the next stiffener on either side. Maximum bending stress in the stiffener is thus:

$Wly/12I$ lbf.in

where W is total distributed load (lb) and l is stiffener length; y is neutral-axis to extreme-fibre distance and I the stiffener section second moment of area. The deflection of the stiffener is given by:

$Wl^3/384EI$

Standard bending and torsion formuli given in earlier sections can then be used to calculate overall body structural loads while either the Moment Distribution or Unit Load methods, discussed in earlier chapters, can be used to calculate the effect of the stiff joints between the 'rib' members on the magnitude and distribution of internal forces. Fig 57 shows the initial bending stress calculation phase for a 20 tonne capacity tipping semitrailer.

Recommendations for floor loadings and dimensions have been provided by British Aluminium[12]; The table in Fig 58 reproduced here is based on the criterion of a maximum deflection of 1/200th the supported span while the table in Fig 59 is to withstand a working stress of 10,000 lb/in^2 — both using BA25WP grade aluminium-alloy sheet. Choice of crossbearer design depends on floor-loading, crossbearer-spacing and method of body attachment. The table in Fig 60 gives some recommended maximum distributed loads for a selection of crossbearer sections. To ensure that the body sidewalls contribute to the longitudinal strength of a tipping body, the crossbearer must also be adequately jointed to the side-pillar ribs as well as these being supported by substantial top-rails and corner posts, Fig 61.

In the case of all-welded bodies, the handbook suggests the corresponding joints shown in Fig 62. This is for a typical body type, for an 8.5 yd^3 (6.5 m^3) capacity, fabricated entirely from 1/4 and 3/16 in (6.35 and 4.76 mm) BA28 plate using MIG welding. Unless warranted by volume, stiffeners would be formed on a press-brake. Folding in of the lower side panels by 45 degrees permits strengthening of the connection between side-pillars and crossbearers. In this case, the body internal width tapers by two inches (50 mm) increase from bottom to top. Subframe longitudinals are of zed section 1/4in (6.35 mm) plate slotted for allowing crossbearers to pass through.

When a hopper-discharge body is employed, it

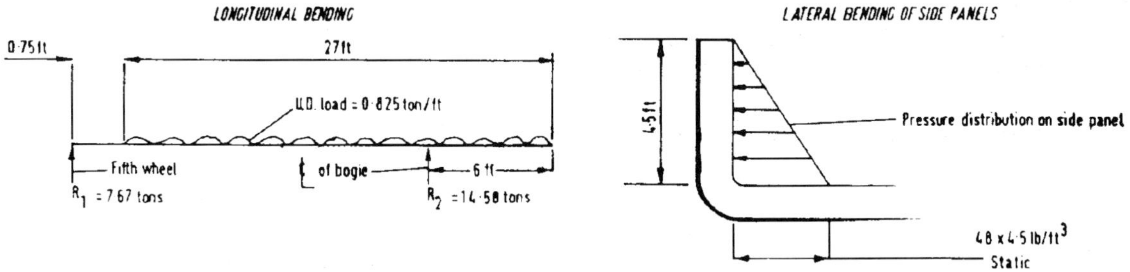

Max. bending moment = 41·5 ton-ft
Second moment of area = 21 520 in^4
Neutral axis 43·3 in below top rail
Max. static stress = 2250 lb/in^2 (compression)
Dynamic load factor = 3
Safety factor = 5
Fully factored stress = 15 × 2250 = 33 750 lb/in^2
Flexural strength of material = 65 000 lb/in^2

Max. bending per stanchion = 1312 lb-ft
Dynamic factor = 3
Second moment of area = 8·34 in^4
Neutral axis = 2·82 in from outside profile
Max. static stress = 16 000 lb/in^2 (compression)
Safety factor = 4
Fully factored stress = 64 000 lb/in^2

Fig 57: Bending stress calculation

Cross-bearer Spacing	Floor Loading (lb./ft²) Uniformly Distributed					
	50	75	100	150	200	250
15″	(0·128″)	(0·128″)	(0·128″)	(0·128″)	(0·144″)	(0·160″)
18″	(0·104″)	(0·128″)	(0·144″)	(0·160″)	(0·188″)	(0·188″)
21″	(0·128″)	(0·144″)	(0·160″)	(0·188″)	(0·219″)	(0·219″)
24″	(0·144″)	(0·160″)	(0·188″)	(0·219″)	(0·250″)	(0·250″)
27″	(0·160″)	(0·188″)	(0·219″)	(0·250″)	(0·281″)	(0·281″)

Fig 58: Recommended flooring thickness based on deflection

Cross-bearer Spacing	Floor Loading (lb. ft.²) Uniformly Distributed					
	50	75	100	150	200	250
15″	10 s.w.g.	10 s.w.g.	10 s.w.g.	10 s.w.g.	10 s.w.g.	8 s.w.g.
18″	10 s.w.g.	10 s.w.g.	10 s.w.g.	8 s.w.g.	8 s.w.g.	3/16″
21″	10 s.w.g.	10 s.w.g.	10 s.w.g.	8 s.w.g.	3/16″	7/32″
24″	10 s.w.g.	10 s.w.g.	8 s.w.g.	3/16″	7/32″	7/32″
27″	10 s.w.g.	8 s.w.g.	8 s.w.g.	7/32″	1/4″	1/4″

Fig 59: Recommended flooring thickness based on stress

Section	Weight (lb. per ft.)	Side Overhang	Cross-bearer Spacing				
			15″	18″	21″	24″	27″
	1·52	26″	256	213	183	160	142
		28″	192	160	137	120	107
		30″	147	122	105	92	81
	1·42	26″	314	262	224	196	174
		28″	235	196	167	147	131
		30″	181	151	129	113	100
	1·81	26″	319	266	228	199	177
		28″	239	199	170	149	133
		30″	184	153	131	115	102
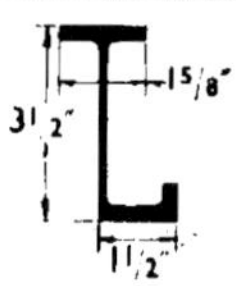	1·60	26″	380	317	271	237	211
		28″	285	237	203	178	158
		30″	219	182	156	137	121
	2·38	26″	774	644	552	483	430
		28″	581	485	415	363	323
		30″	446	372	319	279	248

Fig 60: Recommended maximum distributed load (lb/ft²)

is suggested that hopper panelling of 10 or 12 swg BA25WP plate is used rising to 3/16th in (4.76 mm) in the case of particularly abrasive materials. The hopper sides normally slope 5-15 degrees more than the internal friction angle of the material. Fig 63 is one design approach.

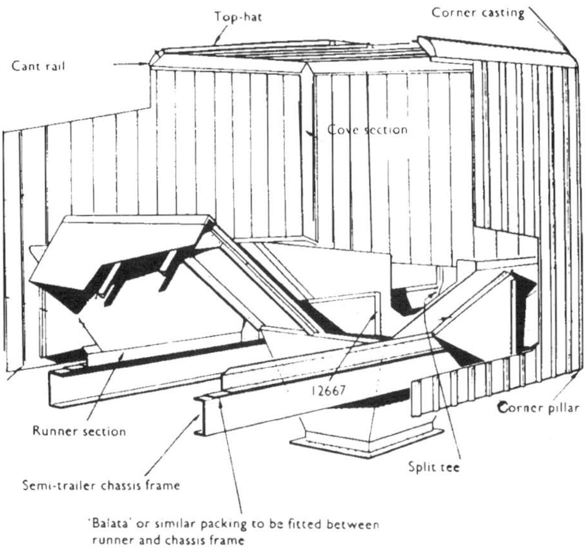

Fig 63: Hopper-discharge body

Powder bulker charge/discharge

The successful design of bulk powder vehicles depends as much on the equipment for loading and unloading the vehicle as on the container for carrying the material in transit.

According to Dr N Mainwaring of Simon Macawber, the physical properties of the bulk solid material are critical to the successful understanding of pneumatic conveying equipment. He sees conveying equipment as split into two groups, positive and negative pressure systems, depending respectively on blowing and suction — the latter being restricted to relatively low product throughputs and short conveying distances. Positive pressure systems can then be sub-divided into dilute and dense phase flows.

While the former are more versatile in terms of the wide range of products that can be conveyed, more recently, dense phase flow systems have increased in use due to their ability to reduce power consumption, minimise particle degradation and lessen the incidence of erosive wear on pipe bends. Development effort is directed towards making dense phase systems suit a wider range of products.

In defining the two main systems the criterion

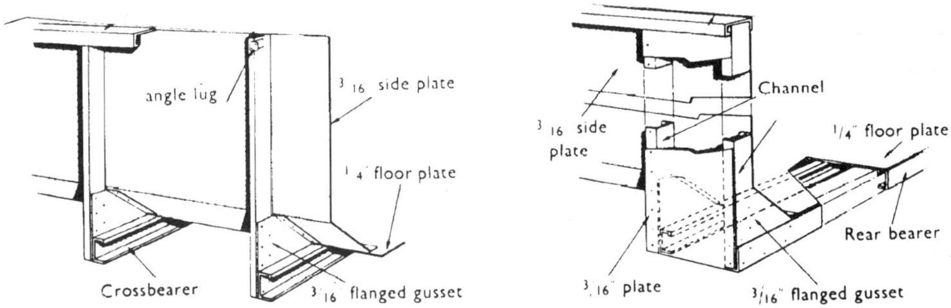

Fig 61: Side-pillar to crossbearer joint

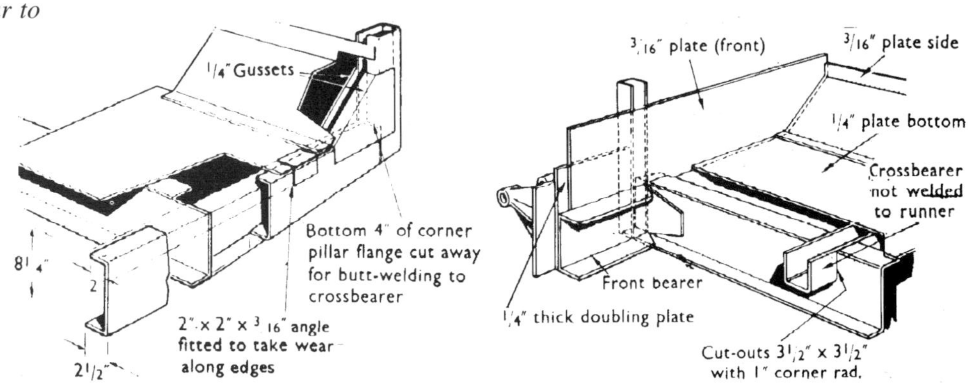

Fig 62: Welded-body joints

of amount of product continuously held in suspension is increasingly used, dense phase systems involving a significant fall-out of product over the conveying distance. Several modes of flow are involved depending on the type of product and the method of feeding air, and product, into the pipeline, Fig 64. Reducing air velocity from the dilute flow regime at (a) leads to the situation at (b) where a small strand of material slides along the bottom of the pipeline. Further reduction in velocity lads to 'moving bed' flow (c), typically occurring at air velocities of 8-10 metres/sec.

Often the moving bed part of the flow involves a pulsating stop-start regime. By reducing velocity to 3- 6 metres/sec, the flow pattern of (d) occurs, when the material in the line briefly fills the entire cross-section and any further velocity reduction would lead to a blocked pipe. Type (d) flow seems to suit material exhibiting good air retention, the author argues, such as cement, pulverised coal and fly-ash. Products with very poor air-retention can, when air velocity is reduced from the flow (a) situation, cause a plug of material to form which will then move at high speed around the pipeline and possibly cause excessive vibration.

If, instead of reducing air velocity from the flow (a) situation, flow commences with a very low air velocity, which is subsequently increased, then different flow modes are experienced. Initially a shearing type flow is experienced as shown in Fig 65(a), then slugs progress along the pipe above the shear line. Slugs progress along the pipe by dropping material off the back of the slug and picking it up from the front of the next one, maintaining constant depth to the stationary bed. Increasing air velocity results in transition to a full bore plug flow, (b). Moving plugs occupy the complete cross-section and again material is swept up at the front and dropped off at the rear. Air velocities are in the order of 0.5-2.0 metres/sec and plugs are separated by an air-space, with a thin stationary layer still existing. This sort of flow is suited to coarse permeable products. With such materials there is an unstable transition between dense and dilute phase flows when excessive vibration is encountered.

A material's suitability to be conveyed in a particular manner is most influenced by particle size distribution, shape distribution, permeability and air retention. The first can be readily measured and

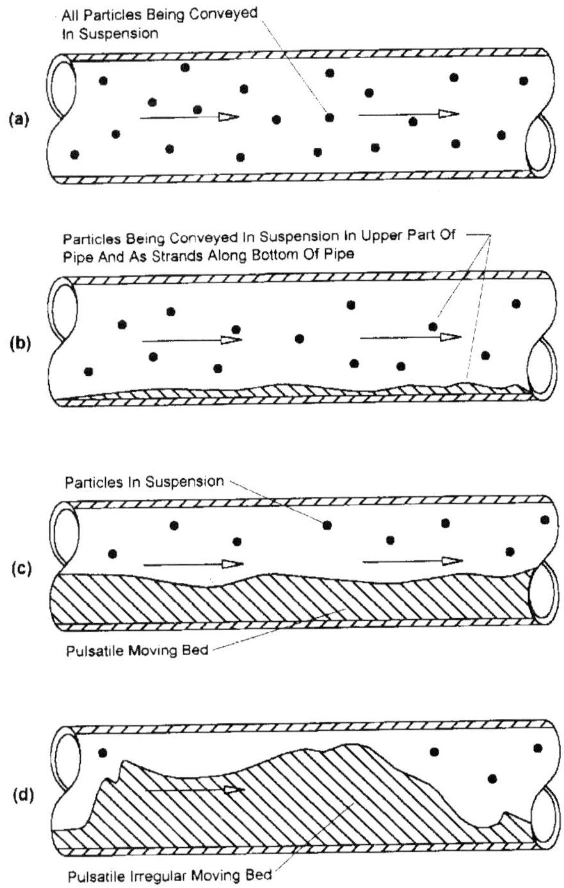

Fig 64: Dilute flow systems

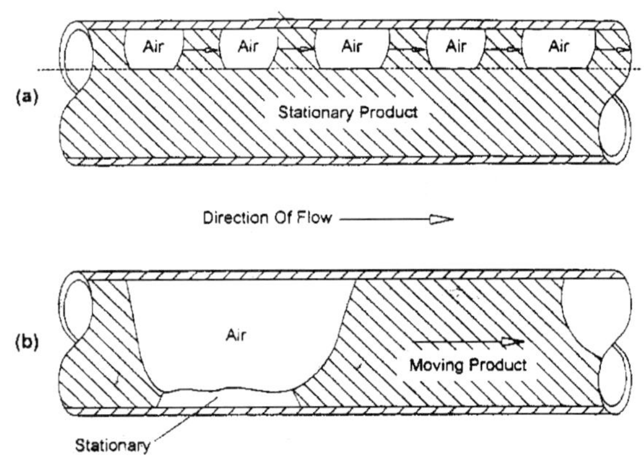

Fig 65: Dense flow systems

generally materials with wide particle size distribution are more problematic than, say, cement. With narrow size distribution materials there is also a direct particle size effect. Those of 150-200 microns are more difficult to convey than those of typically 3000 microns, where a dense plug mode is used. Fig 66 is a size analysis provided by the author. Particle shape distribution is usually qualitatively assessed and its effect associated with size distribution. Permeability can be measured with equipment described in the paper; it is expressed as the relationship between superficial velocity of gas passing through a continuous bed, of such a product, and the pressure drop that results in the direction of the flow, Fig 67. Pressure drop is directionally proportional to air velocity over a certain range, corresponding to flow through a 'packed' bed. Another test described in the paper can be used to determine air-retention and a result shown in Fig 68 is obtained. If material is sufficiently permeable as expressed by being above the line in Fig 69 it is likely to adopt the dense phase mode whereas if it is not permeable, but sufficiently air-retentive expressed by being above the line in Fig 70, the same effect is obtained.

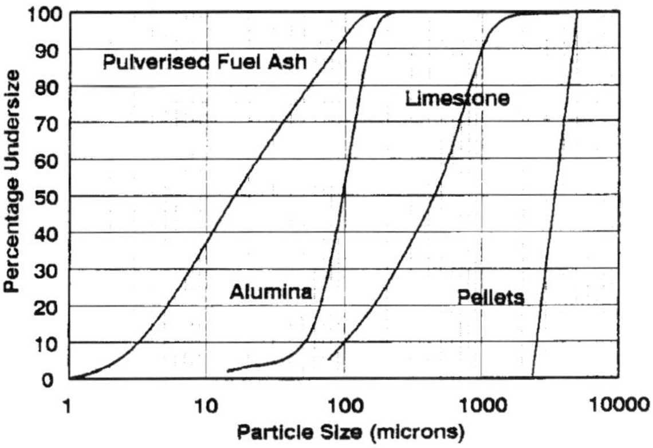

Fig 66: Particle size analysis

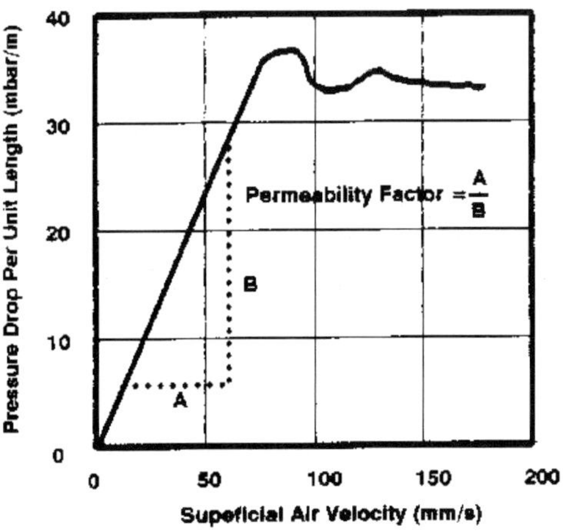

Fig 67: Permeability

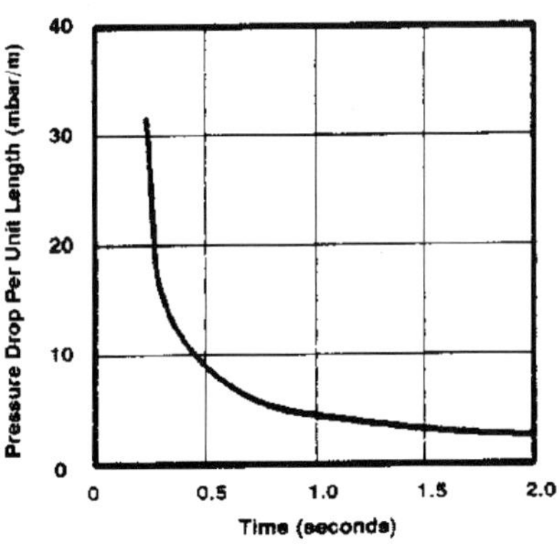

Fig 68: Pressure drop

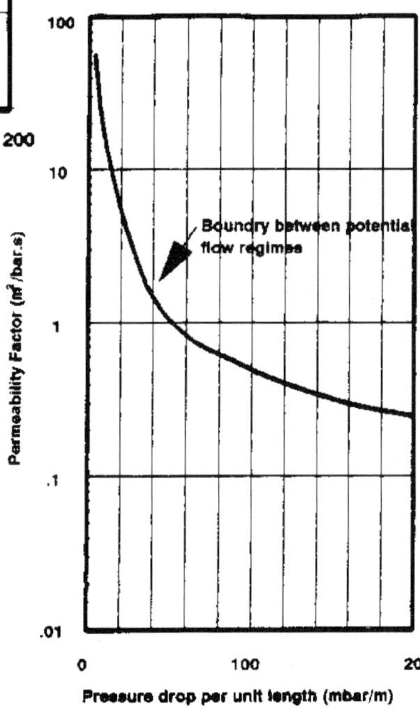

Fig 69: Permeability factor

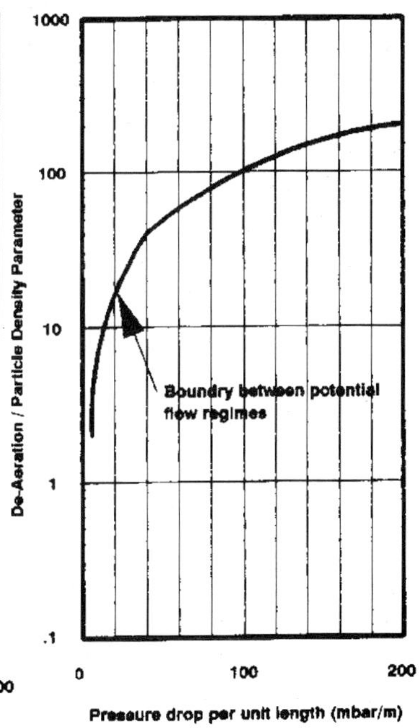

Fig 70: Air-retention factor

Powder conveying systems

Most commonly used dilute phase systems will utilise a rotary valve as a feeding device to control flow of material into the conveying line and to provide a pressure barrier to the source of the material. Generally a pressure differential of 1 bar is the rule. For dense phase conveying, several alternative systems are available; generally these involve a pressure vessel into which a batch of bulk solid is admitted. The material inlet valve is then closed and air continuously fed in. When blowtank pressure reaches a sufficient level the product will pass into the conveying line resulting in the blowtank contents gradually discharging. Use of two instead of one blowtank can lead to continuous rather than batch flow. By proportioning air supply between the blowtank and a supplementary air inlet position, control of feed rate can be obtained. Fig 71 shows two possible arrangements; use of two air input positions results in a particularly versatile system with which, by proportioning the air, it can convey in both dilute and dense phase modes.

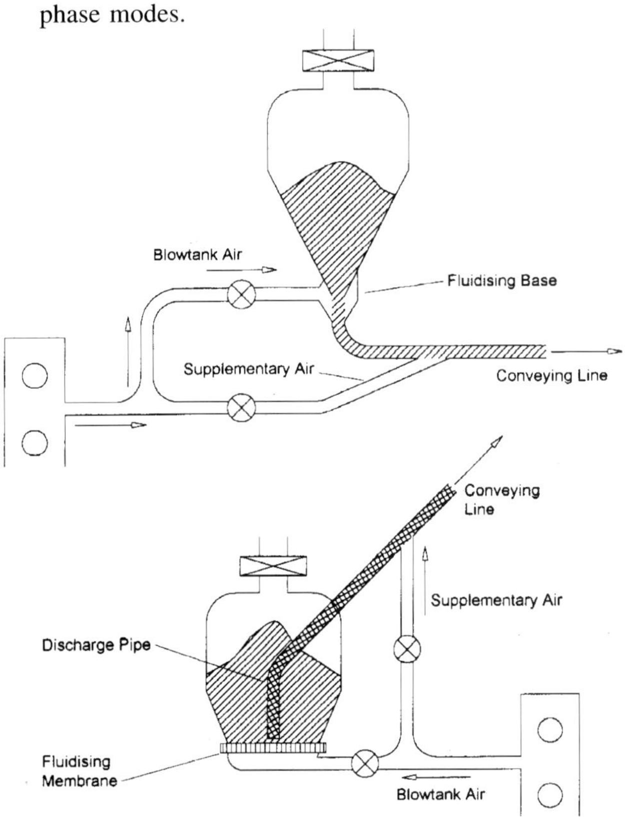

Fig 71 : Alternative systems for proportioning flow

Roll-over stability of articulated tippers

University of Nottingham researchers[15] have now completed a computer modelling technique that allows prediction of tipper stability at the design stage. They describe a theoretical method for predicting the behaviour of a tipping trailer operating on an incline. A computer program has been written which allows a finite-element stiffness matrix for a particular trailer chassis and tractor coupling to be entered, along with data on tyres and suspension, for solution with any combination of ground slope, payload, wind load and tipping angle. In their studies of previous work on this problem the authors make little reference to the tipping problem but some reference to cornering stability of laden tippers. They argue that improvements in handling stability could be made by altering tyre and suspension construction, increasing chassis torsional stiffness and increasing body rigidity.

The body (assumed to be rigid) of the tipping trailer is loaded by the payload P and wind load W. As shown in Fig 72, the equilibrium of the body is maintained by forces H_L at the left-hand hinge, H_g at the right-hand hinge and R at the bottom of the ram (which is considered as part of the body). The positions of these reactions change under load, because H_L, H_R and R are supported by the flexible chassis. A stiffness matrix for the flexible parts of the vehicle (chassis, tyres and suspension and so on) has to be used to obtain the displacements of the connecting 'points' between the body and chassis (Fig 73). A similar iterative procedure allows the true displacements to be determined. The distributed weight of the chassis, wheels and suspension of the trailer, W_c, and the weight of the tractor, W_t, are significant loads on the chassis, which have to be included in the calculation of the displacements. The suspension airbags also exert forces which have to be determined. The three airbags on each side are connected so that they exert six identical forces A_L on the left and six other identical forces A_R on the right, as shown in Fig 74. However, there are four possible operating conditions which exist for the airbag suspension system. Under normal conditions, the airbag pressure is continuously changed to maintain 'height control' H; that is the distance between each side of the central wheel axle and the chassis is maintained at a constant value. If the pressure in one of the sets of airbags reaches a maximum value, a cut-off valve operates and polytropic compression of the fixed mass of gas *C* controls the airbag force and hence the deformations of the suspension on this side, while the other side can still be height controlled. Also, if the gauge pressure becomes zero for one of the sets of airbags, no further control is exerted on that set of airbags *F*. Hence the airbags will be effectively removed from one side while the other side may be either height controlled or controlled by the polytropic compression of a fixed mass of gas. For rolling to the right, the four possible airbag conditions are as follows.

The left and right side airbag pressures are both adjusted to give height control on both sides (that is LH/RH). The left airbag pressure is adjusted to give height control on this side while the right side is controlled by the polytropic compression of a fixed mass of gas (that is LH/RC). The left airbags are inoperative (gauge pressure of zero) while the right airbag pressure is adjusted to give height control on this side (that is LF/RH). The left airbags are inoperative while the right side is controlled by the polytropic compression of a fixed mass of gas (that is LF/RC).

The basis of the analysis is fully described in the paper for these four suspension cases. The computer program for solving the analysis allows ground slope to be incremented in one degree steps from 0 to 10 degrees and the ram extension to be incremented in 0.5 metre intervals over its full stroke, for a given ground slope and wind loading. Although the overall problem is highly non-linear, only small-deformation linear elastic FE analyses are required and a PC program, which uses the results of these as input data, deals with all non-linear effects. It can be used with a fixed or flowing payload and a variety of suspension settings. The computer model is therefore a valuable design tool for predicting stability effects before commitment to metal.

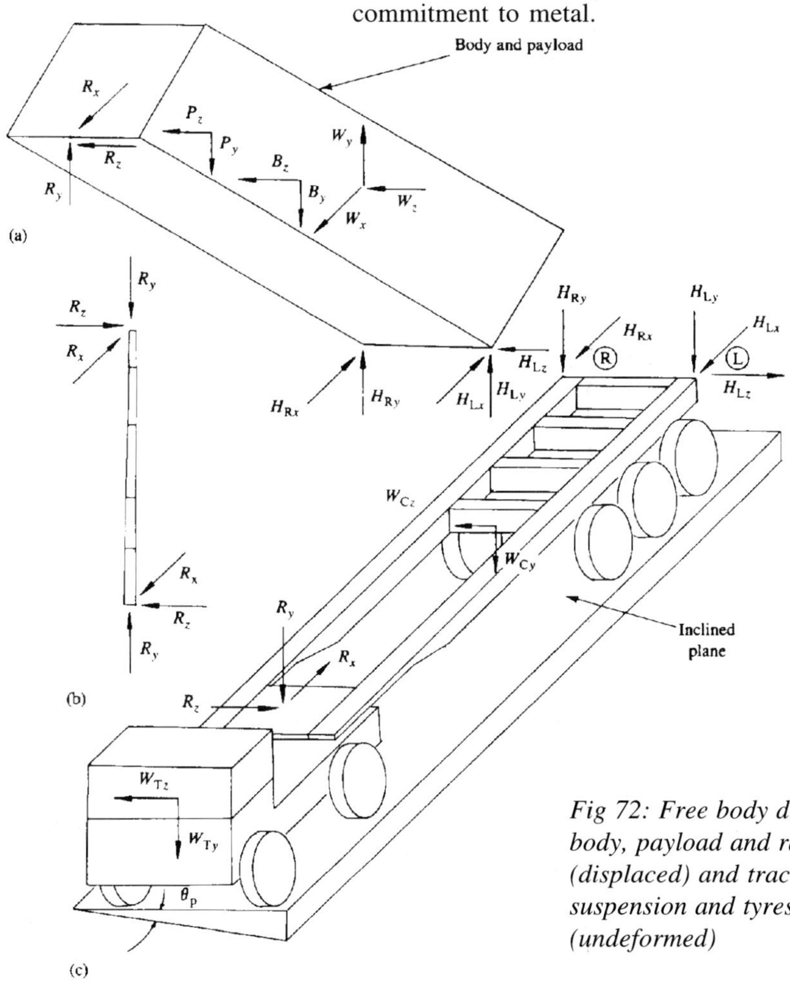

Fig 72: Free body diagram of body, payload and ram (displaced) and tractor, trailer, suspension and tyres (undeformed)

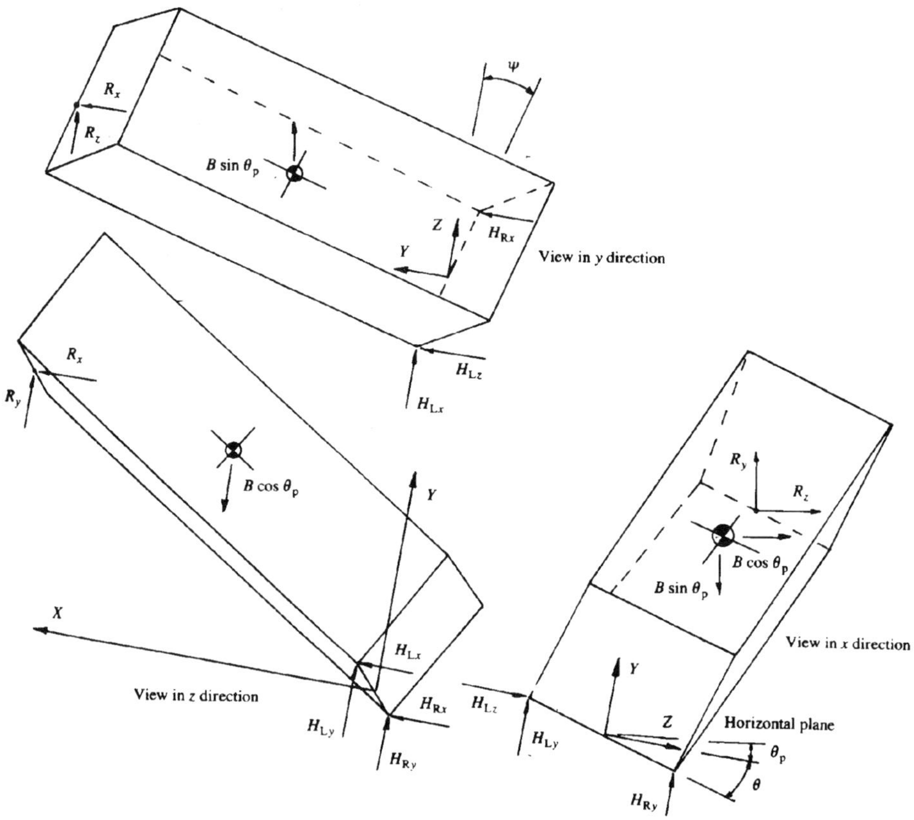

Fig 73: Free body diagram of body and trailer in tipped position, with flexible chassis

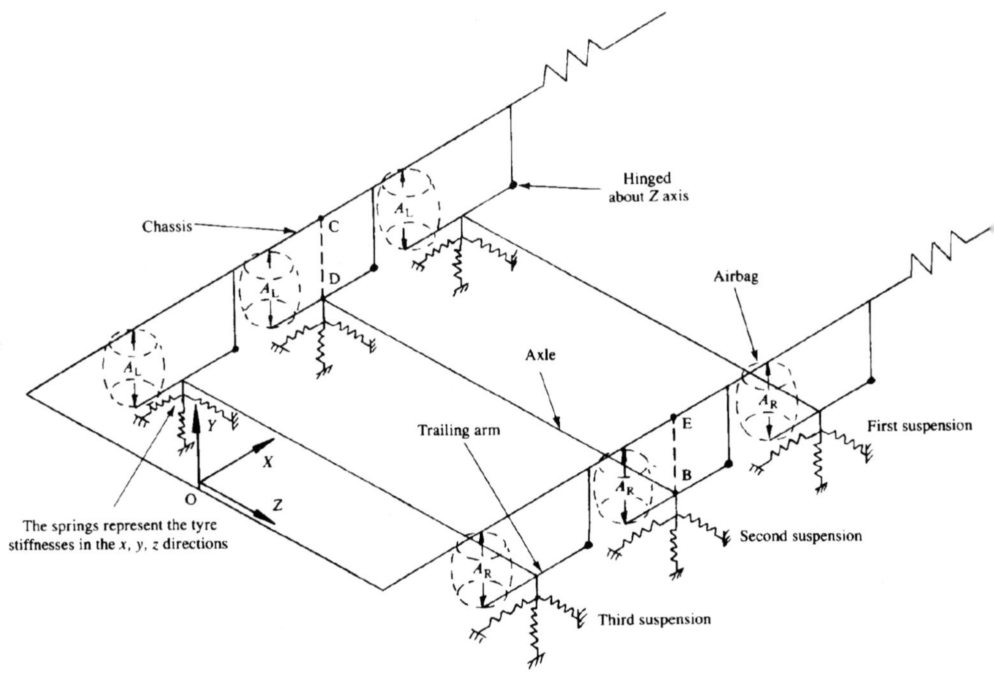

Fig 74: Schematic of rear chassis, suspension and tyres

Calculating tipper stability

The IRTE tipper stability code has focused new attention on designing for the best possible resistance to overturning when long bodied tippers discharge on uneven ground. The problem is also being examined by university researchers who are demonstrating that a rigorous analysis can throw useful light on the problem. Manufacturers are also responding with such techniques as gusseted box section crossmembers, Fig 75.

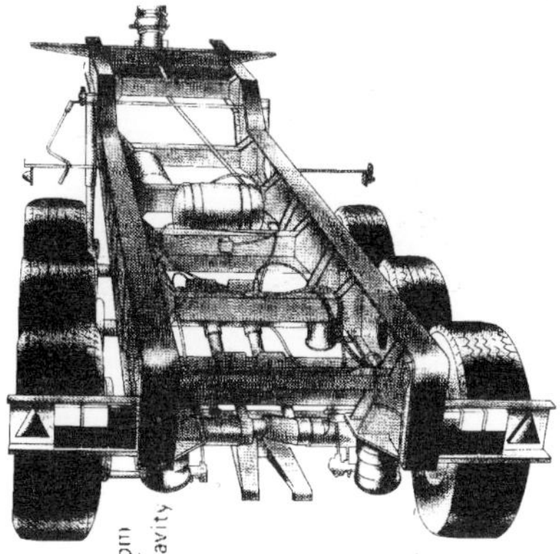

Fig 75: Tipper trailer chassis with gusseted crossmembers

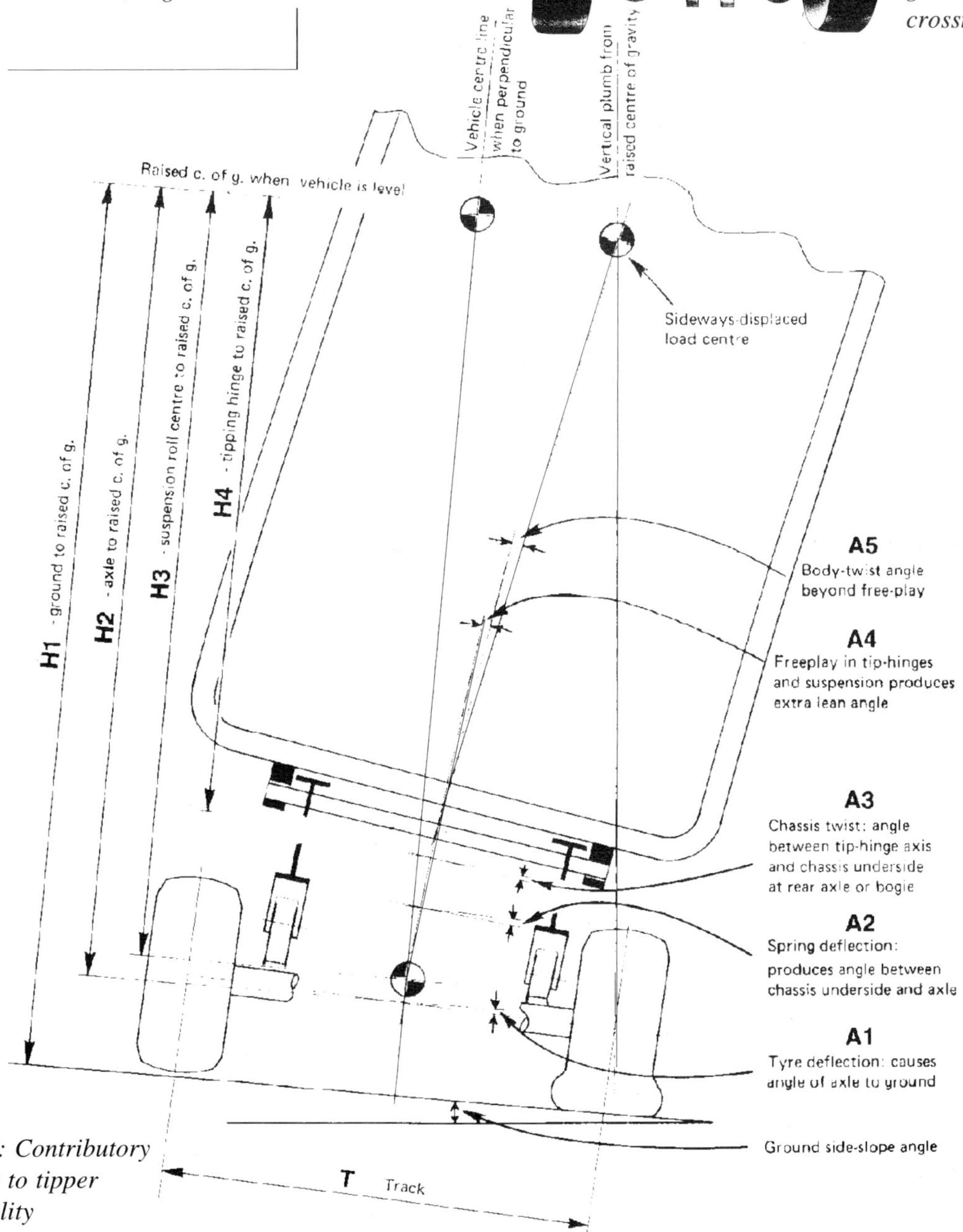

Fig 76: Contributory factors to tipper instability

Improved stability for tipper trucks is the objective of the code-of-practice recently published by the Institute of Road Transport Engineers. The institute drew on several independent expert consultants including Dr George Tramal of the Bristol Faculty of Engineering and former Crane Fruehauf engineering chief Norman Simmons who now operates as NCS Engineering Consultancy Services. The last company was responsible for the diagram of contributory factors shown as Fig76.

This is drawn up on the basis of the commonest type of accident, falling over sideways when discharging load. While the situation can be worsened by such factors as soft ground, sticking payload or a side-wind gust, tolerance of such adversities can be built in to the design of the vehicle and observance of the basic parameters illustrated gives the tipper the best chance for stability under most conditions.

Two categories of minimum stability are recommended by the code: Category B covers the ability to discharge on hard level surfaces and Category A covers the unlevel off-road situation. The criterion of stability is the sideslope angle which the vehicle can withstand without wheel lift when the body is fully raised and fully loaded — which can be both measured and calculated. Because of the sticking load situation it is necessary for the vehicle to have its body fully raised with the load in place and then withstand uneven discharge.

For Category B operating conditions the minimum acceptable sideslope stability for any tipping vehicle must be five degrees. This increases to seven degrees for Category A operation and for both cases it is recommended that the main safety points in operating the vehicle should be marked on it. For tipper manufacturers to have a reliable basis for assessment of stability the chassis maker should supply, on request, the torsional resistance of the chassis frame from hinge point to rear suspension centre-line; the roll-stiffness of the suspension at the point of wheel lift on one side; the effective spring-base dimension; roll-centre height for the suspension and mounting guidelines for both bodies and tipping-gear.

General guidelines in design are for the centre-of-gravity to be as low as possible; high suspension roll-stiffness at the point of overturning; torsional stiff chassis between hinge and suspension; minimum free-play in suspension and hinges (which must be as widely spaced as possible) also close as possible distance between hinge and suspension. Add-on features such as sideways restraint frames between chassis and body can also be beneficial.

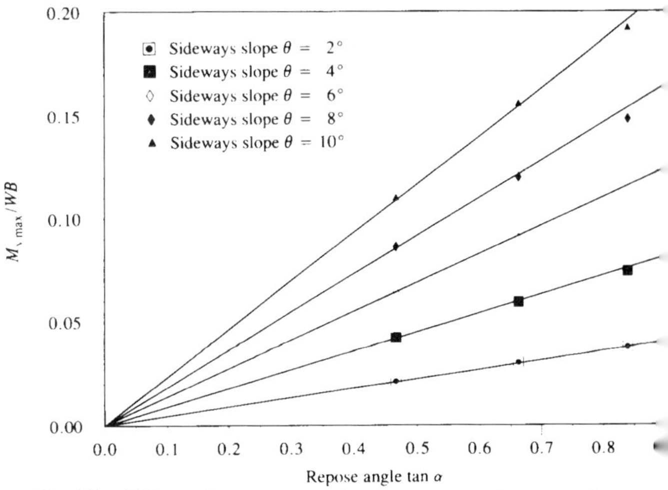

Fig 78: Effect of repose angle on overturning couple

Researchers at Nottingham University[15] have also been involved in tipper stability. This has led to method for calculating roll-over couples, for granular materials with different repose angles, when tipping from different sideslopes. Fig 77 gives a representation of the tipping body and payload which is assumed to be free-flowing and create no friction with the body walls. The researchers have examined the shape of the sectional plane IJKA through the payload and used co-ordinate geometry in three planes to calculate its shape. A computer program has also been written to relate a shape factor with payload volume, mass and centre-of-gravity co-ordinates. So far the work has been restricted to analysis of the body and payload only but further work is envisaged that will include factors pertaining to a large articulated vehicle.

For the body and payload maximum value of roll-over couple due to a free-flowing payload has been found as:

$$0.023q \tan a/WB$$

for the corresponding tipping angle: α - *12* where W is weight of full payload and B is body width; θ is sideways slope of body and α is angle of repose. Fig 78 shows the effect of repose angle on overturning couple.

The various factors that go into the build-up of overbalancing angle, which equates, in a cumulative formula:

$$\sin^{-1}\left\{\frac{(T \div 2)-(H2 \times \sin A1)-(H3 \times \sin A2)-(H3 \times \sin A3)-(H4 \times \sin A4)-(H4 \times \sin A5)}{H1}\right\}$$

This will not give a precise prediction of the overturning angle but will normally be good enough for comparison purposes. Typical values are given below for a tri-axle trailer.

TYRES

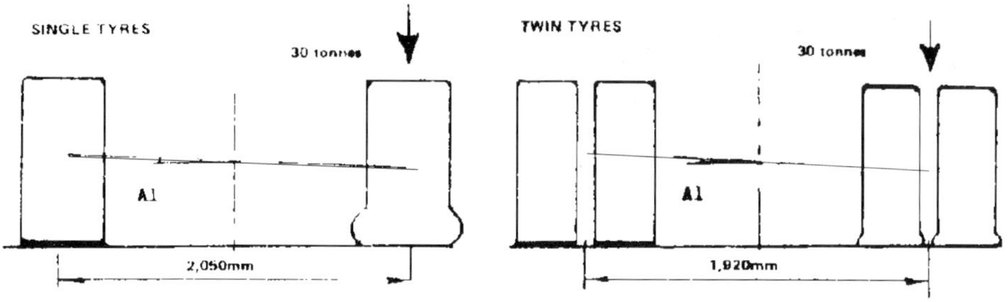

Deflection = 120mm Angle A1 = 3.35° Deflection = 95mm Angle A1 = 2.8°

SPRINGS

Deflection = −17mm 67mm

Angle A2 = 4°

Deflection = −47mm 73mm

Angle A2 = 7°

Air suspension A2 is less than 2°

FREEPLAY

1,200mm SPRING CENTRES	1,000mm SPRING CENTRES
For 30mm play, Angle A3 = 1.43°	For 30mm play, Angle A3 = 1.7°
For zero play, Angle A3 = 0	For zero play, Angle A3 = 0

BODY STIFFNESS

Assume Angle A5 = 2°

CHASSIS STIFFNESS

For a torsionally flexible chassis, Angle A4 approximately 3° (few, or channel or I-section cross-members).

For a torsionally stiff chassis, Angle A4 approximately 1° (tubular or box cross-members).

It is possible to get significant variations in these values. These figures are by way of example only.

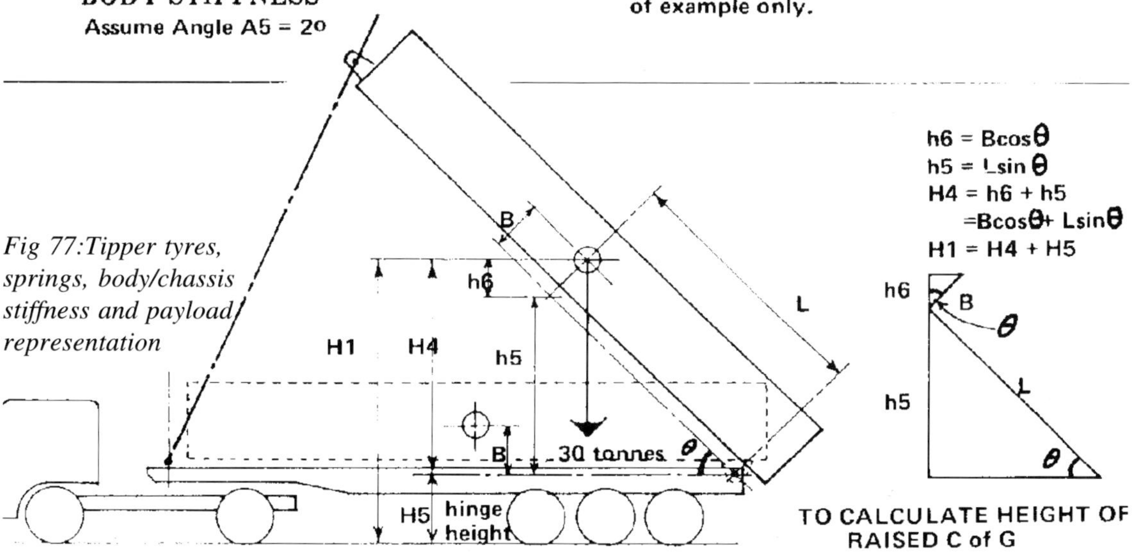

Fig 77:Tipper tyres, springs, body/chassis stiffness and payload representation

Chapter 11: Specialist commercial vehicles

Refrigerated vehicles; refuse/cleansing vehicles; power-take-offs; hydraulic and gear drives; handling trucks; light van design; optimizing the delivery vehicle; vehicles for the disabled; ambulance design; van design, new initiatives; electric vans; EV structures

Refrigerated vehicles

There is intense competition between refrigerated vehicle builders to produce a thermally efficient, lightweight vehicle with maximum internal cube. This section looks at some of the techniques of analysis and construction used to wring out that final margin of performance over the competition.

The basic principles of heat transfer are a prerequisite to an understanding of body thermal efficiency — the ability to hold low temperatures over a complete journey with minimum expenditure of energy. Heat flows by conduction through a material; by radiation from one surface to another, via the intervening space, or by natural/forced convection in which the heat is carried by moving air (fluid) currents from one surface to another. As a vehicle moves along an open road the body will be subject to varying degrees of solar radiation which warm its outer surface. In general, heat will flow by conduction through the thickness of the wall and forced air from the refrigeration plant helps to induce convection currents on to its cooling coils or plates. Conversely when a door is open to atmosphere into the cool chamber cold air will 'fall' out of the body by natural convection and be replaced by warm air from the atmosphere.

Heat flow, being analogous to electrical current, follows the line of least resistance from the atmospheric ambient of, say, +15 C to the body interior which may be held down as low as 18 C for such products as frozen food and ice cream. A principal concern of body design and build is thus to avoid heat paths between outer and inner surfaces; high conductivity metal bridges between inner and outer skins are obviously something to avoid. So too is it necessary to seal against moisture ingress as water also has high conductivity relative to insulating materials. Sandwich panels are, of course, widely used as these have no structural pillars or rails penetrating the insulant. Fig 1 shows Concargo's Permacold sandwich panel body kit which includes pre-engineered door assembly and interlinking corner rebates.

A typical foam plastic insulant will have thermal conductivity of 0.025 kcal m/hr m^2 C. This will vary with density and structural strength and from material to material. Expanded PVC is one of the strongest materials related to weight and thermal insulation. Neoprene is an adhesive which is suitable for bonding expanded-PVC cores to metal skins and Fig 2 shows the kind of vacuum table which can be used to fabricate the panels from slabstock insulant and metal sheets — the latter should be degreased

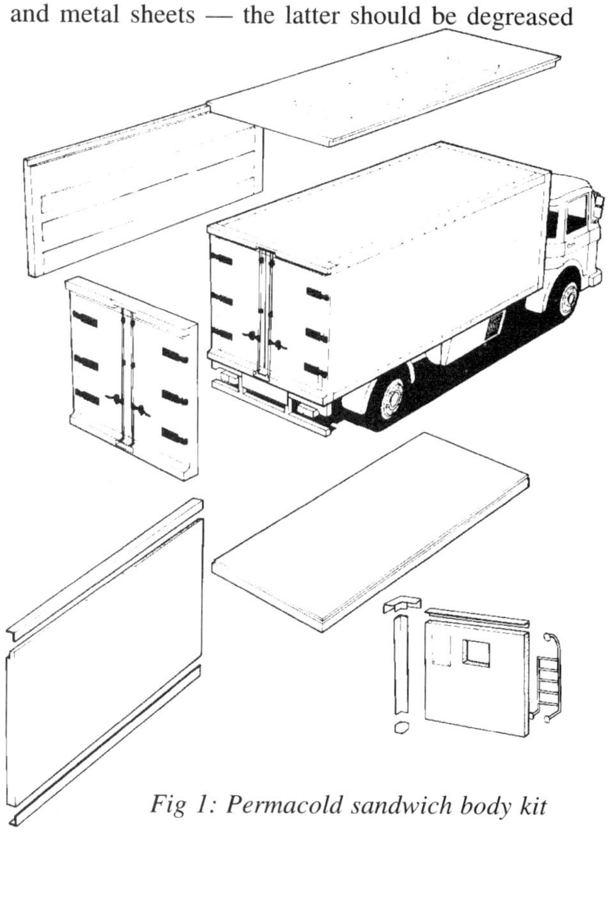

Fig 1: Permacold sandwich body kit

with a suitable solvent. Another common technique is foamed in-situ expanded polyurethane.

Insulated vehicle practice

For large refrigerated articulated trailers, Schmitz Cargobull have moulding presses installed for producing 40 ft (12.2 m) long sandwich panels using expanded-PVC insulant bonded panel-to-panel by polyurethane gap-filling adhesive on the edges. Expanded polyurethane is used as a roof-panel insulant to protect against the higher skin temperatures. PVC foams of 40 and 55 kg/m^3 density are used according to the structural requirements of the panel. Normally the exterior skin is prepainted white aluminium alloy, in 4 ft (1.22 m) wide sheets positioned vertically, but 10% of vans go out with electro-galvanised steel skins. Each is bonded to the cores by epoxy resin. Exhausting by 10 lb/in^2 (69 kN/m^2) in the vacuum bag is equivalent to a side thrust of 250 tonnes. Resulting side-panels are 45, 60 and 80 mm thick.

The floor sandwich comprises a 1 mm thick galvanised under-skin, adjacent to a 15 mm thick plywood layer. The rubbing plate above the kingpin is just the 1 mm thick under-skin but the plywood inner panel is overlayed with aluminium alloy above the kingpin frame, which is embedded in the insulant.

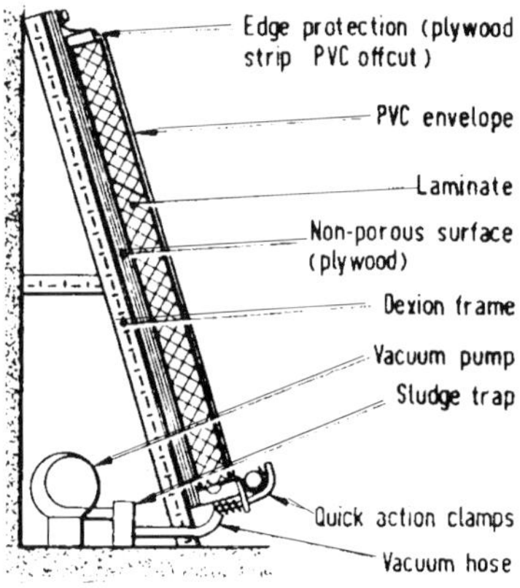

Fig 2: Vacuum table for assembling sandwich panels

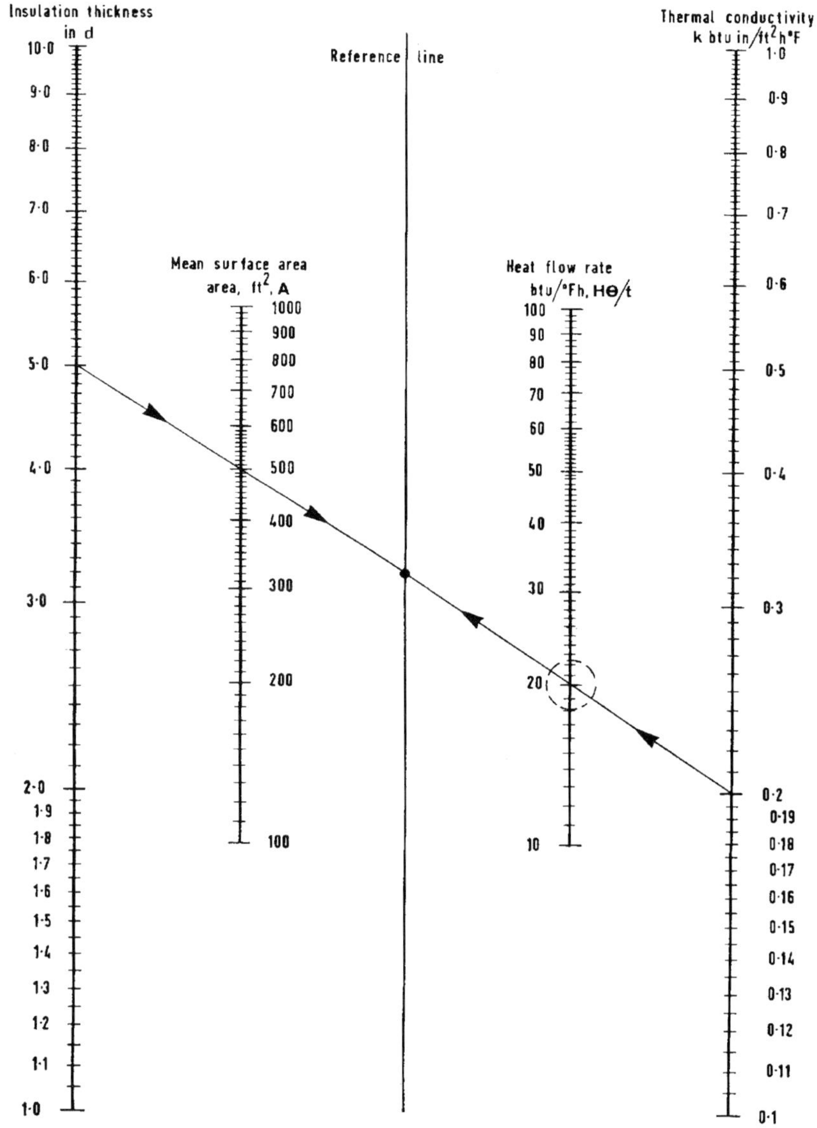

Fig 3: Insulated panel selection nomograph

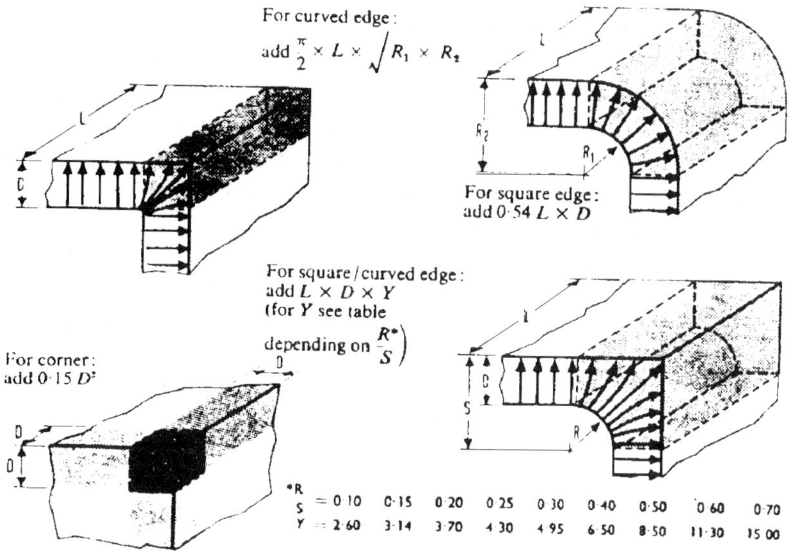

Fig 4: Shape factors for edges and corners

In final body assembly the precision moulded panels are self-jigging and no mechanical fasteners are used to supplement the twin-pack adhesive. Corners are, however, capped with aluminium alloy angle sections over a vapour-sealing PVC tape. The one-piece stainless steel rear frame is fabricated from channel sections which cap the rear edges of all four main body panels. Doors have no lap joint so that either can

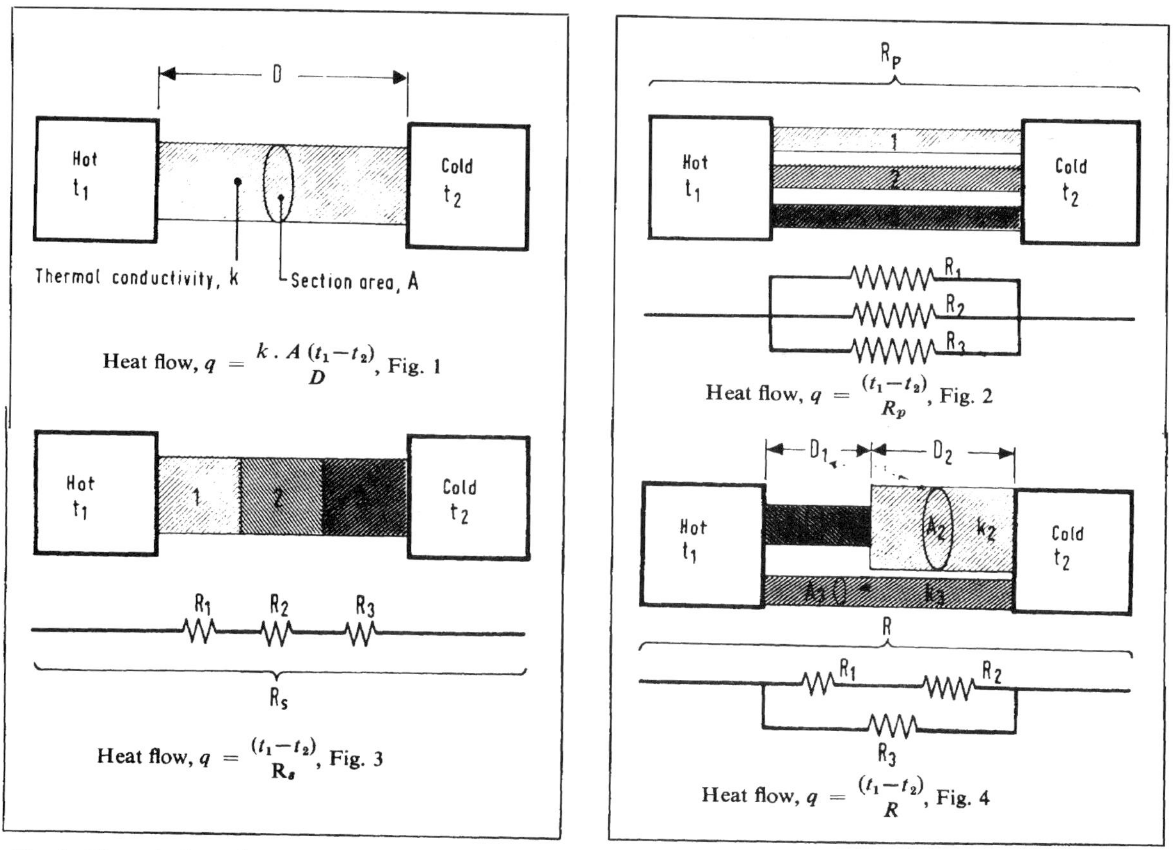

Fig 5: Electrical analogues for heat flow

be opened independently.

Interior body joints are also capped with angles, and sealed with mastic, except the bottom joints which are sealed when a layer of polyester resin is floated over the whole floor — which is impregnated with carborundum grit and topped by a sealing screed. Having an integrated kingpin frame and rubbing plate results in a 4.5 in (114 mm) gain in interior height to 2.5 metres — on a 48 in (1.21 m) high fifth-wheel coupling turntable and within the four metres continental overall height restriction. With interior length of 12.2 metres, and interior width of 2.45 metres, internal cube is just below 75 cubic metres. Unladen weight of the integral trailer on tandem axles is below 14000 lb (6.35 tonnes) — for operation at 38 tonnes GTW.

Analysis of insulated bodies

First assessment of insulation thickness required to achieve a given thermal performance by a sandwich panel can be made using the nomograph in Fig 3 which quotes thermal conductivity in Imperial units.

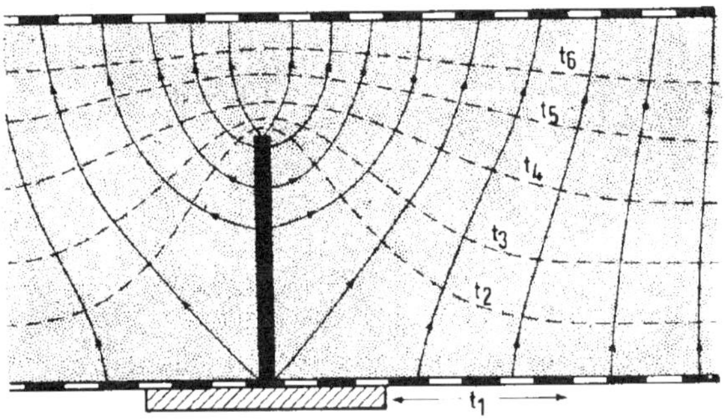

Fig 6: Heat flow from rib

$\frac{n}{l}$	0·333	0·5	1·0	2·0	3·0	∞
For $\frac{m}{l} = \frac{1}{4}$	0·507	0·573	0·609	0·611	0·612	0·611
For $\frac{m}{l} = \frac{1}{2}$	0·1834	0·207	0·2201	0·2205	0·2208	0·2206

Fig 7: Shape factors for rib

Values of X/l for various shapes

m/l	$n/l = 0$ ≡ Rib	$n/l = \infty$ Buttress	T-rib
0·1	0·0079	0·0122	0·0115
0·25	0·050	0·065	0·060
0·5	0·221	0·250	0·221
0·75	0·611	0·650	0·537
0·9	1·181	1·222	0·929
a	*b*	*c*	*d*

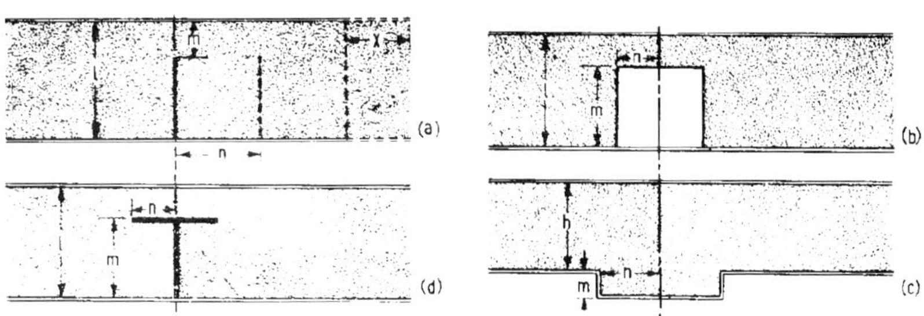

Fig 8: Shapes tested for heat flow

To estimate the thickness *d* of insulation to give minimum heat leakage *H* through panel surface area *A* with insulant conductivity *k*, the equation

$$H/\theta t = kA/d$$

applies which is represented by the nomograph shown — in which a worked example shows intersects drawn for d = 5 in (127 mm), A = 500 ft^2 (46.45 m^2) and k = 0.2 BTU in/ft^{2}h F.

In calculating the heat flow through a complete box body, additional surface area to allow for edges and corners must be allowed for in calculation. These 'shape factors' are suggested in Fig 4 and are an estimate of the two and three dimensional flows which occur at these locations. Fig 5 shows how the electrical analogy to heat flow can be put to effect in calculation — the heat paths being considered as electrical resistances. The basic conduction equation can only be applied in case (a) whereas in other cases it is useful to draw up a thermal circuit to calculate an approximate heat flow for series, parallel and combined networks. Note, of course, that $R_s = R_1 + R_2 + R_3$ whereas $R_p = (1/R_1) + (1/R_2) + (1/R_3)$. Also the actual, and not 'specific', thermal resistances must

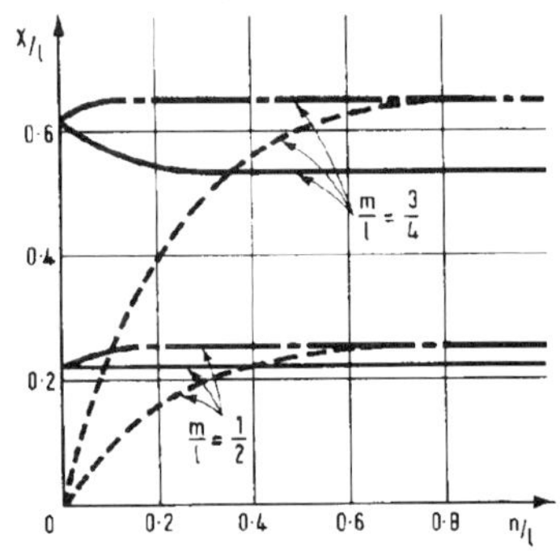

Fig 9: Shape factors for other sections

be used in calculation so that areas and lengths of heat paths are accounted for. Thus in case (d), $R_l = [D_l/k_lA_l](R2, R3, etc.)$ and $(1/R) = [1/(R_l + R_2)] + 1/R_3$. Only complete accuracy can be obtained where the two and three dimensional heat flow effects are accounted for. These can be predicted by finite element techniques but it may be preferable to use physical models of the shapes in question and test them in electrical flow analogue experiments.

In the case, say, of a high conductivity rib extending into an insulated wall, the parallel lines of heat flow are disturbed around the locality of the rib as shown in Fig 6. Pioneer work by the Ship Owners Cargo Research Organisation obtained shape factors for such sections and the table of Fig 7 shows extra thickness of wall, X as a fraction of plain wall thickness l for rib depth m and spacing n. Other shapes shown in Fig 8 were found to have shape factors obtainable from the graph in Fig 9. The heat flow through door frames can be estimated by such techniques and it is a comparatively simple matter to have computer programs written to simulate thermal performance of the vehicle prior to build, as a summation of all the local heat flows. As an illustration, the door-frame and seal section in Fig 10 is accompanied by its electrical resistance analogue for which the elemental resistances can be evaluated by the methods just outlined.

'Insulated' rather than 'refrigerated' vehicles, for the carriage of perishable rather than frozen goods, are sometimes termed 'half-isothermal' vans and may involve limited thermal bridges between inner and outer skins. Such construction is typified by frame pillars in expanded-PVC enclosed by expanded polyurethane insulation as in Fig 11. The importance of the vapour film is vital as it is possible for a large box van body to accumulate as much as 1 tonne of water in a year's operation — which can markedly affect thermal efficiency.

On-board refrigeration plant

Isothermal vans are characterised by on-board refrigeration systems which may be eutectic plates, mechanical units and/or liquid-nitrogen/solid-carbon-dioxide evaporative cooling. Eutectic plates are filled with refrigerant which is only circulated, through a land-based refrigeration plant, when the vehicle is 'in dock'. The plates act as storage coolers in transit and though somewhat heavy when made in metal, new

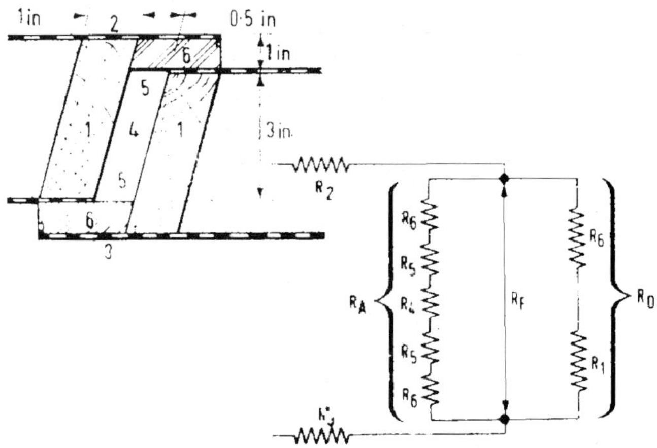

Fig 10: Door frame/seal with its electrical analogue

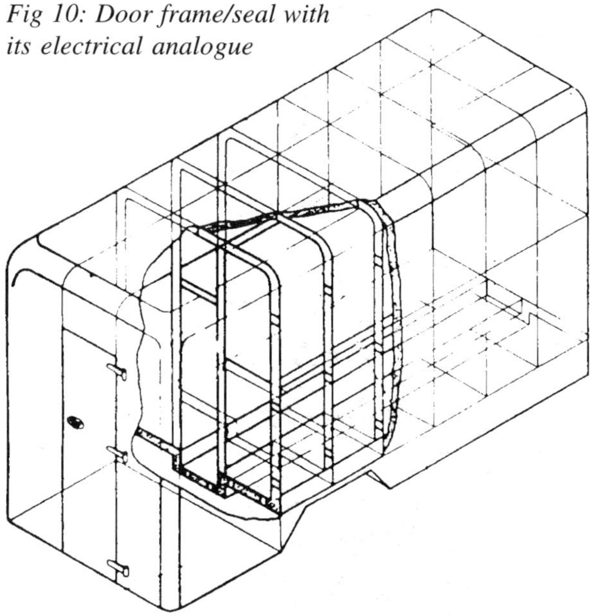

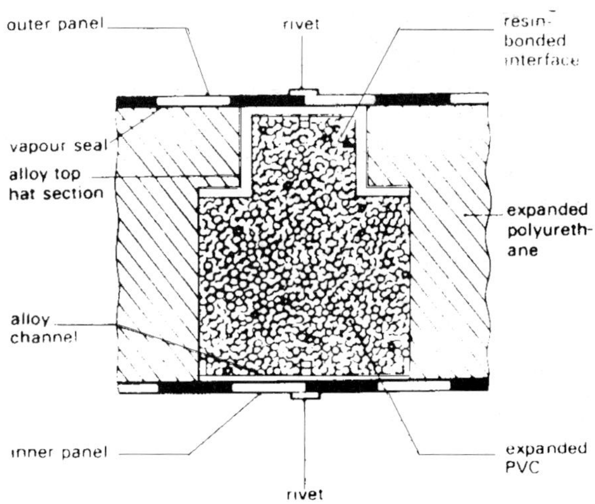

Fig 11: Bridging of inner and outer skins

plastic tubular versions are proving to be acceptable to operators.

Mechanical systems, of course, give maximum mobility to the operator but their weight, too, is not insignificant. A large reefer can involve a plant weighing 0.75 tonne which is usually mounted on the leading face of the trailer front panel. The mounting will usually involve thermal bridges in the frontpanel reinforcement frame which must be allowed for in the thermal efficiency calculation. Direct evaporation systems such as those for liquid nitrogen involve a container to which a series of small bore copper pipes are connected and installed usually at the upper extremities of the van, running from front to rear. A series of jets at predetermined centres spray the liquid over the cargo during initial temperature 'pulldown'. This can take only three minutes, from initial ambient temperature, as the liquid boils at 196 C. Low weight of the equipment is an attraction — being in the order of 50 kg for a 4.8 metre long box body. Fig 12 shows an early system from BOC alongside an early ICI evaporator system for solid carbon dioxide (dry ice), Fig 13. The latter occupied 18 x 21 in (457-533 mm) of floor space and was 30 in (762 mm) high. It weighed 62 lb (28 kg) empty and contained 120 lb (54.5 kg) of dry ice. Cooling capacity up to 20 800 BTU/hr for short periods could be obtained with continuous running at 2000-6000 BTU/hr.

Performance of refrigeration systems depends upon a well sealed body with minimum cold air fall-out. Leakage rates of nine cubic metres per hour have been measured on reefers with poor rear door sealing. An interesting door sealing development by Wilsdon is shown in Fig 14 in which the thermal and weather-proofing duties of the seal are combined. Door jambs are moulded integrally with the wall panels and a close tolerance is held on shut clearance by jigging techniques. The sealing gasket works in compression against the wall-panel thus overcoming conventional shear drag.

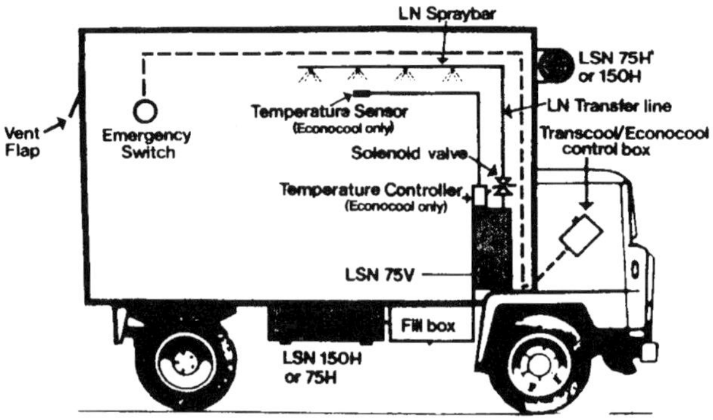

Fig 12: BOC liquid nitogen system

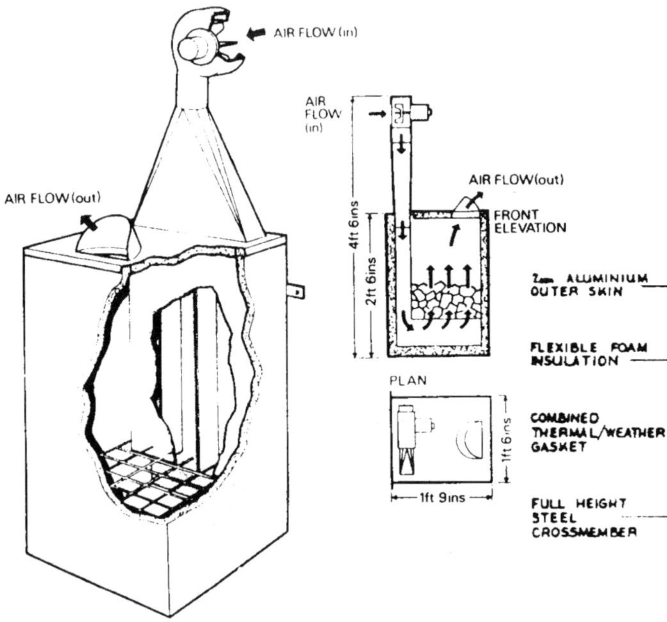

Fig 13: ICI dry ice system

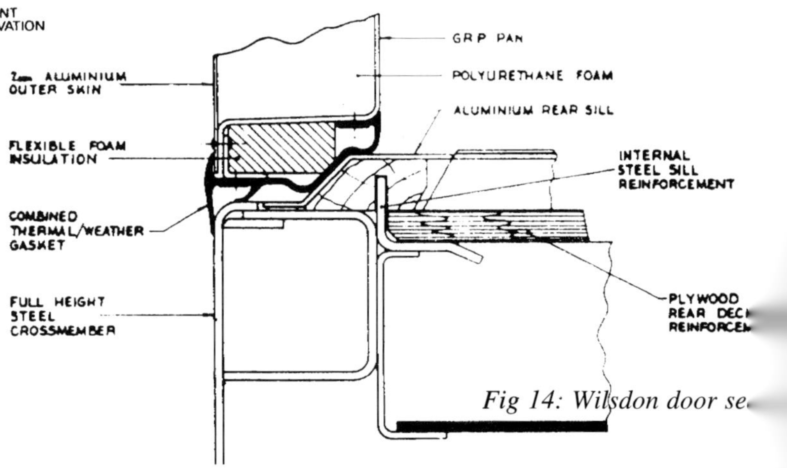

Fig 14: Wilsdon door se

Refuse/cleansing vehicles

Refuse vehicles encompass a variety of technologies and call for considerable creative skill on behalf of design engineers who must successfully integrate them. Some new and unusual approaches to integration are reviewed here alongside consideration of two prevailing technologies, power-take-offs and hydrostatics.

The move from converted truck chassis to small purpose-built sweeper vehicles probably began in the 1970s with an unusual adaptation of the Reliant three-wheeler chassis for sweeper duty in Melford Engineering's Precinct model, Fig 15. Hydrostatic drive was used on the front steered wheel so that the vehicle could drive forward into a blind 90 degree corner and sweep the ground surface in one pass. A total steering angle of 200 deg was achieved enabling the two rotating brushes at the front corner of the machine to be traversed out of a corner in such a way that the only area unswept was in the very corner beyond the 28 in (711 mm) diameter of the brush.

The Schmidt SK152 model, Fig 16, has also advanced the state of the art; based on the Design Council Award winning SK150, it has special steering geometry on the oscillating front axle, and hydrostatic power steering, which allows this 4-wheeler machine also to sweep out a 90 deg internal corner. A patented water circulation system allows water injected into the suction nozzle and tube to be recycled and ensures a dust-free environment during sweeping. An electro-hydraulic anti-stall system operates between the vehicle transmission and the sweeper drive.

The Hestair Eagle Griffon was an interesting development in truck-mounted sweepers, Fig 17, whose considerable suction capacity comes from a

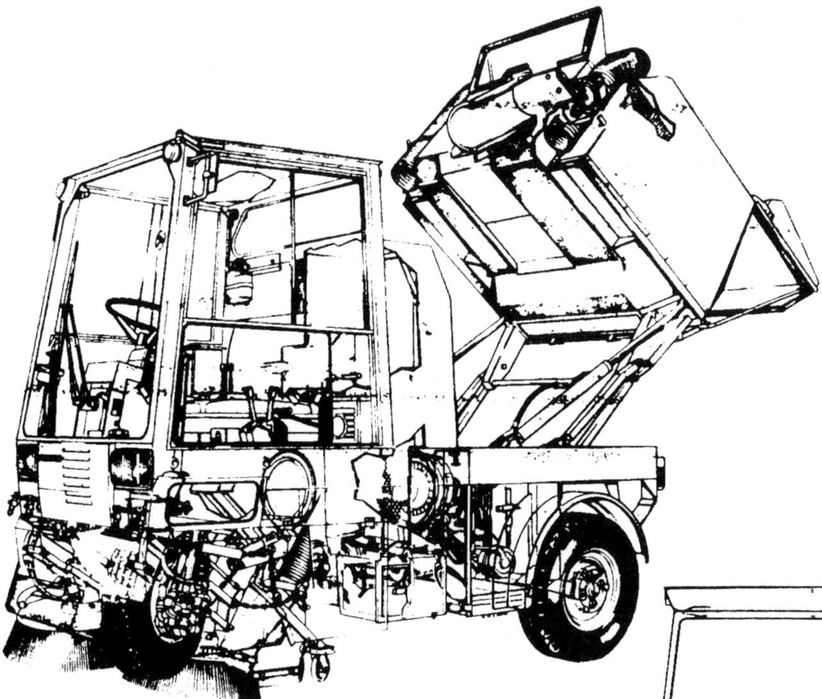

Fig 15: Melford Precinct sweeper

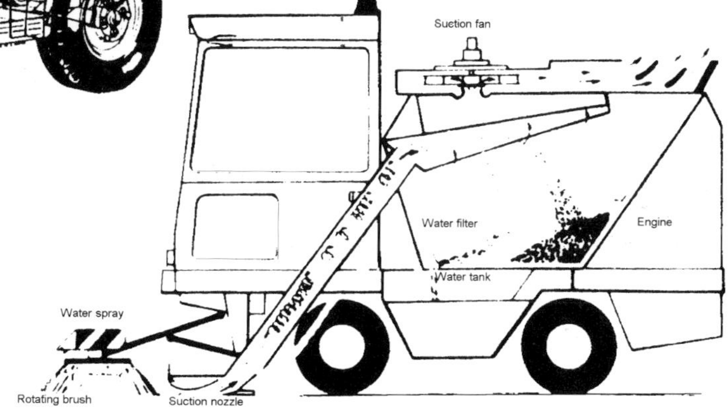

Fig 16: Water circulation system

Fig 17: Eagle Griffon

diesel auxiliary engine driven centrifugal fan (1). The water tank (2) and five cubic metres refuse compartment (3) are constructed in curved high tensile steel panels to withstand the suction vacuum. Shown at (4) is the air-discharge silencer alongside brush-gear (5) and subframe (6); at (7) is the automatic rear door and at (8) the full-width filter screen. Twin inboard fuel tanks (9) are featured also deflectors (10) to improve refuse distribution.

An interesting departure in refuse collection vehicles is the Laird Rotopress (Fig 18) which favours a screw packer rather than the hydraulic one typified by traditional RCVs. On the Dempster Dumpmaster, Fig 19, another alternative is the curved tailgate which is designed to rotate the refuse as the compaction cycle proceeds, Figs 20/21. At (1) the packer head drives refuse (C) up and over; (2) as the packer head retracts, compacted refuse (C) in the upper body falls into empty head space; (3) new load (D) is compacted against (C) and packer head drives

Fig 18: Laird Rotopress

Fig 19: Dempster Dumpmaster

Fig 22: Vector Combi

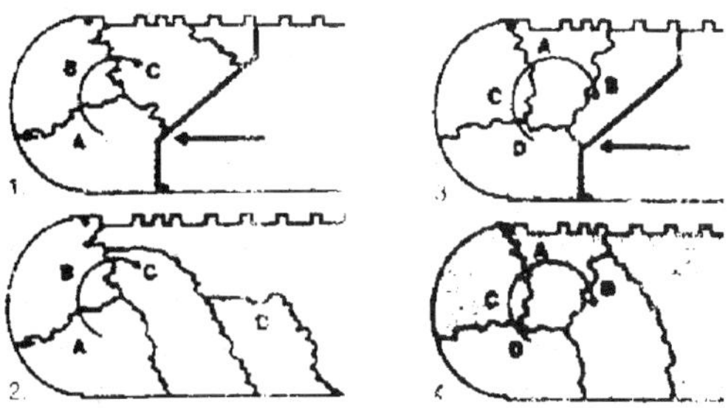

Fig 20: Curved tailgate for compaction and Fig 21 (below): for ejection

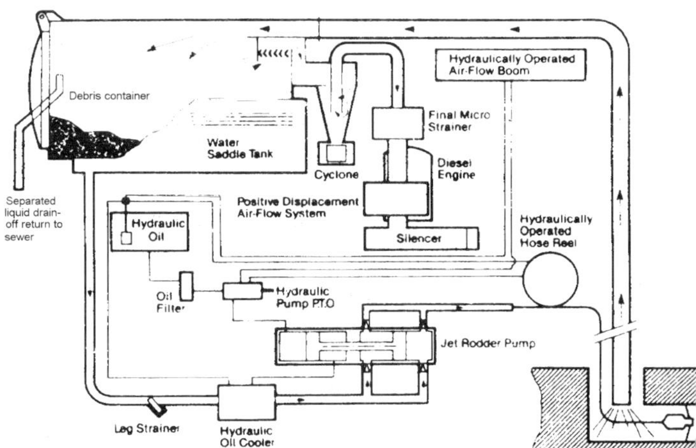

Fig 23: Combi operation

Figs 24/25: Dennis Elite RCV

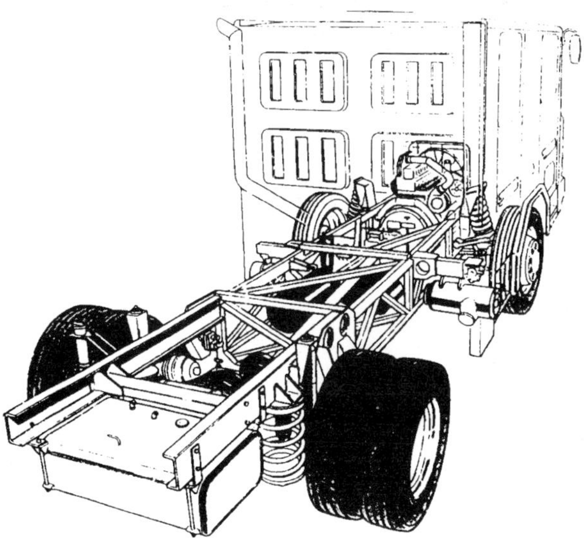

Fig 26: Dennis Rapier

refuse (C) up and around again; (4) refuse (B) falls into empty space left by the packer head. Gully emptier and jetter vehicle design has been advanced by the Peabody Holmes Vactor Combi, Fig 22, whose operating system is shown at Fig 23. The air flow system is driven by an independent diesel engine and the variable rate jetting pump is hydraulically driven from the engine PTO.

New image for RCVs

A move to up-grade the image of refuse-collection vehicles has been taken by Dennis Eagle whose new Elite vehicle has been given a flush-glazed cab with low floor and wide door-openings. It is made from stainless steel and has a low entrance step forward of the front wheels as well as a lowered front chassis-frame which gives a flat-floor. Its tilt-cab is considerably more spacious on the inside as well as providing better visibility than its predecessor. Figs 24/25 show the similarity with the designer's concept renderings. Floor height is 27 inches on 305/70R tyres while interior headroom is over six feet. The engine is set back some 860 mm than that of its predecessor and is mounted 200 mm lower. The chassis frame steps down 340 mm at the front end and channel section side rails have given way to a deep zed section with outwardly angled top flange reinforced by a length of 200 x 50 mm box section. They are spaced apart by the same amount for wide (2.49 metres) and narrow (2.20 metres) versions of the cab. Between-kerbs turning circle has been reduced by up to two metres on some models and when laden on 11R 22.5 tyres, front angle of approach is 15 degrees for crew-cabs and 19 degrees for single-cab models. There is also a variant of the vehicle designed for one-man working, with the driver standing in a well 425 mm above the ground on the additional near-side control position, his controls being arranged for US-style stand-up driving. On all variants the gear-lever is set well forward and the handbrake mounted in the fascia.

Much of the technology has been taken from the Dennis Rapier fire-appliance vehicle, with space-frame chassis, Fig 26. Dramatic improve-ment in ride and handling is achieved by Rapier over the conventional ladder-frame and leaf-spring layout — with the availability of a more rigid frame to mount a suspension of greater sophistication. Over a tonne of weight has been pared from the conventional structure and a low deck height has also been achieved — with a particularly low-slung engine. The latter has permitted the cab to be mounted 250 mm lower than usual, removing the need for entry steps. The entire crew cab tilts in one piece and the lightweight body is made in corrugated aluminium alloy sheet over an extruded section frame — incorporating water tanks made from plastic. Substantial coil springs suspend both front and rear wheels.

Environmentally-friendly RCVs

A special adaptation of the Pietsch K1500 sweeper for the Romanian market, Fig 27, has been described by researchers at the Inar Brasov Research Institute. Protection screens and panels are used for noise attenuation on the principal noise emitters of engine, transmission, exhaust system and fan drive. Another environmental measure is aimed at saving the water used for sprinkling the sweeping zone for dust prevention. Mixed water and dust is collected by absorption then separated and recycled into a buffer tank. The author also described a means of minimising fuel usage and exhaust emission by a mixed drive system having IC-engine drive for the tractor and electric drive for the semi-trailer. The latter drive is used in refuse collection while the former is used in journeys to the discharge tip.

Jetter adapted for water circulation Kaiser of Liechtenstein have built the specially adapted municipal vehicle for water recycling shown in Fig 28. Its rear hose reel has 180 degree hydraulic slew and hydraulics are also used for the security ring closer for the tank's full diameter rear door and the swing open divider wall within it. In front of the tank is the container with vacuum pump for sucking out water for re-cycling. This will tip with the tank as it is raised for discharge purposes. The KDU high-pressure pump is of the pressure regulating variable displacement type and powers a double acting piston to pressurise the water. Pressure up to 200 bar and flow up to 340 litre/min can be accommodated.

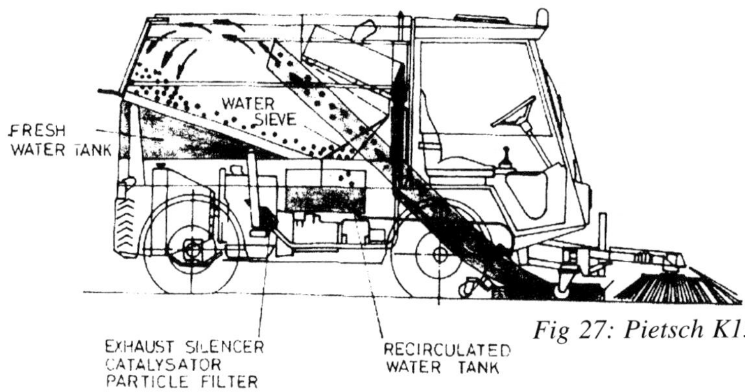

Fig 27: Pietsch K1500 sweeper

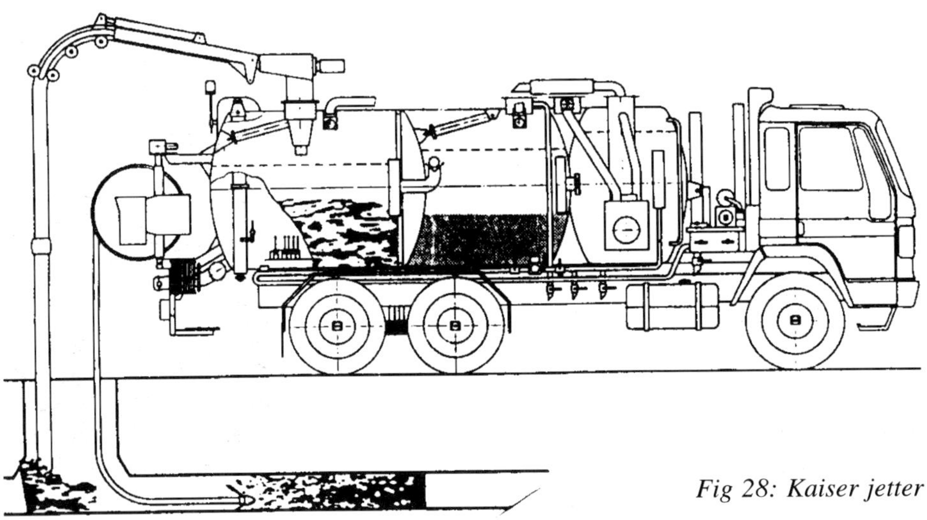

Fig 28: Kaiser jetter

Power-take-off selection

A distinctive characteristic of refuse vehicles is the reliance put on engine and transmission power-take-offs to drive such items as water pumps, suction fans, waste packers and sweeping brushes.

Openings provided in the top, side or bottom of transmission gearboxes are used to access constantly meshing gears for PTO drives to either hydraulic pumps or drive shafts. Typical sidemounting configurations, due to Edbro, are shown in Fig 29 employing the SAE standard four or six bolt facings; eight-bolt facings also exist for heavy duty drives. As well as these layshaft gear drives, other take-off points are front of engine crankshaft; sandwich drives between engine and gearbox and split propshaft drives. Rear mounted layshaft PTOs are another possibility usually involving a splined dog engagement; often step-up gearing is provided in this type of drive, Fig 30. Various adapter plates are also available, Fig 31, for such functions as directional reversing and even replacing a gearbox side plate with another containing PTO opening. Torque throughputs are typically 140 lb ft (190 Nm) for six-bolt facings while eight-bolt facings can have up to full engine powers transmitted through them. In the case of automatic transmissions, having no layshaft, the drive gear is usually large, being mounted on the outer gear of the planetary train: thus requiring reduction gearing. PTO speed available varies in relation to the load imposed by the driven device with this type. The split driveshaft type of PTO, between gearbox and rear axle, Fig 32, might involve a belt drive in certain cases.

Selection procedure involves putting the questions: how many speeds are required and in which direction of rotation; what shaft powers are needed corresponding to these speeds; what type of transmission gearbox is fitted and is it capable of the required PTO powers? If, for example a 25 bhp blower is required to run at 1000 rpm for optimum efficiency from a PTO which only allows 500 rpm, then step-up gearing will be required which will double the torque throughput on the PTO. Generally the higher the pitch line speed of the transmission gearbox PTO gear, the higher the potential power take off capacity. Speed of the pick-off gear with relation to engine speed is also needed: generally pitch line velocities should not fall below 400 and 650 ft (122 and 198 m)/min for 15 and 25 bhp PTOs respectively. Gaskets

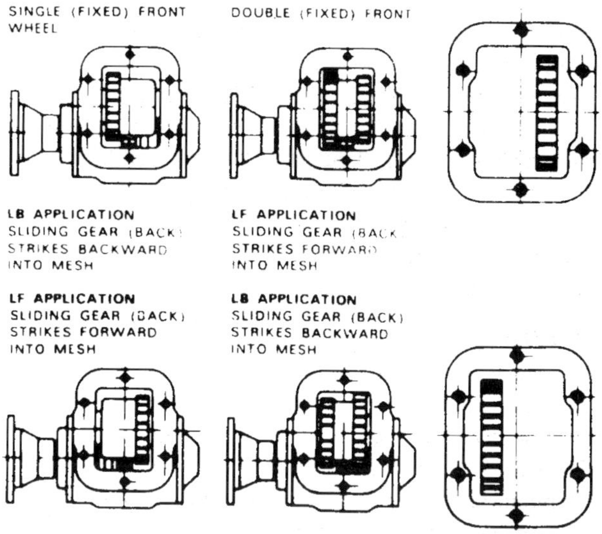

Fig 29: Edbro PTO configurations

Fig 30: Step-up gearing

Fig 31: Adapter plates

Fig 32: Split driveshaft PTO

should be chosen carefully for the PTO face to obtain optimum 0.008 to 0.012 in (0.2-0.3 mm) of gear backlash with fastening bolts tightened to 20 lb ft (27 Nm). PTO control systems vary from direct-lever arrangements to flexible cable remote controls — or even air shifting in and out of gear. With air shift, it is usual to spring-load the PTO gears out of mesh as a fail-safe measure in the event of losing air supply. Electrical solenoid controls are also available, with microprocessor switching.

Use of elementary fluid-mechanics theory is valuable in assessing the power required for the various functions of refuse vehicles. Emptying gullies and cesstanks involves forces and pressures obtained from the basic laws of hydrostatics. Liquids are usually assumed to be incompressible in the analysis and the weight of water is taken as 62.4 lb/ft^3 (1 tonne/m^3). Pressure of water is normal to the wall retaining it so it is useful to remember, with a cylindrical vessel, that total thrust force is calculated on the projected area rather than the total area of the vessel wall. In calculating the force applied by a simple plunger pump required to raise the level of water in a vertical cylindrical tank the mechanical advantage is simply the ratio to tank area to plunger area. Pumping power can thus be calculated on this basis. The pressure head of a liquid is another useful concept in calculation — based on a liquid being subjected to pressure due to its own weight. Pressure equals head (or depth) times density. In Fig 33, intensity of pressure due to weight in a vertical direction is also reflected by an equal horizontal pressure as shown by the triangular pressure profile. This profile affects the position of the centre of pressure on a submerged liquid-retaining hatch.

For low speed flow in pipes, the frictional forces can be neglected and streamline-flow conditions are assumed. For the general flow condition of Fig 34, velocity at (aa) times section area equals velocity at (bb) times its section area. To find the work done in overcoming liquid pressure in a tank, the set-up in Fig 35 can be considered. If water is forced into the tank at velocity v against the internal pressure of p, work done is WH where W is weight of water injected per second and H the equivalent static head in feet of water. If water then flows through the same pipe under the action of the weight of water in the tank $WH = Wv^2/2g$ which is the kinetic energy of the water ejected; the 'static head' then becomes the 'velocity head'. In pumps with a long suction pipe and a low delivery head, pressure due to inertia of the column of water in the suction pipe will be large compared with the pressure on the outside of the delivery valve. Fig 36 shows a work diagram for a pump, the vertical ordinate being plunger pressure either side of a double acting pump. Line (f.e) represents atmospheric pressure while (d.c) is cylinder pressure during the suction stroke and (a.b) cylinder pressure during delivery. Corresponding works done on the suction and delivery strokes are given by dcef and abef provided the pump is operated at low speed.

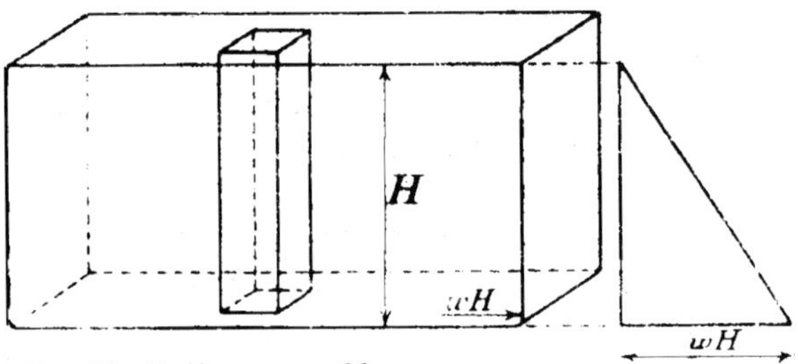

Fig 33: Pressure profile

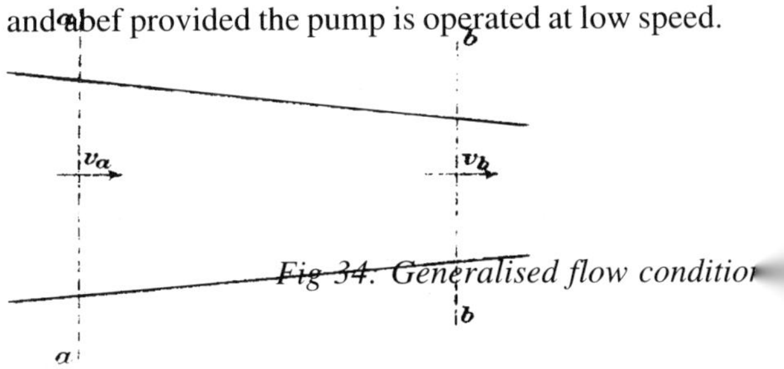

Fig 34: Generalised flow conditio

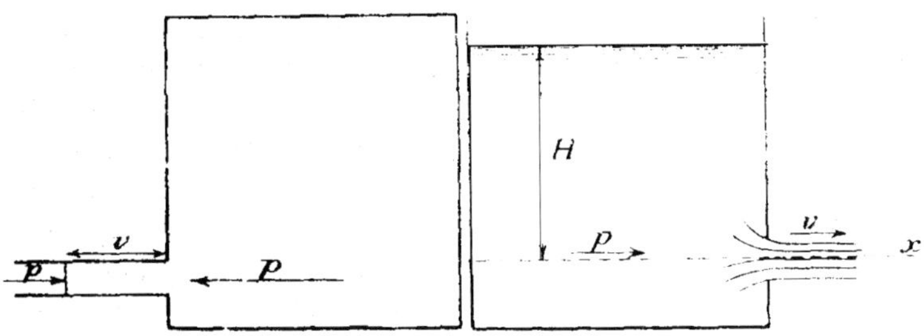

Fig 35: Work done in overcoming pressure

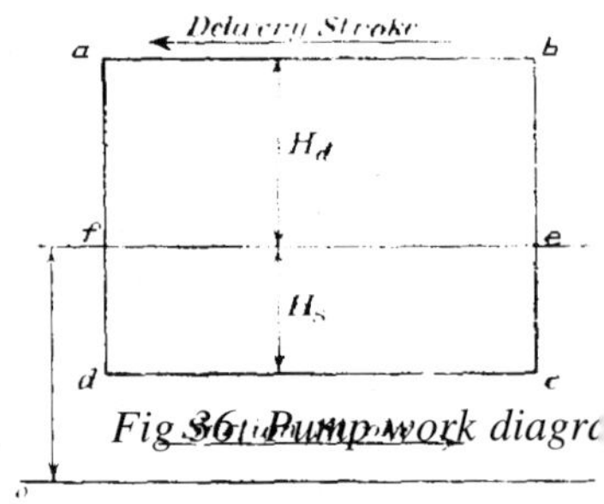

Fig 36: Pump work diagr

Hydraulic systems: avoiding failures

Refuse collection vehicles also place great reliance on hydraulic systems so ways of improving hydraulic system performance and predicting causes of unreliability are considered here.

According to the Fluid Power Centre at the University of Bath[2], the main reason for specifying a hydraulic system is that large forces can be generated within a small installed volume and for minimum weight. Additionally, the medium of hydraulics allows flexible control of motion and forces, whilst achieving a high transmission stiffness. However, the main disadvantage of hydraulic systems is that they require hydraulic fluid. This fluid is apt to leak, propagate noise and to become contaminated with dirt from the environment and wear particles from within the system. The problem of contamination is a major issue and it is often suggested that the majority of hydraulic systems fail due to damage or operational malfunction caused by the presence of contamination in the system. On the positive side, hydraulic systems have a remarkably good safety record, the author argues.

Design In some industrial contexts, the author maintains, hydraulic systems are built by assembling various off-the-shelf components into one of a number of familiar circuit configurations, and hoping for the best. Unfortunately, it is not as easy to build hydraulic circuits in this way due to the non-linear nature of hydraulic components and complex ways in which they can interact with each other. Even carefully designed circuits may show unacceptable dynamic behaviour, under operating conditions, that is very hard to cure. A common problem is the unstable oscillation of counterbalance valves used to prevent runaway of assisted loads.

For a given application, there will be a number of available alternative circuit arrangements that will meet the specified operational and functional requirements. A relatively common cause of faults in user-installed hydraulic circuits is components being fitted the wrong way round. This is particularly easy to do if the port connections on the component are all identical. Diagnosing such faults is often quite straightforward; however, there is a danger that components forming part of a safety feature, such as an unloading circuit, may be incorrectly fitted. In this case, the faulty installation may not be noticed until too late.

Reliability Contamination of the hydraulic fluid is probably the largest single cause of failure in hydraulic systems. Typical contaminants include casting sand, metal burrs and chips. The main effect of contamination is increased wear of components, especially if the contamination includes hard particles; if there are sensitive components, such as spool valves, where clearances are of the same order as the size of particulate, then increased friction and seizure is likely. In time, hydraulic oil will oxidise to form sludges and gums. These may accumulate if a component is used only periodically, causing increased levels of friction. The problem of fluid oxidation may be severe if the hydraulic system is run at high power level with insufficient cooling.

Computer simulation is now increasingly used within industry as a tool for analysis of fluid power systems. Using commercially available packages, the performance of a given circuit configuration and choice of components can be evaluated for a wide range of operating conditions, and potential weaknesses revealed before committing the circuit to hardware. It is common to find components joined by more bolts than is strictly necessary for strength or sealing reasons. This is an example of mechanical redundancy to improve reliability in the event of incomplete assembly or a bolt becoming loose during operation.

Reliability assessment is often in the form of a Failure Modes and Effects Analysis (FMEA) and a Fault Tree Analysis (FTA). The first is a bottom-up process whereby the effects of faults at component level are propagated up the system to determine the top-level event associated with the component fault. In contrast, a fault tree analysis is a top-down process concerned with finding the possible component failures that could lead to a given top-level event. Combining FMEA with FTA gives a 'belt and braces' reliability assessment which hopefully covers all possibilities.

Work currently underway at the Fluid Power Centre at the University of Bath is progressing towards automating the process of FMEA for hydraulic systems. The goal is to improve the efficiency of the task through automation of the extensive calculations involved. However, the more creative parts of the assessment, involving engineering judgement and experience, will still call upon the skill of a designer.

Systems and components An engineer at Vickers Systems Division[3] neatly defines the function of a hydraulic system as conversion of the motions produced by the prime mover into movements at remote locations, controlling both force and position to a precision generally uneconomical by other means. Actuators have extremely favourable power to size and weight ratios and speed can be infinitely variable by simple flow control strategies. Reversibility is easily achieved and protection from overload/overspeed can be accurately set.

Problems of leakage and noise can often be overcome by judicious extra expenditure at the commissioning stage. Reduction in shock loads by using proportional control, and attention to interconnection devices, are the approaches to take. In cases where the hydraulic pump is the principal noise source it is sometimes possible to locate this unit within a structural enclosure, to isolate noise. Where feasible, pump/motor packages enclosed in their own shroud can also be specified. The integration of electronics into valve controls could also be exploited to provide servo-assist proportionality. Fig 37 shows how valves positioned between motor/pump set and actuator can be used to provide protection as well as versatility in position, speed and direction.

An engineer at Denison Hydraulics[4] has outlined hydraulic pump and motor types in current use,

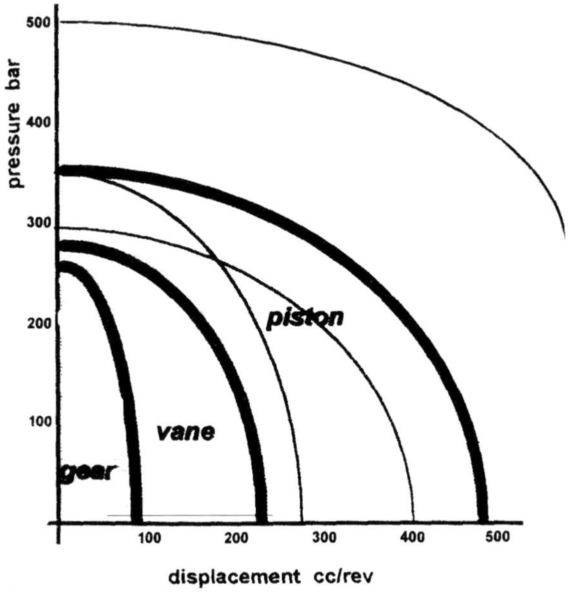

Fig 38: Use of different pump types

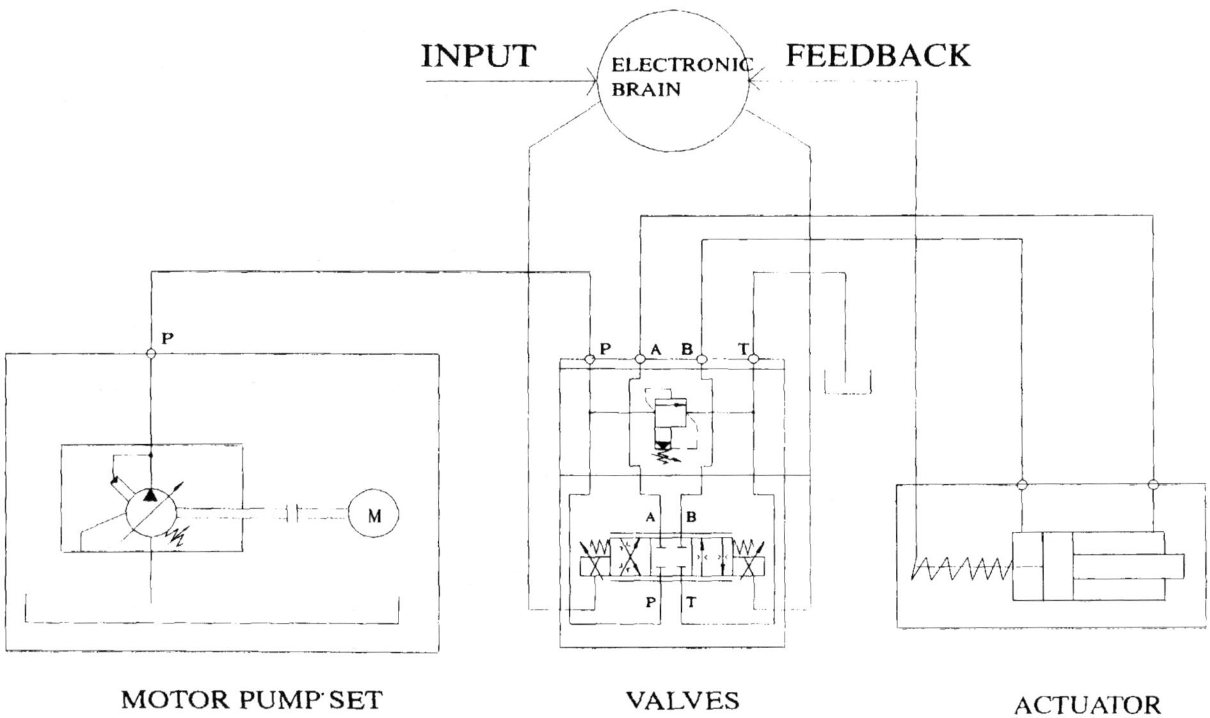

Fig 37: Using valves to provide protection and positioning versatility

to the seminar. He focused on external gear designs in which meshing gears determine flow output according to their face width. These can deliver 0.2 to 400 cm^3/rev when rotating between 500-600 r/min and have 90 % overall efficiency when new. Mobile applications form the largest market for these designs where high efficiency and low noise is demanded. With the trend in the market to higher power requirements, in construction plant and material handling vehicles, for example, larger size units are being specified which leads to an increase in flow or pressure ripples, and so greater noise, thus considerable engineering effort is being made to reduce this effect. One of the techniques in use is harmonic cancellation. helical gear design with relatively longer gear teeth were also being tried. Dual gear units were also being employed to effect cancellation of flow pulses. Other trends noted were to `packaged' pumps with built-in controls, such as load sensing, use of Teflon-coated bearing faces — also a move to higher pressures so as to compete with the bottom end of the piston pump market. The way in which the market is split between gear, piston and vane types is seen, in general terms, in Fig 38.

In discussing hydraulic valves, an engineer from Mannesmann Rexroth[5] points out that the considerable difference in appearance between valves for the mobile and industrial markets lies in the relatively high volume requirements of the former category together with the need for low weight and compactness. Compared with industrial ratings, service life was also comparatively short so that flows and pressures could be much higher for a given design envelope.

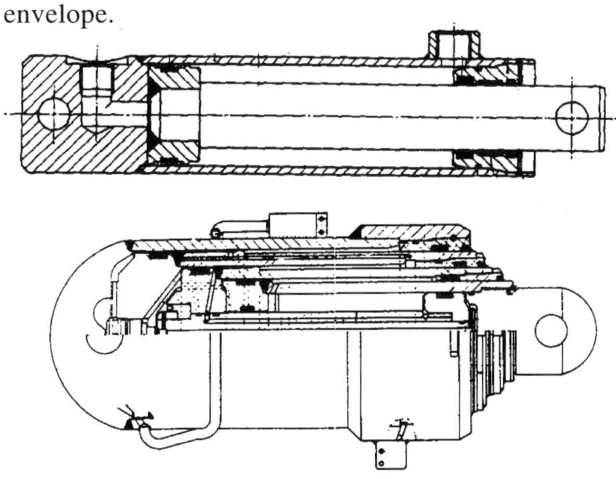

Fig 39: Cylinders for mobile applications

There were three stages in pressure control for a typical hydraulic circuit, pump pressure control, pressure relief and pressure reduction, the last two involving valves. Directional control can also be by pump or valves, the latter usually being electrically-controlled direct action solenoid valves of the so-called CETOP 3 or DIN 6 size. All other directional valves are two-stage, piggy-back, valves with the size 6 as the pilot stage, so that the solenoid can be limited to an economical size. A wide variety of available spool types allows fluid flow in various directions when the valve is in neutral, often critical in reducing hydraulic shocks.

When two or more end-motions are involved in the system, it is necessary to prevent all the pump flow from passing to one of the motions; a simple throttle is generally used for this. For speeding up operation of systems, proportional control may be introduced, in which valve function is controlled in proportion to the electrical control signal. Functions of directional control and flow control can then be combined in one valve.

Actuators, cylinders, hoses and couplings have been considered by engineers at Parker Hannifin[6]. Considerations in cylinder applications included choice of piston rod diameter to avoid rod buckling and ensuring the cushion energy for the application must be appropriate to the cylinder chosen. Calculated energy value should include deceleration of the load, work done due to orientation of the cylinder and work done by the hydrostatic piston force. Energy rating can be critical at the annular end, when the cylinder is extending, as cushion pressure intensification is equal to the ratio of full bore area to the cushion area multiplied by the supply pressure. With cyclic loading situations the fatigue limit of the piston rod thread must be considered and the cylinder aligned to the load when fully retracted so that the rod alignment remains within that specified, for the full stroke.

Fig 39 compares a welded construction cylinder for earthmoving vehicle and fork-tuck applications with an example of a telescopic, three-stage, cylinder construction for application in off-road vehicles and fire-engines. Fig 40 offers installation dos and donts for hose assemblies.

PTO geared drives

According to engineers at Renold Power Transmission[7], the advent of the `forgive and forget' culture is a key feature of present day construction of smaller gearboxes. Key factors in designing for a given power density in a helical gear pair are choice of gear material and gear accuracy. Selection of case hardened and ground gears, instead of through-hardened, will increase the rating by a factor of 3 to 5. Increasing accuracy from BS436 grade 8 to grade 6 would increase rating by a factor of 1.5 to 2.

Gears for high volume fixed ratio applications such as wheelchair drives and windscreen-wiper/window-winder drives are often moulded plastic or sintered metal and have no need for finishing operations but high tooling costs are involved.

In the case of geared electric motors (gearboxes are specified on standard four-pole motors instead of using prohibitively expensive multi-pole motors) gearboxes are becoming increasingly integrated with the electric motor, to reduce parts count. In the case of wormgear drives, the wormshaft end is often bored out and internally keyed to accept the motor shaft; the motor flange is then bolted directly to the gearbox case.

For small gearboxes generally, Fig 43, deep-grooved ball-bearings are the norm since they have adequate capacity for the radial and relatively light axial loads associated with helical gearing. No adjustment across the span is necessary and sometimes assembly can be simplified by specifying them in conjunction with cylindrical roller bearings. Where very high loads are involved taper-roller bearings become the norm, having greater axial load capacity, but adjustment devices are required. In the case of wormgear boxes, angular contact bearings with contact angles in the range 25-40 are needed for the necessary axial load capacity. Circlip location is common for the bearings while gears are usually keyed to the shafts in small gearboxes. Nitrile rubber

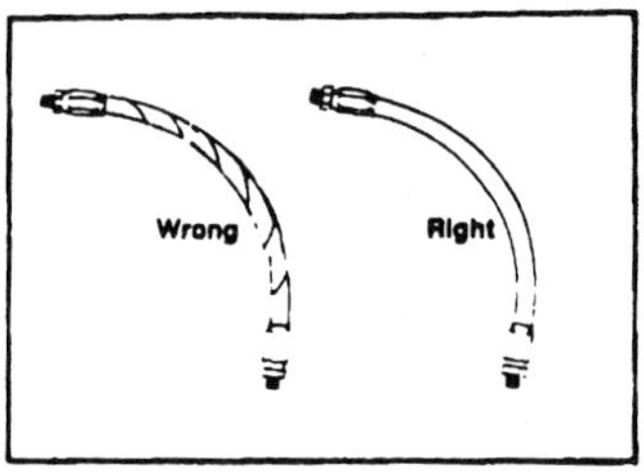

Hose is weakened when installed in a twisted position. Also, pressure in twisted hose tends to loosen fitting connections.

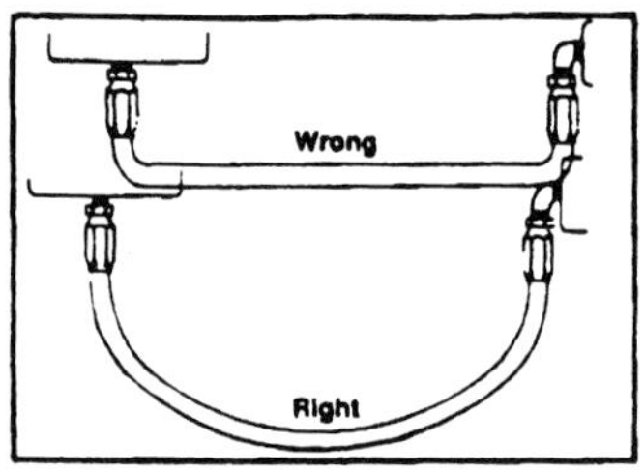

Ample bend radius should be provided to avoid collapsing of line and restriction of flow. Exceeding minimum bend radius will greatly reduce hose assembly life.

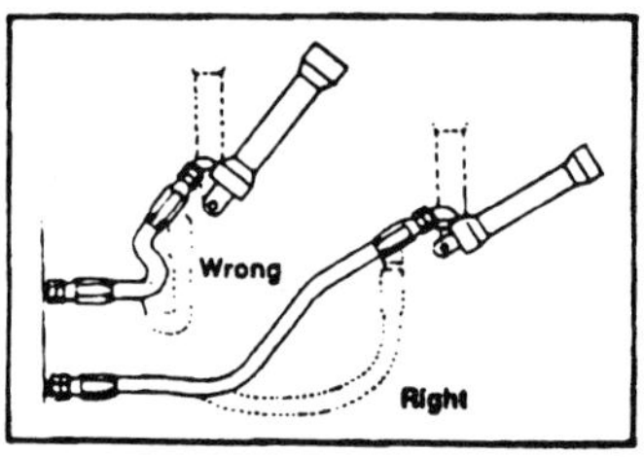

When hose assembly is installed in a flexing application, remember that metal hose fittings are not part of flexible portion. Allow ample free length for flexing.

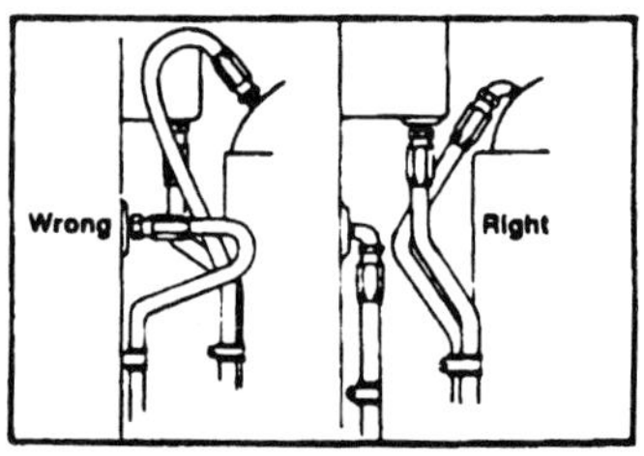

Use elbows or other adapters as necessary to eliminate excess hose length and to ensure neater installation for easier maintenance.

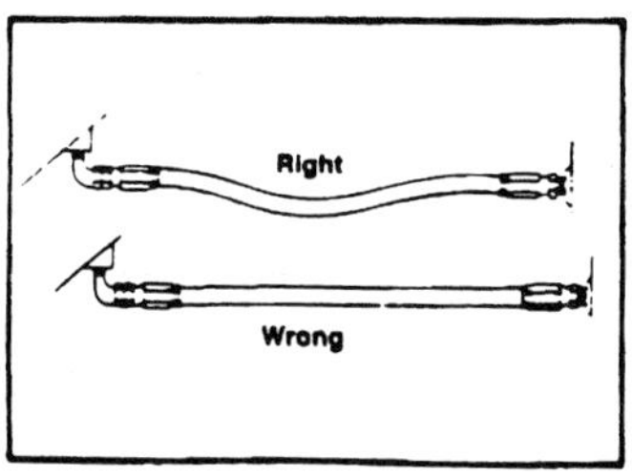

Pressure can change hose length as much + 2%. Provide slack in line to compensate for hose length changes.

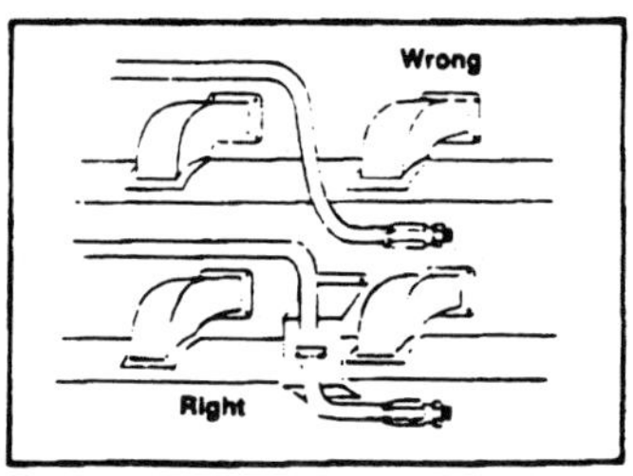

Avoid installing hose line close to exhaust manifold or any other hot section. If this is not possible, isolate hose with fire proof sleeve or other protective means.

Fig 40: Hints for hose assemblies

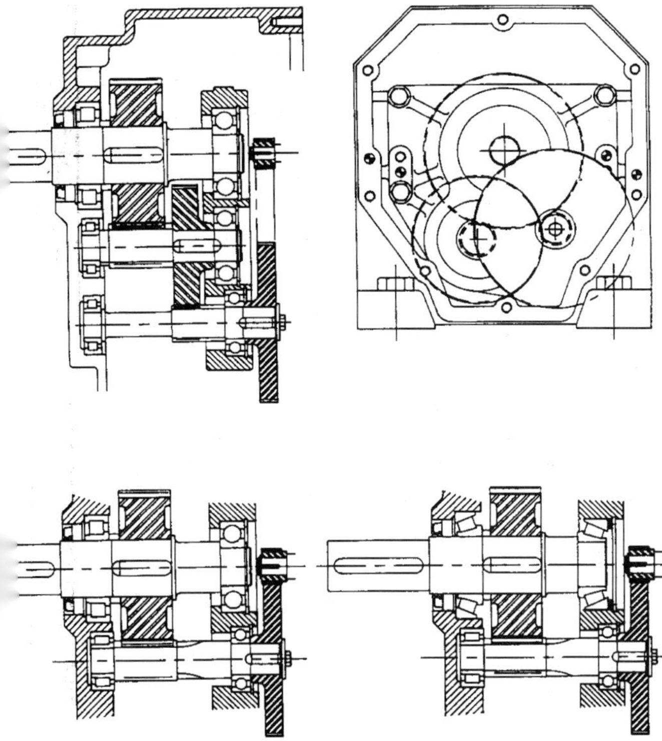

Fig 43: Typical ancillary drive gearbox showing alternative bearing types

lip-type oil seals are normally specified; the material is tolerant of oil temperatures up to 80 C and pressures up to 0.5 bar.

Gear materials and hardening Researchers at Birmingham University[8] have distinguished between surface heat treatment and surface coating techniques for raising hardness of components such as gearwheels. Heat treatment included thermal hardening, carburizing (with its derivative carbonitriding), nitriding (with its derivative nitrocarburising), boriding and metalliding with either vanadium or niobium. Surface coating, on the other hand, encompassed electrochemical treatments, chemical treatments, chemical vapour deposition, physical vapour deposition, spraying processes and welding processes. Increasingly, however, duplex surface engineering is used where improved fatigue and bending strength needs to be combined with low friction, low wear and superior load carrying. This involves the application of two surface modification processes to obtain combinations of properties. Figs 44 and 45 show the benefits of such approaches. The first shows rolling-sliding wear test results for PVD TiN/nitrided steel, plasma nitrided and TiN/nitrided systems while the second shows projected gear load capacity (highest torque without failure) for PVD TiN/non-nitrided steel, untreated steel, plasma nitrided steel and TiN/nitrided systems.

Bearings Researchers at Newcastle University have used[9] the data in Fig 46a to show how the specification of high strength case-carburised and nitrided gears has resulted in significant reductions in

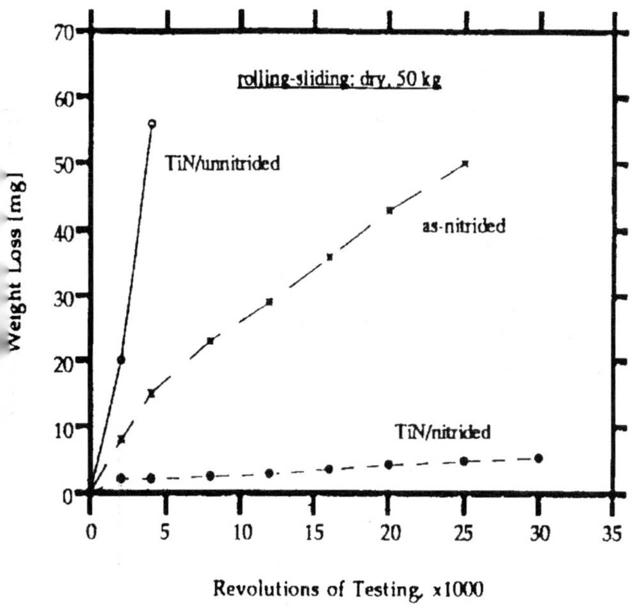

Fig 44: Wear curves for nitrided, TiN coated and duplex treated En40B

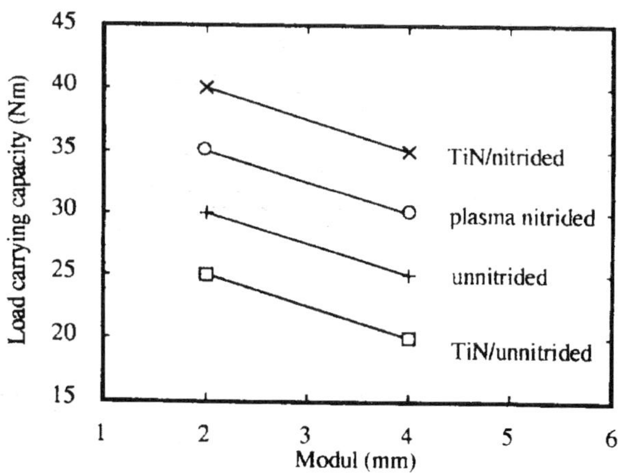

Fig 45: Gear load carrying capacity

gearbox size for a given power. He then argues that since bearing loads increase in proportion to decrease in gear centres, bearing load capacity can often limit gearbox rating. The table in Fig 46b is used to compare the characteristics of different power transmission gearbox bearing types on an infinite life' basis for a 120 mm shaft diameter setup. The roller bearing has the highest specific load capacity, with p_e = 10.3 N/mm^2, a typical hydrodynamic bearing having about one third of this load capacity, with the hydrostatic bearing falling between. For larger shaft diameters, disadvantage of normal rolling element bearings is the relatively low speed capability and potentially limited fatigue life.

The author concludes, however, that rolling element bearings are the appropriate choice for most gearing applications. Bearing life calculations to ISO, where L_{10} bearing life is $(C/P)^p$, can predict the finite life of most systems. However, very well sealed bearings operating in clean oil of appropriate viscosity can exceed this and be calculated using coefficients to the above formula, available from bearing manufactures.

Gear Material	En 24	En 19	En 36
Heat Treatment	H & T 280 HBN	Induction hardened	Case carburised
Finishing	Hobbed	Hobbed and lapped	Ground
Centre Distance Module	650mm 10	470mm 14	390mm 10
Gearcase Size	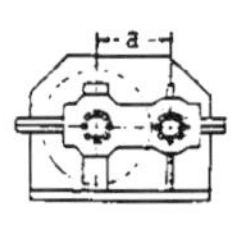	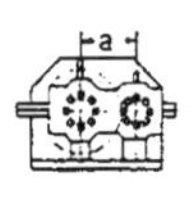	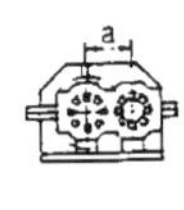
Bearing Weight	95 Kg	105 Kg	120 Kg
Gearbox Weight	4860 Kg	2390 Kg	1581 Kg
% Gearbox Weight	100%	49%	33%
% Cost	100%	66%	63%
Pitting S.F. S_H Bending S.F. S_F	1.3 5.7	1.4 2.3	1.6 2.3

Fig 46a: Gearbox size comparison for different gear materials: pinion torque 28000 Nm at 500 r/min; 3:1 speed ratio

Bearing Comparison for 120mm dia. shaft, Radial Load only	Cylindrical Roller Bearing	Hydrodynamic Bearing	Hydrostatic Bearing
Bearing dimensions, OD x L, mm	215 x 58	162 x 90	160 x 90
Fatigue Load Limit, P_u, N	72000	40000	65000
Maximum Speed rpm	3000	→ 15000	→ 10000
Weight Kg	8.3	5.6	5.5
Typical diametral clearance μm	N 140 ±40 C_2 70 ±40	200 ±40	80 ±20
Specific Load Capacity: $p_e = P_u \div d \,.\, L$ N/mm^2	10.3	3.7	6.0
Effective Coefficient Friction μ	.001/.002	0.005/0.01	.005/.01

Fig 46b: Comparison of bearing types

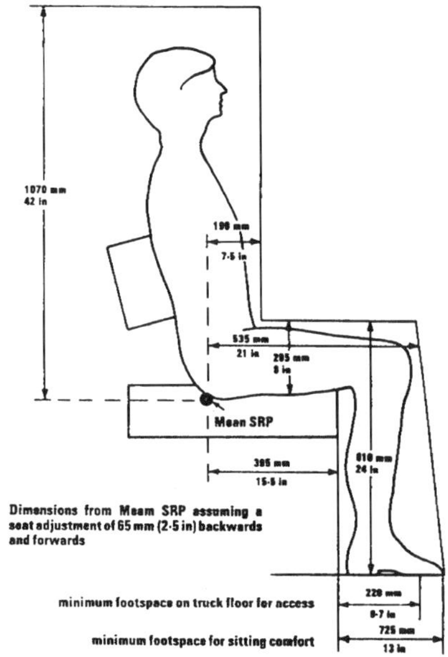

Fig 47: Optimum seat dimensions

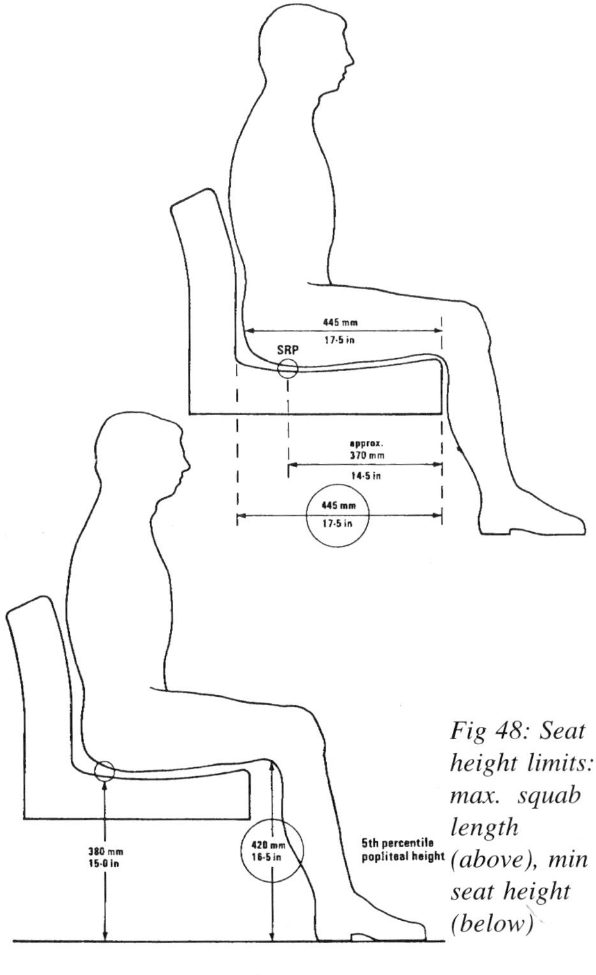

Fig 48: Seat height limits: max. squab length (above), min seat height (below)

Handling trucks

Design of the industrial fork-lift truck is governed to a major extent by the compactness of the aisles in which it has to work and manoeuvre. Ergonomics of the driver's work area must not be compromised, however, otherwise work cycle efficiency will be reduced by the effects of driver fatigue. The stability of the truck is also a crucial factor for a truck which carries its payload outside its wheelbase — as well as the drive motor, batteries and controller.

Work at Birmingham University's Department of Engineering Production has shown that operator fatigue can be considerably reduced in fork-trucks by careful attention to seat design and control positioning — which can dictate the basic design layout of the machine. The optimum seat dimensions recommended by ergonomists, Fig 47, were found by the Birmingham team to have often been exceeded in commercially available trucks. Height limits for seats have been reportedly exceeded by as much as 3.5 inches, resulting in fatigueing under-thigh pressures for drivers in the lower percentiles of stature, Fig 48. Overlong seat squabs can cause drivers to adopt an unnecessarily slumped position and too high a back causes restriction of the shoulders which require freedom for both steering and reversing.

A recommended space envelope around the median plane of the body is often intruded into by the steering wheel — again causing discomfort. The need for the driver to have his lower leg vertical, as shown, is also important to ease frequent entry and egress. Common difficulties are forward projections from beneath the seat, arising from the designer's belief that a leg posture akin to a car driving position would be possible. Another common form of discomfort is overheating of the lower leg caused by proximity to a poorly insulated engine compartment. Finally it is necessary for the safety of the driver for the seat reference point (SRP) to be at least 20 in (508 mm) from the truck side so that no part of the driver's body can project over the edge of the truck.

Hand reach limits, Figs 49/50, worked out by the Birmingham team follow the approach made by classical researchers such as Rebiffe, and described in an earlier chapter. On many commercial designs hand brake and horn were found to be out of reach as, on occasions, were gear and mast control levers. A special problem arises with fork-trucks in that full

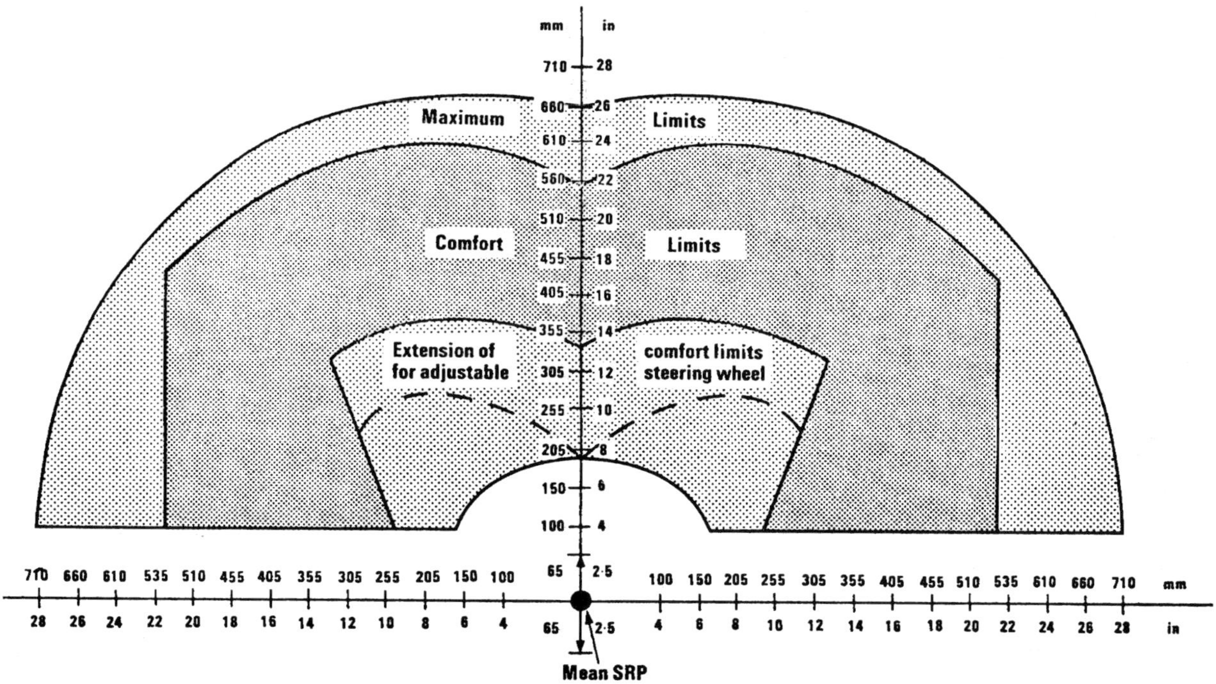

Dimensions from Mean SRP assuming seat adjustment of 65 mm (2·5in) backwards and forwards

Fig 49: Hand-reach limits in plan view

single handed rotation of the steering wheel is often required which can cause problems if an unusually large diameter wheel has been used to bring it within the normal reach envelope for both hands. A well sited and adjustable-position wheel of smaller diameter, coupled with power-assisted steering, is often the answer. Pedal location limits, defined by foot reach envelopes, are unique to the upright driving position required of fork-trucks.

Stability factors of fork-lifts

The other principal constraint on fork-truck layout is that due to static stability of a vehicle whose inherent design involves carriage of its payload outside the wheelbase. This problem has been examined by the US Towmotor Corporation for the worst case of a truck decelerating down a slope, Fig 51. By equating resolved forces parallel to the slope and arranging terms so that a stability influencing factor can be defined as the reciprocal of moments due static level + those due to slope + vehicle acceleration + load acceleration. As this is a cumbersome equation to manipulate and if no slope criteria are specified, then level ground stability is conveniently determined as a first step, as shown in Fig 52. Here the stability factor is $L_f W / L_q Q$.

The constant relationship between the load Q and moment arm L_q is shown by the hyperbolic curves in the figure which apply to an example of truck empty weight of 12,200 lb (5.5 tonnes) with cg location at $L_f = 46$ and H = 30 in (1.17 and 0.76 m). Nominal load capacity is 11,000 lb (5 tonnes) at L_q = 40 in (1.03 m) and the curves represent a stability margin range of zero, 25 % and 50 %.

For the nominal capacity of this truck example there is a stability margin of 27 %. The effect of gradient θ can be found by substituting in:

$$\tan\theta_{max} = (L_f W - L_q Q)/(H_q Q + HW)$$

for the example truck in which cg height of the raised and lowered loads is taken as 100 and 40 in (2.54 and 1.02 m), for which maximum grades work out as 8.2 and 15%. The same expression, transposed, can be used to calculate the maximum translatory deceleration as 0.082 g and 0.15g in the respective cases.

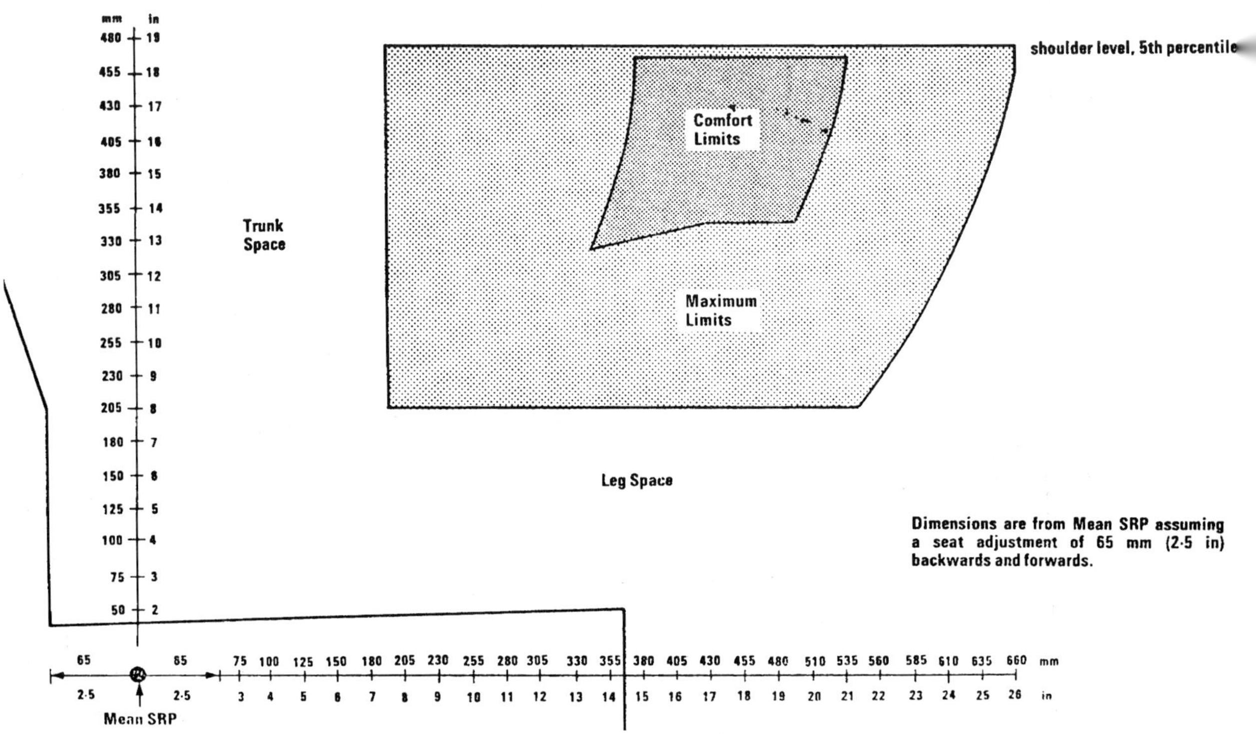

Fig 50: Hand-reach limits in side view for non-adjustable steering wheel

Electric drive of handling trucks

The wide use of electric drive for inside factory and warehouse operations has made the fork truck one of the major EV categories. DC motors are the rule for these applications. Very high starting currents are required for moving off under load. Around 800 amps is common and silicon controlled rectifiers have been used for a number of years to regulate current feed to the traction motors. Sometimes a single electric motor powers the hydraulic systems of both steering and lift/tilt functions. By means of SCR pulse control, lift speed can be controlled electrically rather than hydraulically. Typically such a motor will have three speeds: say, 500 rpm for steering, 1000 rpm for tilt and slow lift then 3000 rpm for rapid lift.

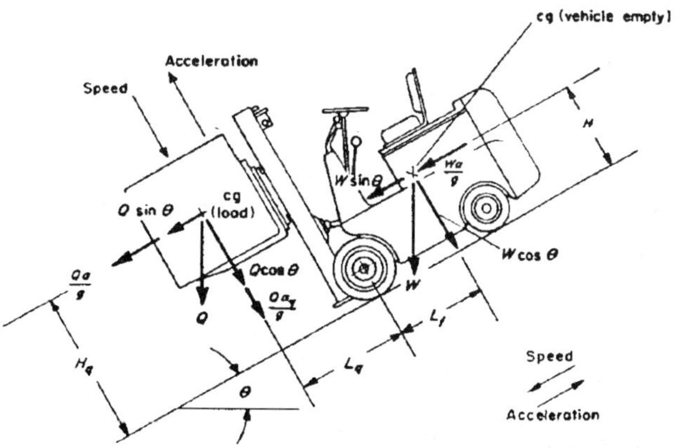

Fig 51: Stability on slope

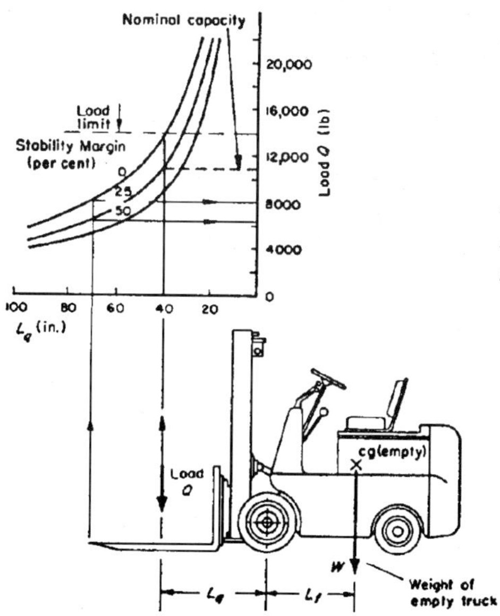

Fig 52: Level-ground stability

The principal of pulse width modulation is shown in Fig 53. Typically a complete switching cycle will occupy 1/120th of a second, applied voltage being either full battery output, say 36 V, and then zero. During the 'on' pulse, motor current rises on a curve governed by inductance or electrical inertia of the windings. At switch-off the current does not stop immediately, but continues flowing through a by-pass 'one-way' diode, dying away on another curve until the transistor switch closes for the next pulse, causing the resultant sawtooth waveform. Varying pulse length alters relative rise-and-fall times, longer pulses permitting higher peaks. Average value can be raised by lifting the whole saw-tooth up or down the graph by moving the trailing edge of the switching pulse sideways within each cycle. Resulting current characteristics are shown in Fig 54. A typical electronic circuit is described in Chapter 15.

Researchers at Imperial College produced a useful method for calculating the range of a multi-stop EV, applicable to categories including fork-trucks, in terms of the drive motions only, not the work motions. To some extent the latter are easier to calculate separately by computing potential and kinetic energy changes in lifting, tilting and swivelling

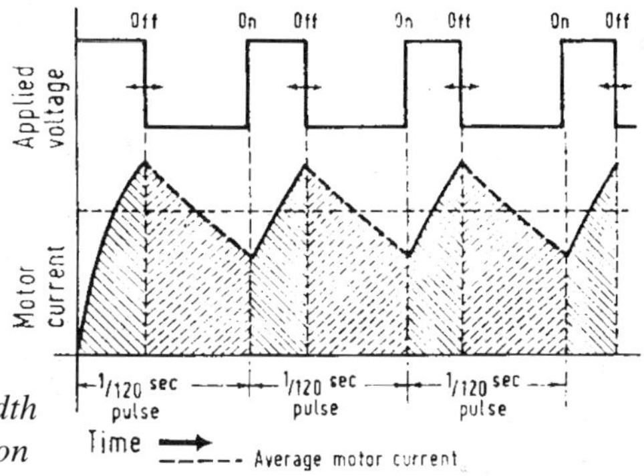

Fig 53: Pulse-width modulation

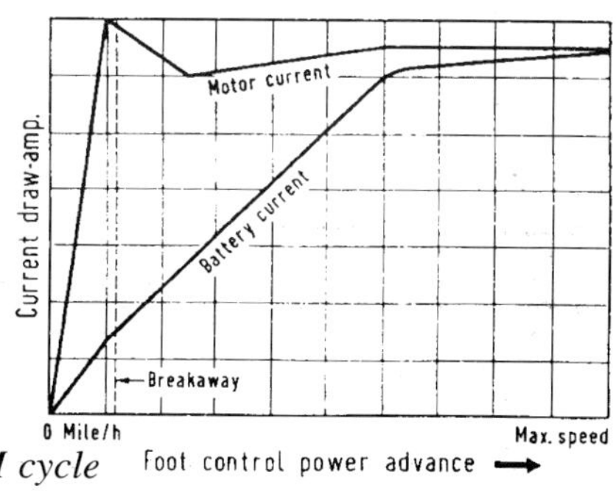

Fig 54: PWM cycle

the load for different duty cycles.

By first converting the energy density of the battery typically 25 watt-hours/kg for its weight at about a five-hour rate of discharge, a figure of torque per kg battery weight can be computed (since 1 Wh/sec = 1 Nm), division of which by the tractive resistance would give a distance travelled D if the percentage of vehicle weight taken up by the battery is 100.

Fraction of battery energy used in moving itself S km against T N/kg tractive resistance is $f_s=ST/D$. If this energy were used entirely in giving up kinetic energy to the battery, the speed obtained would be given by v in $(v^2/2)=D$. Another value would be obtained by factoring down this energy by a percentage representing the starting energy losses. Fraction of battery energy f_u used in accelerating the truck to speed v is thus $(v/u)^2$. If then the battery occupies a fraction f_b of the vehicle weight the energy fraction above becomes $(v/u)^2/f_b$. And if the load on the battery be such that it is a fraction B of that due to a five hour rate the expression becomes $(v/u)^2/f_bB$ (B in fact is unity for those vehicles with intermittent duties which take several hours to discharge the battery). When motor and transmission efficiencies are also accounted for, the two expressions become:

$$f_s=ST/Df_bBE \quad \text{and} \quad f_u=(v/u)^2/f_bBE$$

The duty of the vehicle will be specified as P starts per km to a speed of v km/h and for such a duty the total fraction f of battery capacity required per km will be $f=(Pf_u+f_s)$ where f_s corresponds to 1 km. Total range is thus:

$$(1/f)=f_bBE/[P(v/u)^2 + T/D]$$

Development history of fork trucks The past three decades have seen the development of fork trucks into their present state of the art. The 1960s saw the integral monocoque-type frame, exemplified by the Yale KBE, all-welded from plate. As a 4-wheeler this had a rear wheel steering system representative of then current practice, shown as Fig 55. A 1970s introduction was the Coventry Climax R-15 reach truck with full free lift mast operation permitting unusually good forward vision. The model, Fig 56, was noted for its ergonomic attention to layout of the work-motion levers, involving a staggered fan formation. In diesel engine designs, the Still R-70 is representative of current practice, Fig 57. Another variation is the articulated chassis truck, exemplified by the Translift Bendi, Fig 58.

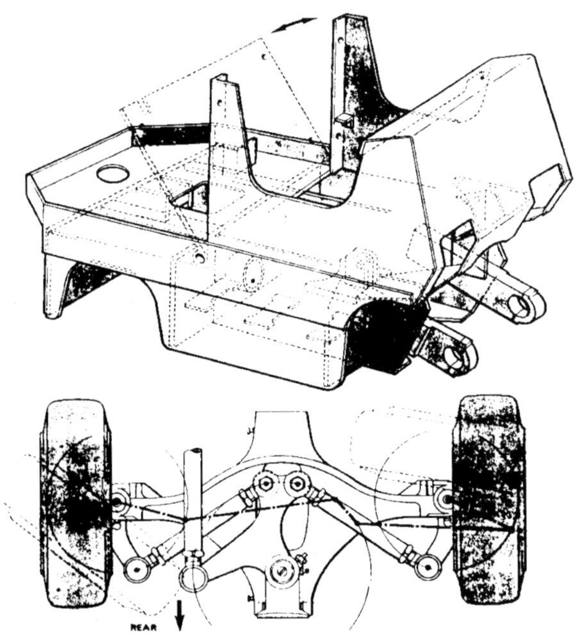

Fig 55: Yale KBE frame and rear-steering

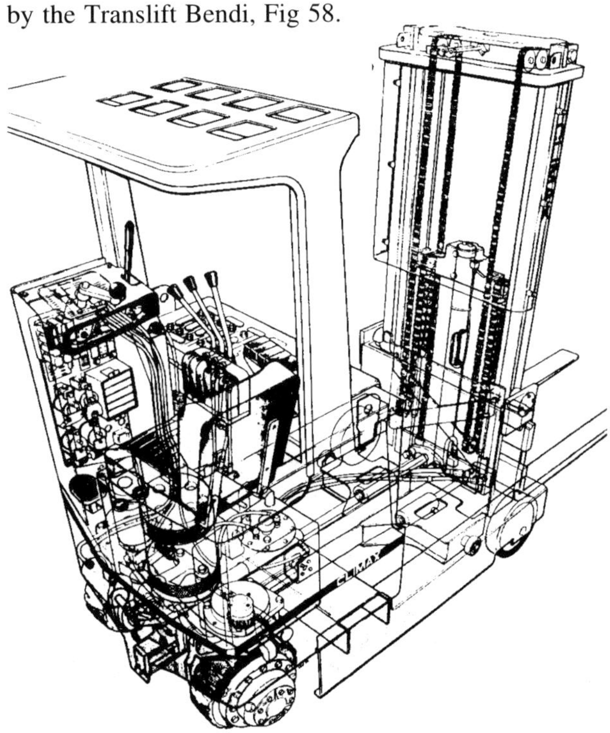

Fig 56: Climax R-15

Fig 57: Still R 70

Fig 58: Translift Bendi

Fig 59: Reliant 850 Pick-up

Light van design

The light van sector is affected considerably by the volume vehicle builders as much of the market is served by volume-car-derived vans at lower gross vehicle weights. At higher GVWs the volume sector are, of course, also heavily into panel vans — but right across the weight range there are many important niches to be filled by the specialists.

The mini-sized volume car is often not available as a van derivative and volume panel vans can be restrictive in terms of optimum access to both cab and load space; opportunities also exist for specialist bodybuilders in the weight bracket between top-of-the-range panel vans and the smallest chassis-mounted custom box vans. The demise of the Rover Minivan in the UK presented the opportunity for the Reliant Fox, recently superceded by a three-wheeler pick-up version, Fig 59, which the company have hinted might soon get the Fox's hard-top box. This is a vehicle which offers outstanding economy by virtue of its very low unladen weight and modest engine capacity. The base for the vehicle is the chassis in Fig 60. Extensive use of aluminium alloy in the power-train and GRP for the bodywork helped to achieve the below 406.5 kg (8 cwt) target of the early vehicle to meet the three-wheeler taxation requirements.

Beyond the top end of the panel van market, an interesting example is the novel integral van made by Navistar in the USA — which overlaps with the large delivery van sector covered in the next section. The International Metro II version, Fig 61 is available in body lengths of 14 to 24 ft (4.3-7.3 m) and has inside dimensions 85 in high x 93 in wide (2.16x2.36 m). This also has a separate chassis frame but the integral cab/body is factory-produced. The cab is 'walk-through' and the body can be accessed from it. Side walls are of GRP skins bonded to a tough core and are said to be considerably more scratch resistant than those of conventional aluminium alloy box vans.

Car-derived vans The 7-10 cwt (356-508 kg) weight-capacity car-derived van, with volume-capacities typically 60 to 100 ft^3 (1.67-2.83 m^3), are subject to legislation detailed in Fig 62a/b which affects the overall design. Countries such as Sweden have additional legislation to be met. Vision, package layout, petrol tank location, rear end configuration, lamps and bumpers are among the items affected. Obviously dimensions are related to those of the base car and roof heights, too, are governed by the need for an aesthetically, and aerodynamically, smooth blend between car and expanded body surface. Suspension system is crucial to the inter wheel arch clearance and appropriate re-design can bring substantial benefits over the base car system — probably sacrificing ride comfort for improved cargo space. Rear access to the space can be improved, too, by appropriate choice of closure between half-lift/half-drop gates, lift tail gate, single/double rear doors or even roller shutter.

Generally, front wings and doors are carried over from the base car to the derived van. New doors could solve the frequent problem found on volume-built vans of the door tumblehome adversely affecting the shape and so the volume capacity, of the van body. Fig 63 shows five suggested ways of assembling the body side to an existing front-end. The most economic solution is that of (d) whereby the existing body side-panel is trimmed along the window line and thus only a new upper panel is required. At (e), the A-post and front cant rail is built up from small panels and a rear side panel assembled to the front-end just forward of the body centre-pillar.

Panel vans

Design techniques used in notable volume panel vans can often be exploited by the specialist builder. A look at some key points is thus worthwhile. The Leyland EA was an early panel van which successfully exploited the tilted engine concept to obtain good walk-through ability on a forward control layout. The low-slung engine-gearbox necessitated a down swept boxed crossmember, Fig 64, at the front, with 40 degree inclined rubber bobbin engine mounts. The tail of the gearbox was carried on twin conical

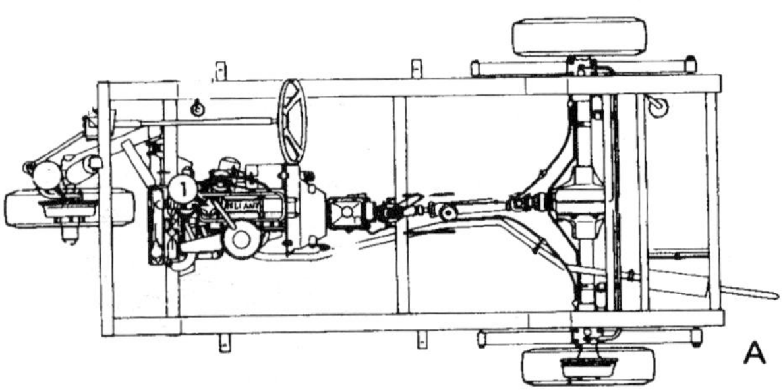

Fig 60: Reliant chassis

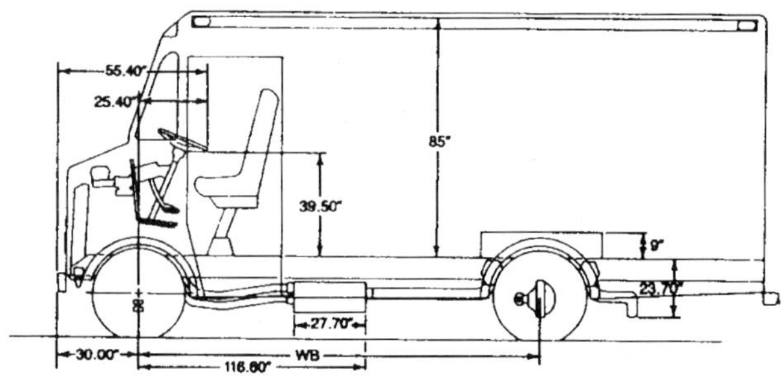

Fig 61: International Metro van

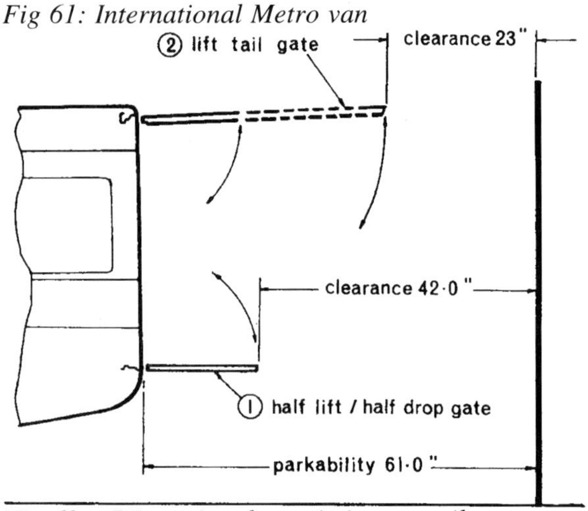

Fig 62a: Dimensional restrictions to tail-gate

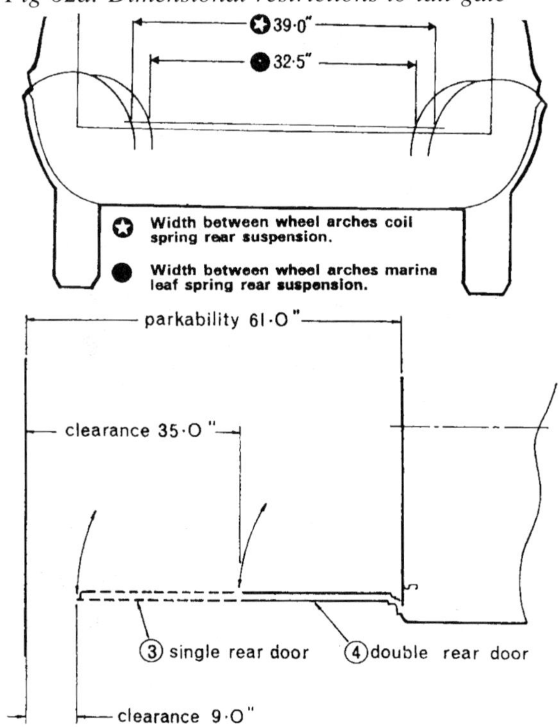

Fig 62b: Dimensional restrictions to rear doors

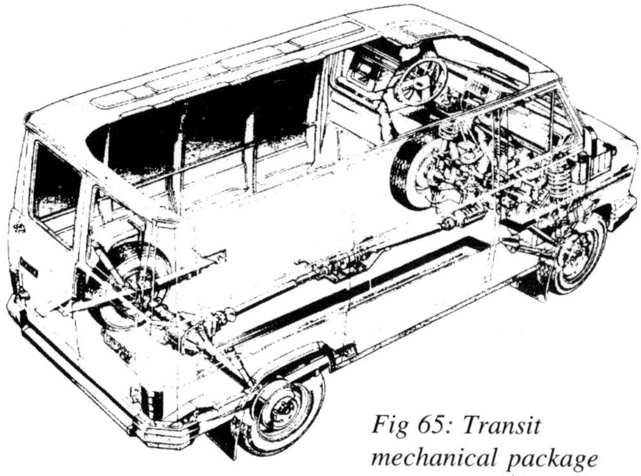

Fig 65: Transit mechanical package

mounts. State-of-the-art is represented by such vans as the Ford Transit. The layout and structure of the base vehicle, Fig 65, contains some important design features. The MacPherson strut IFS is arranged to clear the footwells and a semi forward control gives a nice compromise between available cube, access and driving comfort. Cab floor is a continuation of that of the load space so that good front access to the body is available.

Multipurpose vans

One of the greatest recent excitements in small van design is the multi-convertible Concept '92 vehicle of Hawtal Whiting, Fig 66. This vehicle also doubles as a fun-car and people-carrier. The chassis is a 3-dimensional semi- monocoque which is common base for the variants upon which partially-structural and non-structural cladding can be attached. The optional petrol/electric/lpg power unit fits into a self-contained structural sub-frame which also reacts the suspension loads. The assembly becomes a removable pod for remote servicing of the engine as re-

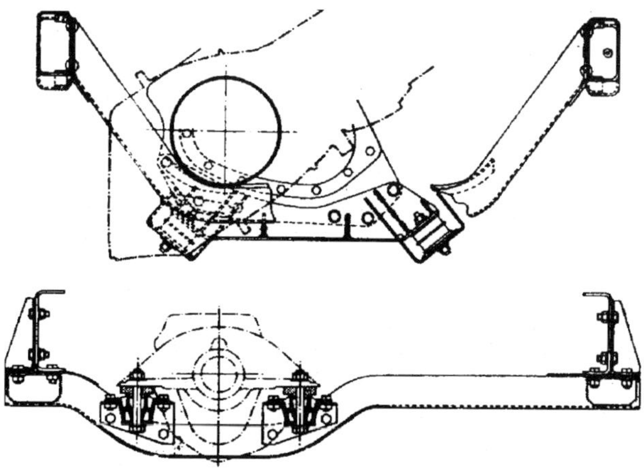

Fig 64: Low-slung engine suspension

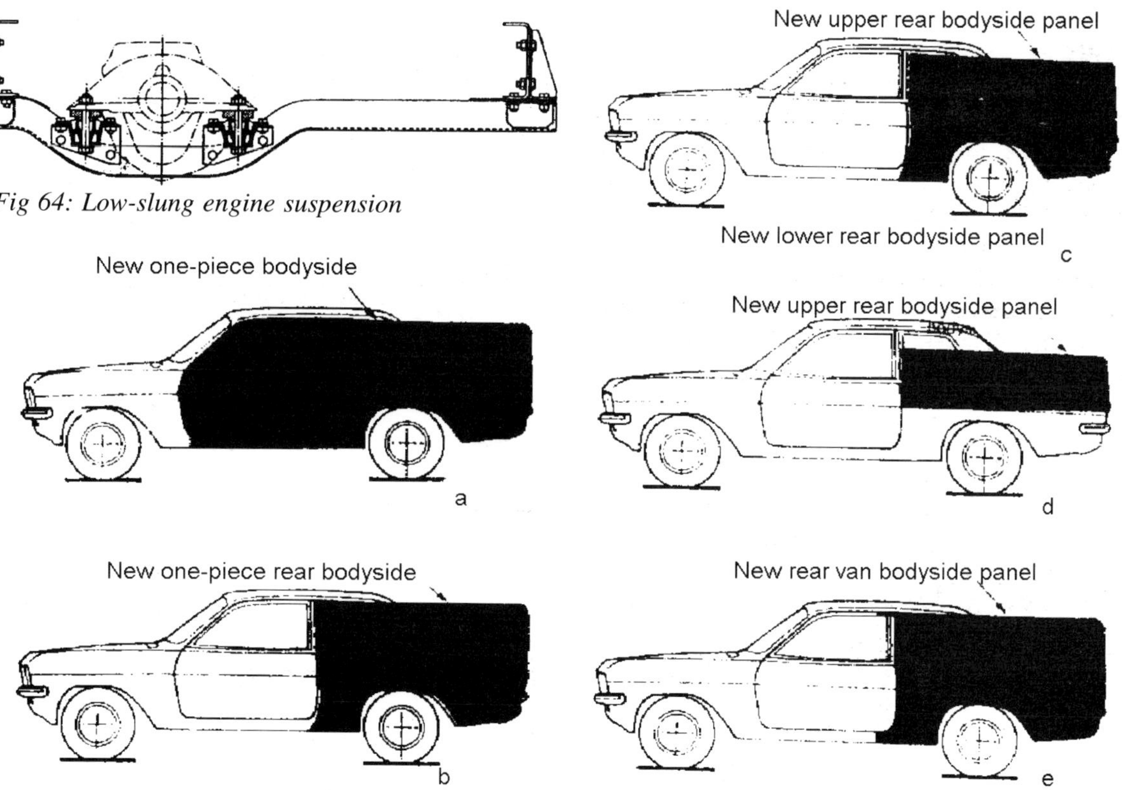

Fig 63: Body integration alternatives

quired. The three metre length is the same as the Rover Mini but considerably greater interior space has been achieved. The vehicle is taller than a Mini but not ungainly so in view of the relatively wide track. A three-across bench seat has therefore resulted and wide doors allow the ingress of wheel chairs. The two front seats are located slightly inboard to clear the wheel arches and can rotate through 45 degrees to improve access to the rear seat.

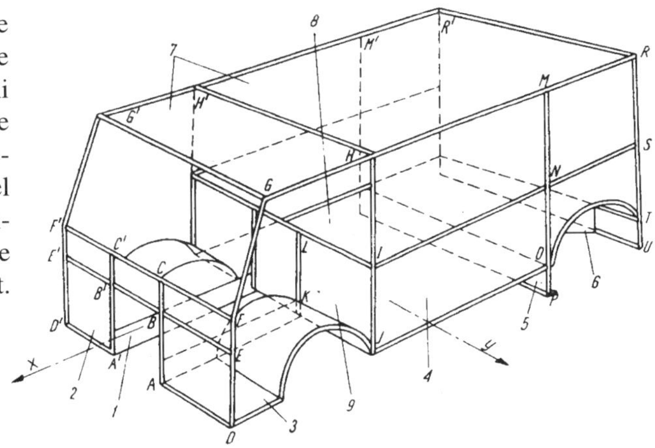

Fig 68: Integral-van proposal

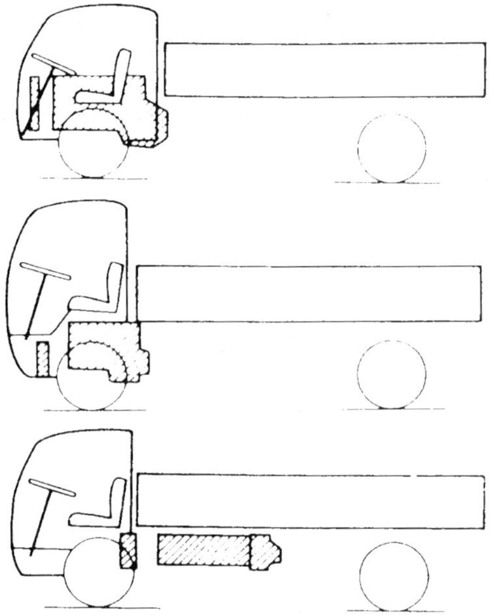

Fig 67: CV engine/drive layouts

Fig 69: Albion Claymore chassis

Fig 66: Hawtal Whiting Concept 92 with its body variants

Optimising the delivery vehicle

This section considers the multistop delivery vehicle of 3.5 tonnes GVW and over with reference to the improvements in fitness-for-purpose which could arise if conceived as a specialist vehicle in its own right.

A considerable challenge exists for engineers to design and build a high volume van which is also suitable for multi-stop delivery work. Panel vans between around 0.75 and 1.5 tonne carrying capacity are generally well arranged for this work but above this payload, the multi-purpose chassis-cab unit of 3.5 tonne GVW upwards tends to be the base for the larger multi-stop delivery vehicle. While cab design for light trucks has improved in respect of accessibility and general convenience, the requirements of the delivery vehicle operator can easily be compromised in favour of the tipper operator, or whomsoever, within the truck chassis' many functions.

The general-purpose cab has inevitably to be designed around the chassis frame and drive-train layout favoured for general duty. With the basic box-tube shape of a van body, it would be possible, as in the case of the light panel van, to use a more flexible integral-construction approach which avoided a high chassis front-end preventing a walk-through layout of the cab floor. There could be similar benefits in the adoption of a mid-engine layout, Fig 67, to ease the cab floor space problems. An integral structure proposal, due to Pawlowski, for a large van is shown in Fig 68. Engine support beams (1) are provided between the wheel-arches in this case. Other structural elements are front-wall (2), cab-floor and wheel-arch (3), side-wall (4), crossmember (5), rear-spring top beam (6), roof (7), floor-plate (8) and bulkhead (9).

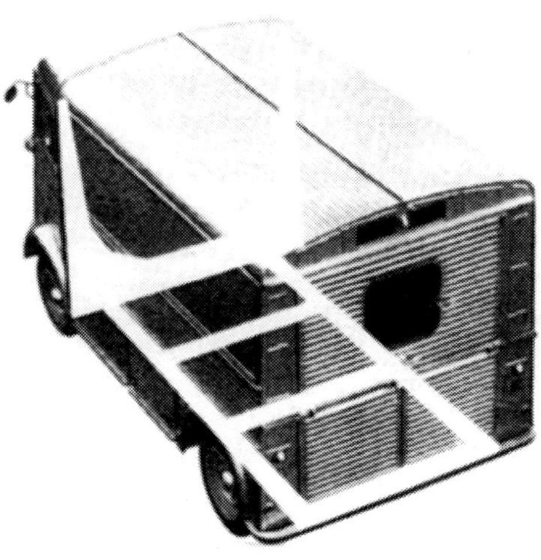

Fig 70:Citroen H-van

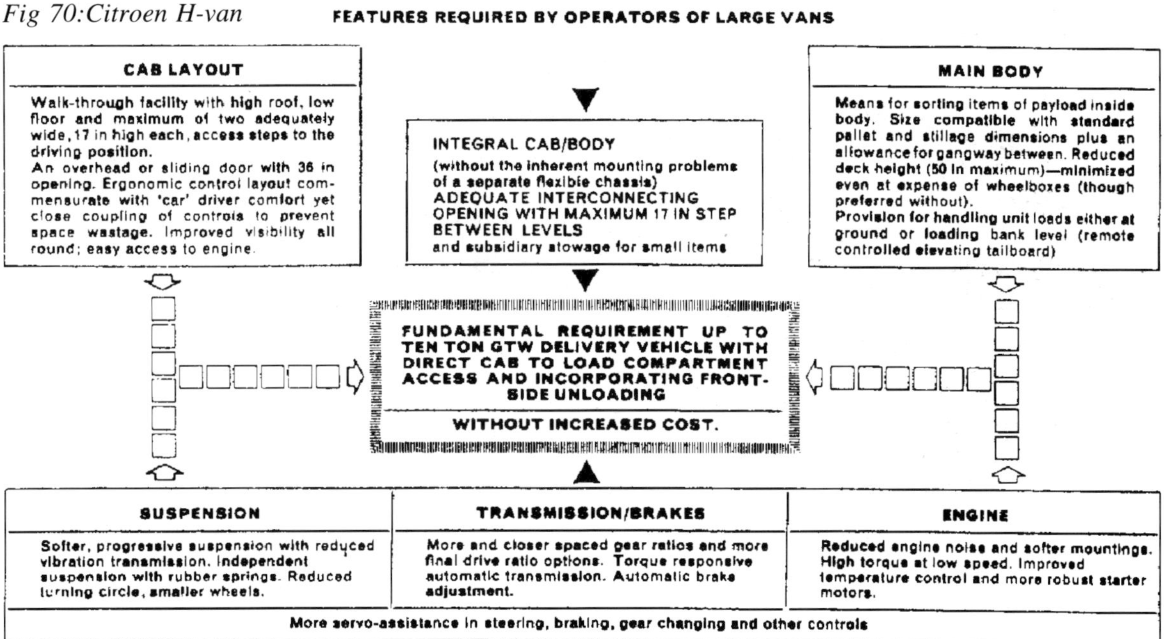

Fig 71: Requirements for a large delivery vehicle

The structure is considered integral if the structural surfaces which form the windscreen frame (F-G-G'-F'), the cab doors (D-G-H-J) and the rear doors (R-U-U'-R') are able to resist shear deformation.

A radical proposal for delivery van layout was made in the earlier chapter on CV packaging, in terms of general layout, which is examined here in greater detail. The concept of a purpose-built large delivery van is not new, of course, as seen in the Albion Claymore of Fig 69, an early 1960s model from a successful medium size company prior to absorption into the ill-fated BLMC. Even this vehicle was compromised by attempts to make it a 'Jack of all trades' but its flat-four mid engine and low drop-front frame were clear pointers in the direction of construction for purpose and a good access was obtained into an otherwise conventional cab. At the same period, the innovative Citroen company was in production with its H van, Fig 70, which showed how front-wheel-drive could also benefit van layout, particularly in respect of the rear cargo area whose low floor was additionally accessed from a mid-forward side door. The H-frame structure showed the way, disposing of the conventional chassis frame in favour of a base from which cargo and cab enclosures could be cantilevered.

The various requirements of fleet operators for large van specifications are summarised in the table of Fig 71, major aspect of which is the cab to cargo area walk-through facility which is absent in many large vans. Fig 72 is a more detailed picture of the above mentioned new-concept integral van, previously illustrated in diagrammatic form. The Albion Claymore flat-four engine envelope is again used in this proposal with drive taken back to a transfer box and then forward to a proprietary steer-drive axle mounted on rubber torsion springs — but with A-frame location from the transfer box. The H-frame in

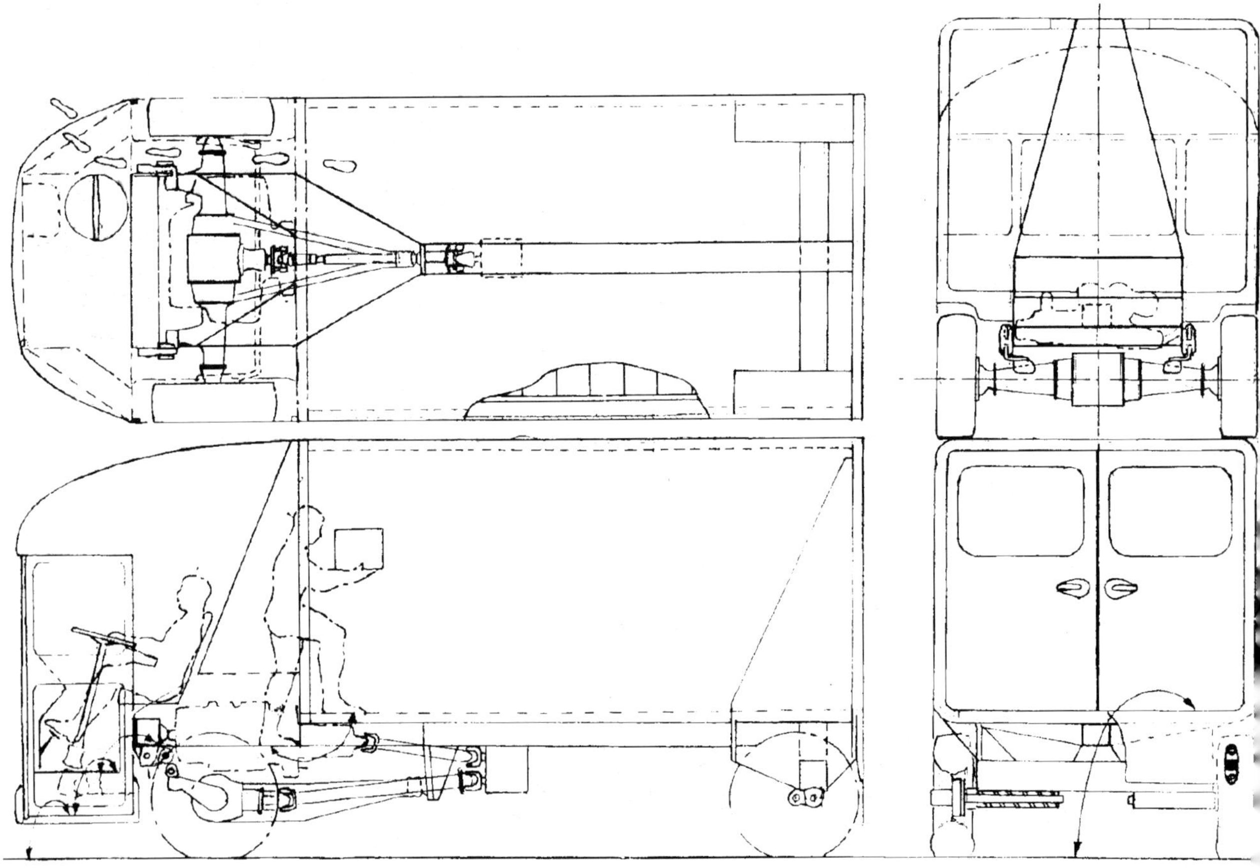

Fig 72: New-concept delivery van

this case cantilevers a sandwich panelled box tube for the cargo area, the floor panel of which acts like a simple truss to transfer floor loading out to the sidewalls. A ring-frame at the rear is tapered forward to support the torsion rubber rear-suspension and provide support for the rear door hinges.

The foam-plastic core material of the sandwich panel would provide structural stability for the skin panels as well as giving good thermal insulation for the payload. The box tube is of such a symmetry as to readily provide optional lengths of vehicle while being relatively simple to analyse as a beam, for strength and rigidity under different load conditions. The structure would be virtually rattle-free compared with conventional separate chassis/body arrangements used for integral box-van conversions. The cab ahead of engine position, and well-aft placement of both axles combine with the use of single-formation wide tyres to make front-wheel drive a practical proposition. Better braking will also result from more regular front rear weight distribution whether laden or unladen. The compaction of mechanical components with FWD also means a general space saving on the vehicle. Modest step heights result from ground-level through cab-floor to that of the cargo area.

Integral bodies on large vans date back to such vehicles as the Bedford-Hawson Easy Access van, Fig 73, used extensively for parcels delivery work. This vehicle would have benefitted considerably from the more recent initiative by Bruce on the Scottish & Newcastle Brewery delivery-vehicle, Fig 74, whose imaginative front drive conversion allowed a usefully low rear deck height. The Leyland TX450 concept distribution vehicle (Fig 75) was also a very important step forward. While not precisely aimed at the local-delivery market, it is nonetheless probably one of the most sophisticated integral van designs yet built.

Fig 73: Hawson Easy-Access van

Fig 74: Bruce delivery vehicle chassis

Fig 75: Leyland TX450 concept

Vehicles for the sick/disabled

With some 10 % of the population having a permanent mobility disability, and a substantial proportion at any one time suffering temporary illness or accident injury, this special vehicle category is of substantial importance. The specification and conversion of vehicles in the medium-sized car to panel-van size category is examined here (the design of individual invalid carriages [wheel chairs] is considered in a later section).

The Institute of Consumer Ergonomics at Loughborough University of Technology have published a guide called 'The ins and outs of car choice' which found a number of key entry/exit problems with conventional cars. Problem areas for elderly and arthritically handicapped people centre around: low seat height; low door frame height; high door sill height; wide door sills and deep floor-wells — also narrow door opening apertures. Other problems are door handles, supports, seat adjustment, seat belts, centre console, hand-control positioning and height of the boot sill. Fig 76 shows the supports commonly grasped in entry and exit with the percentages applying to each. These were observed in carefully controlled trials on high volume produced cars which led to dimensional recommendations as follows: minimum sill height above ground 270 mm; maximum footwell depth 60 mm; recommended seat height 500-550 mm above ground, 250 mm above sill; minimum A-post to seat front distance 440 mm; optimum door opening angle 70 degrees; minimum cant rail height from ground 1370 mm; maximum distance from seat edge to sill outer edge 160 mm; minimum distance from A-post to B-post 980 mm.

Servo-controls for disabled drivers

The expansion of on-vehicle electronics for conventional cars has also some promise for the disabled in the availability of improved servo-controls. Workers at Birmingham University have produced a system adaptable to a wide range of disabilities. A prototype vehicle has been built with a two-axis joystick. The

Fig 78: Elswick Envoy

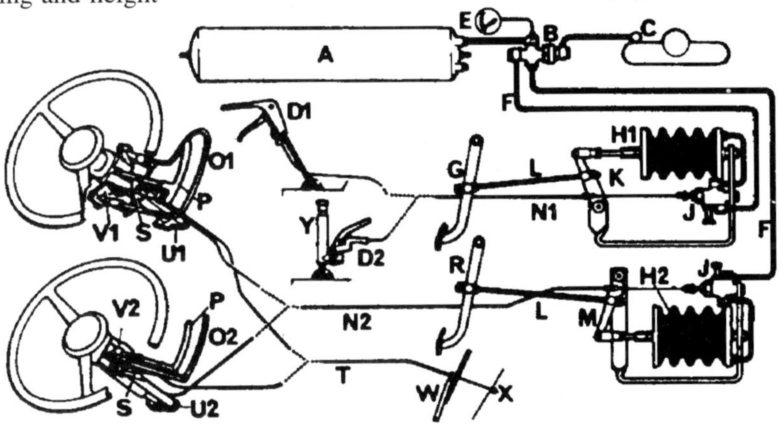

A Vacuum reservoir B Non-return valve C Induction fitting D1 Clutch handle std D2 Clutch handle tee E Vacuum gauge F Vacuum hose 5 in. I.D. G Clutch pedal H1 No. 1 Power unit H2 No. 2 Power unit J Control valves K Clutch relay lever L Rods C/W sliding links M Brake relay lever N1 Clutch control cable N2 Brake control cable O1 Brake handle Std O2 Brake handle slimline P Accel. control handle R Brake pedal S Accel. Pre-set lever T Accel. control cable U1 Column clamp block Std U2 Column clamp block "U" pin V1 Pivot claw V2 Pivot block slimline W Accel pedal X Accel cable retainer Y Clutch handle sleeve

Fig 77: Servo-assisted hand-actuated main brake

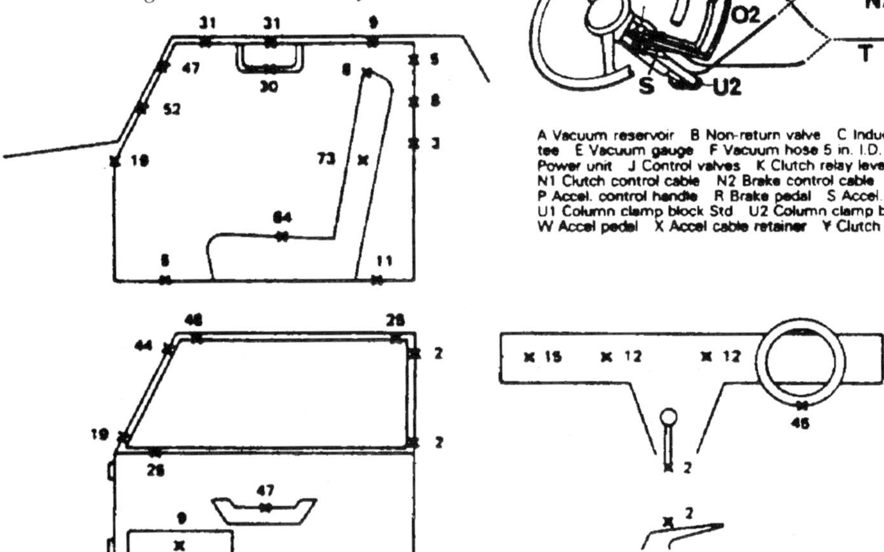

Fig 76: Study of supports grasped in entry/exit

search for low cost electric actuators led the researchers to use a high performance printed-armature DC motor to drive throttle mechanism and automatic gearbox linkages directly. For controlling the brake circuits a high pressure hydraulic system is regulated by a simple modulated solenoid valve. The most difficult part of the project, conducted in the Low Cost Automation Centre, is the steering servo-control for which response has to vary according to conditions and the rate of response has to be adaptive (these variables also need to be programmable). A small gearbox fitted on to the standard steering column of the adapted Ford Escort operate the servomechanisms.

Proprietary control systems are also available from Feeny & Johnson Ltd who have historically been in the business of vacuum servo foot brake assistors. The company's hand control conversion system for cars and commercial vehicles is shown in Fig 77. Throttle and brake levers have handles which form concentric arcs within easy reach of a hand on the steering wheel. The company were also involved in a project led by Reading University and the Transport and Road Research Laboratory in which only electric connection is provided between hand control and steering mechanism. A gearbox on the steering column contains three feedback potentiometers driven by anti-backlash gears. Two electric motors turn the column through worm gears and a 100 mm control handle is moved through a 40 degree arc to control them. On moving the handle, the potentiometers send signals to the motor. A photo pulse generator behind the vehicle's speedometer provides change of gain to the control handle as the speed increases.

Car and taxicab conversions

An alternative to modifying the controls of a conventional car is to provide access into the car for a wheelchair which is then clamped into the driving position so that the user never has to leave his/her seat. This approach was used on the Elswick Envoy shown in Fig 78; for severe disability cases; this could be accompanied by a push-pull mechanism as shown in Fig 79 which achieves braking and acceleration from one handle. A variety of alternative devices are shown in Fig 80 from the catalogue of Petri + Lehr GmbH.

A number of efforts have been made to modify the standard London taxicab to make the vehicle more acceptable to the wheelchair-bound. Already the regulations governing basic layout fit the vehicle

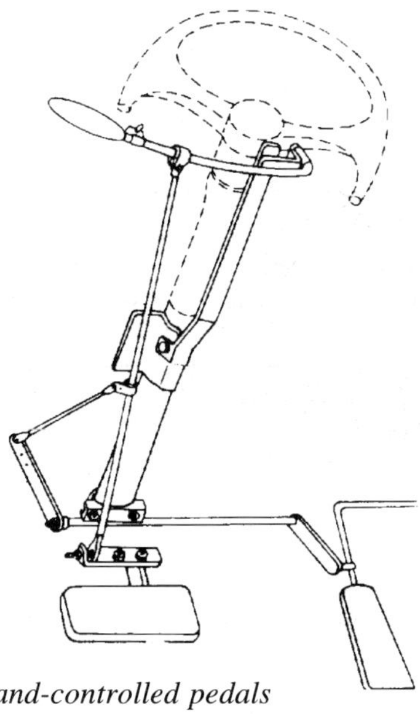

Fig 79: Hand-controlled pedals

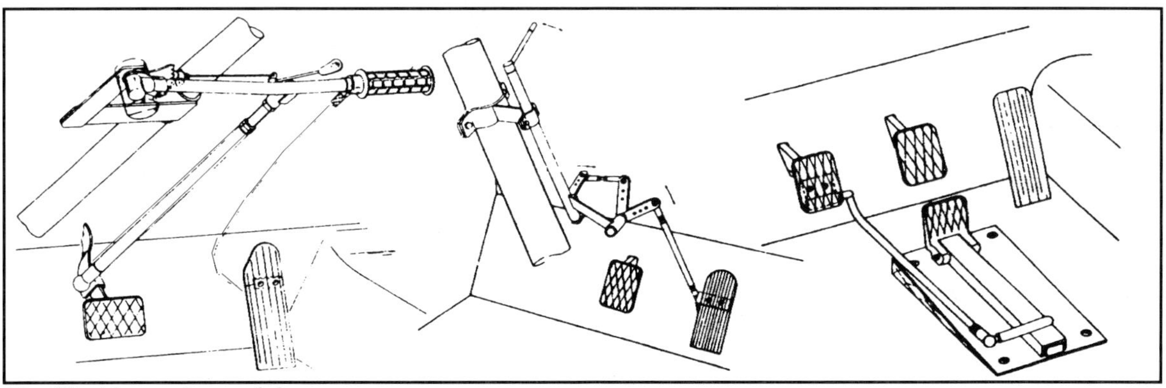

Fig 80: Alternative pedal movers

well for entry and exit by the less severely disabled. The Carbodies Ltd London Cab (Fig 81) now has many updated features which make conversion attractive. The vehicle features five speed all-syncromesh overdrive gearbox and has four wheel hydraulic full servo braking; key dimensions shown demonstrate the potential for conversion.

Ambulance design

Vehicles for the temporarily disabled have been the subject of considerable controversy in their history. Opinion differs in the medical profession as to whether ambulances should provide a heavier 'mobile hospital' type vehicle or a lighter minimally equipped one that would speed the injured between the scene of the accident, or serious illness, to the hospital. In earlier years this contrast could be seen in France's converted cars compared with Britain's converted trucks. More recently there has been a drawing together as designers strive for the optimum arrangement. Fig 82 shows a typical contemporary ambulance which has a Taurus body on an Iveco Ford 35.8 chassis. The 'sitting-case' model illustrated has twin 'one-third' doors at the rear and up to 10 people can be carried. Fig 83 shows the inside of a car-derived van conversion, for a stretcher case, by KFB Extramobile of West Germany. Both of these contrast strongly with an early 1970s design proposal shown in Fig 84. This was by Dr Bothwell of the Birmingham University Accident Research Unit who believed that as first-aiders are not accident surgeons, high speed easy-access ambulances were needed to cut down the journey time from accident location to operating table. Key features of the Bothwell design were a front mounted underfloor flat engine, pitch-free front/rear coupled suspension with automatic levelling and kneel facility; front wheel drive and wishbone rear wheel location to maximise rear-end access; monocoque chassis-less construction for optimum interior space utilisation; up-and-over rear door in conjunction with concertina heat-retention curtains.

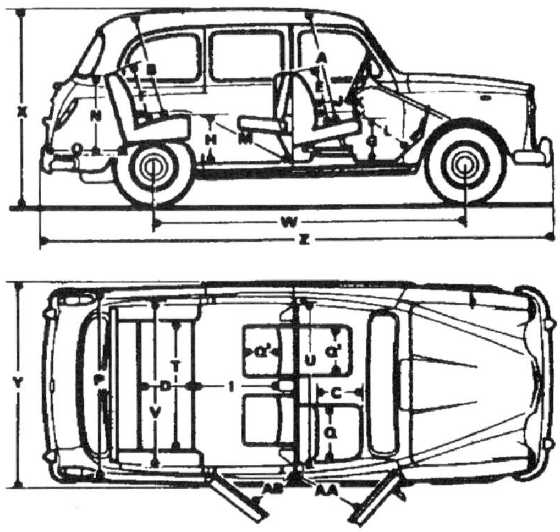

*A (Max.) 36½" (0.93 m.)	*A (Min.) 32" (0.81 m.)	*B 38" (0.97 m.)	*C 16" (0.41 m.)	*D 17½" (0.44 m.)	*E 18" (0.46 m.)
*F 20" (0.51 m.)	*G (Max.) 17" (0.43 m.)	*G (Min.) 13½" (0.34 m.)	*H 14" (0.36 m.)	*I 31¼" (0.79 m.)	*J (Max.) 15½" (0.39 m.)
*J (Min.) 10" (0.25 m.)	*K (Max.) 8½" (0.22 m.)	*K (Min.) 4½" (0.11 m.)	*L (Max.) 40" (1.02 m.)	*L (Min.) 35" (0.89 m.)	*M 57" (1.45 m.)
N 26½" (0.67 m.)	O 25½" (0.65 m.)	P 65" (1.65 m.)	*Q 20" (0.51 m.)	*†Q2 17" (0.43 m.)	*†Q3 13" (0.33 m.)
*T 45½" (1.15 m.)	U 55" (1.40 m.)	V 55½" (1.41 m.)	W 110⅝" (2.81 m.)	X 69½" (1.77 m.)	Y 68½" (1.74 m.)
Z 180½" (4.58 m.)	AA 25½" (0.65 m.)	AB 24" (0.61 m.)		Track – front 56" (1.42 m.) Track – rear 56" (1.42 m.) Turning circle 25 ft. (7.62 m.)	

† These measurements apply to all tip-up seats
* Approximate measurements

Fig 81: London cab and its key dimensions

Fig 82: CV-based ambulance

Fig 83: Car-based ambulance

Paramedic ambulance design

Heavy ambulances of former times are being replaced by small light-weight units suitable for paramedic operations. Considerable reworking of both front and rear drive layouts is required to accomodate intensive-care facilities, Fig 85.

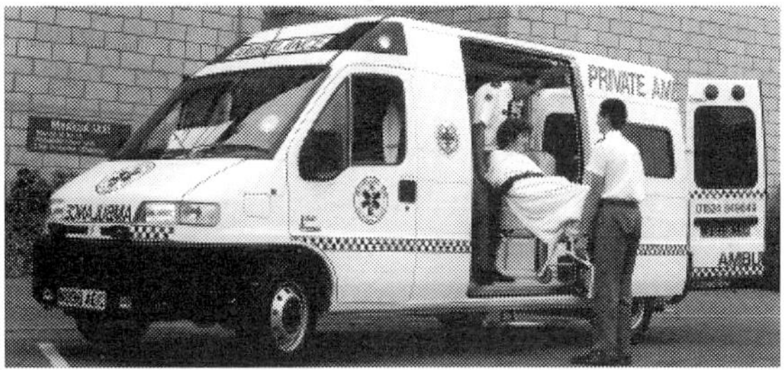

Fig 85: Citroen C25 based vehicle which can accommodate two patients, three crew members, piped oxygen, defibulators and automatic resuscitators.

Fig 86: Spacebus ambulance

According to Dr Alan Hope of the Glasgow Clinical Shock Study Centre some 10 000 inter-hospital transfers, about 60 % with head injuries, take place in the UK each year and create a demand for specialist ambulance design. Another important trend, in transport from the scene of an accident, follows the philosophy of 'taking the hospital to the patient' rather than the patient to the hospital'. Even where it is not possible to convey the appropriate equipment skilled medical management during transfer can be crucial to a patient's survival. The two above factors are relate because the majority of inter-hospital transfers involve moving people to hospitals where intensive therapy unit (ITU) equipment is available.

A conventional transport vehicle cannot provide full ITU facilities; a typical ambulance provides about six square metres of floor space, only 30 % of the recommended ITU bed-space. Selective specification of medical equipment is therefore the rule but interior ergonomic design is crucial in best exploiting restricted space. Key clearance volumes must be defined for on-board staff to, for example manage a cardiac arrest and provide easy access to the patients airway, intravenous and pulse sites, also both sides of the chest wall for massage. This is best achieved by the provision of a side-mounted trolley, rather than central one, with under-trolley ventilator, seating to the side and at the head of the patient. Standing headroom must also be provided. A patient suspension system, in addition to that provided by the vehicle's own system, is also recommended.

Firm securing of interior equipment such as oxygen cylinders is also crucial, to minimise damage

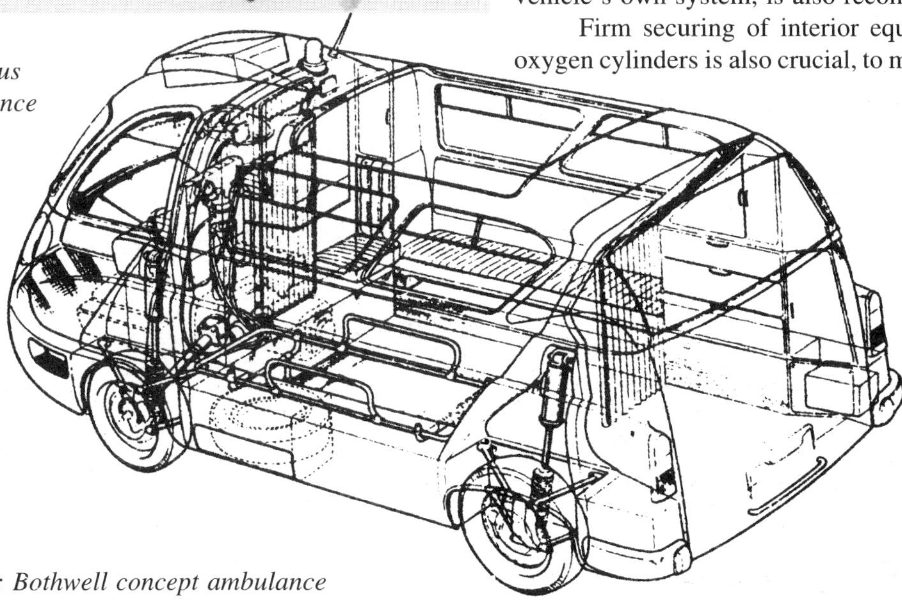

Fig 84: Bothwell concept ambulance

in an accident situation and 209 restraint systems have been proposed. Armrests for side-facing seats are also crucial. Three categories of ambulance emerge: Type A, one equipped for transport of patients not required to need treatment by trained personnel; Type B, one equipped for treatment and monitoring of non-critically ill patients; and Type C, that discussed above for the advanced treatment and monitoring of patients. The latter category differs in the UK from its continental counterparts because of the tradition of staffing the ambulance with paramedic rather than nursing staff. The proposed CEN specification is leading to the view that a specialist vehicle should be constructed rather than a van conversion. Design of storage cupboards should involve clear doors for the contents to be visually inspected and their positioning such as to avoid significant movement by the paramedic for access. Power doors for the main cabin of the ambulance are a valuable feature since they prevent onboard staff from becoming infected during operation of exterior handles; better draught-sealing could be a side benefit.

Ambulance chassis specification in the UK has traditionally involved 3.5 tonne GVW Ford transit, LDV 400 and Mercedes 300 rearwheel drive designs. Front drive vehicle such as Citroen C25 and Renault Master offer the potential of considerably lower chassis heights and greater diversity of body design with the removal of rear axle and prop-shaft. Body design is likely to be affected by the recent Manual Handling Operations Regulations which require less physical effort on the part of staff handling both walking and stretcher patients. An advance over the traditional mildsteel framed ambulance body is seen in the Mastermax body by Locomotors based on a Renault Master chassis. It employs a ferritic stainless steel tubular construction, considerably lighter than normal. The original design, Fig 86, was due to Spacebus and has light composite skinning providing internal height of 73 inches and width of 80 inches. The greater interior cube gives scope for stowage of ambulance equipment at floor level rather than overhead lockers which tend to compromise headroom. Floor height can be down to 475 mm. Even lower floors are possible with chassis conversions such as those by Al-ko, Fig 87. The CRD Sparshatt company have achieved a 200 mm height reduction by this method. There is also the possibility to use tandem axles and thus increase GVW from 3.5 to 4.6 tonnes. The design also provides 200 mm interval holes along the chassis to vary the axle position or insert crossmembers on the all-bolted construction frame; sidemembers are 180 x 50 x 5 mm.

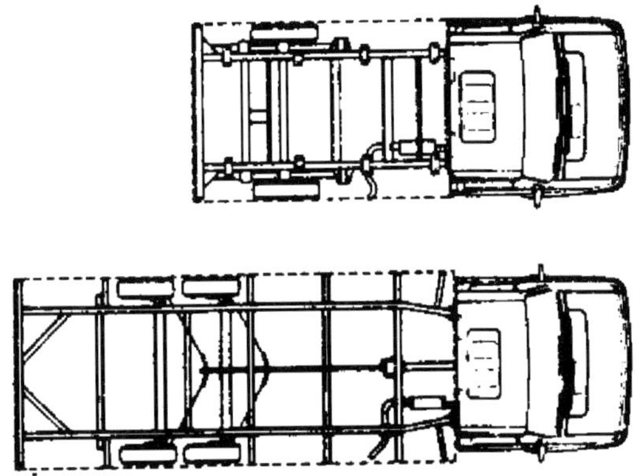

Fig 87: Al-ko conversions have been carried out to Peugeot Talbot Express chassis-cabs to produce parametric ambu/ances for Wadham Stringer. A platform height of just 300 mm from ground level has been achieved. Conversions with three axles are also possible. For the later Sevel chassis-cabs five additional body lengths become available with single axles and six more wth tandems. The rear axles are torsion bar suspended and have progressive rate springing. The chassis frame extension is made from C-section galvanised steel members and resultant weight saving can produce up to 37 % improvement in load carrying capacity over the standard chassis

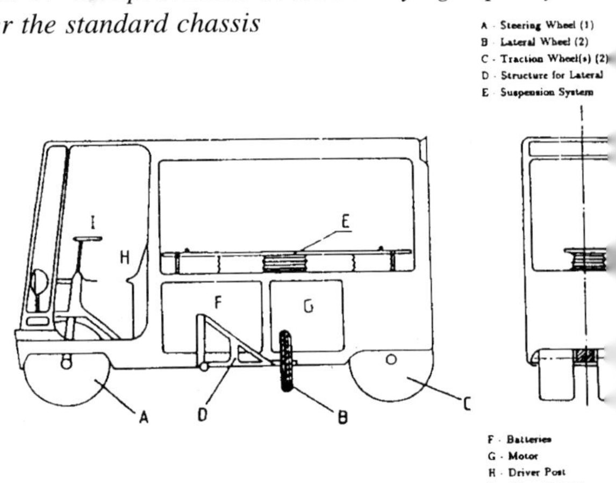

Fig 88: Lisbon Technical University ambulance proposal

Rethinking the ambulance Emergency transport of accident victims in heavy traffic was the impetus behind an interesting vehicle to handle urban emergencies discussed at the recent Autotech congress. Researchers at Lisbon Technical University suggest[12] that the key design factor was provision of an efficient life condition support for the patient in a vehicle with exceptional traffic mobility. That involved a highly compact layout yet a comfortable posture and space for medical assistance to the patient in transit, Fig 88. The short journeys inherent in the duties of the vehicle make it an attractive proposition for electric traction and in the proposed design the patient would be positioned above the battery container. Twin rear wheels mounted on an oscillating axle, the 'banking' of which would be controlled by the driver, or 'rider' in a semi motor-cycle situation, behind the single steering front wheel of the vehicle. Retractable outrigger wheels would also be used to provide cornering stability in less dense traffic conditions. Translucent panelling around the patient compartment would reduce the claustrophobic effect as would careful colour selection for the interior.

Van design: new initiatives

Electric drive and gas propulsion are under scrutiny by a number of makers and some of the projects are discussed here.

Concept electric van: a detailed specification An electric vehicle with equivalent performance to an IC-engined one, and yet based on a conventional Escort van platform, Fig 89, has been achieved by Ford with the use of a sodium/sulphur battery. The battery is enclosed in a hermetically-sealed metal casing, which contains sodium and sulphur on each side of a porous ceramic separator. In order to react on a molecular level, the sodium and sulphur must be free to flow through the separator in liquid form. This requires maintaining an operating temperature of between 290 and 350 C. The Ecostar's battery pack is formed of a double-skinned, stainless steel case, with a glassfibre insulated cavity which has been evacuated to form a vacuum flask. Inside the case, 480 individual sodium/sulphur cells are embedded in a sand matrix to prevent movement and contain the elements in the event of a severe impact.

Each cell is encased in a corrosion-resistant aluminium cylinder and produces about two volts. They are linked to provide a 330 volt power supply and a steady-load current of 80 amps. Power of up to 50kW is available for short periods. The battery pack is mounted under the vehicle floor, ahead of the rear

Fig 89: Ecostar electric van

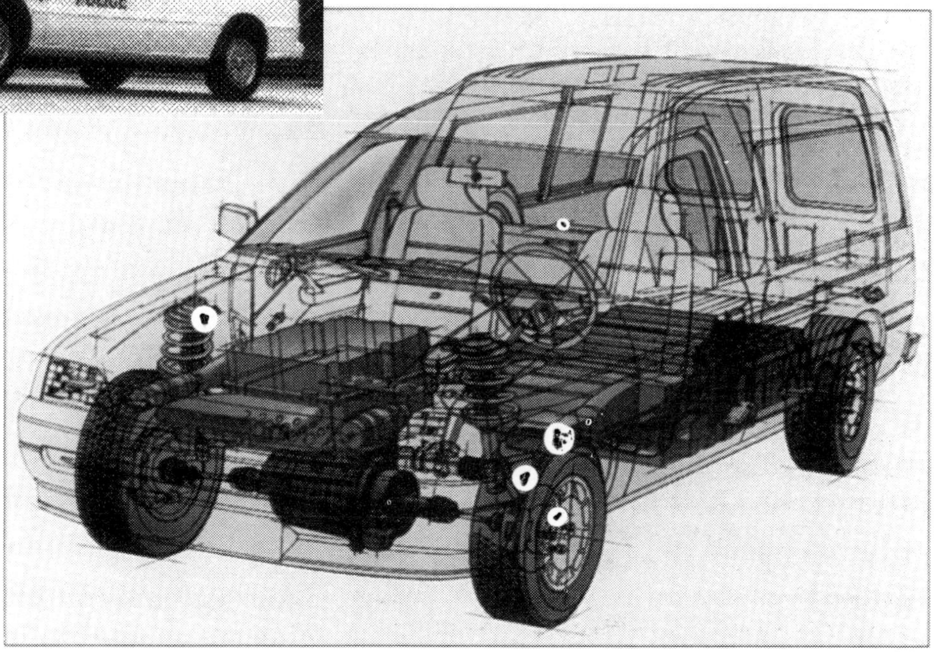

Fig 90: Ecostar package

axle, Fig 90. The vehicle is fitted with an intelligent on-board charging system that automatically adjusts to the supply voltage from the mains. It can be used for overnight recharging and temperature control of the sodium/sulphur battery. A full recharge takes between five and seven hours. In the event of electrical malfunction, the battery pack is isolated by replacing external 200 amp fuses at the battery connection box. The batteries are also equipped with internal fuses that isolate the electrical supply from the vehicle if there is any kind of internal overload.

The Ecostar is powered by a high-speed, three-phase alternating current (AC) motor, rated at 56kW (75hp). It is directly coupled to the transaxle by a single-speed planetary reduction gear set. The motor operates at 330 volts, provided by the sodium/sulphur battery pack. Compared to equivalent DC systems, the AC motor works at a higher rate of efficiency, is smaller, lighter, lower in cost, easier to cool, more reliable and generally more durable. Maximum speed is 13,500 rpm and maximum torque 193 Nm at zero rpm. Regenerative braking comes from the motor acting as a generator, so recharging the battery.

The supply from the sodium/sulphur battery is fed along special heavy-duty cables to the Power Electronics Centre (PEC), which is housed in the engine compartment. The candy-striped, red-on-black cables are strikingly marked to avoid confusion with any other wiring in the van. Encased in aluminium, the PEC incorporates the battery charging electronics and inverters that convert the 330-volt DC supply to AC power. It also includes a transformer which enables a 12-volt auxiliary battery to be recharged from the high-voltage traction batteries. The electrical supply to the PEC is connected and isolated by power relays inside a contactor box, controlled by the ignition key and the electronic modules.

On-board microprocessors are linked by a multiplex database that allows synchronous, high-speed communication between all the vehicle's systems. The Vehicle System Controller (VSC) acts as the user/ vehicle interface and is operated by electrical signals from the accelerator pedal. This 'drive-by-wire' system has no mechanical connection to the speed controller. It is supplemented by a battery controller that monitors the sodium/sulphur operating temperature, the state of charge and recharging. The control system also incorporates a diagnosticdata recorderwhich stores information from all the on-board electronic systems. This locates any operational malfunctions quickly and precisely.

A fault detection system (known as the Power Protection Centre (PPC), has also been built into the battery controller to monitor continuously the main electrical functions. Every four seconds it checks for internal and external leakage between the high-voltage system, the vehicle chassis and the battery case. Should a leak be detected in any wire, a warning light warns the driver that service action should be taken. The vehicle can then be driven safely for a short distance so that repairs can be made. If leakage is detected from both battery leads, the vehicle system cuts off the power to the motor and illuminates a red warning light. The auxiliary power supply is maintained to operate the battery cooling system. An inertia switch is also fitted, which is activated in the event of a vehicle collision and isolates power from the main battery pack. Auxiliary power to the battery coolant pump is also cut off to reduce the risk of hot fluids escaping. The vehicle incorporates a small amount of 'creep' whereby slight brake pressure is required to prevent the vehicle from moving forwards (or backwards when in reverse gear). This results in easier manoeuvring and smoother transmission of power. Auxiliary vehicle systems are pow-

Table 1: Ecostar specification

Powertrain Three-phase, AC Inducbon Motor
Transmission Single-speed integrated front-wheel dnve
Power 56kW
Maximum Torque (Nm) 193
Maximum Speed (rpm) 13 500
Battery Type Sodium/Sulphur
Energy capacity ratMg at 80% DOD 30 kWh
Power ratings: peak intermmittent (kW/hp) 50/70
max continuous (kW/hp) 30/40
On board charger (120/240 V) with 2 metre charging cord on reel.
Required 240 V at 30 amp AC single phase for max'm charging rate
Maximum vehicle weight (kg) 1 851
Kerb weight (kg) 1,406
Payload (kg) 400-463
Rated top speed (governed) 70 mph
Rated 0-50 mph acceleration 12 seconds
Range (Federal Urban Dnving Schedule) 100 miles
Lightweight 14 inch aluminium alloy wheels with specially developed P195/70R14 low resistance tyres

ered by a standard automotive 12-volt lead/acid battery exception the electrically-driven cabin air conditioning system, which is powered directly from the main sodium/sulphurtraction batteries, via a special AC inverter in the PEC module. The climate control unit handles both the air conditioning and a highly efficient 4.5kW ceramic element PTC heater. The heater elements are made from barium titanate with a multilayer metallic coating on each side, impregnated with special chemical additives. The material has low resistance at lowtemperature for a very fast warm-up, while at higher temperatures the power supply is automatically regulated to save electrical energy. A strip of solar panels across the top of the windscreen supply power to a supplementary extractor fan that ventilates the cabin when the vehicle is parked in direct sunlight. This relieves the load on the air conditioning system when the journey is resumed. Some lightweight materials have been used in the Ecostar to offset the 350kg weight of the battery pack. Elimination of the clutch, torque converter and additional gearing is supplemented by a magnesium transmission casing, aluminium alloy wheels, air conditioning compressor and power electronics housing. Plastic composite materials have been used for the rear suspension springs, load floor and rear bulkhead. Use of these materials has helped keep the Ecostar's kerb weight to between 1,338 and 1,452kg, which is 25 per cent heavier than a standard diesel-powered van. The vehicle also has a useful load carrying capacity of up to 463kg and retains similar loadspace dimensions to the standard Escort van. The Ecostar has been developed for optimum performance in urban conditions, where it is expected to be driven most frequently. Its top speed is restricted to 70 mph, whilst 0 to 50 mph takes approximately 12 seconds. Because of the high torque at low speed, acceleration from standing to 30 mph is quicker than diesel and petrol driven vehicles of the same size. The average vehicle range between charges to date has been 94 miles, with a maximum recorded range of 155 miles.

Production electric van: Bradshaw Envirovan
Fig 91 shows the Envirovan built in conjunction with US collaborator Taylor-Dunn. This can carry 1500 lbs on a 3.55 square metre platform at speeds up to 32.5 mph and is aimed specifically at city deliveries. The vehicle relies on 12 six-volt deep-cycle, rechargeable lead-acid batteries for a total of 72V. All accessories, such as internal lights, windscreen wipers, and gauges, run off the 72V system through a DC-DC converter, which steps the power down to 12V, so that all batteries discharge equally. This distributes power requirements evenly across all 12 batteries and prevents one or two of the batteries from draining prematurely. A battery warning indicator shows the current percentage of battery power available, with avisual warning when battery charge is below 20 %. An on-board battery charger, featured on the Envirovan, can be used to recharge the battery packs simply by plugging it into any standard 240V AC power socket. The entire 72V system requires approximately 9.5 hours to fully charge the batteries from an 80 % state of discharge (20 % remaining charge). The battery pack provides approximately 1000 recharging cycles before replacement is required. Battery packs are available for less than £1000 which equates to less than 2p per mile. Recharging costs add an additional 1p per mile giving a total cost per mile of 3p. A 20 bhp General Electric motor has been designed for Envirovan. A range of eight hours/50 miles is available for the vehicle which measures 4.21 metres long x 1.65 wide. It can accelerate to 25 mph in 6 seconds and its controller can generate up to 28 bhp for quick response.

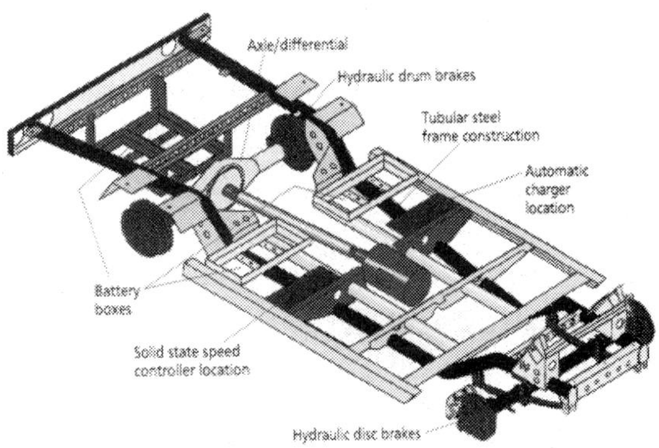

Fig 91: Bradshaw Envirovan

Half car, half van/pick-up

Available in single-cab, club-cab and double-cab forms, Fig 92, Mitsubishi's 1-ton pick-up truck, Fig 93, is aiming to satisfy the different demands of goods and passenger vehicle markets in one package.

Double walling of the cargo box, Fig 93a/b, is achieved by using single sheet inner and outer panels. U-shaped reinforcement members have been added between the panels and a redesigned tailgate and opening has resulted in a recessed handle and useful rave-level securing hooks help the carriage of long-loads with an open tailgate. Petrol-engined versions feature a 4-valve per cylinder design with roller-bearinged rockers, the 2.4 lire variant developing 97 kW. Twin counter-balancer shafts, Fig 94, claim to bring six-cylinder smoothness to the four-cylinder layout. Highest powered diesel variant, a turbocharged and intercooled unit, develops 73kW and 240 Nm torque. As well as choice of manual or automatic transmission there is on-the-fly part-time four-wheel drive and a rear differential lock is provided. The latter is electronically-controlled and may be engaged on the move at speeds under 12 kph. A limited slip differential is also provided. Front disc brakes have been introduced, front and rear track widths increased and a load sensing proportional valve is specified. The chassis frame has had buckling-inducing shapes applied to the front longitudinal inner members to achieve controlled collapse on frontal impact. The body structure has undergone computer modelling to optimise its occupant-protection features. The steering column features a pressure-fit ball which allows the upper column to collapse in on itself and a one-way breakaway capsule, in the upper mount, releases the column so that it falls on to the driver's lap. A fuel-cut-off valve operates in the event of unduly high lateral acceleration, or in roll-over, to prevent leakage from the fuel tank and a screw-in plastic filler cap incorporates a ratchet mechanism to ensure security

PSA electric vans Peugeot's J5 electric van, Fig 95, uses a DC motor developing 25 kW at 1500 rpm in conjunction with a five speed manual gearbox. Range on one charge is 70 km for the 1200 kg vehicle which has a rolling resistance of 12 kg/tonne. Frontal area is 1.8 square metres and drag co-efficient 0.34. The Citroen C15 EV, Fig 96, uses a 16 cell 96 V battery and DC electric motor with electronic con-

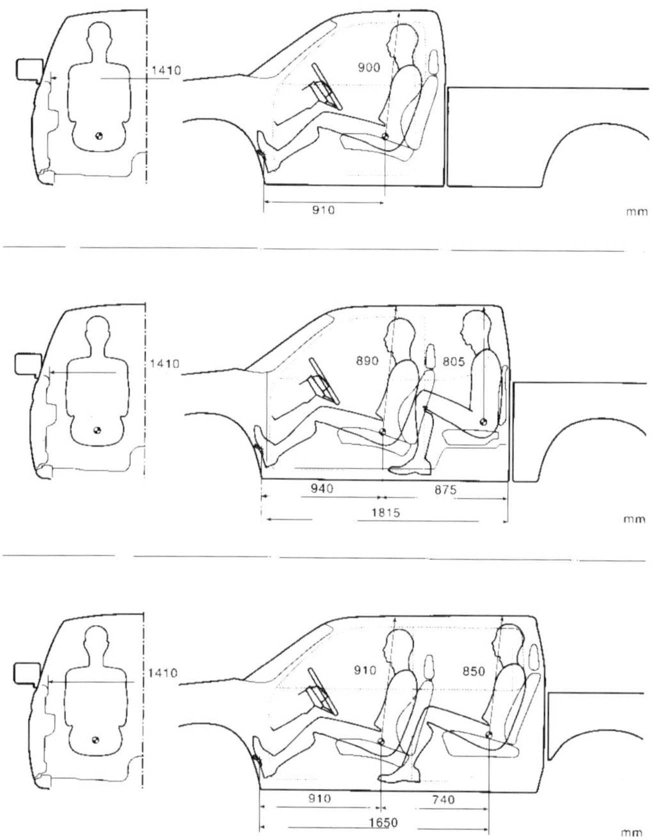

Fig 92: Caboptions

Fig 93a: Mitsubishi 1 tonne pick-up

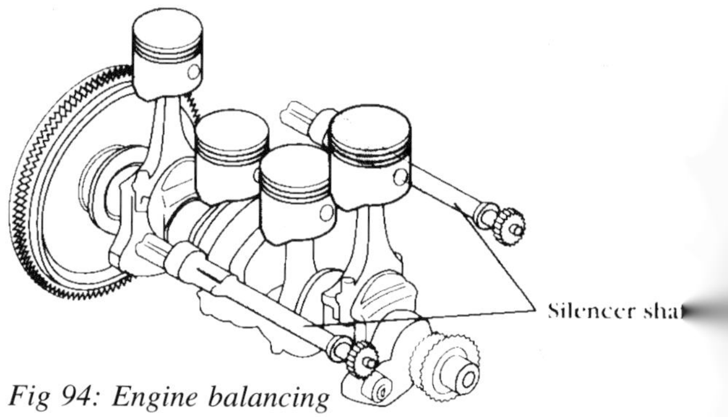

Fig 94: Engine balancing

trol. The Leroy Somer motor is powered by Dryfit batteries which weigh 31.5 kg and have gelled sulphuric acid electrolyte with three year life expectancy. Torque developed is 127 Nm at 1800 rpm; maximum current intensity is 250 A. Gearbox is the same as that used in the petrol-driven car-derived van. Kerbside weight is 1392 kg for the 3.99 metres long vehicle. Range is 70 km and top speed 90 kph.

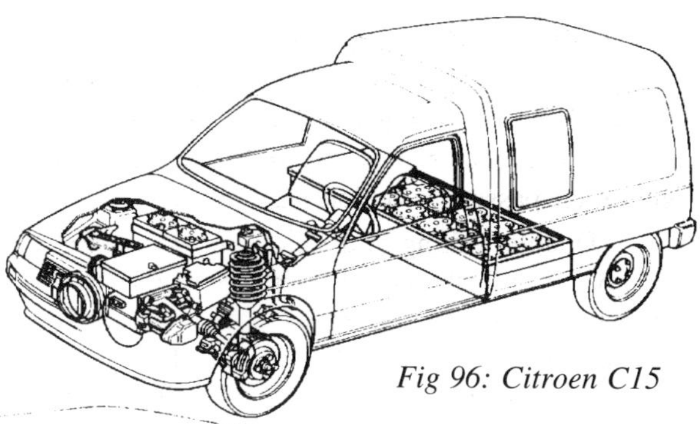

Fig 96: Citroen C15

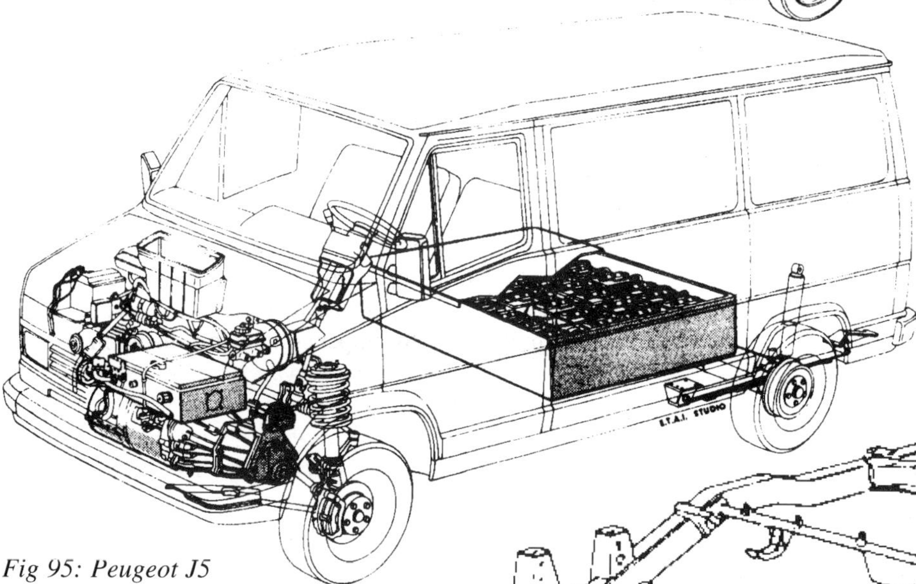

Fig 95: Peugeot J5

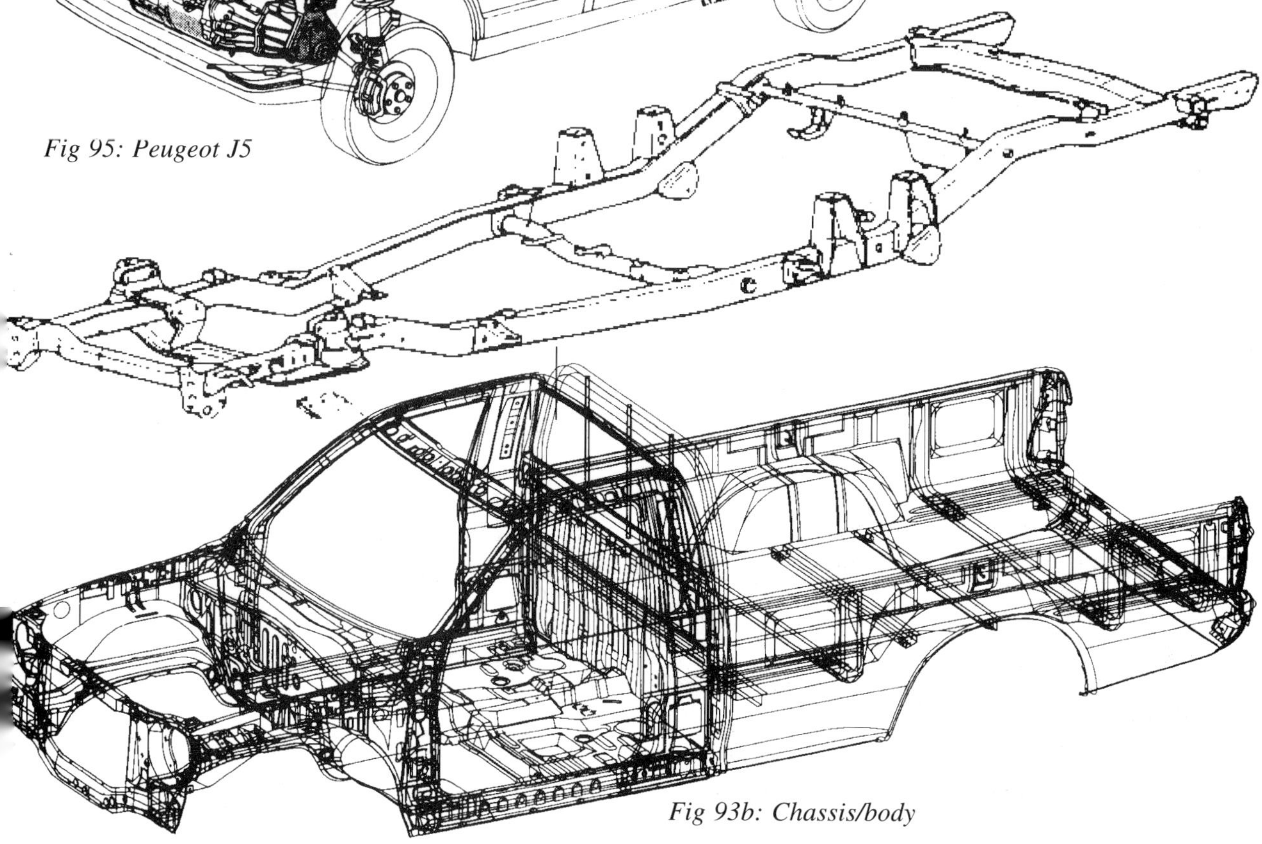

Fig 93b: Chassis/body

Fuel-cells for electric CVs

According to RL Hodkinson of Nelco Electric,within a few years PEM (Proton Exchange Memory) fuel cell powered electric vehicles will be developed. Hodkinson also suggests that the the gas turbine is a serious alternative for powering hybrid electric vehicles and is an efficient solution for buses and trucks where powers over 100 kW are called for.

The PEM cell has a fuel conversion efficiency potential of 83 %; already 58 % has been achieved and 70 % is in prospect for the end of the decade. Costs are gradually reducing, he argues, and £200/kW has already been reached using hydrogen-air fuel cells. The technique of thermal catalytic reforming developed by Hydrogen Power Corporation/ Englehard is the key to obtaining adequate quantities of hydrogen fuel. The chemical reactions involved are:

$$3Fe + 4H_20 = Fe_30_4 + 4H_2$$

with a catalyst at 130 C and hydrogen stored in a hydride fuel tank:

$$Fe_30_4 + CH_4 = 3Fe + H_20 + CO_2$$

the iron being reduced in a central facility, some 1kg of iron being required for a small car to travel 6km on a fuel-cell and a high proportion of hydrocarbon heat energy is converted into hydrogen.

The overall conversion efficiency, hydrocarbon to hydrogen, fuel-cell hydrogen to electricity and electricity to mechanical power is estimated to be within reach of 48 % . Hydrogen would be reformed on demand at fuel stations and the waste heat used to generate electricity for the grid. Natural gas is a particularly attractive hydrocarbon source as a low sulphur content avoids poisoning the catalyst and the pipeline grid avoids the need to distribute conventional fuels by road.

For on-board storage iron titanium hydride is considered to be one solution but US experiments are in train with cryogenic storage that are likely to reduce weight and cost penalties. Fig 97 shows the hydrogen distribution scheme. A fuel-cell hybrid vehicle would have the motive power provided by the fuel cell and a booster battery providing peak power for acceleration. The fuel cell would recharge the battery when the vehicle is stationary. Emissions would be reduced by a factor of 10 compared with IC engines running at stochiometric air-fuel ratios. The PEM cell has graphite electrodes sandwiched by membranes, and a gas-tight seal. Each is 6 mm thick and produces 1 volt off-load and 0.7 volts on-load at

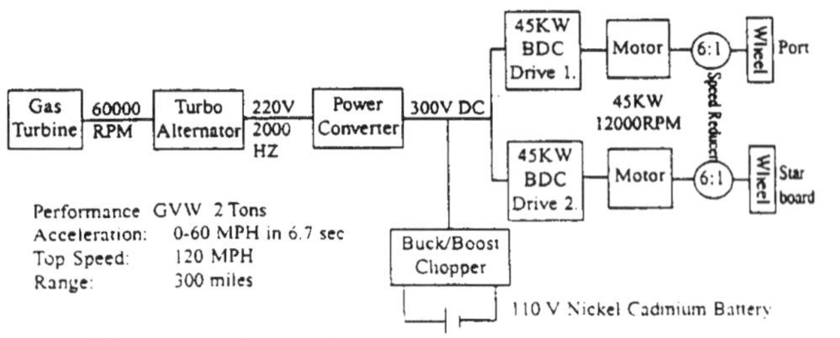

Fig 99: Series hybrid proposal

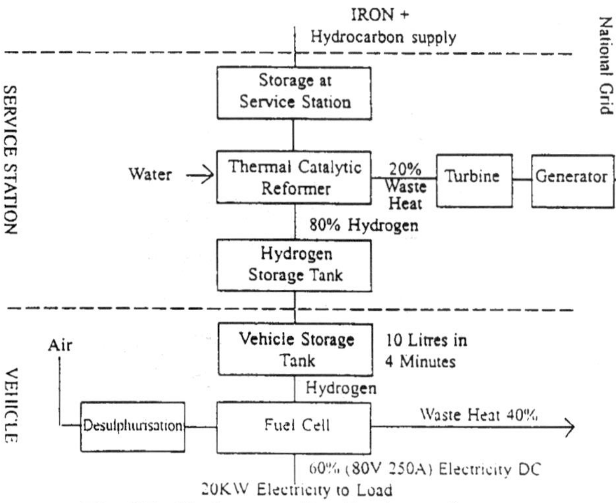

Fig 97: Hydrogen distribution scheme

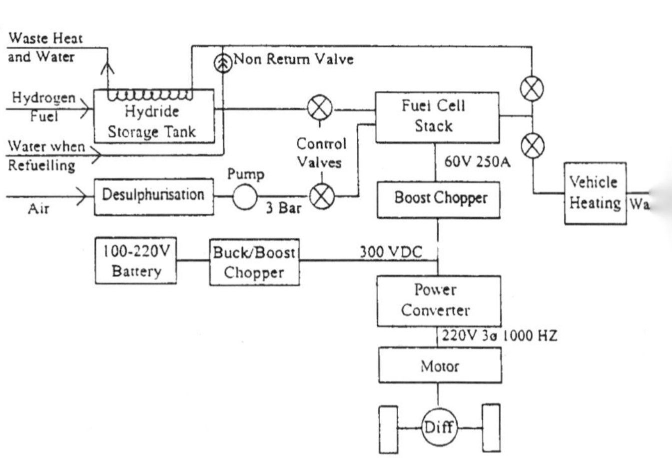

Fig 98: Scheme for safe connectoins

a current of 250 amps — a cell for 15 kW average power would produce about 60-70 volts DC in a size of about 200mm square and 600 mm long. Operating temperature is 80 C and when cold it would give 50\% power, instantly, rising to full power after about three minutes. Fig 98 shows a scheme recommended by Hodkinson for safe connection on vehicles. A low-cost two-pole 220V, 300A remote-control circuit-breaker needs to be developed which will act as a battery isolator with 5kA short-circuit capacity. Fig 99 shows a scheme for a 90 kW series hybrid vehicle based on gas-turbine technology. The turbine would operate through a PWM invertor stage to feed 300-500 volts DC into the main bus. Two separate drives would power one rear wheel each of the vehicle. To reduce weight, motors would be designed for 1200 rpm operation and gearboxes would reduce this to 1800 rpm at the wheel, corresponding to a road speed of 120 mph. The turbine would operate over a 2:1 speed range to optimise efficiency.

As fuel-cells are low-voltage/heavy-current devices, the power conversion scheme of Fig 100 is recommended. Here a 100-stage fuel cell is integrated with a 216 volt battery to give a stabilised 300 volt DC rail with motor controller operating from it — and currents are significantly reduced on the 100 volt system. This system serves another purpose: as the vehicles are fitted with booster batteries for acceleration, the DC link will change significantly according to load and the conversion stabilises this variation.

The mature CNG van The shape of things to come in natural gas vehicles was forecast Tickford who suggest a front traction-module incorporating encapsulated powertrain, isolated cab and gas storage cassette; a mono-fuel CNG engine turbocharged to better ULEV tailpipe emissions; exhaust system with insulated manifold/downpipe and methane oxidising catalytic converter with fast light-off; continuously variable transmission; the gas cylinder cassette to be load-bearing, as a platform deck; integral body and underframe using tailored gauge panels in various materials; alternative superstructure options; all-round levelling air-suspension with raise/lower facility; low rolling-resistance tyres, differentially sized front/rear.

Low-loading van An independent loading module is proposed in a design for a low-loading van by the Dutch companies DSM Resins and Treberg Specials. The idea is that the loading module is re-used when the driving module is replaced. Independent rear-wheel air suspension for the loading module, which is made from reinforced Synolite resin, allows the deck to be lowered to street height. The vehicle is under current evaluation by the Dutch Post Office.

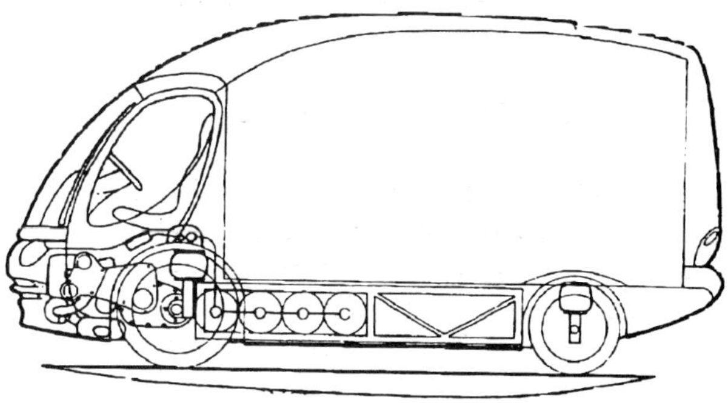

Fig 101: Mature CNG van

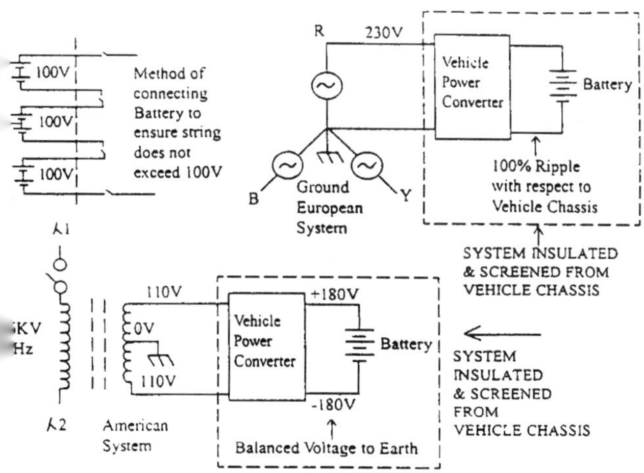

Fig 100: Power conversion scheme

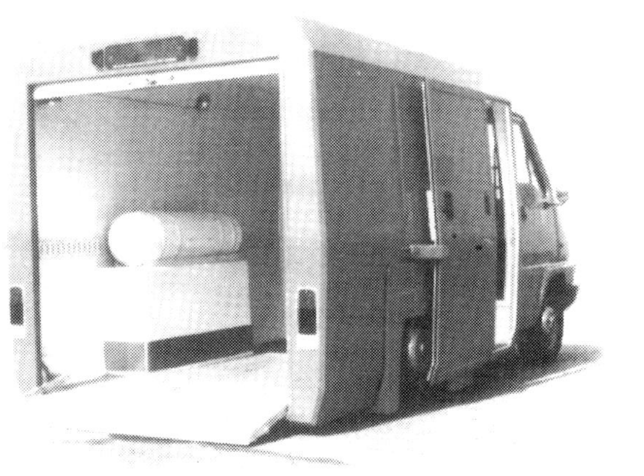

Fig 102: Separate loading module

Structurally optimizing electric vehicles

Daewoo's DEV3 electric vehicle has used structural analysis techniques applied by the Institute for Advanced Engineering in the design of its space frame[13]. The frame comprises welded extruded aluminium alloy members of identical cross-section and the design was optimized by setting up design variables for each member in differential element thicknesses. The structure incorporates a large battery tray over the floor area, the design of which was also optimized such that a light weight EV build was possible even complete with fitted batteries. Its kerb weight of 1187 kg compares with 2100 for the pioneering Lucas Hybrid EV, for example; while battery weight for the DEV3's NiMH units was 373 kg against 550 kg for the Lucas lead-acid units.

For the floor tray, upper and lower skins over the extruded frame members formed a strong sandwich construction, with a box frame around its periphery. Material used is AL 6061-T6, of 310 MPa ultimate tensile strength, joined by TIG welding. Target weight for the space frame was 130 kg and a torsional stiffness of 7940 Nm/deg was sought to give first torsional frequency of 25 Hz and 140 MPa stress under service loads. Fig 103 compares the completed structure with its finite element model and in testing the physical structure a torsional stiffness of 8682 Nm/deg was obtained which was 31 % stiffer than that of the GM Impact vehicle against which it was compared, considered to be the effect of integrating the tray structure; torsional frequency was 26.5 Hz. Service stresses exceeded the target values, however, for 3g bump loading of both-together and individual front wheels, necessitating thickening of the strut member supporting the front of the battery tray.

Of the 34 extruded members comprising the space frame, those in Fig 104 were chosen, in addition to the tray members, Fig 105, for the purpose of optimizing the structure by thickness variation from a base size of 2 mm. The cross section of the tray frame member is shown in Fig 106. A table of the combined members chosen is in Fig 107 and the results of the sizing optimization tabled in Fig 108, leading to a final weight of 96.39 kg for the space-frame.

Quantifying foam as a box beam core material
According to Foamseal Urethane Technology, ITW

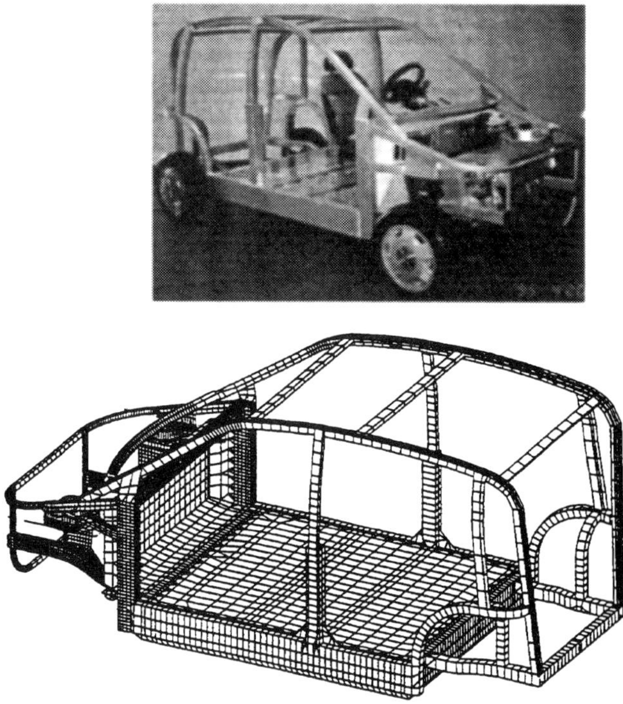

Fig 103: FE model alongside test structure, above

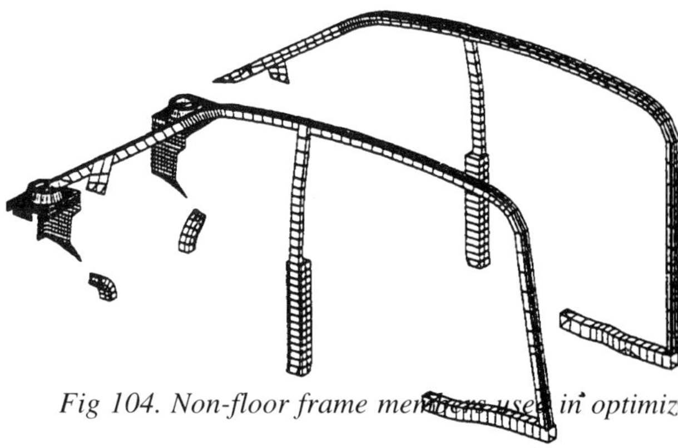

Fig 104. Non-floor frame members used in optimiz

No.	Label	Part Name	Geometry
1	strut	strut	3mm panel
2	A	A-pillar	2mm panel
3	support	front longi. support column	w50 x h50 x t3 extrusion
4	upper B	upper B-pillar	w50 x h30 x t3
5	lower B	lower B-pillar	w80 x h70 x t3
6	rear	rear longi. frame	w60 x h70 x t3
7	roof	side roof rail	t2 extrusion
8	tray1	tray frame	t3
9	tray2	tray frame	t4
10	tray3	battery seperator	t5
11	tray4	floor supporter	t8 panel
12	tray5	tray floor	t3 panel

Fig 107: Members chosen for optimization

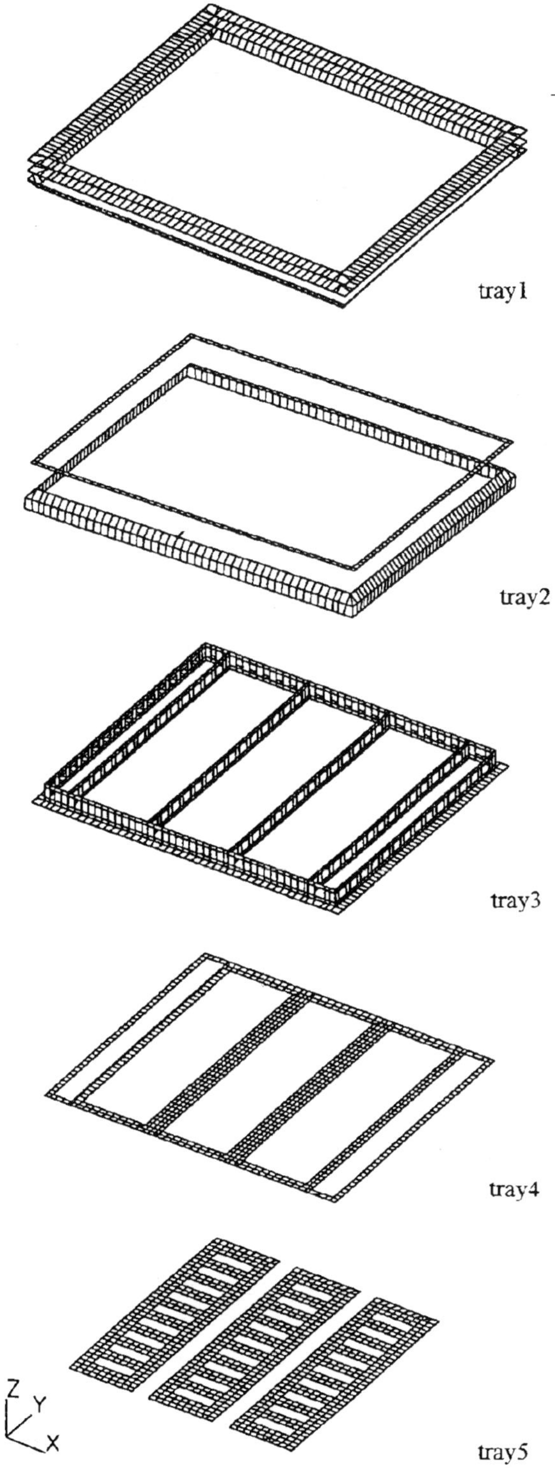

Fig 105: Floor members used in optimization

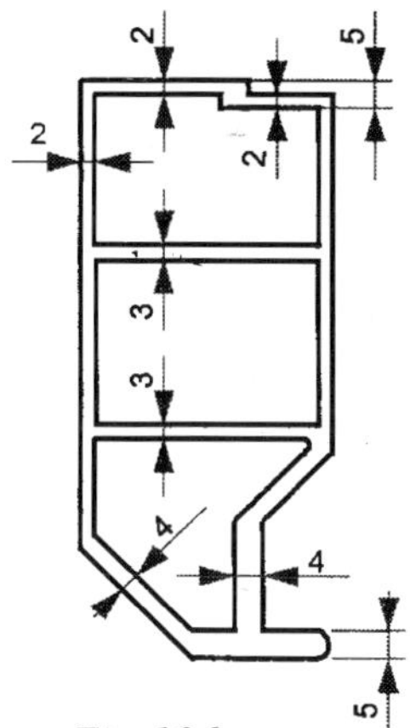

Fig 106: Cross section of floor members

No.	Initial thickness (mm)	Lower bound (mm)	Upper bound (mm)	Final thickness (mm)
1	3	2	5	4.11
2	2	1.5	5	1.50
3	3	2	5	2
4	3	2	5	2
5	3	2	5	2
6	3	2	5	2
7	2	1.5	5	1.55
8	3	2	5	3.03
9	4	2	5	2
10	5	2	7	2
11	8	2	10	2
12	3	2	5	2

Fig 108: Optimization results

and EASI Engineering[14], the results of four-point bending tests replicated by FE analysis, on foam filled box beams, are showing that optimization based on filled areas and foam densities can achieve given weight, cost, strength and stiffness improvement targets in design. In designing for improved roof crush resistance, it is shown that foam filling can reduce pillar sections, reduce metal thickness and sometimes remove reinforcements, that can compensate for the increased weight of the foam. The authors also demonstrate that the FMVSS 201 Head Impact Protection upgrade can be met with a proposed design concept involving foam-filling.

Fig 109 shows stress/strain curves for various densities of PUR foam between two and 30 lb/ft^3. In tests on B-pillar to sill joints, carried out by Ford researchers in conjunction with steel-industry engineers, 5 lb/ft^3 foam increased torsional stiffness by 250 % and a 30 lb/ft^3 foam increased it by 500 % over the unfilled figure. Furthermore, cyclic testing has shown the use of 25 lb/ft^3 foam in this joint was able to delay the onset of fatigue cracks from 10 to 110 % of the life cycle in a typical design.

The filling technique has been in use, particularly for NVH applications on volume production vehicles, since 1982 and excellent adhesion properties is claimed for both electro-coated and painted surfaces, with no foam degradation over the life of the vehicles. Now the use of foams for structural purposes is under active consideration in recent production vehicles have used foam filling of the A and B pillars to preserve roof strength but the relatively high density foams involved have required optimization to ensure weight limits are met.

In the four-point bending test set-up of Fig 110, steel tubes were tested with three foam densities against the base line of an unfilled tube, with the results confirming the effectiveness of the foam in stabilising the thin-walled tube sections against buckling of the skins. Entrapment of the foam by the closed section is seen as an important criterion and the authors doubt whether the foam would be any like as effective in the case of open section beams.

The transient non-linear code LS-DYNA3D was used in the numerical FE analysis with Material Number 57 solid elements adopted for the modelling. End section deformed shapes of the filled and unfilled tubes, obtained in the simulation, are shown in Fig 111 while the force/deflection curves for the test and simulation results are shown in Fig 112. Increase in bending strength and bending stiffness was found to increase almost linearly with increase in foam density.

The B-pillar and rear roof header was modelled for a pick-up truck cab to simulate the FMVSS 2116 roof crush test, Fig 113 showing the undeformed and deformed shapes of the unfilled structure, alongside the model of the cab frame. Several simulations were carried out in order to evaluate foams of different densities and identify critical areas of foam application so as to establish the effect of using reduced

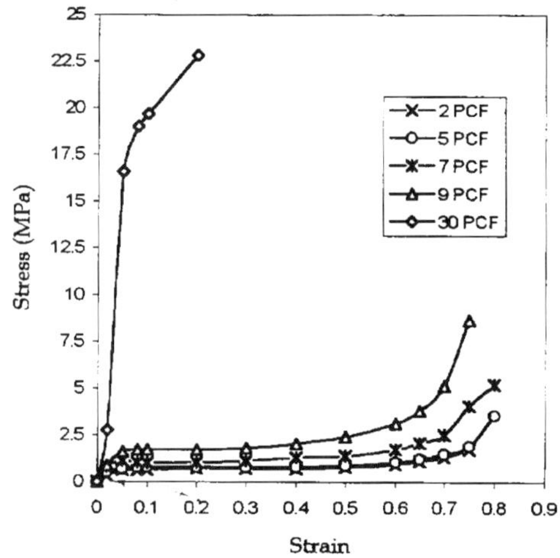

Fig 109: Stress strain curves for different density foams

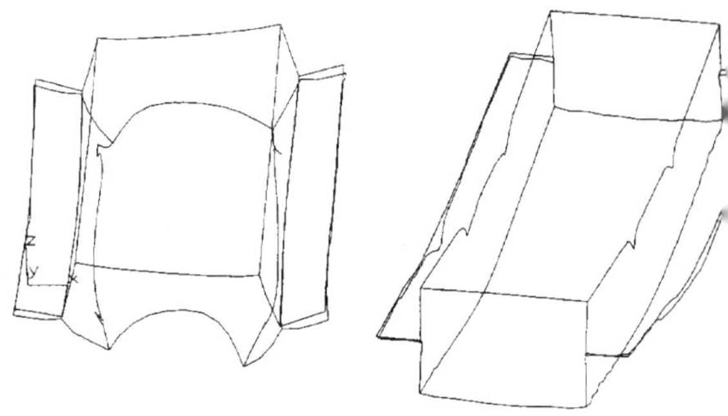

Fig 111: Distorted shape for unfilled, left, and filled structure, right

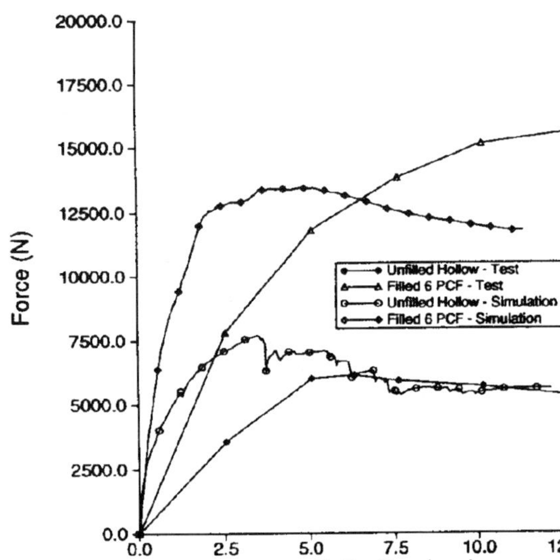

Fig 112 : Deflection curves for unfilled and filled tubes, right

Design Configuration	Maximum Roof Crush Load (N)	Weight of B-pillar (excluding foam) (Kg)	Weight of Foam Filling in each B-Pillar (Kg)	Total Weight of each B-pillar (Kg)
1. Original B-Pillar Section (Unfilled)	51,745	11.18		11.18
2. Original B-Pillar Section; B-Pillar Ring Filled (with 30 pcf Foam)	89,170	11.18	4.3	15.48
3. Original A-Pillar Section (with 5pcf) ; Original B-Pillar Section	52,780	11.18	0.16	11.34
4. Original B-Pillar Section (Filled with 5pcf)	59,200	11.18	0.62	11.80
5. Reduced Section B-Pillar (Filled with 5 pcf)	49,160	10.02	0.38	10.40
6. Reduced Section B-Pillar (Filled with 9 pcf)	52,800	10.02	0.66	10.68

Fig 116 (above): Improvements with foam filling

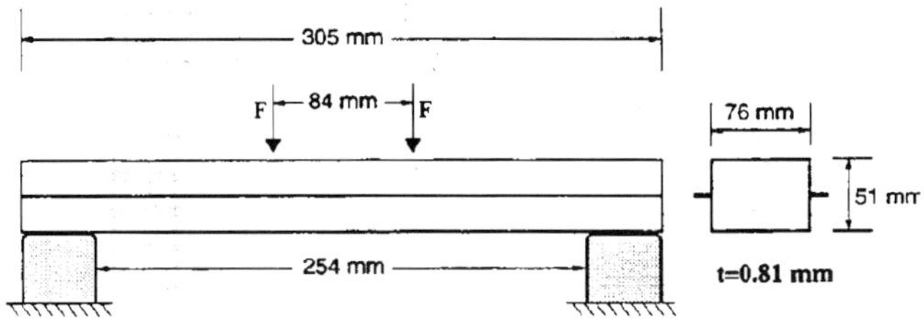

Fig 110: Bending test setup

foam density in certain regions. The effect of a reduced upper pillar section, Fig 114, was also tested and found to outperform the original unfilled section, when filled with 9 lb/ft^3 foam. This allowed energy absorbent foam padding to be added to the pillar interior without loss of interior survival space in the cab, against impact, Fig 115. Improvement to roof crush strength generally is shown in Fig 116.

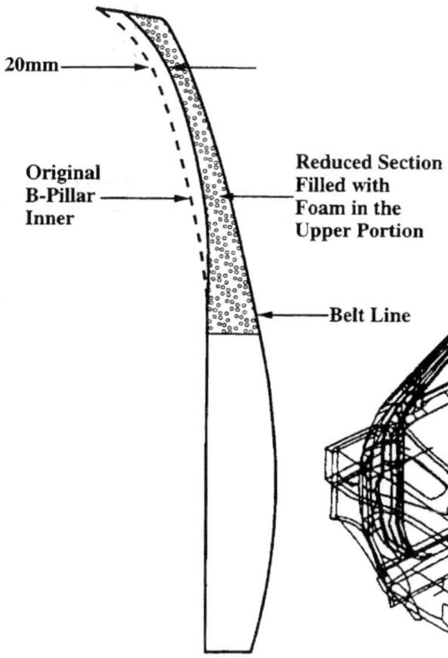

Fig 114: Reduced upper B-pillar section

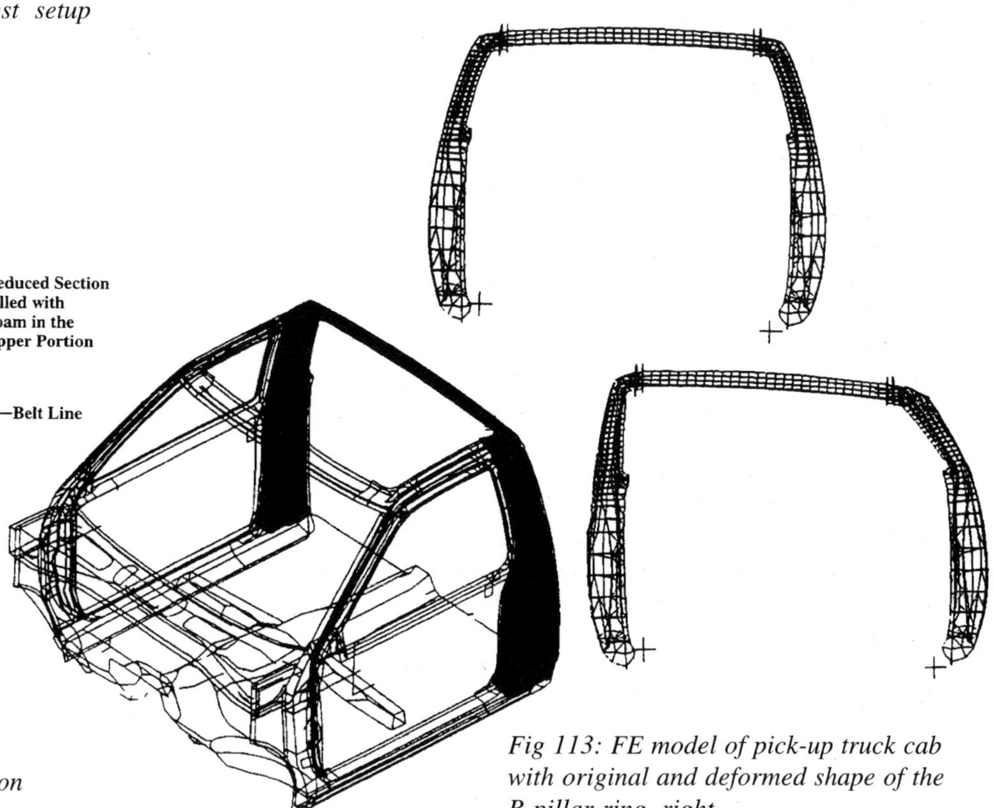

Fig 113: FE model of pick-up truck cab with original and deformed shape of the B-pillar ring, right

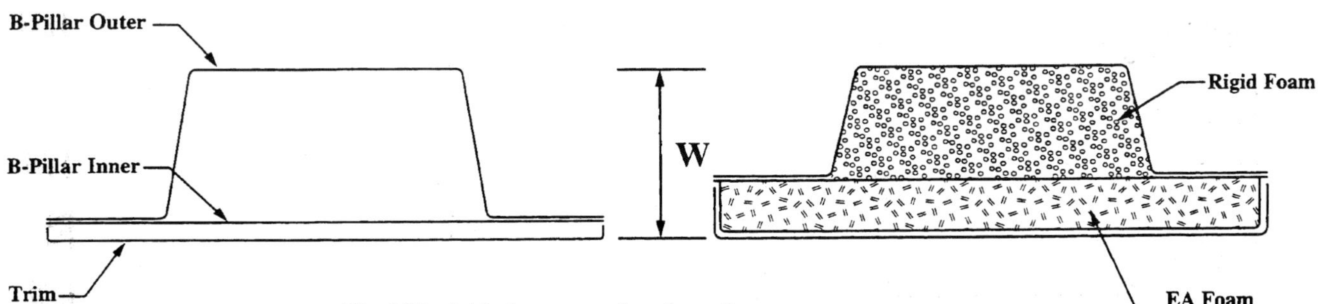

Fig 115: Added energy-absorbent foam

Chapter 12: Specialist passenger vehicles

Buses and coaches; PSV design; PSV systems; low-floor and articulated buses; Advanced bus designs; three-wheeled vehicles; motor cycle engineering; motorcycle developments; developing-country vehicles; light trailers

Buses and coaches

Design of buses and coaches is largely influenced now by the requirement to provide a better passenger environment for new-prospect users of PSVs and touring coaches. The influence of weight reduction and greater safety requirement is also affecting structural configurations.

A number of attempts to produce buses and coaches on a high volume basis have faltered because of the less than enthusiastic reception by operators for 'standardised' designs. The preference by fleet engineers for vehicles designed for local operating conditions, and other specifics of fleet use, ensures a strong future for buses and coaches in the specialist vehicle sector. This lack of volume base also ensures a continuance of separate chassis structures being made alongside integral designs — particularly so with the requirement for separate sizes of body to suit seating capacities for mini-, midi- and full size service buses as well as meeting the separate requirements of double-deckers and touring coaches.

The de-regulation of UK bus services brought about by the 1985 Transport Act has also affected the demands on design, as has the relatively low UK total industry volume. The effect of the single European market is beginning to counteract this particular constraint and running-gear suppliers are increasingly able to tailor mechanical units for the PSV market. There is a realisation at last that buses need no longer be aimed at the bottom of the market, as continental designs have shown how good design can attract passengers right across the social spectrum. Much needed improvements in the service infra-

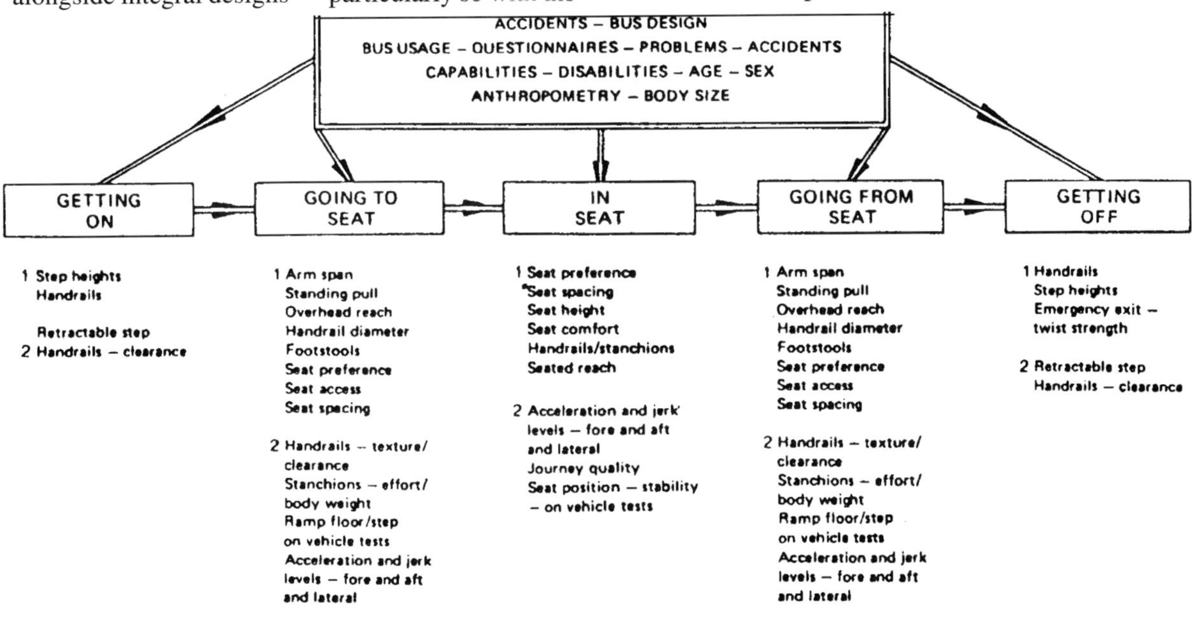

Fig 1: Systemised design approach (TRL)

structure will also yield more passengers and these can be matched by well designed vehicle systems in the areas of entry/exits, seat layouts and ticket issuing systems. Coach comfort, too, is set to be revolutionised by continental practice.

The traditional 'light-weight' sector of the coach market has also virtually disappeared with the 'heavies' (premium coaches) occupying some 90 % of the market. Roll-over protection has also affected design with respect to the large 'picture' windows that were once in vogue; these designs are also less favoured for touring coaches operating in semi-tropical conditions. High floors are likely to become more widespread with the requirement for better passenger visibility and increased underfloor luggage accommodation. In the sub-sector of minicoaches, an increased luxury specification is the norm with the trend away from van conversions as purpose built minibuses are put out to hire at off-peak periods.

Typical understructure construction naturally depends on whether design is chassised or integral. In the former, steel sidemembers of 6 to 8 mm are usually involved while in the latter, 3 to 6 mm would be the norm, at usually not much more than 400 MN/m^2 — to prevent difficulty in forming, drilling and punching operations. For integral underframes of thinner gauge, a zinc or similar coating is usually applied, such coated steels being deep-drawable up to 5 mm thickness. An 18 micron thickness of zinc would require a 50 micron thick layer over it to provide long life protection. Where aluminium alloy underframes are involved, typical alloys are BS 1474 HE 15 and 30 from the 6000 series while for body parts the NS3 alloy from the 3000 series would be typical.

Consultants Booz, Allen and Hamilton, have carried out some key specificational comparisons for recent service buses in which they have highlighted the trend towards making buses more accessible to the aged and disabled. This factor is likely to be one of the main driving influences on future bus design. Some valuable studies by the Transport and Road Research Laboratory have systemised this design approach as shown in Fig 1.

Drive layout for PSVs

Interior layout for full-size service buses can be optimised when a grouped rear engine/drive system leaves an uninterrupted central section. Transmission makers ZF have recently made alternative proposals for engine/drive grouping dependent on their available range of gearboxes, drive-axles and angle drives, Fig 2. A particularly low floor arrangement is seen at (h) for which the engine is seen at (1), the gearbox at (2), rear doors at (3) and angle/spur gear drive at (5,6). The 80 degree angle drive has a 217 mm offset spur gear drive mounted directly to the Ecomat automatic transmission. A drop-axle, type AV130, is featured and the clearance requirements for the other units are provided by this supplier.

These layouts by no means exhaust the possible drive configurations for buses and coaches as evidenced by Fig 3 which shows some recent designs

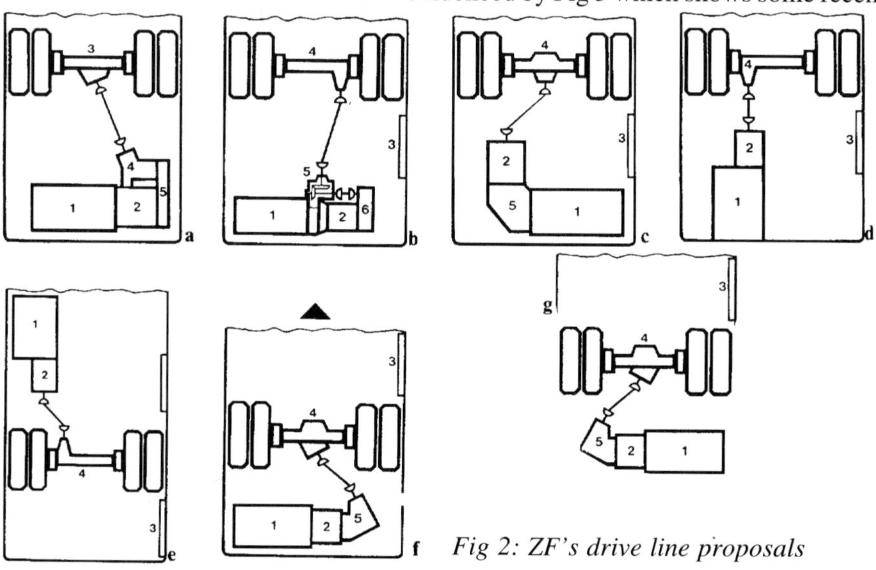

Fig 2: ZF's drive line proposals

in current operation. At (a) Leyland National integral single-decker with fore-aft parallel drive and 'north-south' engine; (b) Leyland Olympian with shallow angle drive, 'east-west' engine and independent front suspension; (c) Mercedes Benz 0405 city bus, again with north-south engine and drive; (d) mid-engined Leyland Tiger coach chassis; (e) MCW Metrobus again with modest angle drive; (f) Volvo's early

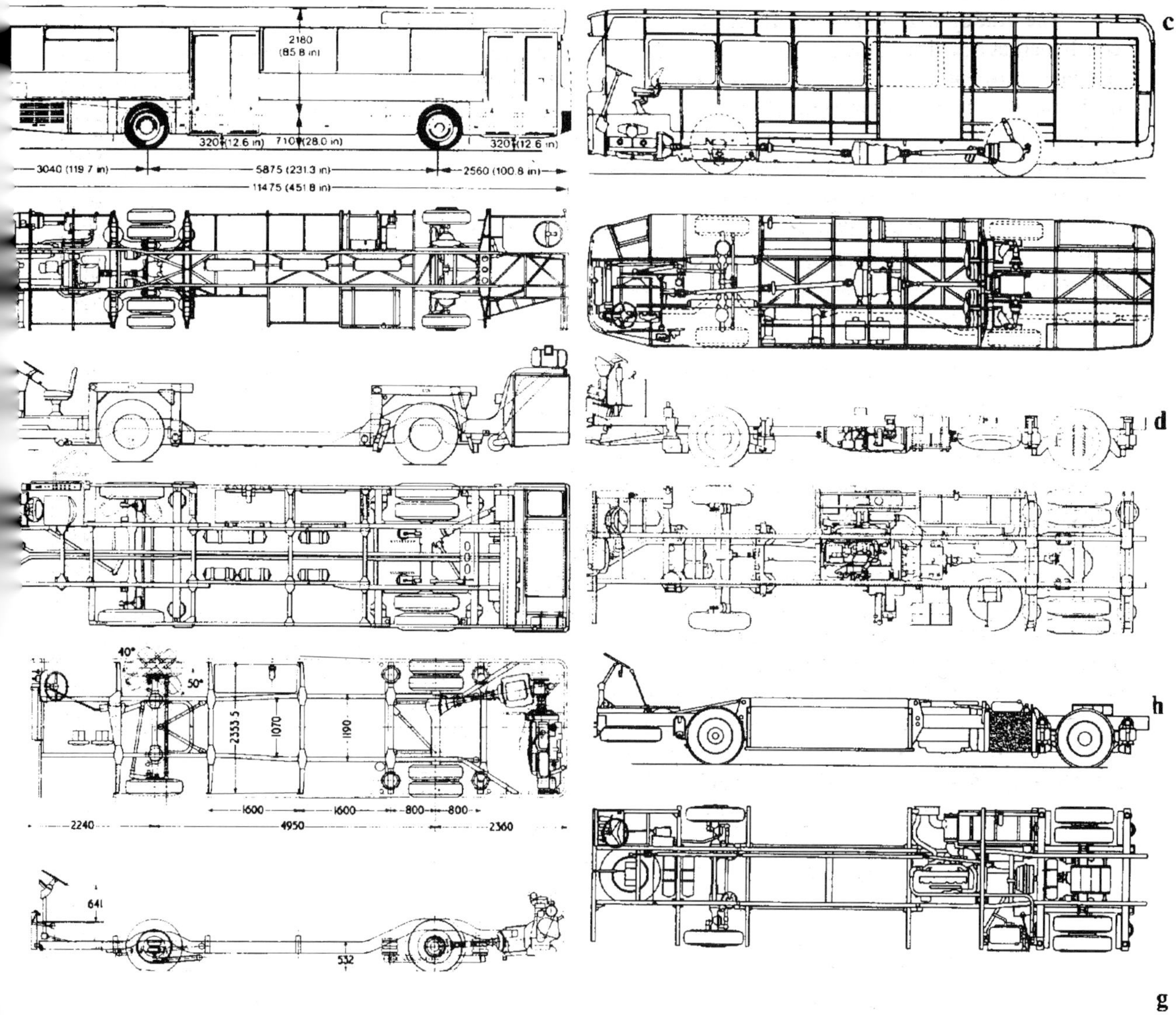

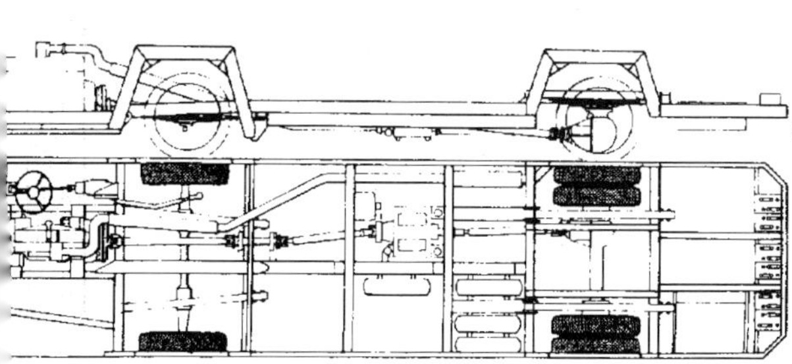

ig 3: Recent PSV configurations

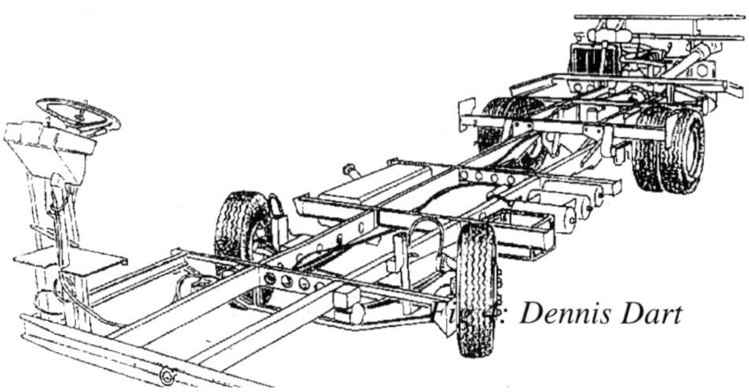

g 4: Dennis Dart

Ailsa bus with front engine and rear drive; (g) the integral structured Saviem SC-10 and (h) the mid to rear engined Dennis Javelin coach with arched chassis. The midi-bus is exemplified by the Dennis Dart, Fig 4, which is of semi-integral construction with a stainless steel frame providing space for 39 seated and 15 standing passengers in its 9 metre overall length.

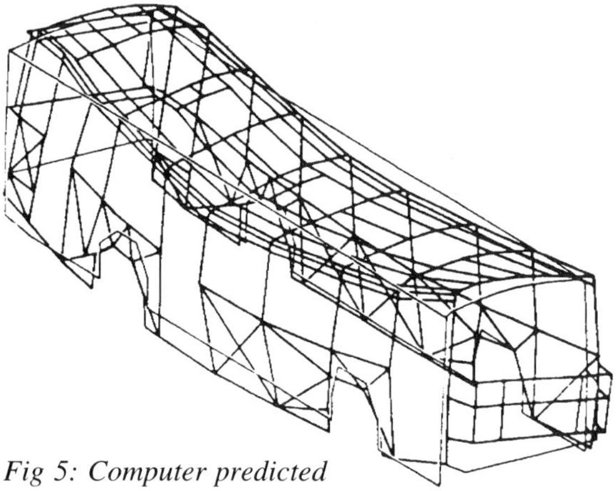

Fig 5: Computer predicted structural vibration

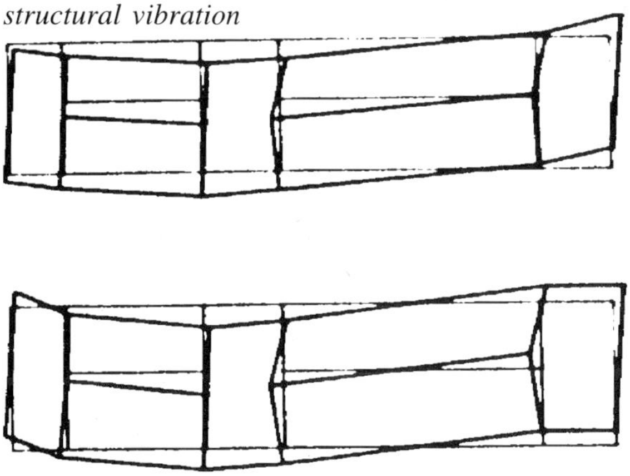

Fig 6: Measured and calculated sideframe deflections

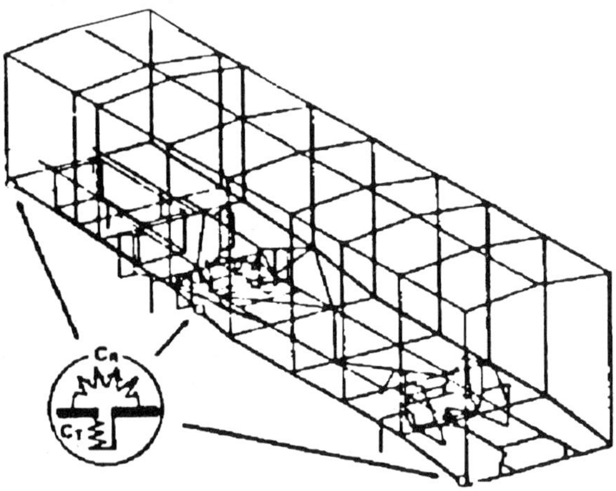

Fig 7: Modelling doorway stiffness

Structural design of PSVs

The construction of semi-integral buses and coaches requires a new level of co-operation between chassis and body builder. DAF Bus have gone as far as formalising a written agreement between the two parties. This has led the chassis-maker into an in-depth programme of structural design involving finite element predictive techniques which allow computer simulation of even the vibration of the structure, Fig 5. The difference in vibration mode shape for different sides of the bus is of particular interest — arising from the effect of door cut-outs in the structure. Fig 6 shows the good conformity between measured and predicted distortion mode of the side frame. To model the stiffness of the doorways, Fig 7, vertical and rotational spring representation were used. Fig 8 shows the magnitude of the bending moment in the roof panel for different frequencies corresponding to speeds driven over a rough road. Fig 9 shows two sideframe structural alternatives and the stress patterns superimposed. In the case of the maximum stress above the rear door, this is reduced from 21.2 to 8.4 N/mm^2 by fitting bonded rather than rubber-mounted windows.

An extension of Pawlowski's Simple Structural Surface method is also available for the design analysis of sideframes. The stiffness of the luggage compartment of a high-floor coach can be seen to effect in the SSS model of Fig 10. The structure is shown here under torsional loading — the load

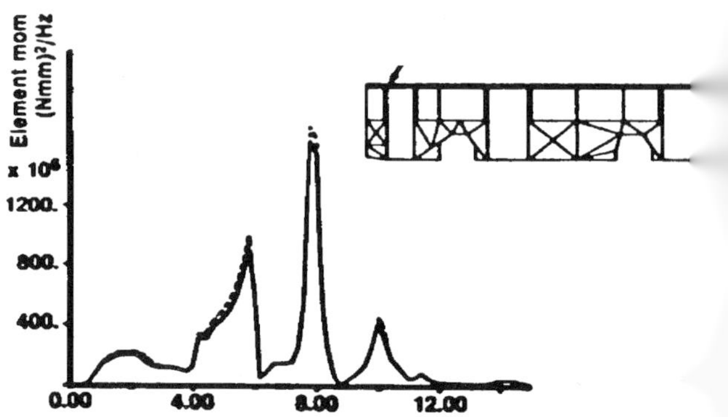

Fig 8: Roof-panel bending moment with respect to frequency in Hz

couple from the front suspension being in equilibrium with the reaction R_f in the sidewall and K_{26} in the front panel, similarly with R_r and $K_{88\text{-}89}$ in the rear wall. High bending moments occur on the window posts in this mode.

A number of bus body builders, including Optare, use the Alusuisse System M 5438 for body construction in aluminium alloy. This has standards for welded and bolted joints, Fig 11. Specially extruded sections, Fig 12, are available in widths up to 0.8 metre. The adaption of this system by Optare to the DAF SB220 chassis resulted in an unladen weight of 970 kg.

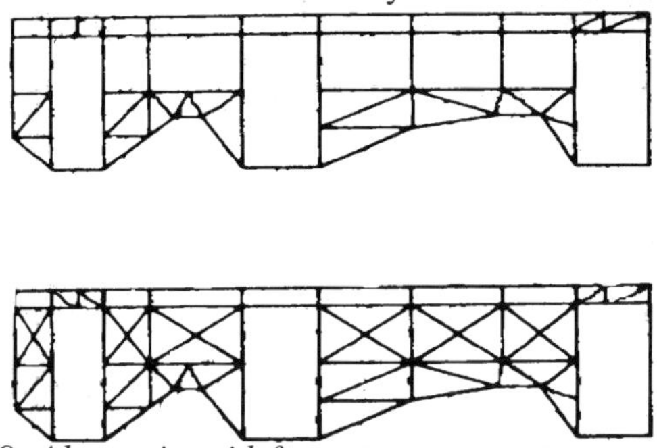

Fig 9: Alternative sideframe structures: above — rubber-mounted windows; below — bonded ones

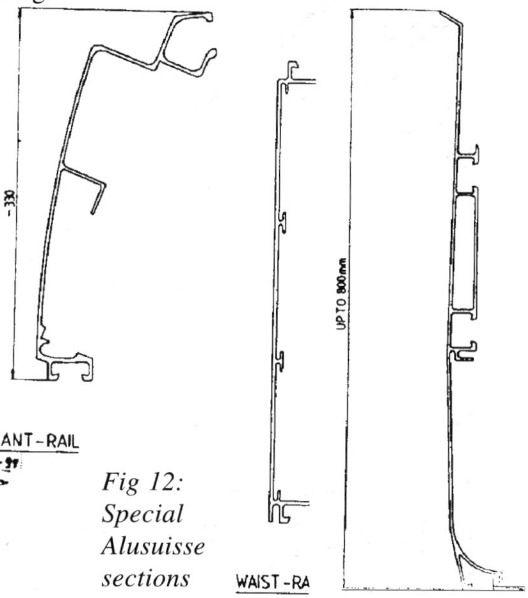

Fig 12: Special Alusuisse sections

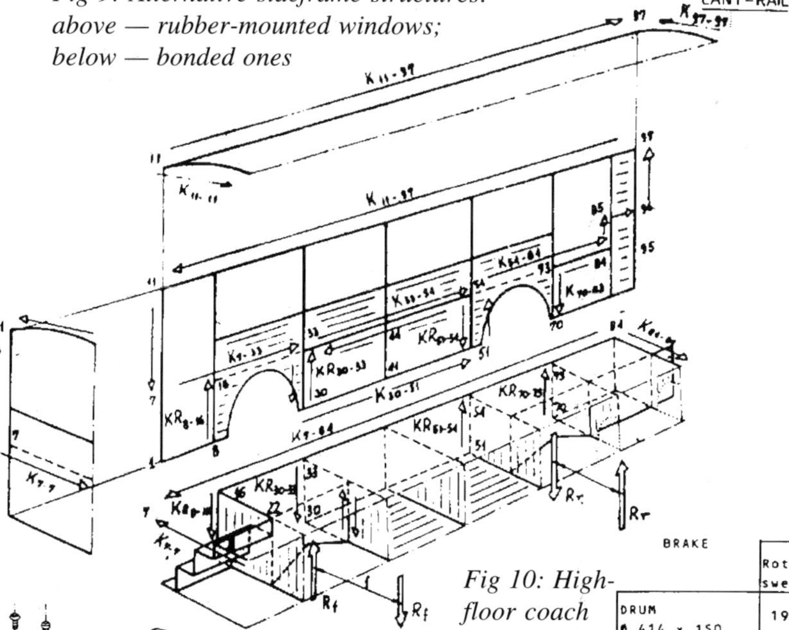

Fig 10: High-floor coach idealisation

Fig 13: Valeo twin-disc brake

BRAKE	SURFACE (cm2) Rotor swept (a)	Linings (b)	Net swept (c)	Total rotor (d)	Net rotor (e)	ROTOR MASS (Kg.)(f)
DRUM Ø 414 x 150	1950	960	990	~4635	~3675	45
DRUM Ø 410 x 220	2833	1728	1105	~8239	~6511	72
SINGLE DISC Ø 392 x 48	1625	374	1251	~4450	~4076	28,5
VALEO TWO-DISC (Ø417 x 30)x2	3306	500	2806	~7916	~7416	38
SINGLE DISC (WHEEL 24,5") Ø 440 x 50	2136	500	1636	~5767	~5267	40

Fig 14: Brake thermal performance compared

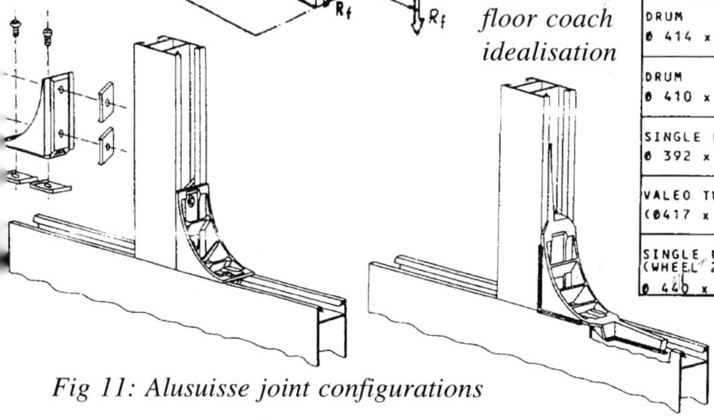

Fig 11: Alusuisse joint configurations

Bus mechanical systems

Braking of large buses also requires specific systems — to meet the EEC directive requiring deceleration of 5 metres/ second^2 for buses weighing more than 5 tonnes or with more than eight passengers. Loss of efficiency up to 20 % of this minimum value is allowed with hot brakes. A number of car categories only involve decelerations of 7-8 metres/second^2 — for comparison. The Valeo company developed their twin-disc brake for such duties, Fig 13, which was designed for homologation on both front and rear axles having load capacities from 6-13 tonnes. Compared with the disc brake installation of a car the problem is one of dissipating heat within a wheel diameter less than twice that of a car but with braking loads up to 10 times higher.

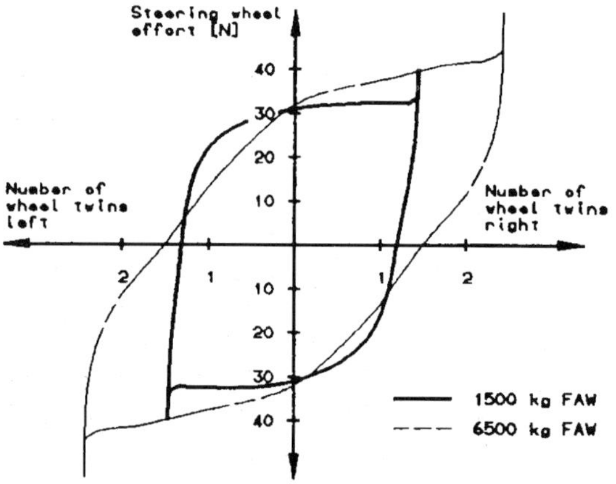

Fig 15: Steering efforts for small and large buses

The company tabulated the characteristics of existing brake systems against their twin-disc design in terms of thermal performance, Fig 14. They gauged thermal performance in terms of the area of swept surface not facing the linings (c). Also considered was the total useful surface for heat exchange from the rotor (e) and the ratio of useful rotor surface to rotor mass (e/f).

Buses also put special demands on steering systems and steering gear makers ZF have shown in Fig 15 the difference in measured effort between a small PSV with power steering from lock-to-lock at manoeuvring speed (full lines) and with the dotted lines a double deck bus illustrating the problem of sensing road feed-back — light and heavy vehicles requiring similar manual work being done in straight ahead driving. As vehicle speeds increase beyond manoeuvring speed, the forces needed to steer the vehicle generally reduce resulting in a yet smaller range of efforts within which the driver can evaluate the road contact. Fig 16 shows road feed-back when driving the light vehicle at low and high speeds straight ahead. This of course is opposite to the ideal situation and leads to consideration of steering-gears which can enhanced in this way.

Newly developed systems even avoid the need to continually adjust gears for reducing backlash and the avoidance of excess energy-transfer into the hydraulic power system. This is achieved by using higher operating pressures and reducing flow rate as well as using suction-control with a radial piston pump, Fig 17.

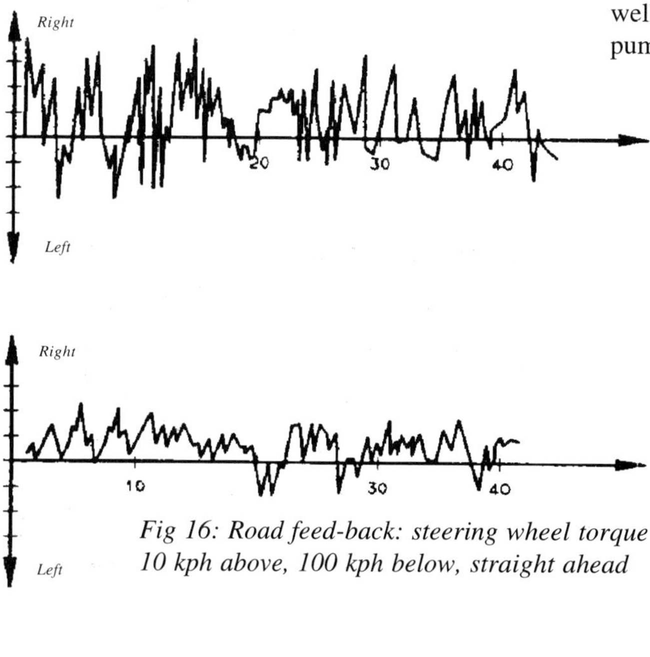

Fig 16: Road feed-back: steering wheel torque (Nm), 10 kph above, 100 kph below, straight ahead

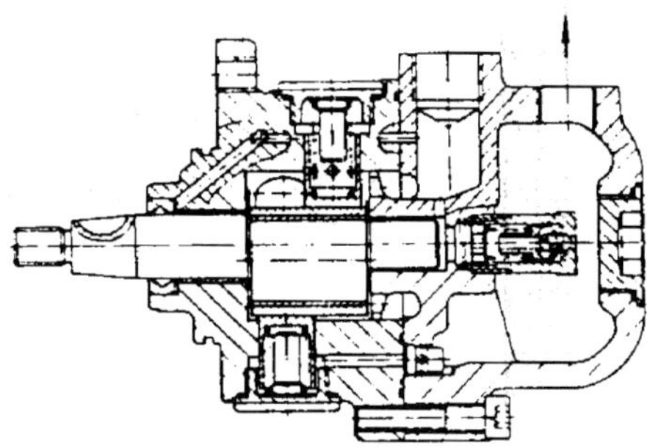

ump

Design peculiarities of PSVs, according to ZF, centre round installation conditions involving long and tortuous input linkages — which can sometimes cause self-centring problems. There are also suspension design effects, both factors contributing to high forces on the steering systems which must be reacted by the vehicle structure without undue compliance. Fig 18 shows a bus gear of the integral power type involving recirculating ball and nut drive. It features hydraulic reaction, with cut-off and limiter valves — the latter being activated only at full lock to reduce loads on the vehicle structure. A vane-type pump, Fig 19, would typically run at 140 bar. An ideal valve characteristic would be as in Fig 20.

The Servotronic variant, Fig 21, first developed for cars helps to solve the feed-back problem by providing electronic sensed and modulated speed-proportional steering. The driver feels safer, and is safer, when the ratio of input to output torque increases with speed, Fig 22 showing the degree of change to the steering characteristic possible.

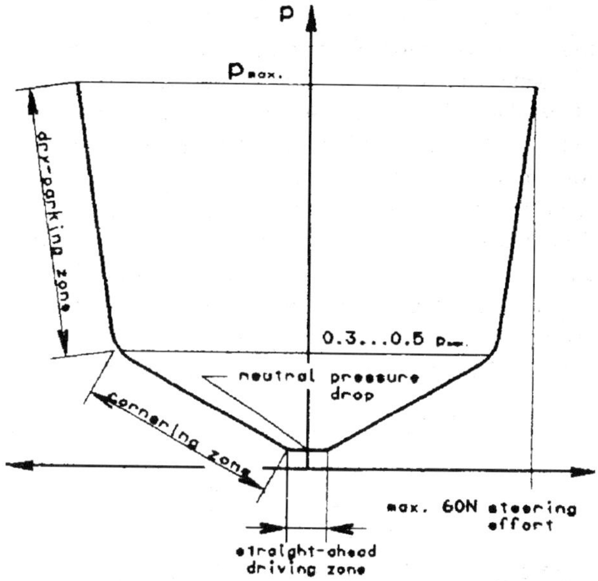

Fig 20: Ideal valve characteristic

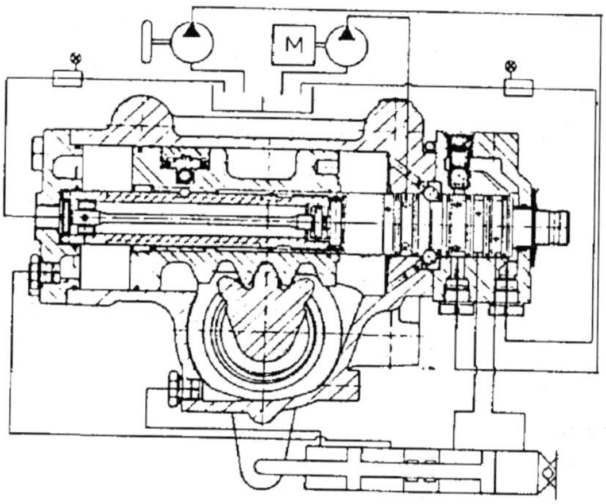

Fig 18: Integral power steering gear

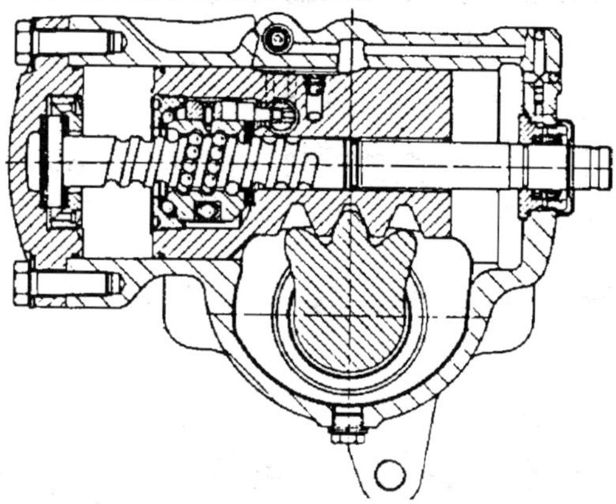

Fig 21: Servotronic steering gear

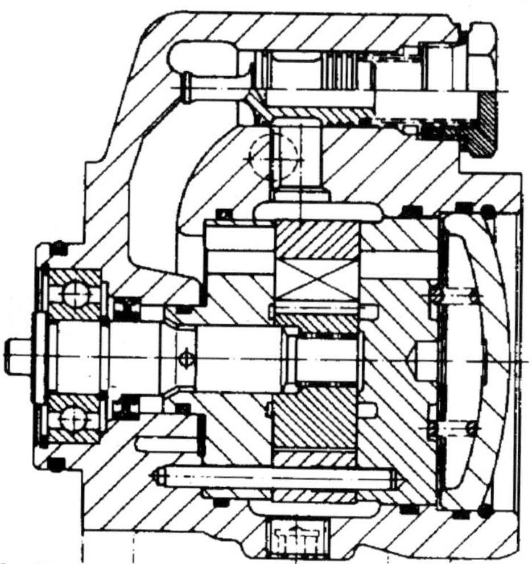

Fig 19: Vane-type pump

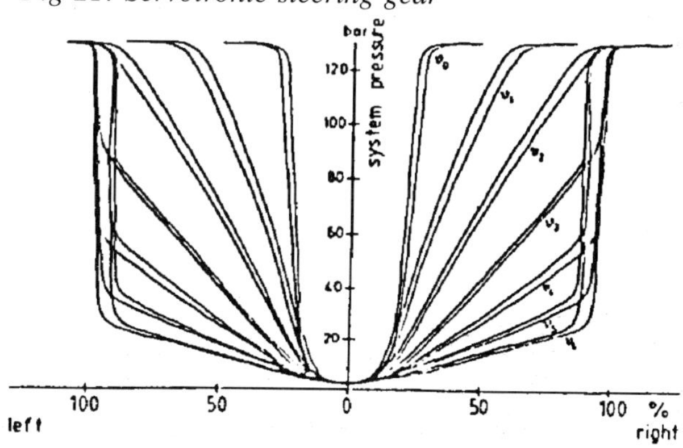

Fig 22: Modification to steering characteristic

Developing low-floor and articulated buses

Despite initial hesitation by UK and other more conservative operators the impetus for low-floor bus production, which began in Germany, is extending to manufactures in many countries worldwide. The latest designs involve electric drive-lines and share common design platforms with trolley buses.

The Union of Public Transport (UITP) has defined the low-floor bus as one which, between at least doors 1 and 2, has a floor low and one which is level enough to remove the need for steps in the aisle, both between these doors and in the vicinity of the doors, with step heights as defined in Figs 23-25. Historically solutions were based on buses having low-floor area between entrance and exit doors, with horizontal engines on the left hand side of the vehicle rear-end, Fig 23. To enter the rear part of the bus passengers had to negotiate either an additional step or floor sloping at around 12.5%, the seats being placed on a podium. Another solution was to place the engine between the axles as seen in Fig 24. While the low-floor extends over the full length of the vehicle the seats on the left side are mostly podium-mounted and aisle widths are too narrow for manoeuvring a wheelchair. Third solution was to place the engine and drive (diagonally) at the rear of the vehicle which results in both loss of passenger space and driveline complication. Low-floor buses are, of course, also complicated by having larger wheel-arch intrusions around which seats have to be fitted.

Future developments include removal of the rear axle by incorporating electric drive-lines and the design of three separate modules for steering, passengers and drive. Either the driver's cab is in the conventional position on the steering module or wheels are positioned at the front corners of the vehicle with the cab between them, which complicates passenger flow with respect to the driver. There are also proposals to place the drive module, containing engine, generator, wheels and hub-motors, at the rear of the vehicle but the resultant extra-long wheelbase makes meeting a normal (12.5 metre) turning circle more difficult.

However, despite these difficulties, sales in Germany of low-floor buses have already reached 80% of the total market. Beyond Germany the products are receiving wider acceptance, too, despite cost

Dennis Dart SLF is the best selling UK low-floor bus, seen here with a 42 seat East Lancs Spryte body

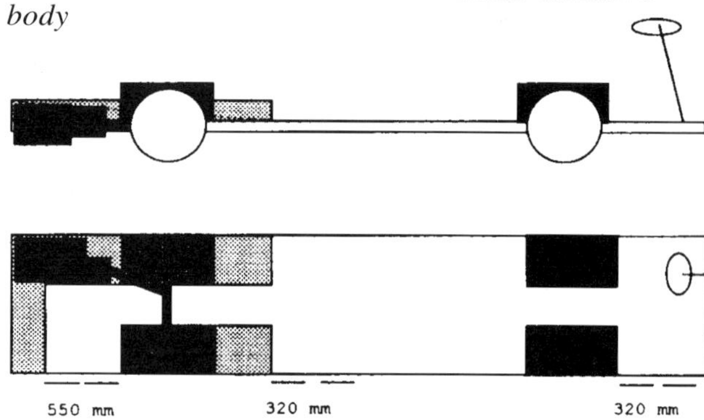

Fig 23: Rear longitudinal engine

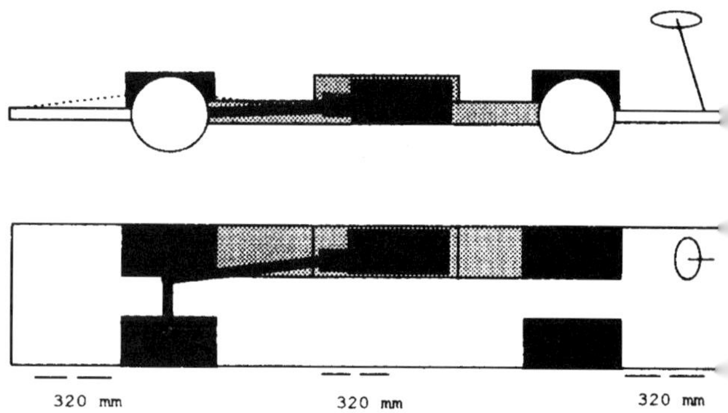

Fig Central longitudinal engine

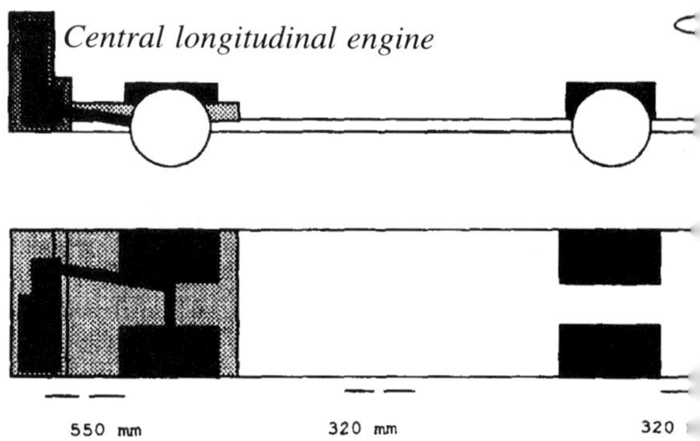

Fig 25: Rear transverse engine

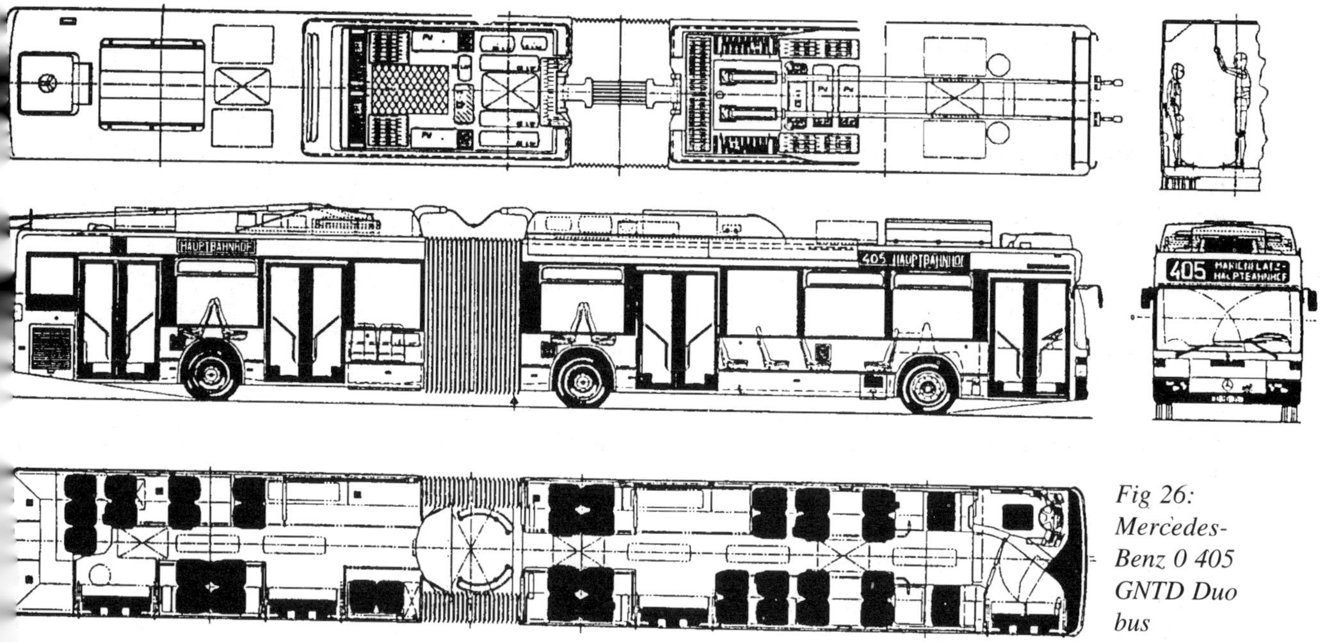

Fig 26: Mercedes-Benz 0 405 GNTD Duo bus

fears. MAN is now supplying its NL202 underframe for bodying by Berkof in Holland and a low-floor range has been introduced in Austria via the MAN subsidiary OAF Graf & Stift. Now many indigenous manufacturers are active.

Electric drive duo buses

According to researchers at Mercedes-Benz[1], a move from central electric motor drives to wheel motors has considerably improved the packaging of the company's dual-drive buses. Passengers previously had to surmount a step in the gangway to clear a central motor with 710 mm ground clearance. Because the drive systems can be used with both duo (rail or overhead contact-line) and hybrid drive buses, wheel motors can enjoy production economies of scale. The spread of drive throughout many axles on multi-wheeler and/or articulated buses means safer handling can be realised since reduced tractive force per wheel increases the tyres ability to develop side force in cornering. Jack-knifing tendency is also reduced for articulated configurations and a simpler

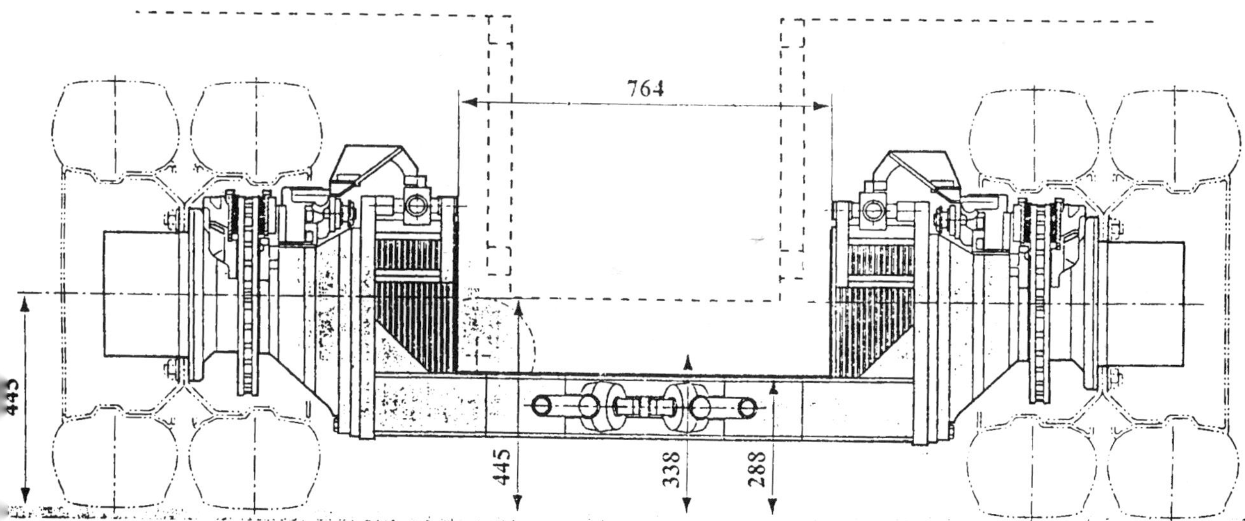

Fig 27: ZF EE wheel drive and AV132 axle

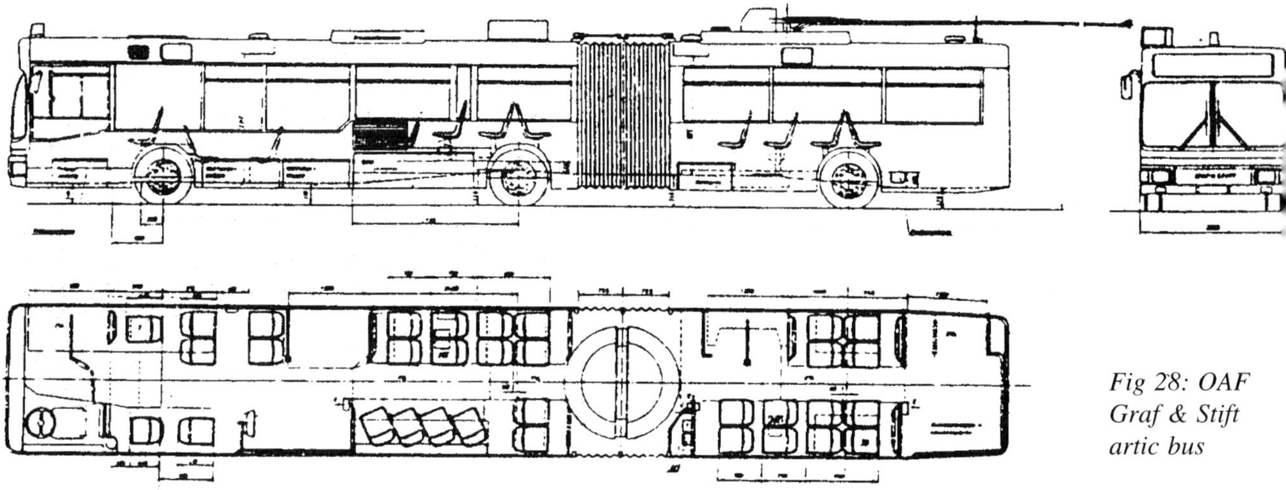

Fig 28: OAF Graf & Stift artic bus

swivel joint is possible. In using asynchronous motors there is a wide band of constant maximum torque over the speed range to assist in hill-climbing. At a certain point in the torque speed curve a region of constant power is manifested with decreasing torque as speed rises, reducing the need for change-speed gearboxes and avoiding ratio steps.

The company's 0 405 GNTD low-floor articulated duo bus, Fig 26, which can carry 156 people, has a flat floor over the length of the vehicle at the door step height of 370 mm. Hub motors of 50 kW are connected to the wheels of the drive axles at the mid and rear stations. Planetary gears reduce the 11 000 rpm motor speed down to wheel speed. The driving heads carry emergency disk brakes which allow the vehicle to be halted in the event of failure of the motor or its regenerative braking system. The twin driving wheels are flange-mounted to the drive motors, Fig 27, which are connected to one another across the vehicle by axles which contain the attachments for the suspension mountings. Power supply to the motors comes either from the 220 kW, at 2200 rpm, diesel generator set, on the vehicle, or via contact line through an automatic current collector on the vehicle roof. All auxiliaries are electric driven so no PTOs are required on the diesel engine.

Low-floor trolley-buses

The Austrian trolley-bus maker OAF-Graf & Stift has tackled the special problem of designing articulated buses, Fig 28, for operation in ancient cities having narrow roads and tight bends between them. They showed at the last EAEC conference that artic buses of 2.5 metres wide and from 16 to 18 metres long can be made to suit such conditions with the help of steered third axles. Fig 29 shows the improvement obtainable with such systems against conventional layouts. Positioning the traction motor ahead of the driven second axle. However, artic buses are usually arranged as pusher vehicles, providing minimum loss of passenger area in low-floor designs, a criterion which had to be matched by the new bus with steerable back axle. Another criterion was the accommodation of all electrical equipment on the chassis so that it could be accessed through side-flaps in the bodywork, instead of the normal roof-mounted arrangements. This resulted in the overall configuration seen in in the figure. Automatic double doors have width of 1350 m on this 138 passenger vehicle for operation in Salzburg. Floor is level from first to second doors but step heights are 317 and 340 mm for first and second doors respectively. Between front axle and second door all seats are at floor level, without step; from the second door the floor rises slightly over the second axle. Between second and third axle it descends to low-floor level at the fourth door. Between second and fourth doors the seats are on small steps. At the third door, the only one with two steps, which are 340 and 200 mm.

MAN running gear is used generally; all axles are air-sprung and the drop centre front axle is located by drag links and a panhard rod. The drive axle is derived from that on 4-wheel low floor buses by MAN but fitted in reverse configuration so its flange faces the drive motor concealed under the right podium behind the second door. It is suspended by four widely spaced air springs . The trailer axle is steerable like the front axle but its angle is constrained by the articulation angle between front and rear parts of the bus. The steering rod of the third axle is connected by a drop arm, and from there by a second steering rod to a linkage point fixed to the rearward most crossmember of the towing vehicle. The gearing of steered wheel motions between the two axles can be altered by varying the drop arm length to suit the operation, Fig 30.

A hydraulic pump for steering power is fitted at the front of the drive motor and alongside the motor is a large pulse-width-modulator used to regulate power output of the motor, by controlling voltage and frequency. A transformer to convert the 600 V at the traction motor to the 24 V required of the vehicle subsystems, is in a podium just ahead of the third axle, on the trailer portion of the vehicle. Electric air heaters beneath the seats are used to warm the passenger areas.

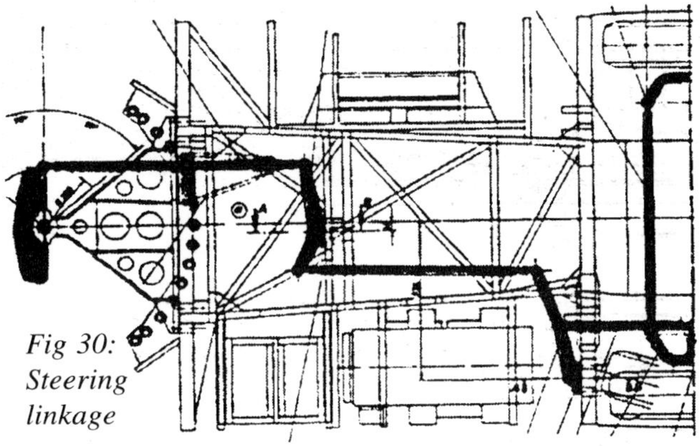

Fig 30: Steering linkage

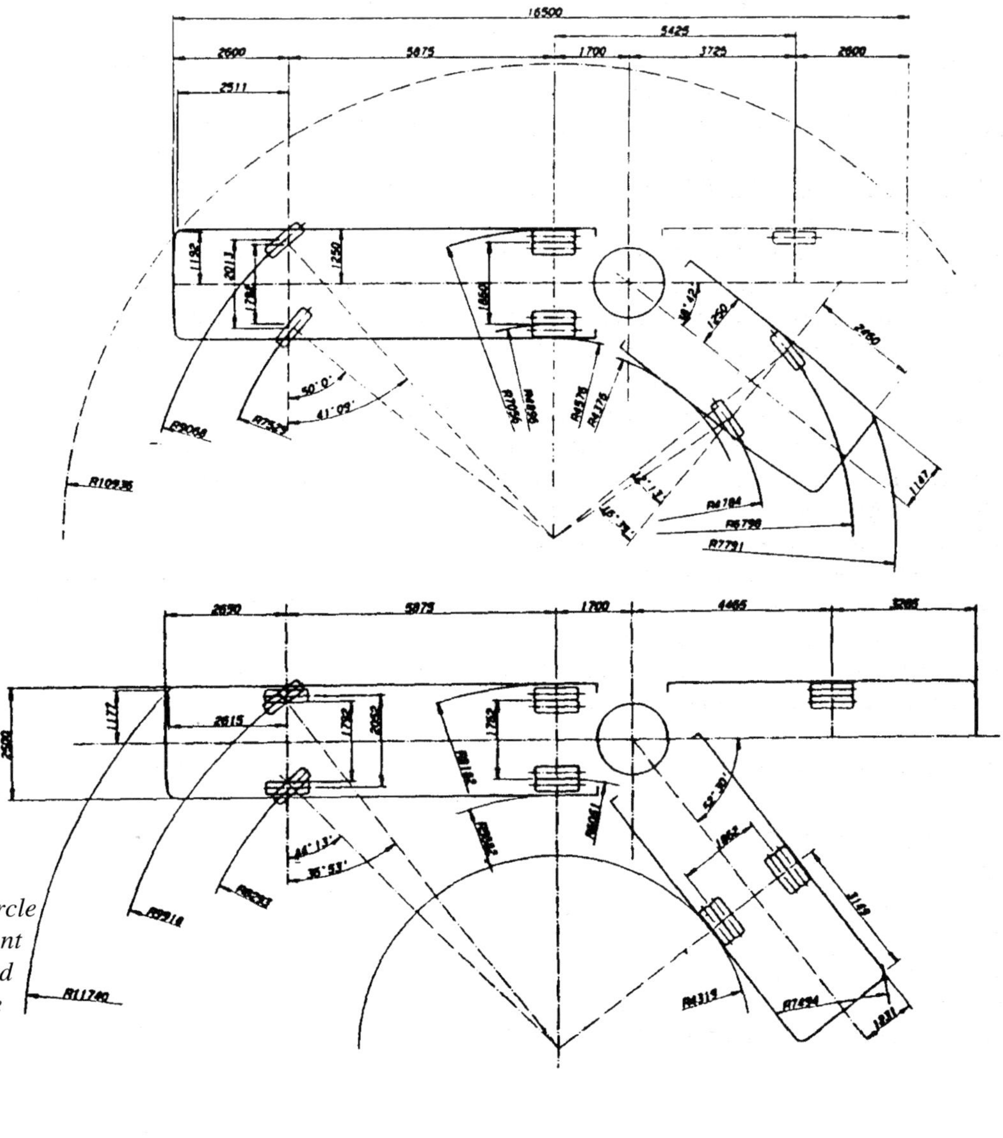

Fig 29: Turning circle improvement with steered trailer axle

Developing a current rigid/artic bus family

One of the first fruits of the combined engineering strengths of the former Leyland Bus product development team with the manufacturing/marketing strengths of Volvo was the B10L. The chassis is used for both two-axle and articulated low-floor city buses, solid modelling, Fig 31, being exploited to refine the basic requirements set out by UITP for these types of vehicle. According to Volvo[2], step entry was fixed at 320 mm for both first and second doors of the rigid model as well as the third door of the artic. The height could be reduced to 230 mm by the kneeling facility provided by the air suspension. The rear exit door, however, constrained by the use of a mechanical driveline required the use of twin steps to reach the 550 mm deck height, which could be lowered to 460 mm by suspension kneeling. The layout required approach and departure angles to be kept at a maximum of seven degrees, using 275/70R 22.5 tyres. Good passenger flow dictated a 950 mm minimum gangway width at the front and 550 mm at the rear. Target height for the turntable on the artic version was 410 mm, with maximum slopes from it of 3 % lateral and 10 % fore-aft.

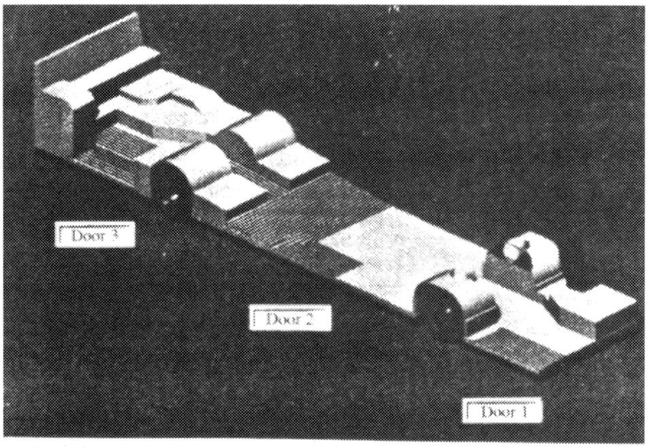

Two-axle

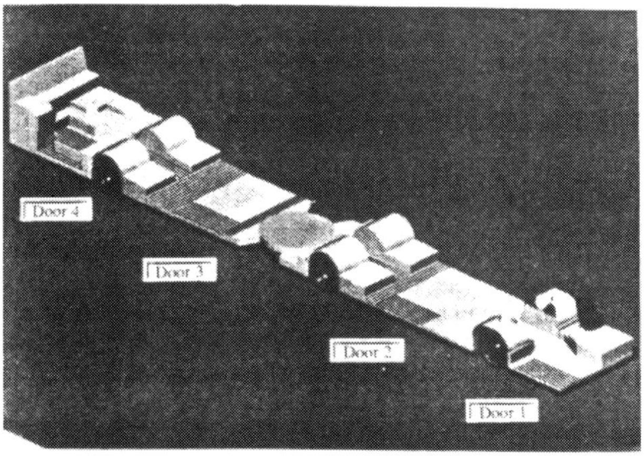

Artic

Fig 31: Solid modelling

Table 1

Exhaust constituent	CNG g/kW h	Diesel Euro II g/kW h
Nitrogen oxides (NO_x)	2.50	7.00
Particulates	0.05	0.15
Carbon monoxide (CO)	0.30*	4.00
Non-methane hydrocarbons (NMHC)	0.20*	
Methane (CH_4)	1.80	
Total hydrocarbons (THC)	2.00*	1.10

* With a catalytic converter.

Engine and automatic transmission are placed behind the rear axle in horizontal in-line position, with the cooling system at the right hand corner of the vehicle, Fig 32, and two moulded plastic fuel tanks in front of the front-wheelarches (or in the bays immediately behind the coupling in the case of the artic). Front air-suspension now features an independent layout, with disk brakes, while the rear involves four axle-mounted suspension beams located by upper and lower parallel radius rods. A special turntable was designed for the articulated version by Hymer and Volvo. Turntable articulation angle is controlled electronically by a system which monitors angular velocity and acceleration of the turntable together with vehicle speed and braking levels, and optimizes the damping for all conditions.

Chassis frame structure, Fig 33, was designed in two modules for the two-axled vehicle, the front one being a combination of formed sections and tubes together with structural wheelarches. The rear module also incorporates shear panels; between them and adjustable length centre section is fitted by the bodybuilder to meet operational wheelbase requirements. Five modules are involved in the articulated configuration, the middle three differing from those of the rigid vehicle, in accommodating the centre axle, turntable and rear turntable abutment.

Future development of the chassis could incorporate multiplexed wiring and adaptive diesel engine controls to suit urban operation. Natural gas engines have also been developed for the B10L which give the emissions benefits over diesels seen in Table 1.

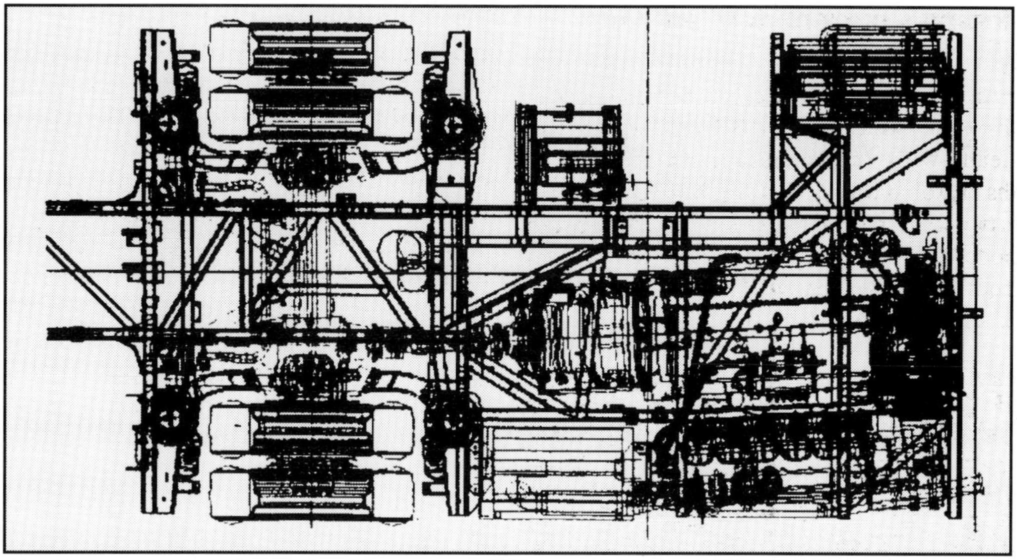

Fig 32: Cooling system location

Advanced bus designs

A joint venture between MAN and Voith has resulted in the NL 202 DE low floor concept city bus, Fig 34, designed to carry 98 passengers at a maximum speed of 70 kph. No steps are involved at any of the entrances which lead directly to a completely level deck height of between 317 to 340 mm. The rear-mounted horizontally positioned diesel engine allows fitment of a bench seat at the rear of the bus; it drives a generator with only electrical connection to the Voith transverse-flux wheel motors which drive the wheels through two-stage hub-reduction gear sets, Fig 35. The diesel is rated at 127 kW and the generator at 135 kW; the controller, Fig 36, is of the IGBT converter type and also developed by Voith. It provides a differential action to the wheel motors on

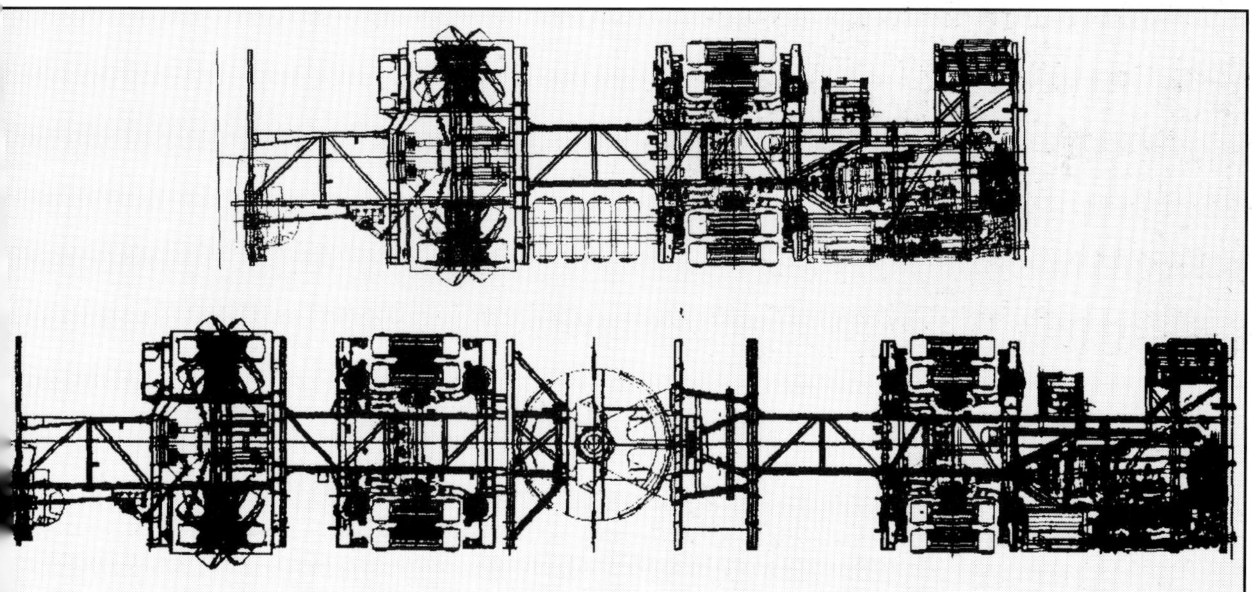

g 33: Chassis structure comparison

cornering. Permanent magnet synchronous wheel-motors are rated at 57 kW and have a maximum speed of 2500 rev/min; see Table 2. The bus is 12 metres long and has water-cooling for its generator, converters and wheel motors. As well as providing virtually jerk-free acceleration, the drive system is seen by MAN as providing the possibility of four-wheel-drive on future articulated buses to improve traction and stability in slippery road conditions. The term transverse flux motor refers to the means used to guide the magnetic flux in the stator; this is new to inverter-supplied PM types and involves a novel collector configuration. Double-sided magnetic force generation is also new and involves a patented double air gap construction having high idling inductances and force densities up to 120 kNm/m^2, with relatively low losses. A new control process permits operation of the motor in a field-weakening type mode, in spite of PM excitation. The generator is almost identical in concept but involves no field weakening. Each has concentric construction of permanent magnets, rotor/stator soft-iron elements and stator winding; *see below*. Armature elements are U-shaped cut strip-wound core sections, embedded in the ring-shaped supporting structures of inner and outer stators. Each core surrounds the windings and forms a stator pole with its cut surfaces facing the rotor. The latter is pot-shaped and positioned between poles of the outer and inner stators. In the stator pole region it comprises magnet and soft iron element while in the winding region a ring of GRP serves as the connecting element. The inverters supply the motors with sinusoidal currents and voltages until the nominal operating point is reached; operating frequency is 10 kHz. In field-weakening mode the induced voltage exceeds intermediate circuit voltage and only square wave voltages are supplied to the motor. Power output then remains constant and the operating frequency equals the fundamental motor frequency. A large speed ratio, 1.5:1, is thus possible.

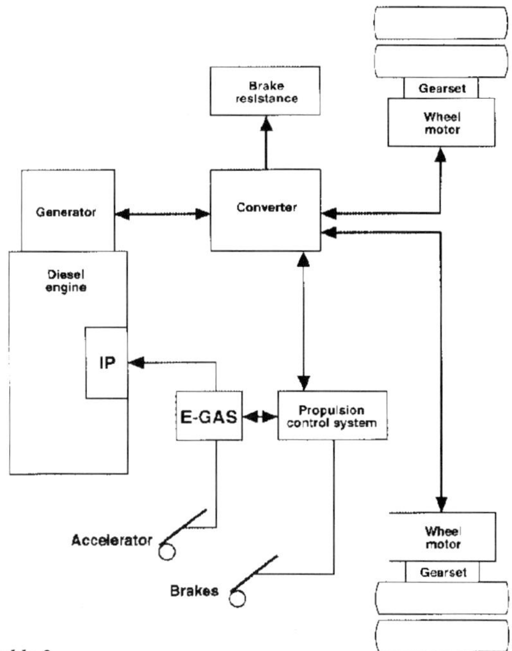

Table 2

	TFM wheel motor	TFM generator
Power	57 kW	135 kW
Rated speed	735 rev./min.	1,750 rev./min.
Approx. max. speed	2,500 rev./min.	2,400 rev./min.
Max. fundamental frequency of stator	1,350 Hz	-
Rated torque	740 Nm	740 Nm
Approx. max. torque	1,050 Nm	740 Nm
Approx. torque conversion	1:3	
Power/weight ratio	1.8 kg/kW	0.9 kg/kW

Fig 34: MAN/ Voith concept city bus

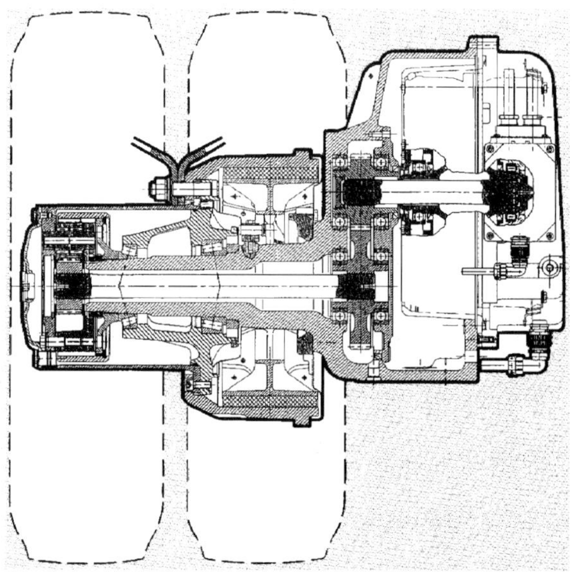

Fig 35: Wheel motor

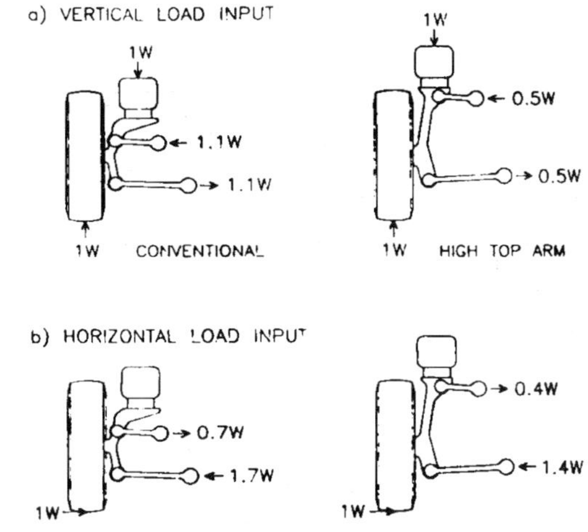

Fig 37: Wishbone spacing

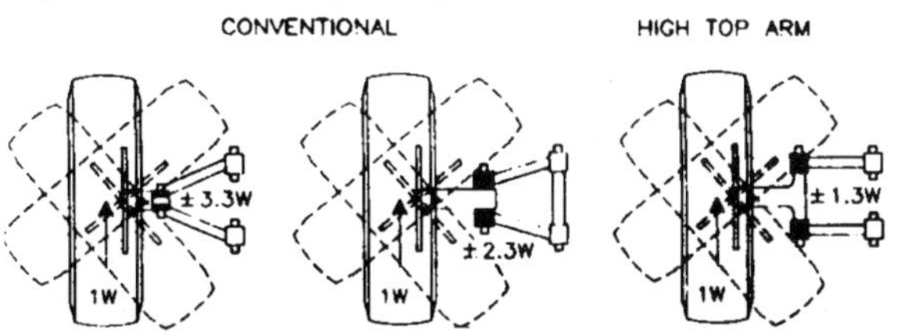

Fig 39: Torque development at the king-post

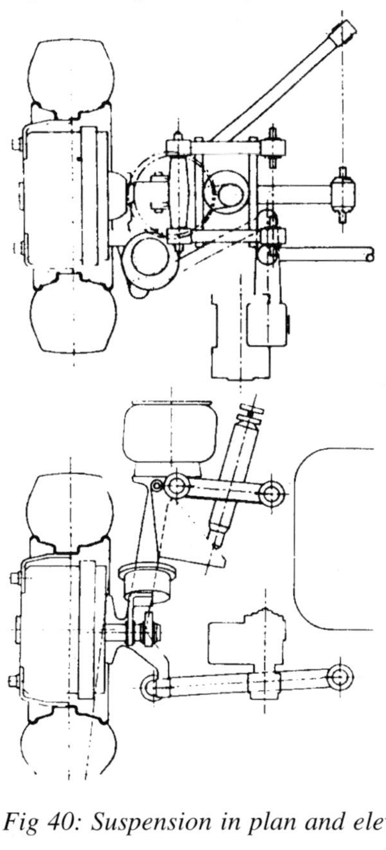

Fig 40: Suspension in plan and elevation

Bus system developments

Consultants Capoco Design[3] have provided a particularly useful examination of independent suspension systems, questioning the apparent lack of read-over from car to bus technology. A review of Capoco's involvement with the Duple 425 integral coach showed some of the advantages to be gained by intelligent use of IFS. Wide separation of the control arms for the wishbone layout was a key aspect, Fig 37. In achieving good handling and high inherent roll stiffness, reduced anchorage bush loadings was another key feature, Fig 38, affecting vertical and horizontal loads of braking and cornering. The ability to thus use softer bushes meant that, as well as reducing 'parasitic' spring rate, improved vibration isolation resulted. Further advantage came in the steer control of the king-post, Fig 39. This is because, for any system within the wheel, any move inboard of the outer bushes is accompanied by a matching increase in offset, from the disturbing effect of the braking load; in this case, due to the 90 degree sector, the effect is self limiting. Since the suspension bushes are spread above the tyre on lock, loads and stiffnesses can also be optimised, the author argues.

Wide spacing of the mounts also means direct assembly on the vehicle frame is possible without shimming, caster and camber are tolerant of slight frame build variations. Soft mounts are particularly useful for designing in longitudinal compliance, one lesson learned from car technology. Responses down to 8 Hz have been reached in car suspension compliance, multi-link geometry being used to careful control wheel motions at this low flexibility. On the 425 a level of 15 Hz was reached as the higher wheel out-of-balance forces on heavier vehicles risked the possibility of otherwise passing through horizontal resonance. Front and plan views of the suspension, Fig 40, show the anti-dive geometry built in. Thus

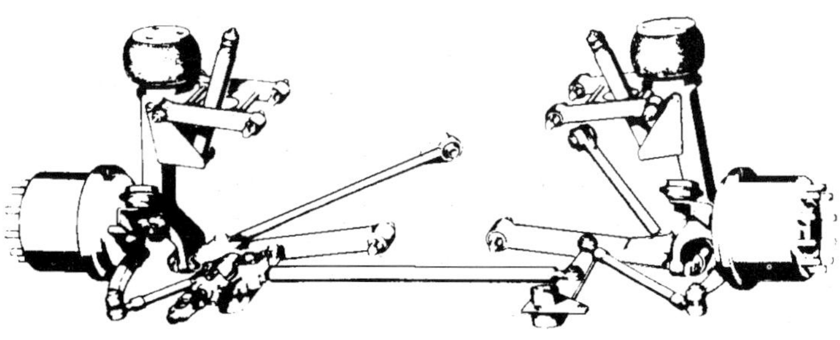

Fig 38: Anchorage bush loadings

far, the author reports, records have shown tyre mileages equal to those obtained with beam axles.

Researchers at Renault Vehicules Industiels have given[4] interesting design background to the R 312 city bus concept, Fig 41, from their company. A fully flat low-floor accessed by three wide entrances characterises the layout. A 1530 mm long single floor module is 570 mm high. The body comprises seven equi-length 'bays' involving seven glazed modules on one side matching four glazed and seven door modules on the other; these form part of 12 metre long sideframe modular assemblies of which are handled in single units. Similar modules exist for roof, floor and ends, as in Fig 42. The so-called rear power package unit, Fig 43, involves a central symmetrical driveline to a portal axle suspended by trailing wishbones.

The authors argue that angle drives, and conventional H-frame suspension layouts, otherwise inevitably cause ramped rear floors — also causing asymmetry in the major modules between RH and LH drive variants. A T-drive transfer gearbox is used to obtain the necessary drive configuration for the inclined central propeller shaft. This is the lowest point of the package — which determines the 8 degree departure angle. The engine is conventionally connected to the automatic gearbox but a special drop-gear unit is flanged to the gearbox casing, a short external prop-shaft linking its output to the transfer box. An aspect of the very low rear floor is the inability to provide upper vee-brackets to stabilise a conventional rear suspension configuration. Hence the use of the innovative trailing arm layout. A special linkage withstands side and longitudinal loading. This is made up from two L-shaped members mounted by anchorage bushes to the frame crossmember, Fig 44.

Fig 42: Structure modules

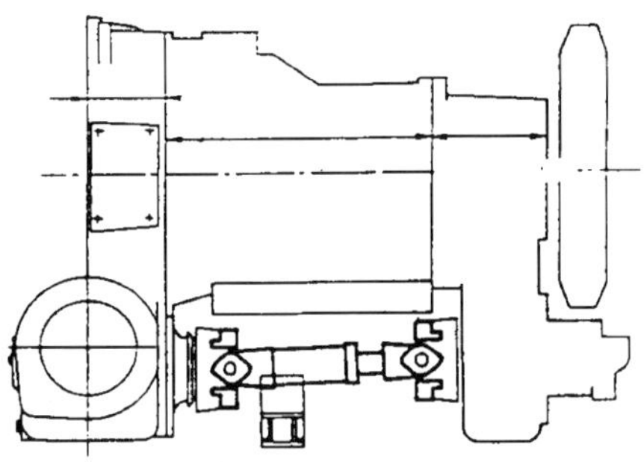

Fig 43: Drive unit

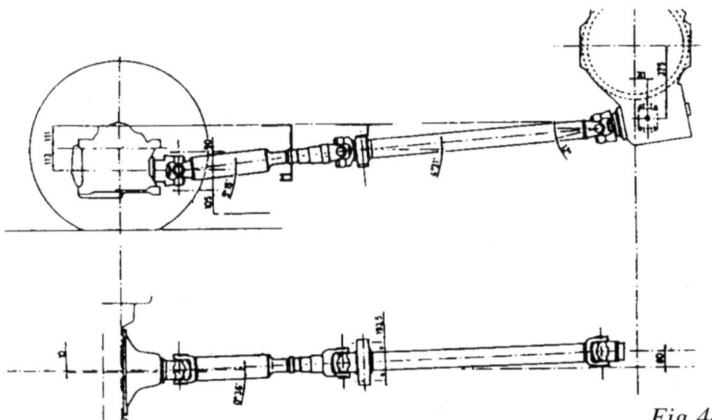

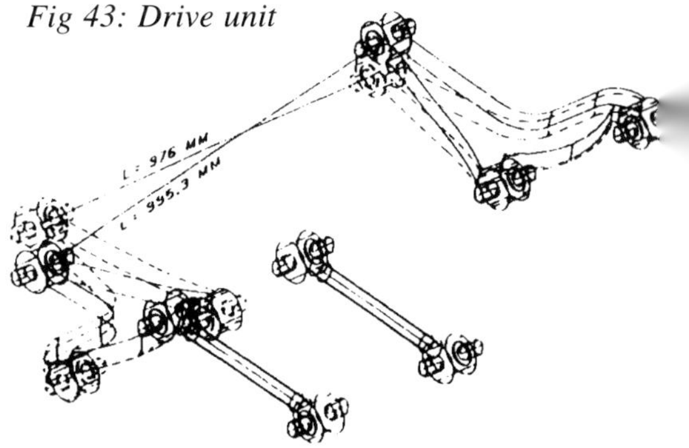

Fig 44: T-drive and axle control arms

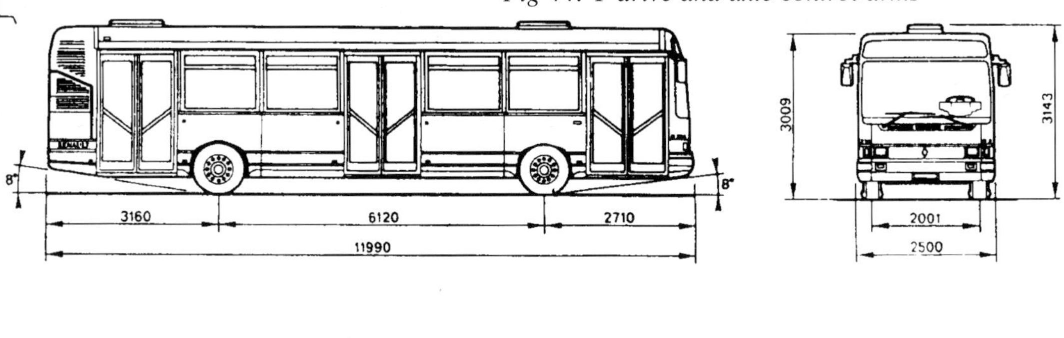

Fig 41: R 312 city bus concept

Advanced integral coach A major step forward in PSV technology is marked by the Mercedes Benz 0 404 coach, Fig 45. A premium specification includes ventilated disc brakes on all wheels, operated by compressed air. Independent front suspension is of elasto-kinematic design, Fig 46, automatically compensating for the effect of side forces on the wheels in cornering. Additional safety comes from the incorporation of ABS and ASR and good road adhesion results from air suspension, front and rear. Drag coefficient is less than 0.4, creditable for a coach body.

Engine encapsulation has reduced drive-by noise level to 80 dB(A) and interior silence is ensured by double-glazed side-windows and installation of the ventilation blowers outside the passenger compartment. Sandwich construction of the raised floor is another benefit in this respect. Body structural members are formed by a fluid cell press. These pressed members replace the tubes of the earlier models. While largely of space frame construction, the structure has become partially monocoque with the structural incorporation of a number of underbody panels. Contoured side panelling and a virtually smooth front panel characterise the exterior and has helped to reduce the Cd figure some 30 % below that of the 0 303.

Interior floor and roof panels form radiation surfaces for the heating system, interlaced with hot water piping throughout the length of the vehicle. These provide an unusually even distribution of heat and the avoidance of annoying cold spots on side panels which many coaches suffer from. Redesign of the airconditioning system has involved fitting the evaporator within the ventilation system and an automatic temperature control system in the rear roof section; the condenser is also stowed at the rear. The driver's compartment is in a separate climate zone to that of the passengers. A convenience for the driver is the specification of an electrically controlled windscreen blind.

Engines up to 380 bhp are now available and the six-speed gearbox has geometrically spaced ratios, with only small stepping between fifth and sixth gears. Sixth is a direct drive gear and in the axle a new higher ratio has been added, of 3.07:1. With this combination engine speed is 1600 rpm at a road speed of 100 kph.

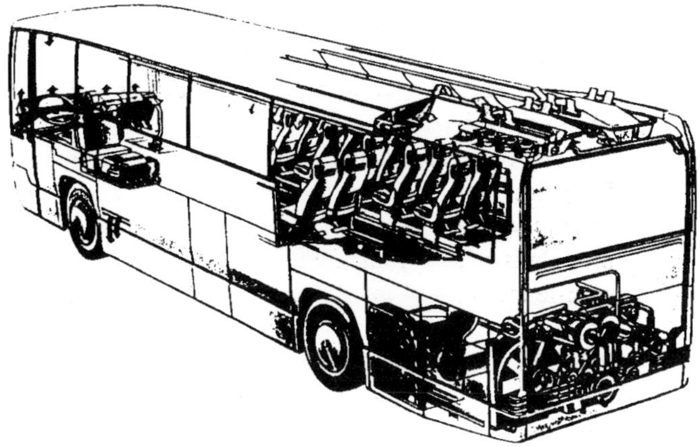

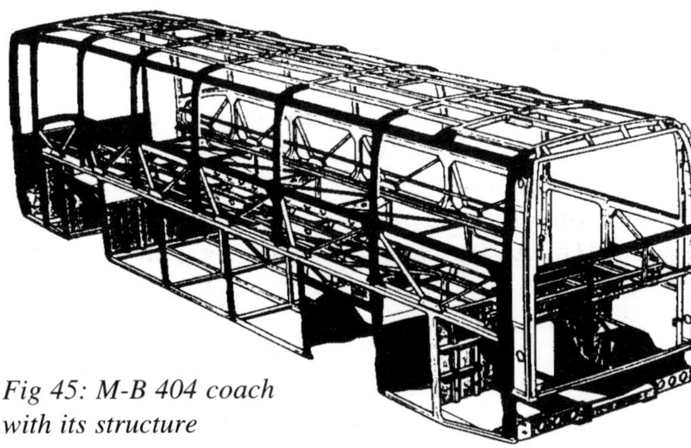

Fig 45: M-B 404 coach with its structure

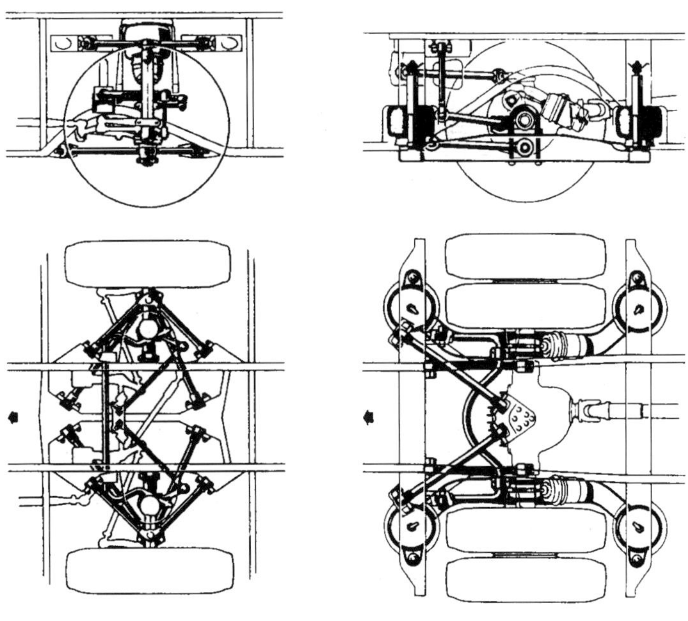

Fig 46: Suspension layout

Diesel-engine/hydraulic accumulator hybrid In Fig 47, the Mitsubishi experimental city bus uses a hydraulic reservoir to accumulate braking energy as stored pressure for subsequent use in braking and driving after conversion through a hydraulic pump/motor system and final-drive gearing. These systems are controlled electronically while letting the driver feel he is at the controls of a normal bus. System specifications are shown in the tables of Fig 48. Accumulative capacities of the equipment were gauged so as to save all braking energy by excluding the operation resistance and energy loss in the drive system based on vehicle kinetic energy at 40 kph.

A high pressure pump/motor system rated at 34.3 MPa was chosen to reduce the size of the equipment. To take power off the pump/motor a through shaft is connected to the final drive and reduction gearbox, incorporated, to allow for the high speed working of the high-pressure hydraulics. An electro-magnetic clutch in this reduction box allows on-off operation of the pump/motor to reduce torque loss when the pump is unloaded in normal running mode. The graph shows a typical driving sequence, such that at start-up (or any change of state effected by the driver), the change-over valve releases hydraulic fluid from the accumulator and the motor develops torque. Accelerating beyond 10 kph, the hydraulic motor starts to assist the engine. on braking the pump activates the regenerative brake before the service brakes and the accumulator is pressurised .

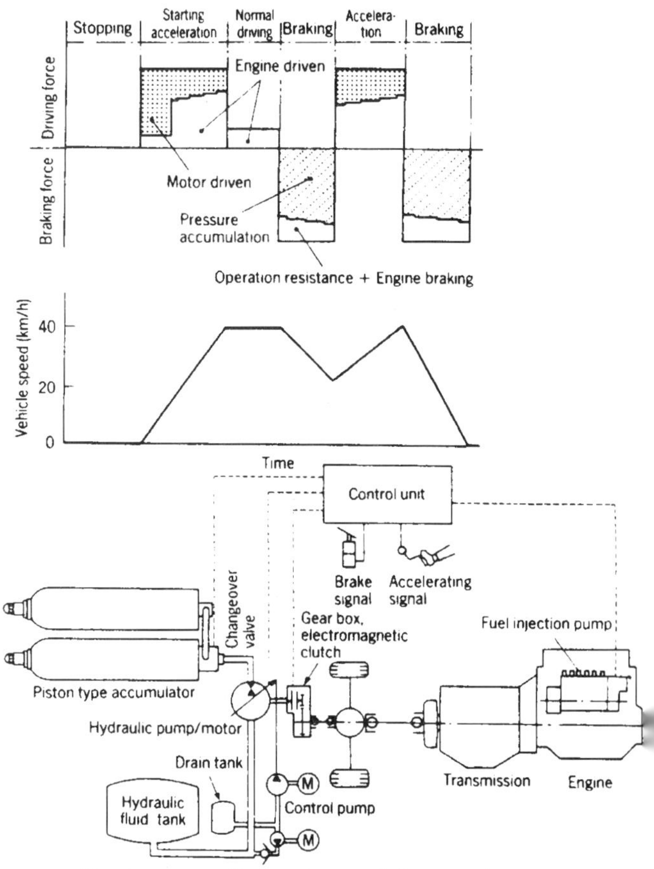

Fig 49: Drive configuration and duty cycle

Vehicle model		U-MP218M
Dimention (mm)	Overall length	10,640
	Overall width	2,490
	Overall height	3,115
Gross vehicle weight (kg)		14,060
Seating capacity		87 passenger + 1 driver
Engine	Type	6D22 (L6, DI)
	Bore × Streke (mm)	130 × 140
	Total displacement (L)	11.149
	Maximum output (kW/min^{-1})	165/2,200
	Maximum torque (Nm/min^{-1})	765/1,400
Tramsmission	Type	M8S5
	Gear ratio	5.431 – 3.463 – 1.747 – 1.000 – 0.741 – 5.677
Final reduction	Type	D8HT (with through shaft)
	Gear ratio	5.571

Hydraulic pump/motor	Structure	Swash plate type axial plunger pump
	Capacity	0 – 90cc/rev variable
	Working pressure range	20.6 – 34.3 MPa
	Theoretical torque	290 – 487 Nm
	Control system	Electromagnetic proportioning solenoid
Accumulator	Type	Piston type
	Capacity	48 L × 2
	Pressurization of nitrogen gas	18.6 MPa

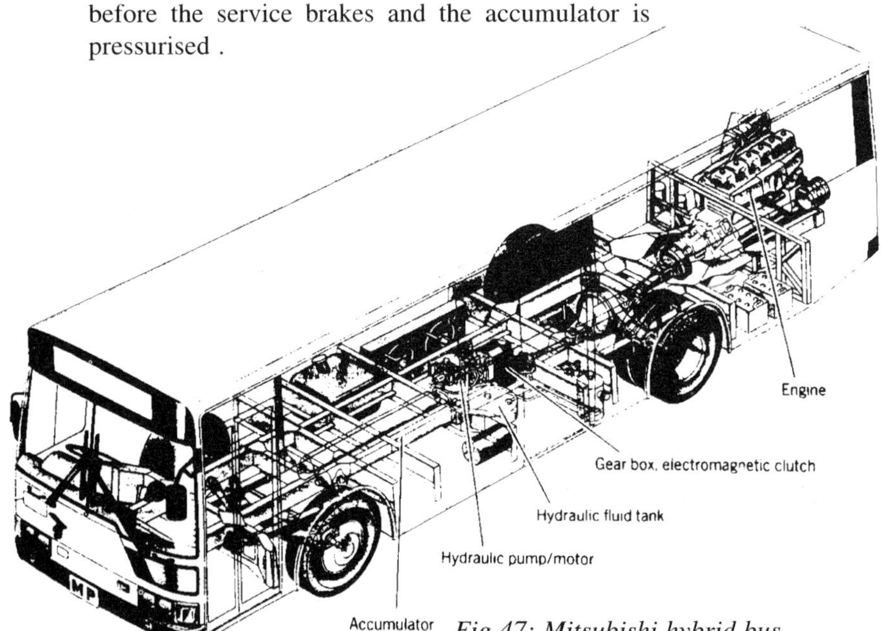

Fig 47: Mitsubishi hybrid bus

Fig 48: System specification

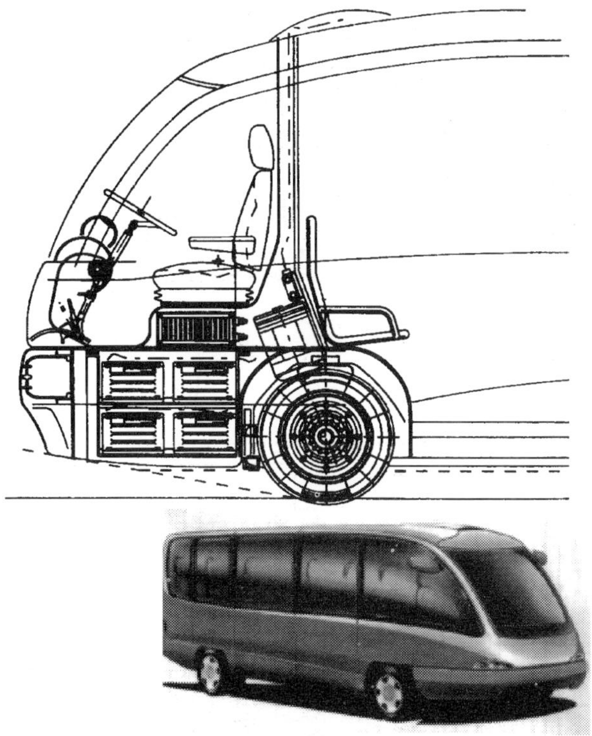

Fig 50: Capoco Centro concept

Electric city-bus designed for low drag The Centro concept midibus from Capoco Design, Fig 50, is intended to show the practicality of 'available now' electric drive technology for city centre operation. The 7.3 metre long vehicle would operate at up to 40 kph and have a range of 40 miles. Energy consumption is reduced by the vehicle's drag coefficient of just 0.31 and low rolling-resistance tyres of modest section profile. A light weight structure comprises mechanically fastened aluminium alloy extrusions covered by a single-piece roof moulding. In full standee form up to 56 passengers can be carried and deck height is just 290 mm. Shown below is the economical packaging of batteries and drive motor, with near vertical drive axis. Width of the vehicle is 2380 mm and height 2550 mm, on a 4050 mm wheelbase.

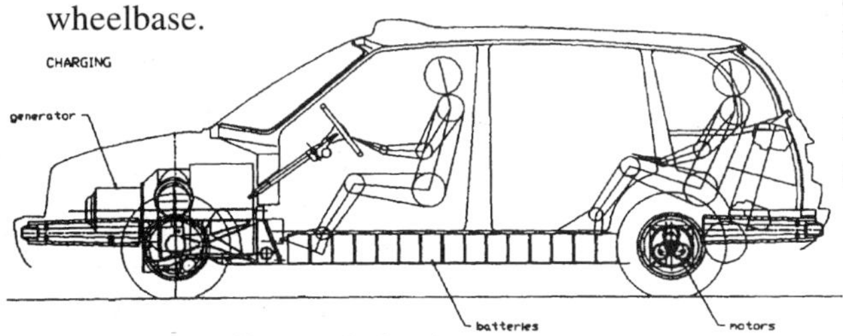

Fig 51: Randle taxi hybrid

Gas-turbine hybrid The Randle design proposal for a hybrid-drive taxi, Fig 51, fits a vehicle category suited to hybrid-drive employing gas-turbines. This is because the on-board electric generator, if subject to deep discharge cycles, only rarely is started up — so the slow time-to-speed of a gas turbine would not be a problem. Ability to start at full GVW on a 1:3 hill would determine the motor size while ability to maintain 70 mph up a 1:20 hill would define generator size. A 65 kW generator was found appropriate to a London taxi duty while average charge rate operating to a ECE-15 cycle would be 56.48 kW. Three operating modes would encompass generator charging batteries, the batteries powering electric drive motor and a combination of the first two with turbine driving wheels for the high performance mode.

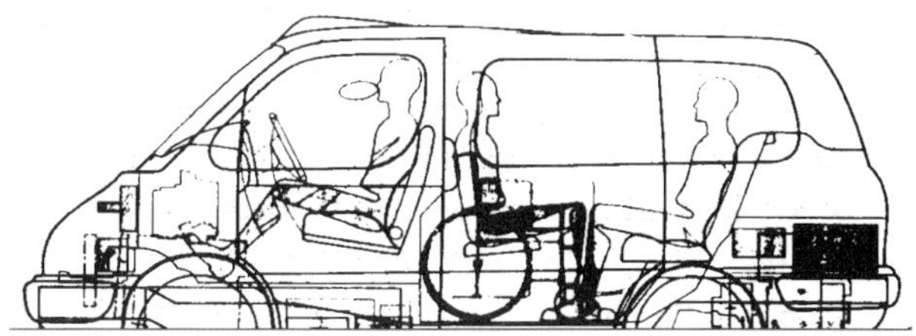

Fig 52: Eurotaxi

Electric taxi The IAD Eurotaxi is based on electric/diesel hybrid running units mounted in a box-section steel chassisframe with fully independent suspension. Sodium sulphur batteries of 50kWh capacity power the 50kW AC induction motor, which is water-cooled, via an inverter system with integral battery charger and a DC-DC converter. The auxiliary power unit is a three-cylinder diesel generator set with output controlled to 25 kW. Regenerative braking is employed and the vehicle is said to have a battery range of 100 miles, with maximum speed 65 mph. An interesting substitute for a high-level brake light is a progressive strip light system which indicates rate of retardation to following vehicles. The high-mounted, cab-forward, driving position provides high standards of visibility while enabling a large rear passenger-compartment — giving easy access for wheelchairs.

Three-wheeled vehicles

Steepening oil prices and the desire for maximum manoeuvrabilty could bring a renewed interest in the three wheeler configuration for both commuter and budget sportscars as long as they are not marketed as any more than motor tricycles.

While the earliest three-wheelers had their configuration derived from carriages such as the pony-and-trap, those to come may yet result from the need to minimise fuel consumption and maximise manoeuvrability. The world's first vehicle to move under its own power, the French steam-powered Cugnot of 1769, was a tricycle and despite the market failure of the Sinclair-C5, its low rolling-resistance configuration is one of its aspects that its many detractors paid tribute to. By the time Carl Benz made the first motor car in 1885, his Motorwagen probably derived its layout from pedal-driven tricycles of the time, motor engineering as a whole developing from cycle engineering at the end of the last century. The Benz's single wheel at the front configuration was particularly suited to front entry into the mid-wheel-base bench seat, an ergonomic factor that could affect the design of a two-seater C5-type vehicle which has often been mooted.

It has been argued that it was mainly misinformed motoring-press comment that changed the majority of light 'cycle-cars' from three to four wheel configuration, between 1905 and 1910; few could foresee the coming of balloon tyres with well-based rims which would remove the danger of a single wheel 'burst' not uncommon with beaded-edge tyres. Morgan, of course, persisted and success in competition from 1910 onwards brought market success even up until the early post World-War 2 years, backed by a tax concession which was later to benefit Reliant in attracting custom from former riders of motor cycle combinations. About the same time, economy 'bubble' cars such as those of Heinkel and Isetta rose in popularity. High oil prices forecast for 2000 could again bring popularity to such vehicles.

Fig 53: Sinclair C5

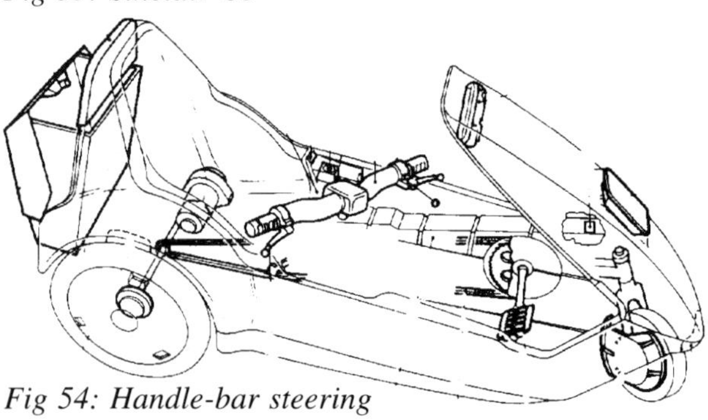

Fig 54: Handle-bar steering

A closer look at C5 technology reveals how a very low cost commuter tricycle could be obtained given the lessons learned in hindsight about the vulnerability of very low profile, poor weather protection and inadequate battery capacity. The vehicle's greatest detraction seem to be an over exuberant marketing campaign that allowed the motoring press to interpret the vehicle as a 'car' rather than cycle derivative. A more acceptable reception might have followed from the selling of the vehicle through cycling and motor-cycling outlets and indeed configuring the vehicle around the higher silhouette of a conventional pedal cycle with power-assistance to a much smaller initial market but one with greater long-term growth potential.

The prone riding position, Fig 53, had ergonomic advantages for applying pedal assistance as did the handlebar steering for accurate handling control, without the expense of conventional gear and linkage control, Fig 54. Conventional size cycle wheels of 26-30 inch diameter would have raised the silhouette substantially above that obtained with the 12.5 - 16 inch ones used, albeit at the cost of greater overall length. While the thermoplastic body moulding was at the time the largest single-piece moulding produced, current technology could no doubt achieve the larger size needed to enshroud a 28-inch wheel machine. The Lotus-designed backbone chassis (Fig 55) was heralded as a neat solution to a low-cost rigid base structure — with flexural compliance built into its rear arms providing a degree of suspension travel at the rear. Performance-wise, an important feature was the achievement of less than one per cent, the 45 kg unladen weight, rolling resistance from the use of narrow section tyres, a technique since exploited by a number of solar-powered vehicles. The aerodynamic drag figure of $CdA = 0.23$ metre^2 would inevitably rise with the higher body profile, of course. The epicyclic reduction gear was an extension of cycle practice, the two-stage reduction, with 13:1 overall ratio, being achieved with glass reinforced nylon gear teeth. While a high castor angle was used on the front steered wheel, a modest castor trail of 11 mm was used. Front calliper and rear drum brakes also lent on cycle technology. Transfer drive to the rear axle was via tooth belt and the permanent magnet motor was rated at 250 W and worked in conjunction with a custom-chip centred electronic controller which, in conjunction with an LED display encouraged the rider to guard against motor overheating or undue battery drain. By far the most important lesson of the C5's abortive launch was that the occupant should have been seen as a 'rider' and not a 'driver' (Fig 56), the latter category expecting no compromise from the 'room on wheels' concept with which the contemporary car is perceived. It is for this reason above most that the interest in three-wheeler car-derivatives from such companies as Morgan, originally, and more recently, Reliant, is on the wane. The famous two-speed Super Sports Morgans which won racing acclaim 60 years ago could no longer expect to generate the size of market for road derivatives which the early vehicles enjoyed. There is a significant minority market, however, which the nostalgia for such cars has kept alive. This is well catered for by such kit-cars as the Lomax 223, Fig 57. Here the Citroen 2CV is cleverly converted as the donor chassis, with an about-face application of a single rear wheel suspension being adapted for the steerable front wheel. The normally fan cooled engine is 'stripped bare to the wind' in the Lomax application, in the manner of the exposed JAP engines of the two-speed Morgans. A somewhat similar theme marked the design of Reliant's Bond Bug, Fig 58, but some say this suffered from lack of the classic configuration of the two-speed Morgan. The body shell was an appealing departure but relatively high cost and somewhat conservative mechanical specification probably limited the appeal.

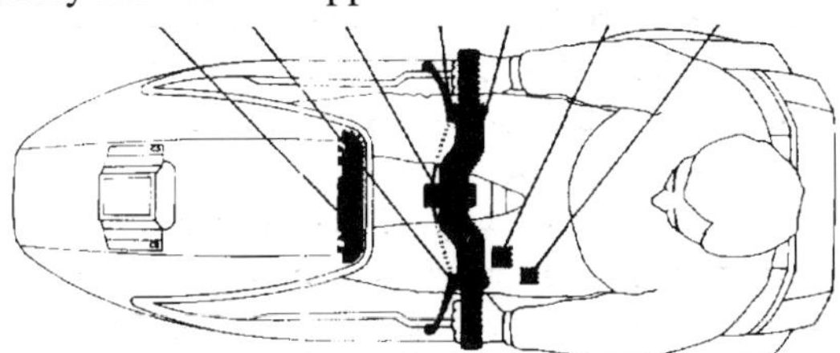

Fig 56: Rider or driver?

Fig 57: Lomax 223 and its modified 2CV front end

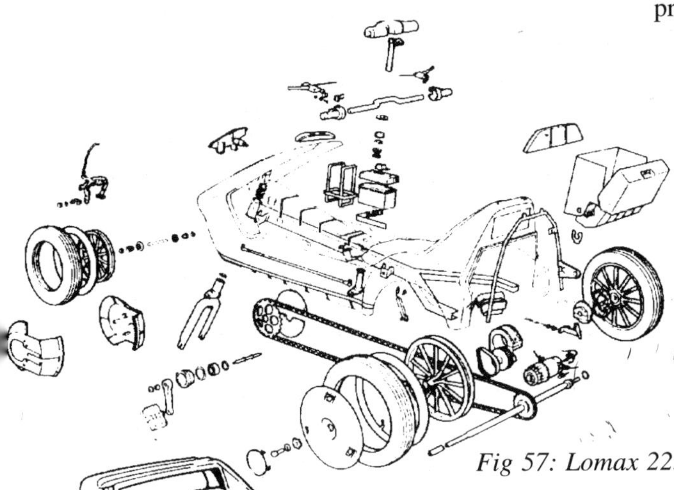

Fig 55: Chassis and running gear

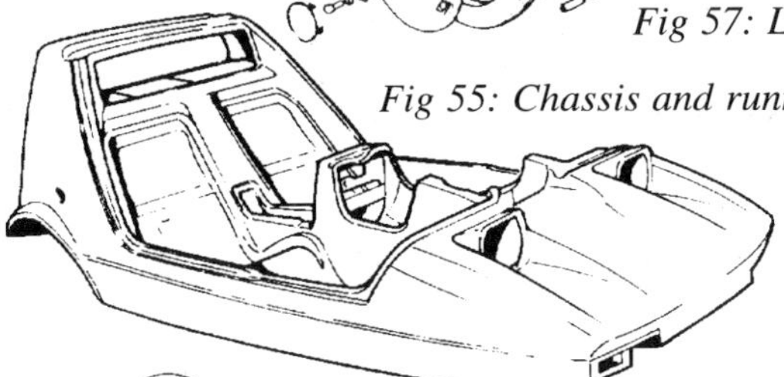

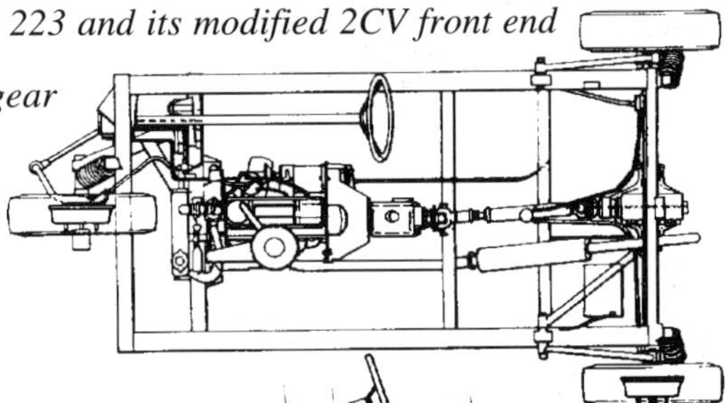

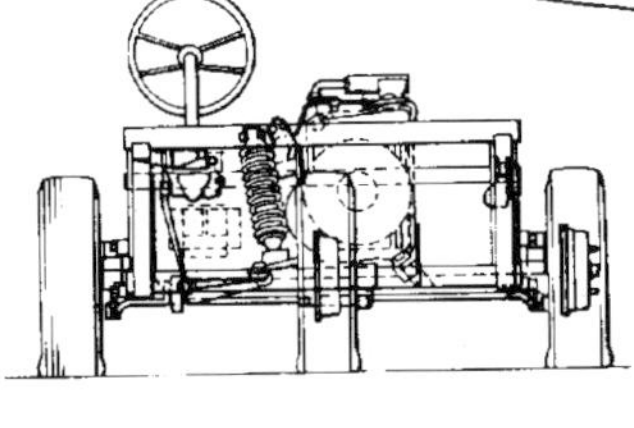

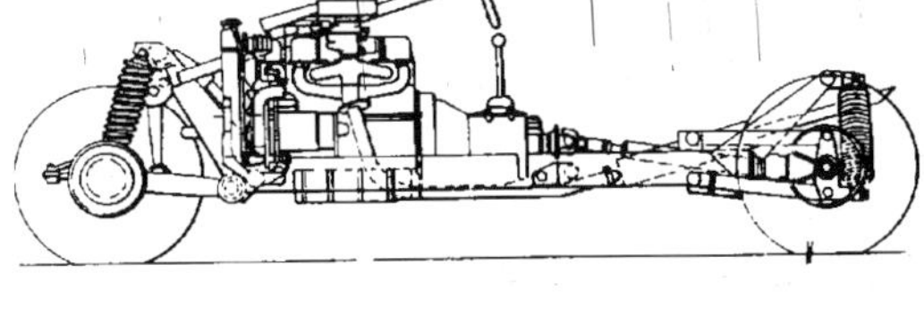

Fig 58: Bond Bug

Motor-cycle engineering

The design of single-track vehicles has a number of distinct approaches relating to the stability of the machine. This section examines the basic force systems involved as well as observing some developing trends in this specialist market.

While the chief virtue of motor-cycles was once perceived as economy, the effect of sport has turned motor-cycling into a pleasure pursuit. The bar-graph in Fig 59 is based on a 1991 Gallup poll and shows that while over two-thirds of motorists seldom drive for pleasure, three quarters of motor-cyclists claim to ride purely for pleasure. There is also an upsurge in urban use for both commuting and courier work. A key characteristic of single-track vehicles is that cornering is performed with a roll-angle adapted to lateral acceleration. The motor-cycle tyre has its maximum of circumferential force, without simultaneous lateral force, at a very small slip-angle — around 0.06-0.08 radians — and lateral forces are mainly developed by roll-angles. Knowledge of power transfer between tyre and road is thus a prime issue, Fig 60. For a given lateral acceleration, wheel slip is far from constant and depends upon curve radius, friction potential, driving speed and tyre condition. Acceleration and braking torques also have a substantial effect. In tests carried out by Darmstadt Technische Hochschule, the roll angle force was found to be caused by the tendency of the cambered tyre to rotate around the point of intersection of the steering axis with the road. Because the tyre is forced on to a much larger radius, however, it is necessary from a balance standpoint to keep tyre slip angle as low as possible when cornering. Tests on a 130/90 V17 tyre held at a constant inflation pressure of 2.5 bar, showed in Fig 61 that dynamic tyre radius depends considerably on roll-angle. Friction maximum is obtained at about 0.8 % slip, Fig 62, which shows the typical steep rise at the lower slip range when riding without both slip- and roll- angles. Steep descent of the curve after reaching 100 % braking results from the change from rolling friction to sliding friction. Above 100 % slip there is another change to rolling friction.

Corresponding data for a roll angle of 45 degrees show the change in characteristics. Friction maximum is not as distinct and is reached only between 20-30 % braking slip. At higher slip ranges

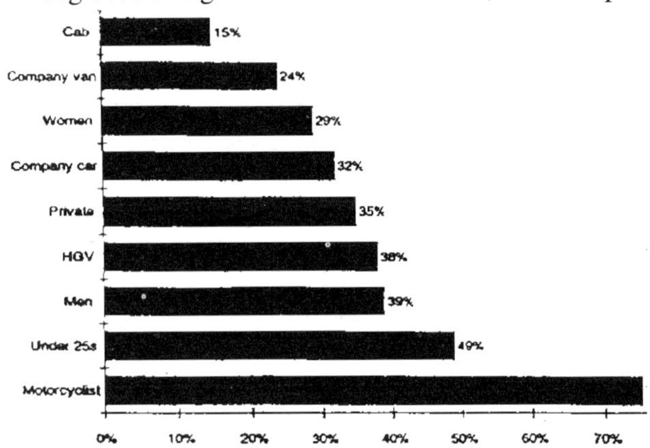

Fig 59: `Driving for pleasure' survey

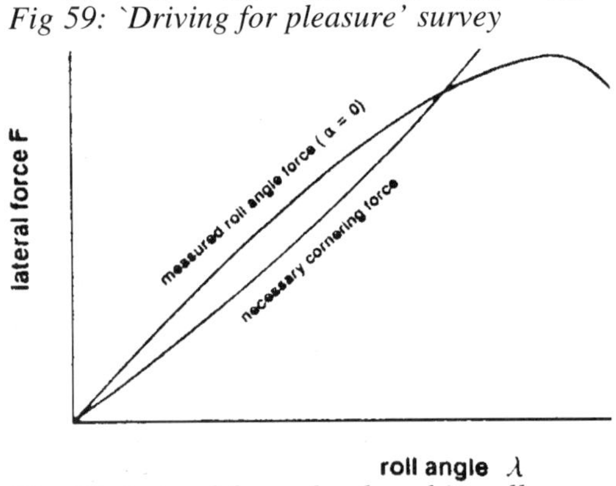

Fig 60: Lateral force developed in roll

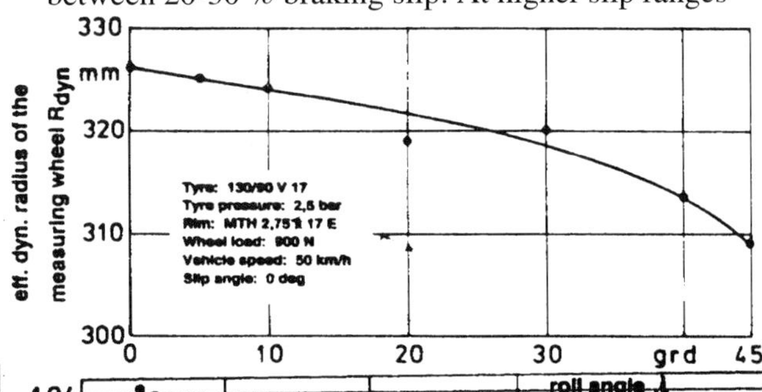

Fig 61: Dynamic tyre radius

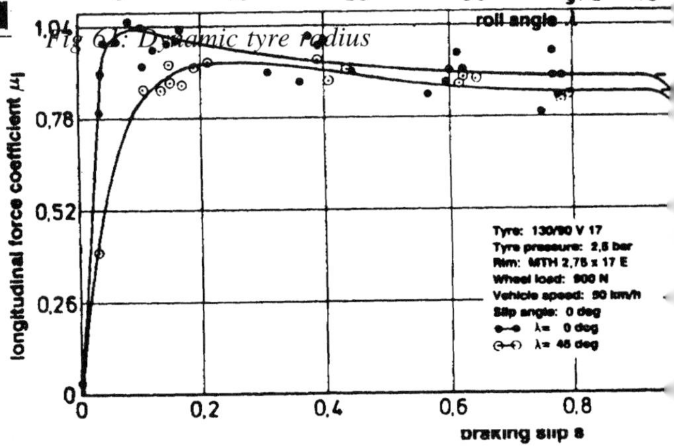

Fig 62: Friction force development

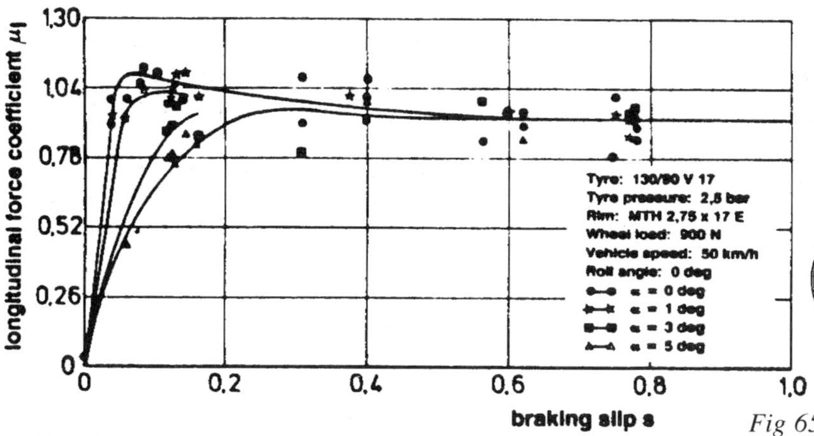

Fig 63: Slip and roll observed simultaneously

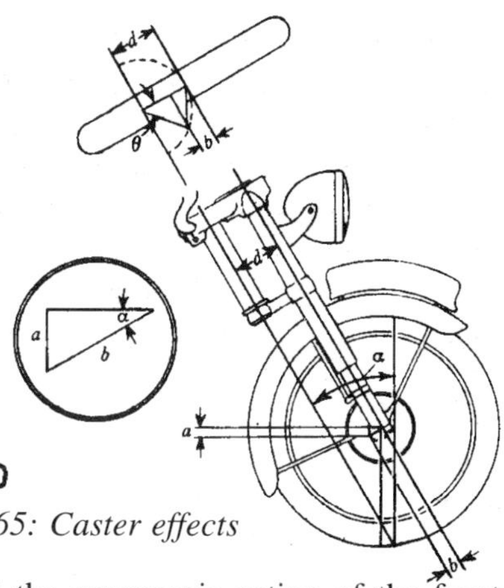

Fig 65: Caster effects

friction levels are only slightly lower than at a roll angle of 0 degrees. When observing slip- and roll-angles at the same time, Fig 63, the roll-angle influence on transferable force predominates in the lower and negative slip-angle range, decreasing with increasing slip angle.

The steering and stability of single-track vehicles was examined in a classic study of the same title by R A Wilson-Jones[5] of the former Enfield company. It was clear that the idea that stability of the rider in 'dynamic' balance on the cycle did not just arise from the gyroscopic effects of the wheels. In fact this only occurs at high speeds and lower down the speed range it is the tendency as the machine tends to fall. That is from the wheels steered ahead position — and the gyroscopic action of the front wheel merely turns the steering in the direction of the fall. This same front-wheel gyroscopic effect also provides self-centering torque for the wheel when it is steered from the straight-ahead when the machine is at speed. The way in which the resultant of the weight of the machine plus centrifugal force acts in the plane of the wheel is clear in view (c) of Fig 64. When the machine corners at slower speeds in the near upright condition of (d) the point of action of AB is behind the steering axis by the extent of the caster trail which provides the self-centring action. These self-centring effects can, of course, be dangerously delayed by excessive friction in the steering head. The effects can also be tuned by fitment of steering dampers, altering the inertia of the steered assembly and by altering gyroscopic torque by changing wheel weight. Steering head angle is another important consideration which can aid self-centring due to the effect of the vehicle weight being lifted as the steering is turned. This is a comparatively small effect, raising the machine by:

$$d \sin\alpha(1-\cos\theta).$$

A second effect is that it tends to increase the angle of lean of the front wheel when the steering is turned and at high steer angles, a point at the side of the tyre contacts the ground, thus lowering the steering head. More important is that the head angle produces the necessary caster trail without negating the initial effect of lifting the machine.

CG height of machine and driver has two opposing effects. As the trail causes the contact point of tyre and road to move left when the steering is

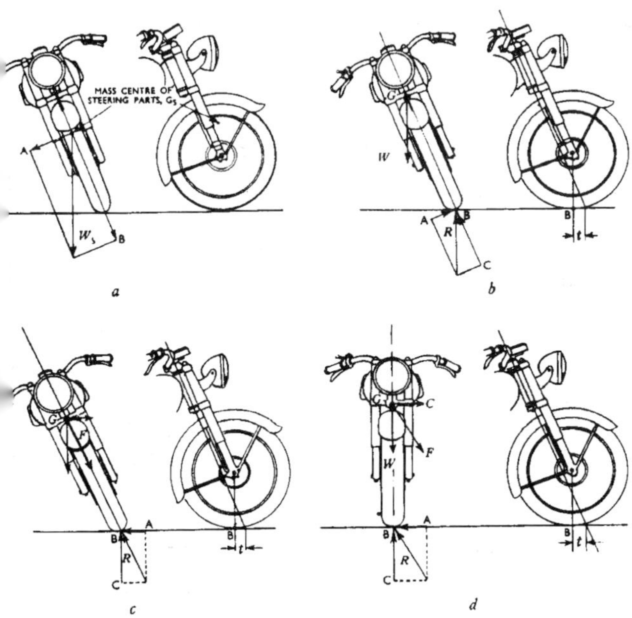

Fig 64: Resolving weight and centrifugal force

turned left, and vice versa, the front end of the line joining these points is deflected to one side of the machine, on steering, Fig 65, and a high CG machine is required to lean over a greater distance. However, due to the width and curvature of the tyre tread, the line of contact with the ground is also displaced to one side as the machine leans and the effects oppose one another.

Typifying recent trends, the new departure for BMW (who now feature ABS on some machines, Fig 66) of the F 650 Euro bike developed in conjunction with Aprilia and Bombardier-Rotax, uses the latter company's 650 cc engine. Designed to combine the attributes of fun-bike and endurance machine, the F650 has engine outputs in either 34 or 48 bhp levels, a weight of just 417 lb (190 kg) and modest seat height of 32 in (813 mm). The engine is a liquid-cooled four valve unit developing up to 57 Nm torque. The gearbox is integrated within the aluminium alloy engine housing which is split vertically in two halves. The covers for the engine housing, having clutch at rear-left and alternator rear-right, are also made in aluminium alloy while the chain pinion cover is in plastic. Cylinder liners are coated in low-friction nickel-silicon while the cast piston is in 'box' configuration, cooled on its underside by splash oil.

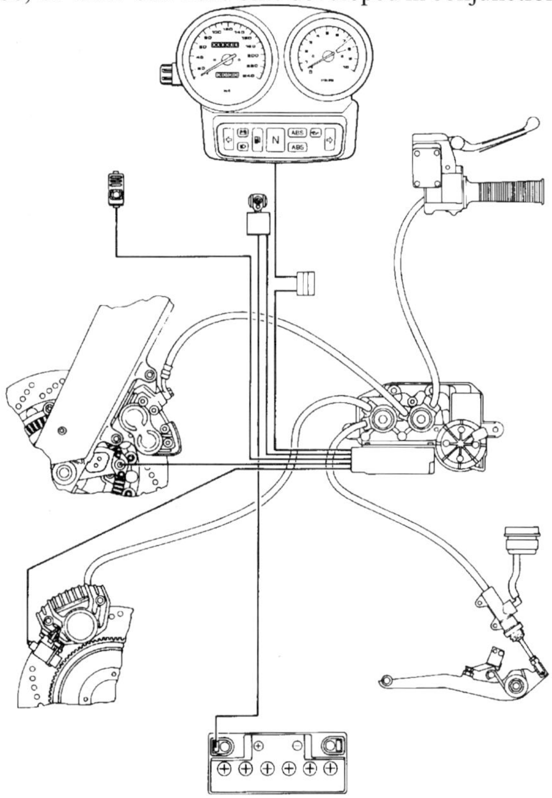

Fig 66: ABS II by BMW

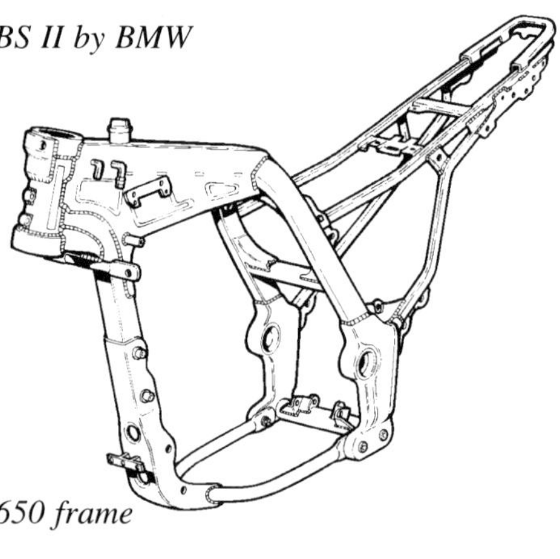

Fig 67: F650 frame

To minimise vibration on this large single-cylinder engine, a gear-driven compensation shaft is provided which runs at crankshaft speed. The engine has dry-sump lubrication and obtains its oil from a 1.5 litre reservoir incorporated in the upper part of the frame. Fuel is supplied through two Mikuni constant-vacuum, automatic choke, carburettors of 33 mm choke size while air is fed to them via a 6.5 litre silencer/filter. The mapped capacitor-ignition system uses two sparking plugs in order to shorten flame travel and minimise pre-ignition, as well as improving part-load combustion. The stainless steel exhaust system has a 7.5 litre expander volume. Transmission of power is via a multi-plate clutch and dog-type five-speed gearbox with gears lubricated by splash oil. An interesting aspect is the recyclability, by registered BMW dealers, of the optional exhaust catalyser to preserve the precious metal. Otherwise the machine is serviced on a 10,000 km interval basis.

The frame of the machine, Fig 67, has a pressed steel section single loop, which also incorporates square-section tubing, further reinforcement being by bolted-on round-section tubing. While the front wheels run in conventional telescopic forks, the rear wheel is located by a delta shaped boxed swinging-arm. Front and rear spring struts are by the fork maker Showa and are adjustable for spring pre-tension and damper-rate. Progressive rate is obtained at the rear by a lever mounting system for the rear strut. Overall wheel travels are 170 mm front and 165 rear. The rear wheel is driven by an O-ring chain, with permanent lubrication between the rollers, running at 1:2.93 transmission ratio. Rubber cushion supports are built in to the rear drive sprocket.

Evolution of motorcycle stability

The configuration of motorcycles has evolved over the years by largely empirical means such that entirely safe designs are now the norm and few design variations threaten stability. However, the revolution in lightweight structures and novel propulsion systems such as hybrid battery-electric/ic-engine drives could result in configurations that might require a more fundamental reappraisal of any trend to instability. It is interesting therefore to look at some of the earliest analytical procedures that were once too cumbersome to apply on an iterative basis, before electronic computation, to see if they now have value in conceptual design calculation should a program be written to optimise the factors concerned based on equations given in early papers.

Pearsell[6] considered the motions of a free-wheeling hoop rolled at steady speed on a smooth surface and concluded that the effect of gravity causing lean of the hoop, and subsequent motion in a curved path, to be counteracted by centrifugal force causing the hoop to correct its lean. Because the hoop is a gyrostat, either gravity or centrifugal force also produce precessional couples giving a tendency to turn to the right, say, if gravity leans the hoop to the right.

By taking moments about the ground contact point in both elevation and plan views, Fig 68, for a two-wheeler vehicle:

$$I_3\ddot{\theta} = I\frac{v}{r}\dot{\varphi}$$

since gain in angular momentum in plan is equal to gain in precession:

$$Mgr\varphi - \frac{Mv^2r}{R} = \frac{Iv}{r}\dot{\theta} + I_2\ddot{\varphi}$$

since the gravity couple increasing lean less the restoring centrifugal couple equals the precessional and leaning gains in angular momentum.

As the instantaneous centre of rotation lies on the axis of the hoop so the rate of change of the hoop's angular position, in plan, must be the ratio of velocity to path curvature radius

$$\dot{\theta} = v/R \qquad \text{such that}$$

$$\dot{\theta}\left(\frac{Iv}{r} + Mvr\right) = Mgr\varphi - I_2\ddot{\varphi}$$

By writing $A\ddot{\theta} = B\dot{\varphi}$

$C\dot{\theta} = D\varphi - E\ddot{\varphi}$

then $\ddot{\varphi} + \frac{BC - AD}{AE}\varphi = \frac{C}{AE} \times \text{const.}$

By examining this equation alongside the author's appendix, seen in Table 3, this shows that if speed is above the critical value at which the coefficient of the angle becomes positive, the motion is a continuous lateral oscillation without damping. Its frequency increases with speed, mean lean angle decreases but amplitude is unaffected.

It is now possible to represent the individual assemblies of the motorcycle to test their effect on the machine's behaviour. By adding frame and rear wheel as a single mass then equation 2 has h substituted for r in its left hand side terms. The path curvature is not changed from that of the hoop and only a difference in swings of the wheels is introduced. As these are relatively small, path radius is considered constant for all parts of the motorcycle. Moment of inertia of the front wheel about the vertical is taken as that of the complete assembly including forks and handle bars, about a steering head which is assumed vertical.

When the steering head is castored, however, there is a centrifugal force component forming an aligning torque for the wheel, but reduced by the torque required to accelerate the rotation of the whole machine about a vertical axis through the rear wheel. Thus equation 1 requires the following terms to be added to its left-hand side:

$$Mv^2nc/\breve{R} + Mn^2lc\ddot{\Upsilon}$$

As the front wheel is steered, the ground-contact moves outside the frame of the plane so the gravity torque is increased by the front wheel load acting with this displacement and equation 2 therefore needs the following term added to its left hand side:

$$Mgnc\varphi$$

Furthermore, as the motorcycle leans over, a similar torque appears which increases the steer angle of the front wheel and the following term is added to the right hand side of equation 1:

$$Mgnc\theta$$

The rear wheel is then considered as a gyroscope, with the precession, in plan, opposing the gravity torque so that the following term is added to the right hand side of equation 2:

$$I_b (v/r)\dot{\gamma}$$

while precession, in elevation, increases rate of increase in front wheel steer angle requiring the following term to appear on the right hand side of equation 1:

$$I_b (v/r)\dot{\varphi}$$

Finally the rake of the steering head has to be considered, by introducing resolved motions, about vertical and horizontal axes to account for lift and fall of the machine when steer angle is applied. Those about the horizontal axes can be added to the equations even though they are formed on the assumption of a vertical steering head. The following term thus appears on the right hand side of equation 2:

$$kI_3\ddot{\theta}$$

and the following term to the right hand side of equation 1:

$$(Ivk/r)\dot{\theta}$$

The author further manipulates the equations in the original paper and draws some useful conclusions which including that weight should be well forward on the front wheel and that the castor angle should be preferably between

$$c = Ik/Mnr \quad \text{and} \quad Mnc/(I_3 + I_4 c^2/l^2)$$

Table 4: Notation

I	Polar moment of inertia of front wheel.
I_b	Polar moment of inertia of rear wheel and engine.
I_2	Moment of the whole about the ground line.
I_3	Moment of the front-wheel system in the head.
I_4	Moment of the whole about the vertical through the rear whee
c	Castor.
l	Wheel-base.
r	Wheel radius.
M	Total mass.
h	Height of centre of gravity.
v	Velocity of cycle.
n	Proportion of weight on the front wheel.
k	Tangent of inclination of head to vertical.
ϕ	Angle of lean.
θ	Angle of deviation of front wheel.
γ	Angle of deviation of rear wheel.
R	Radius of path of cycle.
ρ	Radius of tyre tread.

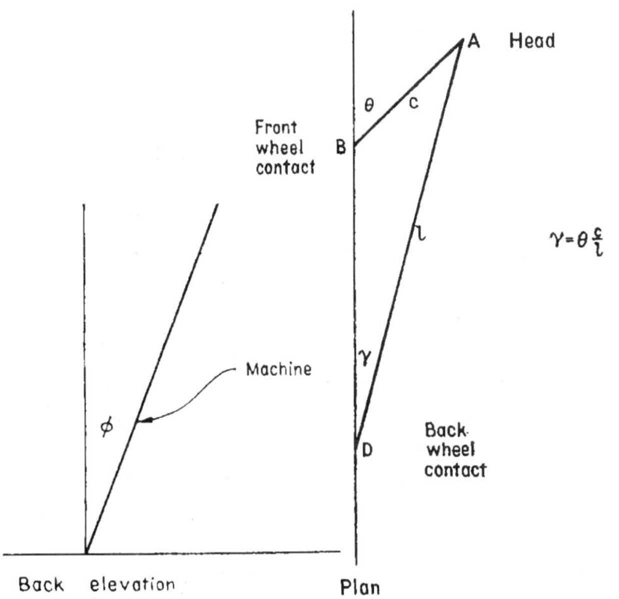

Fig 68: Elevation and plan views

Table 3

If a displacement y varies with time according to the equation

$$y = e^{-pt} \sin qt$$

the motion is a sine form of decreasing amplitude, and the corresponding differential equation is

$$\ddot{y} + 2p\dot{y} + (p^2 + q^2)y = 0$$

or

$$\ddot{y} + P\dot{y} + Qy = 0$$

so that an equation of this form expresses a sine wave

Fig 69: BMW Personal Commuter

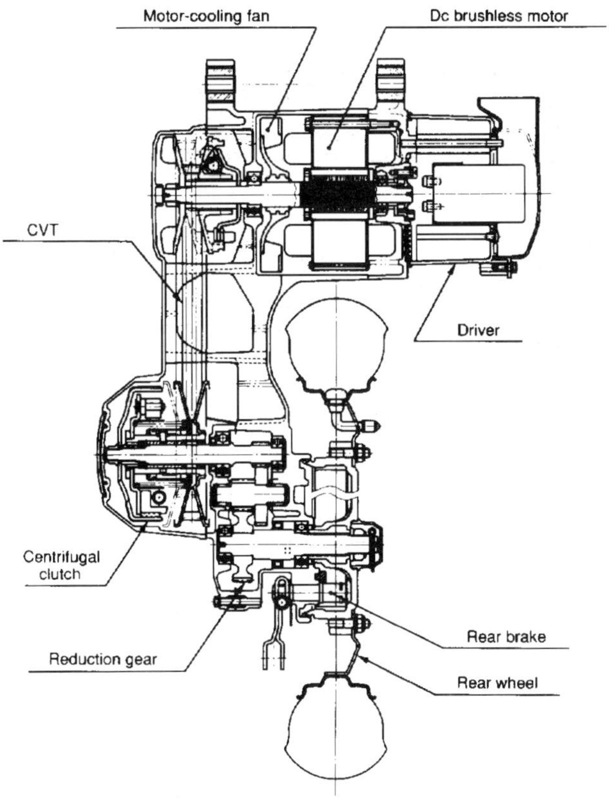

Fig 70: Honda electric drive

Motorcycle developments

Riding a two-wheeler without crash helmet BMW's C1 Personal Commuter, Fig 69, is to be built by Bertone from 2000 and sets out to combine car safety with motorcycle agility and low-cost operation. Power is provided by a 15 bhp 125 cc 4-stroke engine featuring digital engine management and three-way catalytic converter. Front-suspension is the company's Telelever anti-dive configuration and braking the ABS introduced on BMW motorcycles in 1988. However, shaft-drive has given way to a CVT belt drive. The aluminium ally body frame doubles as safety cage and roll-over bar while a deformable front mudguard gives a degree of controlled collapse in frontal impact, the rider having the additional protection of a two-point, and one three-point, belts, obviating against the need for a crash helmet. The 76 cm wide GRP bodywork incorporates windscreen and wiper. The machine weighs 170 kg, can reach 62 mph and can deliver 80 mpg.

Potential for electric scooters The Honda CUV ES uses 12 sealed 1.2V 20 Ah nickel/cadmium batteries, mounting and cooling system for which are shown on the front cover. Thermostatically controlled fan cooling for the batteries maintains temperature below 40 C. A brushless DC motor is employed, developing 8.8 Nm at 3200 rpm; this drives through a CVT transmission, Fig 70, with electronic management. The scooter weighs 130 kg and has a range of 61 km at 20 kph. It is interesting to note that Taiwan, which has over 10 million registered scooters and small motorcycles, has mandated that 5 % of its motorcycles be electric by 1998.

Hybrid Scooter Piaggio have introduced an 'easy-rider' style scooter, the Hexagon (Fig 71), a fully-automatic transmission machine which carry two people in a relaxed position as well as providing helmet stowage. Engine is a 125 cc liquid-cooled single-cylinder two-stroke. Electronic, capacity discharge, ignition and electric starting are featured. The 1.4 metre wheelbase machine weighs 125kg. Another major novelty is the launch of the 'Zip-and-Zip', probably the world's first hybrid drive vehicle to be available on the market. This has both 50cc petrol engine and alternative electric drive which can provide 15 mph for 75 minutes. The petrol engine drives the machine at up to 30 mph, the limit for mopeds. Drive is transferred from one mode to the other by the action of a handlebar switch. Battery can be recharged in eight hours from the on-board charger by connecting to a 220V supply. The electric drive comes from a 650 W permanent-magnet motor connected to two 12V lead-acid batteries, each of 40Ah.

State-of-the-art Yamaha's XJ900S Diversion, Fig 72, has an 892 cc air-cooled 4-cylinder in-line engine and shaft drive. With an eight litre air intake plenum the slant cylinder layout allows the use of downdraught carburettors and virtually straight inlet tracts. There is a digitally mapped ignition, and throttle position sensor, specified and the engine is tuned to provide high torque at low speed. Piston undercrowns are sprayed by oil jets and there is a large oil cooler positioned in the airstream above the cylinder head. There is also an exhaust pollution control system in the form of a secondary air supply from the intake plenum controlled by valves to feed the exhaust ports. Resulting recombustion in the exhaust results in 40 % CO and 30 % HC reduction. The frame is notable for widely spaced upper rails of 48.6 mm diameter, to accommodate the eight litre plenum. Rear suspension is a link-type rising rate design providing 110 mm of vertical wheel travel. Front suspension travel is 140 mm and front brakes comprise a triple disc system of 320 mm diameter with 2-pot callipers. Triumph's new Thunderbird, Fig 73, combines traditional styling with contemporary technology and is launched particularly for the USA market. Engine is a liquid-cooled 885 cc unit. The steel spine frame has also been redesigned for this 1.55 metre wheelbase machine which weighs 220 kg and whose engine develops 70PS at 8000 rpm. Retro styling for crankcase, cover and cylinder head will evoke the marque's famous twins of the 1960s.

Aluminium-frame Welbro have produced an aluminium-alloy framed motorcycle, Fig 74, of striking style, based on specially made Alcan extrusions including a triple box-section in 7020 alloy.

Fig 71: Piaggio Hexagon

Fig 72: Yamaha Diversion

Fig 73: Triumph Thunderbird

Fig 74: Welbro aluminium frame

Developing-country vehicles

Application of an 'appropriate-technology' philosophy is necessary in design of vehicles for operation in very poor countries where conventional manufacturing facilities are simply not available. A combination of the intelligent use of systems from donor vehicles, or bin-picking from major volume builders' overseas plants, with the choice of the most rudimentary engineering solutions, is the usual approach.

The Intermediate Technology Development Group have been prominent in promoting the view that "appropriate" technology should be applied to third world transport needs. Low cost motorised vehicles are particularly seen as the solution to poor areas where an intermediate range vehicle, between that of human porterage and conventional motor vehicles, is appropriate. Opportunities exist, it is argued, for transfer of low-cost vehicle technologies from regions of use to those of non-use.

The Group focus on utilitarian load-carriers for goods or several passengers. In rural India, for example, it is reported that three-wheeled vehicles operating on level ground with diesel engines of 4.5 kW carry loads up to 1000 kg and those with engines of 8.25kW up to 2000 kg. Travel speeds are 20-25 kph. Four wheeled vehicles having 6-8.25 kW diesel engines have payload capacity of 1500-2000 kg. In farming areas, using single-axled tractors powered by 4.5-8.25 kW diesels, trailers of 750-1500 kg capacity are also used for transport tasks. There are also four-wheeler conversions of motorcycles of the type shown in Fig 75, in use in such Far-Eastern locations as the Phillipines, and three-wheelers made from chassis extensions of motorcycle frames, Fig 76. In the latter case the centre of the rear load platform will be just ahead of the rear axle. Weight bias of the load should not exceed 60 % in front of the axle, however. Stability will be interdependent on wheel size and fore-aft CG position, the Group suggest. To provide a simple differential drive unit for the rear axle, former car axles are often converted to chain drive in India. A sprocket is attached in place of the crown-wheel and the rear of the housing cut away to accommodate the drive chain and alternative lubrication methods provided! The axle is reversed to allow the pinion drive flange to face rearwards. Where used-car drive axles are unavailable, a simplified free and fixed hub beam axle of the type shown in Fig 77 may be fabricated. Heavy gauge tube of some 50 mm diameter forms the beam which is machined at either end to receive bearings. A final-drive sprocket is bolted to a boss on the tube.

For purpose-built three-wheelers the Group favours a single driven wheel at the front to overcome the differential problem, despite inferior traction and load-carrying. The engine is recommended to pivot

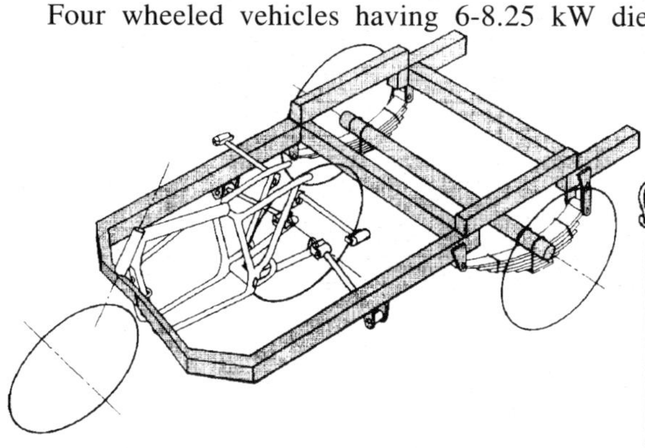

Fig 75: Four-wheeler conversion

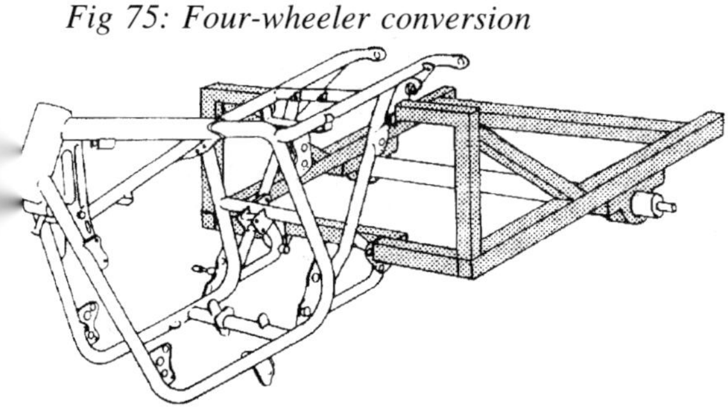

Fig 76: Three-wheeler conversion

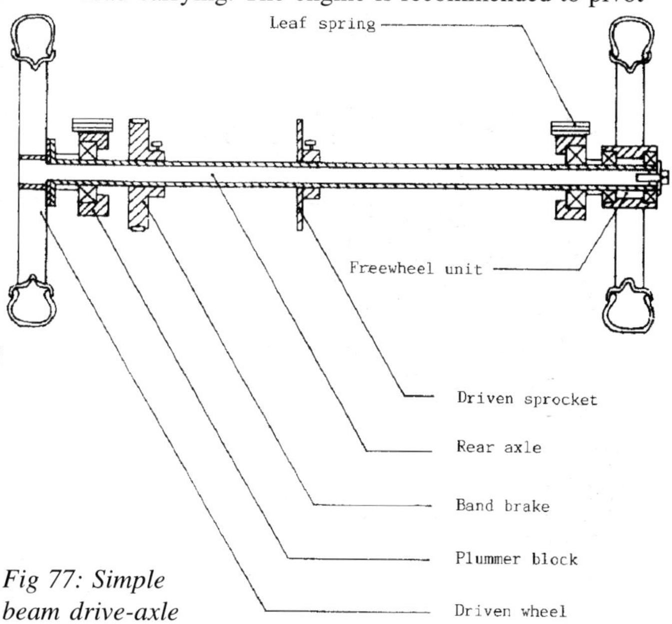

Fig 77: Simple beam drive-axle

with the front steered wheel, to simplify the drive-line, and handlebar control is used. A tensioned-belt clutch may be used as a simpler solution than the conventional plate clutch. Here a twin-pulley vee-belt drive is arranged with the belt set deliberately slack. A spring-loaded idler is used to take up the slack in the belt and the drive disengaged by releasing idler pressure and vice versa. Power engagement is less smooth than with a friction clutch but belt flexibility will take up some of the shock. Where the engine and clutch are mounted independently on the chassis, a flexible coupling should be used to take up some of the inevitable misalignment. Where a speed reduction is required to the final drive, a chain-drive with different sprocket sizes may be the easiest way to achieve this and gross misalignments can be thus better tolerated.

When a change-speed mechanism is required this may be most easily achieved with a stepped-pulley belt-drive system. Fig 78 shows the arrangement designed for the Trag vehicle. Here the engine has a factory-equipped step-down drive of 6:1 while the variable speed portion of the transmission is made up of an upper and lower set of three pulleys (of 1:2, 1:1 and 2:1 ratios) with equal length v-belts positioned over them. Three corresponding levers are positioned on the handle-bars; when one is pulled, the spring-loaded idler-wheel on the desired belt pushes the belt tight and causes it to grip the pulley. The arrangement is reported to have given good service in many parts of the world.

The main modification required to the chassis frame of a three-wheeled vehicle is the attachment of a steering head, a steel tube supporting the front wheel in bearings, Fig 79. Diagonal members are also often needed to brace the rectangular frame. Engine mounting for a stationary-type unit usually involves standing the feet on a suitable base-plate or framed-platform of sufficient rigidity. Practice has proved the suitability of 38 x 8 mm angle sections for this purpose, with engines up to 6 kW. Slotted holes in the fore-aft direction on the platform are a convenient way of adjusting the engine to the drive system before tightening down the mounting bolts.

Suspension systems for Third-World vehicles are usually adaptations of factory-made systems but proprietary caravan suspensions can be a useful standby. The Indespension system of Fig 80 has a square-section member inserted into a hollow axle at

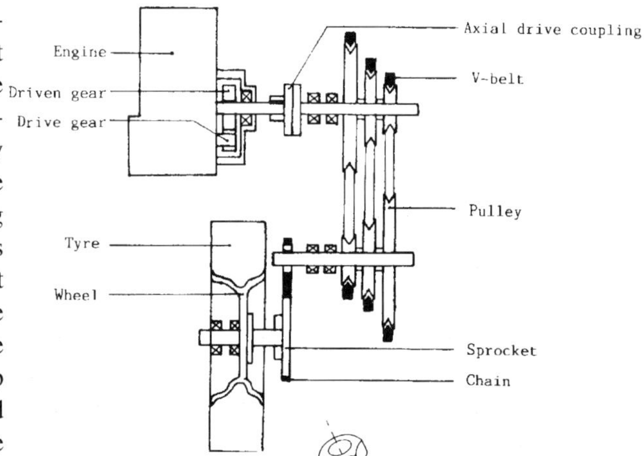

Fig 78: Trag change-speed mechanism

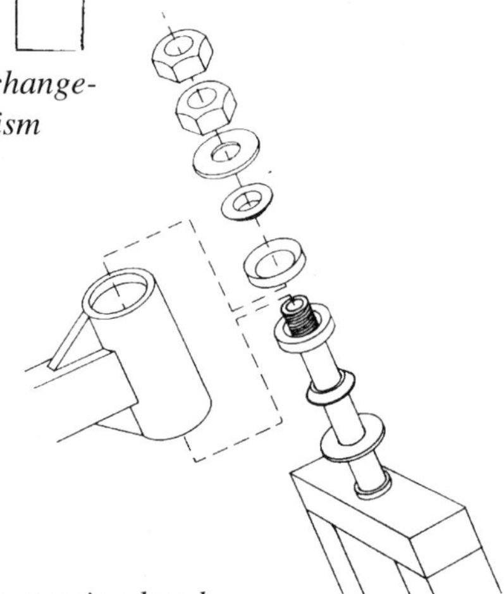

Fig 79: Simple steering head

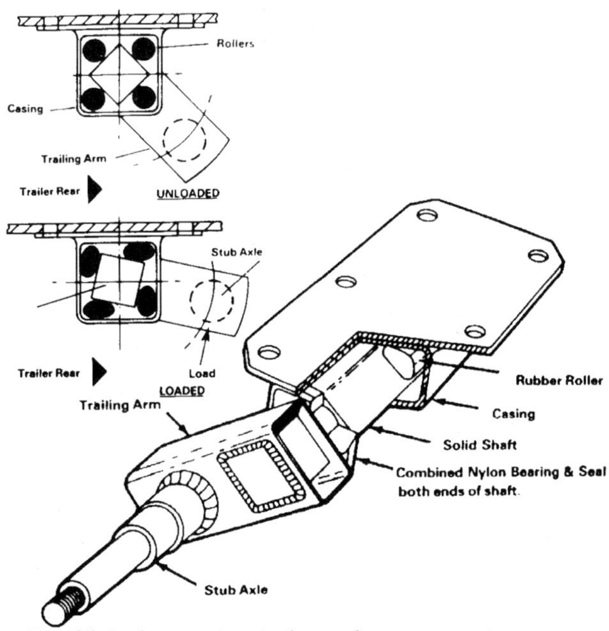

Fig 80:Indespension independent suspension

an interior face angle of 45 degrees and a rubber roller located in the resulting triangular section openings. Rotation of the radius arm causes compression and shear of the rubbers to create the suspension springing.

The front suspension must accommodate steering of the wheels whereas the rear must be capable of providing continuous drive to the wheels. For the three-wheeler, leading-link forks from a suitable motor-cycle are likely to be the easiest solution. For the rear wheels an effective but somewhat complex solution is to use chain drive with driven wheels mounted independently on trailing arms, in the manner of the Bajaj layout shown in Fig 81. The drive chains run from sprockets on an intermediate shaft mounted parallel to the pivoting axle of the trailing arms; the chains lie parallel to the trailing arms with the driven sprockets affixed to the rear wheels. A differential unit is incorporated within the intermediate shaft.

Stability is substantially affected by the castor trail on the front forks of three-wheelers. Two-wheelers require more trail because they are inherently less stable, of course, than three-wheelers so there is likely to be a reduction required in castor trail when mounting motorcycle forks to three-wheelers. However, the amount of castor angle required is likely to be less, due to the heavier front-end weight; but a greater angle is required if high-speed operation on poor surfaces is the aim of the design.

Where the function of the vehicle is extended to agricultural as well as transport operations a more studied approach to the design is necessary, as exemplified by the Tinkabi tractor which has operated successfully in Swaziland. Here the requirement was to replace the use of oxen with a vehicle having a platform to carry a minimum 500 kg while having under-clearance to perform row-crop work. The resulting dimensions are shown in Fig 82. A payload of 1000 kg was achieved, 70 % of the unballasted weight being over the rear wheels. Oxen-style implements are clamped to the rear drawbar which is hand-raised by a spring-assisted lever. Pumps, circular-saws and sprayers can also be drawbar-mounted and belt-driven from a pulley attached to the engine flywheel.

The tractor is powered by a Kirloskar 2-cylinder, air-cooled, direct-injection diesel which is hand-started and produces 12kW at 2000 rpm. Power is transmitted to the rear wheels by a hydrostatic transmission with variable-displacement pump and hy-

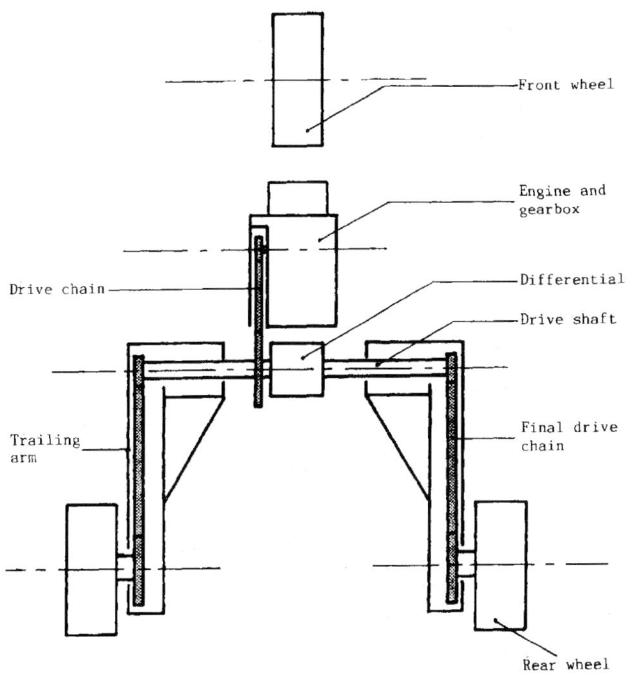

Fig 81: Bajaj rear wheel drive

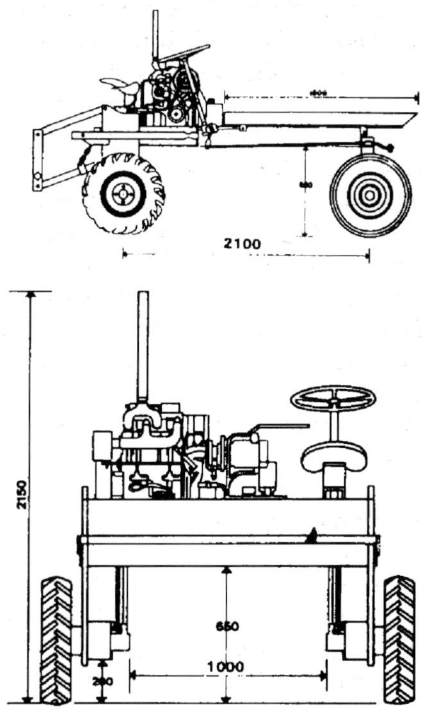

Fig 82: Tinkabi tractor

draulic motors on the wheel mountings. The pump is gear-driven from the engine crankshaft at 10 % step-up speed and has a displacement of 75 litres/min at 2000 rpm. The motors are bi-directional and drive the wheels through 4:1 epicyclic reduction gears. Each has a maximum power output of 11 kW. A single lever control actuates the swashplate mechanism of the pump to control speed direction and braking of the tractor. Maximum drawbar pull of 818 kg is obtainable and speed is infinitely variable up to a maximum of 10 kph. The tractor will pull a plough, capable of turning a single furrow 300 mm wide by 200 mm deep, at a speed of 6.4 kph. Fig 83 shows tables of component and production costs at the time of the design of the machine.

A yet more sophisticated vehicle known as the BTV (Basic Transport Vehicle), Fig 84, was the subject of a study between Lola Cars and Specialised Mouldings Ltd, with Chrysler UK. By applying structural expertise to the base-frame of the vehicle, a more efficient Third-World workhorse results which also has pleasing body-style, with the aid of GRP mouldings. Key structural elements are floor, box-section sidemembers extending from front to rear, dash-bulkhead and two crossmembers, all of which can be made on a press-brake from guillotined panels. The floor-pan comprises two sub-assemblies, and a cross-tube, while the dash is a simple folded panel with flanges turned over at the edges and contoured to fit the sidemembers. The rear floor is basically an angle section, consisting of heel-board and floor, cut out from the wheel-arches. These are reinforced by Z-section transverse members above the rear axle. A tube beneath the floor reacts the loads from the rear-spring front-hangers. Forward of the floor there is a pair of box-sectioned longitudinals comprising top-hat sections and closing plates. At mid-span of these the towers for the front suspension struts are fixed with their bases just forward of the triangular extensions from the dash-panel.

	Cost per unit	
	$US CIF Swaziland	$US FOB ex works
Diesel engine	756	630
Hydraulic pump	258	241
Hydraulic motors	515	471
Steering unit	63	55
350 kg mild steel	122	
Tyres 2 × 14 in	82	
Rims 2 × 15 in	64	
Oil (50 litres SAE 20)	29	
Bearings/fasteners	30	
Sundry items	20	
	1939	
Implement materials	120	
Fasteners/sundries	20	
Total per set	140	

Fig 83: Tinkabi costings

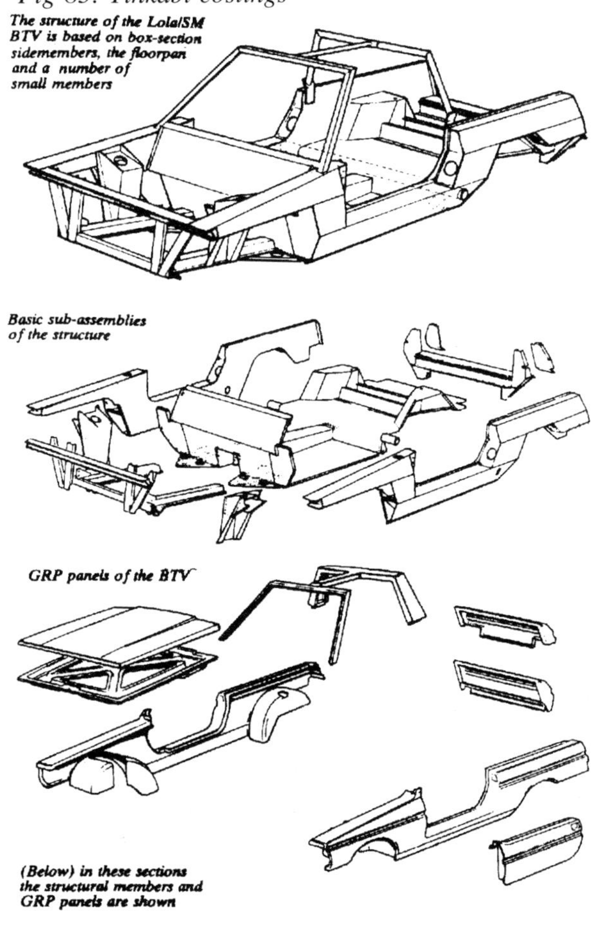

Fig 84: BTV and its structure

Light trailers: new regulations

EC Directive 7516 for light trailers affects most of the specialist vehicle industry — passenger cars, commercial and industrial vehicles. In an effort to raise safety and construction standards type-approval will be required in future. Trailers up to 750 kg are affected with separate requirements for the braked and unbraked categories.

It will be a requirement to tow only half the weight of the towing vehicle and both the maximum permitted drawbar load plus allowable towable mass

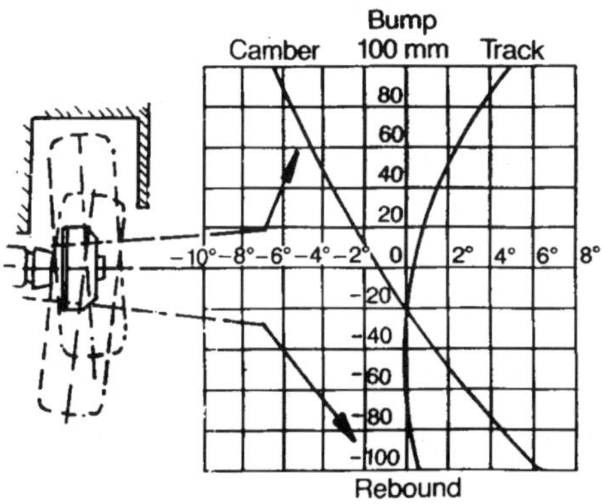

Fig 85: Computer simulation of snaking

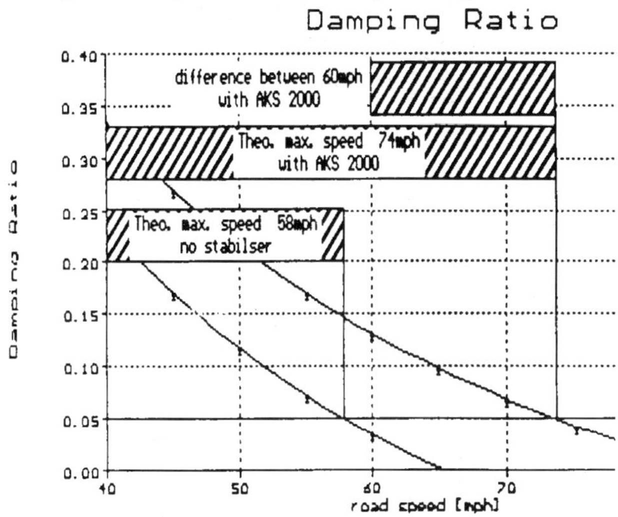

Fig 86: Damping effect of coupling

Fig 87: trailer with Delta axle

will have to be marked on the coupling. Off-road vehicles designed to 70/156/EEC will be able to tow 1.5 times the GVW of the towing vehicle. There will be coupling legislation requiring the fitment of a 40 mm DIN eye with an onion shape jaw for those trailers entering the Mainland Europe market.

In a liberalising part of the legislation others apart from OEMs will be allowed to fit towing brackets and the OEMs will be required to make brackets and fixing point information generally available. Parking brake provision will be a legal requirement and there will be a number of additional braking provisions that affect trailers for `hire-and-reward' that do not affect `own use' operation. OEMs are likely to be going to require test equipment for overrun brakes in order to certify some installations.

For rear lighting, trailers will require the same cluster as on the towing vehicle. There will also be new requirements for side marker-lighting. Here there will be special exemptions for such operations as boat trailers used only for launching and recovery. There will also be conformity certification applying to such items as light lenses.

The need for light-trailer legislation on safety grounds arises, it is argued, from the changed specifications of towing vehicles, particularly cars, in recent years. Twenty-five years ago a tow-car might typically weigh 1.5 tons and its 60 bhp engine would automatically limit tow speeds for caravan combinations to 40 mph; present-day power to weight ratios allowed speeds well in excess of the 60 mph legal limit to be exceeded.

Parking brakes will be required to operate both brake-shoes at each wheel station and the linkage mechanism must provide the correct mechanical advantage. For rear-lighting, trailers will require the same cluster as on the towing vehicle. There will also be new requirements for side marker-lighting. Here there will be special exemptions for such operations as boat trailers used only for launching and recovery. There will also be conformity certification applying to such items as light lenses. The need for light-trailer legislation on safety grounds arises, it is argued, from the changed specifications of towing vehicles, particularly cars, in recent years. Twenty-five years ago a tow-car might typically weigh 1.5 tons and its 60 bhp engine would automatically limit tow speeds for caravan combinations to 40 mph; present-day power to weight ratios allowed speeds well in excess of the

60 mph legal limit to be exceeded.

Parking brakes will be required to operate both brake-shoes at each wheel station and the linkage mechanism must provide the correct mechanical advantage. For rear-lighting, trailers will require the same cluster as on the towing vehicle. There will also be new requirements for side marker-lighting. Here there will be special exemptions for such operations as boat trailers used only for launching and recovery. There will also be conformity certification applying to such items as light lenses. The need for light-trailer legislation on safety grounds arises, it is argued, from the changed specifications of towing vehicles, particularly cars, in recent years. Twenty-five years ago a tow-car might typically weigh 1.5 tons and its 60 bhp engine would automatically limit tow speeds for caravan combinations to 40 mph; present-day power to weight ratios allowed speeds well in excess of the 60 mph legal limit to be exceeded.

Al-Ko Kober have devised a computer program for representing the polar moments of towing and trailer vehicles, Fig 85, in demonstrating the effectiveness of the model 2000 stabiliser, Fig 86. The required damping ratio for the stabiliser can be calculated and the snaking oscillation of the trailer predicted. The company have introduced the Delta CA (compound angle) axle for trailers, Fig 87, which also gives a stability advantage. Automatic control of camber and track variation is achieved. The existing rubber torsion bar independent trailing arm axle is now cranked at the centre of the torsion tube to form a bow-shape configuration which is mounted at an angle to the trailer. Wheel attitude responds to the centrifugal forces generated in cornering. to provide a progressive response and improve tyre contact, particularly in braking.

The tube is cranked backwards 12 degrees and mounted at a 3.5 degrees compound angle upwards from the chassis longitudinals. The resulting arcs of motion for the wheel are said to make tyres better follow the towing vehicle as well as accommodating greater variations in surface irregularities. Camber and track response of the axle, tested at Brunswick University, are shown in Fig 87.

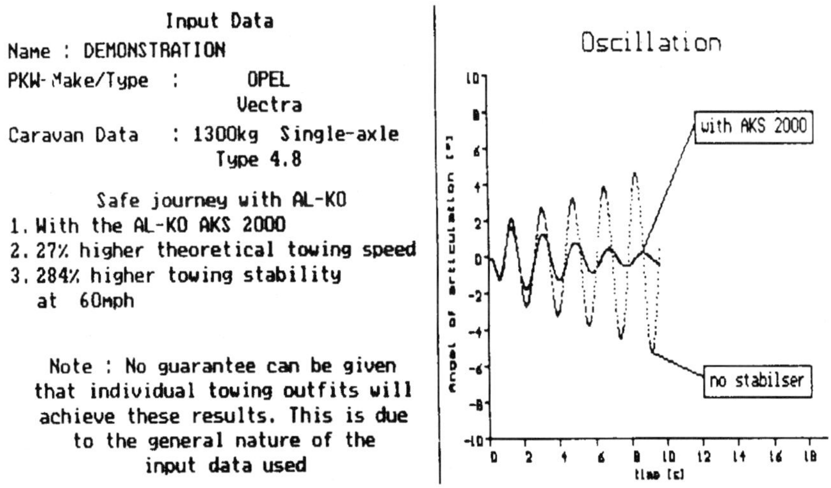

Fig 87: Tested response of axle

References

Chapter 1
1 De Bono, E, A five day course in thinking
2 Fielden, G, Engineering Design, Dept Sci & Ind Res, HMSO, 1963
3 Fores, M, Technik: the missing concept, Higher Education review, Spring 1979
4 Wilson, A, Eminent Victorians, BBC, 1989
5 Sparke, P, Consultant Design: the history and practice of the designer in industry, Pembridge Press, 1983
6 Papanek, V, Design for the real world, Thames and Hudson, 1984
7 Siemens Review, No 6/81, pages 29-32
8 Clausen, H, The case for the engineering designer, *Automotive Design Engineering*, Jun 1965
9 Flurscheim, C, Industrial design in engineering, The Design Council, London, 1983.
10 Whitford, F, Bauhaus, Chapters 10, 12, Thames and Hudson, 1988
11 Thiel, P, Visual awareness and design, University of Washington Press, 1983
12 *Illuminating Engineering*, USA, Vol 57, number 12
13 Kennington, W, The utilitarian aesthetics of automobile body design, Proc, Institute of Automobile Engineers, 1934-5,
14 Ford, B, Modern Britain, Cambridge University Press
15 Yarwood, A, An introduction to AutoCAD Release 13, Longman, 1995
16 Ward, P, Surface modelling, Paper 85062, Proc ISATA, 1985
17 Medland, A, The computer-based design process, Chapman & Hall, 1992
18 Vittorio & Mazzanti, A new surface modelling technique, Paper 86039, Proc ISATA, 1986
Further reading
Design and the economy, The Design Council, 1983
The car programme, Boilerhouse Project exhibition guide, The Conran Foundation
Pugh, S, Total design, Addison-Wesley, 1991
Nahum, A, Alec Issigonis, The Design Council, 1988
Curved surfaces in engineering, CADCentre conference report, IPC Science & Technology, 1972
Effective CADCAM, IMechE conference report 9, 1987
Ranger, T, Surface reappoximation for automotive body panels, *Automotive Engineer*, Apr/May 1987, pp 25-27
McClelland, M, Computers and the design process, *Materials and Design*, Vol 5, Jun/Jly 1984
Karp, H, CAE applications in the automotive industry, SAE paper 850446
Ostrowsky, O, Engineering drawing with CAD applications, Edward Arnold, 1989.
Karp, H, CAE applications in the autootive industry, SAE Paper 850446
Levens, Graphics: analysis and conceptual design, Wiley

Chapter 2
1 Weichenrieder and Haldenwanger, The best function for the seat of a passenger car, Paper 850484 in SAE publication PT-64 Automotive interior trim, 1996
2 Murrell H, Control and instruments, design procedure, *Automotive Design Engineering*, vol 2, no. 14, 1963
3 Gurram and Vertiz, The role of automotive seat cushion deflection in improving ride comfort, SAE Paper 970596, 1997
4 Shen and Vertiz, Redefining seat comfort, SAE Paper 970597, 1997
5 Lewis and Griffin, The transmission of vibration to the occupants of a car seat with a suspended back-rest, Proc IMechE Vol 210, Part D, 1996
6 Rachel, T, Split frame seating, Paper in Materials for automobile interiors, SAE Special Publication, 1992
7 Rebiffe *et al*, Efficient vehicle packaging with front-wheel drive, SAE paper 780131
Further reading
Ergonomics and safety in motor car design, IMechE symposium report, 1966
Woodson, W, Human engineering guide for equipment designers, Berkeley Univ of Cal Press, 1954
Everden, I, Designing for comfort and convenience, *Automotive Design Engineering*, Jun 1963
Ashley, C, Work vehicle suspension seating, Bostrom report, 1976
Mattia et al, Thin Line seating for small cars, SAE paper 820405, 1982
Tilbury, Body interior dimensioning, *Automotive Design Engineering*, Apr 1964
Design of commercial vehicle cabs in relation to driver environment, IMechE conference report, 1973
Davey, J, Looking after the lorry driver, *Design*, April 1965
Karen, T, Ergonomics from the driver's seat, Design Engineering Show conference, 1986
Muller, G, TECHNA, a concept-car to challenge automotive engineers, SAE paper 690267.
Stikeleather et al, A study of vehicle vibration spectra as related to seating dynamics, SAE Paper 720001
Lansdown and Fowkes, An investigation into evaluation of driver information systems, MIRA internal report

Chapter 3
1 Costin F, The aerodynamics of the modern car, *Automotive Engineer*, vol 1, no 1, 1975
2 White, R, A rating method for assessing vehicle drag coefficients, Motor Industry Research Association Report, 1967/9
3 Harvey, Aerodynamics fundamentals, *Automotive Design Engineering*, July 1964
4 Scibor-Rylski, A, Aerodynamic characteristics of the

wedge shaped racing car, *Automotive Design Engineering*, July 1969
5 Research Report, Jim Clark Foundation, 28 Charlotte Square, edinburgh, 1969
6 Scibor-Rylski, A, Road vehicle aerodynamics, Pentech Press, 1984
7 Katz and Dykstra, Study of an open-wheel racing car's rear-wing aerodynamics, SAE Paper 890600
8 Cooper, K, The effect of front-edge rounding and rear edge shaping on the aerodynamic drag of bluff vehicles in ground proximity, SAE paper 850288
9 Assessment of proprietary devices for reducing the aerodynamic drag of commercial vehicles, MIRA REport K.23049, 1978
10 Garry, K, A review of commercial vehicle aerodynamic drag redcurion techniques, Proc IMechE, Vol 199, No D3
11 Hanna, Dr, Paper C498/36/244, Autotech 1995
12 Domini, R, Paper C498/36/181, Autotech, 1995
13 Graysmith and Baxendale, Paper C498/36/160, Autotech 1995
14 Howell and Goodwin, Paper C498/36/158, Autotech 1995
15 Barnard, R, Road Vehicle Aerodynamic Design, Longman
16 Paper C498/36/042, Autotech 1995
17 Anderson, D, Airflow to engine coolant, *Automotive Design Engineering*, March 1969

Further reading

Ed Dorgham, M, Impact of aerodynamics on vehicle design, Inderscience Enterprises, 1993
Domini, J & RG, Aerodynamic influences on the performance of the Grand Prix racing car, ProcIMechE, Vol 198, No 12
Katz, Aerodycnamic model for wing-generated down force on open-wheel-racing-car configurations, SAE Paper 860218, 1986
Garry, K, A review of commercial vehicle aerodynamic drag reduction techniques, Proc IMechE, Vol 199, D3, 1985
Sumitani and Yamada, Development of Aero Slit, SAE paper 890372, 1989
MacDonald, K, Automotive air conditioning, Howard Sams, 1978

Chapter 4

1 Lynam, N, Smart windows for automobiles, SAE Paper 900419
2 Evans, R, Paper C498/8/197, Autotech 1995
3 Materials Design Manual, RFI Shielding, Warner Drive, Springwood Industrial Estate, Rayne Road, Braintree, CM7 7YW, UK
4 Autotech paper C427/40/108, 1995

Further reading

Bedford, A, Development of a headlining, Proc IBCAM Boditek 1987
Lee, W, Effect of loading geometry on stress-strain behaviour of plastic foams, SAE Paper 710040
Monk and Morgan, The manufacture of a new generation of headlamp reflectors, Proc IMechE, Vol 199, B3, 1985.
Lindae, G, Improvements of low-beam pattern by use of polyellipsoidal (PES) headlamps, SAE Paper 850228
Steffens, E, Electrical conductive plastic compounds for EMI shielding, Proc SITEV, 1990
Hicks, H, A fresh approach to an idea for seeing at night, SAE paper 700088, 1970
Watts, A, Automotive seating with dual-hardness foams, *Automotive Engineer*, Oct/Nov 1984
Molian, S, Some key problems in the theory of plane mechanisms, Borough Polytechnic Mechanisms conference, 1969.
Ed Giles, J, Vehicle Equipment, Chapter 3 on Mechanisms, Iliffe, 1969
Lorenzo et al, SAE papers 960151/2, 1996

Chapter 5

1 Washino, S, Automobile electronics, Gordon and Breach scientific Publishers, 1988
2 Ed Adler, U, Automotive electric/electronic systems, Robert Bosch GmbH (distributed by the SAE)
3 Ribbens and Mansour, Understanding automotive electronics, Texas Instruments Leaning Center, Dallas
4 Legg and Webb, Automotive lamp integrity checking system, AB Electronic Products group report
5 Jones, T, 'Commercial Vehicle Electronics' 1986 SAE Ray Buckendale lecture
6 Stevens, SD, The development and testing of an integrated systems demonstrator vehicle, Autotech 95 paper C498/35/157
7 van Zanten et al, Simulation for the development of the Bosch VDC, SAE paper 960486
8 Leffler et al, The new automatic stability control ASC of the 3-Series BMW diesel passenger cars, Autotech 95 paper C498/35/056
9 Soja, R, Microcoded timing coprcessor for enhanced engine management, IEE/IMechE Automotive Electronics conference, 1991
10 Kerslake and Oxley, An enabling technology for automotive systems - LinASIC', IEE/IMechE Automotive Electronics conference, 1991
11 Chidlow and Hollins, Cost-effective multiplexing, IEE/IMechE Automotive Electronics conference, 1991
12 McLaughlin and Khoh, A feasibility study of CAN technology in body electronic control systems, Autotech paper C498/35/028, 1995

Further reading

Bromley, P, Advanced automotive electronics, IFS Publi-

cations, 1989
Miller, G, Introduction to microprocessors, SAE paper 790234
Knowles, D, The automotive computer, Prentice-hall, 1987
Phail, F, Controller Area Network, an in-vehicle network solution, SAE paper 880588
Electronics for mechanical engineers, *OEM Design*, Oct 1971-Oct 1972
Elementary AC Theory, *Automotive Design Engineering*, Jun 1967
Embrey et al, A non multiplexed microprocessor based driver interface switching system for automotive application, AB Electronic Products
Valentine, R, MUX wiring: digital control for vehicular electrical systems, Report AR154S, Motorola Inc
Anderson and Morgan, High reliability automotive interconnections, SAE paper 870552, 1987

Chapter 6

1 Hodgkiss, J, On-board electronics in commercial vehicles, *Automotive Engineer*, April/May 1991
2 11 Maue, HW, Tailgate multiplexed node, Paper 960390, Multiplexing Technology, SAE Special Publication SP-1137, 1996
3 Meyer *et al*, Paper SIA 9506EI8, Proc EAEC, 1995
4 Hoess *et al,* Paper 961009, Proc SAE, 1996
5 Winner *et al*, Paper 961010, Proc SAE, 1996
6 Angermuller *et al,* Paper 960530, Proc SAE, 1996
7 Lissel *et al,* Paper C498/8/166, Autotech 1995.
8 SITEV conference 1990
9 Franzen, S, SITEV conference 1990
10 Thomas, J, SITEV conference 1990
11 Bragas and Elke, SITEV conference 1990
12 Blake, L, The automatic taxi public transport system, IMechE convention on guided transport, 1966
13 Boyd et al, An advanced door-to-door system for inter-urban transportation, SAE paper 690170, 1969
14 Autotech paper C498/1/184, 1995
15 Autotech paperC498/1/224, 1995
Further reading
Hillyer, M, Using motors in automotive multiplex systems, SAE paper 860394
Anderson and Morgan, High reliability automotive interconnectins, SAE paper 870552, 1987

Chapter 7

1 Lovins and Barnett, Supercars: the coming light-vehicle revolution, Rocky Mountain Institute, Summer Study, European Council for an energy-efficient economy, Denmark, 1993

Chapter 8

1 Dominey, Aerodynamics of Grand prix cars, Proc IMechE, Vol 206,
3 Goran *et al*, Paper (C389/418), FISITA Congress, 1992
Further reading
Fothergill et al, Computer-aided concept design of a sports car chassis system, IMechE conference on Vehicle Structures, no 7, 1984
The design and manufacture of low-cost motorised vehicles, Intermediate Technology Publications, 1988
Spicer, J, Car aperture design for easy access, vehicles for the disabled session, Autotech 1987
Dunn, M, The Bentley Turbo R, Proc IMechE, vol 201, no. 95, 1987
Steeds, W, Mechanics of road vehicles, Iliffe, 1960
British Standard BS AU 246: 1992

Chapter 9

1 BR Schneider and P Hofmann, Paper C389/341, FISITA 1992
2 Paper 925071, FISITA, 1992
3 Dare-Bryan, V, Paper presented at the MIRA electric vehicles seminar, 1992
4 Sheet Steel Handbook, Swedish Steel Srip Products, SSAB Tunnplat AB, S-78184 Borlange, Sweden.
Commercial vehicles, engineering and operation, IMechE conference report, 1967
Ed Montgomerie, G, Vehicle engineering handbook, Freight Transport Association, 1990

Chapter 10

5 ALCOA publication, Notes on the design of aluminium monocoque semi-trailer transport tanks
6 ALCAN publication, Semi-trailer road tanker design and details
7 Formulas for stress & strain, McGraw Hill
8 Experimental Station Bulletins 108 and 123, University of Utah
9 Reed, A, Opening papaer, IMechE symposium on Hoppers and Silos (S308 Bulk Materials Handling Committee)
10 Draughtsman and Allied Technicians Association Information Sheet Number Nine
11 Engineering Designer, *December 1966*
12 British Aluminium publication, Commercial vehicle body-building in aluminium
13 Hyde et al, Proc IMechE vol 208 part D3, 1994
8. Proc IMechE, Vol 207, No D4, 1993.
9. IMechE symposium on Hoppers and Silos (S308 Bulk Materials Handling Committee)
Further reading
Bulk materials handling towards the year 2000, IMechE conference report, no 12, 1991
Jenike, A, Gravity flow of bulk solids, Storage and flow of

solids, bulletins 108 and 123 of University of Utah Engineering Experimental Station
Spencer et al, New tools from research make bins, hoppers flow, *Materials Handling Engineering*, June 1965
Rees, C, Design of tanks and cylindrical fluid containing vessels, *The Engineering Designer*, Dec 1966
D'Arcy-Smith, F, Bulk handling of powders in the chemical industry, Institute of Materials Handling conference, 1961
Rotter, J, Strength criteria for steel silo structures, IMechE conference paper C418/035

Chapter 11

1 Toganel, G, Paper E1617, FISITA Congress, 1996
2 Crowther, W, Faulty hydraulics: causes, effects and solutions, Industrial hydraulics practice today IMechE Tribology Group Seminar, 1995
3 Balmer, C, IMechE Tribology Group Seminar, 1995
4 Newnham, P, IMechE Tribology Group Seminar, 1995
5 French, D, IMechE Tribology Group Seminar, 1995
6 Tovey, P, IMechE Tribology Group Seminar, 1995
7 Fuller, D, IMechE Tribology Group Seminar, Gearing and gearbox practice today, June 1995
8 Mao, K, IMechE Tribology Group Seminar, Gearing and gearbox practice today, June 1995
10 Hofmann, D, IMechE Tribology Group Seminar, Gearing and gearbox practice today, June 1995
12 Da Silva and Miranda, Autotech Congress, 1991
13 Tae-Eun *et al*, Structural design of aluminium space frame with battery tray, SAE Paper 960524
14 Lilley and Mani, Roof-crush strength improvement using rigid polyurethane foam, SAE Paper 960435

Further reading

Dickson-Simpson, J, Era of articulation, Transport Press services, 1968
Ed Hoelze et al, Utility vehicle design handbook, SAE, 1991
Evaluation of unit load handling equipment, IMechE conference report, no. 1, 1973
Fenton, J, New concept for delivery vehicles, *Automotive Design Engineering*, Aug 1968
Hales, F, The lateral stability of a simplified articulated vehicle, IMechE symposium on Control of vehicles during braking and cornering, 1963
Coupling articulated vehicles, IRTE Code of Practice, 1986

Chapter 12

1 Niemann, K, IMechE Bus Conference, 1995
2 McKnight, The development of a low floor ciybus — Volvo B10L, Proc IMechE Vol 209 Pt D, 1995
3 Ponsford, A, IMechE Bus 92 conference, 1992
4 Baguelin and Thomas, IMechE Bus 92 conference, 1992
5 Wilson-Jones, R, Steering and stability of single-track vehicles, Proc IMechE, 1952
6 Pearsell, R, Proceedings of the Institute of Automobile Engineers, 1922-23, Part 1.

Further reading

Daniels *et al*, Future transit bus designs, SAE paper 780058
Recommended specification for buses used to operate local services, Disabled Persons Transport Advisory Committee (DIPTAC), DoT, 1988
Brooks et al, Passenger problems on moving buses, TRRL/ Leyland Vehicles report
Bus 92, The expanding role of buses towards the 21st century, Proc IMechE, no. 2, 1992
Webster and McGowan, Planning for the bus, TRRL conference Moving people in cities, 1973
McHugh and Tidbury, The design of touring coach bodywork for safety, IMechE conference paper C200/73
Bretherton et al, The Leyland Titan, ProcIMechE, Vol 192, No 7, 1978
Copus, A, DC traction motors for electric vehicles, EVA conference paper, 1991
Hodkinson, R, Waste heat recovery, EVA conference paper, 1991

Index

G

H

I

J

K

L

M

U

V

W

Y